TODAY'S TECHNICIAN™

SHOP MANUAL

For Automotive Electricity and Electronics

SEVENTH EDITION

Barry Hollembeak

CENGAGE

Australia • Brazil • Mexico • Singapore • United Kingdom • United States

Today's Technician™: Shop Manual for Automotive Electricity and Electronics, Seventh Edition

Barry Hollembeak

SVP, GM Skills & Global Product Management: Jonathan Lau

Product Director: Matthew Seeley

Senior Product Manager: Katie McGuire

Senior Director, Development: Marah Bellegarde

Senior Product Development Manager: Larry Main

Senior Content Developer: Mary Clyne

Product Assistant: Mara Ciacelli

Vice President, Marketing Services: Jennifer Ann Baker

Marketing Manager: Andrew Ouimet

Senior Production Director: Wendy Troeger

Production Director: Andrew Crouth

Senior Content Project Manager: Cheri Plasse

Senior Art Director: Jack Pendleton

Production Service/Composition: SPi Global

Cover image(s): Umberto Shtanzman/ Shutterstock.com

Library of Congress Control Number: 2017951497

Shop Manual ISBN: 978-1-3376-1901-1
Package ISBN: 978-1-3376-1899-1

Cengage
20 Channel Center Street
Boston, MA 02210
USA

Cengage is a leading provider of customized learning solutions with employees residing in nearly 40 different countries and sales in more than 125 countries around the world. Find your local representative at **www.cengage.com**.

Cengage products are represented in Canada by Nelson Education, Ltd.

To learn more about Cengage Learning, visit **www.cengage.com**

Purchase any of our products at your local college store or at our preferred online store **www.cengagebrain.com**

Notice to the Reader

Publisher does not warrant or guarantee any of the products described herein or perform any independent analysis in connection with any of the product information contained herein. Publisher does not assume, and expressly disclaims, any obligation to obtain and include information other than that provided to it by the manufacturer. The reader is expressly warned to consider and adopt all safety precautions that might be indicated by the activities described herein and to avoid all potential hazards. By following the instructions contained herein, the reader willingly assumes all risks in connection with such instructions. The publisher makes no representations or warranties of any kind, including but not limited to, the warranties of fitness for particular purpose or merchantability, nor are any such representations implied with respect to the material set forth herein, and the publisher takes no responsibility with respect to such material. The publisher shall not be liable for any special, consequential, or exemplary damages resulting, in whole or part, from the readers' use of, or reliance upon, this material.

Printed in the United States of America
Print Number: 01   Print Year: 2017

CONTENTS

PHOTO SEQUENCES

JOB SHEETS

Thanks to the support the *Today's Technician*™ series has received from those who teach automotive technology, Cengage, the leader in automotive-related textbooks and learning solutions, is able to live up to its promise to provide new editions of the series every few years. We have listened and responded to our critics and fans and have presented this new updated and revised seventh edition. By revising this series on a regular basis, we can respond to changes in the industry, in technology, in the certification process, and the ever-changing needs of those who teach automotive technology.

We have also listened to instructors when they said that something was missing or incomplete in the previous edition. We have responded to those and have included the results in this seventh edition.

The *Today's Technician*™ series features textbooks and digital learning solutions that cover all mechanical and electrical systems of automobiles and light trucks. The individual titles correspond to the ASE (National Institute for Automotive Service Excellence) certification areas and are specifically correlated to the 2017 standards for Automotive Service Technicians, Master Service Technicians, as well as to the standards for Maintenance and Light Repair.

Additional titles include remedial skills and theories common to all of the certification areas and advanced or specific subject areas that reflect the latest technological trends. *Today's Technician: Automotive Electricity & Electronics, 7e* is designed to give students a chance to develop the same skills and gain the same knowledge that today's successful technician has. This edition also reflects the most recent changes in the guidelines established by the National Automotive Technicians Education Foundation (NATEF).

The purpose of NATEF is to evaluate technician training programs against standards developed by the automotive industry and recommend qualifying programs for certification (accreditation) by ASE. Programs can earn ASE certification upon the recommendation of NATEF. NATEF's national standards reflect the skills that students must master. ASE certification through NATEF evaluation ensures that certified training programs meet or exceed industry-recognized, uniform standards of excellence.

The technician of today and the future must know the underlying theory of all automotive systems, and be able to service and maintain those systems. Dividing the material into two volumes, a Classroom Manual and a Shop Manual, provides the reader with the information needed to begin a successful career as an automotive technician without interrupting the learning process by mixing cognitive and performance learning objectives into one volume.

The design of Cengage's *Today's Technician*™ series was based on features that are known to promote improved student learning. The design was further enhanced by a careful study of survey results, in which respondents were asked to value particular features. Some of these features can be found in other textbooks, while others are unique to this series.

Each Classroom Manual contains the principles of operation for each system and subsystem. The Classroom Manual also contains discussions on design variations of key components used by different vehicle manufacturers. It also looks into emerging technologies that will be standard or optional features in the near future. This volume is organized to build upon basic facts and theories. The primary objective of this volume is to allow the reader to gain an understanding of how each system and subsystem operates. This understanding is necessary to diagnose the complex automobiles of today and tomorrow.

Although the basics contained in the Classroom Manual provide the knowledge needed for diagnostics, diagnostic procedures appear only in the Shop Manual. An understanding of the underlying theories is also a requirement for competence in the skill areas covered in the Shop Manual.

A spiral-bound Shop Manual delivers hands-on learning experiences with step-by-step instructions for diagnostic and repair procedures. Photo Sequences are used to illustrate some of the common service procedures. Other common procedures are listed and are accompanied with fine line drawings and color photos that allow the reader to visualize and conceptualize the finest details of the procedure. This volume also contains the reasons for performing the procedures, as well as when that particular service is appropriate.

The two volumes are designed to be used together and are arranged in corresponding chapters. Not only are the chapters in the volumes linked together, but the contents of the chapters are also linked. The linked content is indicated by marginal callouts that refer the reader to the chapter and page where the same topic is addressed in the companion volume. This valuable feature saves users the time and trouble of searching the index or table of contents to locate supporting information in the other volume. Instructors will find this feature especially helpful when planning the presentation of material and when making reading assignments.

Both volumes contain clear and thoughtfully selected illustrations, many of which are original drawings or photos specially prepared for inclusion in this series. This means that art is a vital part of each textbook and not merely inserted to increase the number of illustrations.

—Jack Erjavec

HIGHLIGHTS OF THE NEW EDITION— CLASSROOM MANUAL

The text, photos, and illustrations in the seventh edition have been updated throughout to highlight the latest developments in automotive technology. In addition, some chapters have been combined. Although chapter 16 covers details associated with alternative powered vehicles, all pertinent information about hybrid vehicles is included in the main text that concerns relative topics. For example, the discussion of batteries in Chapter 5 includes coverage of HEV batteries and ultra-capacitors. Chapter 6 includes AC motor principles and the operation of the integrated starter/generator. Chapter 7 includes the HEV charging system, including regenerative braking and the DC/DC converter.

The flow of basic electrical to more complex electronic systems has been maintained. Chapters are arranged to enhance this flow and reduce redundancy.

Chapter 1 introduces the student to the automotive electrical and electronic systems with a general overview. This chapter emphasizes the interconnectivity of systems in today's vehicles, and describes the purpose and location of the subsystems, as well as the major components of the system and subsystems. The goal of this chapter is to establish a basic understanding for students to base their learning on. All systems and subsystems that are discussed in detail later in the text are introduced, and their primary purpose is described. Chapter 2 covers the underlying basic theories of electricity and includes discussion of Ohm's and Kirchhoff's laws. This is valuable to the student and the instructor because it covers the theories that other textbooks assume the reader knows. All related basic electrical theories are covered in this chapter.

Chapter 3 applies those theories to the operation of electrical and electronic components, and Chapter 4 covers wiring and the proper use of wiring diagrams. Emphasis is on using the diagrams to determine how the system works and how to use the diagram to isolate the problem.

The chapters that follow cover the major components of automotive electrical and electronic systems, such as batteries, starting systems and motor designs, charging systems, and basic lighting systems. This is followed by chapters that detail the functions of the body computer, input components, and vehicle communication networks. From here the student is guided into specific systems that utilize computer functions.

Current electrical and electronic systems are used as examples throughout the text. Most of these systems are discussed in detail. This includes computer-controlled interior and exterior lighting, night vision, adaptive lights, instrumentation, and electrical/electronic accessories. Coverage includes intelligent wiper, immobilizer, and adaptive cruise control systems. Chapter 15 details the passive restraint systems currently used.

HIGHLIGHTS OF THE NEW EDITION—SHOP MANUAL

Like the Classroom Manual, the Shop Manual is updated to match current trends. Service information related to the new topics covered in the Classroom Manual is also included in this manual. In addition, several new Photo Sequences are added. The purpose of these detailed photos is to show students what to expect when they perform the same procedure. They also help familiarize students with a system or type of equipment they may not encounter at school. Although the main purpose of the textbook is not to prepare someone to successfully pass an ASE exam, all the information required to do so is included in the textbook.

To stress the importance of safe work habits, Chapter 1 is dedicated to safety, and includes general HEV safety. As with the Classroom Manual, HEV system diagnosis is included within the main text. This provides the student with knowledge of safe system diagnosing procedures so they know what to expect as they further their training in this area. Included in this chapter are common shop hazards, safe shop practices, safety equipment, and the legislation concerning and the safe handling of hazardous materials and wastes.

Chapter 2 covers special tools and procedures. This chapter includes the use of isolation meters and expanded coverage of scan tools. In addition, a section on what it entails to be an electrical systems technician is included. This section covers relationships, completing the work order, and ASE certification. Another section emphasizes the importance of proper diagnostic procedures.

Chapter 3 leads the student through basic troubleshooting and service. This includes the use of various test equipment to locate circuit defects and how to test electrical and electronic components. Chapter 4 provides experience with wiring repairs along with extended coverage and exercises on using the wiring diagrams.

The remaining chapters have been thoroughly updated. The Shop Manual is cross-referenced to the Classroom Manual by using marginal notes. This provides students the benefit of being able to quickly reference the theory of the component or system that they are now working with.

Currently accepted service procedures are used as examples throughout the text. These procedures also serve as the basis for new job sheets that are included in the Shop Manual chapters.

CLASSROOM MANUAL

Features of this manual include the following:

Cognitive Objectives

These objectives outline the chapter's contents and identify what students should know and be able to do upon completion of the chapter. *Each topic is divided into small units to promote easier understanding and learning.*

Terms to Know List

A list of key terms appears immediately after the Objectives. Students will see these terms discussed in the chapter. Definitions can also be found in the Glossary at the end of the manual.

A Bit of History

This feature gives the student a sense of the evolution of the automobile. This feature not only contains nice-to-know information but also should spark some interest in the subject matter.

Margin Notes

The most important terms to know are highlighted and defined in the margins. Common trade jargon also appears in the margins and gives some of the common terms used for components. This helps students understand and speak the language of the trade, especially when conversing with an experienced technician.

Author's Notes

This feature includes simple explanations, stories, or examples of complex topics. These are included to help students understand difficult concepts.

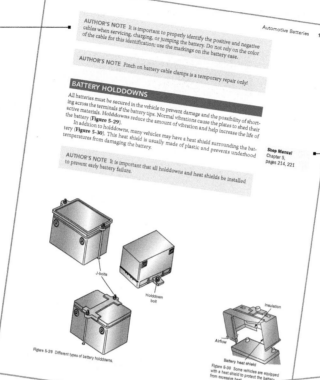

AUTHOR'S NOTE It is important to properly identify the positive and negative cables when servicing, charging, or jumping the battery. Do not rely on the color of the cable for this identification; use the markings on the battery case.

AUTHOR'S NOTE Pinch on battery cable clamps is a temporary repair only!

BATTERY HOLDDOWNS

All batteries must be secured in the vehicle to prevent damage and the possibility of shorting across the terminals if the battery tips. Normal vibrations cause the plates to shed their active materials. Holddowns reduce the amount of vibration and help increase the life of the battery (**Figure 5-29**).

In addition to holddowns, many vehicles may have a heat shield surrounding the battery (**Figure 5-30**). This heat shield is usually made of plastic and prevents underhood temperatures from damaging the battery.

AUTHOR'S NOTE It is important that all holddowns and heat shields be installed to prevent early battery failure.

Shop Manual
Chapter 5,
pages 214, 221

J-bolts

Holddown bolt

Figure 5-29 Different types of battery holddowns.

Insulation

Airflow

Battery heat shield
Figure 5-30 Some vehicles are equipped with a heat shield to protect the battery from excessive heat.

Cross-References to the Shop Manual

Reference to the appropriate page in the Shop Manual is given whenever necessary. Although the chapters of the two manuals are synchronized, material covered in other chapters of the Shop Manual may be fundamental to the topic discussed in the Classroom Manual.

Summary

Each chapter concludes with a summary of key points from the chapter. The key points are designed to help the reader review the chapter contents.

Review Questions

Short-answer essays, fill in the blanks, and multiple-choice questions are found at the end of each chapter. These questions are designed to accurately assess the students' competence in the objectives stated at the beginning of the chapter.

374   Chapter 12

SUMMARY

- Through the use of gauges and indicator lights, the driver is capable of monitoring several engine and vehicle operating systems.
- The gauges include speedometer, odometer, tachometer, oil pressure, charging indicator, fuel level, and coolant temperature.
- The most common types of electromechanical gauges are the d'Arsonval, three coil, two coil, and air core.
- Computer-driven quartz swing needle displays are similar in design to the air-core electromagnetic gauges used in conventional analog instrument panels.
- All gauges require the use of a variable resistance sending unit. Styles of sending units include thermistors, piezoresistive sensors, and mechanical variable resistors.

- Digital instrument clusters use digital and linear displays to notify the driver of monitored system conditions.
- The most common types of displays used on electronic instrument panels are: light-emitting diodes (LEDs), liquid crystal displays (LCDs), vacuum fluorescent displays (VFDs), and a phosphorescent screen that is the anode.
- A head-up display system displays visual images onto the inside of the windshield in the driver's field of vision.
- In the absence of gauges, important engine and vehicle functions are monitored by warning lamps. These circuits generally use an on/off switch–type sensor. The exception would be the use of voltage-controlled warning lights that use the principle of voltage drop.

REVIEW QUESTIONS

Short-Answer Essays

1. What are the most common types of electromagnetic gauges?
2. Describe the operation of the piezoresistive sensor.
3. What is a thermistor used for?
4. What is meant by *electromechanical*?
5. Describe the operation of the air-core gauge.
6. What is the basic difference between conventional analog and computer-driven analog instrument clusters?
7. Describe the operating principles of the digital speedometer.
8. Explain the operation of IC chip–type odometers.
9. Describe the operation of the electronic fuel gauge.
10. Describe the operation of quartz analog speedometers.

Fill in the Blanks

1. The purpose of the tachometer is to indicate _____.
2. A piezoresistive sensor is used to monitor _____ changes.

3. The most common style of fuel level sending unit is _____ variable resistor.
4. The brake warning light is activated by _____ pressure in the brake hydraulic system.
5. In a three-coil gauge, the _____ produces a magnetic field that bucks or opposes the low-reading coil. The _____ coil and the bucking coil are wound together, but in opposite directions. The _____ coil is positioned at a 90° angle to the low-reading and bucking coils.
6. A _____ circuit completes the warning light circuit to ground through the ignition switch when it is in the START position.
7. Digital instrument clusters use _____ and _____ displays to notify the driver of monitored system conditions.
8. _____ is a calculation using the final drive ratio and the tire circumference to obtain accurate vehicle speed signals.

SHOP MANUAL

To stress the importance of safe work habits, the Shop Manual dedicates one full chapter to safety. Other important features of this manual include the following:

Performance-Based Objectives

These objectives define the contents of the chapter and define what the student should have learned upon completion of the chapter. These objectives also correspond with the list of required tasks for ASE certification. *Each ASE task is addressed.*

Although this textbook is not designed to simply prepare someone for the certification exams, it is organized around the ASE task list. These tasks are defined generically when the procedure is commonly followed and specifically when the procedure is unique for specific vehicle models. Imported- and domestic-model automobiles and light trucks are included in the procedures.

Terms to Know List

Terms in this list are also defined in the Glossary at the end of the manual.

Special Tools List

Whenever a special tool is required to complete a task, it is listed in the margin next to the procedure.

Cautions and Warnings

Throughout the text, warnings are given to alert the reader to potentially hazardous materials or unsafe conditions. Cautions are given to advise the student of things that can go wrong if instructions are not followed or if an unacceptable part or tool is used.

Margin Notes

The most important terms to know are highlighted and defined in the margins. Common trade jargon also appears in the margins and gives some of the common terms used for components. This feature helps students understand and speak the language of the trade, especially when conversing with an experienced technician.

Basic Tools List

Each chapter begins with a list of the basic tools needed to perform the tasks included in the chapter.

CHAPTER 5
BATTERY DIAGNOSIS AND SERVICE

Upon completion and review of this chapter, you should be able to:

- Demonstrate all safety precautions and rules associated with servicing the battery.
- Perform a visual inspection of the battery, cables, and terminals.
- Correctly slow and fast charge a battery, in or out of the vehicle.
- Describe the differences between slow and fast charging and determine when either method should be used.
- Perform a battery terminal test and accurately interpret the results.
- Perform a battery leakage test and determine the needed corrections.
- Test a conventional battery's specific gravity.
- Perform an open circuit test and accurately interpret the results.

- Test the capacity of the battery to deliver both current and voltage and to accurately interpret the results.
- Perform a 3-minute charge test to determine if the battery is sulfated.
- Perform a conductance test of the battery and accurately interpret the results.
- Perform a battery drain test and accurately determine the causes of battery drains.
- Remove, clean, and reinstall the battery properly.
- Jump-start a vehicle by use of a booster battery and jumper cables.
- Determine the cause of HV battery system failures.
- Measure HV battery module voltages with a DMM.

Basic Tools
Basic mechanic's tool set
Service information

Terms To Know

Battery ECU
Battery leakage test
Battery terminal test
Capacity test
Charge
Charge rate

Conductance
Fast charging
Hydrometer
Jump assist
Open circuit voltage test
Parasitic drains

Refractometer
Slow charging
Stabilize
State of charge
Sulfation

INTRODUCT...

216   Chapter 5

enough to require hospital treatment. Keep sparks, flames, and lighted cigarettes away from the battery. Also, do not use the battery to lay tools on. They may short across the terminals and result in the battery exploding. Always wear eye protection and proper clothing when working near the battery. Also, most jewelry is an excellent conductor of electricity. Do not wear any jewelry when performing work on or near the battery. Do not remove the vent caps while charging. Do not connect or disconnect the charger leads while the charger is turned on.

CHARGING THE BATTERY

Classroom Manual
Chapter 5, page 118

To charge the battery means to pass an electric current through the battery in an opposite direction than during discharge. If the battery needs to be recharged, the safest method is to remove the battery from the vehicle. The battery can be charged in the vehicle, however. If the battery is to be charged in the vehicle, it is important to protect any vehicle computers by removing the negative battery cable.

Special Tools
Safety glasses
Battery charger
Voltmeter
Fender covers

When connecting the charger to the battery, make sure the charger is turned off. Connect the cable leads to the battery terminals, observing polarity. Attempting to charge the battery while the cables are reversed will result in battery damage. For this reason, many battery chargers have a warning system to alert the technician that the cables are connected in reverse polarity. Rotate the clamps slightly on the terminals to assure a good connection.

Depending on the requirements and amount of time available, the battery can be either slow or fast charged. Each method of charging has its advantages and disadvantages.

Caution
If the battery is to be removed from the vehicle, disconnect the negative battery cable first. Lift the battery out with a carrying tool (Figure 5-5).

WARNING Before charging a battery that has been in cold weather, check the electrolyte for ice crystals. A discharged battery will freeze at a higher temperature than a fully charged battery. Do not attempt to charge a frozen battery. Forcing current through a frozen battery may cause it to explode. Allow it to warm at room temperature for a few hours before charging.

Slow Charging

Slow charging means the charge rate is between 3 and 15 amps (A) for a long period of time. Slow charging the battery has two advantages: it is the only way to restore the battery to a fully charged state and it minimizes the chances of overcharging the battery. Slow

Figure 5-5 Always use a battery carrier to lift the battery.

... = 1.5 amperes

...ent flow through the branches together to get the total current flow:

...amperage = 3 + 2 + 1.5 = 6.5 amperes

Since this is a 12-volt system and total current is 6.5 amperes, total resistance is

R_T = 12 volts/6.5 amps = 1.85 Ω

This method can be mathematically expressed as follows:

$R_T = V_T / A_T$

Series-Parallel Circuits

The equivalent series load, or equivalent resistance, is the equivalent resistance of a parallel circuit plus the resistance in series and is equal to the equivalent resistance of a single load in series with the voltage source.

The series-parallel circuit has some loads that are in series with each other and some that are in parallel (Figure 2-26). To calculate the total resistance in this type of circuit, calculate the equivalent series loads of the parallel branches first. Next, calculate the series resistance and add it to the equivalent series load. For example, if the parallel portion of the circuit has two branches with 4 Ω resistance each and the series portion has

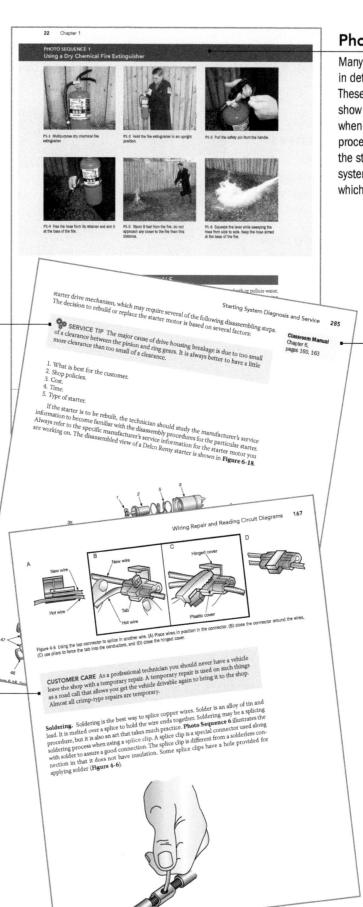

Photo Sequences

Many procedures are illustrated in detailed Photo Sequences. These detailed photographs show students what to expect when they perform particular procedures. They can also provide the student a familiarity with a system or type of equipment, which the school may not have.

Service Tips

Whenever a shortcut or special procedure is appropriate, it is described in the text. These tips generally describe common procedures used by experienced technicians.

Cross-References to the Classroom Manual

Reference to the appropriate page in the Classroom Manual is given whenever necessary. Although the chapters of the two manuals are synchronized, material covered in other chapters of the Classroom Manual may be fundamental to the topic discussed in the Shop Manual.

Customer Care

This feature highlights those little things a technician can do or say to enhance customer relations.

Case Studies

Beginning with Chapter 3, each chapter ends with a case study describing a particular vehicle problem and the logical steps a technician might use to solve the problem.

ASE-Style Review Questions

Each chapter contains ASE-style review questions that reflect the performance-based objectives listed at the beginning of the chapter. These questions can be used to review the chapter as well as to prepare for the ASE certification exam.

ASE Challenge Questions

Each technical chapter ends with five ASE challenge questions. These are not mere review questions; rather, they test students' ability to apply general knowledge to the contents of the chapter.

CASE STUDY

The vehicle owner complains that the brake lights do not light. He also says the dome light is not working. The technician verifies the problem and then checks the battery for good connections and tests the fusible links. All are in good condition.

A study of the wiring diagram indicates that the brake light and dome light circuits share the same fuse. It is also indicated that the ignition switch illumination light circuits share these two circuits. A check of the ignition switch illumination light shows that it is not operating either. The technician checks the fuse that is identified in the wiring diagram. It is blown. When a replacement fuse is installed, the dome and brake lights work properly for three tests, and then the fuse blows again.

Upon further testing of the shared circuits, an intermittent short to ground is located in the steering column in the ignition switch illumination circuit. The technician solders in a repair wire to replace the damaged section. After all repairs are completed, a final test indicates proper operation of all circuits.

ASE-STYLE REVIEW QUESTIONS

1. Splicing copper wire is being discussed.
   Technician A says it is acceptable to use solderless connections.
   Technician B says acid core solder should not be used on copper wires.
   Who is correct?
   A. A only
   B. B only
   C. Both A and B
   D. Neither A nor B

2. Use of wiring diagrams is being discussed.
   Technician A says a wiring diagram is used to help determine related circuits.
   Technician B says the wiring
   exact location

4. Repairs to a twisted/shielded wire are being discussed.
   Technician A says a twisted/shielded wire carries high current.
   Technician B says because a twisted/shielded wire carries low current, any repairs to the wire must not increase the resistance of the circuit.
   Who is correct?
   A. A only
   B. B only
   C. B   d B
          nor B
          sed.
          are
          both

426   Chapter 9

ASE CHALLENGE QUESTIONS

1. *Technician A* says when diagnosing intermittent faults, it is good practice to substitute control modules to see if the problem goes away.
   Technician B says a circuit performance fault indicates that the continuity of the circuit is suspect.
   Who is correct?
   A. A only
   B. B only
   C. Both A and B
   D. Neither A nor B

2. The scan tool displays 5 volts for the ambient
   ature sensor. This indicates:
          or return circuit

4. *Technician A* says the MAP sensor reading with the key on, engine off should equal barometric pressure.
   Technician B says when the engine is started, the MAP sensor signal voltage should increase.
   Who is correct?
   A. A only
   B. B only
   C. Both A and B
   D. Neither A nor B

5. A vehicle with four-wheel ABS has a problem with the right rear wheel locking during heavy braking.
   Technician A says this could be caused by a bad speed sensor mounted at the wheel.
   Technician B says the speed sensor mounted at
   ould cause this problem.

Name _____

SHOP SAFETY SURVEY

As a professional technician, safety should be one of your first concerns. This job sheet will increase your awareness of shop safety items. As you survey your shop and answer the following questions, you will learn how to evaluate the safety of any workplace.

Date _____

Safety   41

JOB SHEET
1

Procedure

Your instructor will review your work at each Instructor Response point.

1. Before you begin to evaluate

Task Completed

Job Sheets

Located at the end of each chapter, job sheets provide a format for students to perform procedures covered in the chapter. A reference to the ASE task addressed by the procedure is included on the Job Sheet.

Basic Electrical Troubleshooting and Service   161

DIAGNOSTIC CHART 3-6
PROBLEM AREA: Relays.
SYMPTOMS: Component fails to turn on.
POSSIBLE CAUSES: Faulty terminals to relay.
Open in relay coil.
Shorted relay coil.
Burned relay high-current contacts.
Internal open in relay high-current circuit.

DIAGNOSTIC CHART 3-7
PROBLEM AREA: Relay.
SYMPTOMS: Component fails to turn off.
POSSIBLE CAUSES: Short across the relay terminals.
Stuck relay high-current contacts.

DIAGNOSTIC CHART 3-8
PROBLEM AREA: Stepped resistors.
SYMPTOMS: Component fails to turn on.
POSSIBLE CAUSES: Faulty terminal connections to the stepped resistor.
Open in the input side of the resistor.
Open in the output side of the resistor.

DIAGNOSTIC CHART 3-9
PROBLEM AREA: Stepped resistors.
SYMPTOMS: Component fails to operate at certain speeds or brightness.
POSSIBLE CAUSES: Excessive resistance at the terminal connections to the stepped resistor.
Open in one or more of the resistor's circuit.
Short across one or more of the resistor's circuit.

DIAGNOSTIC CHART 3-10
PROBLEM AREA: Variable resistors.
SYMPTOMS: Component fails to turn on.
POSSIBLE CAUSES: Faulty terminal connections to the variable resistor.
Open in the input side of the resistor.
Open in the output side of the resistor.
Open in the return circuit of the resistor.
Short to ground in the output circuit.
Short to ground in the input circuit.

Diagnostic Charts

Some chapters include detailed diagnostic charts that list common problems and most probable causes. They also list a page reference in the Classroom Manual for better understanding of the system's operation and a page reference in the Shop Manual for details on the procedure necessary for correcting the problem.

SUPPLEMENTS

Instructor Resources

The *Today's Technician* series offers a robust set of instructor resources, available online at Cengage's Instructor Resource Center and on DVD. The following tools have been provided to meet any instructor's classroom preparation needs:

- An Instructor's Guide including lecture outlines, teaching tips, and complete answers to end-of-chapter questions.
- PowerPoint presentations with images and animations that coincide with each chapter's content coverage.
- Cengage Learning Testing Powered by Cognero® provides hundreds of test questions in a flexible, online system. You can choose to author, edit, and manage test bank content from multiple Cengage Learning solutions and deliver tests from your LMS, or you can simply download editable Word documents from the DVD or Instructor Resource Center.
- An Image Gallery includes photos and illustrations from the text.
- The Job Sheets from the Shop Manual are provided in Word format.
- End-of-Chapter Review Questions are provided in Word format, with a separate set of text rejoinders available for instructors' reference.
- To complete this powerful suite of planning tools, correlation guides are provided to the NATEF tasks and to the previous edition.

MINDTAP FOR AUTOMOTIVE ELECTRICITY & ELECTRONICS, 7E

NEW! The MindTap for *Automotive Electricity & Electronics, seventh edition,* features an integrated course offering a complete digital experience for the student and teacher. This MindTap is highly customizable and combines an enhanced ebook with videos, simulations, learning activities, quizzes, job sheets, and relevant DATO scenarios to help students analyze and apply what they are learning and help teachers measure skills and outcomes with ease.

- *A Guide:* Relevant activities combined with prescribed readings, featured multimedia, and quizzing to evaluate progress, will guide students from basic knowledge to analysis and application.
- *Personalized Teaching:* Teachers are able to control course content by removing, rearranging, or adding their own content to meet the needs of their specific program.
- *Promote Better Outcomes:* Through relevant and engaging content, assignments and activities, students are able to build the confidence they need to succeed. Likewise, teachers are able to view analytics and reports that provide a snapshot of class progress, time in course, engagement, and completion rates.

REVIEWERS

The author and publisher would like to extend special thanks to the following instructors for their contributions to this text:

Brett Baird
Salt Lake Community College
Riverton, UT

Timothy Belt
University of Northwestern Ohio
Lima, OH

Brian Brownfield
Kirkwood Community College
Cedar Rapids, IA

Michael Cleveland
Houston Community College
Houston, TX

Jason Daniels
University of Northwestern Ohio
Lima, OH

Kevin Fletcher
Asheville-Buncombe Technical
Community College
Asheville, NC

Jose Gonzalez
Automotive Training Center
Exton, PA

Jack D. Larmor
Baker College
Flint, MI

Gary Looft
Western Colorado Community College
Grand Junction, CO

Stanley D. Martineau
Utah State University
Price, UT

William McGrath
Moraine Valley College
Warrenville, IL

Tim Mulready
University of Northwestern Ohio
Lima, OH

Gary Norden
Elgin Community College
Elk Grove Village, IL

Michael Stiles
Indian River State College
Pierce, FL

David Tapley
University of Northwestern Ohio
Lima, OH

Omar Trinidad
S. Illinois University Carbondale
Carbondale, IL

Cardell Webster
Arapahoe Community College
Littleton, CO

David Young
University of Northwestern Ohio
Lima, OH

In addition, the author would like to extend a special thank-you and acknowledgment to Jerome "Doc" Viola and Arapahoe Community College in Littleton, Colorado, along with Phil Stiebler from Pro Chrysler Jeep Dodge Ram located in Denver, Colorado, for their valuable assistance with this textbook.

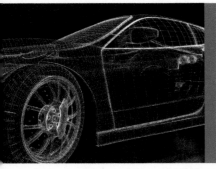

CHAPTER 1
SAFETY

Upon completion and review of this chapter, you should be able to:

- Explain how safety is a part of professionalism.
- List and describe personal safety responsibilities.
- List the different types of eye protection devices and explain the proper application of each.
- Lift heavy objects properly.
- Inspect power tools before use.
- Raise a vehicle using a floor jack and safety stands.
- Raise a vehicle using a hoist.
- Demonstrate the ability to properly run the engine in the shop.
- Classify fires and fire extinguishers.
- Locate, identify, and inspect fire extinguishers in the shop.
- Explain the proper use of the fire extinguisher.
- Define hazardous materials.

- Explain the right-to-know law or workplace hazardous materials information systems (WHMIS).
- Describe the responsibilities of the employer and the employee concerning hazardous materials.
- Determine what constitutes hazardous waste and how to properly dispose of it.
- Describe the basic safety rules of servicing electrical systems.
- Work around batteries safely.
- Explain the safety precautions associated with charging and starting systems.
- List the safety precautions associated with servicing the air bag system.
- Explain the safety precautions that are necessary when servicing the antilock brake system.
- Explain the safety precautions necessary when servicing hybrid electric vehicles (HEVs).

Terms To Know

Air bag module
Air bag system
Antilock brake systems (ABS)
Asbestos
BAT
Carbon monoxide (CO)
Caustic
Conductors
Face shields
Fire blanket
Fire extinguishers
Flammable

Floor jacks
Hand tools
Hazard Communication Standard
Hazardous materials
Hazardous waste
High-efficiency particulate-arresting (HEPA)
High-voltage service plug
Hoists
Material safety data sheets (MSDS)
Occupational safety glasses

One-hand rule
Pneumatic tools
Power tools
Resource Conservation and Recovery Act (RCRA)
Right-to-know laws
Safety goggles
Safety stands
Vehicle lift point
Volatile
Workplace hazardous materials information systems (WHMIS)

INTRODUCTION

Being a professional technician is more than being knowledgeable about automotive systems; it is also an attitude. Being a professional technician includes having an understanding of all the hazards that may exist in the workplace. One of the most obvious traits of a professional is the ability to work productively and safely. This is where knowledge becomes very important. You need it to be productive and you need it to ensure your own safety and the safety of others. This chapter discusses the safety concerns associated with working in an automotive repair shop and the safety concerns associated with the vehicle's fuel and emission systems. In addition to basic shop safety, working on the vehicle's fuel and emission systems presents many special concerns.

Safety is everyone's concern. However, never assume the person working next to you is as conscientious as you are. You must be aware of what is going on around you at all times. As a professional technician, you must perform your work in a manner that protects not only you, but others in the workplace as well.

PERSONAL SAFETY

Personal safety encompasses all aspects of preventing injury, including awareness, attitudes, and dress. All three of these are manifested through neat work habits. Cleaning up spills, keeping tools clean, and organizing the tools and materials in the shop help prevent accidents. Rushing to complete a job may result in a lack of consideration for personal safety and may ultimately cause an accident. Taking time to be neat and safe is rewarded by fewer accidents, higher customer satisfaction, and better pay.

Dress and Appearance

Nothing displays professional pride and a positive attitude more than the way you dress (**Figure 1-1**). Customers demand a professional atmosphere in the service shop. Your appearance instills customer confidence, as well as expresses your attitude toward safety and your chosen profession. Wearing proper and neat clothing can prevent injuries.

Loose-fitting clothing, or clothing that hangs out freely, can cause serious injury. Long-sleeve shirts should have their cuffs buttoned or rolled up tightly. Shirttails should

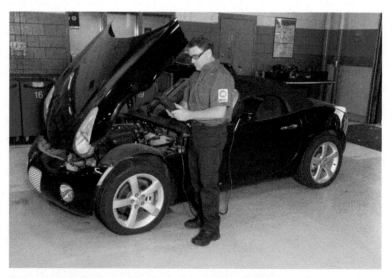

Figure 1-1 Wearing clean and properly fitting clothes along with proper safety equipment is an indication of how serious you are about safety.

be tucked in at all times. Some job positions within an automotive repair facility may require the employee to wear a necktie. If a necktie is worn in the shop area, it should be tucked inside your shirt. Clip-on ties are recommended if you must wear a tie.

Long hair is a serious safety concern. Very serious injury can result if hair gets caught in rotating machinery, fan belts, or fans. If your hair is long enough to touch the bottom of your shirt collar, it should be tied back and tucked under a hat.

Jewelry has no place in the automotive shop. Rings, watches, bracelets, necklaces, earrings, and so forth can cause serious injury. The gold, silver, and other metals used in jewelry are excellent **conductors** of electricity. Your body is also a good conductor. When electrical current flows through a conductor, it generates heat. The heat can be great enough to cause severe burns. Jewelry can also get caught in moving parts, causing serious cuts. Necklaces can cause serious injury or even death if they get caught in moving equipment.

You should wear shoes or boots that will protect your feet in the event something falls, or you stumble into something. It is a good idea to wear safety shoes or boots with steel toes and shanks. Most safety shoes also have slip-resistant soles. Tennis and jogging shoes provide little protection and are not satisfactory footwear in the automotive shop. Never wear any type of open toe or open top shoes, such as sandals.

Smoking, Alcohol, and Drugs in the Shop

Due to the potential hazards, never smoke when working in the shop. A spark from a cigarette or lighter may ignite flammable materials in the workplace. If the shop has designated smoking areas, smoke only in those areas. As a courtesy to your customers, do not smoke in their vehicles. Nonsmokers may not appreciate cigarette odor in their vehicles.

The use of drugs or alcohol must be avoided while working in the shop. Even a small amount of drugs or alcohol affects reaction time. In an emergency, slow reaction time may cause personal injury. If a heavy object falls off the workbench and your reaction time is slowed by drugs or alcohol, you may not get out of the way quickly enough to prevent injury. Also, you are a hazard to your coworkers if you are not performing at your best. Penalties for failure to adhere to your employer's policies concerning smoking, alcohol, and drugs may include termination of employment and possible legal actions.

Eye Protection

The importance of wearing proper eye protection cannot be overemphasized. Every working day there are more than 1,000 eye injuries in the United States. Many of these injuries result in blindness. Sadly, almost all of these are preventable. The safest and surest way to protect your eyes is to wear proper eye protection any time you enter the shop. At a minimum, wear eye protection when grinding, using power tools, hammering, cutting, chiseling, or performing service under the vehicle. In addition, wear eye protection when doing any work that can cause sparks, dirt, or rust to enter your eyes, and when you are working around chemicals. Remember, just because you are not doing the work yourself does not mean you cannot suffer an eye injury. Many eye injuries are caused by a coworker. Wear eye protection any time you are near an eye hazard.

The key to protecting your eyes is the use of *proper* eye protection. All eye protection must meet the newest requirements of the American National Standards Institute (ANSI). Currently this standard is identified as ANSI Z87.1. Regular prescription glasses do not provide adequate protection. Regular glasses are designed to impact standards that are far below those that are required in the workplace. The lens may stop a flying object, but the frame may allow the lens to pop out and hit your face, causing injury. In addition, regular glasses do not provide side protection. Prescription glasses that meet ANSI Z87.1 are available.

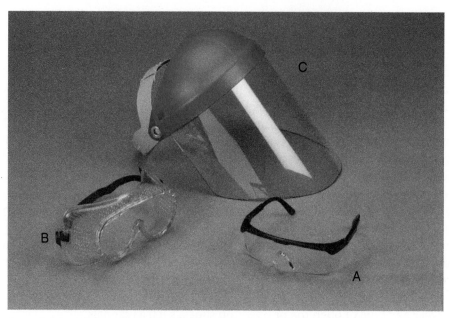

Figure 1-2 Examples of some of the different types of eye protection: (A) occupational safety glasses with side shields; (B) safety goggles that may be worn over prescription glasses; and (C) a face shield that is worn over safety glasses or goggles and protects the face.

There are many types of eye protection (**Figure 1-2**). One of the best ways to protect your eyes is to wear **occupational safety glasses**. These glasses are light and comfortable. They are constructed of tempered glass or safety plastic lens and have frames that prevent the lens from being pushed out upon impact. They have side shields to prevent the entry of objects from the side. Occupational safety glasses are available in prescription lens so they can be worn instead of regular corrective lens glasses.

Safety goggles fit snugly around the area of your eyes to prevent the entry of objects and to provide protection from liquid splashes. The force of impact on the lens is distributed throughout the entire area where the safety goggles are in contact with your face and forehead. Safety goggles are designed to fit over regular glasses.

Face shields are clear plastic shields that protect the entire face. They are used when there is potential for sparks, flying objects, or splashed liquids, which can cause neck, facial, and eye injuries. The plastic is not as strong or impact resistant as occupational safety glasses or safety goggles. If there is a danger of high-impact objects hitting the face shield, it is a good practice to wear safety glasses under the face shield.

Safety glasses provide little or no protection against chemicals. When working with chemicals, such as battery acid, refrigerants, and cleaning solutions, safety goggles should be worn. Full-face shields are not intended to provide primary protection for your eyes. They are designed to provide primary protection for your face and neck and should be worn in addition to eye protection.

Before removing your eye protection, close your eyes and lean forward. Pieces of metal, dirt, or other foreign material may have accumulated on the outside. These could fall into your eyes when you remove your glasses or shield.

Eyewash Fountains

Eye injuries may occur in various ways in an automotive shop. The following are some common types of eye injuries:

1. Thermal burns from excessive heat.
2. Irradiation burns from excessive light, such as from an arc welder.
3. Chemical burns from strong liquids, such as gasoline or battery electrolyte.

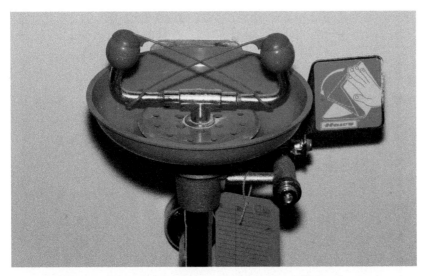

Figure 1-3 An eyewash fountain is used to remove chemicals and dirt from the eyes.

4. Foreign material in the eye.
5. Penetration of the eye by a sharp object.
6. A blow from a blunt object.

Wearing safety glasses and observing shop safety rules will prevent most eye accidents. If a chemical gets into your eyes, it must be washed out immediately to prevent a chemical burn. An eyewash fountain is the most effective way to wash the eyes, and every shop should be equipped with some eyewash facility (**Figure 1-3**). Be sure that you know the location of the eyewash fountain in the shop.

Hearing Protection

When performing any work or machining operation that will generate loud noise, wear some form of hearing protection (**Figure 1-4**). Ear plugs work well for providing up to 30 dB reduction in noise levels. Another option is to use specially designed hearing protection earmuffs.

Figure 1-4 Ear plugs or earmuffs provide the necessary hearing protection.

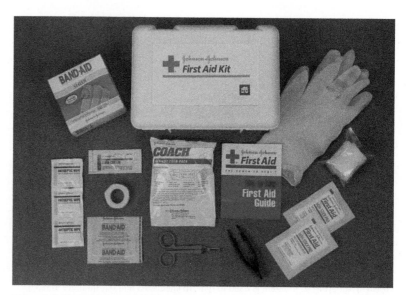

Figure 1-5 First-aid kit and typical contents.

First-Aid Kits

First-aid kits should be clearly identified and conveniently located (**Figure 1-5**). These kits contain such items as bandages and ointments required for minor cuts. All shop personnel must be familiar with the location of first-aid kits. At least one of the shop personnel should have basic first-aid training, and this person should oversee administering first aid and keeping the first-aid kits filled.

Hand Protection

Good hand protection is often overlooked (**Figure 1-6**). A scrape, cut, or burn can seriously impair your ability to work for many days. A well-fitting pair of heavy work gloves should be worn while grinding, welding, or when handling high-temperature components. Special rubber gloves are recommended for handling **caustic** chemicals. Caustic chemicals can destroy or eat through something and are considered extremely corrosive.

Many technicians wear latex, vinyl, or nitrile gloves to help protect their hands and to keep them clean. These are similar to the type of gloves worn by doctors and dentists during examinations. Latex gloves are inexpensive and provide good hand protection; however, some people are allergic to latex. If you wear latex gloves and develop a rash or redness on your hands, discontinue use. Vinyl gloves are also available and provide good resistance to tears and many nonaggressive liquids. Also, vinyl gloves are latex-free, so

Figure 1-6 Different types of gloves that are typically worn in the automotive service shop.

those who are allergic to latex can wear them. At a higher cost, nitrile gloves are latex-free synthetic rubber gloves that are superior to latex or vinyl in puncture resistance. In addition, nitrile gloves resist a wide range of chemicals that are harmful to either latex or vinyl.

Latex, vinyl, or nitrile gloves should be worn if you have an open cut or other injury on your hand, to prevent infection and the spread of diseases. In addition, these gloves should be worn if you are required to render first aid or medical assistance to someone who is injured. Because of the serious nature of blood-borne pathogens (disease- and infection-causing microorganisms carried by blood and other potentially infectious materials), it is important that you take every precaution to protect yourself regardless of the perceived status of the individual you are assisting. In other words, whether or not you think the blood/body fluid is infected with blood-borne pathogens, you treat it as if it is.

Rotating Belts and Pulleys

Many times the technician must work around rotating parts such as fans, generators, power steering pumps, air pumps, water pumps, and air conditioner compressors (**Figure 1-7**). Other rotating equipment or components of concern include tire changers, spin balancers, drills, bench grinders, and drive shafts. Always think before acting. Be aware of where you are placing your hands and fingers at all times. Do not place rags, tools, or test equipment near moving parts. In addition, make sure you are not wearing any loose clothing or jewelry that can get caught.

Electric Cooling Fans

Be very cautious around electric cooling fans. Some of these fans will operate even if the ignition switch is turned off. They are controlled by a temperature-sensing unit in the engine block or radiator and may turn on any time the coolant temperature reaches

© iStockphoto/sdbower

Figure 1-7 The accessory drive system on an engine can present serious safety concerns.

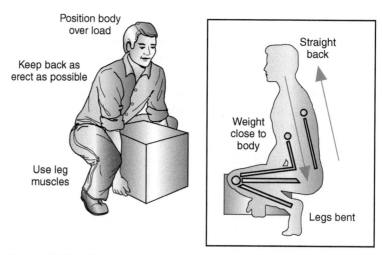

Figure 1-8 When lifting a heavy object, keep your back straight and lift with your legs.

a certain temperature. Before working on or around an electric cooling fan, you should become familiar with its operation, and, if necessary, you should disconnect the electrical connector to the fan motor or the negative battery cable.

Lifting

Back injuries are one of the most crippling injuries in the industry, yet most of them are preventable. Most occupational back injuries are caused by improper lifting practices. These injuries can be avoided by following a few simple lifting guidelines:

1. Do not lift a heavy object by yourself. Seek help from someone else.
2. Do not lift more than you can handle. If the object is too heavy, use proper equipment to lift it.
3. Do not attempt to lift an object if there is not a good way to hold onto it. Study the object to determine the best balance and grip points.
4. Do not lift with your back. Your legs have some of the strongest muscles in your body. Use them.
5. Place your body close to the object. Keep your back and elbows straight (**Figure 1-8**).
6. Make sure you have a good grip on the object. Do not attempt to readjust the load once you have lifted it. If you are not comfortable with your balance and grip, lower the object and reposition yourself.
7. When lifting, keep the object as close to your body as possible. Keep your back straight and lift with your legs.
8. While carrying the object, do not twist your body to change directions. Use your feet to turn your whole body in the new direction.
9. To set the load down, keep the object close to your body. Bend at the knees and keep your back straight. Do not bend forward or twist.
10. If you need to place the object onto a shelf or bench top, place an edge of the object on the surface and slide it into place. Do not lean forward.

TOOL AND EQUIPMENT SAFETY

Technicians would not be able to do their job without tools and equipment. Most injuries caused by tools and equipment are the result of improper use, improper maintenance, and/or carelessness.

Figure 1-9 Assortment of hand tools.

Hand Tools

Hand tools use only the force generated from the body to operate (**Figure 1-9**). They multiply the force received through leverage to accomplish the work. Here are some very simple steps that you can take to help assure safe hand tool use:

1. Do not use tools that are worn out or broken.
2. Do not use a tool to do something that it was not designed to do. Use the proper tool for the job.
3. Keep your tools clean and in good condition.
4. Point sharp edges of tools away from you.
5. Do not hold small components in your hands while using tools such as screwdrivers. The tool may slip and cause injury to your hand.
6. Examine your work area for things that can cause injury if a tool slips or a fastener breaks loose quickly. Readjust yourself or the tool to avoid injuries.
7. Do not put sharp tools in your pockets.

Power Tool Safety

 **WARNING** **Always wear safety glasses when operating a power tool.**

Power tools use forces other than those generated by the body (**Figure 1-10**). They can use compressed air, electricity, or hydraulic pressure to generate and multiply force. Many times a technician will be required to use power tools when performing electrical service.

Figure 1-10 Examples of power tools.

Pneumatic tools are
often called *air tools*.

Drills and hole saws will be used to install new accessories onto the vehicle or to drill holes for wiring to pass through. Grinders, drill presses, and hydraulic presses may be used to help fabricate or modify components. **Pneumatic tools** are powered by compressed air and are used to remove or fasten components. All of these tools can cause injury if not used properly. Use the following guidelines when working with power tools:

1. Ask your instructor if you are not sure of the correct operation of a tool.
2. Always wear proper eye protection when using power tools.
3. Check that all safety guards and safety equipment are installed on the tool.
4. Before using an electrical tool, check the condition of the plug and cord. The plug should be a three-pronged plug. Never cut off the grounding prong. Do not use the tool if the wires are frayed or broken. Plug the tool only into a grounded receptacle.
5. Before using an air tool, check the condition of the air hose. Do not use the tool if the hose shows signs of weakness such as bulges or fraying. Also, the tool should be properly oiled.
6. Before using a hydraulic tool, check the condition of all hoses and gauges. Do not use the tool if any of these are defective.
7. Make sure other people are not in the area when you turn the tool on.
8. Do not leave the area with the tool still running. Stay with the tool until it stops. Then disconnect it.
9. Make all adjustments to the tool before turning it on.
10. If the tool is defective or does not pass your safety inspection, put a sign on it and report it to your supervisor or instructor.

Inspecting Power Tools Improper use of power tools is the cause of many accidents each year. Many of these accidents could have been avoided if the tool was checked before it was used.

Use this checklist for the electric power tool(s) that you are inspecting. Do not plug the electrical cord into the socket until the inspection is completed. Consider the following:

1. Name of tool.
2. Is the tool clean?
3. Are safety guards in place?
4. Does the tool appear to be in good condition?
5. Is there a ground terminal on the plug?
6. Is the electrical cord in good condition?

If you answered *no* to any of the previous questions, tag the tool and report the defect to your instructor.

Use this checklist for the pneumatic tool(s) that you are inspecting. Do not connect the tool to the air outlet until the inspection is completed. Consider the following:

1. Name of tool.
2. Is the tool clean?
3. Are safety guards in place?
4. Does the tool appear to be in good condition?
5. Is the air hose in good condition?

If you answered *no* to any of the previous questions, tag the tool and report the defect to your instructor.

Compressed Air Safety

Compressed air is used in an automotive shop to do many things. However, cleaning off your clothes is not one of them. Dirt and other objects blown off your clothes can cause serious injury to you and others. There may be dirt in the nozzle or hose that will be

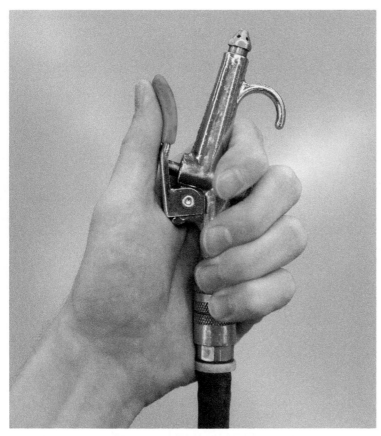

Figure 1-11 Only approved safety air nozzles should be used.

ejected at a high rate of speed and can be forced into someone's skin or eyes. In addition, most shops have compressed air systems equipped with automatic oilers. The pressure can push the oil and air bubbles through your skin and into your blood.

Use only approved safety nozzles when using compressed air to dry parts that have been cleaned (**Figure 1-11**). Safety nozzles have a relief passage that prevents high pressures from being expelled directly out the front. It is best not to use compressed air to dry parts; however, there are instances where air must be used to dry small passages. In these cases, use only air pressure that has been regulated to about 25 psi.

Check the air hoses for signs of wear. Do not use them if they are bulging and frayed or if the couplers are damaged.

LIFTING THE VEHICLE

Many service procedures require lifting the vehicle. There are two basic methods of lifting the vehicle from the floor: floor jack and safety stands, and hoists. Each requires the technician to follow certain safety rules to prevent injury and vehicle damage.

Floor Jack and Safety Stand Use

Floor jacks are used to lift a vehicle a short distance off the floor or when only a portion of the vehicle needs to be raised (**Figure 1-12**). Before using a floor jack, check it for signs of hydraulic fluid leaks and for damage that would compromise its safe use. Before lifting the vehicle, place wheel blocks in front of and behind one of the tires that will remain on the ground.

Many jack manufacturers and service information provide illustrations for the proper **vehicle lift points** on a vehicle (**Figure 1-13**). Vehicle lift points are the areas

Special Tools
Floor jack
Safety stands
Wheel blocks
Service information

Figure 1-12 Floor jacks are used to raise a vehicle a short distance off the floor.

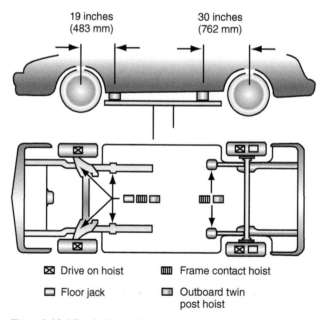

Figure 1-13 Lift point illustrations are usually provided in the service information.

the manufacturer recommends for safe vehicle lifting. These areas are structurally strong enough to sustain the stress of lifting. If this information is not available, always place the floor jack on major strength parts. These areas include the frame, cross member, and differential. If you are in doubt about the proper lift point, ask your instructor. Never lift on sheet metal or plastic parts.

The floor jack is to be used only to lift the vehicle off the floor. It is not intended to support the vehicle while someone is under it. Use **safety stands** to support the vehicle (**Figure 1-14**). Use one safety stand for each quarter of the vehicle that is lifted

Safety stands are also referred to as floor stands or jack stands.

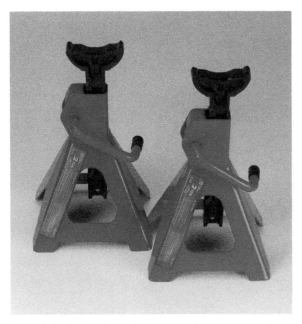

Figure 1-14 Safety stands should be used to support the vehicle after it has been lifted by a floor jack.

(**Figure 1-15**). Place the safety stand under the frame or a major support component of the vehicle. When the vehicle is lowered onto the stands, make sure that they do not tilt.

Before using the floor jack, make sure that it has a sufficient rating to lift and sustain the weight of the vehicle. Next, inspect it for proper lubrication and hydraulic fluid level. Check the operation of the jack while looking for signs of hydraulic fluid leaks. If the jack does not pass any one of these inspections, tag it and notify your instructor immediately.

To lift the entire vehicle, begin by placing the vehicle's transmission into PARK. Place it in first gear if the vehicle has a manual transmission. Set the parking brake and place wheel blocks around the rear wheels (**Figure 1-16**). Position the floor jack under the front of the vehicle at a location strong enough to support the weight. The jack

Figure 1-15 Support the vehicle by safety stands located at each corner that is lifted. Do not rely on the floor jack to support the vehicle.

Figure 1-16 Before lifting the vehicle with floor jacks, block the wheels to prevent them from rolling.

should be centered between the front tires and positioned so that the lift will be straight up and down (**Figure 1-17**).

⚠ WARNING **If you are lifting only one wheel of the vehicle, be careful not to lift it so high that it can slip off the jack saddle.**

Operate the jack until the jack saddle contacts the vehicle lift point. Check for good contact. If things look good, lift the front of the vehicle a couple of inches off the floor. Recheck the position of the jack. Continue to check the jack position throughout the lifting procedure. If the vehicle or jack begins to lean, lower the jack and reset it. Lift the vehicle to the required height. Do not lift higher than is necessary.

Do not get under a vehicle that is supported only by a floor jack. Place safety stands under the vehicle in locations that will support the weight. Use two safety stands to

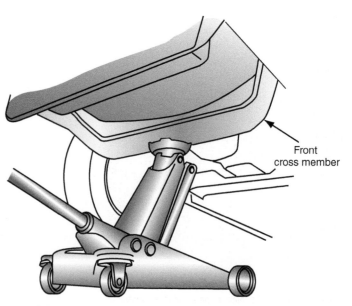

Front cross member

Figure 1-17 Example of correct lifting location for raising the front of the vehicle.

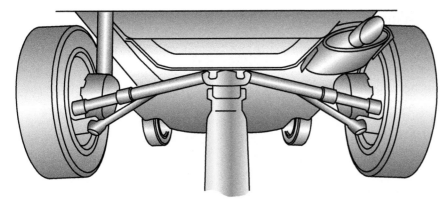

Figure 1-18 Example of proper lift procedure for raising the rear of the vehicle.

support the front of the vehicle. Make sure that the safety stands are located where they will not lean or slip. Slowly lower the vehicle onto the stands.

Place the floor jack under the rear of the vehicle (**Figure 1-18**). Follow the same procedure to raise the rear of the vehicle. Use two safety stands to support the rear of the vehicle.

When the vehicle is properly lifted and supported by safety stands, it is safe to work under the vehicle.

Use the same lift points to lower the vehicle. Once one end of the vehicle is on the floor, place wheel blocks around those wheels. Then lower the other end.

Hoist Safety

Hoists are used when the entire vehicle needs to be raised, usually high enough for the technician to stand underneath the vehicle (**Figure 1-19**). When a vehicle is placed on the hoist, it must be centered. The balance of the vehicle must be taken into consideration, as well as the effects on balance if a heavy component is removed from the vehicle.

Locate the correct lift points in the service information for the vehicle you are working on. Other sources of lift point location information include the Automotive Lift Institute and lift manufacturers. Pay close attention to any warnings or special considerations listed there.

Center the vehicle over the hoist; keep in mind the vehicle's center of gravity and balance point. Locate the hoist pads under the lift points (**Figure 1-20**). Adjust the pads so that the vehicle will be lifted level. Vehicles that have had their suspensions modified to

Special Tools
Frame contact hoist
Service information

Figure 1-19 One type of hoist that may be used in the automotive shop. The hoist is used to raise the entire vehicle.

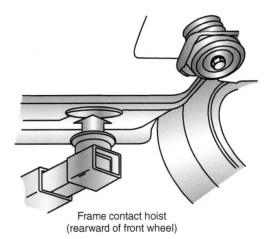

Frame contact hoist
(rearward of front wheel)

Figure 1-20 Locate the hoist pad at the correct lift point for the vehicle you are lifting.

raise or lower the vehicle's height may require special lifting considerations. In addition, vehicle accessories (such as running boards) may present obstacles to safe lifting of the vehicle. In these instances, use of special lift arm adapters may be required. If the vehicle is not level on the pads, or the pads are not in the proper position, lower the vehicle and readjust as needed. Never get under a vehicle that is not sitting properly on the hoist.

Lift the vehicle a few inches off the floor. Confirm good pad contact by shaking the vehicle while observing for signs of any movement. If the vehicle is not secure on the hoist, or unusual noises are heard while lifting, lower it to the floor and reset the pads.

Once the vehicle is at the desired height, lock the hoist (**Figure 1-21**). Do not get under the vehicle until the hoist locks have been set.

To lower the vehicle, release the locks and put the control valve into the lower position (**Figure 1-22**). Once it is returned to the floor, push the contact pads out of the path of the tires.

Figure 1-21 Make sure the hoist's locks are properly engaged before continuing to do any other tasks on or around the vehicle.

Figure 1-22 Release the hoist locks and lower the vehicle.

Figure 1-23 Testing the ventilation system before using it.

Figure 1-24 Connect the ventilation system hose to the vehicle's exhaust before starting the engine. Make sure the hose is secure and there are no kinks in the hose.

RUNNING THE VEHICLE WHILE IN THE SHOP

Many times it will be necessary to run the engine while the vehicle is in the shop. This presents the possibility of **carbon monoxide (CO)** poisoning if it is not done safely. Carbon monoxide is an odorless, colorless, and toxic gas produced as a result of the combustion process. All shops should be equipped with a ventilation system that will remove the vehicle exhaust. If the shop is not equipped with a ventilation system, a hose must be routed from the vehicle's exhaust to outdoors.

⚠ WARNING **Some of the early warning signs of carbon monoxide poisoning are headaches, dizziness, and blurred vision. If you experience any of these while in the shop, notify your instructor immediately and get some fresh air. If symptoms persist, seek medical attention.**

Check the operation of the shop's ventilation system before connecting to the vehicle. Turn on the ventilation motor and place your hand over the hose (**Figure 1-23**). You should feel a strong, consistent vacuum. A weak vacuum indicates a restriction or a leak in the ventilation system. If you encounter a problem, notify your instructor.

Connect the ventilation hose to the vehicle's exhaust (**Figure 1-24**). Place wheel blocks around a tire and place the transmission in PARK or NEUTRAL. Set the parking brake. The engine is now ready to be started.

Be aware of the warning signs of carbon monoxide poisoning. Report any symptoms to your instructor immediately.

⚠ WARNING **Be careful when working around electric engine cooling fans. These fans are controlled by a thermostat and can come on without warning, even when the engine is not running. Whenever you must work around these fans, disconnect the electrical connector to the fan motor before reaching into the area around the fan. Make sure you reconnect the connector before you return the car to the customer.**

FIRE HAZARDS AND PREVENTION

Fires are classified by the types of materials that are involved (**Table 1.1**). Technicians should be able to locate the correct fire extinguisher to control all the types of fires they are likely to experience (**Figure 1-25**). Technicians must also be able to fight a fire in an emergency.

Labels on the fire extinguisher will indicate the types of fires that it will put out (**Figure 1-26**). Become familiar with the use of a fire extinguisher.

TABLE 1.1   A Guide to Fire Classification and Fire Extinguisher Types

	Class of Fire	Typical Fuel Involved	Type of Extinguisher
Class A Fires (green)	For Ordinary Combustibles Put out a class A fire by lowering its temperature or by coating the burning combustibles.	Wood Paper Cloth Rubber Plastics Rubbish Upholstery	Multipurpose dry chemical
Class B Fires (red)	For Flammable Liquids Put out a class B fire by smothering it. Use an extinguisher that gives a blanketing, flame-interrupting effect; cover whole flaming liquid surface.	Gasoline Oil Grease Paint Lighter fluid	Carbon dioxide Halogenated agent Standard dry chemical Purple K dry chemical Multipurpose dry chemical
Class C Fires (blue)	For Electrical Equipment Put out a class C fire by shutting off power as quickly as possible and by always using a nonconducting extinguishing agent to prevent electric shock.	Motors Appliances Wiring Fuse boxes Switchboards	Carbon dioxide Halogenated agent Standard dry chemical Purple K dry chemical Multipurpose dry chemical
Class D Fires (yellow)	For Combustible Metals Put out a class D fire of metal chips, turnings, or shavings by smothering or coating with a specially designed extinguishing agent.	Aluminum Magnesium Potassium Sodium Titanium Zirconium	Dry power extinguishers and agents only

Figure 1-25 Know the location and types of fire extinguishers that are available in the shop.

Figure 1-26 Be sure to know what type of fires the extinguisher will put out.

Gasoline

Gasoline is so commonly found in automotive repair shops that its dangers are often forgotten. A slight spark or an increase in heat can cause a fire or explosion. Gasoline is a very explosive liquid and is very powerful. One exploding gallon of gasoline has a force equal to 14 sticks of dynamite. The expanding vapors from gasoline are extremely dangerous, and these vapors are present even in cold temperatures. Gasoline vapors are heavier than air; therefore, when an open container of gasoline is sitting about, the vapors spill out

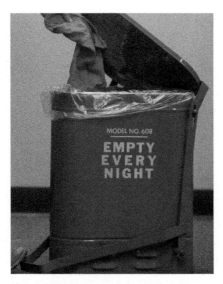

Figure 1-27 Dirty rags and towels must be stored in approved containers.

over the sides of the container onto the floor. These fumes are more **flammable** (support combustion) than liquid gasoline and can easily explode.

Never smoke around gasoline since even the droppings of hot ashes can ignite the gasoline. If an engine has a gasoline leak or you have caused a leak by disconnecting a fuel line, stop the leak and clean up the spilled gasoline immediately. While stopping the leak, be extra careful not to cause sparks. If any rags are used in the cleanup of the gasoline, they must be placed in an approved container (**Figure 1-27**). Due to extreme fire hazards, it is important to immediately clean up any gasoline spilled on the floor. Also, many of the compounds in petroleum are toxic, especially if they are in high concentrations.

The chemicals in petroleum that do not evaporate quickly are biodegradable. Optimum degradation occurs if the gasoline is diluted and there is enough air, water, and nutrients for the microbes to eat up the chemicals. These properties of gasoline are an advantage in the cleanup and disposal of small spills. Spreading absorbent material such as kitty litter, sand, ground corncobs, straw, sawdust, woodchips, peat, synthetic absorbent pads, or even dirt can stop the flow and soak up the gasoline. Keep in mind that the absorbent does not make gasoline nonflammable.

Brooms can be used to sweep up the absorbent material and put it into buckets, garbage cans, or barrels. Remember to control ignition sources. Be aware of local laws concerning gasoline spills. Some states or municipalities require notification of any gasoline spill larger than 5 gallons.

Gasoline should always be stored in approved containers (**Figure 1-28**) and never in glass containers. If the glass container is knocked over or dropped, a terrible explosion can occur. Approved gasoline storage cans have a flash-arresting screen at the outlet. These screens prevent external ignition sources from igniting the gasoline within the can while the gasoline is being poured.

Follow these safety precautions regarding gasoline containers:

1. Always use approved gasoline containers that are painted red for proper identification.
2. Do not fill gasoline containers completely. Always leave the level of gasoline at least 1 inch (25 mm) from the top of the container. This action allows expansion of the gasoline at higher temperatures. If gasoline containers are completely full, the gasoline will expand when the temperature increases. This expansion forces gasoline from the can and creates a dangerous spill.

Figure 1-28 Gasoline must be stored in approved containers. Store any type of combustible materials in an approved safety cabinet.

3. If gasoline containers must be stored, place them in a well-ventilated area such as a storage shed. Filled gasoline containers must be stored within an approved safety cabinet.
4. When a gasoline container must be transported, be sure it is secured against upsets. Do not transport or fill gasoline containers on plastic truck bed liners. Static electricity can be generated and ignite the vapors.
5. Do not store a partially filled gasoline container for long periods of time because it may give off vapors and produce a potential danger.
6. Never leave gasoline containers open except while filling or pouring gasoline from the container.
7. Do not prime an engine with gasoline while cranking the engine.
8. Never use gasoline as a cleaning agent.

Diesel Fuel

A **volatile** substance easily vaporizes or explodes. Although diesel fuel is not as volatile as gasoline, it should still be stored and handled in the same way as gasoline. It is also not as refined as gasoline and tends to be a "dirty" fuel. It normally contains impurities, including active microscopic organisms that can be highly infectious. If diesel fuel happens to enter an open cut or sore, thoroughly wash it immediately. If it gets into your eyes, flush them immediately and seek medical help.

Solvents

Cleaning solvents are not as volatile as gasoline, but they are still flammable. They should be treated and stored in the same way as gasoline. Whenever using solvents, wear eye and hand protection.

Rags

Oily and greasy rags can also cause fires. Used rags should be stored in an approved container and never thrown out with normal trash. Like gasoline, oil is a hydrocarbon and can ignite with or without a spark or flame.

Fire Extinguisher Use

Fire extinguishers are portable apparatuses containing chemicals, water, foam, or special gas that can be discharged to extinguish a small fire. Tour the shop area and become familiar with the location of the fire extinguishers. Use a report sheet to record the locations.

Also indicate the type of each extinguisher and what kinds of fires it will extinguish. Check the gauge and record the state of charge for each extinguisher.

⚠ WARNING **The following is not intended to be used as a lab exercise unless expressly directed by your instructor. Use the photo sequence as a guide to become familiar with fire extinguisher use. You must be willing to fight a fire in the shop, if the occasion arises.**

⚠ WARNING **Do not risk your life fighting a fire. If it is evident that the fire is out of control, get out. Always be aware of where you are and the location of the nearest exit. Do not open garage doors in the event of a fire because the extra oxygen will intensify the flames.**

The proper use of a fire extinguisher is very important. It is possible to deplete an extinguisher and still not put out even the smallest of fires if the extinguisher is used improperly. Procedures vary depending on the agents used. Technicians must become familiar with all the extinguishers equipped in the shop. **Photo Sequence 1** illustrates the proper use of a multipurpose dry chemical extinguisher. This type of extinguisher is the most widely used in the automotive shop.

Fire Blankets

A **fire blanket** is a safety device that can be used to extinguish small fires (**Figure 1-29**). These blankets are made of nonflammable materials such as fiberglass or aramid fibers. Fires that can be put out by a fire blanket include grease/oil fires and electrical fires. Also, fire blankets are useful for putting out clothing fires since they do not stick to fire-damaged skin.

Fire blankets work by preventing the supply of oxygen to a fire. Wrapping something that is burning in a blanket smothers the flames. To get the benefits of your fire blanket, you must use it properly. It is important to read the instructions and become familiar with the proper procedures prior to having to use them in an actual emergency.

When using a fire blanket, you must protect your hands from the fire. Wrap your hands into the top edge of the blanket as you put the blanket on the flame.

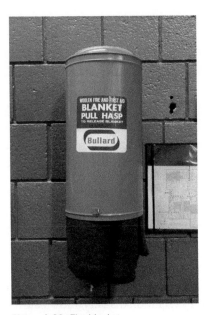

Figure 1-29 Fire blanket.

PHOTO SEQUENCE 1
Using a Dry Chemical Fire Extinguisher

P1-1 Multipurpose dry chemical fire extinguisher.

P1-2 Hold the fire extinguisher in an upright position.

P1-3 Pull the safety pin from the handle.

P1-4 Free the hose from its retainer and aim it at the base of the fire.

P1-5 Stand 8 feet from the fire; do not approach any closer to the fire than this distance.

P1-6 Squeeze the lever while sweeping the hose from side to side. Keep the hose aimed at the base of the fire.

HAZARDOUS MATERIALS

Hazardous materials are materials that can cause illness, injury, or death or pollute water, air, or land. Many solvents and other chemicals used in an automotive shop have warning and caution labels that should be read and understood by everyone who uses them. Many service procedures generate what are known as hazardous wastes. Examples of hazardous waste are used or dirty cleaning solvents and other liquid cleaners.

Right-to-Know Laws

In the United States, **right-to-know laws** concerning hazardous materials and wastes protect every employee in a workplace. The general intent of these laws is for employers to provide a safe working place as it relates to hazardous materials. The right-to-know laws state that employees have a right to know when the materials they use at work are hazardous. The right-to-know laws started with the **Hazard Communication Standard** published by the Occupational Safety and Health Administration (OSHA) in 1983.

This document was originally intended for chemical companies and manufacturers that required employees to handle hazardous materials in their work situation. At the present time, most states have established their own right-to-know laws. Meanwhile, the federal courts have decided to apply these laws to all companies, including automotive service shops.

Under the right-to-know laws, employers have three responsibilities regarding their employees' handling of hazardous materials. The first responsibility concerns employee training and providing information. All employees must be trained about the types of hazardous materials they will encounter in the workplace. All employees must be informed about their rights under legislation regarding the handling of hazardous materials. In addition, information about each hazardous material must be posted on **material safety data sheets (MSDS)** available from the manufacturer (**Figure 1-30**). These sheets contain extensive information and facts about hazardous materials. In Canada, MSDS are called **workplace hazardous materials information systems (WHMIS)**.

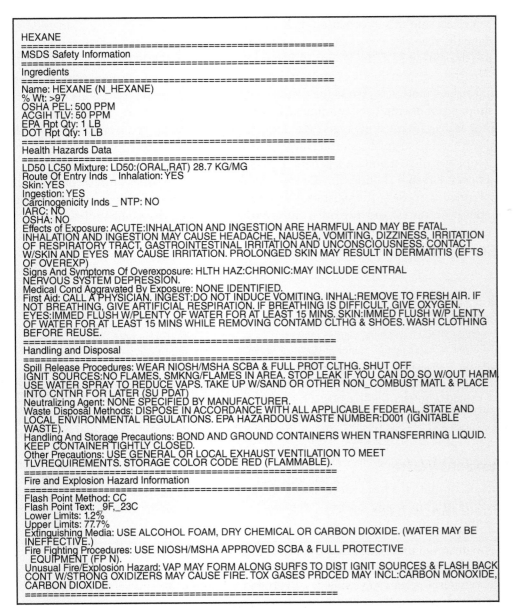

```
HEXANE
=============================================================
MSDS Safety Information
=============================================================
Ingredients
=============================================================
Name: HEXANE (N_HEXANE)
% Wt: >97
OSHA PEL: 500 PPM
ACGIH TLV: 50 PPM
EPA Rpt Qty: 1 LB
DOT Rpt Qty: 1 LB
=============================================================
Health Hazards Data
=============================================================
LD50 LC50 Mixture: LD50:(ORAL,RAT) 28.7 KG/MG
Route Of Entry Inds _ Inhalation: YES
Skin: YES
Ingestion: YES
Carcinogenicity Inds _ NTP: NO
IARC: NO
OSHA: NO
Effects of Exposure: ACUTE:INHALATION AND INGESTION ARE HARMFUL AND MAY BE FATAL.
INHALATION AND INGESTION MAY CAUSE HEADACHE, NAUSEA, VOMITING, DIZZINESS, IRRITATION
OF RESPIRATORY TRACT, GASTROINTESTINAL IRRITATION AND UNCONSCIOUSNESS. CONTACT
W/SKIN AND EYES  MAY CAUSE IRRITATION. PROLONGED SKIN MAY RESULT IN DERMATITIS (EFTS
OF OVEREXP)
Signs And Symptoms Of Overexposure: HLTH HAZ:CHRONIC:MAY INCLUDE CENTRAL
NERVOUS SYSTEM DEPRESSION.
Medical Cond Aggravated By Exposure: NONE IDENTIFIED.
First Aid: CALL A PHYSICIAN. INGEST:DO NOT INDUCE VOMITING. INHAL:REMOVE TO FRESH AIR. IF
NOT BREATHING, GIVE ARTIFICIAL RESPIRATION. IF BREATHING IS DIFFICULT, GIVE OXYGEN.
EYES:IMMED FLUSH W/PLENTY OF WATER FOR AT LEAST 15 MINS. SKIN:IMMED FLUSH W/P LENTY
OF WATER FOR AT LEAST 15 MINS WHILE REMOVING CONTAMD CLTHG & SHOES. WASH CLOTHING
BEFORE REUSE.
=============================================================
Handling and Disposal
=============================================================
Spill Release Procedures: WEAR NIOSH/MSHA SCBA & FULL PROT CLTHG. SHUT OFF
IGNIT SOURCES:NO FLAMES, SMKNG/FLAMES IN AREA. STOP LEAK IF YOU CAN DO SO W/OUT HARM.
USE WATER SPRAY TO REDUCE VAPS. TAKE UP W/SAND OR OTHER NON_COMBUST MATL & PLACE
INTO CNTNR FOR LATER (SU PDAT)
Neutralizing Agent: NONE SPECIFIED BY MANUFACTURER.
Waste Disposal Methods: DISPOSE IN ACCORDANCE WITH ALL APPLICABLE FEDERAL, STATE AND
LOCAL ENVIRONMENTAL REGULATIONS. EPA HAZARDOUS WASTE NUMBER:D001 (IGNITABLE
WASTE).
Handling And Storage Precautions: BOND AND GROUND CONTAINERS WHEN TRANSFERRING LIQUID.
KEEP CONTAINER TIGHTLY CLOSED.
Other Precautions: USE GENERAL OR LOCAL EXHAUST VENTILATION TO MEET
TLVREQUIREMENTS. STORAGE COLOR CODE RED (FLAMMABLE).
=============================================================
Fire and Explosion Hazard Information
=============================================================
Flash Point Method: CC
Flash Point Text: _9F,_23C
Lower Limits: 1.2%
Upper Limits: 77.7%
Extinguishing Media: USE ALCOHOL FOAM, DRY CHEMICAL OR CARBON DIOXIDE. (WATER MAY BE
INEFFECTIVE.)
Fire Fighting Procedures: USE NIOSH/MSHA APPROVED SCBA & FULL PROTECTIVE
   EQUIPMENT (FP N).
Unusual Fire/Explosion Hazard: VAP MAY FORM ALONG SURFS TO DIST IGNIT SOURCES & FLASH BACK
CONT W/STRONG OXIDIZERS MAY CAUSE FIRE. TOX GASES PRDCED MAY INCL:CARBON MONOXIDE,
CARBON DIOXIDE.
=============================================================
```

Figure 1-30 An example of a material safety data sheet (MSDS).

The employer has a responsibility to place MSDS where they are easily accessible by all employees. The MSDS provide extensive information about the hazardous material such as the following:

1. Chemical name.
2. Physical characteristics.
3. Protective equipment required for handling.
4. Explosion and fire hazards.
5. Other incompatible materials.
6. Health hazards such as signs and symptoms of exposure, medical conditions aggravated by exposure, and emergency and first-aid procedures.
7. Safe handling precautions.
8. Spill and leak procedures.

The second responsibility of the employer is to make sure that all hazardous materials are properly labeled. The label information must include health, fire, and reactivity hazards posed by the material, as well as the protective equipment necessary to handle the material. The manufacturer must supply all warning and precautionary information about hazardous materials, and this information must be read and understood by the employee before handling the material. Pay great attention to the information on the label. By doing so, you will use the substance in the proper and safe way, thereby preventing hazardous conditions.

The third responsibility of the employer is for maintaining permanent files regarding hazardous materials. These files must include information on hazardous materials in the shop, proof of employee training programs, and information about accidents such as spills or leaks of hazardous materials. The employer's files must also include proof that employees' requests for hazardous material information, such as MSDS, have been met. The employer must maintain a general right-to-know compliance procedure manual.

There are responsibilities for the employees as well. Employees must be familiar with the intended purpose of the substance, the recommended protective equipment, accident and spill procedures, and any other information regarding the safe handling of hazardous materials. This training must be given annually to employees and provided to new employees as part of their job orientation.

 A BIT OF HISTORY

During the 1960s, disabling injuries increased 20% and 14,000 workers were dying on the job each year. In pressing for prompt passage of workplace safety and health legislation, Senator Harrison A. Williams Jr. called attention to the need to protect workers against such hazards as noise, cotton dust, and asbestos. Representative William A. Steiger also worked for passage of a bill to protect workers. On December 29, 1970, President Richard M. Nixon signed The Occupational Safety and Health Act of 1970, also known as the Williams-Steiger Act.

Hazardous Waste

Waste is considered a **hazardous waste** if it is on the Environmental Protection Agency (EPA) list of known and harmful materials or if it has one or more of the following characteristics:

1. Any material that reacts violently with water or other chemicals is considered hazardous.
2. If a material releases cyanide gas, hydrogen sulfide gas, or similar gases when exposed to low-pH acid solutions, it is hazardous.
3. If a material burns the skin or dissolves metals and other materials, it is considered hazardous.

4. Materials are hazardous if they leach one or more of eight heavy metals in concentrations greater than 100 times the primary drinking water standard. These materials are considered toxic.

5. A liquid is hazardous if the temperature at which the vapors on the surface of the fuel will ignite when exposed to an open flame is below 140°F (60°C), and a solid is hazardous if it ignites spontaneously.

A complete list of EPA hazardous wastes can be found in the Code of Federal Regulations. It should be noted that no material is considered hazardous waste until the shop is finished using it and is ready to dispose of it. New oil is not a hazardous waste; however, used oil is. Once you drain oil from an engine, you have generated hazardous waste and now become responsible for its proper disposal. There are many other wastes that need to be handled properly after you have removed them, such as batteries, brake fluid, transmission fluid, and engine coolant.

No fluids drained from a vehicle should be allowed to enter sewage drains. Some fluids, such as coolant, can be captured and recycled in the shop with special equipment. Filters for fluids (transmission, fuel, and oil filters) also need to be handled in designated ways. Used filters need to be drained and then crushed or disposed of in a special shipping barrel. Most regulations demand that oil filters be drained for at least 24 hours before they are disposed of or crushed.

Federal and state laws control the disposal of hazardous waste materials. It is the responsibility of the employer and the employee to ensure that everyone in the shop is familiar with these laws. Hazardous waste disposal laws include the **Resource Conservation and Recovery Act (RCRA)**. This law basically states that hazardous waste generators are responsible for the waste from the time it becomes a waste material until the proper waste disposal is completed. Therefore, the user must store hazardous waste material properly and safely and be responsible for the transportation of this material until it arrives at an approved hazardous waste disposal site where it is processed according to the law. A licensed waste management firm normally does the disposal. The hazardous waste coordinator for the shop should have a written contract with the hazardous waste hauler.

The RCRA controls these types of automotive waste:

1. Paint and body repair products waste.
2. Solvents for parts and equipment cleaning.
3. Batteries and battery acid.
4. Mild acids used for metal cleaning and preparation.
5. Waste oil, engine coolants, or hydraulic fluids.
6. Air-conditioning refrigerants.
7. Engine oil, transmission, and fuel filters.

NEVER, under any circumstances, use these methods to dispose of hazardous waste material:

1. Pour hazardous wastes on weeds to kill them.
2. Pour hazardous wastes on gravel streets to prevent dust.
3. Throw hazardous wastes in a dumpster.
4. Dispose of hazardous wastes anywhere but an approved disposal site.
5. Pour hazardous wastes down sewers, toilets, sinks, or floor drains.
6. Bury hazardous wastes in the ground.

Handling and Disposing Shop Wastes

To assure that you are following the required procedures concerning the handling and disposal of hazardous waste, the following sections are provided as guidelines. Be sure to familiarize yourself with federal, state, and local regulations in your area.

Oil Disposal. Used oils can be recycled, properly disposed, or used to fuel waste oil burners. In all cases, proper collection of used oil is essential. A drip table or screen table is commonly used to collect the oil into a designated collection bucket. These tables will collect oil as it drips off parts. Be sure not to mix oils unless approved by the recycle company or allowed by the waste oil furnace.

Oil filters must be drained for at least 24 hours before disposal. Once properly drained, they are crushed and then set to be properly disposed of.

Any hazardous waste that is stored on the premises must be properly labeled. Also, be sure to adhere to the regulations as to the maximum quantity that may be stored on site.

Solvents. If possible, reduce the use of solvents by utilizing less toxic alternatives. For example, water-based cleaning solvents should be used instead of petroleum-based solvents. Used solvents that have become too dirty to continue their use must be disposed of properly. Usually, this requires a hazardous waste management company to come to the shop on a regular basis and service the equipment. However, some equipment can be serviced in-house. When storing solvents, be sure they are in approved containers and are tightly sealed. Evaporation of the solvent may have an environmental impact.

Properly label used solvents and store them in containment areas. Store solvents with other compatible wastes only.

Asbestos Exposure

Asbestos. It describes a number of naturally occurring fibrous materials and is classified as a carcinogen. **Asbestos** has been identified as a health hazard and has been shown to cause diseases that result in lung cancers known as mesothelioma. When asbestos fibers are breathed in, they cause scarring of the lungs and damage the air passages. The scars result in holding locations for the asbestos fibers that the body is unable to expel. Asbestos has been/is used for clutch discs, brake pads and shoes, and gaskets. Because asbestos is a health hazard, all OSHA and EPA regulations must be adhered to when disposing of parts that contain this material.

The EPA does not regulate the removal of brake pads and clutches unless more than 50% of the shop's work involves grinding or debonding of the asbestos material. If the shop is above this 50% requirement, then the asbestos materials are regulated as a hazardous waste and must be handled according to the set regulations.

Even if your shop is below the 50% rule, it is very important that as a technician you protect yourself from the dangers associated with inhaling asbestos dust. The EPA recommends that the shop capture asbestos dust into a separate container. The OSHA-preferred method of servicing components consisting of asbestos dust is to use a low-pressure/wet-cleaning method. Never use compressed air to blow off the dust from the parts.

The wet-cleaning method is done by mixing water with an organic solvent. The solvent is flowed over the brake parts prior to removal of the brake drum or around the brake disc prior to brake disassembly. A catch container is positioned to trap the contaminated solvent. During disassembly of the brakes, the solvent is reapplied to assure all asbestos dust is removed.

Another asbestos cleaning method is to use a **high-efficiency particulate-arresting (HEPA)** vacuum cleaner (**Figure 1-31**). The HEPA vacuum cleaner captures the asbestos dust in a special filter. The brake or clutch parts are covered with a special tent. Built-in gloves in the tent allow the technician to clean the parts while viewing through a window. This procedure prevents direct contact with the asbestos and draws the dust particles into the container.

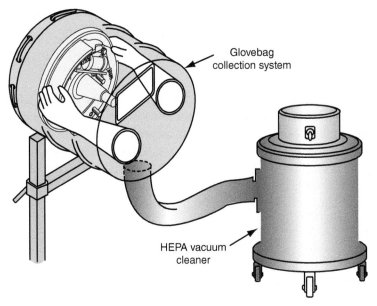

Figure 1-31 A high-efficiency particulate-arresting (HEPA) vacuum cleaner used to capture asbestos fiber dust.

When the HEPA filter is full, it is wetted with a mist of water and then removed. The filters must be placed into an impermeable container, properly labeled, and properly disposed of in accordance with all laws.

Another procedure that is allowed only if the shop performs less than five brake or clutch services a week is the wet wipe method. The technician uses a spray bottle or low-pressure air-charged container to spray a mist of water and detergent onto all brake and clutch components. The components are then wiped with a clean cloth. To reduce the risks of asbestos poisoning, the following guidelines should be followed:

- Never smoke while working around asbestos.
- Wash your hands, arms, and face before eating.
- Take showers at the completion of your work shift.
- Do not wear your work clothes home. Change into your work clothes when you arrive at work and change out of them when leaving.

AUTHOR'S NOTE More information concerning work environment safety can be found by contacting the United States EPA Office of Compliance at http://es.inel.gov or the Coordinating Committee for Automotive Repair (CCAR)-Greenlink at http://www.ccar-greenlink.org.

ELECTRICAL SYSTEM SAFETY

There are many safety requirements that must be followed when working on the vehicle's electrical system. In addition to personal safety, there is the concern of damaging the electrical system with improper service techniques. The following are a few of the safety rules.

Battery Safety

Before attempting to do any type of work on or around the battery, you must be aware of certain precautions. To avoid personal injury or property damage, follow these precautions:

1. Battery acid is very corrosive. Do not allow it to come into contact with your skin, eyes, or clothing. If battery acid should get into your eyes, rinse them thoroughly with clean water and seek immediate medical attention. If battery acid comes into contact with your skin, wash with clean water. Baking soda added to the water will neutralize the acid. If the acid is swallowed, drink large quantities of water or milk followed by milk of magnesia and a beaten egg or vegetable oil.
2. When making connections to a battery, be careful to observe polarity, positive to positive and negative to negative.
3. When disconnecting battery cables, always disconnect the negative (ground) cable first.
4. When connecting battery cables, always connect the negative cable last.
5. Avoid any arcing or open flames near a battery. The vapors produced by the cycling of a battery are very explosive. Do not smoke around a battery.
6. Follow the manufacturer's instructions when charging a battery. Charge the battery in a well-ventilated area. Do not connect or disconnect the charger leads while the charger is turned on.
7. Do not add electrolyte to the battery if it is low. Add only distilled water.
8. Do not wear any jewelry while servicing the battery. These items are excellent conductors of electricity. Severe burns may result if current flows through them by accidental contact with the battery positive terminal and a ground.
9. Never lay tools across the battery. They may come into contact with both terminals, shorting out the battery and causing it to explode.
10. Wear safety glasses and/or a face shield when servicing the battery.

Batteries are very dangerous components of the vehicle. It is important that you be able to demonstrate the ability to work around the battery in a safe manner. Throughout this manual there will be many instances where you will be required to perform a task involving the battery. Chapter 5 covers the subject of removing and testing the battery. Do not perform any tests or disconnect the battery until you have completed that chapter. The purpose of this section is to assist you in becoming more familiar with the battery to allow you to work safely around it.

Point out the following components of the battery to your instructor:

1. Negative terminal.
2. Positive terminal.
3. Vents.

Answer the following questions concerning battery safety (answer written or orally per your instructor):

1. Why be concerned about battery acid?
2. What should be done if battery acid splashes into your eyes?
3. What should you do if battery acid gets onto your skin?
4. What is meant by polarity? Why is it a concern when connecting a battery?
5. Which terminal must be disconnected first?
6. When connecting battery cables, which cable is to be connected last?
7. Why is wearing jewelry discouraged around the battery?
8. Why is smoking not allowed around the battery?
9. Why are tools not to be laid across the top of the battery?

10. What safety protection should be worn while servicing or working around the battery?
11. What other safety precautions must be observed?

If you do not understand any of the safety precautions associated with working on or around the battery, ask your instructor.

Starting System Service Safety

Before testing or servicing the starting system, become familiar with these precautions that should be observed:

1. Refer to the recommendations given in the service information for correct procedures for disconnecting a battery. Some vehicles with on-board computers must be supplied with an auxiliary power source to maintain computer memories.
2. Disconnect the battery ground cable before disconnecting any of the starter circuit's wires or removing the starter motor.
3. Be sure the vehicle is properly positioned on the hoist or on safety stands.
4. Before performing any cranking test, be sure the vehicle transmission is in PARK or NEUTRAL and the parking brakes are applied. Put wheel blocks in front of and behind one tire.
5. Follow the service information procedures for disabling the ignition system.
6. Be sure all test leads are clear of any moving engine components.
7. Never clean any electrical components in solvent or gasoline. Clean with denatured alcohol, or wipe only with clean rags.

Charging System Service Safety

The following are some general rules for servicing the generator and the charging system:

1. Do not run the vehicle with the battery disconnected. The battery acts as a buffer and stabilizes any voltage spikes that may cause damage to the vehicle's electronics.
2. When performing charging system tests, do not allow output voltage to increase over 16 volts (V).
3. If the battery needs to be recharged, disconnect the battery cables while charging.
4. Do not attempt to remove electrical components from the vehicle with the battery connected.
5. Before connecting or disconnecting any electrical connections, the ignition switch must be in the OFF position, unless directed otherwise in the service information.
6. Avoid contact with the **BAT** terminal of the generator while the battery is connected. BAT is the terminal identifier for the conductor from the generator to the battery positive terminal. Battery voltage is always present at this terminal.

Air Bag Safety

The **air bag system** is designed as a supplemental restraint that, in the case of an accident, will deploy a bag out of the steering wheel or passenger-side dash panel to provide additional protection against head and face injuries. An air bag system demands that the technician pay close attention to safety warnings and precautions when working on or around it. Most air bags are deployed by an explosive charge. Accidental deployment of the air bag can result in serious injury. When working on or around the steering wheel or **air bag module**, be aware of your hands and arms. The air bag module is the air bag and inflator assembly together in a single package (**Figure 1-32**). Do not place your arm over the module. If the air bag deploys, injury can result.

Caution

Always double-check the polarity of the battery charger's connections and leads before turning the charger on. Incorrect polarity can damage the battery or cause it to explode.

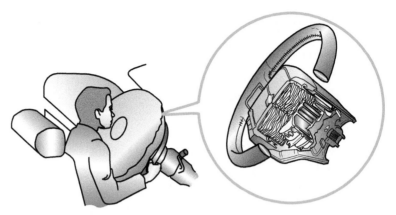

Figure 1-32 Typical air bag module.

Air bag systems contain a means of deploying the bag even if the battery is disconnected. This system is needed in the event the battery is damaged or disconnected during an accident. The reserve energy can be stored for over 30 minutes after the battery is disconnected. Follow the service information procedures for disabling the system.

When carrying the air bag module, carry it so that the bag and trim are facing up and away from your body (**Figure 1-33**). In the event of accidental deployment, the charge will be away from you. Do not face the module toward any other people.

When you place the module on the bench, face the bag and trim up (**Figure 1-34**). This provides a free space for the bag to expand if it deploys. If the module will be stored for any period of time, it must be stored in a cool dry place. Store the module with the trim up and do not place anything on top of the module.

While troubleshooting the air bag system, do not use electrical testers such as battery-powered or AC-powered voltmeters, ohmmeters, and so on, or any other equipment except those specified in the service information. Do not use a test light to troubleshoot the system.

When it is necessary to make a repair or replace a component in the air bag system, always disconnect the negative battery cable before making the repair. It is a good practice

Figure 1-33 Proper method of carrying an air bag module.

Figure 1-34 Place the air bag module on the bench with trim facing up.

to insulate the terminal with tape or a rubber hose to prevent it from coming into contact with the battery post. Some manufacturers recommend that the air bag inflator module(s) be disconnected in addition to the negative battery cable.

AUTHOR'S NOTE Be sure to follow all manufacturer's warnings, cautions, and special service notes when working on or around the air bag system.

Although it is unlikely that an air bag module will inflate on its own, it is possible to ignite it while performing service. To prevent injury, be aware of where the module(s) is located in the vehicle. The most common location is in the center of the steering wheel. However, all late-model vehicles also have a passenger-side air bag (**Figure 1-35**). Some manufacturers are installing air bag systems for the back-seat passengers as well.

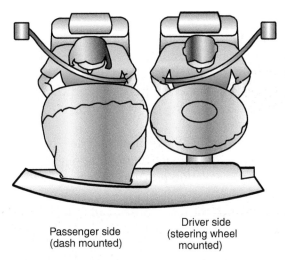

Passenger side
(dash mounted)

Driver side
(steering wheel
mounted)

Figure 1-35 An air bag system with both a driver and a passenger-side air bag module.

Figure 1-36 The air bag wiring is usually identified with a bright yellow harness and/or connector.

Figure 1-37 Some manufacturers identify air bag circuits in their service information with a warning symbol.

⚡ WARNING **Obey all of the warnings in the service information when working on or around the air bag system. Failure to follow these warnings may result in air bag deployment and injury.**

To prevent accidental deployment of the air bag system, disconnect the negative battery cable before disconnecting or connecting any electrical connectors in the system. It is important to be able to recognize the components of the air bag system. Most manufacturers place the wiring of the air bag system into a bright yellow harness tube or use bright yellow insulation or connectors (**Figure 1-36**). The wires are usually tagged to alert the technician. Walk around an air bag–equipped vehicle with your instructor. Your instructor will point out the components of the system and review the necessary safety precautions.

Some manufacturers denote air bag–related circuitry in their service information with a warning symbol (**Figure 1-37**).

Antilock Brake System Safety

Antilock brake systems (ABS) automatically pulsate the brakes to prevent wheel lockup under panic stop and poor traction conditions. ABS is available on most of today's vehicles. There are many different systems used, and each has its own safety requirements regarding servicing the system. Become familiar with the warnings and cautions associated with the system you are working on by studying the service information before performing any service.

Certain components of the ABS are not intended to be serviced individually. Do not attempt to remove or disconnect these components. Only those components with approved removal and installation procedures in the service information should be serviced.

Some operations require that the tubes, hoses, and fittings be disconnected. Many earlier ABS used high hydraulic pressures (up to 2,800 psi [19,305 kPa]) and an accumulator to store this pressurized fluid. Before disconnecting any lines or fittings, the accumulator must be fully depressurized. Many late-model ABS do not use an accumulator; therefore, these systems do not require depressurizing. However, always refer to the correct service information before servicing a brake system. The following is a common method of depressurizing the ABS. However, follow the service information procedures for the vehicle you are working on:

1. Place the ignition switch in the OFF position.
2. Pump the brake pedal a minimum of 20 times.
3. There should be noticeable change in pedal feel when the accumulator is discharged.

GENERAL HYBRID ELECTRIC VEHICLE SAFETY

Since the hybrid electric vehicle (HEV) system can use voltages in excess of 300 volts (both DC and AC), it is vital that the service technician be familiar with, and follow, all safety precautions. Failure to perform the correct procedures can result in electrical shock, battery leakage, explosion, or even death. The following are some general service precautions to be aware of:

- Test the integrity of the insulating gloves prior to use.
- Wear high-voltage insulating gloves when disconnecting the service plug.
- Do not attempt to test or service the system for 5 minutes after the high-voltage service plug is removed. At least 5 minutes is required to discharge the high-voltage capacitors inside the inverter module.
- Never cut the orange high-voltage power cables (**Figure 1-38**). The wire harnesses, terminals, and connectors of the high-voltage system are identified by orange. In addition, high-voltage components may have a "High Voltage" caution label attached to them (**Figure 1-39**).
- Never open high-voltage components.
- Use insulated tools when required.
- Do not wear metallic objects that may cause electrical shorts.
- Follow the service information diagnostic procedures.
- Wear protective safety goggles when inspecting the high-voltage (HV) battery.
- Before touching any of the high-voltage system wires or components, wear insulating gloves, make sure the high-voltage service plug is removed, and disconnect the auxiliary battery.
- Remove the service disconnect plug prior to performing a resistance check.

Special Tools

DMM capable of reading 400 VDC
Insulating gloves
Insulating tape

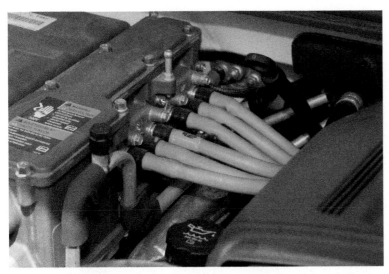

Figure 1-38 High-voltage cables are identified with bright orange conduit or insulation.

Figure 1-39 High-voltage warning label. Be sure to familiarize yourself with all warnings associated with service on EV and HEV vehicles.

- Remove the service plug prior to disconnecting or reconnecting any connectors or components.
- Isolate any high-voltage wires that have been removed with insulation tape.
- Properly torque the high-voltage terminals.

⚠ WARNING **The technician must verify that the system remains powered down when performing any repairs that involve contact with high-voltage or hybrid components or systems.**

⚠ WARNING **The high-voltage checkout procedure must be performed to ensure that the vehicle is properly powered down.**

⚠ WARNING **Whenever the vehicle has been left unattended, recheck that the service disconnect has not been reinstalled.**

⚠ WARNING **Prior to performing any diagnostic or service procedure, you must thoroughly read and follow all applicable high-voltage safety procedures.**

⚠ WARNING **Be sure to utilize proper safety equipment when working on any high-voltage system. Failure to do so may result in serious injury or death.**

⚠ WARNING **Wait a minimum of 5 minutes after performing the high-voltage battery disconnect procedure before accessing the high-voltage system. Failure to do so may result in serious injury or death.**

High-Voltage Service Plug

The HEV is equipped with a **high-voltage service plug** that disconnects the HV battery from the system. Usually, this plug is located near the battery (**Figure 1-40**). Prior to disconnecting the high-voltage service plug, the vehicle must be turned off. Some manufacturers also require that the negative terminal of the auxiliary battery be disconnected. Once the high-voltage service plug is removed, the high-voltage circuit is shut off at the intermediate position of the HV battery and there is no high voltage in the vehicle's systems.

Figure 1-40 The high-voltage service plug is usually located near the HV battery.

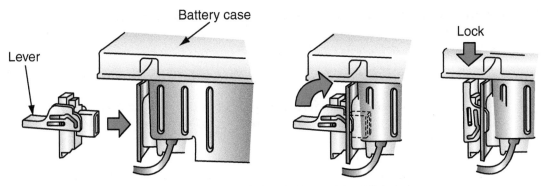

Figure 1-41 To install the service plug, make sure that the lever is down and then fully locked.

⚙️ **SERVICE TIP** Diagnostic trouble codes will be erased once the batteries are disconnected. Prior to disconnecting the system, be sure to check and record any DTCs.

The high-voltage service plug assembly contains a safety interlock reed switch. The reed switch is opened when the clip on the high-voltage service plug is lifted. The open reed switch turns off power to the service main relay (SMR). The main fuse for the high-voltage circuit is inside the high-voltage service plug assembly.

However, never assume that the high-voltage circuits are off. The removal of the high-voltage service plug does not disable the individual HV batteries. Use a DMM to verify that 0 volts is in the system before beginning service. When testing the circuit for voltage, set the voltmeter to the 400 VDC scale.

After the high-voltage service plug is removed, a minimum of 5 minutes must pass before beginning service on the system. This is required to discharge the high voltage from the condenser in the inverter circuit.

To install the high-voltage service plug, make sure the lever is locked in the DOWN position (**Figure 1-41**). Slide the plug into the receptacle, and lock it in place by lifting the lever upward. Once it is locked in place, it closes the reed switch.

Preparing HEV for Service

When working on an HEV, always assume the HV system is live until you have proven otherwise. If the vehicle has been driven into the service department, you know that the HV system was energized since most HEVs do not move without the HV system operating.

It is critical that proper tools be used when working on the HV system. These include protective hand tools and a digital multimeter (DMM) with an insulation test function. The meter must be capable of checking for isolation up to 1,000 volts and measuring resistance at over 1.1 megaohms. In addition, the DMM insulation test function is used to confirm proper isolation of the HV system components after a repair is performed.

Always remove the service disconnect plug and perform the HV checkout procedure to prove that the HV system has been powered down. Never perform repairs until you have performed this procedure to ensure the system is safe to work on. Follow **Photo Sequence 2** as a guide to the safety procedures that must be followed whenever servicing the HEV's high-voltage systems.

Special Tools

Safety goggles
Insulating gloves
Digital multimeter
(DMM)

PHOTO SEQUENCE 2
HEV High-Voltage Isolation

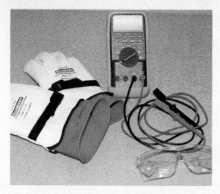

P2-1 Tools required to perform this procedure include safety goggles, 1,000-volt-rated insulating gloves, and a digital multimeter (DMM).

P2-2 Remove the key from the ignition.

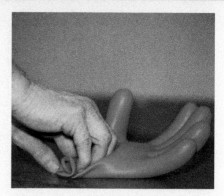

P2-3 Test the integrity of the insulating gloves prior to use. Wear the gloves until the high-voltage isolation procedure is complete.

P2-4 Put on your eye protection.

P2-5 If equipped, disconnect the 12-volt inline connector to isolate the 12-volt battery. If the vehicle does not have the in-line connector, disconnect the negative (–) terminal of the auxiliary (12-volt) battery. Always disconnect the auxiliary battery prior to removing the high-voltage service plug.

P2-6 Remove the high-voltage service plug and put it into your tool box or somewhere it cannot be accidentally reinstalled by someone else.

P2-7 Cover the high-voltage service plug receptacle with insulation tape.

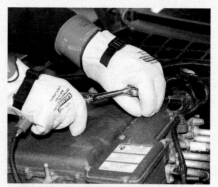

P2-8 After waiting at least 5minutes for the high-voltage condenser inside the inverter to discharge, remove any covers necessary to access the isolation test locations.

P2-9 Use a DMM to confirm that high-voltage circuits have 0 volts before performing any service procedure.

Figure 1-42 Follow the one-hand rule while testing the high-voltage system of an HEV.

Whenever possible use the **one-hand rule** when servicing the HV system (**Figure 1-42**). The one-hand rule means working with only one hand while servicing the HV system so that in the event of an electric shock the high voltage will not pass through your body. It is important to follow this rule when performing the HV checkout procedure since confirmation of HV system power down has not been proven yet.

Insulated Glove Integrity Test

The insulating gloves that the technician must wear for protection while servicing the HV system must be tested for integrity before they are used. If there is a leak in the gloves, high-voltage electricity can travel through the hole to the technician's body. The insulating gloves must meet Class "0" requirements of a rating of 1,000 volts (**Figure 1-43**). In addition, the technician should wear rubber-soled shoes, cotton clothing, and safety glasses with side shields. Remove all jewelry and make sure metal zippers are not exposed. Always have a second set of insulation gloves available and let someone in the shop know their

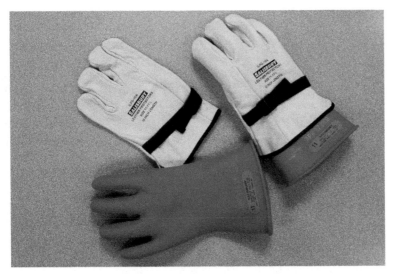

Figure 1-43 Class "0" insulation gloves must be worn when working on high-voltage systems.

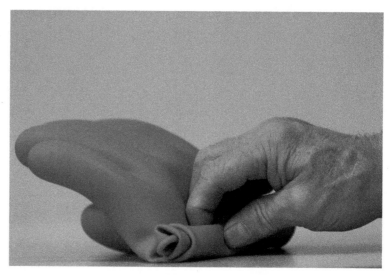

Figure 1-44 Test the insulating gloves for leaks before each use.

location. When preparing to work on an HEV, let others know in the event they must come to your aid.

To test the insulating gloves:

1. Remove the gloves from leather protectors and inspect for any tears or worn spots.
2. Blow air into the gloves to inflate them and seal the opening by folding the base of the glove.
3. Slowly rotate the base of the glove toward the fingers to increase the pressure (**Figure 1-44**).
4. Look and feel for pinholes.

Repeat the steps for each glove. This procedure must be performed for each new set of gloves and before each use.

> **⚡ Caution**
>
> Once the high-voltage service plug is removed, do not operate the power switch. Doing so may damage the hybrid vehicle control ECU.

 WARNING **Do not use the gloves if they fail the leak test or are damaged.**

SUMMARY

- Being a professional technician means more than having knowledge of vehicle systems. It also requires an understanding of all the hazards in the work area.
- As a professional technician, you should work responsibly to protect yourself and the people around you.
- Technicians must be aware that it is their responsibility to prevent injuries in the shop, and that their actions and attitudes reflect how seriously they accept that responsibility.

- Long-sleeve shirts should have their cuffs buttoned or rolled up tightly, and shirttails should be tucked in at all times. Neckties should be tucked inside your shirt; ideally, only clip-on ties should be worn.
- Long hair should be tied back and tucked under a hat.
- Jewelry has no place in the automotive shop.
- The safest and surest method of protecting your eyes is to wear proper eye protection anytime you enter the shop.

- When working around rotating pulleys and belts, be aware of where you are placing your hands and fingers at all times.
- Most occupational back injuries are caused by improper lifting practices.
- Most injuries caused by tools and equipment are the result of improper use, improper maintenance, and carelessness.
- Never use compressed air for cleaning off your clothes.
- The floor jack is to be used only to lift the vehicle off the floor. Use safety stands to support the vehicle after it is lifted.
- Fires are classified by the types of materials involved. Fire extinguishers are classified by the type of fire they will extinguish.
- Batteries can cause serious injury if all safety rules are not followed when working on or around them.
- The air bag system demands that the technician pay close attention to safety warnings and precautions when working on or around them.
- Air bag systems contain a means of deploying the bag even if the battery is disconnected. The reserve energy can be stored for over 30 minutes after the battery is disconnected.

- Since the hybrid system can use voltages in excess of 300 volts (both DC and AC), it is vital that the service technician be familiar with, and follow, all safety precautions.
- Remove the high-voltage service plug and put it where it cannot be accidentally reinstalled by someone else.
- Do not attempt to test or service the high-voltage system of an HEV for 5 minutes after the high-voltage service plug is removed.
- Use a DMM to confirm that high-voltage circuits have 0 volts before performing any service procedure on the high-voltage system of an HEV.
- The insulating gloves that are worn to protect the technician while servicing the high-voltage system of an HEV must be tested for integrity before use.
- The HEV is equipped with a high-voltage service plug that disconnects the HV battery from the system.
- Never assume that the high-voltage circuits of an HEV are off.
- Do not use the insulating gloves if they fail the leak test or are otherwise damaged.

ASE-STYLE REVIEW QUESTIONS

1. Which of the following is *not* included in proper gasoline spill cleanup?

   A. Spread absorbent material to stop the flow and soak up the gasoline.

   B. Immediate containment of the spill.

   C. Use a soap and water solution to wash the gasoline into a container.

   D. Be aware that the absorbent does not reduce flammability concerns.

2. *Technician A* says the right-to-know laws require employers to train employees regarding hazardous waste materials.

   Technician B says the right-to-know laws have no provisions requiring proper labeling of hazardous materials.

   Who is correct?

   A. A only   C. Both A and B

   B. B only   D. Neither A nor B

3. All of the following statements concerning hybrid high-voltage system safety are true, EXCEPT:

   A. Disconnect the motor generators prior to turning the ignition off.

   B. Test the insulating gloves for leaks prior to use.

   C. Do not attempt to test or service the system for 5 minutes after the high-voltage service plug is removed.

   D. Turn the power switch to the off position prior to performing a resistance check.

4. *Technician A* says electrical fires are extinguished with Class A fire extinguishers.

   Technician B says gasoline is extinguished with Class B fire extinguishers.

   Who is correct?

   A. A only   C. Both A and B

   B. B only   D. Neither A nor B

5. *Technician A* says material safety data sheets (MSDS) explain employers' and employees' responsibilities regarding handling and disposal of hazardous materials.

   Technician B says material safety data sheets (MSDS) contain specific information about hazardous materials.

   Who is correct?

   A. A only   C. Both A and B

   B. B only   D. Neither A nor B

6. *Technician A* says a solid that ignites spontaneously is considered a hazardous material.

   Technician B says a liquid is considered a hazardous material if the vapors on the surface will ignite when exposed to an open flame whose temperature is below 140°F (60°C).

   Who is correct?

   A. A only   C. Both A and B

   B. B only   D. Neither A nor B

7. *Technician A* says safety glasses are sufficient when working with battery acid and refrigerants.

   Technician B says full-face shields are designed to provide protection for the face, neck, and eyes of the technician.

   Who is correct?

   A. A only   C. Both A and B

   B. B only   D. Neither A nor B

8. Which of the following statements on HEV safety is correct?

   A. Always place the high-voltage service plug where someone will not accidentally reinstall it.

   B. Use a voltmeter set on 400 VDC to determine if the high-voltage system voltage is at 0 volts before servicing.

   C. Test the integrity of the insulating gloves prior to use.

   D. All of the above.

9. Which of the following statements concerning gasoline storage is true?

   A. Always use approved gasoline containers that are painted blue for proper identification.

   B. Prevent air from entering the container by filling gasoline containers completely full.

   C. Always transport gasoline containers on plastic bed liners.

   D. Do not store a partially filled gasoline container for long periods of time.

10. *Technician A* says once the service plug is disconnected there is no high voltage in the vehicle systems.

    Technician B says prior to disconnecting the high-voltage service plug, the vehicle must be turned off and the auxiliary battery may need to be disconnected.

    Who is correct?

    A. A only   C. Both A and B

    B. B only   D. Neither A nor B

Name _____   Date _____

SHOP SAFETY SURVEY

As a professional technician, safety should be one of your first concerns. This job sheet will increase your awareness of shop safety items. As you survey your shop and answer the following questions, you will learn how to evaluate the safety of any workplace.

Procedure

Task Completed

Your instructor will review your work at each Instructor Response point.

1. Before you begin to evaluate your workplace, evaluate yourself. Are you dressed for work?   ☐ Yes   ☐ No

   If yes, why? _____

   _____

   If no, what must you correct to be properly dressed? _____

   _____

2. Do your safety glasses meet all required standards (ANSI)?   ☐ Yes   ☐ No

   Do they have side shields? _____

3. Carefully inspect your shop, noting any potential hazards._____

NOTE: A hazard is not necessarily a safety violation but is an area of which you must be aware.

4. Are there safety areas marked around grinders and other machinery?

   ☐ Yes ☐ No

5. What is the air pressure in your shop? _____

6. Where are the tools stored in your shop? _____

   Are they clean and neatly stored?   ☐ Yes   ☐ No

7. If you could, how would you improve the tool storage?

   _____

   _____

   _____

8. What kind of hoist is used in your shop?

   _____

   _____

9. Ask your instructor to demonstrate hoist usage.

10. Where is the first-aid kit in your shop?

    _____

11. Where is the main power shutoff or emergency shutoff controls located?

    _____

    _____

12. List the location of the exits.

_____

_____

_____

13. Describe the emergency evacuation procedures.

_____

_____

_____

14. Where are the hazardous materials stored?

_____

_____

_____

15. What is the procedure for handling hazardous waste?

_____

_____

_____

16. What is the procedure to be followed in your shop in case of an accident?

_____

_____

17. Have your instructor supply you with a vehicle make, model, and year. Using the   ☐
appropriate shop manual, draw an illustration showing the lifting points on the given
vehicle.

Instructor's Response

_____

_____

_____

Name _____   Date _____

FIRE HAZARD INSPECTION

Fire is always a danger in any automotive shop. The very nature of automotive work involves the use of many highly flammable chemicals. Because of this, a technician must be very careful. Watch for and immediately correct all fire hazards.

Procedure

Task Completed

1. Are there any flammable liquids stored in your shop? _____

    Are they stored properly?   ☐ Yes   ☐ No

    Why or why not? _____

    _____

2. Where are the fire extinguishers located in your shop?

    _____

    _____

    Against what class fires can each extinguisher be used?

    _____

    _____

3. Explain to your instructor how to use each fire extinguisher in your shop.   ☐

4. Does your shop have a fire blanket?   ☐ Yes   ☐ No

    If so, where is it kept?

    _____

    _____

5. Where are the fire alarms located?

    _____

    _____

6. Where are the fire exits located?

    _____

    _____

7. Where are the fire escape routes posted?

    _____

    _____

8. Where are the fireproof cabinets located?

    _____

    _____

9. Where are dirty shop towels to be disposed?

_____

_____

Instructor's Response

_____

_____

_____

Name _____   Date _____

HYBRID SAFETY

Upon completion of this job sheet, you should be familiar with the critical safety procedures involved in servicing a high-voltage hybrid system.

NATEF Correlation ——————————————————————————————————————

This job sheet addresses the following **MLR/AST/MAST** task:

B.7.   Identify safety precautions for high voltage systems on electric, hybrid-electric, and diesel vehicles.

NATEF

Tools and Materials

- HEV
- Service information
- Insulating gloves
- Eye protection

Describe the vehicle being worked on:

Year _____ Make _____ Model _____

VIN _____ Engine type and size _____

Procedure

Task Completed

1. Use the service information data and determine the location on the vehicle for the high-voltage service plug.

   _____

   _____

   _____

2. What must be done prior to disconnecting the service plug?

   _____

   _____

3. How long must you wait after the plug is disconnected before servicing the system?

   _____

   _____

4. Access the 12-volt auxiliary battery and remove the negative terminal.   ☐

5. Test the insulating gloves for leaks. Are the gloves safe to use?   ☐ Yes ☐ No

   If no, inform your instructor.

6. Put on the insulating gloves and eye protection.   ☐

7. Remove the service plug. What device(s) are integrated into the service plug assembly?

   _____

   _____

8. Reinstall the service plug.   ☐

9. Review your observations with your instructor.   ☐

Instructor's Response

_____

_____

_____

CHAPTER 2
SPECIAL TOOLS AND PROCEDURES

Upon completion and review of this chapter, you should be able to:

- Explain the proper use of jumper wires.
- Explain the proper use of a test light.
- Explain the proper use of a logic probe.
- Explain the proper use of analog volt/amp/ohmmeters.
- Explain the proper use of digital volt/amp/ohmmeters.
- Describe when to use the different types of multimeters.

- Explain the proper use of a digital storage oscilloscope.
- Use Ohm's law to determine electrical values in different types of circuits.
- Locate service information.
- Explain the concepts of working as an electrical systems technician.

Basic Tool

Basic mechanic's tool set

Terms To Know

Ammeter	Flat rate	Pulse width
Average responding	Frequency	Root mean square (RMS)
Backprobe	Glitches	Scan tool
Captured signal	Hertz	Sine wave
Cycle	Jumper wire	Sinusoidal
D'Arsonval movement	Logic probe	Square wave
Diagnostic trouble codes (DTCs)	Multimeter	Test light
	Noise	Voltage drop
Digital multimeter (DMM)	Ohmmeter	Voltmeter
Duty cycle	Open circuit testing	Work order

INTRODUCTION

This chapter covers many of the typical shop procedures that the electrical systems technician may encounter. This includes proper troubleshooting procedures, the use of special test equipment, the use of service information, and workplace practices.

Also, this chapter covers what it means to work as an electrical systems technician. This includes compensation programs, the importance of workplace and customer relations, communication, and certification.

BASIC ELECTRICAL TROUBLESHOOTING

Classroom Manual
Chapter 2, page 20

To be able to properly diagnose electrical components and circuits, you must be able to use many different types of electrical test equipment. In this chapter, you will learn when and how to use the most common types of test equipment. You will also learn which test instrument is best to use to identify the cause of the various types of electrical problems.

> **AUTHOR'S NOTE** In the past voltmeters, ammeters, and ohmmeters were often separate test equipment. Today these meters are included in a single digital multimeter (DMM), which is also called a digital volt/ohmmeter (DVOM). The DMM is discussed later in this chapter.

Classroom Manual
Chapter 2, page 21

Classroom Manual
Chapter 2, page 23

Classroom Manual
Chapter 2, page 21

Electrical current is a term used to describe the movement or flow of electricity. The greater number of electrons flowing past a given point in a given amount of time, the more current the circuit has. This current, like the flow of water or any other substance, can be measured. The unit for measuring electrical current is the ampere. The instrument used to measure electrical current flow in a circuit is called an **ammeter**.

When any substance flows, it meets resistance. The resistance to electrical flow can be measured by an instrument called an **ohmmeter**. As current flows through a resistance, there is less voltage left over. The amount of voltage drop is a direct result of the amount of resistance. The ohmmeter sends a small current through the leads and the component being tested. It then measures the voltage drop and calculates the resistance value in ohms.

Voltage is electrical pressure. Voltage is the force developed by the attraction of electrons to protons. The more positive one side of the circuit is, the more voltage is present in the circuit. Voltage does not flow; rather, it is the pressure that causes current flow.

To have electricity, some force is needed to move the electrons between atoms. This force is the pressure that exists between a positive and negative point within an electrical circuit. This force, also called electromotive force, is measured in units called volts. One volt is the amount of pressure (force) required to move one ampere of current through a resistance of one ohm. Voltage is measured by an instrument called a **voltmeter**.

The amount of current that flows in a circuit is determined by the resistance in that circuit. As resistance goes up, the current goes down. The energy used by a load is measured in volts. Amperage stays constant in a circuit, but the voltage is dropped as it powers a load. Measuring voltage drop determines the amount of electrical energy changed to another form of energy by the load.

Troubleshooting electrical problems involves using meters, test lights, and jumper wires to determine if any part of the circuit is open or shorted, or if there is unwanted resistance.

To troubleshoot a problem, always begin by verifying the customer's concern. Operate the affected system along with other systems to get a complete understanding of the problem. Often there are other problems, which are not as evident or bothersome to the customer, which will provide helpful information for diagnostics. Obtain the correct wiring diagram for the car and study the circuit that is affected. From the diagram, you should be able to identify testing points and probable problem areas. Then test and use logic to identify the cause of the problem.

An ammeter and a voltmeter connected to a circuit at the different locations shown in **Figure 2-1** should give readings as indicated when there are no problems in the circuit. An open exists whenever there is not a complete path for current flow. If there is an open

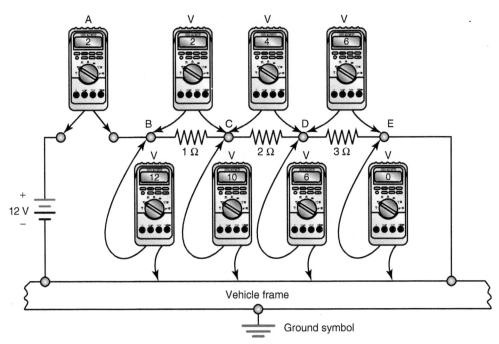

Figure 2-1 A basic circuit being tested with an ammeter and a voltmeter.

anywhere in the circuit, the ammeter will read zero current. If the open is in the 1-ohm resistor, a voltmeter connected from C to ground will read zero. However, if the resistor is open and the voltmeter is connected to points B and C, the reading will be 12 volts. The reason is that the battery, ammeter, voltmeter, 2-ohm resistor, and 3-ohm resistor are all connected together to form a series circuit. Because of the open in the circuit, there is only current flow in the circuit through the meter, not the rest of the circuit. This current flow is very low because the meter has such high resistance. Therefore, the voltmeter will show a reading of 12 volts, indicating no voltage drop across the resistors.

To help you understand this concept, look at what happens if the 2-ohm resistor is open instead of the 1-ohm resistor. A voltmeter connected from point C to ground would indicate 12 volts. The 1-ohm resistor in series to the high resistance of the voltmeter would have little effect on the circuit. If an open should occur between point E and ground, a voltmeter connected from point B, C, D, or E to ground would read 12 volts. A voltmeter connected across any one of the resistors, from B to C, C to D, or D to E, would read 0 volts, because there will be no voltage drops if there is no current flow.

A short would be indicated by excessive current and/or abnormal voltage drops. These examples illustrate how a voltmeter and ammeter can be used to check for problems in a circuit. An ohmmeter also can be used to measure the values of each component and compare these measurements to specifications. If there is no continuity across a part, it is open. If there is more resistance than called for, there is high internal resistance. If there is less resistance than specified, the part is shorted.

⚡ WARNING **Because the human body is a conductor of electricity, observe all safety rules when working with electricity.**

⚠ **Caution**

Any broken, frayed, or damaged insulation material requires replacement or repair to the wire. Exposed conductor material from damaged insulation can result in a safety hazard and damage to circuit components.

TEST EQUIPMENT

Since electricity is an invisible force, the proper use of test tools will permit the technician to "see" the flow of electrons. Knowing what is being looked at and being able to interpret various meter types will assist in electrical system diagnosis. To diagnose and repair

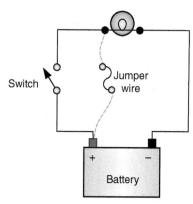

Figure 2-2 Using a fused jumper wire to bypass the switch.

electrical circuits correctly, a number of common tools and instruments are used. The most common tools are jumper wires, test lights, voltmeters, ammeters, and ohmmeters.

Jumper Wires

One of the simplest types of test equipment is the **jumper wire**. Connecting one end of the jumper wire to battery positive will provide an excellent 12-volt power supply for testing a component. If the component does not operate when it is in its own circuit, but does operate when battery voltage is jumped to it, the component is good and the fault is within the circuit (**Figure 2-2**).

> **AUTHOR'S NOTE** The term *battery positive* refers to any portion of the circuit from the battery positive terminal to the load (insulated side) that has battery voltage.

Jumper wires can be used to bypass individual wires, connectors, components, or switches (**Figure 2-3**). Bypassing a component or wire helps to determine if that part is faulty. If the problem is no longer evident after the jumper wire is installed, the part being bypassed is usually faulty.

To protect the circuit being tested, it is recommended the jumper wire be fitted with an in-line fuse holder or circuit breaker. This will allow the quick changing of fuses to correctly protect the circuit and help prevent damage to the circuit if the jumper wire is connected improperly.

⚡ WARNING **Never connect a jumper wire across the terminals of the battery. The battery could explode, causing serious injury.**

Test Lights

There are two types of test lights commonly used in diagnosing electrical problems: non-powered and self-powered. A **test light** is used when the technician needs to look for electrical power in the circuit. A typical nonpowered test light has a transparent handle that contains a light bulb. A sharp probe extends from one end of the handle while a wire with a clamp extends from the other end (**Figure 2-4**). If the circuit is operating properly, clamping the lead of the test light to ground and probing the circuit at a point of voltage should light the lamp (**Figure 2-5**).

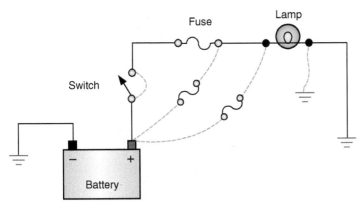

Figure 2-3 Examples of locations a fused jumper wire can be used to bypass a portion of the circuit, and to test the ground circuit. Anytime you bypass the circuit fuse, use a jumper wire that is fitted with a fuse. Never bypass the load component.

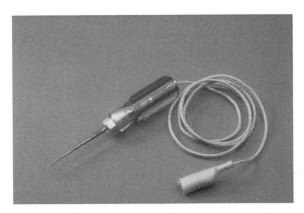

Figure 2-4 A typical 12-volt test light used to probe for voltage in a circuit.

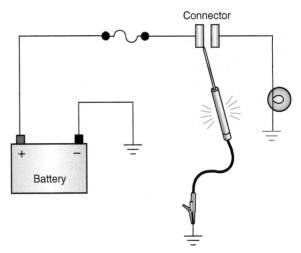

Figure 2-5 If voltage is present, the test light will illuminate.

A test light is limited because it does not display how much voltage is at the point of the circuit being tested. However, by understanding the effects of voltage drop, the technician will be able to interpret the brightness of the test light and relate the results to the expectations of a good circuit. If the lamp is connected after a voltage drop, the lamp will light dimly. Connecting the test lamp before the voltage drop should light the lamp brightly. The light should not illuminate at all if it is probing for voltage after the last resistance.

> ⚠ **Caution**
>
> Do not use a test light to test for power in an electronic circuit, such as a computer or printed circuit board. The increased current draw of the test light may damage the system components.

> ⚙ **SERVICE TIP** A headlight bulb fitted with wires and clips will make a good test light for checking circuits that carry 5 amps or more of current. Do not use this lamp to test circuits that carry less than 5 amps of current.

It must be understood that just because the test light lights brightly, that does not mean the circuit may not have problems. If the light is bright, that means the circuit can handle the amount of current needed to light the test light. Most test lights require very

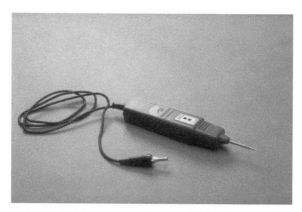

Figure 2-6 Typical self-powered continuity tester.

little amperage to turn on (e.g., 250 mA). If testing a circuit that needs to handle 8 amps of current, you should use a test light that will require that amount of current.

Another type of circuit tester is the self-powered continuity tester (**Figure 2-6**). The continuity tester has an internal battery that powers a light bulb. With the power in the circuit turned off or disconnected, the ground clip is connected to the ground terminal of the load component. By probing the feed wire, the light will illuminate if the circuit is complete (has continuity). If there is an open in the circuit, the lamp will not illuminate.

⚠ WARNING   **Never use a self-powered test light to test the air bag system. The battery in the tester can cause the air bag to deploy.**

Logic Probes

Many computer-controlled systems use a pulsed voltage to transmit messages or to operate a component. A standard or self-powered test light should not be used to test these circuits since they may damage the computer. However, a **logic probe** (**Figure 2-7**) can be used. A logic probe looks something like a test light except it contains three different-colored LEDs. The red LED will light if there is high voltage at the point in the circuit being tested. The green LED will light to indicate the presence of low voltage. The yellow LED lights to indicate the presence of voltage pulses. If the voltage is a pulsed voltage from a high level

Figure 2-7 A typical logic probe.

to a low level, the yellow LED will be on and the red and green LEDs will cycle, indicating the change in voltage. Logic probes can be used only on digital circuits. Analog circuits change voltage over time; the logic probe works on only pulsed voltages.

MULTIMETERS

The **multimeter** is one of the most versatile tools used to diagnose electrical systems. It can be used to measure voltage, resistance, and amperage. In addition, some types of multimeters are designed to test diodes and measure frequency, duty cycle, temperature, and rotation speed. Multimeters are available in analog (swing needle) and digital display.

Analog Meters

Analog meters use a sweeping needle and a scale to display test values (**Figure 2-8**). All analog meters use a **D'Arsonval movement** (**Figure 2-9**). A D'Arsonval movement is a small coil of wire mounted in the center of a permanent horseshoe-type magnet. A pointer or needle is mounted to the coil. When taking a measurement, current flows through the coil and creates a magnetic field around the coil. The coil rotates within the permanent magnet as its magnetic field interacts with the magnetic field of the permanent magnet. The amount of rotation is determined by the strength of the magnetic field around the coil. Since the needle moves with the coil, it reflects the amount of coil movement and its direction.

Digital Meters

With modern vehicles incorporating computer-controlled systems, it is mandatory for today's technician to know how to properly use a **digital multimeter (DMM)** (**Figure 2-10**). Digital multimeters (also called DVOMs) display values using liquid crystal displays instead of a swinging needle. They are basically computers that determine the measured value and display it for the technician. Computer systems have integrated circuits (ICs) that operate on very low amounts of current. Analog meters can overload computer circuits and burn out the IC chips since they allow a large amount of current to flow through the circuit. On the other hand, most digital multimeters have very high input resistance (impedance), which prevents the meter from drawing current when connected to a circuit. Most DMMs have at least 10 megohms (10 million ohms) impedance. This reduces the risk of damaging computer circuits and components.

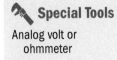
Special Tools
Analog volt or
  ohmmeter
DMM
Backprobing tools

⚡ **Caution**

Do not use an analog meter on a computer-controller circuit unless expressly directed to do so in the service information. Damage to the circuit or computer may result.

Digital multimeters (DMM) are also referred to as DVOMs (digital volt/ohmmeters).

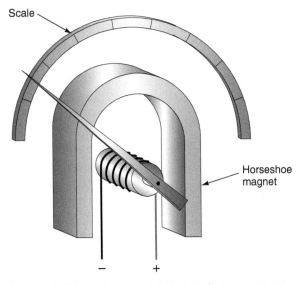
Figure 2-9 D'Arsonval movement is the basis for the movement of an analog meter.

Figure 2-8 An analog meter.

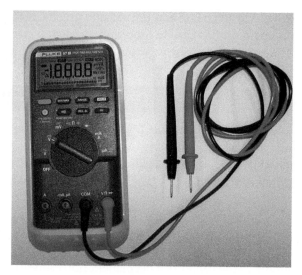

Figure 2-10 Digital multimeter.

⚠ **Caution**

Not all DMMs are rated at 10 megohms of impedance. Be sure of the meter you are using to prevent electronic component damage.

Special Tool
DMM

Classroom Manual
Chapter 2, page 21

Classroom Manual
Chapter 2, page 24

Digital meters rely on electronic circuitry to measure electrical values. The measurements are displayed with LEDs or on a liquid crystal display (LCD). Digital meters tend to give more accurate readings and are certainly much easier to read. Rather than reading a scale at the point where the needle lines up, digital meters simply display the measurement in a numerical value. This also eliminates the almost certain error caused by viewing an analog meter at an angle.

Voltmeters

A voltmeter can be used to measure the voltage available at the battery. It can also be used to test the voltage available at the terminals of any component or connector. In addition, a voltmeter tests voltage drop across an electrical circuit, component, switch, or connector.

A voltmeter has two leads: a red positive lead and a black negative lead. The red lead should be connected to the positive side of the circuit or component. The black lead should be connected to ground or to the negative side of the component. A voltmeter is connected in parallel with a circuit (**Figure 2-11**).

Figure 2-12 shows how to check for voltage in a closed circuit. The voltage at point A is 12 volts positive. There is a drop of 6 volts over the 1-ohm resistor and the reading is 6 volts positive at point B. The remaining voltage drops in the motor load and the voltmeter reads 0 at point C, indicating normal motor circuit operation.

When reading voltage in the same circuit that has an open (**Figure 2-13A**), 12 volts will be indicated at any point ahead of the open. This is indicated at points A, B, and C, but not through X. Since the circuit is open, and there is no electrical flow, there is no voltage drop across a resistor or load.

The loss of voltage due to resistance in wires, connectors, and loads is called **voltage drop**. Voltage drop is the amount of electrical energy converted to another form of energy. For example, to make a lamp light, electrical energy is converted to heat energy. It is the heat that makes the lamp light.

Testing for voltage by disconnecting a connector and then measuring for applied voltage is called **open circuit testing**. This can be misleading since virtually no current is flowing through the voltmeter and the circuit is not loaded. Thus, the meter may display full battery voltage although there is resistance in the circuit. This is why voltage drop testing under load is the preferred method.

To measure voltage drop across each load, it must be determined what point is the most positive and what point is the most negative in the circuit. A point in the circuit can be either positive or negative, depending on what is being measured. Referring to

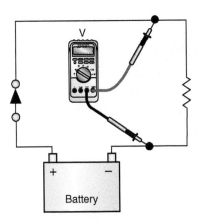

Figure 2-11 Connecting a voltmeter in parallel to the circuit.

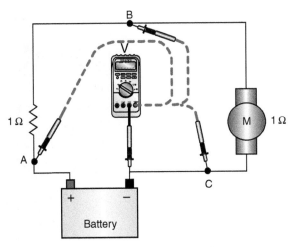

Figure 2-12 Checking voltage in a closed circuit.

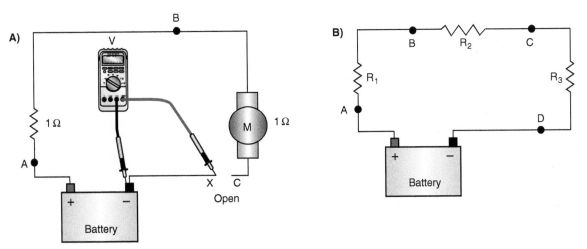

Figure 2-13 Using a voltmeter to (A) find an open circuit and (B) to check voltage drop.

Figure 2-13B, if voltage drop across R_1 is being measured, the positive meter lead is placed at point A and the negative lead at point B. The voltmeter will measure the difference in volts between these points. However, to measure the voltage drop across R_2 the polarity of point B is positive and point C is negative. The reason is that point B is the most negative for R_1, yet it is the most positive point for R_2. To measure voltage drop across R_3, point C is positive and point D is negative. The positive lead of the voltmeter should be placed as close as possible to the positive side of the battery in the circuit. The procedure for measuring a voltage drop is shown in **Photo Sequence 3**.

> ⚙ **SERVICE TIP** Using a voltmeter to measure voltage drop is one of the easiest ways of testing a circuit's ability to carry current under load.

Often, voltage drop testing requires the technician to **backprobe** a connector to access the terminal. Backprobing requires the use of a backprobing tool (**Figure 2-14**) that will slide into the back side of a connector. The tool should be inserted between the wire's insulation and the connector's seal until it contacts the terminal (**Figure 2-15**). When performed properly the wire, connector, or terminal will not be damaged. It is important to use as small a backprobing tool as possible.

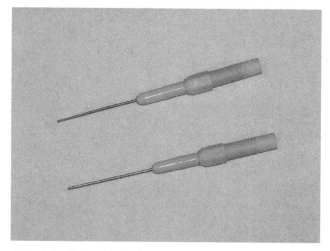

Figure 2-14 Backprobing tools.

Figure 2-15 Backprobing tool properly inserted into the back of a connector.

PHOTO SEQUENCE 3
Performing a Voltage Drop Test

P3-1 The tools required to perform this task include a voltmeter, backprobing tools, and fender covers.

P3-2 Set the voltmeter to its lowest DC voltage scale.

P3-3 To test the voltage drop of the whole insulated side of the circuit, connect the red (positive) voltmeter test lead to the battery positive (+) terminal.

P3-4 Use the backprobing tool to connect the black (negative) voltmeter test lead into the low-beam terminal of the headlight socket. Make sure that you are connected to the input side of the headlight.

P3-5 Turn on the headlights (low beam) and observe the voltmeter readings. The voltmeter will indicate the amount of voltage that is dropped between the battery and the headlight.

All wiring must have resistance values low enough to allow enough voltage to the load for proper operation. The maximum allowable voltage loss due to voltage drops across wires, connectors, and other conductors in an automotive circuit is 10% of the system voltage. Therefore, in a 12-volt automotive electrical system, this maximum total loss is 1.2 volts.

Figure 2-16 shows two headlights connected to a 12-volt battery by two wires. Each wire has a resistance of only 0.05 ohm. Each headlight has a resistance of 2 ohms. The two headlights are wired in parallel and have a total resistance of 1 ohm. If the wires had no resistance, the total resistance of the circuit would be 1 ohm and the current flowing through the circuit would be 12 amps. However, the wires have resistance and the total resistance of the circuit is 1.1 ohms (1 ohm + 0.05 ohm + 0.05 ohm). Therefore, the circuit current is 10.9 amps (I = E / R, I = 12 V / 1.1 ohms). The voltage drop across the bulbs would be 12 volts if there was no resistance in the wires. Now the voltage drop across the bulbs is only 10.9 volts (E = I × R, E = 10.9 amps × 1 ohm). Although it may not be very noticeable, the light bulbs will not be as bright as they should be because of the decreased current and the decreased voltage drop.

A voltmeter can also be used to check for proper circuit grounding. If the voltmeter reading indicates full voltage at the lights, but the bulbs are not illuminated, the bulbs or sockets could be bad or the ground connection is faulty. An easy way to check for a defective bulb is to replace it with a known good bulb.

If the bulbs are good, the problem lies in either the light sockets or the ground wires. Connect the voltmeter to the ground wire and a good ground. If the light socket is defective, the voltmeter will read 0 volts. If the socket is not defective, but the ground wire is broken or disconnected, the voltmeter will read very close to battery voltage. In fact, any voltage reading would indicate a bad or poor ground circuit. The higher the voltage, the greater the problem.

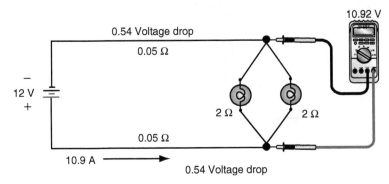

Figure 2-16 Measuring the voltage drop across a lamp. Notice the wires have some resistance.

Thus far, we have discussed using a voltmeter on a direct current (DC) circuit. The voltmeter can also measure alternating current (AC). There are two methods used to display AC voltage: **root mean square (RMS)** and **average responding**. If the AC voltage signal is a true **sine wave**, both methods would display the same value. However, most automotive sensors do not produce a pure sine wave signal. The technician must know how the different meters will display the AC voltage reading under these circumstances. The type of display the voltmeter uses can be found in the meter's operating manual.

Ohmmeters

In contrast to the voltmeter, which uses the voltage available in the circuit, an ohmmeter is battery powered. The circuit being tested must be electrically open or isolated. If the power is on in the circuit, the ohmmeter may be damaged.

The two leads of the ohmmeter are placed across or in parallel with the circuit or component being tested. The red lead is placed on the positive side of the circuit and the black lead is placed on the negative side of the circuit (**Figure 2-17**). The meter sends current through the component and determines the amount of resistance based on the voltage dropped across the load. The scale of an ohmmeter reads from 0 to infinity (∞). A 0 reading means there is no resistance in the circuit and may indicate a short in a component that should show a specific resistance. For example, a coil winding should have a high resistance value; a 0-ohm reading would indicate the coil windings are being bypassed. An infinity reading indicates a number higher than the meter can measure. This usually is an indication of an open circuit.

AUTHOR'S NOTE There is not a correlation between the ohmmeter reading and continuity. The ohmmeter reads the resistance value determined during the test. For example, a reading of 0 ohms does not mean there is 0 continuity. It means there is no resistance between the test points. An ohmmeter reading of 250 ohms does not mean there are 250 units of continuity, but 250 ohms of resistance.

The test chart, shown in **Table 2-1**, illustrates the readings that may be expected from an ohmmeter or voltmeter under different conditions. It is important to become familiar with these examples in order to analyze circuits.

Root mean square (RMS) meters convert the AC signal to a comparable DC voltage signal.

Average responding meters display the average voltage peak.

A **sine wave** is a waveform that shows voltage changing polarity from positive to negative.

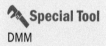

Special Tool

DMM

Classroom Manual Chapter 2, page 23

Caution

Since the ohmmeter is self-powered, never use an ohmmeter on a powered circuit.

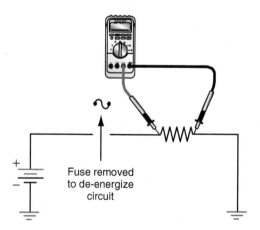

Figure 2-17 Measuring resistance with an ohmmeter. The meter is connected in parallel with the component being tested after power is removed from the circuit.

TABLE 2-1   CIRCUIT TEST CHART

Type of Defect	Test Unit	Expected Results
Open	Ohmmeter	∞ infinite resistance between conductor ends
	Test light	No light after open
	Voltmeter	Ø volts at end of conductor after the open
Short to ground	Ohmmeter	Ø resistance to ground
	Test light	Lights if connected across fuse
	Voltmeter	Generally not used to test for ground
Short	Ohmmeter	Lower than specified resistance through load component Ø resistance to adjacent conductor
	Test light	Light will illuminate on both conductors
	Voltmeter	A voltage will be read on both conductors
Excessive resistance	Ohmmeter	Higher than specified resistance through circuit
	Test light	Light illuminates dimly
	Voltmeter	Voltage will be read when connected in parallel over resistance

Ohmmeters are also used to trace and check wires or cables. Assume that one wire of a four-wire cable is to be found. Connect one probe of the ohmmeter to the known wire at one end of the cable and touch the other probe to each wire at the other end of the cable. Any evidence of resistance indicates the correct wire. Using this same method, you can check a suspected defective wire. If low resistance is shown on the meter, the wire is intact. However, this does not mean that testing of the wire under load is not required. If infinite resistance is measured, the wire is defective (open). If the wire is okay, continue checking by connecting the probe to other leads. Any indication of resistance indicates that the wire is shorted to one of the other wires and that the harness is defective.

Ammeters

An ammeter measures current flow in a circuit. Circuit problems can also be identified by using an ammeter. An ammeter must be placed in series with the circuit being tested (**Figure 2-18**). Normally, this requires disconnecting a wire or connector from a component and connecting the ammeter between the wire or connector and the component. The red lead of the ammeter should always be connected to the most positive side of the connector and the black lead should be connected to the least positive side.

Most handheld multimeters have a 10-ampere protection device. This is the highest amount of current flow the meter can read. When using the ammeter, start on a high scale and work down to obtain the most accurate readings.

Classroom Manual
Chapter 2, page 22

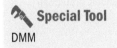 **Special Tool**
DMM

⚠ **Caution**

Never place the leads of an ammeter across the battery or a load. This puts the meter in parallel with the circuit and will blow the fuse in the ammeter or possibly destroy the meter.

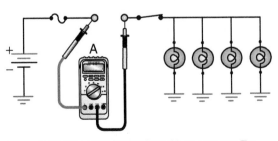

Figure 2-18 Measuring current flow with an ammeter. The meter must be connected in series with the circuit.

Figure 2-19 An ammeter with an inductive pickup. The inductive pickup eliminates the need to connect the meter in series with the circuit.

It is much easier to test current using an ammeter with an inductive pickup (**Figure 2-19**). The pickup clamps around the wire or cable being tested. The inductive clamp has an arrow on it to indicate proper attachment to the conductor. The arrow indicates current flow direction using the conventional flow theory of positive to negative. These ammeters measure amperage based on the magnetic field created by the current flowing through the wire. This type of pickup eliminates the need to separate the circuit to insert the meter leads.

Because ammeters are built with very low internal resistance, connecting them in series does not add any appreciable resistance to the circuit. Therefore, an accurate measurement of the current flow can be taken.

For example, assume that a circuit shown in Figure 2-18 normally draws 5 amps and is protected by a 10-amp fuse. If the circuit constantly blows the fuse, a short exists somewhere in the circuit. Mathematically, each light should draw 1.25 amperes (5/4 = 1.25). To find the short, disconnect all lights by removing them from their sockets. Then, close the switch and read the ammeter. With the load disconnected, the meter should read 0 amperes. If there is any reading, the wire between the fuse block and the socket is shorted to ground.

If 0 amps was measured, reconnect each light in sequence; the reading should increase 1.25 amperes with each bulb. If, when making any connection, the reading is higher than expected, the problem is in that part of the light circuit.

⚙ **SERVICE TIP** Most handheld DMMs have the option for an inductive pickup that will measure amperage between 10 and 20 amps (depending on the model). This makes measuring amperage on some circuits easier since it eliminates the need to connect the test leads in series.

Additional DVOM Functions

Some multimeters feature additional functions besides measuring AC or DC voltage, amperage, and ohms. Many multimeters are capable of measuring engine revolutions per minute (rpm), ignition dwell, diode condition, distributor condition, frequency, and temperature.

Some meters have a MIN/MAX function that displays the minimum, maximum, and average values received by the meter during the time the test was being recorded. This feature is valuable when checking sensors, output commands, or circuits for electrical **noise**. Noise is an unwanted voltage signal that rides on a signal. Noise is usually the result of radio frequency interference (RFI) or electromagnetic induction (EMI). The noise causes slight increases and decreases in the voltage signal to, or from, the computer. Another definition of noise is an AC signal riding on a DC voltage. The computer may attempt to react to the small changes in the signal as a result of the noise. This means the computer is responding to the noise and not the voltage signal, resulting in incorrect component operation.

Also, some multimeters may have the capabilities to measure **duty cycle**, pulse width, and frequency. Duty cycle is the measurement of the amount of "on" time as compared to the time of a **cycle** (**Figure 2-20**). A cycle is one set of changes in a signal that repeats itself several times. The duty cycle is displayed in a percentage. For example, a 60% duty cycle means the device is on 60% of the time and off 40% of the time of one cycle.

Pulse width is similar to duty cycle except that it is a measurement of time the device is turned on within a cycle (**Figure 2-21**). Pulse width is usually measured in milliseconds.

As previously mentioned, some multimeters can measure **frequency** (**Figure 2-22**). Frequency is a measure of the number of cycles that occur in one second. The higher the frequency, the more cycles that occur in a second. Frequency is measured in **hertz**. If the cycle occurs once per second, the frequency is 1 hertz. If the cycle is 300 times per second, the frequency is 300 hertz.

To accurately measure duty cycle, pulse width, and frequency, the meter's trigger level must be set. The trigger level tells the meter when to start counting. Trigger levels can be set at certain voltage levels or at a rise or fall in the voltage. Normally, meters have a built-in trigger level that corresponds with the voltage setting. If the voltage does not

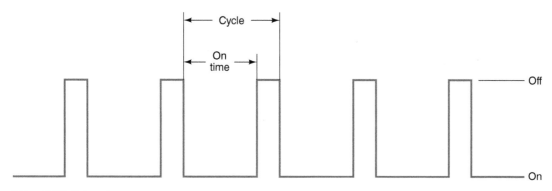

Figure 2-20 Duty cycle.

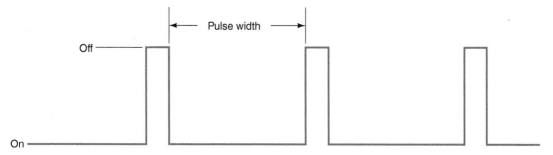

Figure 2-21 Pulse width.

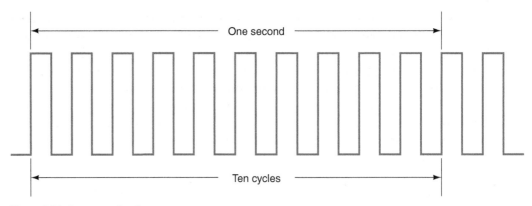

Figure 2-22 Frequency signal.

reach the trigger level, the meter will not begin to recognize a cycle. On some meters, you can select between a rise or fall in voltage to trigger the cycle count. A rise in voltage is a positive increase in voltage. This setting is used to monitor the activity of devices whose power feed is controlled by a computer. A fall in voltage is negative voltage. This setting is used to monitor ground-controlled devices.

Reading the DVOM

With the increased use of the DVOM, it is important for the technician to be able to accurately read the meter. By becoming proficient in the use of the DVOM, technicians will have confidence in their conclusions and recommended repairs. This will eliminate the replacement of parts that were not faulty. There are deviations between the different DVOM manufacturers as to the way the display is presented, but most follow the method described here.

Multimeters either have an "auto range" feature, in which the meter automatically selects the appropriate scale or must be manually set to a particular scale. Either way, to designate particular ranges and readings, meters display a prefix before the reading or range. Meters use the prefix because they cannot display long numbers. Values such as 20,200 Ω cannot be displayed as a whole number. As a result, scales are expressed as a multiple of tens or use the prefix K, M, m, and μ (**Table 2.2**). The prefix K stands for kilo and represents 1,000 units. For example, a reading of 10K equals 10,000. An M stands for mega and represents 1,000,000 units. A reading of 10M would represent 10,000,000. An m stands for milli and represents 0.001 of a unit. A reading of 10m would be 0.010. The symbol μ stands for micro and represents 0.000001 of a unit. In this case a reading of 125.0μ would represent 0.000125.

TABLE 2-2   COMMON PREFIXES USED ON DIGITAL MULTIMETERS

Prefix	Symbol	Relation to Basic Unit
Mega	M	1,000,000
Kilo	K	1,000
Milli	m	0.001 or $\dfrac{1}{1,000}$
Micro	μ	0.000001 or $\dfrac{1}{1,000,000}$
Nano	n	0.000000001
Pico	p	0.000000000001

If the display has no prefix before the unit being measured (V, A, Ω), the reading displayed is read directly. For example, if the reading was 1.243 V, the actual voltage value is 1.243. However, if there is a prefix displayed, then the decimal point will need to be floated to determine actual readings. If the prefix is M (mega), then the decimal is floated six places to the right. For example, a reading display of 2.50 MΩ is actually 2,500,000 ohms. A reading display of 0.250 MΩ is actually 250,000 ohms.

A prefix of K (kilo) means the decimal point needs to move three places to the right. For example, a display reading of 56.4 KΩ is actually 56,400 ohms. A reading of 1.264 KΩ is actually 1,264 ohms.

If the prefix is m (milli), the decimal is floated three places to the left. For example, a reading of 25.4 mA represents 0.0254 amperes. A display of 165.0 mA is actually 0.165 amperes.

Finally, if the prefix is a μ (micro), the decimal is floated six points to the left. A reading displayed as 125.3μ would represent 0.0001253 amperes while a reading of 4.6 μA is actually 0.0000046 amperes.

When using the ohmmeter function of the DVOM, make sure power to the circuit being tested is turned off. Also, be sure to calibrate the meter before taking measurements. This is done by holding the two test leads together. Most DVOMs will self-calibrate while others will need to be adjusted by turning a knob until the meter reads zero. Connect the DVOM in parallel to the portion of the circuit being tested. If continuity is good, the DVOM will read zero or close to zero even on the lowest scale. If the continuity is very poor, the DVOM will display an infinite reading. This reading is usually shown as a "1.000," a "1," or an "OL."

OL stands for *over limit.*

INSULATION TESTER

An insulation tester can be a stand-alone meter or an additional function of a DVOM (**Figure 2-23**). It is used to measure the insulation resistance on the powered-down HV system when servicing a HEV. Isolation tests are performed only on systems that have the power removed.

To test for insulation leakage of the high-voltage system, begin by inserting the insulation test probe into the + terminal and the ground lead into the − terminal. Do not connect the leads to the volt-ohm terminals. Turn the knob to the "INSULATION" position. In

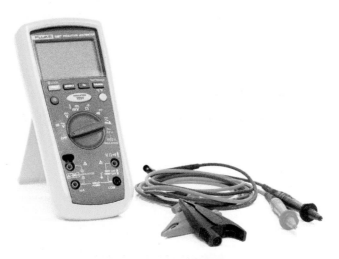

Figure 2-23 An insulation meter is used on many HEVs to determine insulation leaks in the high-voltage system.

this position, a battery level check is performed. Confirm that the battery level test has passed. Press the "RANGE" button to select the desired voltage range. If performing an isolation test on an HEV, this should be set to 500 volts. Next, connect the leads to the circuit you are testing. The meter will automatically detect if the circuit is powered. The meter display should indicate a series of dashed lines. If the meter displays >30 V, then there is voltage greater than 30 volts in the circuit and the test meter will not perform the insulation test. Push and hold the "TEST" button on the red insulation test lead. While the test is in progress, the applied voltage will be displayed in the lower right corner on the screen and the resistance in M ohms or G ohms will be displayed in the center of the screen. Leave the test leads attached to the circuit and release the "TEST" button. The meter will continue to display the resistance reading while the circuit is discharged through the meter.

LAB SCOPES AND OSCILLOSCOPES

Special Tool

Lab scope

Oscilloscopes are commonly called scopes.

An oscilloscope is a visual voltmeter (**Figure 2-24**). For many years, technicians have used scopes to diagnose ignition, fuel injection, and charging systems. These scopes, called "tune-up scopes," were normally part of a large diagnostic machine, although some were stand-alone units. In recent years, an electronic scope, referred to as a lab scope, has become the diagnostic tool of choice for many good technicians.

The oscilloscope is very useful in diagnosing many electrical problems quickly and accurately. Digital and analog voltmeters do not react fast enough to read systems that cycle quickly. The oscilloscope may be considered as a very-fast-reacting voltmeter that reads and displays voltages. The scope allows the technician to view voltage over time (**Figure 2-25**). These voltage readings appear as a voltage trace on the oscilloscope screen.

An upward movement of the voltage trace on an oscilloscope screen indicates an increase in voltage, and a downward movement of this trace represents a decrease in voltage. As the voltage trace moves across an oscilloscope screen, it represents a specific length of time. Most oscilloscopes of this type are referred to as analog scopes or real-time scopes. This means the voltage activity is displayed without any delay.

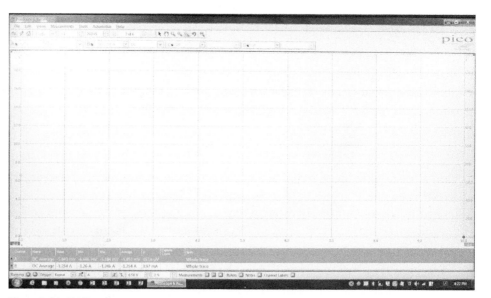

Figure 2-24 Oscilloscope.

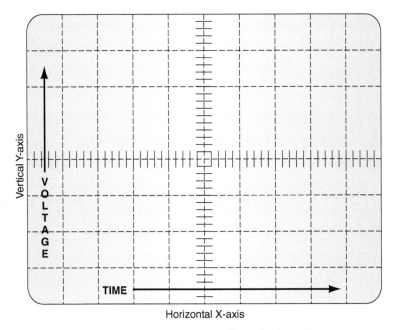

Horizontal X-axis

Figure 2-25 Grids on a scope screen serve as a time and voltage reference.

Today, most technicians use a variation of the oscilloscope called a lab scope. The divisions of the scope screen are set up as a grid pattern. The horizontal movement of the waveform represents time. Voltage is measured with the vertical position of the waveform. Since the scope displays voltage over time, the waveform moves from left (the beginning of measured time) to the right (the end of measured time). The value of the divisions can be adjusted to improve the view of the voltage waveform. For example, the vertical scale can be adjusted so that each division represents 0.5 volt, and the horizontal scale can be adjusted so that each division equals 0.005 (5 milliseconds). This allows the technician to view small changes in voltage that occur in a very short period of time. The grid serves as a reference for measurements.

The lab scope is the best method to use for verifying an input or output signal. Monitoring voltage or amperage provides the ability to measure frequency, pulse width (time), and duty cycle (time ON versus time OFF).

Since a scope displays actual voltage, it will display any electrical noise or disturbances that accompany the voltage signal (**Figure 2-26**). This noise can cause intermittent problems with unpredictable results. When a computer receives a voltage signal with noise, it will try to react to the minute changes. As a result, the computer responds to the noise rather than the voltage signal.

Electrical disturbances or **glitches** are momentary changes in the signal. These can be caused by intermittent shorts to ground, shorts to power, or opens in the circuit. These problems can occur for only a moment or may last for some time. A lab scope is handy for finding these and other causes of intermittent problems. By observing a voltage signal and wiggling or pulling a wiring harness, any looseness can be detected by a change in the voltage signal. This type of testing is commonly referred to as a "wiggle test."

The digital scope, commonly called a digital storage oscilloscope or DSO, converts the voltage signal into digital information and stores it into its memory. Some DSOs send the signal directly to a computer or a printer, or save it to a disk. To help with diagnosis, a technician can freeze the **captured signal** for closer analysis. DSOs also have the ability to capture low-frequency signals. Low-frequency signals tend to flicker when displayed on an analog screen. To have a clean waveform on an analog scope, the signal must be

The squares formed by the grid are called major divisions. These are further defined by smaller divisions called minor divisions.

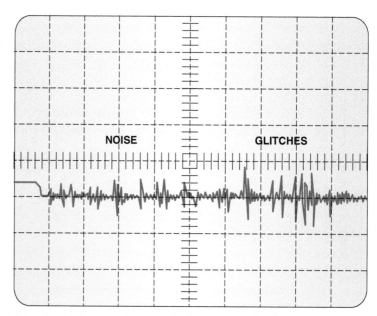

Figure 2-26 RFI noise and glitches showing as voltage signals.

repetitive and occurring in real time. The signal on a DSO is not quite real time. Rather, it displays the signal as it occurred a short time before.

This delay is actually very slight. Most DSOs have a sampling rate of one million samples per second. This is quick enough to serve as an excellent diagnostic tool. This fast sampling rate allows slight changes in voltage to be observed. Slight and quick voltage changes cannot be observed on an analog scope.

Both an analog and a digital scope can be dual-trace scopes (**Figure 2-27**). This means they both have the capability of displaying two traces at one time. By watching two traces simultaneously, you can watch the cause and effect of a sensor, as well as compare a good or normal waveform to the one being displayed.

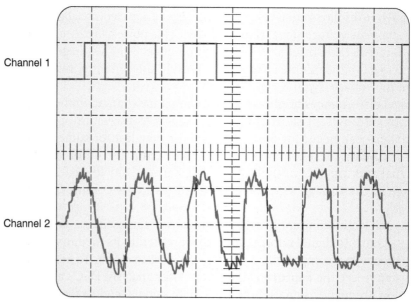

Figure 2-27 A dual-trace scope can show two patterns at the same time.

Waveforms

A waveform represents voltage over time. Any change in the amplitude of the trace indicates a change in the voltage. When the trace is a straight horizontal line, the voltage is constant (**Figure 2-28**). A diagonal line up or down represents a gradual increase or decrease in voltage. A sudden rise or fall in the trace indicates a sudden change in voltage.

Scopes can display AC and DC, one at a time or both, as in the case of noise caused by RFI. The consistent change of polarity and amplitude of the AC signal causes slight changes in the DC voltage signal. A normal AC signal changes its polarity and amplitude over a period of time. The waveform created by AC voltage is typically a sine wave (**Figure 2-29**). One complete sine wave shows the voltage moving from zero to its positive peak, moving down through zero to its negative peak, and returning to zero. If the rise and fall from positive and negative is the same, the wave is said to be **sinusoidal**. If the rise and fall are not the same, the wave is nonsinusoidal.

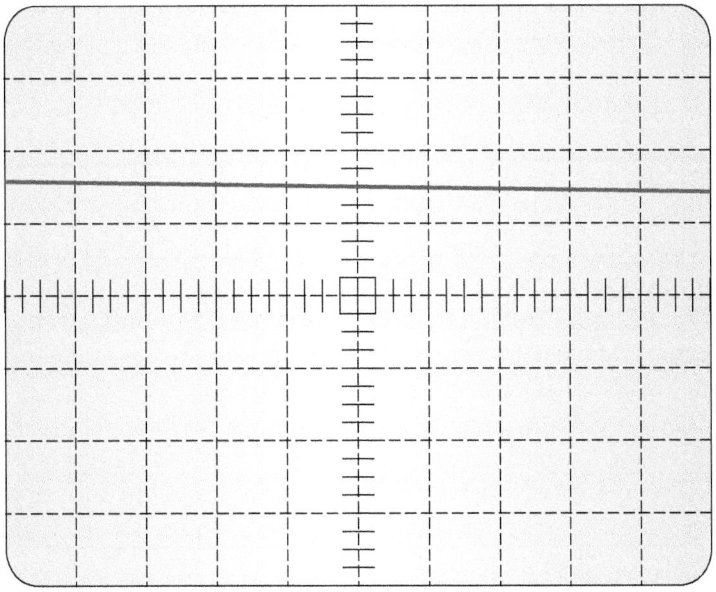

The waveform on a scope is commonly called a trace.

Figure 2-28 A waveform showing a constant voltage.

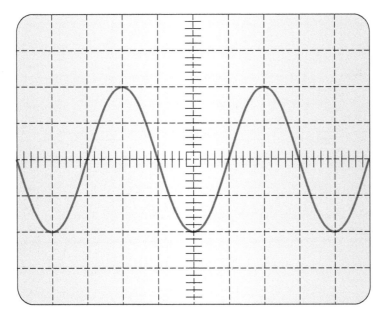

Figure 2-29 An AC voltage sine wave.

One complete sine wave is a cycle. The number of cycles that occur per second is the frequency of the signal. Checking frequency or cycle time is one way of checking the operation of some electrical components. Input sensors are the most common components that produce AC voltage. Permanent magnet voltage generators produce an AC voltage that can be checked on a scope (**Figure 2-30**). AC voltage waveforms should also be checked for noise and glitches. These may send false information to the computer.

DC voltage waveforms may appear as a straight line or a line showing a change in voltage. Sometimes a DC voltage waveform will appear as a **square wave**, which shows voltage making an immediate change (**Figure 2-31**). This type of wave represents voltage being applied (circuit being turned on), voltage being maintained (circuit remaining on), and no voltage applied (circuit is turned off). Of course, a DC voltage waveform may also show gradual voltage changes.

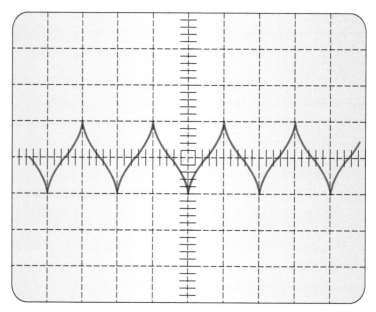

Figure 2-30 An AC voltage trace from a typical permanent magnetic generator-type sensor.

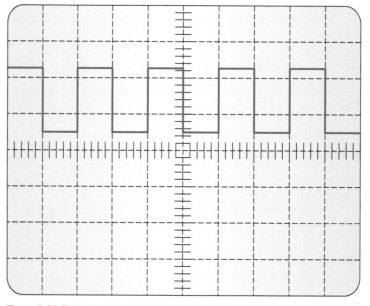

Figure 2-31 Typical square wave.

Scope Controls

Depending on the manufacturer and the model of the scope, the type and number of its controls will vary. However, most scopes have controls for intensity, vertical (Y-axis) adjustments, horizontal (X-axis) adjustments, and trigger adjustments. The intensity control is used to adjust the brightness of the trace. This allows for clear viewing regardless of the light around the scope screen.

The vertical adjustment controls the voltage displayed. The voltage setting of the scope is the voltage that will be shown per division (**Figure 2-32**). If the scope is set at 0.5 (500 millivolts) volt, a 5-volt signal will need 10 divisions. Likewise, if the scope is set to 1 volt, 5 volts will need only five divisions. While using a scope, it is important to set the vertical so that voltage can be accurately read. Setting the voltage too low may cause the waveform to move off the screen, while setting it too high may cause the trace to be flat and unreadable. The vertical position control allows the vertical position of the trace to be moved anywhere on the screen.

The horizontal position control allows the horizontal position of the trace to be set on the screen. The horizontal control is actually the time control of the trace (**Figure 2-33**). Setting the horizontal control is setting the time base of the scope's sweep rate. If the time per division is set too low, the complete trace may not show across the screen. Also, if the time per division is set too high, the trace may be too crowded for detailed observation. The time per division (TIME/DIV) can be set from very short periods of time (millionths of a second) to full seconds.

Trigger controls tell the scope when to begin a trace across the screen. Setting the trigger is important when trying to observe the timing of something. Proper triggering will allow the trace to repeatedly begin and end at the same points on the screen. There are usually numerous trigger controls on a scope. The trigger mode selector has a NORM and AUTO position. In the NORM setting, no trace will appear on the screen until a voltage signal occurs within the set time base. The AUTO setting will display a trace regardless of the time base.

Slope and level controls are used to define the actual trigger voltage. The slope switch determines whether the trace will begin on a rising or falling of the voltage signal (**Figure 2-34**). The level control determines where the time base will be triggered according to a certain point on the slope.

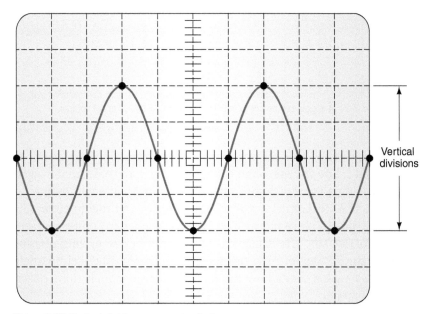

Figure 2-32 Vertical divisions represent voltage.

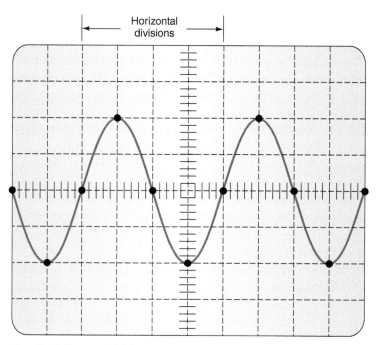

Figure 2-33 Horizontal divisions represent time.

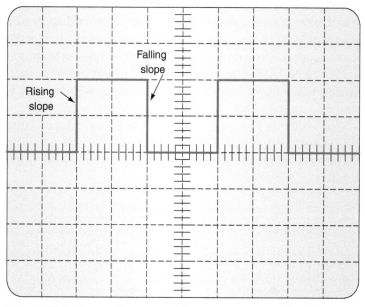

Figure 2-34 A trigger can be set to start the trace with a rise or fall of voltage.

A trigger source switch tells the scope which input signal to trigger on. This can be Channel 1, Channel 2, line voltage, or an external signal. External signal triggering is very useful when observing a trace of a component that may be affected by the operation of another component. An example of this would be observing fuel injector activity when changes in throttle position are made. The external trigger would be voltage change at the throttle position sensor. The displayed trace would be the cycling of a fuel injector. Channel 1 and Channel 2 inputs are determined by the points of the circuit being probed. Some scopes have a switch that allows inputs from both channels to be observed at the same time or alternately.

SCAN TOOLS

The introduction of powertrain computer controls brought with it the need for tools capable of troubleshooting electronic control systems. There are a variety of computer **scan tools** available today that do just that (**Figure 2-35**). Connected to the computer through diagnostic connectors, a scan tool can access DTCs, run tests to check system operations, and monitor the activity of the system. Trouble codes and test results are displayed on an LED screen or printed out on the scanner printer.

A scan tool receives its testing information from one of several sources. Some scan tools have a programmable read-only memory (PROM) chip that contains all the information needed to diagnose specific model lines. This chip is normally contained in a cartridge, which is plugged into the tool. The type of vehicle being tested determines the appropriate cartridge that should be inserted. These cartridges contain the test information for that particular car. A cartridge is typically needed for each make and model vehicle. As new systems are introduced on new car models, a new cartridge is made available.

Scan testers have the capability to test many on-board computer systems such as engine computers, antilock brake computers, air bag computers, and suspension computers, depending on the year and make of vehicle and the type of scan tester. In many cases, the technician must select the computer system to be tested with the scan tester after the tester is connected to the vehicle.

The scan tester is connected to specific diagnostic connectors on various vehicles. Some manufacturers have one diagnostic connector and they connect the data wire from each on-board computer to a specific terminal in this connector. Other vehicle manufacturers have several different diagnostic connectors on each vehicle, and each of these connectors may be connected to one or more on-board computers. A set of connectors is supplied with the scan tester to allow tester connection to various diagnostic connectors on different vehicles.

Special Tool
Scan Tool

The scan tool is also called a *scanner* and a *scan tester.*

Figure 2-35 A typical handheld scan tester.

The scan tester must be programmed for the model year, make of vehicle, and type of engine. With some scan testers, this selection is made by pressing the appropriate buttons on the tester as directed by the digital tester display. On other scan testers, the appropriate memory card must be installed in the tester for the vehicle being tested. Some scan testers have a built-in printer to print test results, while other scan testers may be connected to an external printer.

As automotive computer systems become more complex, the diagnostic capabilities of scan testers continue to expand. Many scan testers now have the capability to store, or freeze, data into the tester during a road test and then play back this data when the vehicle is returned to the shop.

Some scan testers now display diagnostic information based on the fault code in the computer memory. Service bulletins published by the scan tester manufacturer may be indexed by the tester after the vehicle information is entered in the tester. Other scan testers will display sensor specifications for the vehicle being tested.

Scan Tester Features

Scan testers display data and **diagnostic trouble codes (DTCs)** associated with the operation of computer systems and perform many other diagnostic functions. DTCs are fault codes that represent a circuit failure in a monitored system. On many vehicles, scan testers have the capability to diagnose various computer systems such as engine, transmission, antilock brake system (ABS), suspension, and air bag. Scan testers vary depending on the manufacturer, but many scan testers have the following features:

1. A display window displays data and messages to the technician. Messages are displayed from left to right. Most scan testers display at least four readings at the same time.
2. A memory cartridge that plugs into the scan tester. These memory cartridges are designed for specific vehicles and electronic systems. For example, a different cartridge may be required for the transmission computer and the engine computer. Most scan tester manufacturers supply memory cartridges for domestic and imported vehicles.
3. A power cord that connects from the scan tester to the battery terminals or cigarette lighter socket.
4. An adapter cord that plugs into the scan tester and connects to the data link connector (DLC) on the vehicle (**Figure 2-36**). A special adapter cord is supplied with the tester for the diagnostic connector on each make of vehicle.
5. A serial interface for connecting optional devices, such as a printer, terminal, or personal computer.
6. A keypad that allows the technician to enter data and reply to tester messages.

Typical keys that may be on a scan tester include the following:

1. Numbered keys covering digits 0 through 9.
2. Horizontal or vertical arrow keys that allow the technician to move backward and forward through test modes and menus.
3. ENTER keys to enter information into the tester.
4. PAGE BACK key that allows the technician to interrupt the current procedure and go back to the previous modes.
5. "F" keys to allow the technician to perform special functions described in the scan tester manufacturer's manuals.
6. MORE key that allows the technician to obtain additional diagnostic information from the scan tester software.
7. YES and NO keys to allow the technician to select or reject specific procedures.

Figure 2-36 The scan tester connects to the diagnostic link connector (DLC).

⚡ WARNING **The tester manufacturer's and vehicle manufacturer's recommended scan tester diagnostic procedures must be followed while diagnosing computer systems. Improper test procedures may result in scan tester damage and computer system damage.**

The advantages offered by the use of a scan tool include the following:

1. A scan tester provides quick access to data from various on-board computers. Some vehicles have several DLCs to which the scan tester must be connected to access data from a specific computer. For example, some vehicles have separate DLCs to access the powertrain control module and ABS computer data. Vehicles that have on-board diagnostic (OBD II) systems have a central DLC and data links from the various on-board computers to this DLC. Accessing this computer data greatly reduces diagnostic time.
2. A wide variety of modules are available for many scan testers. These modules allow the same scan tester to display data from many vehicles, including imported vehicles. Some scan tester modules access service bulletin information related to engine and transmission problems. This information is available in a book published by the scan tester manufacturer.
3. The vehicle can be driven on a road test with the scan tester connected to the DLC. This allows the technician to observe computer data during various operating conditions when a specific problem may occur. Most scan testers have a snapshot capability that freezes computer data into the scan tester memory for a specific period of time. This data may be played back after the technician returns to the shop.
4. Most scan testers can be connected to a printer, and a copy of the scan tester data can be printed. This allows improved communication between the customer, service writer, and the technician.
5. A scan tester can be connected to a personal computer (PC). This connection allows data to be transferred from the scan tester to the PC. This data may be saved and recalled at a future time. With a computer modem, this information may be transferred to an off-site diagnostic center for analysis.

⚠ **Caution**
Always keep scan tester leads away from rotating parts such as belts and fan blades. Personal injury or property damage may result if scan tester leads become tangled in rotating parts.

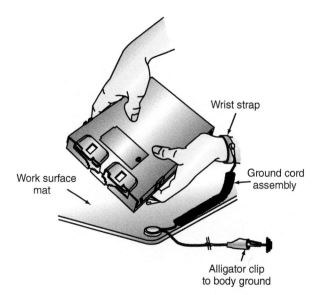

Figure 2-37 Static straps are needed to protect electronic devices.

STATIC STRAP

You are probably familiar with static charges in one form or another. The most common experience with static electricity is when you slide your feet across a carpeted floor and then touch something. You might feel and see a slight static discharge. The action of sliding your feet across the carpet placed a slight electrical charge on you. A change in the number of electrons on you puts you at a different charge level than the objects around you. When you touch them, there is a discharge between you and the object. Although this discharge generally does nothing to you other than to perhaps surprise you, it can do potentially great damage to electronic circuitry. Today's technicians must realize that static electricity will have to be discharged safely before they begin working on an electronic component or processor.

To effectively work on these circuits, some precautions are necessary. Generally, these can be summarized by the statement that you must be at the same electrical potential as the component you are working on and the vehicle you are working in. Many manufacturers suggest the use of static straps that connect the technician, the component, and the ground system of the vehicle together (**Figure 2-37**). The theory behind this is to place all things that will touch at the same electrical potential so that a discharge will not take place. Even if you do not have all the special static straps, run jumper ground wires between the components and the vehicle, and ground yourself to the vehicle before you begin working.

⚠ **Caution**

Reduce electrical load prior to connecting the memory keeper. Close all doors and make sure the ignition is in the OFF position.

MEMORY KEEPERS

Memory keepers, or battery backups, maintain the computer's memory when the battery is disconnected for service. Such components as radio presets, clock settings, memory seat positions, and computer functions are lost when the battery is disconnected. A simple memory keeper (**Figure 2-38**) plugs into the power outlet (cigarette lighter) socket, and a 9-volt battery will keep the settings for up to 4 hours. These devices should not be used if service is being performed to the air bag system.

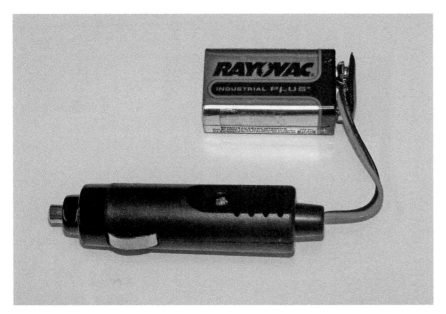

Figure 2-38 A memory keeper will keep the computer memories operational while the battery is disconnected.

SERVICE INFORMATION

With today's complex electrical systems, it would be impossible to repair every customer concern that is brought into the service bay without having the proper service information. The service information (in either paper form or electronic format) is one of the most important and valuable tools for today's technician. It provides information concerning system description, service procedures, specifications, and diagnostic information. In addition, the service information provides information concerning wiring harness connections and routing, component location, and fluid capacities. Service information can be supplied by the vehicle manufacturer or through aftermarket suppliers.

It is also important to stay current with updated service information. This is usually published as Technical Service Bulletins (TSBs). These documents may provide information concerning fixes for a problem, new part numbers to replace a defective unit, corrections to service information, and general information of system operation.

To obtain the correct information, you must be able to identify the engine you are working on. This may involve using the vehicle identification number (VIN). This number has a code for model year and engine. Which numbers are used varies between manufacturers, but the service information will provide instructions for proper VIN usage.

Procedural information provides the steps necessary to perform the task. Most service manuals provide illustrations to guide the technician through the task. To get the most out of the service manual, you must use the correct manual for the vehicle and system being worked on, and follow each step in order. Some technicians lead themselves down the wrong trail by making assumptions and skipping steps.

Torque, end play, and clearance specifications may be located within the text of the procedural information. In addition, specifications may be provided in a series of tables. The heading above the table provides a quick reference to the type of specification information being provided.

Diagnostic procedures are often presented in a chart form (**Figure 2-39**) or a tree. The procedure guides you through the process as system tests are performed. The result

A1 CHECK THE DTCs FROM BOTH THE CONTINUOUS AND ON-DEMAND PAM SELF-TESTS

Check the PAM DTCs from the continuous and on-demand self-tests.

Are any PAM DTCs recorded?

Yes : REFER to DTC CHARTS .

No : GO to A2.

A2 CHECK THE MESSAGE CENTER FOR CORRECT OPERATION

While observing the message center, disable and enable the parking aid system.

Does the message center display REAR PARK AID OFF when the parking aid system is disabled, and display REAR PARK AID ON when the parking aid system is enabled?

Yes : GO to A3.

No : REFER to INFORMATION CENTER - INSTRUMENTATION, MESSAGE CENTER, AND WARNING CHIMES .

A3 CHECK FOR CORRECT REVERSE GEAR INPUT

Apply the parking brake.

Ignition ON.

Select REVERSE.

Enter the following diagnostic mode on the scan tool: PAM DataLogger.

Monitor the PAM TRANSGR PID.

Does the PID read REVERSE (transmission in reverse)?

Yes : GO to A4.

No : GO to A5.

A4 CHECK THE PARKING AID SENSORS FOR CORRECT ALIGNMENT

Figure 2-39 A diagnostic chart leads the technician to the cause of the fault.

of a test then directs you to different steps. Keep following the steps until the problem is isolated.

Since the service manual is divided into several major component areas, a table of contents is provided for easy access to the information. Each component area of the vehicle is covered under a section in the service information. Using the table of contents identifies the section to turn to. Once in the appropriate section, a smaller, more specific table of contents will direct you to the page on which the information is located. Due to the extensive amount of information provided in service manuals, the manual may be divided into several volumes.

Today, most service and parts information is provided through computer services (**Figure 2-40**). This is a popular method since libraries of printed service information can

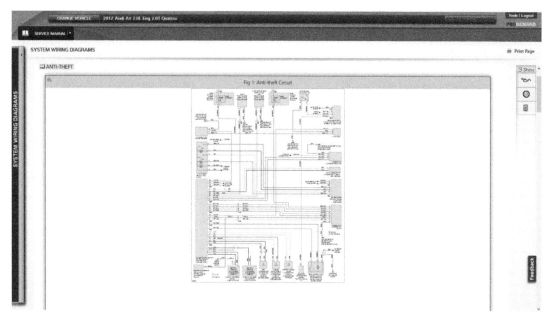

Figure 2-40 Computerized service information and web-based materials retrieval systems have replaced paper service information.

take large areas of space. Computerized systems can have the information stored on disks, or the computer can be connected via the internet to a central database. The computer system helps the technician find the required information quicker and easier than in a book-type manual. Using the computer keyboard, light pen, touch-sensitive screen, or mouse, the technician makes choices from a series of menus on the monitor. If needed, the information can be printed to paper.

Another tool that can be of assistance to the technician is hotline services. There are many companies and organizations that provide online assistance to technicians. The hotline assistants use database information, along with factory service manuals, TSBs, and other sources, to help the technician repair the vehicle.

Also, some companies and organizations provide a service of tracking the service history of vehicles. Manufacturers will generally do this for any of their vehicles that are serviced in their dealerships. Subscription services are available that will provide similar information to aftermarket service centers. This information lets the technician know if service has been performed on the vehicle that may be related to the present problem. Also, some Web site organizations will track if the vehicle has ever been involved in an accident, stolen, or caught in a flood.

WORKING AS AN ELECTRICAL SYSTEMS TECHNICIAN

To be a successful automotive technician, you need to have good training, a desire to succeed, and a commitment to becoming a good technician and a good employee. A good employee works well with others and strives to make the business successful. The required training is not just in the automotive field. Good technicians need to have good reading, writing, and math skills. These skills will allow you to better understand and use the material found in service information and textbooks, as well as provide you with the basics for good communication with customers and others.

Compensation

Technicians are typically paid according to their abilities. Most often, new or apprentice technicians are paid by the hour. While being paid they are learning the trade and the business. Time is usually spent working with master technicians or doing low-skill jobs. As apprentices learn more, they can earn more and take on more complex jobs. Once technicians have demonstrated a satisfactory level of skills, they can go on **flat rate**.

Flat rate is a pay system in which technicians are paid for the amount of work they do. Each job has a flat rate time. Pay is based on that time, regardless of how long it took to complete the job. To explain how this system works, let's look at a technician who is paid $30.00 per flat rate hour. If a job has a flat rate time of 3 hours, the technician will be paid $90.00 for the job, regardless of how long it takes to complete. Experienced technicians beat the flat rate time most of the time. Their weekly pay is based on the time "turned in," not on the time spent. If the technician turns in 60 hours of work in a 40-hour workweek, he or she actually earned $45.00 each hour worked. However, if he or she turned in only 30 hours in the 40-hour week, the hourly pay is $22.50.

The flat rate system favors good technicians who work in a shop that has a large volume of work. The use of flat rate times allows for more accurate repair estimates to the customers. It also rewards skilled and productive technicians.

Workplace Relationships

When you begin a job, you enter into a business agreement with your employer. When you become an employee, you sell your time, skills, and efforts. In return, your employer pays you for these resources. As part of the employment agreement, your employer also has certain responsibilities:

- Instruction and Supervision—You should be told what is expected of you. A supervisor should observe your work and tell you if it is satisfactory and offer ways to improve your performance.
- A Clean, Safe Place to Work—An employer should provide a clean and safe work area as well as a place for personal cleanup.
- Wages—You should know how much you are to be paid, what your pay will be based on, and when you will be paid before accepting a job.
- Fringe Benefits—When hired, you should be told what benefits to expect, such as paid vacations and employer contributions to health insurance and retirement plans.
- Opportunity—You should be given a chance to succeed and possibly advance within the company.
- Fair Treatment—All employees should be treated equally, without prejudice or favoritism.

On the other side of this business transaction, employees have responsibilities to their employers. Your obligations as an employee to the employer include the following:

- Regular Attendance—A good employee is reliable.
- Following Directions—As an employee, you are part of a team, and doing things your way may not serve the best interests of the company.
- Responsibility—You must be willing to answer for your behavior and work. You need to also realize that you are legally responsible for the work you do.
- Productivity—Remember, you are paid for your time as well as your skills and effort.
- Loyalty—Loyalty is expected; by being loyal you will act in the best interests of your employer, both on and off the job.

Customer Relations

Another responsibility you have is good customer relations. Learn to listen and communicate clearly. Be polite and organized, particularly when dealing with customers. Always be as honest as you possibly can.

Look like and present yourself as a professional, which is what automotive technicians are. Professionals are proud of what they do and they show it. Always dress and act appropriately and watch your language, even when you think no one is near.

Respect the vehicles you work on. They are important to the lives of your customers. Always return the vehicle to the owner in a clean, undamaged condition. Remember, a vehicle is the second-largest expense a customer has. Treat it that way. It doesn't matter if you like the car. It belongs to the customer; treat it respectfully.

Explain the repair process to the customer in understandable terms. Whenever you are explaining something to a customer, make sure you do this in a simple way without making the customer feel stupid. Always show the customer respect and be courteous. Not only is this the right thing to do but it also leads to loyal customers.

Communicating with the Customer

Depending on the size of the service center that you are working in, you may or may not talk directly with the customer. A service advisor or manager might be in between the consumer and you, the technician. In either case, getting the correct information from the customer cannot be overemphasized. The customer is likely the person who was driving the vehicle when the problem showed up. The conversation that someone has with the customer can be extremely useful and save hours of fruitless work. A repair ticket that states "drivability problem" will require the technician to figure out the driving conditions that were present when the problem occurred. It is possible that the technician will not be able to duplicate the conditions and not observe the problem. The vehicle is returned to the consumer with the note of "no problem found." Think about how frustrating this could be, especially if the costumer experiences the problem the very next day. The conversation with the customer should have revealed important information for the technician. When did the problem occur? What specifically did the vehicle do or not do? What was the outside temperature? Was the engine warm or cold? What driving conditions produced the problem? Can you duplicate the problem with the technician in the vehicle?

Think about how much more information the technician has in the second example. It is likely that less time will be necessary to fix the vehicle because the technician has a starting point. Once you get out in the field, try to develop your communication skills and especially your ability to listen to the customer. The customer's information, if you have heard it, will save you countless hours of frustration and no doubt result in a better repair. Better repairs bring customers back to the service center the next time they need service.

Completing a Work Order

The **work order** is a contract between the customer and the shop to have specified service performed. In larger shops, a service writer is employed to have contact with the customer and to begin the work order process. In smaller shops, the technician may be required to also write up the work order. Regardless of who is responsible for the write-up, a work order must be written for every vehicle that is brought to the shop for repairs.

The work order must contain certain information (**Figure 2-41**). This includes information about the customer, the customer's vehicle, and the customer's concern or service request. In addition, estimates of cost to perform the service are included on the work order along with estimated date and time the work will be completed.

AUTHOR'S NOTE The work order is also referred to as a Repair Order.

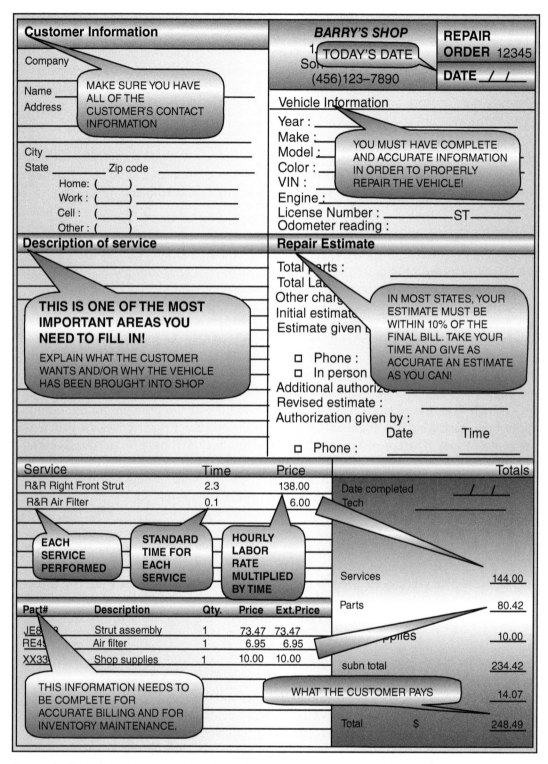

Figure 2-41 The work order must be completed properly.

The work order is a legal document that is an agreement to provide services and legally protects both the customer and the shop. The work order is signed by the customer who gives the shop the authorization to perform the service and accepts all terms that are noted on the work order. Also, the customer is protected against being charged more than the estimate given by the shop. If the costs exceed the original estimate, the customer must authorize the higher amount. Some states allow shops to be within 10% of the estimate, while others hold the shop to the amount that was estimated. Other functions of the work order are for use by payroll and general record keeping.

Today, most service facilities use a form of shop management software. The information for the completion of the work order is input using the computer's keyboard. The software package also helps in the estimation of repair costs. The software also takes information from the work order and saves it in various files, each defined by its purpose.

One of the responsibilities of the technician is to document the cause and correction of the concern. It may not be possible to provide an estimate to the customer at the time of the original write-up since the problem has not been diagnosed yet. At this time the customer agrees to pay the shop to perform the diagnostic routine. Often a limit is agreed upon by the parties. Once the problem has been identified and the root cause is determined, the cost of the repair must be estimated and the customer gives approval. The cause of the complaint needs to be documented in such a manner that the customer and service writer understand what the issue is. The correction actions are documented for purposes of warranty. Whatever the repair entailed is to be recorded. For example, if a wire was repaired, document this repair and the location of the repair. If parts are installed, list the parts.

ASE Certification

An obvious sign of your knowledge and abilities, as well as your dedication to the trade, is ASE certification. The National Institute for Automotive Service Excellence (ASE) has established a voluntary certification program for automotive, heavy-duty truck, auto body repair, and engine machine shop technicians. In addition to these programs, ASE also offers individual testing in some specialty areas. This certification system combines voluntary testing with on-the-job experience to confirm that technicians have the skills needed to work on today's vehicles (**Figure 2-42**). ASE recognizes two distinct levels of service capability—the automotive technician and the master automotive technician.

After passing at least one exam and providing proof of two years of hands-on work experience, you become ASE certified. Retesting is necessary every five years to remain certified. A technician who passes one examination receives an automotive technician

Figure 2-42 ASE certification communicates your pride in your profession to the customers.

shoulder patch. The master automotive technician patch is awarded to technicians who pass all eight of the basic automotive certification exams. You may receive credit for one of the two years by substituting relevant formal training in one, or a combination, of the following:

- High school training. Three years of training may be substituted for one year of experience.
- Post-high school training. Two years of post-high school training in a trade school, technical institute, or community college may be counted as one year of work experience.
- Short courses. For shorter periods of post-high school training you may substitute two months of training for one month of work experience.
- Apprenticeship programs. You may receive full credit for the experience requirement by satisfactorily completing a three- or four-year apprenticeship program.

Each certification test consists of 40 to 80 multiple-choice questions. A panel of technical service experts write the questions, including domestic and import vehicle manufacturers, repair and test equipment and parts manufacturers, working automotive technicians, and automotive instructors. All questions are pretested and quality checked on a national sample of technicians before they are included in the actual test. Many test questions force the student to choose between two distinct repair or diagnostic methods. The test questions focus on basic technical knowledge, repair knowledge and skill, and testing and diagnostic knowledge and skill.

DIAGNOSTICS

The true measure of a good technician is an ability to find and correct the cause of problems. Service manuals and other information sources will guide you through the diagnosis and repair of problems, but these guidelines will not always lead you to the exact cause of the problem. To do this you must use your knowledge and take a logical approach while troubleshooting. Diagnosis is not guessing, and it's more than following a series of interrelated steps in order to find the solution to a specific problem. Diagnosis is a way of looking at systems that are not functioning the way they should and finding out why. It is knowing how the system should work and deciding if it is working correctly. Through an understanding of the purpose and operation of the system, you can accurately diagnose problems.

Most good technicians use the same basic diagnostic approach. Simply because this is a logical approach, it can quickly lead to the cause of a problem. Logical diagnosis follows these steps:

1. Gather information about the problem.
2. Verify that the problem exists.
3. Thoroughly define what the problem is and when it occurs.
4. Research all available information and knowledge to determine the possible causes of the problem.
5. Isolate the problem by testing.
6. Continue testing to pinpoint the cause of the problem.
7. Locate and repair the problem.
8. Verify the repair.

SUMMARY

- A test light is used when the technician needs to "look" for electrical power in the circuit. The test light allows the technician to see if current is at a point in the circuit by lighting the light.
- A logic probe provides a means for testing voltages on electronic circuits without damaging the circuit.
- Digital multimeters (DMM) display values using liquid crystal displays instead of a swinging needle. They are computers that determine the measured value and display it for the technician.
- A voltmeter measures the voltage potential between two points in a circuit.
- An ohmmeter is used to measure the resistance of a circuit or part of a circuit.
- An ammeter is a special meter used to measure current flow in a circuit.
- Electrical noise is an unwanted voltage signal that rides on a signal. Noise is usually the result of radio frequency interference (RFI) or electromagnetic induction (EMI).
- Duty cycle is the percentage of on time the circuit component is turned on as compared to the total time of the cycle.
- A cycle is one set of changes in a signal that repeats itself several times.
- Pulse width is the amount of time, measured in milliseconds, that a component is turned on.
- Frequency is a measure of the number of cycles that occur in one second.
- Hertz is the measurement of frequency.
- A lab scope provides a visual display of electrical waves.
- Glitches may be the result of momentary shorts to ground, shorts to power, or opens in the circuit.
- A sine wave is a waveform that shows voltage-changing polarity from positive to negative.
- Having straight vertical sides and a flat top indicating a fast-acting on-off voltage state identifies a square wave pattern.
- Scan tools interface with the vehicle's computer system to allow the technician to "talk" with the computers.
- The service manual (in either paper form or electronic format) is one of the most important and valuable tools for today's technician.
- The service manual provides information concerning engine identification, service procedures, specifications, and diagnostic information.
- Technical Service Bulletins (TSBs) may provide information concerning fixes for a problem, new part numbers to replace a defective unit, corrections to service manual information, and general information of system operation.
- Service manual procedural information provides the steps necessary to perform the task.
- Since the service manual is divided into several major component areas, a table of contents is provided for easy access to the information.
- Service and parts information can also be provided through computer services.
- Technicians are typically paid according to their abilities. New or apprentice technicians are paid by the hour. Once technicians have demonstrated a satisfactory level of skills, they can go on flat rate.
- When you begin a job, you enter into a business agreement with your employer. When you become an employee, you sell your time, skills, and efforts. In return, your employer pays you for these resources.
- As part of the employment agreement, your employer also has certain responsibilities: instruction and supervision; a clean, safe place to work; wages; fringe benefits; opportunity; and fair treatment.
- Your obligations as an employee to the employer include regular attendance, following directions, responsibility, productivity, and loyalty.
- When communicating with customers, be polite, respectful, organized, and honest.
- An obvious sign of your knowledge and abilities, as well as your dedication to the trade, is ASE certification.
- The work order is a contract between the customer and the shop to have specified service performed.
- The true measure of a good technician is an ability to find and correct the cause of problems.
- Diagnosis is not guessing, and it is more than following a series of interrelated steps in order to find the solution to a specific problem.

ASE-STYLE REVIEW QUESTIONS

1. *Technician A* says a test light is ideal for checking for voltage on a low-current, low-power circuit.
   Technician B says to use an analog multimeter to test computer-controlled circuits.
   Who is correct?
   A. A only
   B. B only
   C. Both A and B
   D. Neither A nor B

2. The use of an ammeter is being discussed.
   Technician A says the ammeter is used to measure current flow.
   Technician B says the ammeter must be connected in parallel to the circuit being tested.
   Who is correct?
   A. A only
   B. B only
   C. Both A and B
   D. Neither A nor B

3. A DVOM is being used to measure current flow. The meter is displaying 85.5 mA.
   Technician A says this represents 0.0855 amperes.
   Technician B says the decimal point needs to be moved six points to the left.
   Who is correct?
   A. A only
   B. B only
   C. Both A and B
   D. Neither A nor B

4. *Technician A* says a 10% duty cycle indicates that the load device is turned on most of the time.
   Technician B says the pulse width is measured in degrees.
   Who is correct?
   A. A only
   B. B only
   C. Both A and B
   D. Neither A nor B

5. A vehicle is being tested for a draw against the battery with the ignition switch in the OFF position. The specifications state the draw should be between 10 and 30 milliamps. The DVOM reads 0.251 amps.
   Technician A says this draw is within the specification range.
   Technician B says the draw is too high.
   Who is correct?
   A. A only
   B. B only
   C. Both A and B
   D. Neither A nor B

6. *Technician A* says an ohmmeter that reads 0 ohms means there is no continuity of the circuit or component being tested.
   Technician B says an infinite reading on the ohmmeter means there is good continuity.
   Who is correct?
   A. A only
   B. B only
   C. Both A and B
   D. Neither A nor B

7. *Technician A* says a voltmeter that is connected in parallel to the load device will indicate the voltage drop across the device.
   Technician B says an ohmmeter reading of 0.00 Ω when connected in parallel to the coil of an A/C compressor indicates the coil is shorted.
   Who is correct?
   A. A only
   B. B only
   C. Both A and B
   D. Neither A nor B

8. *Technician A* says the upward voltage traces on an oscilloscope screen indicate a specific length of time.
   Technician B says the oscilloscope provides accurate voltage over time readings.
   Who is correct?
   A. A only
   B. B only
   C. Both A and B
   D. Neither A nor B

9. The best way of determining if a switch is faulty is being discussed.
   Technician A says to use a jumper wire to bypass the switch and to connect the various circuits controlled by the switch.
   Technician B says to use an ammeter to test continuity through the switch.
   Who is correct?
   A. A only
   B. B only
   C. Both A and B
   D. Neither A nor B

10. *Technician A* says the service manual (in either paper form or electronic format) is one of the most important and valuable tools for today's technician.
    Technician B says technical service bulletins (TSBs) provide information concerning fixes for a problem, new part numbers to replace a defective unit, corrections to service manual information, and general information of system operation.
    Who is correct?
    A. A only
    B. B only
    C. Both A and B
    D. Neither A nor B

ASE CHALLENGE QUESTIONS

1. *Technician A* says a voltmeter measures the electrical potential between two points of the circuit.
   Technician B says an ammeter reading of 0.00 when connected in series to a circuit indicates a short to ground.
   Who is correct?
   A. A only
   B. B only
   C. Both A and B
   D. Neither A nor B

2. A scanner allows a technician to:
   A. measure circuit resistance.
   B. determine load voltage drop.
   C. view computer inputs.
   D. load test the battery.

3. All of the following are true concerning the use of an ohmmeter EXCEPT:
   A. Connect the ohmmeter with the circuit power off.
   B. An "OL" indicates that the reading is over limits.
   C. 0.00 indicates no resistance.
   D. Connect the test leads of the meter in series to the load.

4. The signal output of a 5-volt throttle position sensor is being checked (from idle to the wide-open throttle position) with an oscilloscope. The test lead is set to the $10 \times$ position. Which of the following represents the correct position of the vertical adjustment selector?
   A. 0.1 volt
   B. 0.5 volt
   C. 2 volts
   D. 5 volts

5. *Technician A* says logic probes can be very helpful in testing analog signals.
   Technician B says the nonpowered test lights should not be used to test most computer circuits.
   Who is correct?
   A. A only
   B. B only
   C. Both A and B
   D. Neither A nor B

6. The insulation test function is being discussed.
   Technician A says the test is performed with the circuit powered.
   Technician B says the meter displays the circuit's voltage.
   Who is correct?
   A. A only
   B. B only
   C. Both A and B
   D. Neither A nor B

7. All of the following concerning the use of jumper wires are true EXCEPT:
   A. The jumper wire should be fused.
   B. The jumper wire can be used to bypass a switch.
   C. The jumper wire can be used to bypass the load device.
   D. The jumper wire can be used to supply an alternate ground circuit.

8. When a connector is unplugged and applied voltage is measured, the reading is battery voltage.
   Technician A says this means the circuit is good.
   Technician B says an open circuit voltage test does not indicate if the circuit has resistance.
   Who is correct?
   A. A only
   B. B only
   C. Both A and B
   D. Neither A nor B

9. While using a test light to check for voltage, the light comes on bright.
   Technician A says this indicates that the circuit can handle the load that is required by the test light.
   Technician B says the amperage to turn the light on is dependent on the type and rating of the bulb.
   Who is correct?
   A. A only
   B. B only
   C. Both A and B
   D. Neither A nor B

10. Which of the following statement is most correct?
    A. Voltage drop testing tests the circuit under load.
    B. An ammeter must be connected in parallel to the load.
    C. An ohmmeter must be connected in series to a circuit.
    D. An insulation meter is used to measure high amperage.

Name _____   Date _____

USING OHM'S LAW TO CALCULATE ELECTRICAL PROPERTIES

Upon completion of this job sheet, you should be able to demonstrate knowledge of automotive electrical circuits.

NATEF Correlation ―――――――――――――――――――――――――

This job sheet addresses the following **MLR/AST/MAST** task:

A.2.   Demonstrate knowledge of electrical/electronic series, parallel, and series-parallel circuits using principles of electricity (Ohm's law).

Using Ohm's law, solve the following problems:

Exercise 1—Series Circuit

Refer to the following circuit.

Use Ohm's law to calculate the following values, when $R_1 = 2$ ohms and $R_2 = 4$ ohms:

      Total circuit resistance = _____ ohms

      Circuit current = _____ amps

      Current through R_1 = _____ amps

      Current through R_2 = _____ amps

      Voltage drop across R_1 = _____ volts

      Voltage drop across R_2 = _____ volts

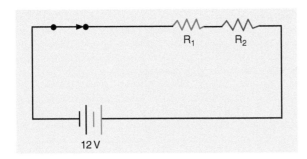

If the resistance of R_1 increases to 8 ohms, what are the new values?

      Total circuit resistance = _____ ohms

      Circuit current = _____ amps

      Current through R_1 = _____ amps

      Current through R_2 = _____ amps

      Voltage drop across R_1 = _____ volts

      Voltage drop across R_2 = _____ volts

Exercise 2—Series Circuit

Refer to the following circuit.

Use Ohm's law to calculate the following values, when $R_1 = 3$ ohms and $R_2 = 6$ ohms:

Total circuit resistance = _____ ohms

Circuit current = _____ amps

Current through R_1 = _____ amps

Current through R_2 = _____ amps

Voltage drop across R_1 = _____ volts

Voltage drop across R_2 = _____ volts

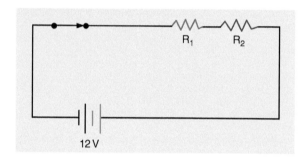

Exercise 3—Parallel Circuit

Refer to the following circuit.

Use Ohm's law to calculate the following values, when $R_1 = 12$ ohms and $R_2 = 12$ ohms:

Total circuit resistance = _____ ohms

Circuit current = _____ amps

Current through R_1 = _____ amps

Current through R_2 = _____ amps

Voltage drop across R_1 = _____ volts

Voltage drop across R_2 = _____ volts

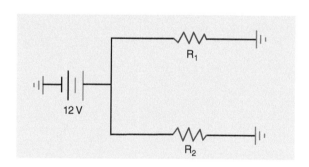

Exercise 4—Parallel-Series Circuit

Refer to the following circuit.

Use Ohm's law to calculate the following values, when $R_1 = 1$ ohm, $R_2 = 3$ ohms, $R_3 = 2$ ohms, and $R_4 = 2$ ohms:

Total circuit resistance = _____ ohms

Circuit current = _____ amps

Current through R_1 = _____ amps

Current through R_2 = _____ amps

Current through R_3 = _____ amps

Current through R_4 = _____ amps

Voltage drop across R_1 = _____ volts

Voltage drop across R_2 = _____ volts

Voltage drop across R_3 = _____ volts

Voltage drop across R_4 = _____ volts

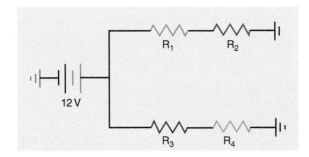

Exercise 5—Parallel-Series Circuit

Refer to the following circuit.

Use Ohm's law to calculate the following values, when R_1 = 1 ohm, R_2 = 3 ohms, R_3 = 2 ohms, and R_4 = 10 ohms:

Total circuit resistance = _____ ohms

Circuit current = _____ amps

Current through R_1 = _____ amps

Current through R_2 = _____ amps

Current through R_3 = _____ amps

Current through R_4 = _____ amps

Voltage drop across R_1 = _____ volts

Voltage drop across R_2 = _____ volts

Voltage drop across R_3 = _____ volts

Voltage drop across R_4 = _____ volts

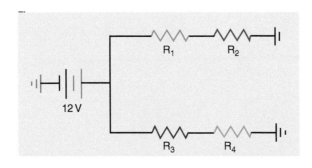

Exercise 6—Parallel Circuit

Refer to the following circuit.

Use Ohm's law to calculate the following values, when $R_1 = 2$ ohms, $R_2 = 3$ ohms, and $R_3 = 6$ ohms:

   Total circuit resistance = _____ ohms

   Circuit current = _____ amps

   Current through R_1 = _____ amps

   Current through R_2 = _____ amps

   Current through R_3 = _____ amps

   Voltage drop across R_1 = _____ volts

   Voltage drop across R_2 = _____ volts

   Voltage drop across R_3 = _____ volts

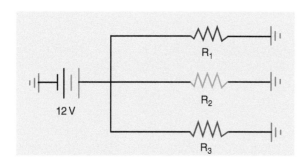

Exercise 7—Series-Parallel Circuit

Refer to the following circuit.

Use Ohm's law to calculate the following values, when $R_1 = 1$ ohm, $R_2 = 2$ ohms, $R_3 = 3$ ohms, and $R_4 = 6$ ohms:

   Total circuit resistance = _____ ohms

   Circuit current = _____ amps

   Current through R_1 = _____ amps

   Current through R_2 = _____ amps

   Current through R_3 = _____ amps

   Current through R_4 = _____ amps

   Voltage drop across R_1 = _____ volts

   Voltage drop across R_2 = _____ volts

Voltage drop across R_3 = _____ volts

Voltage drop across R_4 = _____ volts

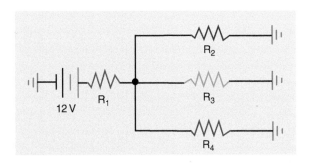

Exercise 8—Series-Parallel Circuit

Refer to the following circuit that has resistance on the ground side.

Use Ohm's law to calculate the following values, when R_1 = 6 ohms, R_2 = 3 ohms, and R_3 = 2 ohms:

Total circuit resistance = _____ ohms

Circuit current = _____ amps

Current through R_1 = _____ amps

Current through R_2 = _____ amps

Current through R_3 = _____ amps

Voltage drop across R_1 = _____ volts

Voltage drop across R_2 = _____ volts

Voltage drop across R_3 = _____ volts

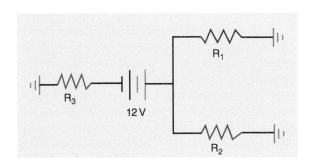

Instructor's Response

_____

_____

_____

Name _____   **Date** _____

METER SYMBOL INTERPRETATION

Upon completion of this job sheet, you should be able to convert commonly found symbols into numeric values.

NATEF Correlation ——————————————————————————————

This job sheet addresses the following **MLR** task:

A.4.   Demonstrate proper use of a digital multimeter (DMM) when measuring source voltage, voltage drop (including grounds), current flow, and resistance.

This job sheet addresses the following **AST/MAST** task:

A.3.   Demonstrate proper use of a digital multimeter (DMM) when measuring source voltage, voltage drop (including grounds), current flow, and resistance.

Convert the following values into the electrical units noted:

1. 2.4 KΩ = _____ Ω

2. 954 mV = _____ V

3. 5. 76 Ω = _____ Ω

4. 2 mA = _____ A

5. 22 KΩ = _____ Ω

6. 4.5 mA = _____ A

7. 456 mA = _____ A

8. 1,024 mV = _____ V

9. 0.786 KΩ = _____ Ω

10. 32 KΩ + 112 Ω = _____ Ω

11. 1,400 Ω = _____ KΩ

12. 0.000235 A = _____ mA

13. 0.987 V = _____ mV

14. 5 KV = _____ mV

15. 123,955 Ω = _____ KΩ

16. 144,000 mA = _____ A

17. 126 mV + 11.874 V = _____ V

18. 320,000 Ω = _____ Ω

19. 0.000045 A = _____ mA

20. 12,600 mV = _____ V

Instructor's Response

_____

_____

_____

Name _____   Date _____

USING A TEST LIGHT

Upon completion of this job sheet, you should be able to properly use a test light to check continuity in electrical circuits.

NATEF Correlation

This job sheet addresses the following **MLR** task:

A.6.   Check operation of electrical circuits using a test light.

This job sheet addresses the following **AST/MAST** task:

A.5.   Demonstrate proper use of a test light on an electrical circuit.

Tools and Materials
- A vehicle
- A test light
- Wiring diagrams for the vehicle

Describe the vehicle being worked on:

Year _____ Make _____ VIN _____

Model _____ Engine type and size _____

Procedure   **Task Completed**

1. Using the proper wiring diagram and your instructor's assistance, determine which wire is the positive feed and which wire is the ground for one of the low-beam headlights on your assigned vehicle. Identify the color of each wire:

   Positive _____ Negative _____

2. Test proper operation of the test light by connecting the negative lead to the negative post of the battery and touching the probe tip to the positive lead of the battery. If the test light is working properly, the light should illuminate. If the light does not turn on, inform your instructor. Did the light work properly?   □ Yes □ No

3. With the headlight switch in the OFF position, disconnect the headlight connector.   □

4. Connect the negative lead of the test light to a good ground. Locate the probe of the test light into the connector cavity identified as positive feed. Did the test light come on?   □ Yes □ No

   Explain your results: _____

   _____

   _____

5. Turn the headlight switch into the ON position and retest the circuit again. Did the test light come on?   □ Yes □ No

6. Turn the headlight switch OFF.   □

7. Move the wire clip of the test light to the positive post of the battery and locate the probe into the cavity identified as the ground. Does the test light come on?
   □ Yes □ No

   Explain the reason: _____

   _____

   _____

Instructor's Response

_____

_____

_____

Name _____   Date _____

USE OF A VOLTMETER

Upon completion of this job sheet, you should be able to measure available voltage and voltage drop.

NATEF Correlation

This job sheet addresses the following **MLR** task:

A.4.   Demonstrate proper use of a digital multimeter (DMM) when measuring source voltage, voltage drop (including grounds), current flow, and resistance.

This job sheet addresses the following **AST/MAST** task:

A.3.   Demonstrate proper use of a digital multimeter (DMM) when measuring source voltage, voltage drop (including grounds), current flow, and resistance.

NATEF

Tools and Materials

- A vehicle
- Wiring diagram for the vehicle
- A DMM
- Basic hand tools

Describe the vehicle being worked on:

Year _____ Make _____ VIN _____

Model _____ Engine type and size _____

Procedure   **Task Completed**

1. Set the DMM to the appropriate scale to read 12 volts DC.   ☐

2. Connect the meter across the battery (positive to positive and negative to negative). What is your reading on the meter? _____ volts

3. With the meter still connected across the battery, turn on the headlights of the vehicle. What is your reading on the meter? _____ volts

4. Keep the headlights on. Connect the positive lead of the meter to the point on the vehicle where the battery's ground cable attaches to the frame. Keep the negative lead where it is.

   What is your reading on the meter? _____ volts

   What is being measured? _____

5. Disconnect the meter from the battery and turn off the headlights.   ☐

6. Refer to the correct wiring diagram and determine what wire at the right headlight delivers current to the lamp when the headlights are on and low beams selected. Color of the wire

   _____

7. From the wiring diagram, identify where the headlight is grounded.

   Place of ground _____

8. Connect the negative lead of the meter to the point where the headlight is grounded.   ☐

9. Connect the positive lead of the meter to the power input of the headlight.   ☐

10. Turn on the headlights.

What is your reading on the meter? _____volts

What is being measured? _____

11. What is the difference between the reading here and the battery's voltage?
_____volts

12. Explain why there is a difference.

_____

_____

_____

_____

Instructor's Response

_____

_____

_____

Name _____   Date _____

USE OF AN OHMMETER

Upon completion of this job sheet, you should be able to check continuity of a circuit and measure resistance on a variety of components.

NATEF Correlation

This job sheet addresses the following **MLR** task:

A.4.   Demonstrate proper use of a digital multimeter (DMM) when measuring source voltage, voltage drop (including grounds), current flow, and resistance.

This job sheet addresses the following **AST/MAST** task:

A.3.   Demonstrate proper use of a digital multimeter (DMM) when measuring source voltage, voltage drop (including grounds), current flow, and resistance.

ASE NATEF

Tools and Materials

- A vehicle
- A DMM
- Wiring diagram for vehicle

NOTE: An ohmmeter works by sending a small amount of current through the path to be measured. Because of this, all circuits and components being tested must be disconnected from power. An ohmmeter must never be connected to an energized circuit; doing so may damage the meter. The safest way to measure ohms is to disconnect the negative battery cable before taking resistance readings.

Describe the vehicle being worked on:

Year _____ Make _____ Model _____

VIN _____ Engine type and size _____

Procedure   **Task Completed**

1. Locate the fuse panel or power distribution box.   ☐

2. With no power to the fuses, check the resistance of each fuse.   ☐

   Summarize your findings.

3. Disconnect the primary wires leading to the ignition coil.

   Now connect the leads of the digital meter across the terminals of the coil.

   Your reading is: _____ ohms

4. Reconnect the wires to the coil.

   Carefully remove one spark plug wire from the spark plug and ignition coil or distributor cap.

   Now connect the leads of the digital meter across the wire.

   Your reading is: _____ ohms

5. Carefully reinstall the spark plug wire.

   Locate the cigar lighter inside the vehicle.

   Now connect the leads of the digital meter from the heating coil to its case.

   Your reading is: _____ ohms

6. Refer to the service manual and find out how to remove the bulb in the dome light. Remove it.

   Now connect the leads of the digital meter across the bulb.

   Your reading is: _____ ohms

7. Reinstall the bulb.

   Remove the rear brake light bulb.

   Now connect the leads of the digital meter across the bulb.

   Your reading is: _____ ohms

8. Reinstall the bulb.

   Disconnect the wire connector to one of the headlights.

   From the wiring diagram, identify which terminals are for low-beam operation.

   Now connect the leads of the digital meter across the low-beam terminals.

   Your reading is: _____ ohms

9. From the wiring diagram, identify which terminals are for high-beam operation. Now connect the leads of the digital meter across the high-beam terminals.

   Your reading is: _____ ohms

10. You measured the resistance across several different light bulbs. On each you should have read a different amount of resistance. Based on your findings, which light bulb would be the brightest and which would be the dimmest? Explain why.

   _____

   _____

   _____

Instructor's Response

_____

_____

_____

Name _____   Date _____

LOCATING SERVICE INFORMATION

Upon completion of this job sheet, you should be able to locate the following service information; service procedures, technical service bulletins, and repair history.

NATEF Correlation

This job sheet addresses the following **MLR/ AST/MAST** task:

A.1.   Research vehicle service information, including vehicle service history, service precautions, and technical service bulletins.

NATEF

Tools and Materials

• Applicable service information

Describe the vehicle being worked on:

Year _____ Make _____ Model _____

VIN _____ Engine type and size _____

Procedure   **Task Completed**

1. What system has your instructor assigned you?

   _____

2. What procedure has your instructor assigned?

   _____

3. Reference the service procedures assigned. Confirm proper selection with your   ☐
   instructor.

4. Are there any precautions or warnings list in the procedure?

   Yes ☐   No ☐

   If yes, what does the first one relate to?

   _____

   _____

5. Are there any TSBs associated with the system you are assigned?   Yes ☐   No ☐

   If yes, list them by TSB number:

   _____

   _____

6. Look up any Repair History associated with the vehicle.

   Is there any Repair History?   Yes ☐   No ☐

   If yes, what is the most recent repair?

   _____

   _____

   _____

Instructor's Response

_____

_____

_____

Name _____   Date _____

LAB SCOPE SETUP

Upon completion and review of this job sheet, you should be able to setup a lab scope to test an input sensor.

NATEF Correlation

This job sheet addresses the following **MAST** task:

A.11.   Check electrical/electronic circuit waveforms; interpret readings and determine needed repairs.

NOTE: The intent of this job sheet is to acquaint you with basic setup procedures of the lab scope to obtain the proper waveform. The best reference is the lab scope manufacturer's User Guide. Also, use your instructor as a resource during this activity.

ASE NATEF

Tools and Materials

- Wiring diagram
- Lab scope
- Lift or jacks with stands

Describe the vehicle being worked on:

Year _____ Make _____ Model _____

VIN _____ Engine type and size _____

Procedure   **Task Completed**

1. What input speed sensor was assigned to you by your instructor?

   _____

2. Referring to the service information, identify the location of the input speed sensor connector. Where is it located?

   _____

   _____

3. Backprobe the signal circuit of the sensor connector.   ☐

4. Connect the lab scope to the backprobe tools and record the initial time and voltage ranges used.

   _____

   _____

5. What tigger are you going to use for the setup?

   _____

   _____

6. Observe the trace while the sensor is providing an input signal.

   General results:

   _____

   _____

   _____

7. Change the time and voltage levels to get the clearest waveform. At what setting is this accomplished?

_____

_____

_____

8. Capture the waveform and play it back. Are there any concerns indicated by the waveform?

_____

_____

_____

9. What is the highest voltage value of the waveform?

_____

_____

Instructor's Response

_____

_____

_____

CHAPTER 3
BASIC ELECTRICAL TROUBLESHOOTING AND SERVICE

Upon completion and review of this chapter, you should be able to:

- Describe how different electrical problems cause changes in an electrical circuit.
- Diagnose and repair circuit protection devices.
- Test switches with a variety of test instruments.
- Test relays and relay circuits for proper operation.
- Identify and test variable and fixed resistors with a voltmeter, ohmmeter, or lab scope.
- Test capacitors.
- Diagnose diodes for opens, shorts, and other defects.
- Diagnose transistors for opens, shorts, and other defects.
- Locate and repair opens in a circuit.
- Locate and repair shorts in a circuit.
- Locate and repair the cause of unwanted high resistance in a circuit.
- Set up a digital storage oscilloscope (DSO) and identify waveforms

Terms To Know

Analog signal
Capacitor
Circuit
Darlington
Digital signal
Diode
Feedback
Fuse

Fusible link
Gauss gauge
Normally open (NO)
Open circuit
Overload Peak
Peak-to-peak voltage
Potentiometer

Pulse width modulation (PWM)
Relay
Rheostat
Short to ground
Shorted circuit
Stepped resistor
Variable resistor

INTRODUCTION

Troubleshooting electrical problems involves the same tools and methods, regardless of which **circuit** has the problem. All electrical circuits must have proper voltage, current, and resistance. Testing for the presence of these, measuring them, and comparing your measurements to specifications are the key to effective diagnosis. To do this, you must have a solid understanding of these basic electrical properties.

Voltage is the electrical pressure that causes electrons to move provided there is a complete path for them to do so. *Current* is the aggregate flow of electrons through a wire and can be defined as the rate of electron flow. *Resistance* is defined as opposition to current flow. An electrical circuit must have resistance in it in order to change electrical energy to light, heat, or movement.

The term **circuit** means a circle and is the path of electron flow consisting of the voltage source, conductors, load component, and return path to the voltage source.

An electrical circuit may develop an open, a short, or an excessive voltage drop that will cause it to operate improperly. An **open circuit** is a circuit in which there is a break in continuity. The open can be on either the insulated side or the ground side. A **shorted circuit** decreases the resistance of the circuit. This happens by shorting across to another circuit or by shorting to ground. When there is a circuit-to-circuit short, one of the circuits is not controlled by its switch. Since the shorted circuit becomes a new parallel leg to the circuit, the entire parallel circuit will turn on and off, with the switch controlling the other circuit. With this type of problem, many strange things can happen. When a circuit is shorted to ground, a new parallel leg is present. This new leg has very low resistance and causes the current in the circuit to increase drastically.

High-resistance problems can occur anywhere in the circuit. However, the effect of high resistance is the same regardless of where it is. Additional or unwanted resistance in series with a circuit will always reduce the current in the circuit and will reduce the amount of voltage drop by the component in the circuit.

TESTING FOR CIRCUIT DEFECTS

Electrical circuits may develop an open, a short, a ground, or an excessive voltage drop that will cause the circuit to operate improperly.

Testing for Opens

It is possible to test for opens using a voltmeter, a DSO, a test light, a self-powered test light, an ohmmeter, or a fused jumper wire. The test equipment used will depend on the circuit being tested and the accessibility of the components.

The technician must determine the correct operation of the circuit before attempting to determine what is wrong. **Figure 3-1** shows the voltmeter readings that should be obtained in a properly operating parallel circuit.

Classroom Manual
Chapter 3, page 80

🔧 Special Tools
DMM
Test light
Self-powered test light
Fused jumper wire
DSO

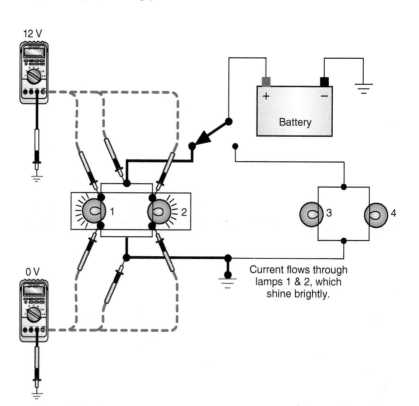

Figure 3-1 Voltmeter readings that would be expected in a properly operating parallel circuit.

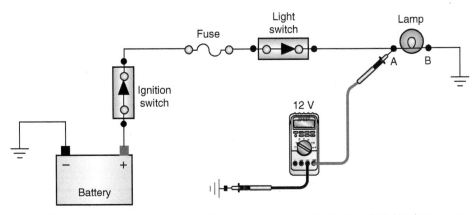

Figure 3-2 Knowing the expected voltage at different locations in the circuit will assist in locating an open.

The easiest method of testing a circuit is to start at the most accessible place and work from there. If the load component is easily accessible, test for voltage at the input to the load (**Figure 3-2**). Remember, to get reliable test results, the load component must remain in the circuit. Do not do any open circuit testing unless directed to do so in the service information. When making a test connection, think ahead of what the expected results should be. Use the following procedure for locating the open:

1. With the switch in the ON position, measure the voltage at point A. If voltage is 10.5 volts or higher, check the ground side (point B). If the voltage at point B is less than 10.5 volts, there is excessive resistance or an open in the ground circuit. If the voltage at point A is less than 10.5 volts, continue testing.
2. Work toward the battery. Test all connections for voltage. If voltage is present at a connection, then the open is between that connection and the previously tested location (**Figure 3-3**). Use a fused jumper wire to bypass that section to confirm the location of the open.

An open circuit will not blow a fuse. However, a blown fuse opens the circuit.

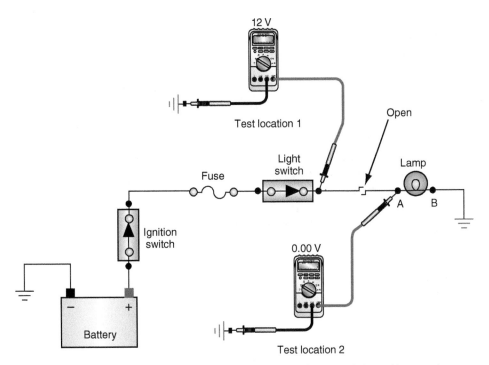

Figure 3-3 Working from the battery positive terminal toward ground, an open is located between the points where voltage was measured and where it was not.

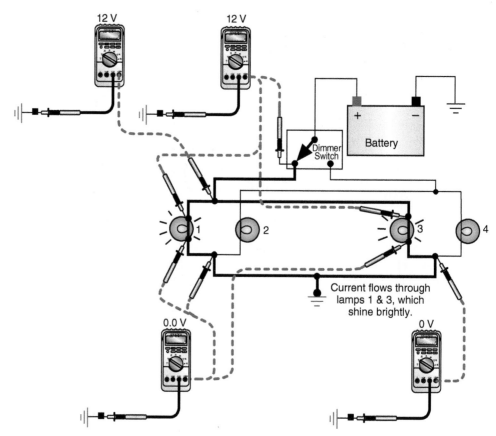

Figure 3-4 Properly operating complex parallel circuit.

3. If battery voltage is present at point B, the open is in the ground circuit. Use a fused jumper wire to connect the ground circuit. Then retest the component.

In more complex circuits, the open may have very different results. In a normally operating circuit, the voltmeter readings would be as indicated in **Figure 3-4**. If an open occurs in the ground side of the circuit, the circuit converts to a series circuit (**Figure 3-5**). This type of circuit defect is a form of **feedback** that results in lamps coming on that are not intended to. If the electrons cannot find a path to ground through the intended circuit, they will attempt to find an alternate path to ground. This may result in turning on components that are in the path. Normal voltage is applied to lamp 3, but lamps 1, 2, and 4 are in series and will illuminate dimmer than normal. The voltmeter will read 12 volts at the locations illustrated in Figure 3-5. However, the voltmeter will not indicate 0 volt on the ground side of bulb 1. Using a fused jumper wire from the ground terminal of the bulb to chassis ground would confirm the open circuit.

Testing for Shorts

Locating a copper-to-copper (conductor-to-conductor) short can be one of the most difficult tasks for a technician. If the short is within a component, the component will operate at less than optimum or not at all. An ohmmeter can be used to check the resistance of the component. If there is a short, then the amount of resistance will be lower than specified. If specifications for the component are not available, it may be necessary to replace the component with a known good unit. Do this only after it has been determined that the insulated and ground side circuits are in good condition.

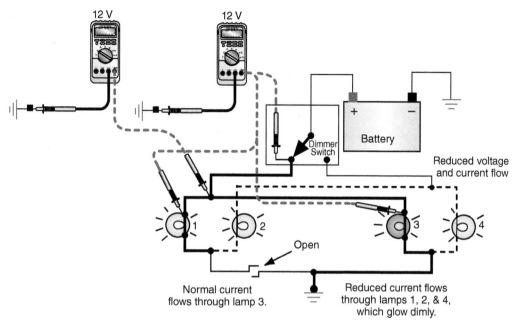

Figure 3-5 An open in the ground circuit can convert the circuit to a series circuit. The dashed line represents the resulting current path to ground.

If the short is between circuits, the result will be components operating when not intended (**Figure 3-6**). Visually check the wiring for signs of chafing, burned insulation, and melted conductors that will indicate a short. Also, check common connectors that are shared by the two affected circuits. Corrosion can form between two terminals of the connector and result in the short.

If the visual inspection does not isolate the cause of the copper-to-copper short, remove one of the fuses for the affected circuits. (If the affected circuits share a common fuse, remove it.) Install a buzzer that has been fitted with terminals across the fuse holder terminals (**Figure 3-7**). Activate the circuit that the buzzer is connected to. In Figure 3-6, if the buzzer is connected to fuse B, then switch 1 would be turned on. Disconnect the

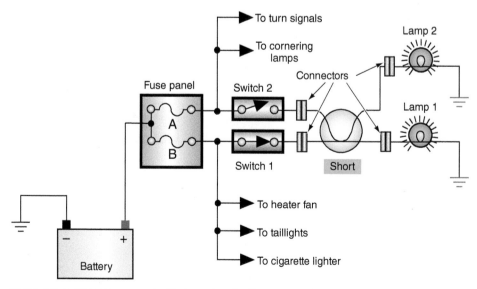

Figure 3-6 A copper-to-copper short between two circuits.

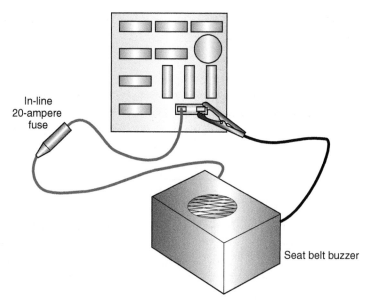

Figure 3-7 The buzzer will sound until the cause of the short is found and corrected.

A short may not blow a fuse depending on the amount of current flowing.

Classroom Manual
Chapter 3, page 82

loads that are supposed to be activated by this switch (lamp 1). Disconnect the wire connectors in the circuit from the load back to the switch. If the buzzer stops when a connector is disconnected, the short is in that portion of the circuit.

Testing for a Short to Ground

A fuse that blows as soon as it is installed indicates a **short to ground**. This condition allows current to return to ground before it has reached the intended load component. If the circuit is unfused, the insulation and conductor will melt. Not all shorts will blow the fuse, however. If the short to ground is on the ground side of the load component but before a grounding switch, the component will not turn off (**Figure 3-8**). If the short to ground is after the load and grounding switch (if applicable), circuit operation may not be affected.

To confirm that the circuit has a short to ground before the load, remove the fuse and connect a test light in series across the fuse connections. If the test light illuminates, the circuit has a short to ground. An ohmmeter can be used by first removing the fuse and then testing for continuity between the output side of the fuse box and chassis ground. If the switch is open, any reading other than infinite or "OL" may indicate a short to ground. Always check the service information since a very high resistance value may be normal

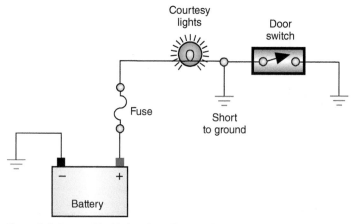

Figure 3-8 A ground in this location will cause the lamp to remain on.

for that circuit, especially if a computer or electronic components are involved. If the switch is closed, the ohmmeter should read the resistance value of the load(s).

> **⚙ SERVICE TIP** By studying the wiring diagram of the circuit carefully, the general location of the short may be determined. With the test light or ohmmeter connected at the fuse box as discussed, pay attention to the light or readings as switch is turned on and off. Also, disconnect the load device and other in-line connectors while observing the test equipment.

When attempting to pin point the location of the short to ground, it may be difficult to use a test light or voltmeter because the fuse blows before any testing can be conducted. To prevent this, connect a cycling circuit breaker that is fitted with alligator clips across the fuse holder (**Figure 3-9**). The circuit breaker will continue to cycle open and closed, allowing you to test for voltage.

Testing for shorts may be complicated if there are several circuits protected by a single fuse and if the ground is located in a section of wire that is not accessible. There are a couple of methods that can be used to locate the fault.

One method is to connect a test light, in series with a cycling circuit breaker, across the fuse holder (**Figure 3-10**). While observing the test light, disconnect individual circuits one at a time until the light goes out. The fault is in the circuit that was disconnected when the light went off.

⚒ Special Tools

Test light
Circuit breaker fitted
    with terminals
Gauss gauge or
    compass
DMM

⚠ Caution

Use a circuit breaker that is rated between 25 and 30 amperes. The use of a circuit breaker rated too high will damage the circuit.

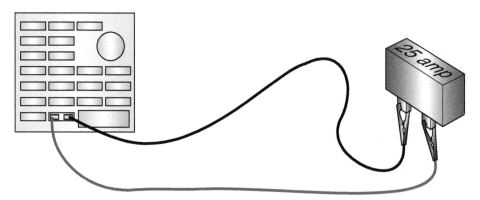

Figure 3-9 Use a circuit breaker to protect the circuit while checking for the short to ground.

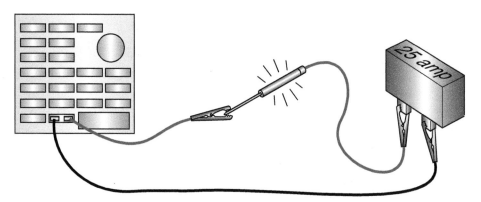

Figure 3-10 The test light will allow technicians to see when they have located the faulty circuit.

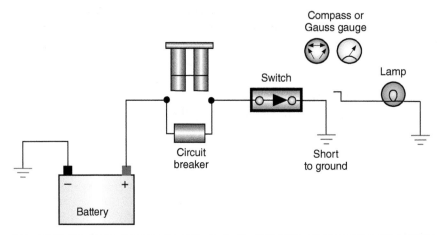

Figure 3-11 The needle of a compass or Gauss gauge will fluctuate over the portion of the circuit that has current flowing through it. Once the short to ground has been passed, the needle will stop fluctuating.

A second method is to use a **Gauss gauge** or a compass to locate the short to ground. A Gauss gauge is a meter that is sensitive to the magnetic field surrounding a wire conducting current. The gauge or compass works on the principle that a magnetic field is developed around a conductor that is carrying current. With a cycling circuit breaker bypassing the blown fuse, trace the path of the circuit with the gauge or compass. The needle will fluctuate as long as the gauge is over the conductor and the circuit breaker is cycling. The needle will stop fluctuating when the point of the short to ground is passed (**Figure 3-11**). This method will work even through the vehicle's trim. It will be necessary to follow all of the circuits protected by the fuse. Consult the wiring diagram for this information.

Commercial short finders are available to assist in locating a short to ground. A transmitter is connected in series to the circuit and a receiver is moved along the wire harness. A short is indicated by LED illumination and audible sounds. Some short finders may identify the direction and distance to the short.

A buzzer can be substituted for the test light.

Testing for Voltage Drop

As discussed in Chapter 2, voltage drop (when considered as a defect) defines the portion of applied voltage that is used up in other points of the circuit rather than that used by the load component. It is a resistance in the circuit that reduces the amount of electrical pressure available beyond the resistance. Excessive voltage drop may appear on either the insulated or the ground return side of a circuit. To test for voltage drop, the circuit must be active (current flowing). The source voltage must be as specified before voltage drop readings can be valid. Whenever voltage drop is suspected, both sides of the circuit must be checked.

Classroom Manual
Chapter 3, page 83

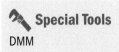 **Special Tools**
DMM

Excessive voltage drop caused by high resistance can be identified by dim or flickering lamps, inoperative load components, or slower-than-normal electrical motor speeds. Excessive resistance will not cause the fuse to blow.

Voltage drop testing is usually preferred over just using an ohmmeter to measure resistance. Remember that an ohmmeter works by sending a low voltage through the conductors to calculate resistance. This low voltage does not provide enough current to see how the circuit will behave once it is under full electrical load. It is possible to have a circuit or component pass the specifications for an ohmmeter test yet fail the voltage drop test. The voltage drop test will test the circuit under load and provide a more accurate representation of the circuit's integrity.

To perform a voltage drop test on any circuit, the positive voltmeter lead must be connected to the most positive portion of the circuit. Follow the instructions shown in Photo Sequence 4 to conduct a voltage drop test. Consult the service information for the maximum amount of voltage drop allowed. When the voltage drop decreases to within specifications, the cause of the excessive resistance is located. In **Photo Sequence 4**, a faulty relay was the cause of the excessive resistance.

When testing the ground side of the circuit, the ground connection terminal of the load component is the most positive location and the battery negative post is the most negative (**Figure 3-12**). Usually more than 0.1 volt indicates excessive resistance in the ground circuit.

According to many manuals, the maximum allowable voltage drop for an entire circuit, except for the drop across the load, is 10% of the source voltage. Although 1.2 volts may be considered the maximum acceptable amount, it is still too much for low-current circuits. Many good technicians use 0.5 volt as the maximum allowable drop. However, there should be no more than 0.1 volt dropped across any one wire or connector. This is the most important specification to consider and remember.

It is possible to calculate voltage drop by testing for available voltage. Use Ohm's law to determine the correct amount of voltage drop that should be across a component. Test for available voltage on both sides of the load component (**Figure 3-13**). Subtract the available voltage readings to obtain the amount of voltage drop across the component.

As mentioned, voltage drop testing is preferred over ohmmeter testing. It is also preferred over open voltage testing. Refer to the circuit illustrated in **Figure 3-14** as an example of how open circuit testing may lead to incorrect troubleshooting results. If the connector at the lamp is disconnected and a voltmeter is used to measure the applied voltage to the connector terminal with the switch closed, it will probably read full battery voltage (**Figure 3-14A**). This would be true even though the measurement is being taken after the resistance. Since the circuit is not loaded, due to the high impedance of the meter, there is no voltage dropped over the resistance. If the test was done by leaving the connector attached to the load and back probing the feed wire, the circuit will be loaded (provided there are no opens). Connecting the voltmeter to measure voltage drop will now indicate a problem with the circuit (**Figure 3-14B**), and the voltmeter will read available voltage at that point in the circuit.

> The higher the circuit current, the higher the allowable voltage drop. The lower the circuit current, the less the allowable voltage drop.

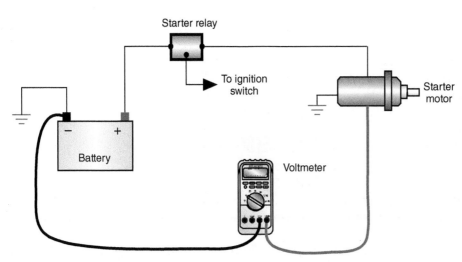

Figure 3-12 Testing the ground side of the starter motor circuit for high resistance by measuring voltage drop. Notice the voltmeter connections have the most positive portion of the circuit at the starter motor housing.

PHOTO SEQUENCE 4
Voltage Drop Test to Locate High Circuit Resistance

P4-1 Tools required to test for excessive resistance in a starting circuit include fender covers, a DMM, and a remote starter switch.

P4-2 Connect the positive lead of the meter to the positive battery post. If possible, do not connect the lead to the cable clamp.

P4-3 Connect the negative lead of the meter to the main battery terminal on the starter motor.

P4-4 To conduct a voltage drop test, current must flow through the circuit. In this test, the ignition system is disabled and the engine is cranked using a remote starter switch.

P4-5 With the engine cranking, read the voltmeter. The reading is the amount of voltage drop.

P4-6 If the reading is greater than specifications, test at the next connection toward the battery. In this instance, the next test point is the starter side of the solenoid.

P4-7 Crank the engine and touch the negative test lead to the starter side of the solenoid. Observe the voltmeter while the engine is cranking.

P4-8 Test in the same manner on the battery side of the solenoid. This is the voltage drop across the positive circuit from the battery to the solenoid.

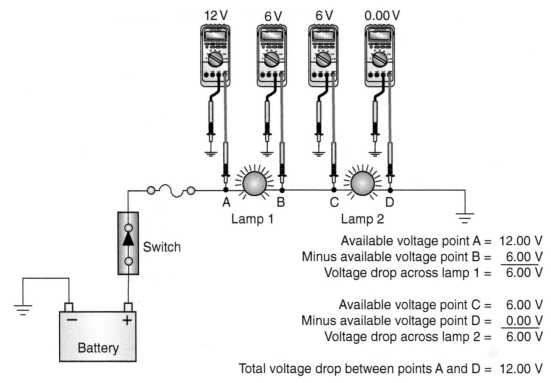

Available voltage point A = 12.00 V
Minus available voltage point B = 6.00 V
Voltage drop across lamp 1 = 6.00 V

Available voltage point C = 6.00 V
Minus available voltage point D = 0.00 V
Voltage drop across lamp 2 = 6.00 V

Total voltage drop between points A and D = 12.00 V

Figure 3-13 Using available voltage to calculate voltage drop over a component. This method is used if the wires of the circuit are too long to test with standard test leads.

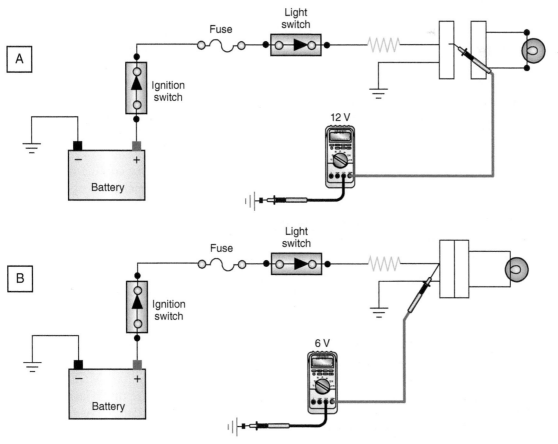

Figure 3-14 Comparison between readings of an open circuit test and a loaded voltage drop test.

Classroom Manual
Chapter 3, page 75

Special Tools

DMM
Test light

⚡ **Caution**

Fuses and other protection devices do not wear out. They fail because something went wrong. Never replace a fuse or fusible link, or reset a circuit breaker, without finding out why it failed.

A blade type fuse is called a spade fuse.

A **fusible link** is commonly called a fuse link.

⚡ **Caution**

Always disconnect the battery ground cable prior to servicing any fusible link.

TESTING CIRCUIT PROTECTION DEVICES

A protection device is designed to turn off the system whenever excessive current or an **overload** occurs. There are three basic types of **fuses** in automotive use: cartridge, blade, and ceramic. A fuse is a replaceable element that will melt should the current passing through it exceed the fuse rating. The cartridge fuse is found on most older domestic cars and a few imports. To check this type of fuse, look for a break in the internal metal strip. Discoloration of the glass cover or glue bubbling around the metal caps is an indication of overheating. Late-model vehicles use blade or spade fuses. To check the fuse, pull it from the fuse panel and look at the fuse element through the transparent plastic housing. Look for internal breaks and discoloration. The ceramic fuse is used on many older European imports. To check this type of fuse, look for a break in the contact strip on the outside of the fuse. All types of fuses can be checked with an ohmmeter or test light. If the fuse is good, there will be continuity through it.

Fuses are rated by the current at which they are designed to blow. A three-letter code is used to indicate the type and size of fuses. Blade fuses have codes ATC or ATO. All glass SFE fuses have the same diameter, but the length varies with the current rating. Ceramic fuses are available in two sizes, code GBF (small) and the more common code GBC (large). The amperage rating is also embossed on the insulator. Codes such as AGA, AGW, and AGC indicate the length and diameter of the fuse. Fuse lengths in each of these series are the same, but the current rating can vary. The code and the current rating are usually stamped on the end cap.

The current rating for blade fuses is indicated by the color of the plastic case (**Table 3-1**). In addition, it is usually marked on the top. The insulator on ceramic fuses is color coded to indicate different current ratings.

Fuses are located in a box or panel, usually under the dashboard, behind a panel in the foot well, or in the engine compartment. Fuses are generally numbered and the main components abbreviated. On late-model cars there may be icons or symbols indicating which circuits they serve. This identification system is covered in more detail in the owner's and service information.

A **fusible link** is a conductor with a special heat-resistant insulation. When there is an overload in the circuit, the link melts and opens the circuit. Fusible links are used in circuits where limiting the maximum current is not extremely critical. They are often installed in the positive battery lead to the ignition switch and other circuits that have power with the key off.

A fusible link is a short length of small-gauge wire installed in a conductor. Since the fusible link is a lighter gauge of wire than the main conductor, it melts and opens the circuit before damage can occur in the rest of the circuit. Fusible link wire is covered with a special insulation that bubbles when it overheats, indicating that the link has melted. If the insulation appears good, pull lightly on the wire. If the link stretches, the wire has melted. Of course, when it is hard to determine if the fusible link is burned out, check for continuity through the link with a test light or ohmmeter.

To replace a fusible link, cut the protected wire where it is connected to the fusible link. Then, tightly crimp or solder a new fusible link of the same rating as the original link. Since the insulation on the manufacturer's fusible links is flameproof, never fabricate a fusible link from ordinary wire because the insulation may not be flameproof.

Many late-model vehicles use maxi-fuses instead of fusible links. Maxi-fuses look and operate like two-pronged blade or spade fuses, except they are much larger and can handle more current. (Typically, a maxi-fuse is four to five times larger.) Maxi-fuses are located in a fuse box in the engine compartment and/or passenger compartment, under the dash or the rear seat.

Table 3-1 Typical color coding of protection devices

Blade Fuse Color Coding	
Ampere Rating	**Housing Color**
4	pink
5	tan
10	red
15	light blue
20	yellow
25	natural
30	light green
Fuse Link Color Coding	
Wire Link Size	**Insulation Color**
20 GA	blue
18 GA	brown or red
16 GA	black or orange
14 GA	green
12 GA	gray
Maxi-fuse Color Coding	
Ampere Rating	**Housing Color**
20	yellow
30	light green
40	amber
50	red
60	blue

SERVICE TIP To calculate the correct fuse rating, use Watt's law: watts divided by volts equals amperes. For example, if you are installing a 55-watt pair of fog lights, divide 55 by the battery voltage (12 volts) to find out how much current the circuit must carry. Since 55 / 12 = 4.58, the current is approximately 5 amperes. To allow for current surges, the correct in-line fuse should be rated slightly higher than the normal current flow. In this case, an 8- or 10-ampere fuse would do the job.

Maxi-fuses are easier to inspect and replace than are fuse links. To check a maxi-fuse, look at the fuse element through the transparent plastic housing. If there is a break in the element, the maxi-fuse has blown. To replace it, pull it from its fuse box or panel. Always replace a blown maxi-fuse with a new one having the same ampere rating.

Some circuits are protected by circuit breakers. Like fuses, circuit breakers are rated in amperes. There are two types of circuit breakers: cycling or those that must be manually reset.

In the cycling type, the bimetal arm will begin to cool once the current to it is stopped. Once it returns to its original shape, the contacts are closed and power is restored. If the current is still too high, the cycle of breaking the circuit will be repeated.

Classroom Manual
Chapter 3, page 79

Two types of noncycling or resettable breakers are used. One is reset by removing the power from the circuit. There is a coil wrapped around a bimetal arm (**Figure 3-15A**). When there is excessive current and the contacts open, a small current passes through the coil. This current through the coil is not enough to operate a load, but it does heat up both the coil and the bimetal arm. This keeps the arm in the open position until power is removed. The other type is reset by depressing a reset button. A spring pushes the bimetal arm down and holds the contacts together (**Figure 3-15B**). When an overcurrent condition exists and the bimetal arm heats up, the bimetal arm bends enough to overcome the spring and the contacts snap open. The contacts stay open until the reset button is pushed, which snaps the contacts together again.

A visual inspection of a fuse or fusible link will not always determine if it has an open (**Figure 3-16**). To accurately test a circuit protection device, use an ohmmeter, voltmeter, or test light.

With the fuse or circuit breaker removed from the vehicle, connect the ohmmeter's test leads across the protection device's terminals (**Figure 3-17**). On its lowest scale, the ohmmeter should read 0 to 1 ohm. If it reads infinite, the protection device is open. Test a fusible link in the same way (**Figure 3-18**). Before connecting the ohmmeter across the fusible link, make sure there is no current flow through the circuit. To be safe, disconnect the negative cable of the battery.

> A circuit breaker is typically abbreviated c.b. in a fuse chart of a service information.

⚙ **SERVICE TIP** Before using a test light, it is good practice to check the tester's lamp. To do this, simply connect the test light across the battery. The light should come on.

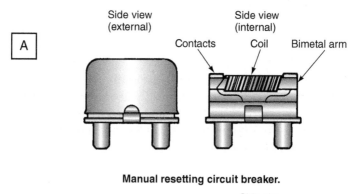

Figure 3-15 The two basic types of circuit breakers.

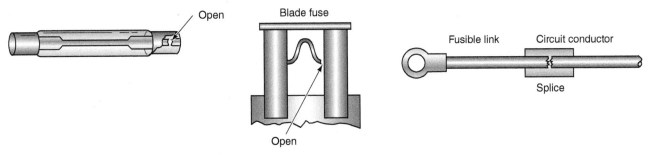

Figure 3-16 A fuse or fusible link can have a hidden fault that cannot be seen by the technician.

Figure 3-17 A good fuse will have zero resistance when tested with an ohmmeter.

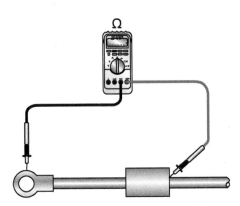

Figure 3-18 A fusible link can be tested with an ohmmeter once it is disconnected from power.

To test a circuit protection device with a voltmeter, check for available voltage at both terminals of the unit (**Figure 3-19A**). If the device is good, voltage will be present on both sides. A test light can be used in place of a voltmeter (**Figure 3-19B**). The lamp should illuminate when each test terminal is touched with the lamp's probe.

Measuring voltage drop across a fuse or other circuit protection device will tell you more about its condition than whether or not it is open. If a fuse, a fuse link, or a circuit breaker is in good condition, a voltage drop of near zero will be measured. If 12 volts are read, the fuse is open. Any reading between 0 and 12 volts indicates some voltage drop. If there is voltage drop across the fuse, it has resistance and should be replaced. Make sure you check the fuse holder for resistance as well.

CUSTOMER CARE Any time you install additional electrical accessories for customers, provide them with information concerning the type and size of fuses installed so they can put this information with their owner's manual.

Diagnosing PTC Circuit Protection Devices

PTCs are rated much like a fuse; so, if it experiences an excessive overload, it can be damaged. When the circuit protected by the PTC experiences a current overload, the PTC resistance increases which reduces the amperage to a safe level. At this point, the PTC

⚠ Caution

Fuses are rated by amperage and voltage. Never install a larger rated fuse into a circuit than the one that was designed by the manufacturer. Doing so may damage or destroy the circuit. Also, do not replace a fusible link with a resistor wire or vice versa.

⚠ Caution

Do not use an unfused jumper wire to bypass the protection device. Circuit damage may result.

Classroom Manual
Chapter 3, page 79

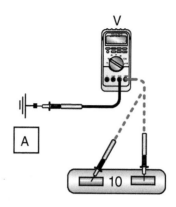

Top view of mini-fuse

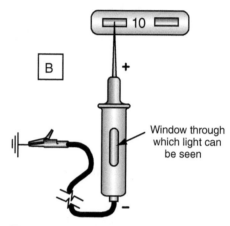

Window through
which light can
be seen

Figure 3-19 (A) Voltmeter test of a circuit protection device. Battery voltage should be present
on both sides. (B) The test light will illuminate on
both terminals if the fuse is good.

acts as the load for the circuit. Normally the PTC will not go to a complete open state
during this time.

Consider the circuit shown in **Figure 3-20** that uses a PTC to protect the power
window motor. If the motor is inoperative and a check of voltage to the motor indicates that 0 volt is present, the problem may be the PTC. Often the PTC is installed
into the fuse box. In many cases it is not serviceable as separate units, but test ports
are generally provided. Because disconnecting the battery to perform resistance tests
with an ohmmeter results in computers resetting, loss of adaptive memories, and
other set values, it is more desirable to use a voltmeter to test the PTC. Simply measure the voltage drop between the input side of the PTC and the output side while
operating the accessory. Typically, there should be less than 0.050 voltage drop across
these points. Because the resistance of the PTC increases as the temperature of the
circuit increases, a high voltage drop indicates the PTC resistance is increasing. This
may be due to a short in the circuit; so, confirm this is not the case by disconnecting
in-line and component connectors while observing the voltmeter. If the voltmeter
reading decreases toward 0 volt when a connector is unplugged, then there is a short
in the circuit.

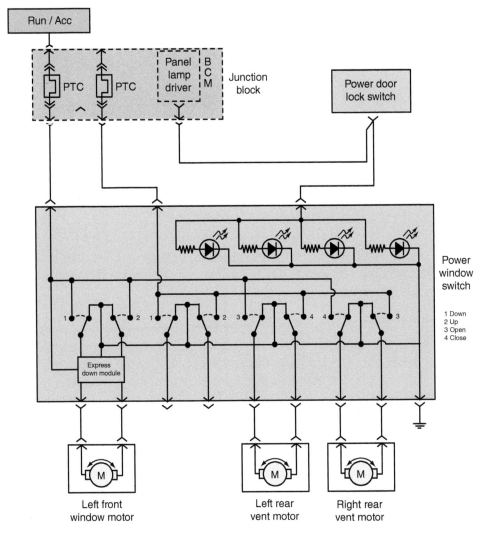

Figure 3-20 PTC-protected circuit.

TESTING AND REPLACING ELECTRICAL COMPONENTS

All electrical components can fail. Testing them is the best way of determining if they are good or bad. For the most part, the proper way to check electrical components is determined by what the component is supposed to do. If we think about what something is supposed to do and how it does it, we can figure out how to test it. Sometimes, removing the component and testing it on a bench is the best way to check it.

Switches

The easiest method of testing a **normally open (NO)** switch is to use a fused jumper wire to bypass the switch (**Figure 3-21**). An NO switch will not allow current flow when it is in its rest position. The contacts are open until they are acted on by an outside force that closes to complete the circuit. If the circuit operates with the switch bypassed, the switch is defective. Voltage drop across switches should also be checked. Ideally, when the switch is closed, there should be no voltage drop. Any voltage drop indicates resistance, and the switch should be replaced.

Classroom Manual
Chapter 3, page 53

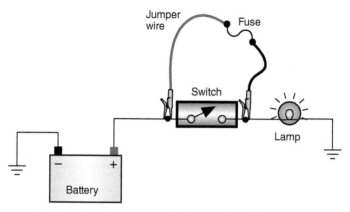

Figure 3-21 Using a fused jumper wire to bypass the switch.

Special Tools

Jumper wires
Test light
DMM

A voltmeter or test light can be used to check for voltage on both sides of the switch (**Figure 3-22**). A faulty NO switch would have voltage present at the input side of the switch but not on the output side when in the ON position.

> **SERVICE TIP** Using an ohmmeter to test a switch is good for determining if the switch works and if the contacts move; however, it may not accurately indicate that a resistance is present. A voltage drop test is better for determining if there is resistance in the switch.

If the circuit being tested is powered through the ignition switch, it must be in the RUN position.

If the switch is removed, it can be tested with an ohmmeter. With the switch contacts open, there should be no continuity between the terminals (**Figure 3-23**). When the contacts are closed, there should be zero resistance through the switch contacts. On

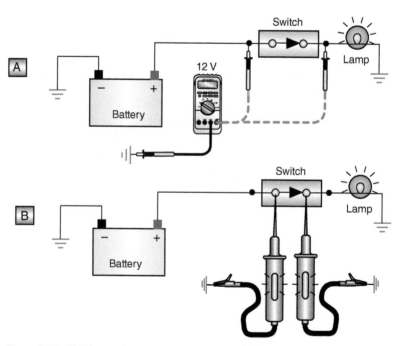

Figure 3-22 (A) Using a voltmeter to test a switch; voltage should be the same on both sides of the switch. (B) Using a test lamp to test a switch; the lamp should illuminate on both sides of the switch.

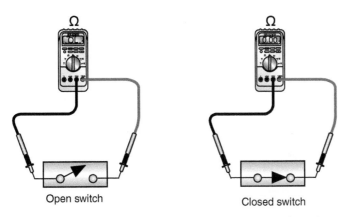

Figure 3-23 The continuity through a switch can be checked with an ohm-meter. With the switch open, there should be infinite resistance. With the switch closed, there should be zero resistance.

Headlamp switch connector
(connector end view)

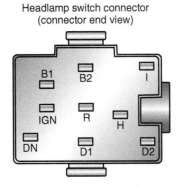

Pin number	Circuit	Circuit function
B1	38 (BK/O)	Power supply to battery
B2	195 (T/W)	Tail lamp switch feed
I	19 (LB/R)	Instrument panel lamp feed
IGN		Not used
R	14 (BR)	Tail lamp and side marker lamps
H	15 (R/Y)	Headlamp dimmer switch feed
DN		Not used
D1	54 (LG/Y)	Interior lamp switch feed
D2	706 (GY)	Battery saver door switch feed

Figure 3-24 Example of headlight switch connector callouts that are used for continuity checks.

complex ganged-type switches, the technician should consult the service information for a continuity diagram (**Figure 3-24**). If there is no continuity chart, use the wiring diagram to make your own chart.

Relays

A **relay** is a device that uses low current to control a high-current circuit. The relay can be either a normally open or a normally closed design. The relay can be checked using a jumper wire, voltmeter, ohmmeter, or test light. If the terminals are easily accessible, the jumper wire and test light may be the fastest method.

Classroom Manual
Chapter 3, page 56

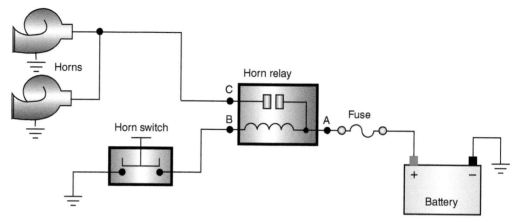

Figure 3-25 A relay circuit with a ground control switch.

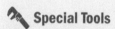
Check the wiring diagram for the relay being tested to determine if the control is through an insulated or a ground switch. Use **Figure 3-25** as a guide to test a ground-switch–controlled relay. Follow these steps:

1. Use a voltmeter to check for available voltage to the battery side of the relay (terminal A). If battery voltage is not present at this point, the fault is in the circuit from the battery to the relay. If battery voltage is present, continue testing.
2. Probe for battery voltage at control terminal B. If voltage is not present at this terminal, then the fault is in the relay coil. If voltage is present, continue testing.
3. Use a fused jumper wire to connect terminal B to a good ground. If the horn sounds, the fault is in the control circuit from terminal B to the horn switch ground. If the horn does not sound, continue testing.
4. Connect the fused jumper wire from the battery positive to terminal C. If the horn did not sound, there is a fault in the circuit from the relay to the horn ground. If the horn sounded, the fault is in the relay.

> **SERVICE TIP** The procedures presented here can be used to test the relay to determine the type of fault it has. However, the easiest way to test a relay is to substitute it with a *known good* relay of the same type. If the circuit operates with the substitute relay, the old relay is the faulty component.

If the relay is controlled by the computer, it is not recommended that a test light be used. The test light may draw more current than the circuit is designed to carry and damage the computer. Refer to **Figure 3-26** for procedures using a voltmeter to test a relay. Use a DMM set as follows:

1. Connect the negative voltmeter test lead to a good ground.
2. Connect the positive voltmeter test lead to the output wire (terminal B). Turn on the ignition switch. If no voltage is present at this terminal, go to step 3. If the voltmeter reads 10.5 volts or higher, turn off the control circuit. The voltmeter should then read 0 volt. If it does, then the relay is good. If the voltmeter still reads any voltage, the relay is not opening and needs to be replaced.
3. Connect the positive voltmeter test lead to the power input terminal (terminal A). The voltmeter should indicate battery voltage. If voltage is below this value, the circuit from the battery to the relay is faulty. If the voltage value is correct, continue testing.

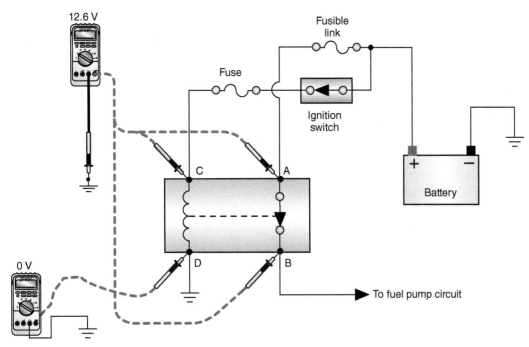

Figure 3-26 Testing relay operation with a voltmeter.

4. Connect the positive voltmeter test lead to the control circuit terminal (terminal C). The voltage should read 10.5 volts or higher. If not, check the circuit from the battery to the relay (including the ignition switch). If the voltage is 10.5 volts or higher, continue testing.

5. Connect the positive voltmeter test lead to the relay ground terminal (terminal D). If more than 0.1 volt is indicated on the meter, there is a poor ground connection. If the reading is less than 0.1 volt, replace the relay.

If the relay terminals are not accessible, remove the relay from its holding fixture and bench test it. Use an ohmmeter to test for continuity between the relay coil terminals (**Figure 3-27**). If the meter indicates an infinite reading, replace the relay. If continuity is indicated, use a pair of fused jumper wires to energize the coil (**Figure 3-28**). Check for continuity through the relay contacts. If the meter indicates an infinite reading, the relay is defective. If there is continuity, the relay is good and the circuits will have to be checked.

Be careful not to touch the coil terminals with the ohmmeter test leads while the coil is energized.

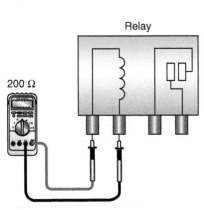

Figure 3-27 Testing the resistance of the relay coil.

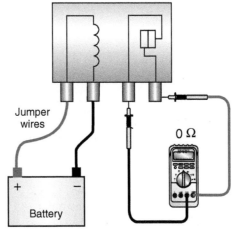

Figure 3-28 Bench testing a relay.

Be sure to check your service information for resistance specifications and compare the relay to them. It is easy to check for an open coil. However, a shorted coil will also prevent the relay from working. Low resistance across a coil would indicate that it is shorted. Too low resistance may also damage the transistors and/or driver circuits because of the excessive current that would result.

Testing Stepped Resistors

A **stepped resistor** has two or more fixed resistor values (**Figure 3-29**). Using an ohmmeter is a good method of testing a stepped resistor. To obtain accurate test results, it is a good practice to remove the resistor and bench test it. Connect the ohmmeter leads to the two ends of the resistor (**Figure 3-30**). Compare the results with manufacturer's specifications. Be sure to place the ohmmeter on the correct scale to read the anticipated amount of resistance. Repeat for each of the resistances.

A stepped resistor can also be checked with a voltmeter or DSO. By measuring the voltage after each part of the resistor block and comparing the readings to specifications, you can tell if the resistor is good or not.

Testing Variable Resistors

A **variable resistor** provides for an infinite number of resistance values within a range. As with the stepped resistor, a good method of testing a variable resistor is with an ohmmeter. However, it is possible to use a voltmeter, DSO, or test light.

A **rheostat** is a two-terminal variable resistor used to regulate the current in an electrical circuit. To test a rheostat, locate the input and output terminals and connect the ohmmeter test leads to them. Rotate the resistor knob slowly while observing the ohmmeter. The resistance value should remain within the specification limits and change in a smooth and constant manner. If the resistance values are out of limits or the resistance value jumps as the knob is turned, replace the rheostat.

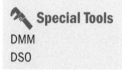

Classroom Manual
Chapter 3, page 60

Special Tools
DMM
DSO

⚠ **Caution**

Operating a blower motor resistor out of the air flow of the heating, ventilation, and air-conditioning (HVAC) housing will cause the resistor to overheat and be damaged.

Classroom Manual
Chapter 3, page 61

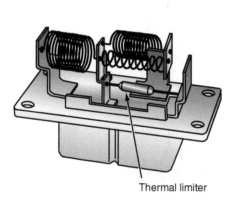

Thermal limiter

Figure 3-29 A stepped resistor used in the heater blower motor circuit.

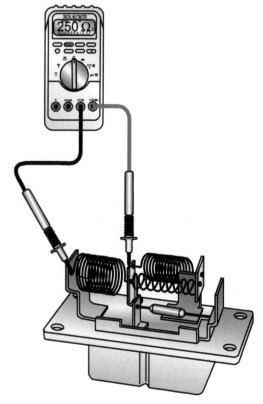

Figure 3-30 Ohmmeter testing of a stepped resistor.

Figure 3-31 Using an ohmmeter to test the continuity between terminals A and C of a potentiometer.

Figure 3-32 Testing continuity between terminals A and B of a potentiometer while the wiper is being moved.

If a voltmeter is used, the voltage drop readings between the two terminals should be smooth and consistent. A test light should change in brightness as the knob is turned; the rheostat is defective if the light blinks at any point.

A **potentiometer** is a three-wire variable resistor that acts as a voltage divider to produce a continuously variable voltage output signal proportional to a mechanical position. To test a potentiometer, connect the ohmmeter test leads to terminals A and C (**Figure 3-31**). Check the results with specifications. Next connect the ohmmeter test leads to terminals A and B (**Figure 3-32**). Check the resistance at the stop and observe the ohmmeter as the wiper is moved to the other stop. The resistance values should be within specification and smooth and constant.

A voltmeter can be used in the same manner. However, jumper wires may need to be used to gain access to the test points (**Figure 3-33**). Because potentiometers are primarily used in computer-controlled circuits, it is not recommended that a test light be used. A DSO is a very good tool to use for variable resistance circuits. The DSO allows for easy reading of the voltage trace as the wiper of the sensor is moved. The voltage indicated by the trace should be within specified limit and should change smoothly as the wiper moves.

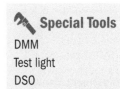

Special Tools

DMM

Test light

DSO

⚠ Caution

Do not pierce the insulation to test the potentiometer. These circuits usually operate between 5 to 9 volts. Piercing the insulation may break some of the wire strands, resulting in a voltage drop that will give errant information to the computer. Even if the conductor is not broken, moisture can enter and cause corrosion.

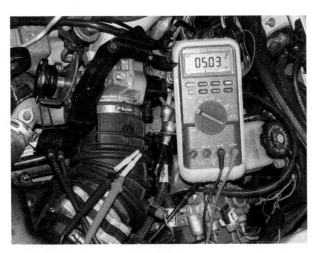

Figure 3-33 It may be necessary to use jumper wires to connect the wire connector to the sensor in order to measure voltage.

Figure 3-34 A capacitor with its rating marks.

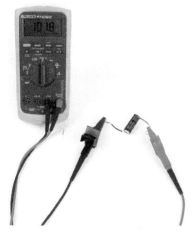

Figure 3-35 Using the capacitor test function of the DMM, the reading should be within 6% of the capacitor's rating.

Testing Capacitors

Classroom Manual
Chapter 3, page 61

Capacitors are used to store and release electrical energy. Because the capacitor stores voltage, it will also absorb voltage changes in the circuit. By providing for this storage of voltage, damaging voltage spikes can be controlled. Capacitors are also used to reduce radio noise. It is possible for the capacitor to fail internally.

Some DMMs have a capacitance test function. Prior to testing the capacitor, make sure power to the capacitor is turned off. Next you must dissipate the stored voltage in the capacitor. This can be done by using a 12-volt test light connected across the terminals of the capacitor. Once the capacitor is discharged, the light will go out. Once the capacitor is discharged, it should be removed from the circuit.

The capacitor will have its rating listed on its shell (**Figure 3-34**). Also, note the tolerance of the rating. If there is no tolerance range listed, use 6% as a general rule.

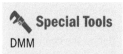

Special Tools

DMM

Based on the rating of the capacitor, select the correct micro-farad setting on the meter. If the meter does not have an auto-range function, select a range that is higher than the rating of the capacitor. Connect the black test lead of the DMM onto the ground terminal of the capacitor (or the case of the capacitor if it does not have a terminal). The red test lead is connected to the power terminal. The reading on the DMM should match the rating of the capacitor (**Figure 3-35**).

TESTING DIODES

Classroom Manual
Chapter 3, page 65

A **diode** is an electrical one-way check valve that will allow current to flow in one direction only. Regardless of the bias of the diode, it should allow current flow in one direction only. A diode that allows current flow in both directions is shorted. An open diode does not allow current flow in either direction.

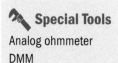

Special Tools

Analog ohmmeter
DMM

You may run into problems when checking a diode with a high-impedance digital ohmmeter. Since many diodes won't allow current flow through them unless the voltage is at least 0.6 volt, a digital meter may not be able to forward bias the diode. This will result in readings that indicate the diode is open, when in fact it may not be. Because of this problem, many digital multimeters are equipped with a special diode-testing feature. This feature allows for increased voltage at the test leads. Connect the meter's leads across the diode (**Figure 3-36**). Observe the reading on the meter. Then reverse the meter's leads and observe the reading on the meter. The value displayed is the voltage required to

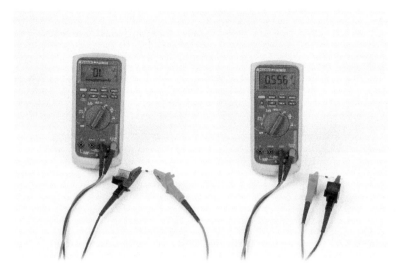

Figure 3-36 Use the diode test function of the DMM to check a diode for an open and short.

forward bias the diode. A silicon diode should read between 400 and 800 mV in the forward direction and open in reverse. For a germanium diode, it is between 200 and 400 mV in the forward direction. If the diode is open, the meter will display "OL" or another reading to indicate infinity or out of range. Some meters during diode check will make a beeping noise when there is continuity.

    Diodes may also be tested with a voltmeter. Using the same logic as when testing with an ohmmeter, test the voltage drop across the diode with the circuit turned on. The meter should read low voltage in one direction and near source voltage in the other direction.

The voltage displayed on the meter is referred to as turn-on voltage or diode drop.

Testing Zener Diodes

If the Zener diode is out of the circuit and you need to diagnose it for an open or short, then test the Zener as described for the standard diode. However, if you desire to measure its Zener voltage level, you will have to build a test circuit (**Figure 3-37**). The power supply voltage should be set to a value slightly higher than the Zener value. For example,

Classroom Manual
Chapter 3, page 67

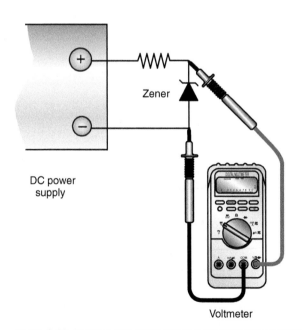

DC power
supply

Zener

Voltmeter

Figure 3-37 Making a test power supply to measure Zener voltage.

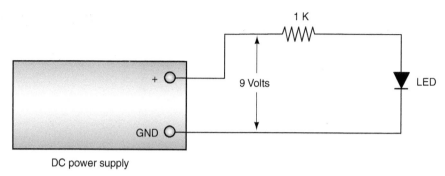

Figure 3-38 Testing an LED with a test circuit.

for a 12-volt diode, the supply voltage should be about 15 volts. This can be made using many styles of "project boxes" from most electronic stores. The value of the resistor (R) should limit the current to about 1 mA. For example, when using 15 volts with a 12-volt Zener, use a 3.3 K resistor.

Once the circuit is built, read the Zener voltage using a digital voltmeter. If the voltmeter indicates 600 mV, the diode is reverse biased and will need to be reinstalled into the circuit.

> ⚙ **SERVICE TIP** The avalanche diode is tested in the same manner as the Zener diode. However, most avalanche diodes used in the AC generator are rated between 20 and 30 volts.

Classroom Manual
Chapter 3, page 68

Testing LEDs

The turn-on voltage of an LED is usually between 1.5 and 2.5 volts. If your DMM has a diode test function, then the LED can be tested in the same manner as a standard diode. The difference will be that the meter will read between 1,600 and 2,500 mV when the diode conducts instead of the 600 mV you read on a standard diode.

It is possible to test an LED without the use of a DMM. To do this, first build the test circuit as shown (**Figure 3-38**). By plugging the LED into the circuit, the LED should light. If the LED doesn't light, then reverse the polarity on the diode. If it still doesn't light, then the LED is faulty.

Classroom Manual
Chapter 3, page 69

Testing Zener Diodes and LEDs in a Circuit

It is not necessary to remove a Zener diode or the LED from the circuit to test it. To test a Zener, use a voltmeter and measure the voltage across it. Connect the negative lead to the anode and the positive lead to the cathode. The meter should read the Zener voltage. If you read 0 volt, the Zener is shorted. This is true if the power and ground circuits to the Zener are confirmed as being good. If the voltmeter reads a voltage that is higher than the Zener's rated voltage, the diode is open.

For an LED that is supposed to be lit but isn't, use a voltmeter to measure the voltage across it. If you measure more than 3 volts, the LED is open.

TESTING TRANSISTORS

Although replacing a transistor that is part of an integrated circuit board is not often done in the automotive repair industry, some bipolar transistors can be easily tested and replaced. In order to test a bipolar transistor, it is first necessary to identify its type

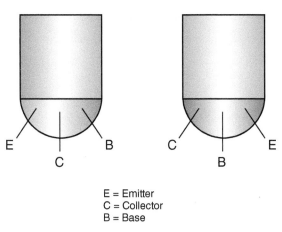 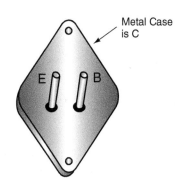

E = Emitter
C = Collector
B = Base

Figure 3-39 There are several configurations of the transistor legs.

(NPN or PNP) and lead arrangement. To perform this test, it is first necessary to identify the base leg of the transistor. There are several configurations of the transistor legs (**Figure 3-39**). This can be done using a digital multimeter. Since transistors behave as back-to-back diodes, the collector and emitter can be identified because the doping for the base-to-emitter junction is always much higher than that for the base-to-collector junction. Therefore, the forward voltage drop will be a couple of millivolts higher on the DMM reading when set on the diode test function. Follow **Photo Sequence 5** to identify the type of bipolar transistor.

If none of the six possible lead connection combinations indicates a pair of low readings, or if more than one combination results in a pair of low readings, the transistor is probably faulty. Keep in mind that the base-to-collector junction voltage drop is always slightly lower than the emitter-to-base junction drop.

Once the type of transistor is identified, it can be tested using the diode function of the DMM. Connect the red lead of the meter to the base of the transistor and the black lead to the emitter. A good NPN transistor will read a voltage of between 450 and 900 mV. A good PNP transistor will read open. With the red lead still on the base, move the black lead to the collector. The reading should be the same as that in the previous test.

Next, reverse the meter leads and repeat the test. With the black lead connected to the base of the transistor and the red lead to the emitter, a good PNP transistor will read a voltage of between 450 and 900 mV. A good NPN transistor will read open. Leave the black lead on the base and move the red lead to the collector. The reading should be the same as that in the previous test.

Finally, place one meter lead on the collector, the other on the emitter. The meter should read open. Reverse your meter leads and the meter should read open. This is the same for both NPN and PNP transistors.

Testing Darlington Transistors

A **Darlington** is a special type of configuration usually consisting of two transistors fabricated on the same chip or mounted in the same package.

Testing is basically similar to that of normal bipolar transistors except that in the forward direction the base-to-emitter reading will be 0.2 to 1.4 volts when reading the DMM on the diode function. This higher voltage is due to the pair of junctions that are in series.

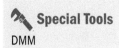 **Special Tools**
DMM

⚡ **Caution**

Do not hold the transistor in your hand while testing it. For every degree the transistor increases in temperature, the base-to-emitter diode drop decreases by 2 mV. This is a significant amount when determining the base-to-emitter and base-to-collector junctions.

Some DMMs cannot read above 1.2 volts, resulting in a good Darlington testing as open. Confirm your DMM is able to read higher than 1.4 volts.

Classroom Manual
Chapter 3, page 73

PHOTO SEQUENCE 5
Identifying Bipolar Transistors

P5-1 Label the pins of the unknown device; 1, 2, and 3.

P5-2 Place the positive probe of the DMM on pin 1 and measure the diode drop to pins 2 and 3.

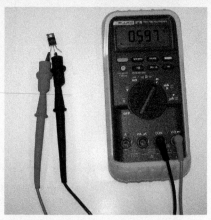

P5-3 If the positive probe is on the base of a good NPN transistor, you should read a low diode drop to pins 2 and 3. The base-to-collector diode drop will be slightly lower than the base-to-emitter reading.

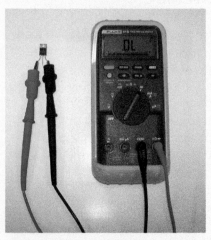

P5-4 If one, or both, measurements to pins 2 and 3 are high, put the positive probe on pin 2 and retest. If still high, put the positive probe on pin 3 and retest.

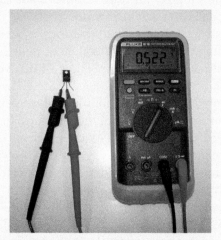

P5-5 If the reading is high when all three pins are tested, repeat the tests with the negative probe as the common pin. A pair of low readings now indicates a PNP transistor.

Digital Storage Oscilloscope (DSO)

The greatest advantage of a digital storage oscilloscope (DSO) is the speed at which it samples electrical signals. Mechanical switching speed of switches and relays is measured in thousands of a second or milliseconds. Electrical/electronic switching speed is measured in millionths of a second or microseconds. Radio frequency interference (RFI) is measured in billionths of a second. The DSO operates at 25 million samples per second. Another method of expressing this operating speed is to say the DSO is capable of sampling a signal in 40 billionths of a second. DSO sampling speed is at least 47,000 times faster than automotive testers such as other engine analyzers.

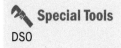

Special Tools

DSO

> ⚙ **SERVICE TIP** Some power transistors have built-in diodes that are reverse biased across the collector-to-emitter junction and resistors between the base-to-emitter junction. The resistance is usually 50 ohms. If not aware of this, you can be confused by the reading if testing the transistor as a standard bipolar transistor. You will need to know the specifications of the power transistor in order to properly test it. Power transistors without internal damper diodes test just about like bipolar transistors.

This sampling speed allows the DSO to provide an extremely accurate, expanded display of input sensors and output actuators compared to multimeters. Such increased speed allows the DSO to display glitches or momentary defects in input sensors and output actuators. The extremely fast sampling of the DSO allows this scope to display a graph of input sensor and output actuator operation. Some DSOs have the capability to display two voltage traces across the screen (**Figure 3-40**). Other DSOs, such as the Simu-Tech, display six voltage traces simultaneously.

DSO Screen

On the DSO screen, voltage is displayed vertically. Voltage change is shown as a vertical movement. Vertical grids on the screen provide a voltage measurement, and the voltage level between the grids is adjustable (**Figure 3-41**). The technician must know the voltage in the circuit being tested in order to select the voltage per division on the DSO that provides the most detail with the signal remaining on the screen.

Horizontal movement on the screen represents time. The milliseconds per division on the horizontal grid are adjustable with the time button on the DSO. Each time a DSO samples a voltage signal, it displays a dot on the screen. The DSO then connects the dots to provide a waveform. When a faster signal is being read, a shorter time base should be selected on the DSO.

If the time base selected is too long and the voltage too high, the waveform is too small to read. Conversely, if the time base is too short and the voltage scale too low, the

> ⚠ **Caution**
>
> While diagnosing computer systems, always place test equipment, such as DSOs or scan testers, in a secure position where they will not fall on the floor or into rotating components. Severe meter damage may occur if the DSO or scan tester is dropped.

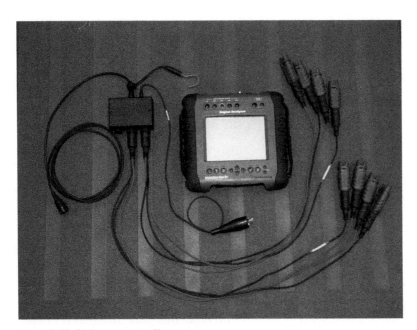

Figure 3-40 Digital storage oscilloscope.

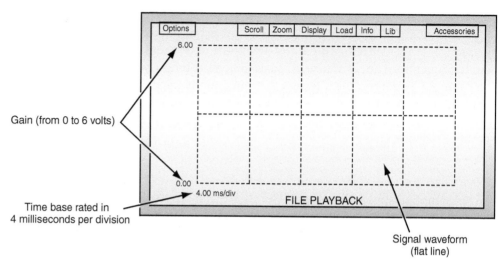

Figure 3-41 DSO vertical and horizontal screen grids.

0.637 represents 63.7% of the peak voltage. Fifty percent is not used because the rate of change in voltage is slower toward the peak than it is nearer the 0 value. The voltage is above 50% for a longer time than it is below it.

waveform is too large for the screen. The technician must select the proper time base for the voltage signal being measured so the waveform is displayed on the screen.

Peak, Average, and Root Mean Square Related to an AC Voltage Waveform

The term **peak** represents the highest point in *one* cycle of an AC voltage waveform. When both the highest and lowest peaks are considered in an AC voltage waveform, the term **peak-to-peak voltage** is the total voltage measured between these peaks. For example, an AC voltage waveform with a 60-volt peak would have a 120-volt peak-to-peak. The average voltage in an AC voltage waveform is calculated by multiplying 0.637 × peak voltage. The average voltage on a 60-volt peak would be $0.637 \times 60 = 38.2$ volts.

In many cases root mean square (RMS) is used to describe AC voltage. For example, if one cycle of an AC voltage waveform from a 120-volt household electrical outlet is divided into four parts at 90° intervals, the instantaneous voltage and current are recorded for each degree in a 90° interval and then averaged. The square root of the average may be calculated by multiplying 0.7071 × the peak voltage. The peak voltage for the average 120-volt household outlet is about 170 volts at 60 hertz (Hz). Therefore, $0.7071 \times 170 = 120.207$ RMS. If this same voltage was measured using the average voltage method, it would be 108.29 volts.

0.7071 is one divided by the square of two.

Selecting DSO Voltage and Time Base

To display a waveform for a 120-volt household electrical outlet, round off the peak voltage of 170 to 200 volts. There are eight vertical voltage divisions on the DSO screen with four divisions above and below the center-line. Select 50 volts per division to display the high and low peaks on the waveform. Assuming the 0-volt position in the waveform is positioned in the center of the screen, the 50 volts per division selection provides 200 volts above and below the screen centerline to display the 170-volt peak voltage above and below the center-line. If the volts per division setting is increased, the peaks appear shorter on the screen.

In a 60-hertz AC voltage, one cycle occurs in approximately 18 milliseconds (ms). Displaying one complete AC cycle requires about 20 ms. The average DSO has 10 horizontal divisions. Since $20 / 10 = 2$ ms per division, this time base selection displays one AC voltage waveform. If 4 ms per division is selected, 2 AC voltage waveforms are displayed.

Increasing the time base displays more AC voltage cycles on the screen. Conversely, decreasing the time base displays fewer AC voltage cycles on the screen, and the waveform appears expanded.

Each time a DSO takes a voltage sample, it displays a dot on the screen and then connects these dots to display a waveform. When the millisecond (ms) time base is too low, the waveform is expanded horizontally and a reduced number of dots are used in the waveform display. This may result in an altered and incomplete waveform display. Ideally, one to three cycles should be displayed on the screen for the best display.

When the DEFAULT button is pressed on some DSOs, a baseline volts per division and ms per division are automatically selected internally. If the volts per division and ms per division selected by the technician are incorrect for the voltage signal being tested, this default mode baseline should provide settings to display a waveform on the screen. Then the volts per division and ms per division may be adjusted to provide the desired display.

Trigger and Trigger Slope

The trigger selection tells the DSO when to begin displaying a waveform. Until the DSO has a trigger level, it doesn't know when to begin the waveform display. When testing an input sensor that operates in a 0-volt to 5-volt range, select a trigger level of one-half this range.

Trigger slope informs the DSO whether the voltage signal is moving upward or downward when it crosses the trigger level. When a negative trigger slope is selected, the voltage signal is moving downward as it crosses the trigger level. Selecting a positive trigger level results in an upward voltage signal trace when it crosses the trigger level. A marker on the left side of the screen indicates the 0-volt or ground voltage position. This marker may be moved with the DSO controls (**Figure 3-42**). A second marker at the top of the screen indicates the trigger location. Since the control buttons on DSOs vary, the technician must spend some time to become familiar with a particular DSO.

Types of Voltage Signals

An **analog signal** is a varying voltage within a specific range over a period of time (**Figure 3-43**). A throttle position sensor (TPS) produces an analog voltage signal each time the throttle is opened (**Figure 3-44**).

A **digital signal** is either on or off. It may be described as one that is always high or low. The leading edge of a digital signal represents an increasing voltage, while the trailing edge of the signal represents a decreasing voltage. If the component is turned on by

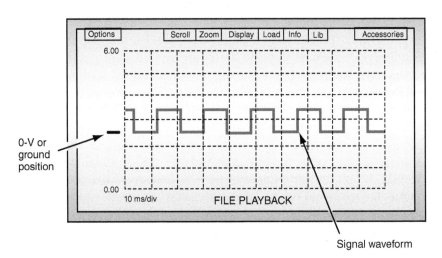

Figure 3-42 Ground-level reference marker on the left side of the screen.

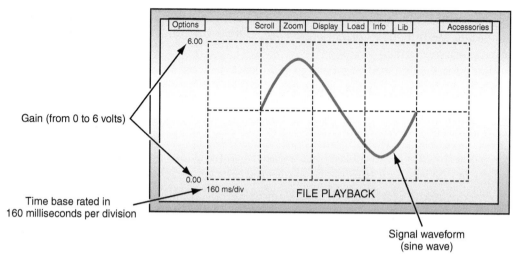

Figure 3-43 Analog voltage signal.

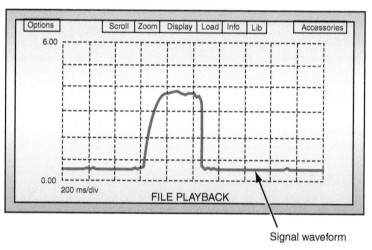

Figure 3-44 TPS analog voltage signal.

insulated side switching, the line across the top of the leading and trailing edge signals represents the length of component on time, which is called pulse width (**Figure 3-45**).

The computer measures the distance between the leading edge of a digital signal and the leading edge of the next signal to determine the frequency of the waveform. The distance between the leading edge of one digital signal and the leading edge of the next digital signal is referred to as one cycle (**Figure 3-46**). The computer counts the number of cycles over a period of time to establish the frequency. For example, if 92 cycles are occurring per second, the frequency is 92 hertz.

The relationship between the on time and off time in a digital signal is called duty cycle. For example, if the component on time and off time are equal, the component has a 50% duty cycle. When the component has a 90% duty cycle, the component is on for 90% of the time in a cycle and off for 10% of the time in a cycle. The computer controls some outputs, such as a carburetor mixture control solenoid, by varying the pulse width

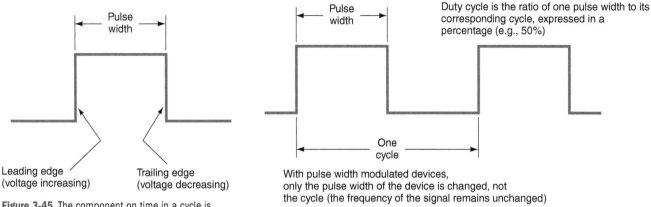

Figure 3-45 The component on time in a cycle is called pulse width.

Figure 3-46 The distance between leading edges is one cycle.

while the frequency remains constant. This type of computer control is referred to as **pulse width modulation (PWM)**. Other outputs, such as fuel injectors, are controlled by varying the frequency and the pulse width.

User-Friendly DSOs

DSOs with simplified, user-friendly controls have recently been introduced to the automotive service industry. In these DSOs, the auto-range function automatically selects the proper voltage and time base for the signal being received. The technician may use the DSO controls to turn off the auto-range function, and manually select the voltage range and time base.

When the menu key is pressed, various menus are displayed and the vertical arrow keys allow the technician to scroll through the menu to select a specific test. Digital readings are displayed on the screen with most waveforms. For example, minimum, average, and maximum millivolt (mV) readings are provided with an O_2 sensor waveform. The DSO automatically adjusts for zirconia or titania O_2 sensors. This DSO has multimeter and ignition waveform capabilities.

Some user-friendly DSOs have a removable application module to help prevent scope obsolescence. A software program card plugs into the bottom of the DSO. This DSO also sets the voltage range and time base automatically for the signal being tested. Four voltage waveforms or six multimeter functions may be displayed on the DSO screen.

CASE STUDY

A customer brings his/her vehicle to the shop because the dash lights are not illuminating. The technician checks the fuses and all are good. She then substitutes a bulb, known to be good, in the printed circuit, but it still does not illuminate. Next, she uses a voltmeter and checks for applied voltage to the panel light circuit. The test indicates that 12.6 volts are present.

The technician then performs a voltage drop test on the ground side of the circuit and the voltmeter indicates 12.6 volts. She concludes that the printed circuit has an open in the ground side of its circuit. Upon receiving written approval from the customer to perform repairs, she replaces the printed circuit. Verification of the repair confirms her diagnosis.

ASE-STYLE REVIEW QUESTIONS

1. Circuit defects are being discussed.

   Technician A says an open can be only on the ground side of the circuit.

   Technician B says an unwanted resistance can result from a corroded connector.

   Who is correct?

   A. A only
   B. B only
   C. Both A and B
   D. Neither A nor B

2. Testing the fuse is being discussed.

   Technician A says sometimes a visual inspection of a fuse or fusible link does not reveal that it is open.

   Technician B says to use a jumper wire to bypass the fuse in order to test the circuit.

   Who is correct?

   A. A only
   B. B only
   C. Both A and B
   D. Neither A nor B

3. Testing of a switch is being discussed.

   Technician A says a switch can be tested with a voltmeter.

   Technician B says to use a fused jumper wire to see if the switch is operating.

   Who is correct?

   A. A only
   B. B only
   C. Both A and B
   D. Neither A nor B

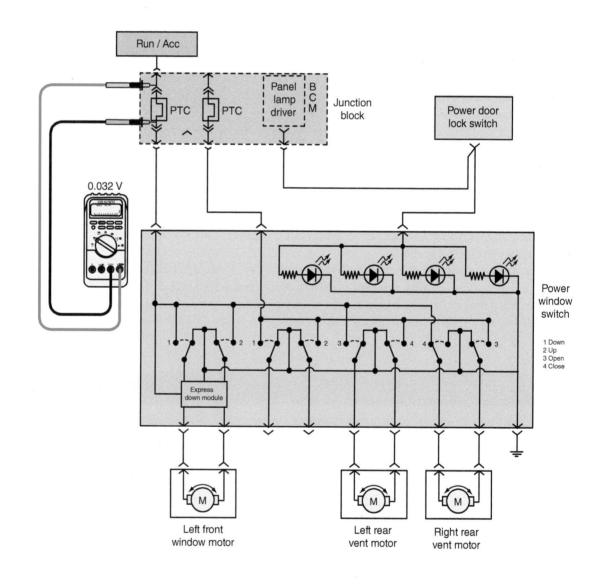

4. The voltmeter reading illustrated in Figure 10-41 indicates:

   A. An open in the circuit between the control module and the motor

   B. A properly operating PTC

   C. An open PTC

   D. None of the above

5. An LED does not light. A voltmeter indicates 0 volt across the LED. All of the following can be the cause **EXCEPT**:

   A. Current-limiting resistor is open.

   B. The LED is open.

   C. Short to ground between the current-limiting resistor and the LED.

   D. Open ground circuit.

6. Voltage drop testing is being discussed.

   Technician A says it is possible to calculate voltage drop by testing for available voltage on both sides of the component.

   Technician B says excessive voltage drop can be on either the power side or the ground side of the circuit.

   Who is correct?

   A. A only   C. Both A and B

   B. B only   D. Neither A nor B

7. The results of conductor-to-conductor shorts are being discussed.

   Technician A says if the windings of an electrical motor are shorted together, then the amount of resistance will be higher than specified.

   Technician B says a short between circuits can result in both circuits operating by closing one switch.

   Who is correct?

   A. A only   C. Both A and B

   B. B only   D. Neither A nor B

8. The testing of a shorted circuit is being discussed.

   Technician A says if a fuse blows as soon as it is installed, this indicates a short to ground.

   Technician B says if the short to ground is on the ground side of the load component but before a grounding switch, the component will not turn off.

   Who is correct?

   A. A only   C. Both A and B

   B. B only   D. Neither A nor B

9. Testing of a transistor is being discussed.

   Technician A says test the base-collector junction and the base-emitter junction as if they were standard diodes.

   Technician B says the resistance between the collector and emitter should read open circuit.

   Who is correct?

   A. A only   C. Both A and B

   B. B only   D. Neither A nor B

10. Testing of a potentiometer is being discussed.

   Technician A says a voltmeter can be used if jumper wires are connected to the wire connector and sensor to gain access to the test points.

   Technician B says the wires can be pierced to test for voltage.

   Who is correct?

   A. A only   C. Both A and B

   B. B only   D. Neither A nor B

ASE CHALLENGE QUESTIONS

1. The fuse for an A/C blower motor circuit fails after a short period of time.

   Technician A says that the blower motor may be binding internally.

   Technician B says that the blower motor ground circuit may have excessive resistance.

   Who is correct?

   A. A only   C. Both A and B

   B. B only   D. Neither A nor B

2. Relay testing is being discussed.

   Technician A says that it is acceptable during the test sequence to bypass the relay control coil terminals with a fused jumper wire.

   Technician B says that the voltage drop across the relay load contact terminals is checked with the relay control circuit de-energized.

   Who is correct?

   A. A only   C. Both A and B

   B. B only   D. Neither A nor B

3. The troubleshooting of a parallel circuit that contains three dimly lit bulbs is being discussed. A voltmeter that is placed across each of the bulbs indicates 7.2 volts.

   Technician A says that the power supply that is common to all three bulbs may be faulty.

   Technician B says that the ground terminal that is common to all three bulbs may have excessive resistance.

   Who is correct?

   A. A only   C. Both A and B

   B. B only   D. Neither A nor B

4. An A/C compressor clutch coil spike suppression diode is being tested with ohmmeter DMM set on diode test mode. The meter reads infinite in both directions.

   Technician A says that the diode is electrically open.

   Technician B says that the use of this diode would result in the failure of the circuit fuse.

   Who is correct?

   A. A only   C. Both A and B

   B. B only   D. Neither A nor B

5. How much total resistance is in a 12-volt circuit that is drawing 4 amperes?

   A. 0.333 ohm   C. 4.8 ohms

   B. 3 ohms   D. 48 ohms

6. A voltmeter that is connected across the input and output terminals of an instrument cluster illumination lamp rheostat indicates 12.6 volts with the switch in the maximum brightness position and the engine off. Which of the following statements is true?

   A. The voltage available to the lamps will be 12.6 volts.

   B. The voltage available to the lamps will be 0 volt.

   C. The rheostat is operating correctly.

   D. More information is required to determine if the lamps will operate correctly.

7. The left-rear and right-rear taillights and the left-rear brake light of a vehicle illuminate dimly whenever the brake pedal is pressed; however, the right-rear brake light operates properly.

   Technician A says the left-rear taillight and brake light may have a poor ground connection.

   Technician B says the brake light switch may have excessive resistance.

   Who is correct?

   A. A only   C. Both A and B

   B. B only   D. Neither A nor B

8. The horn of a vehicle equipped with a horn relay sounds weak and distorted. Which of the following is the least likely cause of this problem?

   A. High resistance in the relay load circuit.

   B. High resistance in the horn ground circuit.

   C. Excessive voltage drop between the relay load contact and the horn.

   D. Excessive voltage drop across the relay coil winding.

9. The power seats do not move fore or aft. The technician notices that the motor attempts to move the seat, then stops. The LEAST likely cause of this is:

   A. A binding track

   B. A short in the motor armature

   C. A faulty PTC

   D. An obstruction under the seat

10. The current draw of a window motor is being measured.

    Technician A says the ammeter can be connected on the power side of the motor.

    Technician B says the ammeter can be connected on the ground side of the motor.

    Who is correct?

    A. A only   C. Both A and B

    B. B only   D. Neither A nor B

Name _____   Date _____

VOLTAGE DROP TESTING VS OPEN CIRCUIT TESTING

Upon completion of this job sheet, you should be able to perform a voltage drop test and an open circuit test and determine the benefits and limitations of each to locating high-resistance faults.

NATEF Correlation ——————————————————————————————

This job sheet addresses the following **MLR** tasks:

A.4.   Demonstrate proper use of a digital multimeter (DMM) when measuring source voltage, voltage drop (including grounds), current flow, and resistance.

A.6.   Check operation of electrical circuits using a test light.

This job sheet addresses the following **AST/MAST** tasks:

A.3.   Demonstrate proper use of a digital multimeter (DMM) when measuring source voltage, voltage drop (including grounds), current flow, and resistance.

A.4.   Demonstrate knowledge of the causes and effects from shorts, grounds, opens, and resistance problems in electrical/electronic circuits.

A.5.   Demonstrate proper use of a test light on an electrical circuit.

NATEF

Tools and Materials
- A vehicle
- Fender covers
- Safety glasses
- DMM
- 35-ohm resistor attached to fused jumper wires
- Test light

Describe the vehicle being worked on:

Year _____ Make _____ Model _____

VIN _____ Engine type and size _____

Procedure   **Task Completed**

1. Measure the voltage across the battery terminals and record your results.

   _____

   _____

2. Connect one end of the fused jumper wire with the 35-ohm resistor to the battery   ☐
   positive (+) terminal. CAUTION: Be sure the other end of the jumper wire does not
   come into contact with ground.

3. With the black lead of the DMM connected to the negative (−) terminal of the battery,
   connect the red lead to the free end of the jumper wire. Record the voltage reading.

   _____

   _____

4. Steps 1 through 3 represented open circuit testing. This would be the same as disconnecting a connector and testing for applied voltage on the feed terminal. The free end
   of the installed jumper wire represents the disconnected connector. Is this reading in
   step 3 the same as in step 1?   ☐ Yes   ☐ No

   Explain why: _____

   _____

5. Measure the voltage across the battery terminals and record.

   _____

   _____

6. Connect the terminal of the test light to the negative (−) post of the battery or a good ground. Connect the probe of the test light to the positive (+) terminal of the battery. Explain the behavior of the test light.

   _____

   _____

7. Move the probe of the test light to the free end of the jumper wire with the resistor installed. How does the test light behave?

   _____

   _____

8. With the test light still connected, connect the black lead of the DMM to ground and the red lead to the tip of the probe. What is the voltage reading?

   _____

   _____

9. Steps 5 through 8 show testing the circuit under load. The test light represents the load. Why is the voltage different in Step 8 than the reading you had in Step 3?

   _____

   _____

10. With the test light still connected as in Steps 6 and 7, place the red DMM test lead at the positive (+) terminal of the battery and the black lead at the probe of the test light. What is the voltage reading?

    _____

    _____

11. What does this voltage represent?

    _____

    _____

12. Explain to your instructor how you would perform the load and voltage drop tests on the ground side of a circuit. Demonstrate this using the jumper wire with the resistor installed, the test light, and the DMM.

    _____

    _____

    _____

Instructor's Response

_____

_____

_____

Name _____   Date _____

TESTING FOR AN OPEN CIRCUIT

Upon completion of this job sheet, you should be able to test a circuit and locate the open.

NATEF Correlation ————————————————————————————

This job sheet addresses the following **MLR** tasks:

A.3.	Use wiring diagrams to trace electrical/electronic circuits.
A.4.	Demonstrate proper use of a digital multimeter (DMM) when measuring source voltage, voltage drop (including grounds), current flow, and resistance.
A.6.	Check operation of electrical circuits using a test light.
A.7.	Using fused jumper wires, check operation of electrical circuits.

This job sheet addresses the following **AST/MAST** tasks:

A.3.	Demonstrate proper use of a digital multimeter (DMM) when measuring source voltage, voltage drop (including grounds), current flow, and resistance.
A.4.	Demonstrate knowledge of the causes and effects from shorts, grounds, opens, and resistance problems in electrical/electronic circuits.
A.5.	Demonstrate proper use of a test light on an electrical circuit.
A.6.	Use fused jumper wires to check operation of electrical circuits.
A.7.	Use wiring diagrams during the diagnosis (troubleshooting) of electrical/electronic circuit problems.

NATEF

Tools and Materials

- A vehicle
- Fender covers
- Safety glasses
- DMM
- Test light
- Fused jumper wire
- Wiring diagram for the vehicle

Describe the vehicle being worked on:

Year _____ Make _____ Model _____

VIN _____ Engine type and size _____

Procedure   Task Completed

1. What is the customer's complaint?

   _____

   _____

2. Can the complaint be verified?   ☐ Yes   ☐ No

   (If no, consult your instructor.)

3. Are there any other related symptoms?   ☐ Yes   ☐ No

   If yes, describe the symptom.

   _____

   _____

4. Following the wiring diagram, use a voltmeter, fused jumper wire, or test light to trace the circuit to locate the open. □

5. Describe the location of the open.

_____

_____

_____

Instructor's Response

_____

_____

_____

Name _____   Date _____

TESTING FOR A SHORT TO GROUND

Upon completion of this job sheet, you should be able to test a circuit and locate the short to ground.

NATEF Correlation

This job sheet addresses the following **MLR** tasks:

A.3.	Use wiring diagrams to trace electrical/electronic circuits.
A.4.	Demonstrate proper use of a digital multimeter (DMM) when measuring source voltage, voltage drop (including grounds), current flow, and resistance.
A.6.	Check operation of electrical circuits using a test light.
A.7.	Use fused jumper wires to check operation of electrical circuits.

This job sheet addresses the following **AST/MAST** tasks:

A.3.	Demonstrate proper use of a digital multimeter (DMM) when measuring source voltage, voltage drop (including grounds), current flow, and resistance.
A.4.	Demonstrate knowledge of the causes and effects from shorts, grounds, opens, and resistance problems in electrical/electronic circuits.
A.5.	Demonstrate proper use of a test light on an electrical circuit.
A.6.	Using fused jumper wires check operation of electrical circuits.
A.7.	Use wiring diagrams during the diagnosis (troubleshooting) of electrical/electronic circuit problems.

ASE NATEF

Tools and Materials

- A vehicle
- Fender covers
- Safety glasses
- DMM
- Test light
- Circuit breaker fitted with alligator clips
- Gauss gauge
- Wiring diagram for the vehicle

Describe the vehicle being worked on.

Year _____ Make _____ Model _____

VIN _____ Engine type and size _____

Procedure   **Task Completed**

1. Which circuit is affected by the short to ground?

   _____

2. Are there any other related symptoms? ☐ Yes ☐ No

   If yes, describe the symptom(s).

   _____

3. Pull the fuse or circuit protection for the affected circuit from the fuse box.   ☐

4. With ignition switch in the RUN position, connect the test light across the fuse terminals in the fuse box. Does the test light illuminate?   ☐ Yes ☐ No

5. What does this test indicate?

   _____

   _____

   _____

6. With the test light connected across the fuse terminals of the fuse box, disconnect components and connectors in the affected circuit that are identified in the wiring diagram.   ☐

7. Did the test light go out when a component was disconnected?   ☐ Yes ☐ No

8. What can be concluded thus far?

   _____

   _____

9. Connect the test circuit breaker across the fuse terminals of the fuse box and use the Gauss gauge to find the location of the short to ground.   ☐

10. Describe the location of the short to ground.

    _____

    _____

    _____

    _____

Instructor's Response

_____

_____

_____

Name _____   **Date** _____

TESTING CIRCUIT PROTECTION DEVICES

Upon completion of this job sheet, you should be able to test circuit protection devices for opens.

NATEF Correlation ────────────────────────────────

This job sheet addresses the following **MLR** tasks:

A.4.   Demonstrate proper use of a digital multimeter (DMM) when measuring source voltage, voltage drop (including grounds), current flow, and resistance.

A.6.   Check operation of electrical circuits using a test light.

This job sheet addresses the following **MLR/AST/MAST** tasks:

A.9.   Inspect and test fusible links, circuit breakers, and fuses; determine necessary action.

This job sheet addresses the following **AST/MAST** tasks:

A.3.   Demonstrate proper use of a digital multimeter (DMM) when measuring source voltage, voltage drop (including grounds), current flow, and resistance.

A.5.   Demonstrate proper use of a test light on an electrical circuit.

ASE **NATEF**

Tools and Materials

- A vehicle equipped with fusible links and PTC circuit protection
- Fender covers
- Safety glasses
- A DMM
- Test light

Describe the vehicle being worked on:

Year _____ Make _____ Model _____

VIN _____ Engine type and size _____

Procedure   **Task Completed**

1. Locate the fuse panel or power distribution center.   ☐

2. Check that the test light is working properly by connecting it across the battery. ☐

3. Connect the negative lead of a test light to a good ground.   ☐

4. Turn the ignition switch to the RUN position.   ☐

5. Touch the probe of the test light onto the metal test tabs on each side of the fuse. ☐

6. Did the test light illuminate on each side of the fuse? ☐ Yes ☐ No Why or why not?

   _____

   _____

   _____

7. Repeat for all fuses in the fuse box. Record your findings.

   _____

   _____

   _____

8. Remove any fuses that failed the test.   ☐

9. Visually inspect the fuse. Record your findings.

_____

_____

_____

10. Use an ohmmeter and test the fuses. Record your findings.

_____

_____

_____

11. Locate a fusible link on the vehicle.   ☐

12. Use a voltmeter and measure the voltage drop over the link. Record your results.

_____

_____

_____

13. Disconnect power to the fusible link and use an ohmmeter to test the link. Record your results.

_____

_____

_____

14. Locate a PTC on the vehicle.

15. Use service information to identify the PTC and its location from the circuit assigned   ☐
to you by your instructor.

_____

_____

16. While operating the accessory, measure the voltage on the input side of the PTC.

_____

17. While operating the accessory, ensure the voltage on the output side of the PTC.

_____

18. Connect the red test lead to the input side of the PTC and the black lead to the output
side of the PTC and operate the system the PTC protects. Record the voltmeter reading.

_____

19. Based on your observations, what can you conclude about the functionality of the PTC?

_____

_____

Instructor's Response

_____

_____

_____

Name _____   Date _____

TESTING SWITCHES

Upon completion of this job sheet, you should be able to test a switch and properly determine needed repairs.

NATEF Correlation

This job sheet addresses the following **MLR** tasks:

A.3.	Use wiring diagrams to trace electrical/electronic circuits.
A.4.	Demonstrate proper use of a digital multimeter (DMM) when measuring source voltage, voltage drop (including grounds), current flow, and resistance.
A.6.	Check operation of electrical circuits using a test light.
A.7.	Use fused jumper wires to check operation of electrical circuits.

This job sheet addresses the following **AST/MAST** tasks:

A.3.	Demonstrate proper use of a digital multimeter (DMM) when measuring source voltage, voltage drop (including grounds), current flow, and resistance.
A.5.	Demonstrate proper use of a test light on an electrical circuit.
A.6.	Using fused jumper wires, check operation of electrical circuits.
A.7.	Use wiring diagrams during the diagnosis (troubleshooting) of electrical/electronic circuit problems.

This job sheet addresses the following **AST** task:

A.10.	Inspect and test switches, connectors, relays, solenoids, solid state devices, and wires of electrical/electronic circuits; determine necessary action.

ASE CERTIFIED NATEF

Tools and Materials

- A vehicle
- Fender covers
- Safety glasses
- DMM
- Test light
- Fused jumper wires
- Wiring diagram for the vehicle

Describe the vehicle being worked on:

Year _____ Make _____ Model _____

VIN _____ Engine type and size _____

Procedure

Task Completed

1. Locate the brake light switch (or another switch as directed by your instructor). ☐

2. Disconnect the switch from the wire harness. ☐

3. Use an ohmmeter to measure the resistance of the switch with the brake pedal released. Record your results.

_____

4. With the ohmmeter still connected across the switch terminal, press the brake pedal and record the ohmmeter reading:

_____

5. Based on your results, is the switch operating properly?   ☐ Yes   ☐ No

Why? _____

_____

6. With the electrical connector to the switch still unplugged, connect a fused jumper wire across the battery feed and brake light circuits.   ☐

7. Do the brake lights come on?   ☐ Yes   ☐ No

If Yes, what is the faulty component?

_____

_____

Instructor's Response

_____

_____

_____

Name _____   Date _____

TESTING RELAYS

Upon completion of this job sheet, you should be able to test a relay and properly determine needed repairs.

NATEF Correlation

This job sheet addresses the following **MLR** tasks:

A.3.	Use wiring diagrams to trace electrical/electronic circuits.
A.4.	Demonstrate proper use of a digital multimeter (DMM) when measuring source voltage, voltage drop (including grounds), current flow, and resistance.
A.6.	Check operation of electrical circuits using a test light.
A.7.	Use fused jumper wires to check operation of electrical circuits.

This job sheet addresses the following **AST/MAST** tasks:

A.3.	Demonstrate proper use of a digital multimeter (DMM) when measuring source voltage, voltage drop (including grounds), current flow, and resistance.
A.5.	Demonstrate proper use of a test light on an electrical circuit.
A.6.	Using fused jumper wires, check operation of electrical circuits.
A.7.	Use wiring diagrams during the diagnosis (troubleshooting) of electrical/electronic circuit problems.

This job sheet addresses the following **AST** task:

A.10.	Inspect and test switches, connectors, relays, solenoids, solid state devices, and wires of electrical/electronic circuits; determine necessary action.

NATEF

Tools and Materials

- A vehicle
- Fender covers
- Safety glasses
- DMM
- Fused jumper wires
- Test light
- Wiring diagram for the vehicle

Describe the vehicle being worked on:

Year _____ Make _____ Model _____

VIN _____ Engine type and size _____

Procedure

Task Completed

1. Locate the low-speed fan relay (or another relay as directed by your instructor).   ☐

2. Disconnect the relay from the relay box.   ☐

3. Check the wiring diagram for the relay being tested.

   Identify the terminal callouts and their purpose.

   _____

   _____

   _____

   _____

   _____

4. Is the control circuit of the relay through an insulated or a ground switch?

   _____

   _____

5. Use a voltmeter to check for available voltage to the two terminals of the relay box that connect to the relay terminals.

   Is voltage available at both terminals?   ☐ Yes   ☐ No

   If no, what would be your next step?

   _____

   _____

6. Connect a test light or DMM across the two terminals of the relay box that connects   ☐
   to the control side of the relay.

7. Use the switch(es) that normally control the relay and turn them ON. Did the test light illuminate or the voltmeter read close to battery voltage?   ☐ Yes   ☐ No

   If No, what does this indicate?

   _____

   _____

   If Yes, what does this indicate?

   _____

   _____

8. Use a fused jumper wire to connect the high-current input terminal of the relay box to the high-current output terminal. Did the fan come on?   ☐ Yes   ☐ No

   If No, what does this indicate?

   _____

   _____

   _____

   If Yes, what does this indicate?

   _____

   _____

   _____

9. Use the service information to locate the specifications for the resistance of the relay coil. Record your findings.

_____

10. Use an ohmmeter to measure the resistance of the relay's coil. Is the reading within specifications?   ☐ Yes   ☐ No

If No, what does the reading indicate the problem is?

_____

_____

11. Use a pair of fused jumper wires to energize the coil. Check for continuity through the relay contacts. What is your reading?

_____

_____

12. Based on your results of this series of relay tests, what are your conclusions?

_____

_____

_____

_____

_____

Instructor's Response

_____

_____

_____

Name _____   Date _____

TESTING A DIODE

Upon completion of this job sheet, you should be able to test a diode for a short or open.

NATEF Correlation —————————————————————————————————

This job sheet addresses the following **MLR** task:

A.4.   Demonstrate proper use of a digital multimeter (DMM) when measuring
             source voltage, voltage drop (including grounds), current flow, and resistance.

This job sheet addresses the following **AST/MAST** task:

A.3.   Demonstrate proper use of a digital multimeter (DMM) when measuring
             source voltage, voltage drop (including grounds), current flow, and resistance.

This job sheet addresses the following **AST** task:

A.10.   Inspect and test switches, connectors, relays, solenoids, solid state devices, and
             wires of electrical/electronic circuits; determine necessary action.

NATEF

Tools and Materials

- An assortment of diodes
- DMM

Procedure

1. Which side is the stripe around the diode on?   ☐ Anode   ☐ Cathode

2. Use the ohmmeter function of the DMM and measure the resistance in both
   directions through the diode. Record your results.

   Forward biased _____

   Reverse biased _____

3. Explain why a DMM ohmmeter is not recommended for testing a diode.

   _____

   _____

   _____

4. Use a DMM to test a second diode using the diode test function. Connect the red test
   lead to the anode and the black lead to the cathode side of the diode. Record the meter
   readings.

   _____

   _____

5. What does this reading represent?

   _____

   _____

6. Reverse the test leads across the diode. What is the reading?

   _____

   _____

7. What does this reading represent?

   _____

   _____

8. What is your conclusion concerning this diode?

   _____

   _____

Instructor's Response

_____

_____

_____

Name _____   Date _____

USING A LAB SCOPE ON SENSORS AND SWITCHES

Upon completion of this job sheet, you should be able to connect a lab scope and observe the activity of various sensors and switches.

NATEF Correlation ——————————————————

This job sheet addresses the following **MLR** tasks:

A.3.   Use wiring diagrams to trace electrical/electronic circuits.

A.4.   Demonstrate proper use of a digital multimeter (DMM) when measuring source voltage, voltage drop (including grounds), current flow, and resistance.

This job sheet addresses the following **AST/MAST** tasks:

A.3.   Demonstrate proper use of a digital multimeter (DMM) when measuring source voltage, voltage drop (including grounds), current flow, and resistance.

A.7.   Use wiring diagrams during the diagnosis (troubleshooting) of electrical/ electronic circuit problems.

This job sheet addresses the following **MAST** task:

A.11.   Check electrical/electronic circuit waveforms; interpret readings and determine needed repairs.

This job sheet addresses the following **MLR/AST/MAST** task:

A.1.   Research vehicle service information, including vehicle service history, service precautions, and technical service bulletins.

This job sheet addresses the following **AST** task:

A.10.   Inspect and test switches, connectors, relays, solenoids, solid state devices, and wires of electrical/electronic circuits; determine necessary action.

ASE CERTIFIED **NATEF**

Tools and Materials

- A vehicle with accessible sensors and switches
- Service information for the assigned vehicle
- Component locator manual for the assigned vehicle
- Fender covers
- Safety glasses
- Lab scope
- DMM

Describe the vehicle being worked on:

Year _____ Make _____ Model _____

VIN _____ Engine type and size _____

Procedure

1. Connect the lab scope test leads across the battery. Make sure the scope is properly set. Observe the trace on the scope. Is there evidence of noise? Explain.

_____

_____

_____

_____

2. Locate the A/C compressor clutch control wires. Connect the DMM to read available voltage. Start and run the engine. Observe the meter, and then turn the compressor on. What happened on the meter?

_____

_____

Now connect the lab scope to the same point with the compressor turned off. Observe the waveform and then turn the compressor on. What happened to the trace?

_____

_____

3. Turn off the engine but keep the ignition on. Locate the throttle position (TP) sensor and identify the purpose of each wire to it. List each wire and describe the purpose of each.

_____

_____

4. Connect the DMM to read reference voltage at the TP sensor. What do you read?

_____

_____

Now move the leads to read the output of the sensor. Starting with the throttle closed, slowly open the throttle until it is wide open. Watch the voltmeter while doing this. Describe your readings.

_____

_____

5. Now connect the lab scope to read reference voltage at the TP sensor. Describe what you see on the trace (is the voltage steady, evidence of noise, EMI, RFI?).

_____

_____

Now move the leads to read the output of the sensor. Starting with the throttle closed, slowly open the throttle until it is wide open. Watch the trace while doing this. Describe your readings.

_____

_____

6. Locate the oxygen sensor and identify the purpose of each wire to it. Connect the DMM to read voltage generated by the sensor. (To do this, you may use an electrical connector for the O_2 sensor that is positioned away from the hot exhaust manifold.) Now run the engine. Watch the meter as the O_2 sensor warms and describe what happened below.

_____

_____

7. Now connect the lab scope to read voltage output from the sensor. Watch the trace and describe what happened below.

_____

_____

_____

8. Explain what you observed as the differences between testing with a DMM and a lab scope.

_____

_____

_____

Instructor's Response

_____

_____

_____

DIAGNOSTIC CHART 3-1

PROBLEM AREA:	Electrical opens.
SYMPTOMS:	Electrical component will not operate.
POSSIBLE CAUSES:	**1.** Broken conductor. **2.** Defective switch. **3.** Defective relay. **4.** Blown fuses. **5.** Burned fusible links. **6.** Burned or defective circuit breakers.

DIAGNOSTIC CHART 3-2

PROBLEM AREA:	Excessive resistance resulting in lowered electrical output.
SYMPTOMS:	Electrical components fail to operate or operate at reduced efficiency.
POSSIBLE CAUSES:	**1.** Excessive resistance. **2.** Corroded or damaged conductor. **3.** Excessive resistance in the switch. **4.** Excessive resistance in the relay. **5.** Improper stepped resistor values **6.** Defective variable resistor.

DIAGNOSTIC CHART 3-3

PROBLEM AREA:	Copper-to-copper short or short to ground.
SYMPTOMS:	No electrical component operation or operation when another control switch is activated.
POSSIBLE CAUSES:	**1.** Broken or burned insulation and/or connectors causing copper-to-copper short. **2.** Broken or burned insulation and/or connectors causing a short to ground.

DIAGNOSTIC CHART 3-4

PROBLEM AREA:	Switch.
SYMPTOMS:	Component fails to turn on.
POSSIBLE CAUSES:	Faulty terminals to switch. Burned switch contacts. Internal open in switch.

DIAGNOSTIC CHART 3-5

PROBLEM AREA:	Switch.
SYMPTOMS:	Component fails to turn off.
POSSIBLE CAUSES:	Short across the terminals to switch. Stuck switch contacts. Internal short in switch.

DIAGNOSTIC CHART 3-6

PROBLEM AREA:	Relays.
SYMPTOMS:	Component fails to turn on.
POSSIBLE CAUSES:	Faulty terminals to relay. Open in relay coil. Shorted relay coil. Burned relay high-current contacts. Internal open in relay high-current circuit.

DIAGNOSTIC CHART 3-7

PROBLEM AREA:	Relay.
SYMPTOMS:	Component fails to turn off.
POSSIBLE CAUSES:	Short across the relay terminals. Stuck relay high-current contacts.

DIAGNOSTIC CHART 3-8

PROBLEM AREA:	Stepped resistors.
SYMPTOMS:	Component fails to turn on.
POSSIBLE CAUSES:	Faulty terminal connections to the stepped resistor. Open in the input side of the resistor. Open in the output side of the resistor.

DIAGNOSTIC CHART 3-9

PROBLEM AREA:	Stepped resistors.
SYMPTOMS:	Component fails to operate at certain speeds or brightness.
POSSIBLE CAUSES:	Excessive resistance at the terminal connections to the stepped resistor. Open in one or more of the resistor's circuit. Short across one or more of the resistor's circuit.

DIAGNOSTIC CHART 3-10

PROBLEM AREA:	Variable resistors.
SYMPTOMS:	Component fails to turn on.
POSSIBLE CAUSES:	Faulty terminal connections to the variable resistor. Open in the input side of the resistor. Open in the output side of the resistor. Open in the return circuit of the resistor. Short to ground in the output circuit. Short to ground in the input circuit.

DIAGNOSTIC CHART 3-11

PROBLEM AREA:	Variable resistors.
SYMPTOMS:	Component fails to operate at certain speeds or brightness.
POSSIBLE CAUSES:	Excessive resistance at the terminal connections to the variable resistor. Excessive internal resistance. Intermittent opens in the resistor.

DIAGNOSTIC CHART 3-12

PROBLEM AREA:	Circuit protection devices.
SYMPTOMS:	Component fails to turn on.
POSSIBLE CAUSES:	Faulty terminal connections to the protection device. Excessive current draw due to electrical short. Excessive current draw due to mechanical restrictions. Incorrect amperage rating of fuse for the circuit.

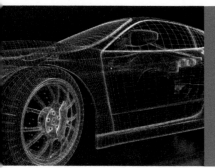

CHAPTER 4

WIRING REPAIR AND READING CIRCUIT DIAGRAMS

Upon completion and review of this chapter, you should be able to:

- Perform repairs to copper wire using solderless connections.
- Solder splices to copper wire.
- Repair aluminum wire according to manufacturer's requirements.
- Repair twisted/shielded wire.
- Replace fusible links.

- Repair and/or replace the terminals of a hard-shell connector.
- Repair and/or replace the terminals of weather and metri-pack connectors.
- Read a wiring diagram to correctly determine the operation of the circuit.
- Use the wiring diagram to diagnose possible causes for the system fault.

 Basic Tool

Basic mechanic's tool set

Service information

Terms To Know

Color codes	Pull to seat	Trouble codes
Crimping	Push to seat	Troubleshooting
Crimping tool	Solderless connectors	Vehicle identification
Ground straps	Splice	number (VIN)
Heat shrink tube	Splice clip	

INTRODUCTION

Many electrical repairs will involve the replacement or repairing of a damaged conductor. To locate the problem area, today's technician must be capable of reading and understanding electrical diagrams and schematics. Once it is determined that the battery is operating correctly, the schematic should always be the starting point in **troubleshooting** an electrical system. Troubleshooting is the diagnostic procedure of locating and identifying the cause of the fault. It is a systematic process of elimination by use of cause and effect. By using the schematic, the technician is able to understand how the circuit should work. This is essential before attempting to determine why it does not work.

The component locator will assist the technician in finding the location of the electrical components and harnesses shown in the schematic. Many times the component locator will also list the locations of connectors, grounds, and splices. Most electronic generated wiring diagrams will provide a hyperlink that will jump to the component locator section.

The process of troubleshooting an electrical concern is as follows:

1. Confirm the concern. Perform a check of the system to gain an understanding of what is wrong. If the on-board computer monitors the faulty system, enter diagnostics to retrieve any **trouble codes**. Trouble codes are the output of the self-diagnostics program in the form of alpha/numeric codes that indicate faulty circuits or components.
2. Study the electrical schematic. This will indicate any shared circuits. Trying to operate the shared circuits will help direct the technician to the problem area. If the shared circuits operate correctly, the problem is isolated to the wiring or components of the problem system. If the shared circuits do not operate, the problem is usually in the power or ground circuit.
3. Locate and repair the fault. By narrowing down the possible causes and taking measurements as required, the fault is located. Before replacing any components, check the ground and power leads. If these are good, then the component is bad.
4. Test the repair. Repeat a check of the system to confirm that it is operating properly.

Classroom Manual
Chapter 4, page 87

WIRE REPAIR

Not all electrical repairs involve removing and replacing a faulty component. Many times the cause of the malfunction is a damaged conductor. The technician must make a repair to the circuit that will not increase the resistance. It should also be a permanent repair. There are many methods to repair a damaged wire. The type of repair used will depend on factors such as:

⚠ Caution

A solid copper wire may be used in low-voltage, low-current circuits where flexibility is not required. Do not use solid wire where high voltage, high current, or flexibility is required, unless solid wire was used by the manufacturer.

1. Type of repair required.
2. Ease of access to the damaged area.
3. Type of conductor.
4. Size of wire.
5. Circuit requirements.
6. Manufacturer's recommendations.

The most common methods of wire repair include wrapping damaged insulation with electrical tape or tubing, crimping the connections with solderless connectors, and soldering splices.

Copper Wire Repairs

Copper wire is the most commonly used primary wire in the automobile. The insulation may break down, or the wire may break due to stress or excessive motion. The wire may also be damaged due to excessive current flow through the wire. Any of these conditions require that the wire be repaired.

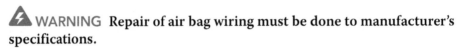 **WARNING Repair of air bag wiring must be done to manufacturer's specifications.**

In some instances, it may be necessary to bypass a length of wire that is not accessible. In this case, cut the wire before it enters the inaccessible portion and at the other end where it leaves the area. Install a replacement wire and reroute it to the load component (**Figure 4-1**). Be sure to protect the wire by using conduit, straps, hangers, and grommets as needed.

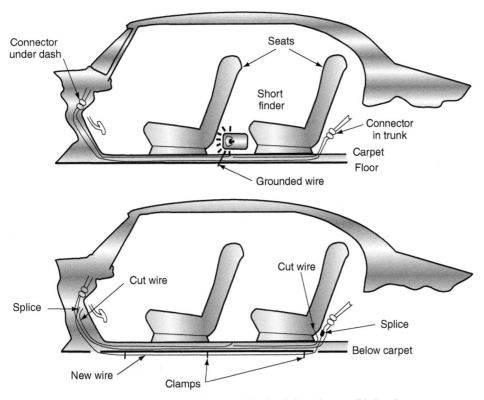

Figure 4-1 Routing a new wire to replace a damaged wire that is in an inaccessible location.

Figure 4-2 Types of solderless connectors and terminals sometimes used for temporary wire repair.

> **Crimping** means to bend, or deform by pinching, a connector so that the wire connection is securely held in place.

The two most common methods of splicing copper wire are with solderless connectors or by soldering.

Crimping. **Crimping** of **solderless connectors** is an acceptable method to **splice** wires that are not subjected to weather elements, dirt, corrosion, or excessive movement. Also, do not use crimped connections in electronic circuits. A poor connection, or corrosion over time, can result in improper electronic control operation of the system. Do the following to make a splice using solderless connections:

> **Solderless connectors** are hollow metal tubes covered with insulating plastic. They can be butt connectors or terminal ends (**Figure 4-2**).

1. Use the correct size of stripping opening on the **crimping tool** to remove enough insulation to allow the wire to completely penetrate the connector. The crimping tool has different areas for performing several functions (**Figure 4-3**). This single tool will cut the wire, strip the insulation, and crimp the connector.

> **Splice** is a term used to mean joining of single wire ends or the joining of two or more electrical conductors at a single point.

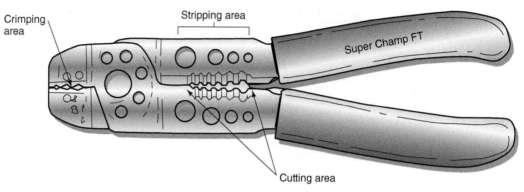

Figure 4-3 A typical crimping tool used for making electrical repairs.

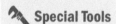
Special Tools

Crimping tool
Electrical tape or heat shrink tube
Solderless connector
Safety glasses
Fender covers

Classroom Manual
Chapter 4, page 92

2. Place the wire into the connector and crimp the connector (**Figure 4-4**). To get a proper crimp, place the open area of the connector facing toward the anvil. Be sure the wire is compressed under the crimp.
3. Insert the stripped end of the other wire into the connector and crimp in the same manner.
4. Use electrical tape or a piece of **heat shrink tube** to provide additional protection. Heat shrink tube is plastic tubing that shrinks in diameter when exposed to heat. Use a controlled heat source such as a heat gun to shrink the tubing. Never expose the tube to open flames.

Adhesive splices provide moisture-proof connections. The adhesive splice has a heat-shrinkable adhesive sleeve that—when heated—melts and flows to fill any voids and create a permanent seal. These are far more reliable than standard solderless connectors.

Another type of crimping connector is the tap splice connector. This type of connector allows for adding an additional circuit to an existing feed wire without stripping the wires (**Figure 4-5**). Although tap connectors make connecting wires easy, these should not be used for permanent repairs. They are unreliable for making and maintaining good connections. Anytime an additional circuit is attached to an existing circuit, make sure the fuse and conductor of the circuit being tapped into has a large enough capacity. Tap connectors are used to add a circuit in parallel with an existing circuit. This causes circuit resistance to decrease and circuit amperage to increase. Also, the wire size of the existing circuit must be capable of carrying the additional amperage.

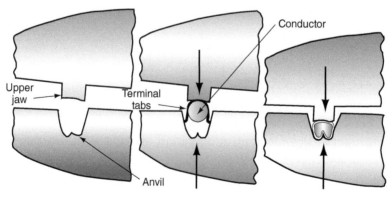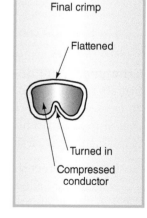

Figure 4-4 Properly crimping a connector.

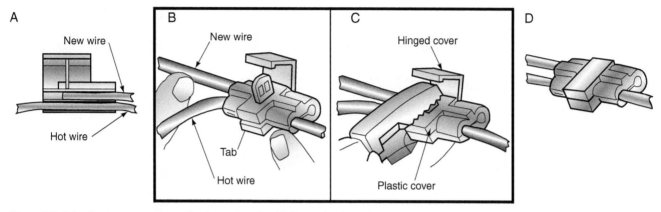

Figure 4-5 Using the tap connector to splice in another wire. (A) Place wires in position in the connector, (B) close the connector around the wires, (C) use pliers to force the tab into the conductors, and (D) close the hinged cover.

CUSTOMER CARE As a professional technician you should never have a vehicle leave the shop with a temporary repair. A temporary repair is used on such things as a road call that allows you get the vehicle drivable again to bring it to the shop. Almost all crimp-type repairs are temporary.

Soldering. Soldering is the best way to splice copper wires. Solder is an alloy of tin and lead. It is melted over a splice to hold the wire ends together. Soldering may be a splicing procedure, but it is also an art that takes much practice. **Photo Sequence 6** illustrates the soldering process when using a **splice clip**. A splice clip is a special connector used along with solder to assure a good connection. The splice clip is different from a solderless connection in that it does not have insulation. Some splice clips have a hole provided for applying solder (**Figure 4-6**).

Figure 4-6 Splice clip. Some splice clips have a hole for applying solder.

PHOTO SEQUENCE 6
Soldering Copper Wire

P6-1 Tools required to solder copper wire: 100-watt soldering iron, rosin core solder, crimping tool, splice clip, heat shrink tube, heat gun, safety glasses, sewing seam ripper, electrical tape, and fender covers.

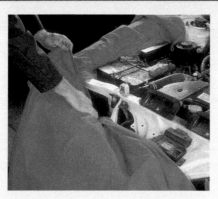

P6-2 Place the fender covers over the vehicle fenders.

P6-3 Disconnect the fuse that powers the circuit being repaired. Note: if the circuit is not protected by a fuse, then disconnect the battery.

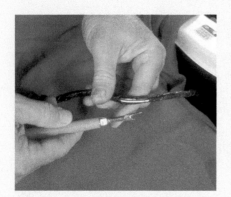

P6-4 If the wiring harness is taped, use a seam ripper to open the wiring harness.

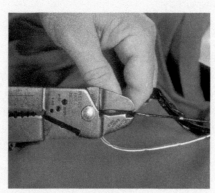

P6-5 Cut out the damaged wire using the wire cutters on the crimping tool.

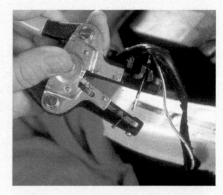

P6-6 Using the correct size stripper, remove about 1/2 inch (12 mm) of the insulation from both wires. Be careful not to nick or cut any of the wires.

P6-7 Determine the correct gauge and length of replacement wire.

P6-8 Using the correct size stripper, remove 1/2 inch (12 mm) of insulation from each end of the replacement wire.

P6-9 Select the proper-size splice clip to hold the splice. Do not install the clip at this time.

PHOTO SEQUENCE 6 (CONTINUED)

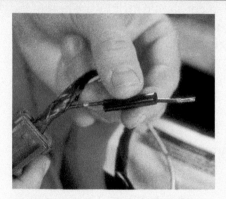

P6-10 Place the correct length and size of heat shrink tube over one of the wires. Slide the tube far enough away from the splice so it is not exposed to the heat of the soldering iron.

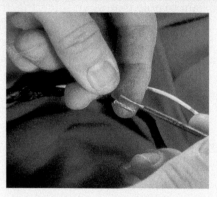

P6-11 Overlap the two splice ends and hold in place with thumb and forefinger.

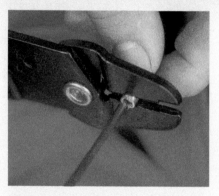

P6-12 Center the splice clip around the wires and crimp in place. Make sure that the copper wires extend beyond the splice clip in both directions. Crimp the clip on both ends.

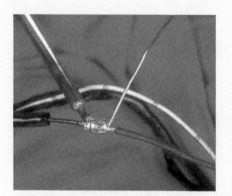

P6-13 Heat the splice clip with the pre-tinned soldering iron while applying solder to the opening in the back of the clip. Do not apply solder to the iron; the iron should be $180°$ away from the opening of the clip.

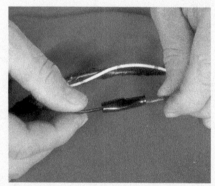

P6-14 After the solder cools, slide the heat shrink tube over the splice.

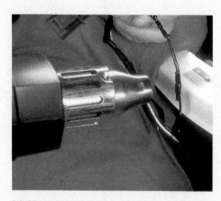

P6-15 Heat the tube with the hot air gun until it shrinks around the splice. Do not overheat the tube.

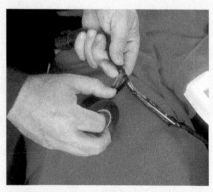

P6-16 Re-tape the wiring harness.

The process of applying solder to the tip of the iron is called *tinning*.

If a splice clip is not used, the wire ends should be braided together tightly using one of the methods illustrated (**Figure 4-7**). Remove about 1 inch (25 mm) of the insulation from the wires and join the wires. Then the splice should be heated with the soldering gun. It is important to note that when soldering, the solder should melt by the heat of the wire splice, not the heat of the soldering tool. The solder should melt and flow evenly among all of the wire strands (**Figure 4-8**). Always use rosin core solder when making electrical repairs. Acid core solder is used for other purposes than electrical repairs and can cause the wire to corrode, which would lead to high resistance. Insulate the splice with heat shrink tube or good-quality electrical tape.

> ⚙ **SERVICE TIP** It is easier to heat the wire with the soldering iron if the solder is first melted onto the tip of the iron. This will transfer the heat from the iron to the wire more quickly.

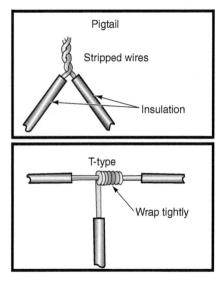

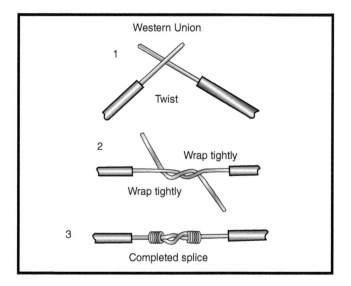

Figure 4-7 Methods for joining wires together.

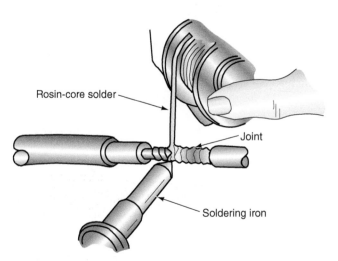

Figure 4-8 When soldering, apply the solder to the joint, not to the tip of the soldering iron.

WARNING Before cutting into a wire to make a splice, look for other splices or connections. Never have two or more splices within 1.5 inches (40 mm of each other. Also, always use wire of the same size or larger than the wire being replaced.

Repairing Aluminum Wire

 WARNING Attempting to solder aluminum wire will damage the conductor.

General Motors has used single-stranded aluminum wire in limited applications where no flexing of the wire is expected. This wire usually has a thick plastic insulator and is placed in a brown harness.

After cutting away the damaged wire, strip all wire of the last 1/4 inch (6 mm) of insulation. Be careful not to nick or damage the conductor. Apply a generous coating of petroleum jelly to the wire and connector (**Figure 4-9**). The petroleum jelly will prevent corrosion from developing in the core.

Crimp the connector in the usual manner. Insulate the splice with heat shrink tube. Do not use electrical tape since it will not stay in place due to the petroleum jelly.

 WARNING Do not connect copper wire to aluminum wire. The two different metals will conduct differently and have different expansion rates that may cause the connection to break and present a fire hazard.

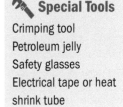

Special Tools

Crimping tool
Petroleum jelly
Safety glasses
Electrical tape or heat shrink tube

Splicing Twisted/Shielded Wire

Twisted/shielded wire is used in many computer communication circuits. It protects the circuit from electrical noise that would interfere with the operation of the computer controls (**Figure 4-10**). These wires may carry as low as 0.1 amp (A) of current. It is important that the splice made in these wires does not have any resistance. The added resistance may give false signals to the computer or actuator. **Photo Sequence 7** illustrates the procedure for splicing shielded wire.

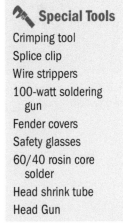

Special Tools

Crimping tool
Splice clip
Wire strippers
100-watt soldering gun
Fender covers
Safety glasses
60/40 rosin core solder
Head shrink tube
Head Gun

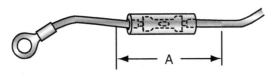

Figure 4-9 Apply petroleum jelly to the areas shown.

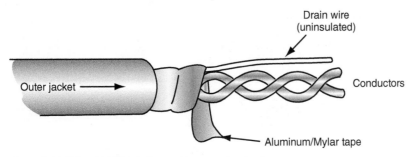

Figure 4-10 Twisted/shielded wire used in computer circuits.

PHOTO SEQUENCE 7
Splicing Shielded Wires

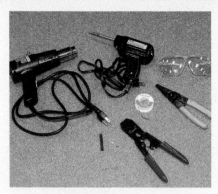

P7-1 Tools required to perform this task include a crimping tool, splice clip, wire strippers, soldering gun, rosin core solder, heat shrink tube, heat gun, and safety glasses.

P7-2 Locate and cut out the damaged section of wire.

P7-3 Being careful not to cut into the Mylar tape or to cut the drain wire, remove about 1 inch (25 mm) of the outer jacket from the ends of each cable.

P7-4 Unwrap the Mylar tape. Do not remove the tape from the cable.

P7-5 Untwist the conductors.

P7-6 Remove a small amount of the insulation from the ends of the conductors. The conductor ends should be staggered so that the splices will not be on top of each other. Also, the heat shrink tube should be located over the cable, but away from the work area.

P7-7 Locate a splice clip over the two ends of the conductors.

P7-8 Solder the splice.

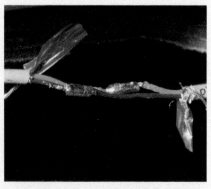

P7-9 Repeat the splice and solder procedure for the second conductor. Make sure that the two splices will not touch each other.

PHOTO SEQUENCE 7 (CONTINUED)

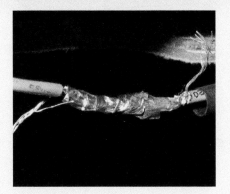

P7-10 Wrap the conductors with the Mylar tape. Do not wrap the drain wire in the tape.

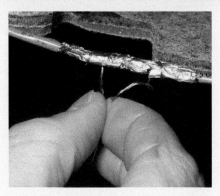

P7-11 Wrap the drain wire around the outside of the Mylar tape.

P7-12 Splice the drain wire with a clip and solder.

P7-13 Locate the heat shrink tube over the splice area and use the heat gun to shrink the tube over the cable.

REPLACING FUSIBLE LINKS

Not all opens within a fusible link are detectable by visual inspection only. Test for battery voltage on both sides of the fusible link to confirm its condition. If the fusible link must be replaced, it is cut out of the circuit and a new fusible link is crimped or soldered into place.

There are two types of insulation used on fusible links: Hypalon and Silicone/GXL. Hypalon can be used to replace either type of link. However, do not use Silicone/GXL to replace Hypalon. To identify the type of insulation, cut the blown link's insulation back. The insulation of the Hypalon link is a solid color all the way through. The insulation of Silicon/GXL will have a white inner core.

When cutting off the damaged fusible link from the feed wire, cut it beyond the splice (**Figure 4-11**). When making the repair, do not use a fusible link longer than 9 inches (228 mm). This length will not provide sufficient overload protection. Splice in the repair link by crimping or soldering.

If the damaged fusible link feeds two harness wires, use two fusible links. Splice one link to each of the harness wires (**Figure 4-12**).

Classroom Manual
Chapter 3, page 77

 **Special Tools**
Crimping tool
100-watt soldering iron
60/40 rosin core solder
Safety glasses
Fender covers

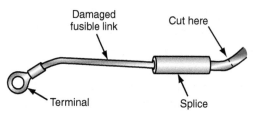

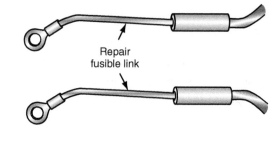

Figure 4-11 Replacing a fusible link.

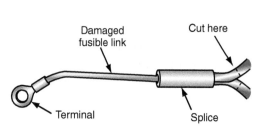

Figure 4-12 Replacing a fusible link that feeds two circuits.

Caution

Disconnect the battery negative cable before performing any repairs to the fusible link.

CUSTOMER CARE If the fusible link or any of the fuses are blown, it is important to locate the cause. Fuses do not wear out. If they blow, it is due to an overload of current in the circuit. By using a cause-and-effect diagnosis approach, the fault can be identified and repaired the first time the vehicle is in the shop. The following illustrates how to use this method:

Effect	Cause
Turn signals not operating	Blown fuse
Blown fuse	Short to ground
Short to ground	Broken wire in trunk
Broken wire	Loose jack handle

Caution

Use the fusible link gauge required by the manufacturer.

This method leads the technician to fix the fault and find the cause.

REPAIRING CONNECTOR TERMINALS

Special Tools

Pick
Crimping tool
Safety glasses
Fender covers

The connector terminal will require repairs due to abuse, improper disconnecting procedures, and exposure potential to the elements. The method of repair depends on the type of connector.

AUTHOR'S NOTE There are many differing connector service procedures, even within the same family of connector types. The following are examples of typical procedures for the different types of connectors. Always refer to the correct service information for the procedure for the connector you are servicing.

MOLDED CONNECTORS

If the connector is a one-piece, molded-type connector, it cannot be disassembled for repairs (**Figure 4-13**). Molded in-line connectors allow for the male and female connector halves to be separated; however, the connector itself cannot be disassembled. If the connector is damaged, it must be cut off and a new connector spliced in.

Classroom Manual
Chapter 4, page 92

Hard-Shell Connectors

⚠ WARNING **Do not place your fingers or body next to the connector. If excessive force is needed to depress the tang, the pick may be pushed out the back and cause injury.**

Classroom Manual
Chapter 4, page 92

Hard-shell connectors usually provide a means of removing the terminals for repair. Use the special tool, or a proper size pick, to depress the locking tang of the connector (**Figure 4-14**). Pull the lead back far enough to release the locking tang from the connector. Remove the pick, and then pull the lead completely out of the connector. Make the repair to the terminal using the same procedures for repairing copper or aluminum wire.

Re-form the terminal locking tang to assure a good lock in the connector (**Figure 4-15**). Use the pick to bend the lock tang back into its original shape. Insert the lead into the back of the connector. A noticeable "catch" should be felt when the lead is halfway through the connector. Gently push back and forth on the lead to confirm that the terminal is locked in place.

Weather-Pack Connectors

The terminals of the weather-pack connector are secured by a hinged secondary lock or a plastic terminal retainer. To perform repairs, first disconnect the two halves by pulling up on the primary lock while pulling the two halves apart (**Figure 4-16**). Unlock the secondary locks and swing them open (**Figure 4-17**).

Classroom Manual
Chapter 4, page 94

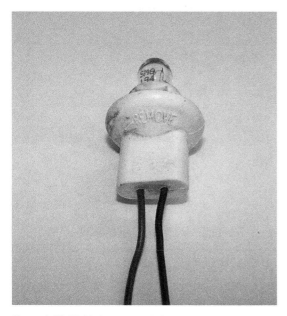

Figure 4-13 Molded connector halves are a one-piece design that cannot be disassembled for repairs.

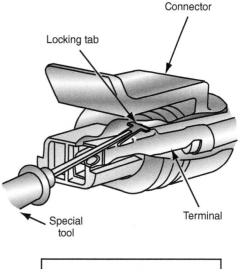

Push narrow pick between terminal and connector body.

Figure 4-14 Depress the locking tang to remove the terminal from the connector.

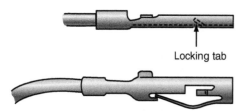

Figure 4-15 Re-form the locking tang to its original position before inserting the terminal back into the connector.

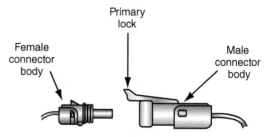

Figure 4-16 The weather-pack connector has two locks. Use the primary lock to separate the halves.

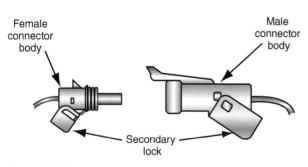

Figure 4-17 Unlock the secondary lock to remove the terminals from the connector.

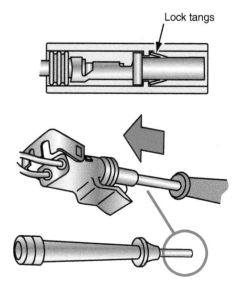

Figure 4-18 Use the recommended special tool to unlock the tang on the terminal.

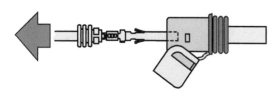

Figure 4-19 After the lock tang has been depressed, remove the lead from the back of the connector.

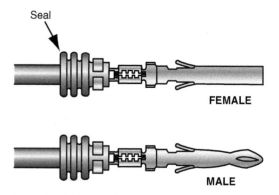

Figure 4-20 Male and female connectors.

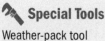
Special Tools

Weather-pack tool
Crimping tool
Safety glasses
Fender cover

Depress the terminal locking tangs using the special weather-pack tool. Push the cylinder of the tool into the terminal cavity from the front until it stops (**Figure 4-18**). Pull the tool out, and then gently pull the lead out of the back of the connector (**Figure 4-19**).

The terminal is either a male or a female connector (**Figure 4-20**). Use the correct terminal for the repair. Feed the wire through the seal and connect the repair lead in the normal manner of crimping and soldering (**Figure 4-21**). Re-form the terminal lock tang by bending it back into its original position (**Figure 4-22**).

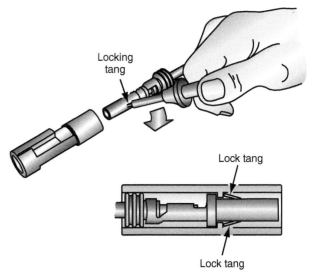

Figure 4-22 Re-forming the locking tangs of the terminal.

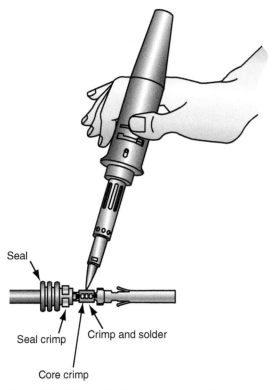

Figure 4-21 Crimp and solder the terminal to the lead.

Insert the lead from the back of the connector until a noticeable "catch" is felt. Gently push and pull on the lead to confirm that it is locked to the connector. Close the secondary locks and reconnect the connector halves.

> **SERVICE TIP** Use a weather-pack repair kit that provides new seals for a complete repair.

Metri-Pack Connectors

There are two types of metri-pack connectors: **pull to seat** and **push to seat**. The push-to-seat terminal removal is illustrated in **Figures 4-23** and **4-24**.

The pull-to-seat terminal is removed by inserting a pick into the connector and under the lock tang (**Figure 4-25**). Gently pull back on the lead while prying up on the lock tang. When the lock tang is free of the tab in the connector, push the lead through the *front* of the connector.

To make the repairs to the terminal, insert the stripped wire through the seal and the connector body (**Figure 4-26**). Crimp and solder the terminal to the wire. Pull the wire lead and terminal back into the connector body until the terminal is locked (**Figure 4-27**).

GROUND STRAPS

Ground straps between the powertrain components and the vehicle's chassis are used to complete the return path to the battery. Ground straps are also used to connect sheet metal parts, even though there is no electrical circuit involved, to suppress electromagnetic induction (EMI) since the sheet metal could behave as a large capacitor and the resulting electrostatic field can interfere with computer-controlled circuits.

Classroom Manual
Chapter 4, page 94

Special Tools
Pick
Crimping tool
Safety glasses
Fender covers

The connectors are called **pull to seat** or **push to seat** to depict the method used to install the terminals into the connector.

Special Tools
DMM
Lab scope

Classroom Manual
Chapter 4, page 90

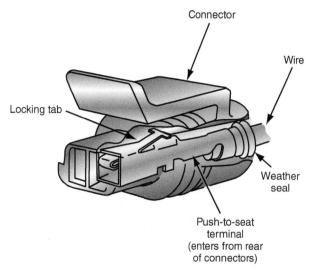

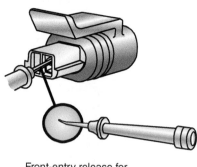

Figure 4-23 Use a pick to unlock the male terminal locking tang.

Front-entry release for push-to-seat terminal

Figure 4-24 Use a pick to unlock the nib of the terminal retainer for female terminals.

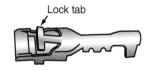

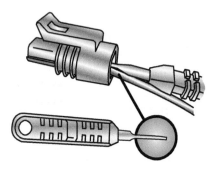

Rear-entry release for pull-to-seat terminal

Figure 4-25 Pull up on the lock tang to release the terminal from the connector.

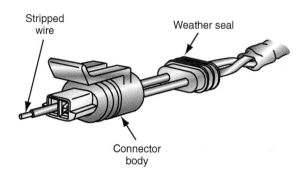

Figure 4-26 The wire lead must be installed into the seal and connector before attaching the terminal.

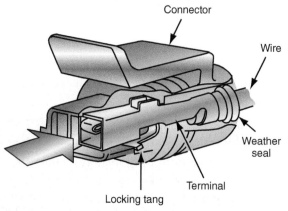

Figure 4-27 Make sure that the terminal locks into the connector body.

Like any other conductor, resistance or opens in the ground strap, or its connections, can cause multiple issues. For example, the engine starter may not engage or crank slow, fuel injectors may not properly energize, and ignition coils may not function.

If electron flow cannot return through the designed ground path, it will attempt to find an alternate path. Sometimes this may be evidenced by pitting of the slip yoke at the rear of the transmission or by pitting of wheel bearings. This pitting is the result of arcing as the current jumps from one component to another and wears away the metal.

Ground straps that are used for electrical circuits can be tested using the voltage drop test. Each connection of the ground strap should have a voltage drop of 0 volt (V). In addition, the ground strap itself should have no voltage drop between its ends. Be sure to confirm that the connection terminals are tight and clean. It is a good practice to use the voltage drop test to confirm proper connections anytime the ground strap is removed and reinstalled.

Ground straps that are used to control EMI generally do not have electrical circuits involved, so a voltage drop test would not be possible. However, a voltmeter or lab scope set on the lowest scale may pick up any voltage that is passing through the strap. Also, an ohmmeter can be used to measure the resistance between connector ends.

> ⚙ **SERVICE TIP** When troubleshooting an intermittent electrical problem, be sure to check the condition of *all* ground straps on the vehicle.

READING WIRING DIAGRAMS

When attempting to locate a possible cause for system malfunction, it is important to have the correct wiring diagram for the vehicle being worked on. There may be a different diagram for each model and even for the same models equipped with different options. Also, diagrams may differ between two- and four-door models. In some cases, it may be necessary to use the date of manufacture and/or the **vehicle identification number (VIN)** to determine the correct diagram to use. The VIN is assigned to a vehicle for identification purposes. The identification plate is usually located on the cowl, next to the left upper instrument panel (**Figure 4-28**). It is visible from the outside of the vehicle. Other locations for the VIN plate are in the door opening, glove box, and engine compartment.

Next, study the method used to identify circuits and **color codes**. Usually this information is provided in the service information. Also, become familiar with the electrical symbols used by the manufacturer.

Classroom Manual
Chapter 4, page 97

Figure 4-28 VIN plate location.

Before you begin to try to use a wiring diagram, you must first have an idea of what you are looking for. It may be a particular component, circuit, or connector. The best way to start the process is by identifying the component or one of the components that doesn't work correctly. Then look in the index for the wiring diagram and find where that component is shown. Computer-driven wiring diagrams may provide a search feature. This makes locating the component within the wiring diagrams much easier and faster.

The electrical section in most service information breaks down the electrical system of the automobile into individual circuits. This approach makes it easier to find a particular component. Of course, you still need to use the index or search function to find the page the component and its circuit is on.

If the printed form of the service information uses a total vehicle wiring diagram, finding the component may be a little trickier. Wiring diagrams are usually indexed by grids. The diagram is marked into equal sections like a street map. The wiring diagram's index will list a letter and number for each major component and many different connection points. If the wiring diagram is not indexed, you can locate the component by relating its general location in the vehicle to a general location on the wiring diagram. Most system diagrams are drawn so the front of the car is on the left of the diagram.

Once you have found the component or part of the circuit you were looking for, identify all of the components, connectors, and wires that are related to that component. This is done by tracing through the circuit, starting at the component. Tracing does not mean taking a pencil and marking on the wiring diagram. Tracing means taking your pencil and drawing out the circuit on another piece of paper. It doesn't have to be pretty to work; it just needs to be accurate. Tracing may also mean taking your finger and following the wires to where they lead. In order for tracing to have any value, you need to identify the power source for the component and/or for the circuit, all related loads (sometimes this involves tracing the circuit back through other pages of the wiring diagram), and the ground connection for the component and for all of the related loads. During this step you should be able to answer the following questions concerning the circuit:

1. Where does the voltage come from?
2. When should voltage be present at any point in the circuit?
3. Are there any other events that must happen in order for the circuit to operate?
4. What voltage and amperage would be expected at any point in the circuit?

After you have studied the circuit, you know how the circuit is supposed to work, describe the problem you are hoping to solve. Ask yourself what could cause this. Limit your answers to those items in your traced wiring diagram. Also, limit your answers to the description of the problem. It is wise to make a list of all probable causes of the problem; then number them according to probability. For example, if no dash lights come on, it is possible that all of the bulbs are burned out. However, it is not as probable as a blown fuse. After you have listed the probable causes in order of probability, look at the wiring diagram to identify how you can quickly test to find out which is the cause. Diagnostics is made easier as your knowledge of electricity grows. It also becomes easier with a good understanding of how the circuit works.

Figure 4-29 is a schematic for a blower control circuit of a heater system. By tracing through the circuit, the technician should be able to determine the correct operation of the system. It cannot be overemphasized that the technician should not attempt to figure out why the circuit is not working until it is understood how it is supposed to work.

In this circuit, battery voltage is applied to the fuse box when the ignition switch is in the RUN position. The 30-amp fuse protects circuit 181 to the motor. Notice that circuit 181 connects to the blower motor resistor block through connector C606. When it leaves

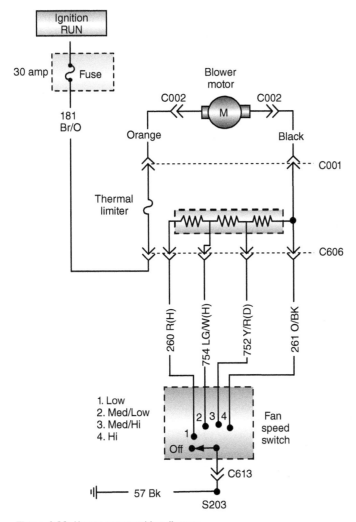

Figure 4-29 Heater system wiring diagram.

the resistor block (after flowing through the thermal limiter), it goes through connector C001 and attaches to the motor through connector C002. Details of these connectors are shown with the schematic.

The circuit then leaves the motor and enters the resistor block through the same connectors. There is a splice in the resistor block that connects this wire to a series of resistors. Connector C606 directs the various circuits out of the resistor block to the blower switch and ground through connector C613. If the switch is placed in the OFF position, the circuit to ground is opened and the blower motor should not operate. If the switch is placed in the LOW position, the current will flow through all of the resistors and the switch through circuit 260. The switch completes the circuit to ground. As the switch is placed in different speed positions, the amount of resistance and the circuit number change. The motor speed should increase as the amount of resistance decreases.

If the customer complains that the heater motor does not work at all, in any speed position, consider the following possible causes:

1. Open fuse.
2. Open in the lead from the ignition switch to the fuse box.
3. Bad ground connection after the switch.
4. Inoperative blower motor.
5. Open in circuit 181.
6. Open thermal limiter.

7. Open in the orange wire between connectors C001 and C002.
8. Open in the black wire between connectors C002 and C001.
9. Disconnected, damaged, or corroded connector C606 or C613.
10. Faulty switch.

However, if the customer complains that the motor does not operate only in the low position, the problem is limited to three possibilities:

1. The third resistor in the series is open.
2. An open in circuit 260 from the resistor block to the blower switch.
3. A faulty switch.

Once the potential problem areas are determined, use the color code to locate the exact wires. Test the leads for expected voltages at different locations in the circuit.

By understanding the way the system is supposed to work, the problem of determining where to look for the problems is simplified. Practice in reading wiring schematics is the only good teacher.

The example just discussed hopefully has given an understanding of the advantages to using a wiring diagram before actually performing diagnostic tests. Today's vehicles have an ever-increasing number of electrical accessories and include the use of computer modules to perform many tasks. When the manufacturers design the vehicle, many of the electrical circuits are constructed in a parallel arrangement. All of this combines to make the electrical circuits more challenging to diagnose. Proper use of the wiring diagram can help you get the diagnosis accomplished intelligently.

To illustrate again how the proper use of the wiring diagram helps to make intelligent diagnostics, consider a simple lighting circuit (**Figure 4-30**). In this circuit, the fuse is wired in series to the three parallel lamps. The common components of this circuit include the fuse and the conductors connecting the battery to the point identified as S1. After the S1 splice location, the three lamps branch off. If lamp A is the only nonfunctioning component of the circuit, by using the wiring diagram you should be able to determine possible fault areas very quickly. If lamps B and C function, the open must be beyond common-point S1. However, if none of the lamps function, the open is probably before the common point. By using the wiring diagram, diagnostic time has been saved by not having to test half of the circuit.

> **AUTHOR'S NOTE** This type of diagnostic procedure of using the wiring diagram to isolate possible fault locations is sometimes referred to as "common-point diagnostics."

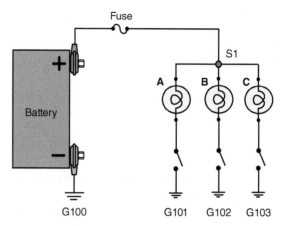

Figure 4-30 Lighting circuit used as an example of using the wiring diagram to locate an open.

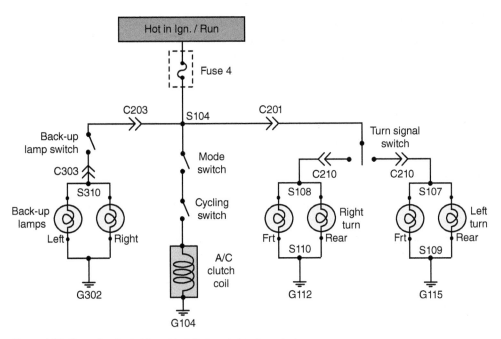

Figure 4-31 Example of using the wiring diagram to locate a short.

In the previous examples of wire diagram usage, we focused mainly on locating opens or faulty components. The wiring diagram is also an excellent tool for locating shorts to ground. Consider the circuit shown in **Figure 4-31** and assume that fuse 4 is blown. Knowing that the fuse blows as a result of excessive current flow in the circuit that can result from a short to ground condition, you need to isolate the location of the short.

Fuse 4 in this circuit protects many parallel circuits that encompass a wiring harness that goes from the front to the rear of the vehicle. To trace out the circuits within the harness would be very time consuming. Using the wiring diagram and common-point diagnostics will identify the most likely location of the short.

The first step is to isolate the faulty circuit. Turn off all loads that are protected by fuse 4. With all circuits turned off, there should be no current flow through the circuit. Remove the fuse and insert a test light across the fuse box terminals. This installs the test light in series with the circuit. Turn the ignition switch to the RUN position and observe the test light. If the test light is illuminated, the short is in the common circuit between the fuse and the switches. In this instance, you would then be looking for a short in the dashboard area after the fuse and before the individual circuit switches. The next step would be to separate the connectors one at a time to see if the test light goes out. This will isolate the circuit further.

If the test light did not illuminate when the ignition switch was placed in the RUN position, the location of the short is after the common circuit and the switches. Replace the fuse with a new properly rated fuse and turn on each circuit separately until the fuse blows. The short would be in the section of the circuit that, when energized, caused the fuse to blow.

Keep in mind that the customer may have brought this vehicle in because the A/C was not working. If all that you did was replace the fuse and it did not blow when the A/C was turned on, you missed the fact that the fuse will blow when the backup lamp switch is closed. Without referencing the wiring diagram, it is possible to make the assumption that there is an intermittent short in the A/C circuit or within the compressor clutch. Remember, one of your diagnostic steps is to determine related symptoms and that the last step is to verify your repair as successful.

CASE STUDY

The vehicle owner complains that the brake lights do not light. He also says the dome light is not working. The technician verifies the problem and then checks the battery for good connections and tests the fusible links. All are in good condition.

A study of the wiring diagram indicates that the brake light and dome light circuits share the same fuse. It is also indicated that the ignition switch illumination light circuit is shared with these two circuits. A check of the ignition switch illumination light shows that it is not operating either. The technician checks the fuse that is identified in the wiring diagram. It is blown. When a replacement fuse is installed, the dome and brake lights work properly for three tests, and then the fuse blows again.

Upon further testing of the shared circuits, an intermittent short to ground is located in the steering column in the ignition switch illumination circuit. The technician solders in a repair wire to replace the damaged section. After all repairs are completed, a final test indicates proper operation of all circuits.

ASE-STYLE REVIEW QUESTIONS

1. Splicing copper wire is being discussed.
   Technician A says it is acceptable to use solderless connections.
   Technician B says acid core solder should not be used on copper wires.
   Who is correct?
   A. A only
   B. B only
   C. Both A and B
   D. Neither A nor B

2. Use of wiring diagrams is being discussed.
   Technician A says a wiring diagram is used to help determine related circuits.
   Technician B says the wiring diagram will give the exact location of the components in the car.
   Who is correct?
   A. A only
   B. B only
   C. Both A and B
   D. Neither A nor B

3. The fuse for the parking lights is open.
   Technician A says the blown fuse may be due to high resistance in one of the parking light bulb connectors.
   Technician B says the fuse probably wore out due to age.
   Who is correct?
   A. A only
   B. B only
   C. Both A and B
   D. Neither A nor B

4. Repairs to a twisted/shielded wire are being discussed.
   Technician A says a twisted/shielded wire carries high current.
   Technician B says because a twisted/shielded wire carries low current, any repairs to the wire must not increase the resistance of the circuit.
   Who is correct?
   A. A only
   B. B only
   C. Both A and B
   D. Neither A nor B

5. Replacement of fusible links is being discussed.
   Technician A says not all open fusible links are detectable by visual inspection.
   Technician B says to test for battery voltage on both sides of the fusible link to confirm its condition.
   Who is correct?
   A. A only
   B. B only
   C. Both A and B
   D. Neither A nor B

6. *Technician A* says troubleshooting is the diagnostic procedure of locating and identifying the cause of the fault.
   Technician B says troubleshooting is a step-by-step process of elimination by the use of cause and effect.
   Who is correct?
   A. A only
   B. B only
   C. Both A and B
   D. Neither A nor B

7. *Technician A* says a replacement fusible link should be at least 9 inches (25 mm) long.

   Technician B says if the damaged fusible link feeds two harness wires, use one replacement fusible link.

   Who is correct?

   A. A only
   C. Both A and B
   B. B only
   D. Neither A nor B

8. Repairing connectors is being discussed.

   Technician A says molded connectors are a one-piece design and cannot be separated for repairs.

   Technician B says to replace the seals when repairing the weather-pack connector.

   Who is correct?

   A. A only
   C. Both A and B
   B. B only
   D. Neither A nor B

9. *Technician A* says all metri-pack connectors use male terminals.

   Technician B says the connectors of the metri-pack are called pull to seat or push to seat to depict the method used to install the terminals into the connector.

   Who is correct?

   A. A only
   C. Both A and B
   B. B only
   D. Neither A nor B

10. *Technician A* says there may be a different wiring diagram for each model of a vehicle.

    Technician B says it may be necessary to use the VIN number to get the correct wiring diagram for the vehicle.

    Who is correct?

    A. A only
    C. Both A and B
    B. B only
    D. Neither A nor B

ASE CHALLENGE QUESTIONS

1. *Technician A* says that acid core solder should be used whenever aluminum wires are to be soldered.

   Technician B says that solderless connectors should not be used if a weather-resistant connection is desired.

   Who is correct?

   A. A only
   C. Both A and B
   B. B only
   D. Neither A nor B

2. Which of the following electrical troubleshooting routines is in the correct sequence?

   A. Study the electrical diagram, confirm the concern, locate and repair the fault, and test the repair.

   B. Study the electrical diagram, locate and repair the fault, confirm the concern, and test the repair.

   C. Confirm the concern, study the electrical diagram, test the repair, and locate and repair the fault.

   D. Confirm the concern, study the electrical diagram, locate and repair the fault, and test the repair.

3. Wire repair procedures are being discussed.

   Technician A says the aluminum wire should never be soldered; it should only be crimped whenever repairs are necessary.

   Technician B says that damaged twisted/shielded wires should be connected with splice clips and then the connections should be soldered.

   Who is correct?

   A. A only
   C. Both A and B
   B. B only
   D. Neither A nor B

4. *Technician A* says that when replacing a fusible link, the gauge size of the replacement fusible link can be decreased but never increased.

   Technician B says that a 14-gauge fusible link can be replaced with an equivalent length of 14-gauge stranded wire.

   Who is correct?

   A. A only
   C. Both A and B
   B. B only
   D. Neither A nor B

5. *Technician A* says that a wire identified as "754 LG/W (H)" on a wiring diagram refers to a wire located in circuit 754 that is light green with white hash marks.

Technician B says that a wire that is labeled 181 BR/O refers to an 18-gauge wire that is brown with an orange stripe.

Who is correct?

A. A only

C. B only

B. Both A and B

D. Neither A nor B

6. Using **Figure 4-32**, what would be the **LEAST LIKELY** cause for fuse 16 blowing?

A. A short to ground between terminal 2 of the fuse and terminal 86 of the relay.

B. A short to ground between terminal 85 of the relay and terminal 47 of the IMP.

C. A short to ground at terminal 30 of the relay.

D. A short to ground in circuit X2.

7. A customer states that his or her tail lamps are not functioning. When testing the lamp operation, it is noted that the tail/stop lamps and the turn signals do not operate. During the diagnostic routine, the technician also notices that the backup lamps are also not functioning. All fuses test good. Using all of the wiring diagrams with indicated test voltages in **Figure 4-33a** through **d**, what is the most likely cause of this condition?

A. G301 is open.

B. G302 is open.

C. Faulty integrated power module.

D. Open in the Z918 circuit.

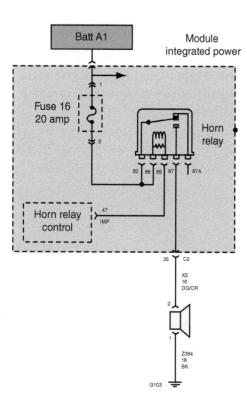

Figure 4-32

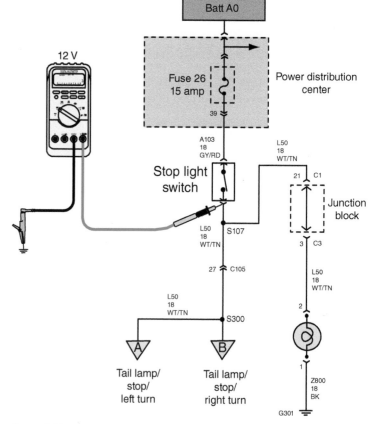

Figure 4-33a

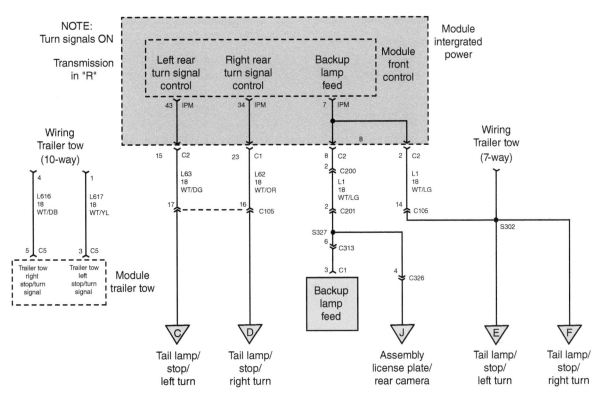

NOTE:
Turn signals ON

Transmission in "R"

Module intergrated power

Left rear turn signal control

Right rear turn signal control

Backup lamp feed

Module front control

43 IPM   34 IPM   7 IPM

8

Wiring Trailer tow (10-way)

4   1

L616 18 WT/DB   L617 18 WT/YL

5 C5   3 C5

Module trailer tow

Trailer tow right stop/turn signal

Trailer tow left stop/turn signal

15 C2   23 C1   8 C2   2 C2

L63 18 WT/DG   L62 18 WT/OR   2 C200   L1 18 WT/LG

L1 18 WT/LG

17   16 C105   2 C201   14 C105

S327   S302

6 C313

3 C1   4 C326

Backup lamp feed

Wiring Trailer tow (7-way)

C   D   J   E   F

Tail lamp/ stop/ left turn

Tail lamp/ stop/ right turn

Assembly license plate/ rear camera

Tail lamp/ stop/ left turn

Tail lamp/ stop/ right turn

Figure 4-33b

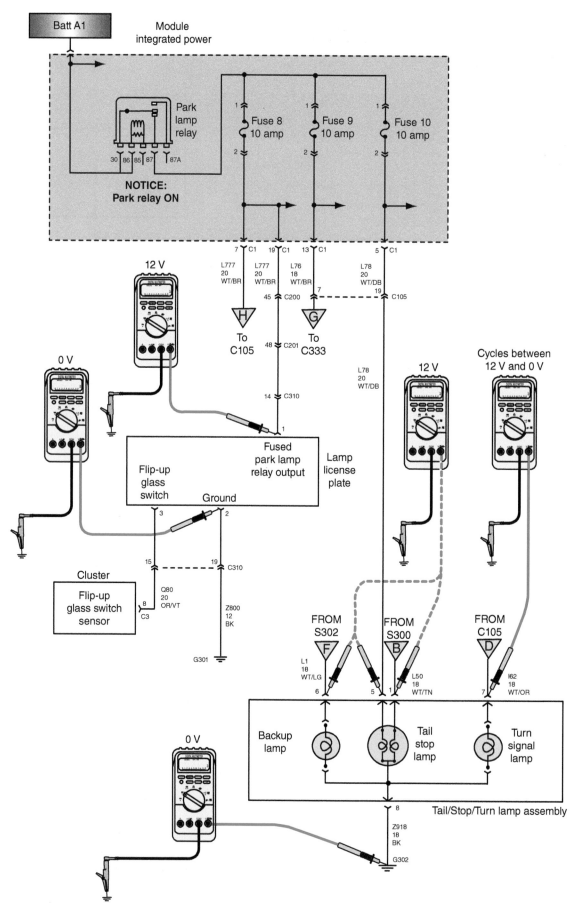

Figure 4-33c

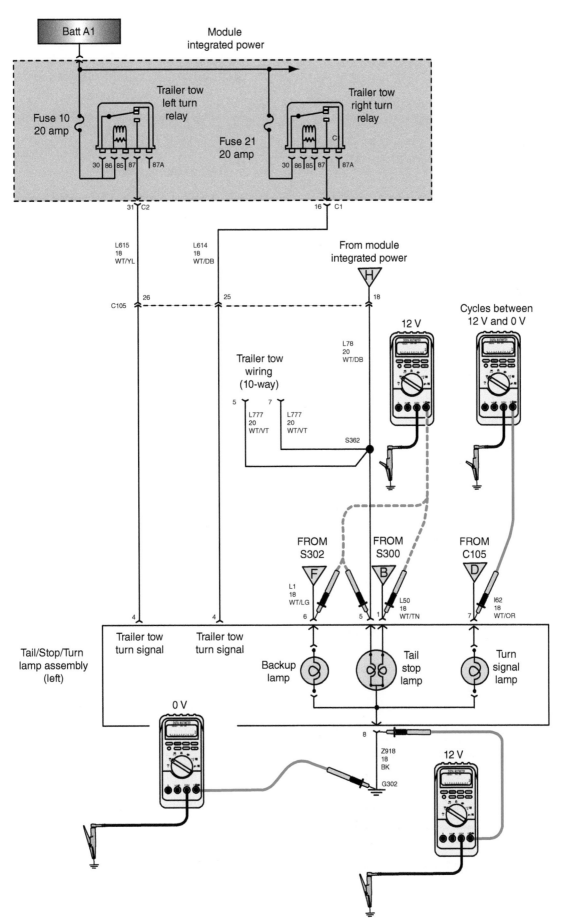

Figure 4-33d

8. Referring to **Figure 4-34**, how is the blower speed controlled?

   A. The A/C–heater controller uses a high-side driver for each speed circuit.

   B. The totally integrated power module uses PWM to control the ground circuit of the motor.

   C. A/C–heater controller uses a low-side driver for each speed circuit.

   D. All of the above.

9. Referring to the wiring diagram illustrated in **Figure 4-35a** through **d**, the fog lights do not work when the headlight switch is in the AUTO position, but work in the HEAD and PARKLAMP positions. What is the most likely cause?

   A. Open in the headlamp switch resistor between the 3 and 6 positions.

   B. Open in the resistor below the 3 position of the headlamp switch.

   C. Short to ground in circuit L80.

   D. Short to voltage at terminal 4 of headlamp switch.

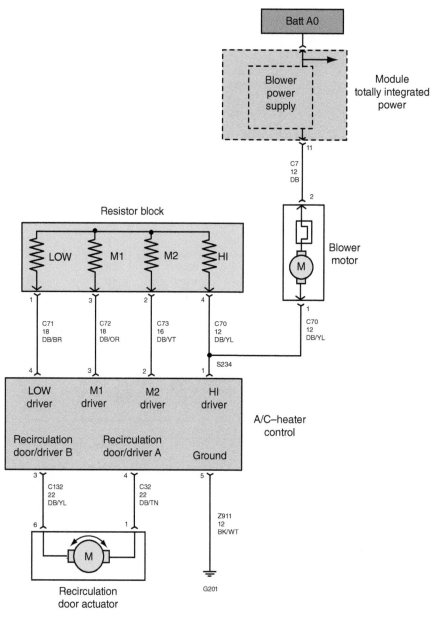

Figure 4-34

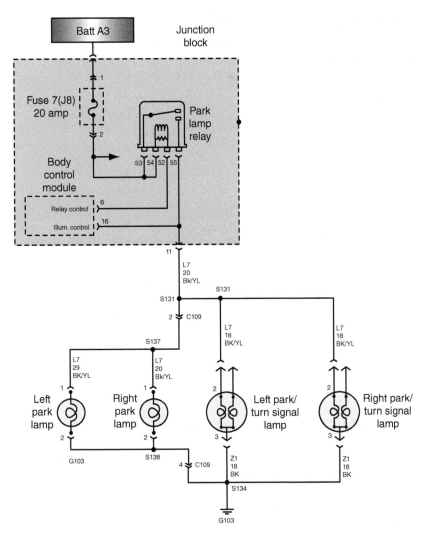

Figure 4-35a

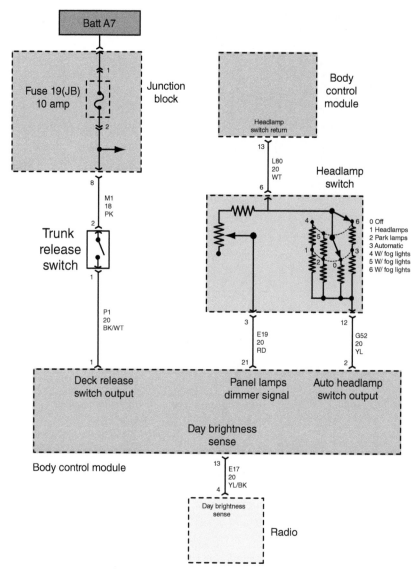

Figure 4-35b

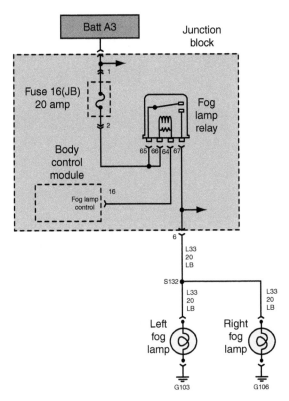

Figure 4-35c

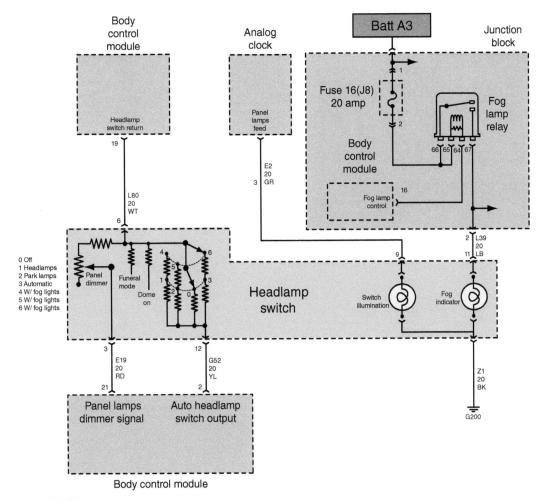

Figure 4-35d

10. Referring to **Figure 4-36a** through **d**, the coil for cylinder 4 and the injector for cylinder 3 both fail to function. What is the possible cause?

A. Poor connection at splice S134.

B. Open in the K13 circuit.

C. Poor connection in the S185 splice.

D. Open in circuit between S184 and PCM C1/38 terminal.

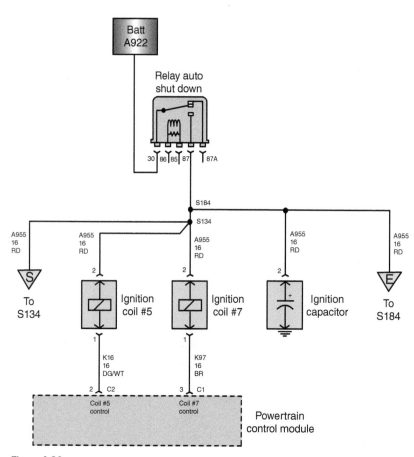

Figure 4-36a

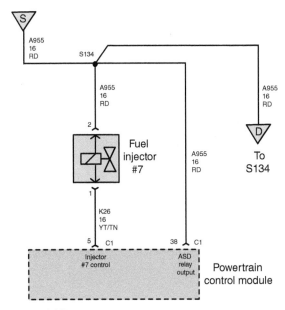

Figure 4-36b

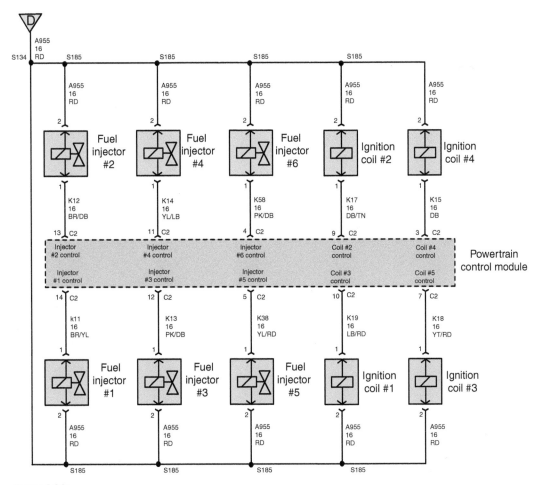

Figure 4-36c

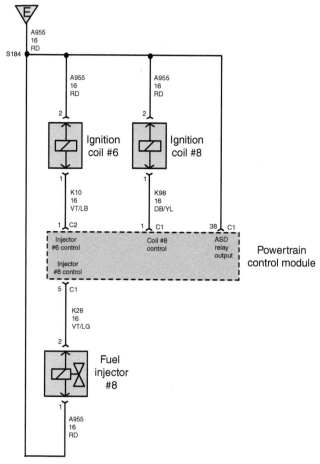

Figure 4-36d

Name _____   Date _____

PART IDENTIFICATION ON A WIRING DIAGRAM

Upon completion of this job sheet, you should be able to locate different parts on a wiring diagram.

NATEF Correlation ————————————————————————

This job sheet addresses the following **MLR** tasks:

A.3.   Use wiring diagrams to trace electrical/electronic circuits.

A.11.   Identify electrical/electronic system components and configuration.

This job sheet addresses the following **AST/MAST** task:

A.7.   Use wiring diagrams during the diagnosis (troubleshooting) of electrical/
electronic circuit problems.

NATEF

Tools and Materials

- A wiring diagram in a service manual or from electronic service information system (assigned by the instructor)

Describe the vehicle being worked on:

Year _____ Make _____ Model _____

VIN _____ Engine type and size _____

Service information used _____

Wiring diagram is found on pages _____ to _____

Procedure   **Task Completed**

Study the wiring diagram; then answer the following questions.

1. Are all circuit grounds clearly marked in the wiring diagrams?

   _____

   _____

2. What is represented by lines that cross each other?

   _____

   _____

3. Are most switches shown in their normally open or normally closed position?

   _____

   _____

4. Do all wires have a color code listed by them?

   _____

5. Is the internal circuitry of all components shown in their schematic drawing?

   _____

6. **List the page and figure numbers** (the location) where the following electrical components are shown in the wiring diagram. Then **draw the schematical symbol** used by this wiring diagram to represent the part.

Component	Location	Drawing
Windshield wiper motor	_____	_____
Dome (courtesy) light	_____	_____
A/C compressor clutch	_____	_____
Turn signal flasher unit	_____	_____
Fuse	_____	_____
Fuel gauge sending unit	_____	_____

7. Attach print out of wiring diagram (if using electronic retrieval system)   ☐

Instructor's Response

_____

_____

_____

Name _____   Date _____

USING A COMPONENT LOCATOR

Upon completion of this job sheet, you should be able to identify components and their locations indicated on the wiring diagram that is used for electrical system diagnosis.

NATEF Correlation

This job sheet addresses the following **MLR** tasks:

A.3.　　　Use wiring diagrams to trace electrical/electronic circuits.

A.11.　　　Identify electrical/electronic system components and configuration.

This job sheet addresses the following **AST/MAST** task:

A.7.　　　Use wiring diagrams during the diagnosis (troubleshooting) of electrical/ electronic circuit problems.

This job sheet addresses the following **MLR/AST/MAST** task:

A.1.　　　Research vehicle service information, including vehicle service history, service precautions, and technical service bulletins.

ASE CERTIFIED NATEF

Tools and Materials

• Service information with component locator capabilities.

Describe the vehicle being worked on:

Year _____ Make _____ Model _____

VIN _____ Engine type and size _____

Service information used _____

Procedure

Using the component locator for the vehicle that has been assigned, the student will find the locations for the taillight, headlight, and stoplight circuits, connectors, connections, plugs, and switches.

1. **Taillight circuit:**

   _____

   Location of source of power: _____

   Location of the switch: _____

   Does the switch control the operation of the taillights directly, or is it an input to a module? DIRECTLY INPUT

   If a module, which one and where is it located?

   _____

   _____

   List the ID numbers of any in-line connectors between the switch (or module) and the taillights: _____

   List the location(s) of the in-line connectors: _____

   _____

   _____

Where is the location of the ground connection for the left taillight?

_____

_____

2. **Headlight circuit:**

Location of source of power: _____

_____

Location of switch: _____

_____

Does the switch control the operation of the headlights directly, or is it an input to a module? DIRECTLY INPUT

If a module, which one and where is it located?

_____

_____

List the ID numbers of any circuit splices between the switch (or module) and the left headlight: _____

List the location(s) of the splices: _____

_____

_____

Where is the location of the ground connection for the left headlight?

_____

_____

3. **Stoplight circuit:**

Location of source of power: _____

Location of switch: _____

Does the switch control the operation of the stoplights directly, or is it an input to a module? DIRECTLY INPUT

If a module, which one and where is it located?

_____

_____

On which side of the vehicle does the wiring harness run between the switch (or module) and the stoplights? _____

Instructor's Response

_____

_____

_____

Name _____ Date _____

USING A WIRING DIAGRAM

Upon completion of this job sheet, you should be able to find the power source, ground connection, and controls for electrical circuits using a wiring diagram.

NATEF Correlation

This job sheet addresses the following **MLR** tasks:

A.3.   Use wiring diagrams to trace electrical/electronic circuits.

A.11.   Identify electrical/electronic system components and configuration.

This job sheet addresses the following **AST/MAST** task:

A.7.   Use wiring diagrams during the diagnosis (troubleshooting) of electrical/ electronic circuit problems.

Exercise 1

Using the diagram in **Figure 4-37**, answer the following (list wire colors):

1. How is the windshield wiper motor circuit powered? _____

_____

_____

_____

_____

2. How is the circuit grounded?

_____

_____

_____

_____

3. How is the circuit controlled?

   a. If it is controlled by a switch, is the switch normally open or closed?

   b. Is the circuit power switched or ground switched?

   c. Is the circuit controlled by a variable resistor?

   d. Is the circuit controlled by a mechanical or vacuum-operated device?

_____

_____

_____

_____

4. How is the circuit protected (identify all fuses)?

_____

_____

_____

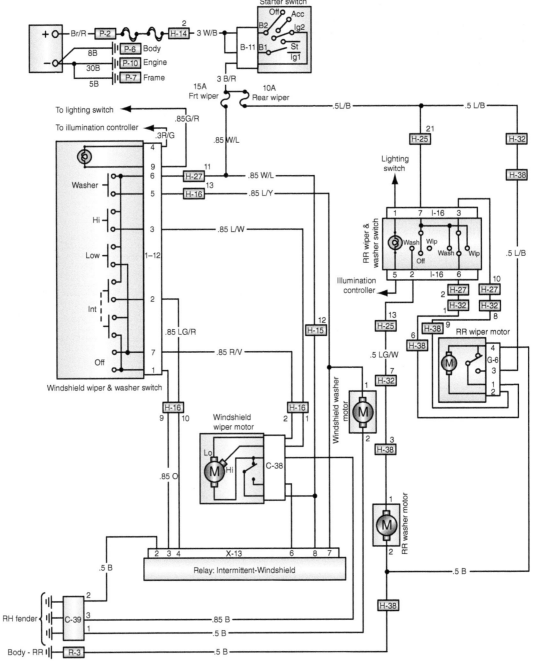

Figure 4-37

Exercise 2

Using the wiring diagram in **Figure 4-38**, answer the following:

1. How is the circuit powered? _____

_____

_____

_____

_____

_____

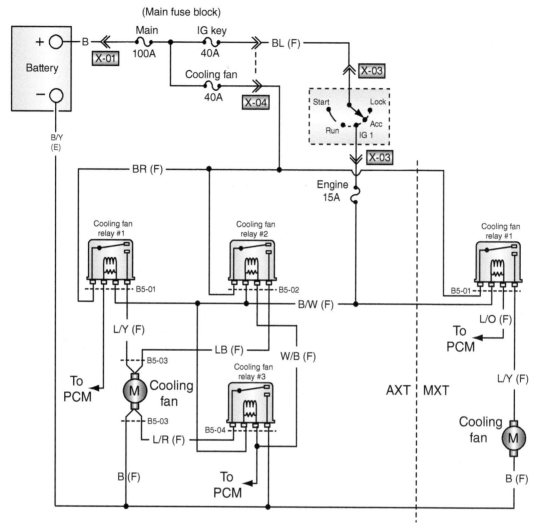

Figure 4-38

2. How is the circuit grounded?

_____

_____

_____

_____

3. How is the circuit controlled?

   a. If it is controlled by a switch, is the switch normally open or closed?

   b. Is the circuit power switched or ground switched?

   c. Is the circuit controlled by a variable resistor?

   d. Is the circuit controlled by a mechanical or vacuum-operated device?

_____

_____

_____

_____

4. How is the circuit protected?

_____

_____

_____

_____

Exercise 3

Using the wiring diagram in **Figure 4-39**, answer the following:

1. How is the circuit powered?

_____

_____

_____

_____

_____

2. How is the circuit grounded?

_____

_____

_____

_____

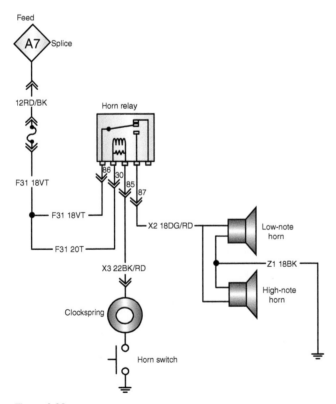

Figure 4-39

3. How is the circuit controlled?

    a. If it is controlled by a switch, is the switch normally open or closed?

    b. Is the circuit power switched or ground switched?

    c. Is the circuit controlled by a variable resistor?

    d. Is the circuit controlled by a mechanical or vacuum-operated device?

_____

_____

_____

_____

4. How is the circuit protected?

_____

_____

_____

_____

Exercise 4

Using the wiring diagram in **Figure 4-40**, answer the following:

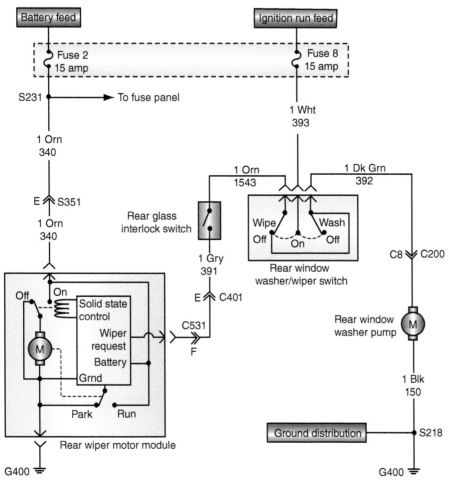

Figure 4-40

1. How is the circuit powered?

_____

_____

_____

_____

2. How is the circuit grounded?

_____

_____

_____

_____

3. How is the circuit controlled?

   a. If it is controlled by a switch, is the switch normally open or closed?

   b. Is the circuit power switched or ground switched?

   c. Is the circuit controlled by a variable resistor?

   d. Is the circuit controlled by a mechanical or vacuum-operated device?

_____

_____

_____

_____

4. How is the circuit protected?

_____

_____

_____

_____

Exercise 5

Using the wiring diagram in **Figure 4-41**, answer the following:

1. How is the circuit powered?

_____

_____

_____

_____

2. How is the circuit grounded?

_____

_____

_____

_____

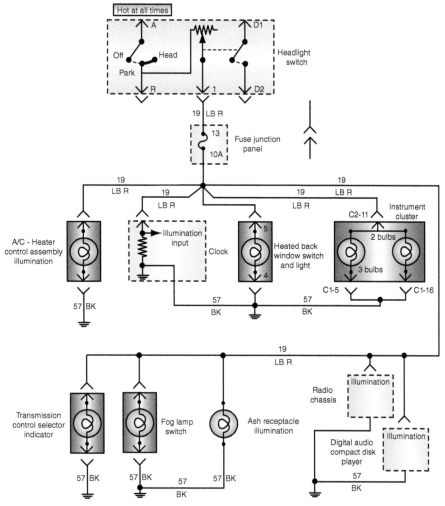

Figure 4-41

3. How is the circuit controlled?

   a. If it is controlled by a switch, is the switch normally open or closed?

   b. Is the circuit power switched or ground switched?

   c. Is the circuit controlled by a variable resistor?

   d. Is the circuit controlled by a mechanical or vacuum-operated device?

   _____

   _____

   _____

   _____

4. How is the circuit protected?

_____

_____

_____

_____

Instructor's Response

_____

_____

_____

Name _____   Date _____

SOLDERING COPPER WIRES

Upon completion of this job sheet, you should be able to properly solder two copper wires.

NATEF Correlation

This job sheet addresses the following **MLR** task:

A.10.   Repair and/or replace connectors, terminal ends, and wiring of electrical/
electronic systems (including solder repair).

This job sheet addresses the following **AST** task:

A.11.   Repair and/or replace components, connectors, terminal ends, and wiring of
electrical/electronic systems (including solder repair).

This job sheet addresses the following **MAST** task:

A.10.   Inspect, test, repair, and/or replace components, connectors, terminals,
harnesses, and wiring in electrical/electronic systems (including solder repair).

NATEF

Tools and Materials

- Two pieces of copper wire
- 100-watt soldering iron
- 60/40 rosin core solder
- Crimping tool
- Splice clip
- Heat shrink tube
- Heating gun
- Safety glasses

Procedure:

Task Completed

1. Using the correct size stripper, remove about 1/2 inch (12 mm) of the insulation from   ☐
   both wires. Be careful not to nick or cut any of the wires.

2. Select the proper size splice clip to hold the splice.   ☐

3. Place the correct length and size of heat shrink tube over the wire. Slide the tube far   ☐
   enough away so the wires are not exposed to the heat of the soldering iron.

4. Overlap the two splice ends and hold in place with thumb and forefinger. While the   ☐
   wire ends are overlapped, center the splice clip around the wires and crimp into place.

5. Heat the splice clip with the soldering iron while applying solder to the opening in   ☐
   the back of the clip. Apply only enough solder to make a good connection. The solder
   should travel through the wire.

6. After the solder cools, slide the heat shrink tube over the splice and heat the tube with   ☐
   the hot air gun until it shrinks around the splice. Do not overheat the tube.

Instructor's Response

_____

_____

_____

DIAGNOSTIC CHART 4-1

PROBLEM AREA	Burned fusible link.
SYMPTOMS	Several electrical components fail to operate.
POSSIBLE CAUSES	1. Short to ground in circuit. 2. Excessive circuit current flow due to high mechanical resistance of actuators.

DIAGNOSTIC CHART 4-2

PROBLEM AREA	Burned, broken, or defective connectors, insulation, or conductors.
SYMPTOMS	Open or short circuits that prevent component operation.
POSSIBLE CAUSES	1. Corroded wiring. 2. Improper or no wire protectors. 3. Corroded connectors. 4. Damaged conductor. 5. Damaged connectors.

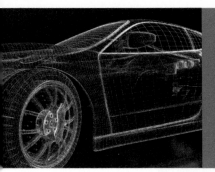

CHAPTER 5
BATTERY DIAGNOSIS AND SERVICE

Upon completion and review of this chapter, you should be able to:

- Demonstrate all safety precautions and rules associated with servicing the battery.
- Perform a visual inspection of the battery, cables, and terminals.
- Correctly slow and fast charge a battery, in or out of the vehicle.
- Describe the differences between slow and fast charging and determine when either method should be used.
- Perform a battery terminal test and accurately interpret the results.
- Perform a battery leakage test and determine the needed corrections.
- Test a conventional battery's specific gravity.
- Perform an open circuit test and accurately interpret the results.

- Test the capacity of the battery to deliver both current and voltage and to accurately interpret the results.
- Perform a 3-minute charge test to determine if the battery is sulfated.
- Perform a conductance test of the battery and accurately interpret the results.
- Perform a battery drain test and accurately determine the causes of battery drains.
- Remove, clean, and reinstall the battery properly.
- Jump-start a vehicle by use of a booster battery and jumper cables.
- Determine the cause of HV battery system failures.
- Measure HV battery module voltages with a DMM.

⚒ **Basic Tools**

Basic mechanic's tool set

Service information

Terms To Know

Battery ECU	Conductance	Refractometer
Battery leakage test	Fast charging	Slow charging
Battery terminal test	Hydrometer	Stabilize
Capacity test	Jump assist	State of charge
Charge	Open circuit voltage test	Sulfation
Charge rate	Parasitic drains	Three-minute charge test

INTRODUCTION

A discharged or weak battery can affect more than just the starting of the engine. The battery is the heart of the electrical system of the vehicle. It is important that it is not overlooked when servicing most electrical problems. Because of its importance, the battery should be checked whenever the vehicle is brought into the shop for service. A battery test series will show the state of charge and output voltage of the battery, which determines if it is good, is in need of recharging, or must be replaced.

Classroom Manual
Chapter 5, page 113

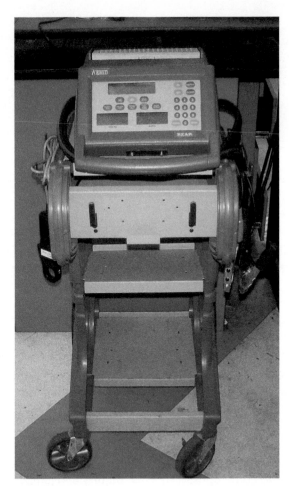

Figure 5-1 A computer-based generator, regulator, battery, and starter tester.

There are many different manufacturers of battery test equipment. Most modern testers are computer based and conduct the tests automatically after a particular test is selected (**Figure 5-1**). Always follow the procedures given by the tester's manufacturer.

GENERAL PRECAUTIONS

Before attempting to do any type of work on or around the battery, the technician must be aware of certain precautions. To avoid personal injury or property damage, take the following precautions:

1. Battery acid is very corrosive. Do not allow it to come in contact with skin, eyes, or clothing. If battery acid gets into your eyes, rinse them thoroughly with clean water and receive immediate medical attention. If battery acid comes in contact with skin, wash with clean water. Baking soda added to the water will help neutralize the acid. If the acid is swallowed, drink large quantities of water or milk followed by milk of magnesia and a beaten egg or vegetable oil.
2. When making connections to a battery, be careful to observe polarity, positive to positive and negative to negative.
3. When disconnecting battery cables, always disconnect the negative (ground) cable first.
4. When connecting battery cables, always connect the negative cable last.

5. Avoid any arcing or open flames near a battery. The vapors produced by the battery cycling are very explosive. Do not smoke around a battery.

6. Follow manufacturer's instructions when charging a battery. Charge the battery in a well-ventilated area. Do not connect or disconnect the charger leads while the charger is turned on.

7. Do not add additional electrolyte to the battery if it is low. Add only distilled water.

8. Do not wear any jewelry or watches while servicing the battery. These items are excellent conductors of electricity. They can cause severe burns if current flows through them by accidental contact with the battery positive terminal and ground.

9. Never lay tools across the battery. They may come into contact with both terminals, shorting out the battery and causing it to explode.

10. Wear safety glasses or face shield when servicing the battery.

11. If the battery's electrolyte is frozen, allow it to defrost before doing any service or testing of the battery. While it is defrosting, look for leaks in the case. Leakage means the battery is cracked and should be replaced.

BATTERY INSPECTION

Before performing any electrical tests, the battery should be inspected, along with the cables and terminals. A complete visual inspection of the battery should include the following items:

1. Battery date code: This provides information as to the age of the battery (**Figure 5-2**).

2. Condition of battery case: Check for dirt, grease, and electrolyte condensation. Any of these contaminants can create an electrical path between the terminals and cause the battery to drain. Also, check for damaged or missing vent caps and cracks in the case. A cracked or buckled case could be caused by excessive tightening of the hold-down fixture, excessive under-hood temperatures, buckled plates from extended undercharged conditions, freezing, or excessive charge rate.

Classroom Manual
Chapter 5, page 115

Special Tools
Fender covers
Safety glasses
Battery filler bottle

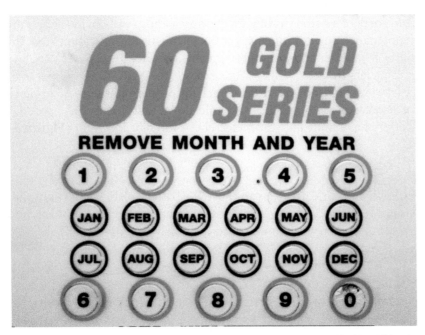

Figure 5-2 The battery sticker will usually have the date the battery was sold, plus additional information.

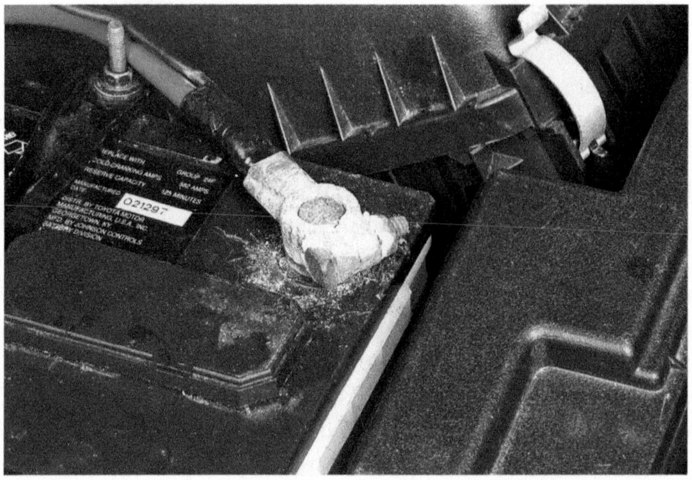

Figure 5-3 Inspect the condition of the battery cables and terminals.

Classroom Manual
Chapter 5, page 134

Classroom Manual
Chapter 5, page 122

3. Electrolyte level, color, and odor: If necessary, add distilled water to fill to 1/2 inch (12 mm above the top of the plates). After adding water, charge the battery before any tests are performed. Discoloration of electrolyte and the presence of a rotten egg odor indicate an excessive charge rate, excessive deep cycling, impurities in the electrolyte solution, or an old battery.

4. Condition of battery cables and terminals: Check for corrosion, broken clamps, frayed cables, and loose terminals (**Figure 5-3**). These conditions result in voltage drop between the battery terminal and the end of the cable. The sulfuric acid that vents out with the battery gases attacks the battery terminals and battery cables. As the sulfuric acid reacts with the lead and copper, deposits of lead sulfate and copper sulfate are created. These deposits are resistive to electron flow and limit the amount of current that can be supplied to the electrical and starting systems. If the deposits are bad enough, the resistance can increase to a level that prevents the starter from cranking the engine.

5. Battery abuse: This includes the use of bungee cords and 2 × 4s for hold-down fixtures, too small of a battery rating for the application, and obvious neglect to periodic maintenance. In addition, inspect the terminals for indications that they have been hit by a hammer and for improper cable removal procedures. Finally, check for proper cable length.

6. Battery tray and hold-down fixture: Check for proper tightness. Also, check for signs of acid corrosion of the tray and hold-down unit. Replace as needed.

7. If the battery has a built-in hydrometer, check its color indicator (**Figure 5-4**).

> **SERVICE TIP** Grid growth can cause the battery plate to short out the cell. If there is normal electrolyte level in all cells but one, that cell is probably shorted and the electrolyte has been converted to hydrogen gas.

One common cause of early battery failure is overcharging. If the charging system is supplying a voltage level over 15.5 volts (V), the plates may become warped. Warping of the plates results from the excess heat that is generated as a result of overcharging. Overcharging also causes the active material to disintegrate and shed off of the plates.

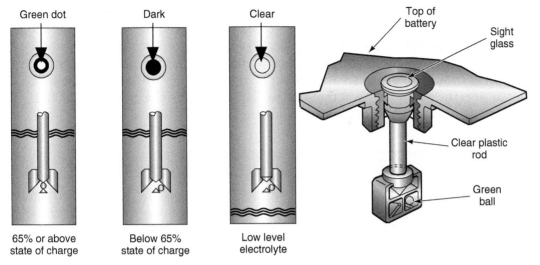

Figure 5-4 Built-in hydrometer used to indicate a battery's state of charge.

If the charging system does not produce enough current to keep the battery charged, the lead sulfate can become crystallized on the plates. If this happens, the sulfate is difficult to remove and the battery will resist recharging. The recharging process converts the sulfate on the plates. If there is an undercharging condition, the sulfate is not converted and it will harden on the plates.

The sulfate that is not converted back to H_2SO_4 hardens on the plates. This results in battery sulfation, which permanently damages the battery.

Some technicians believe that a battery's internal resistance is high when the battery is fully charged; however, this not true. Keep in mind that the lead sulfate on the battery plates acts as an insulator. The more sulfate on the plates, the higher the battery's internal resistance. The increased resistance of a discharged battery allows it to accept a higher rate of charge without gassing or overheating as compared to a battery that is near a full charge. As the battery is charged, the sulfate is removed and there will not be much left to sustain the reverse chemical reaction. The battery's "natural absorption rate" is the level of charge current that can be applied without overheating the battery and breaking down the electrolyte into hydrogen and oxygen. Overcharging occurs when the charge current is greater than the natural absorption rate.

Vibration is another common reason for battery failure. If the battery is not secure, the plates will shed the active material as a result of excessive vibration. If enough material is shed, the sediment at the bottom of the battery can create an electrical connection between the plates. The shorted cell will not produce voltage, resulting in a battery that will have only 10.5 volts across the terminals. With this reduced amount of voltage, the starter usually will not be capable of starting the engine. To prevent this problem, make sure that proper hold-down fixtures are used.

Even during normal battery operation, the active materials on the plates will shed over a period of time. The negative plate also becomes soft. Both of these events will reduce the effectiveness of the battery.

⚠ WARNING **There are many safety precautions associated with charging the battery. The hydrogen gases produced by a charging battery are very explosive. Exploding batteries are responsible for over 15,000 injuries per year that are severe** enough to require hospital treatment. Keep sparks, flames, and lighted cigarettes away from the battery. Also, do not use the battery to lay tools on. They may short across the terminals and result in the battery exploding. Always wear eye protection and proper clothing when working near the battery. Also, most jewelry is an excellent conductor of electricity. Do not wear any jewelry when performing work on or near the battery. Do not remove the vent caps while charging. Do not connect or disconnect the charger leads while the charger is turned on.

CHARGING THE BATTERY

Classroom Manual
Chapter 5, page 118

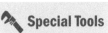

Special Tools

Safety glasses
Battery charger
Voltmeter
Fender covers

Caution

If the battery is to be removed from the vehicle, disconnect the negative battery cable first. Lift the battery out with a carrying tool (**Figure 5-5**).

To **charge** the battery means to pass an electric current through the battery in an opposite direction than during discharge. If the battery needs to be recharged, the safest method is to remove the battery from the vehicle. The battery can be charged in the vehicle, however. If the battery is to be charged in the vehicle, it is important to protect any vehicle computers by removing the negative battery cable.

When connecting the charger to the battery, make sure the charger is turned off. Connect the cable leads to the battery terminals, observing polarity. Attempting to charge the battery while the cables are reversed will result in battery damage. For this reason, many battery chargers have a warning system to alert the technician that the cables are connected in reverse polarity. Rotate the clamps slightly on the terminals to assure a good connection.

Depending on the requirements and amount of time available, the battery can be either slow or fast charged. Each method of charging has its advantages and disadvantages.

WARNING Before charging a battery that has been in cold weather, check the electrolyte for ice crystals. A discharged battery will freeze at a higher temperature than a fully charged battery. Do not attempt to charge a frozen battery. Forcing current through a frozen battery may cause it to explode. Allow it to warm at room temperature for a few hours before charging.

Slow Charging

Slow charging means the charge rate is between 3 and 15 amps (A) for a long period of time. Slow charging the battery has two advantages: it is the only way to restore the battery to a fully charged state and it minimizes the chances of overcharging the battery. Slow

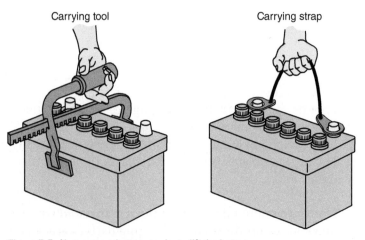

Carrying tool   Carrying strap

Figure 5-5 Always use a battery carrier to lift the battery.

charging the battery causes the lead sulfate on the plates to convert to lead dioxide and sponge lead throughout the thickness of the plate.

Fast Charging

Fast charging uses a high current for a short period of time to boost the battery. Fast charging the battery will bring the state of charge up high enough to crank the engine. However, fast charging is unable to recharge the battery as effectively as slow charging. Fast charging the battery converts only the lead sulfate on the outside of the plates. The conversion does not go through the plates. After the battery has been fast charged to a point that it will crank the engine, it should then be slow charged to a full state.

Charge Rate

The **charge rate** is the speed at which the battery can safely be recharged at a set amperage. The charge rate required to recharge a battery depends on several factors:

1. Battery capacity. High-capacity batteries require longer charging time.
2. State of charge.
3. Battery temperature.
4. Battery condition.
5. Battery type.

Slow charging is the easiest on the battery. However, slow charging requires a long period of time. The basic rule of thumb for slow charging the battery is 1 amp for each positive plate in one cell.

Slow charging of the battery may not always be practical due to the time involved. In these cases, fast charging is the only alternative. To determine the charging time for a full charge based on charge rate amperes, use **Table 5-1**.

An alternative method is to connect a voltmeter across the battery terminals while it is charging. If the voltmeter reads fewer than 15 volts, the charging rate is low enough. If the voltmeter reads over 15 volts, reduce the charging rate until voltage reads below 15 volts. Keeping the voltage at 15 volts will ensure the quickest charge and a safe rate for the battery.

There are three methods of determining if the battery is fully charged:

1. Specific gravity holds at 1.264 or higher after the battery is stabilized.
2. An open circuit voltage test indicates 12.68 or higher after the battery has been stabilized.
3. The ammeter on the battery charger falls to approximately 3 amps or less and remains at that level for 1 hour.

Slow charging is often referred to as "trickle charging."

 Caution

Fast charging the battery requires that the battery be monitored at all times, and the charging time must be controlled. Do not fast charge a battery for longer than 2 hours. Excessive fast charging can damage the battery. Do not allow the voltage of a 12-volt battery to exceed 15.5 volts. Also, don't allow the temperature to rise above 125°F (51.7°C)

TABLE 5-1   table showing the rate and time of charging a battery. Electrolyte temperatures should not exceed 125°F (51.7°C) during charging

Open Circuit Voltage	Battery Specific Gravity	State of Charge (%)	Charging Time of Full Charge at 80°F (267°C)					
			At 60 A	At 50 A	At 40 A	At 30 A	At 20 A	At 10 A
12.6	1.265	100	Full charge					
12.4	1.225	75	15 min.	20 min.	27 min.	35 min.	48 min.	90 min.
12.2	1.190	50	35 min.	45 min.	55 min.	75 min.	95 min.	180 min.
12.0	1.155	25	50 min.	65 min.	85 min.	115 min.	145 min.	280 min.
11.8	1.120	0	65 min.	85 min.	110 min.	150 min.	150 min.	370 min.

⚙ SERVICE TIP If a battery is severely discharged and will not take a slow charge, connect a good battery in parallel (with jumper cables). Fast charge for 30 minutes; then disconnect the good battery and slow charge the discharged battery.

Recharging Recombination Batteries

Classroom Manual
Chapter 5, page 125

Most recombination batteries (including gel cell, valve-regulated, and absorbed glass mat) batteries will accept being recharged very well. This is due, in part, to their low internal resistance. However, overcharging is very harmful to recombination batteries. Since recombination batteries use a special sealing design, overcharging will dry out the electrolyte by forcing the oxygen and hydrogen from the battery through the safety valves.

If a battery is continually undercharged, a layer of sulfate will build up on the positive plate. The sulfate then acts to resist the flow of electrons. This may also result in plate shedding which reduces battery performance and shortens battery life.

When recharging the recombination battery, disconnect the negative battery cable. Charge the battery directly at the battery terminals (do not use any jumper cables or chassis ground locations). It is critical that the charger being used will properly limit the voltage to no more than 14.4 volts and no less than 13.8 volts at 68°F (20°C). This requires special charging equipment designed to recharge recombination batteries. Older-type battery chargers use higher voltages and charging rates and usually cannot be used to recharge recombination batteries. A recombination battery that is charged at too high a voltage may experience shorter battery life because of damage resulting from the battery temperature increasing to over 100°F (37.8°C).

Most modern battery chargers are compatible with recombination-type batteries and have a switch to cycle between flooded and recombination batteries. Be sure to check the documentation supplied with the charger you are using to recharge a recombination cell battery.

It is a good practice to measure the voltage across the recombination battery before recharging it. If the voltage is less than 10 volts, it may take several hours to recharge the battery since the charger will apply very low current to the battery until the battery voltage increases.

Hybrid vehicles with 12-volt auxiliary batteries are recharged in the same manner as other recombination batteries. However, there may be special service procedures that must be performed in preparation of isolating and charging the battery.

MAINTAINING MEMORY FUNCTIONS

Today's vehicles are equipped with several computer control modules that use memories to store values and presets. These include radio station selection, memory seat positions, transmission shift schedule learning, and adaptive fuel strategies. These memories require battery voltage to be maintained and will reset if the battery is disconnected. If the battery is disconnected to perform battery tests or battery service, or to be charged, then a memory keeper discussed in Chapter 2 can be used to maintain the memories. However, if the battery is to be disconnected to perform electrical circuit tests or repairs, then the memory keeper should not be used and the memory function will need to be restored.

BATTERY TEST SERIES

When the battery and cables have been completely inspected and any problems have been corrected, the battery is ready to be tested further. For the tests to be accurate, the battery must be fully charged.

Classroom Manual
Chapter 5, page 132

Battery Terminal Test

The **battery terminal test** checks for poor electrical connections between the battery cables and terminals. Use a voltmeter to measure voltage drop across the cables and terminals. It is a good practice to perform the battery terminal test anytime the battery cable is disconnected and reconnected to the terminals. By performing this test, comebacks, due to loose or faulty connections, can be reduced.

The battery must be fully charged to perform this test. To test the negative terminal connection, connect the positive voltmeter test lead to the cable clamp and connect the negative meter lead to the battery terminal (**Figure 5-6**). Follow the vehicle manufacturer's recommended method to disable the ignition system to prevent the vehicle from starting. This may be done by removing the ignition coil secondary wire from the distributor cap and putting it to ground. Many systems require the removal of the fuel pump or electronic fuel injection (EFI) relay in order to prevent the engine from starting.

Crank the engine and observe the voltmeter reading. If the voltmeter shows over 0.3 volt, there is a high resistance at the cable connection. Remove the battery cable using the clamp puller (**Figure 5-7**). Clean the cable ends and battery terminals (**Figure 5-8**).

Perform the test for the positive terminal in the same manner. Connect the voltmeter leads across the terminal and clamp, observing polarity.

 Special Tools
Safety glasses
Voltmeter
Terminal pliers
Terminal puller
Terminal and clamp
   cleaner
Fender covers

Battery Leakage Test

Battery drain can be caused by a dirty battery. The dirt can actually allow current flow over the battery case. This current flow can drain a battery as quickly as leaving a light on. A **battery leakage test** is conducted to see if current is flowing across the battery case. To

Classroom Manual
Chapter 5, page 118

Figure 5-6 Voltmeter connections for the battery terminal test.

Figure 5-7 Use battery clamp pullers to remove the cable end from the terminal. Do not pry the clamp off.

Figure 5-8 A terminal cleaning tool is used to clean the cable's clamps and the battery's terminals.

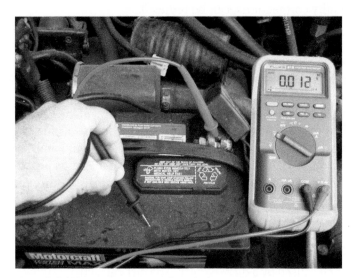

Figure 5-9 Using a voltmeter to perform the battery leakage test.

Special Tools
Voltmeter
Fender covers
Safety glasses

Classroom Manual
Chapter 5,
pages 118, 122

perform a battery leakage test, set a voltmeter to a low DC volt scale. Connect the positive test lead to the positive terminal of the battery. Move the negative test lead across the top and sides of the battery case (**Figure 5-9**). If the meter reads voltage, a current path from the negative terminal of the battery to its positive terminal is being completed through the dirt. Keep in mind that you should not measure voltage anywhere on the case of the battery. If voltage is present, remove the battery. Then use a baking soda and water mixture to clean the case of the battery. When cleaning the battery, don't allow the baking soda and water solution to enter its cells. After the case is clean, rinse it off with clean water.

State of Charge Test

Measuring the **state of charge** is a check of the battery's electrolyte and plates. The state of charge test uses the measurement of the specific gravity to determine the charge of the battery.

To use the hydrometer to test the battery's state of charge:

1. Remove all battery vent caps.
2. Check the electrolyte level. It must be high enough to withdraw the correct amount of solution into the **hydrometer**. A hydrometer measures the specific gravity of a liquid (**Figure 5-10**).

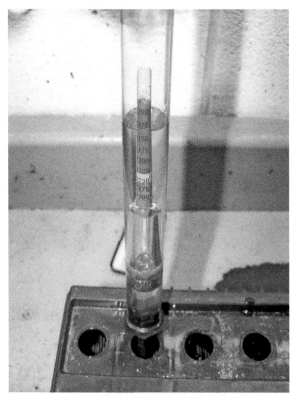

Figure 5-10 A temperature correction hydrometer is used to measure the specific gravity of the electrolyte solution, providing an indication of the battery's state of charge.

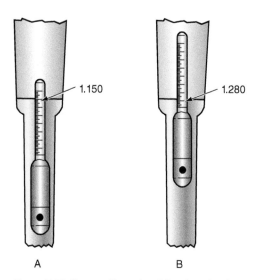

Figure 5-11 The specific gravity of the electrolyte is read at the point where the electrolyte intersects the float. (A) A low reading; (B) a high reading.

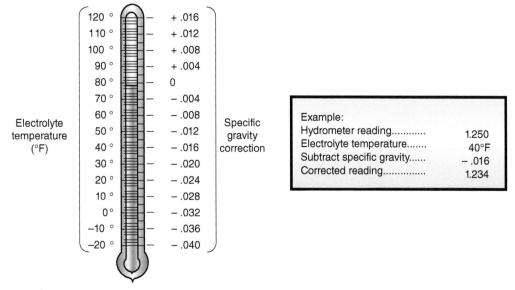

Figure 5-12 Correct the specific gravity reading according to the temperature of the electrolyte.

Example:	
Hydrometer reading............	1.250
Electrolyte temperature.......	40°F
Subtract specific gravity......	– .016
Corrected reading..............	1.234

3. Squeeze the bulb and place the pickup tube into the electrolyte of a cell.
4. Slowly release the bulb. Draw in enough solution until the float is freely suspended in the barrel. Hold the hydrometer in a vertical position.

Special Tool

Hydrometer

The float rises and the specific gravity is read where the float scale intersects the top of the solution (**Figure 5-11**). The reading must also be corrected by compensating for temperatures (**Figure 5-12**).

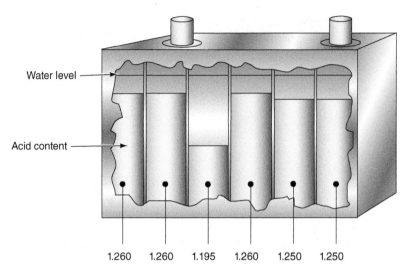

Figure 5-13 A defective cell can be determined by the specific gravity readings.

If the electrolyte level is too low to perform the test, add distilled water to the cell. Do not take hydrometer readings until the water and electrolyte have been mixed by charging the battery.

Test Results. As a battery becomes discharged, its electrolyte has a larger percentage of water. Thus, a discharged battery's electrolyte will have a lower specific gravity number than that of a fully charged battery.

A fully charged battery will have a hydrometer reading near 1.265. Remember, the specific gravity is also influenced by the temperature of the electrolyte and the readings must be corrected to the temperature. If the corrected hydrometer reading is below 1.265, the battery needs recharging or it may be defective.

A defective battery can be determined with a hydrometer by checking every cell. If the specific gravity has a 0.050-point variation between the highest and lowest cell readings, the battery is defective (**Figure 5-13**). When all the cells have an equal gravity, even if all are low, the battery can usually be regenerated by recharging.

Specific gravity tests should not be used as the sole determinant of battery condition. If the cells of the battery do not have the same specific gravity, the battery should be replaced. When the specific gravity of all the cells is good or bad, the voltage of the battery must be considered before coming to a conclusion about the battery's condition. A battery with low specific gravity and acceptable voltage is normally only discharged, perhaps due to a charging system problem. However, a battery with good specific gravity readings but low voltage readings is always bad and needs to be replaced.

Optical Refractometer

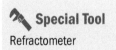

Special Tool
Refractometer

A **refractometer** uses the refractions of light to determine and display a very accurate specific gravity reading. All that is required is a couple of drops of electrolyte to be placed on the glass slide of the refractometer (**Figure 5-14**). While you hold the lens of the meter up to your eye, the light will refract through the sample and display the specific gravity in the window (**Figure 5-15**).

> **SERVICE TIP** If the vehicle has many circuits that place a constant drain on the battery (computer, clock, memory radios, etc.), the negative battery cable should be disconnected before taking a voltmeter reading.

Figure 5-14 Placing a sample of electrolyte onto the refractometer's slide.

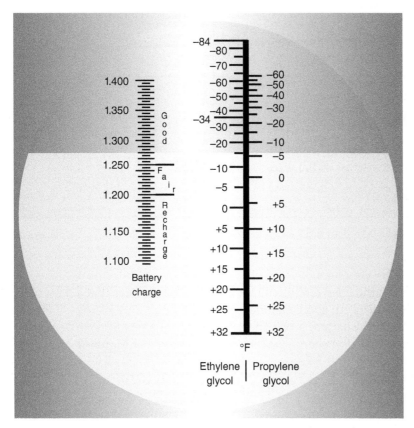

Figure 5-15 The window will indicate the specific gravity.

Open Circuit Voltage Test

The battery **open circuit voltage test** is used to determine the battery's state of charge. It is used when a hydrometer is not available or cannot be used. To obtain accurate open circuit voltage test results, the battery must be stabilized. To **stabilize** the battery means the surface charge is removed by placing a large load on the battery for 15 seconds. If the battery has just been recharged, perform the capacity test, and then wait at least 10 minutes to allow battery voltage to stabilize. Connect a voltmeter across the battery terminals, observing polarity (**Figure 5-16**). Measure the open circuit voltage. Take the reading to the 1/10 volt. A good battery will have an open circuit voltage of 12.6 volts.

Special Tools
Digital voltmeter
Fender covers
Safety glasses

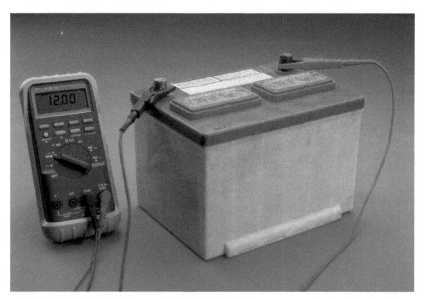

Figure 5-16 Open circuit voltage test using a voltmeter.

Open Circuit Voltage Table	
Open Circuit Voltage	Charge Percentage
11.7 volts or less	0%
12.0 volts	25%
12.2 volts	50%
12.4 volts	75%
12.6 volts or more	100%

Figure 5-17 Open circuit voltage test results relate to the specific gravity of the battery's cells.

To analyze the open circuit voltage test results, consider that a battery at a temperature of 80° F (26.7° C), in good condition, should show about 12.4 volts. If the state of charge is 75% or more, the battery is considered charged. The relationship between open circuit voltage test results and charge percentage is illustrated in **Figure 5-17**.

Capacity Test

Special Tools

VAT-60
Electrolyte thermometer
Safety glasses
Fender covers

Classroom Manual
Chapter 5, page 132

The capacity test is also called the load test.

The **capacity test** provides a realistic determination of the battery's condition. For this test to be accurate, the battery must pass the state of charge or open circuit voltage test. If it does not, recharge the battery and test it again.

During the capacity test, a specified load is placed on the battery while the terminal voltage is observed. A good battery should produce current equal to 50% of its cold-cranking rating (or three times its ampere-hour rating) for 15 seconds and still provide 9.6 volts.

Depending on the equipment used, certain steps need to be followed. If the tester uses a carbon pile to load the battery, follow the general steps outlined in **Photo Sequence 8**.

Once the capacity test is completed and you have recorded the voltage value, the readings need to be corrected to electrolyte temperature (**Figure 5-18**).

If voltage level is below the specifications listed in Figure 5-18, observe the battery voltage for the next 10 minutes. If the voltage rises to 12.45 volts or higher, the battery

Electrolyte Temperature								
°F	70+	60	50	40	30	20	10	0
°C	21+	16	10	4	–1	–7	–12	–18
Minimum Voltage (12-Volt Battery)	9.6	9.5	9.4	9.3	9.1	8.9	8.7	8.5

Figure 5-18 Correcting the readings of the capacity test to temperature readings.

must be replaced. This means that the battery can hold a charge but has insufficient cold-cranking amperes.

If the voltage does not return to 12.4 volts, recharge the battery until the open circuit test indicates a voltage of 12.66 volts. Repeat the capacity test. If the battery fails again, replace the battery.

The capacity test on a 12-volt auxiliary battery that is used on a hybrid vehicle is performed in the same manner as just described. However, be sure to identify the rating of the battery since most of these batteries have a low capacity.

⚠ **Caution**
While performing the load test, do not load the battery longer than 15 seconds.

PHOTO SEQUENCE 8
Performing a Battery Capacity Test

P8-1 Charge the battery to at least 75% and allow the battery to stabilize.

P8-2 Determine the rating on the battery's label and figure the load test specification.

P8-3 Determine the temperature of the electrolyte. If the case is sealed, measure the temperature of the case.

P8-4 Connect the large load leads across the battery terminals, observing polarity.

P8-5 Confirm that the meter reads zero amperes. If needed, adjust ammeter reading to zero.

P8-6 Connect the ammeter's inductive pickup around the negative load tester lead (not the battery's negative cable).

P8-7 Turn the load control knob slowly until the ammeter indicates the amperage amount determined in step 2.

P8-8 Record the voltmeter reading while applying the load for 15 seconds.

If the capacity test readings of a clean and fully charged battery are equal to or above specification, the battery is good. If the battery tests are borderline, perform the 3-minute charge test.

Three-Minute Charge Test

If a conventional battery fails the load test, it is not always the fault of the battery. It is possible that the battery has not been receiving an adequate charge from the charging system. The **three-minute charge test** determines the battery's ability to accept a charge, and for **sulfation**. Sulfation is a chemical action within the battery that interferes with the ability of the cells to deliver current and accept a charge (**Figure 5-19**). A battery must have failed the capacity test to get accurate results from a 3-minute charge test.

Special Tools

Battery charger
Voltmeter
Safety glasses
Fender covers

Classroom Manual
Chapter 5, page 118

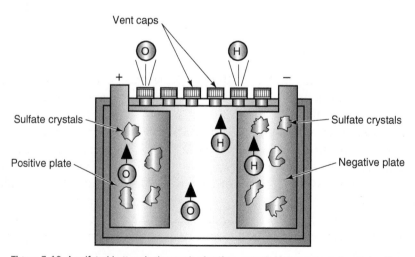

Figure 5-19 A sulfated battery is the result of sulfate crystals that penetrate the plates. The crystals become insoluble and will not allow the battery cell to deliver current nor accept a charge.

To conduct the 3-minute test:

1. Remove the ground cable. The battery must be disconnected from the vehicle's electrical system since the high voltage that is possible during this test can damage the computers.
2. Connect a battery charger to the battery, observing polarity.
3. Connect a voltmeter across the battery terminals, observing polarity.
4. Turn on the battery charger to 40 amps (20–25 for maintenance-free batteries).
5. Maintain this rate of charge for 3 minutes.
6. Check the voltage reading at 3 minutes. If fewer than 15.5 volts, the battery is not sulfated. If the voltmeter reading is above 15.5 volts, the battery is sulfated or there is a poor internal connection.
7. If the battery passes the 3-minute test, slowly recharge the battery and do the capacity test again.
8. If the battery passes the capacity test this time, test the charging system.

⚠ WARNING **Some battery manufacturers, such as Delco, do NOT recommend the 3-minute charge test.**

CUSTOMER CARE One of the best things you can do for your customers is to assist them in choosing the correct battery. Battery selection needs to be based on the make of the vehicle, electrical options on the vehicle, driving habits, and climatic conditions. The largest current capacity rating that can be achieved in a given battery group may benefit some customers but may be a waste of money for others.

Computerized Load-Testing Equipment

Recently, automotive equipment manufacturers have developed battery load testers that use computer technology. With this type of equipment, the technician types in the CCA rating of the battery and the tester automatically performs the test. The results of the test are printed or displayed for the technician. Some systems will plot the voltages on a graph as the test is performed. The tester compares the results from the capacity test with its data for the CCA rating and makes a determination for a proper recommendation to the technician.

Battery Conductance Testing

Conductance is a measurement of the battery's plate surface that is available for chemical reaction. This determines how much power the battery can supply and describes the ability of a battery to conduct current. Conductance testing of the battery has proven to be a reliable test of the battery's capacity (**Figure 5-20**). The higher the conductance value (or lower internal resistance), the better the performance potential of the battery.

🔧 **Special Tools**
Conductance Tester
Battery Charger

Conductance testing is performed by sending a low-frequency AC signal through the battery. A portion of the AC current's returned pulse is then captured and a conductance measurement is calculated. Conductance battery testers have the ability to accurately test batteries that are not fully charged. Battery internal damage is also detected without having to charge the battery. An on-screen display directs the technician through the steps required to perform the battery test.

A fully charged battery will have a conductance reading between 110% and 140% of its CCA rating. As the plate surfaces sulfate, it will lose its conductance. If the battery

Figure 5-20 Handheld conductance tester.

loses its cranking ability, the conductance will drop below the CCA rating. At this point, the battery will need to be replaced. **Photo Sequence 9** illustrates a typical procedure for using the conductance tester for determining a battery's condition.

PHOTO SEQUENCE 9
Using a Conductance Battery Tester

P9-1 The GR8 is a common battery/ starter/ charging system tester that uses conductance.

P9-2 With the tester connected to the power supply, use the arrow keys to select the System Test icon in the Main Menu.

P9-3 Connect the small clamps from the tester to the battery terminals.

P9-4 Select the POST TYPE based on battery design.

P9-5 Press the NEXT soft key to continue.

P9-6 Determine the battery type and make the appropriate selection.

PHOTO SEQUENCE 9 (CONTINUED)

P9-7 Select the battery's rating units.

P9-8 Use the UP or DOWN arrow keys to select the battery rating.

P9-9 The load test will begin.

P9-10 After the test, the tool will display a state of health decision along with an analysis of the battery's state of charge. Use the UP or DOWN arrow keys to scroll through the screens.

P9-11 The results of the battery test series can be printed.

P9-12 Pressing the End key will return to the Main Menu.

BATTERY DRAIN TEST

If a customer complains that the battery is dead when attempting to start their vehicle after it has not been used for a short while, the problem may be a current drain from one of the electrical systems. The most common cause for this type of drain is a light that is not turning off—such as glove box, trunk, or engine compartment illumination lights.

These **parasitic drains** on the battery can cause various driveability problems. With lowbattery voltage, several problems can result; for example:

1. The computer may go into backup mode or "limp-in" mode of operation.
2. The computer may set false trouble codes.
3. To compensate for the lowbattery voltage, the computer may raise the engine speed.

The procedure for performing the battery drain test may vary according to the manufacturer. However, battery drain can often be observed by connecting an ammeter in series with the negative battery cable or by placing the inductive ammeter pickup lead around the negative cable. If the meter reads 30 mA or more, there is excessive drain. Visually check the trunk, glove box, and under-hood lights to see if they are on. If they are, remove the bulb and watch the battery drain. If the drain is now within specifications, find out why the circuit is staying on and repair the problem. If the cause of the drain is not the lights, go to the fuse panel or distribution center and remove one fuse at a time while watching the ammeter. When the drain decreases, the circuit protected by the fuse you removed last is the source of the problem.

Special Tools
Test light
DVOM
VAT-60 or equivalent
Multiplying coil
Terminal pliers
Terminal puller
Cable clamp spreader
Safety glasses
Fender covers

The open circuit voltage reading must be 11.5 volts or higher to perform the battery drain test.

Figure 5-21 Using a test light to prevent the vehicle's computers from powering down.

If the vehicle is equipped with computer-controlled air suspension systems, it may be necessary to disconnect the module to eliminate it from the test.

The following is a typical procedure for determining and locating parasitic drains against the battery. First, it is necessary to determine if a draw is occurring. This can be done by connecting an inductive ammeter clamp around the negative battery cable. The meter must be capable of reading less than 1 amp accurately. With all accessories turned off, the ammeter reading indicates the current draw against the battery. It is normal for some vehicles to have higher than 30 mA draw for up to an hour after the ignition switch is turned off. This allows the computers to perform their "administrative" tasks. Be sure to confirm the normal time-out period with the proper service information. If the ammeter reads higher than allowed current, the cause of the drain must be determined.

If an inductive clamp is not used, then the ammeter needs to be connected in series with the negative battery cable and the battery terminal. In this case it may be advisable to use a test light in series with the negative battery cable and the battery terminal first (**Figure 5-21**). This is done to prevent the vehicle's computers from powering down and then powering back up again when the ammeter is connected. The additional amperage of the computers powering back up may blow the fuse in the ammeter. Prior to disconnecting the clamp from the terminal, connect the test lamp. As the clamp is removed, maintain connection of the lamp with the clamp and terminal. If the test light is on, there is a drain against the battery that is sufficient to light the lamp. After the time period has expired, the test light should go out. If not, then connect the ammeter leads in series between the cable and terminal (with the test light still connected). Once the ammeter is connected, remove the test light and read the amount of current draw.

An alternate method of isolating the circuit that has the excessive drain is to measure the voltage drop across the fuse. Since a fuse may protect several circuits, it is possible to mask the problem when a fuse is pulled. For example, if a fuse protects the power supply circuits to three different computers, then pulling that fuse will cause them all to totally turn off. Usually a computer does not turn off completely when it powers down, but the draw is very slight. If one of the three computers was failing to power down, then excessive drain would be against the battery. If the fuse is pulled and all three computers power down, it is possible that the defective computer will reset and when the fuse is plugged in again there no longer will be a parasitic load. Since pulling the fuse caused the load reading on the ammeter to drop, the circuit is identified. However, once the fuse was plugged back in, the load is no longer indicated and isolating the cause is made more difficult. It would be advantageous to identify the circuit first, and then unplug the computers one at a time until the draw is identified.

To perform this method of circuit identification, use the test tabs on the top of the fuse (maxi-fuses will require removing the lens). Connect a voltmeter set on the millivolt scale across the test points and observe the meter's reading (**Figure 5-22**). A fuse has resistance

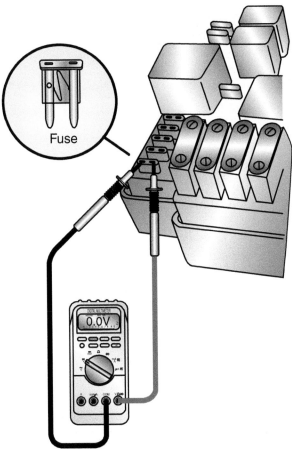

Figure 5-22 Using a voltmeter across a fuse to determine if current is flowing in the circuit.

Fuse Value	Fuse Type	Divide by
5	Mini	16.5
10	Mini	7.5
15	Mini	4.5
20	Mini	3.5
25	Mini	2.5
30	Mini	2.0
20	Cartridge	1.0
30	Cartridge	1.5
40	Cartridge	1.0
50	Cartridge	0.5

Figure 5-23 Determining how much current is flowing through a fuse.

so that it can get hot and burn through if excessive current passes through it; this resistance will provide a voltage drop. If there is no (or very little) current flowing through the circuit, then the voltmeter will read 0 volt. However, if current is flowing through the circuit, then the voltmeter will provide a reading. The reading will indicate how much current is flowing in the circuit, which is a function of the rating of the fuse (**Figure 5-23**). For example, if the voltmeter reads 20 mV when connected across a 15-amp fuse, then the current draw is 20 / 4.5 = 4.4 amps. It is not important to determine the amount of current flow. You are not looking for 1 or 2 mA but more than 30 mA; therefore, any reading displayed on the voltmeter indicates excessive current flow.

AUTHOR'S NOTE Using an ammeter to test for battery drain will not indicate an internal short in the battery.

BATTERY REMOVAL AND CLEANING

It is natural for dirt and grease to collect on the top of the battery. If allowed to accumulate, the dirt and grease can form a conductive path between the battery terminals. This may result in a drain on the battery. Also, normal battery gassing will deposit sulfuric acid as the vapors condense. Over a period of time, the sulfuric acid will corrode the battery

Classroom Manual
Chapter 5, page 135

 **Special Tools**

Baking soda

Cleaning brushes

Terminal pliers
  (Figure 5-24)

Cable clamp spreader

Terminal puller

Terminal and clamp
  cleaning tool

Battery-lifting strap

Putty knife

Protective terminal
  coating

Safety glasses

Heavy rubber gloves

Fender covers

⚡ Caution

Do not allow the baking soda solution to enter the battery cells. This will neutralize the acid in the electrolyte and destroy the battery.

The baking soda and water solution consists of one tablespoon per quart of water.

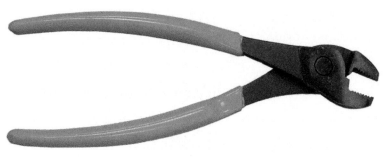

Figure 5-24 Battery terminal pliers.

terminals, cable clamps, and hold-down fixtures. As the corrosion builds on the terminals, it adds resistance to the entire electrical system.

Periodic battery cleaning will eliminate these problems. To be able to clean the battery correctly, it is best to remove it from the vehicle. Removing the battery protects the vehicle's finish and other under-hood components. Follow **Photo Sequence 10** for the procedure for removing the battery from the vehicle and cleaning it. Consult the manufacturer's service information for precautions concerning the vehicle's computer controls.

When replacing the battery into the vehicle, make sure it is properly seated in the tray. Connect the hold-down fixture and secure the battery. Do not overtighten the hold-down fixture. Install the positive cable and tighten; then install the negative cable. Be sure to observe polarity. Perform a battery terminal test to confirm good connections.

CUSTOMER CARE Before installing the battery back into the vehicle, it is a good practice to clean the battery tray. First scrape away any heavy corrosion with a putty knife. Next, clean the tray with a baking soda and water solution **(Figure 5-25)**. Flush with water and allow to dry. After the tray has dried, paint it with rust-resistant paint. After the paint has completely dried, coat the tray with silicone base spray. These extra steps will protect the tray from corrosion.

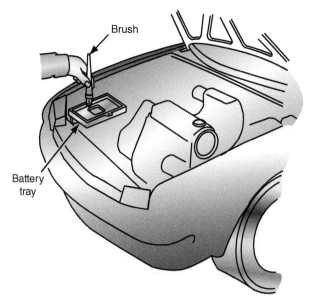

Figure 5-25 After scraping the battery tray clean with a putty knife, use the baking soda mixture to remove all electrolyte residues.

PHOTO SEQUENCE 10
Removing and Cleaning the Battery

P10-1 Tools needed to remove the battery from the vehicle include rags, baking soda, pan, terminal pliers, cable clamp spreader, terminal puller, assorted wrenches, terminal and clamp cleaning tool, battery lifting strap, safety glasses, heavy rubber gloves, rubber apron, and fender covers.

P10-2 Place the fender covers on the vehicle to protect the finish.

P10-3 Loosen the clamp bolt for the negative cable using terminal pliers and wrench of correct size. Be careful not to put excessive force against the terminal.

P10-4 Use the terminal puller to remove the cable from the terminal. Do not pry the cable off of the terminal.

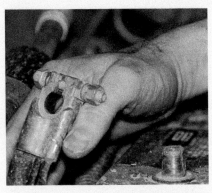

P10-5 Locate the negative cable away from the battery.

P10-6 Loosen the clamp bolt for the positive cable and use the terminal puller to remove the cable. If the battery has a heat shield, remove it.

P10-7 Disconnect the hold-down fixture.

P10-8 Using the battery-lifting clamp, remove the battery out of the tray. Keep the battery away from your body. Wear protective clothing to prevent acid spills onto your hands.

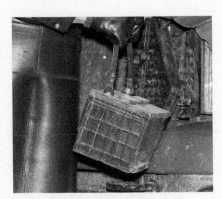

P10-9 Transport the battery to the bench. Keep it away from your clothes.

PHOTO SEQUENCE 10 (CONTINUED)

P10-10 Mix a solution of baking soda and water.

P10-11 Brush the solution over the battery case. Be careful not to allow the solution to enter the cells of the battery.

P10-12 Flush the solution off with water.

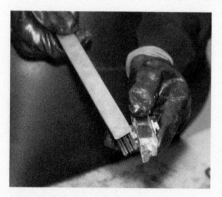

P10-13 Use a wire brush to remove corrosion from the hold-down brackets.

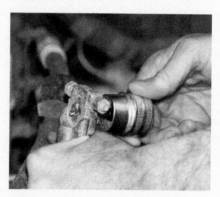

P10-14 Use the terminal cleaning tool to clean the cable.

P10-15 Use the cleaning tool to clean the battery posts.

P10-16 Install the battery into the tray and install the hold-down hardware. The cables are installed positive first, followed by the negative cable.

P10-17 Coat the battery terminals with corrosion-preventive spray. Perform a terminal test to assure tight connections.

Figure 5-26 A protective pad under the battery clamp prevents corrosion of the clamp and terminal.

⚠ WARNING **Be careful not to touch the positive terminal with the wrench when tightening the negative cable clamp.**

Spray the cable clamps with a protective coating to prevent corrosion. A little grease or petroleum jelly will prevent corrosion as well. Also, protective pads that go under the clamp and around the terminal are available (**Figure 5-26**).

JUMPING THE BATTERY

There will be times when you will have to use a boost battery and jumper cables to jump-start a vehicle (**Figure 5-27**). It is important that all safety precautions be followed. Jump-starting a dead battery can be dangerous if it is not done correctly. The following steps should be followed to safely jump-start most vehicles:

1. Make sure the two vehicles are not touching each other. The excessive current flow through the vehicles' bodies can damage the small ground straps that attach the engine block to the frame. These small wires are designed to carry only 30 amps. If the vehicles are touching, as much as 400 amps may be carried through them.

🔧 **Special Tools**

Booster battery
Jumper cables
Jumper box
Fender covers
Safety glasses

⚠ **Caution**

Do not use more than 16 volts to jump-start a vehicle that is equipped with an engine control module. The excess voltage may damage the electronic components.

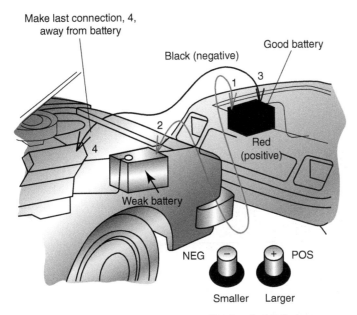

Figure 5-27 Proper jumper cable connections and sequence for jump-starting a vehicle.

2. For each vehicle, engage the parking brake and put the transmission in neutral or park.
3. Turn off the ignition switch and all accessories, on both vehicles.
4. Attach one end of the positive jumper cable to the disabled battery's positive terminal.
5. Connect the other end of the positive jumper cable to the booster battery's positive terminal.
6. Attach one end of the negative jumper cable to the booster battery's negative terminal.
7. Attach the other end of the negative jumper cable to an engine ground on the disabled vehicle.
8. Attempt to start the disabled vehicle. If the disabled vehicle does not start readily, start the jumper vehicle and run at fast idle to prevent excessive current draw.
9. Once the disabled vehicle starts, disconnect the ground-connected negative jumper cable from its engine block.
10. Disconnect the negative jumper cable from the booster battery.
11. Disconnect the positive jumper cable from the booster battery and then from the other battery.

⚡ WARNING **Do not connect this cable end to the battery negative terminal. Doing so may create a spark that will cause the battery to explode.**

⚡ WARNING **A battery that has been rapidly discharged will create hydrogen gas. Do not attach jumper cables to a weak battery if starting the vehicle has been attempted. Wait for at least 10 minutes before connecting the jumper cable and attempting to start the vehicle.**

A jumper box is often used instead of jumping from another vehicle's battery (**Figure 5-28**). The jumper box contains a small 12-volt battery that can supply enough current to start the vehicle. The box is connected to the vehicle battery by clamps similar to that of a battery charger. Polarity must be observed when connecting the jumper box. When the jumper box is not in use, it is plugged into a 120-volt wall socket to keep it charged.

Figure 5-28 A typical jumper box.

HV BATTERY SERVICE

Classroom Manual
Chapter 5, page 127

It is important to remember that most hybrid vehicles have two separate batteries. One is the high-voltage (HV) battery pack; the other is the conventional 12-volt battery. In most hybrids, the HV battery pack provides the electrical power to start the engine and powers the electric motors. The 12-volt battery is used to power the basic electrical system, such as the lights, accessories, and power equipment. When diagnosing a problem with the HV battery, it is important to consider that the 12-volt battery also supplies the power for the electronic controls that monitor and regulate the operation of the hybrid system. If the 12-volt battery source is not operating properly, then the hybrid system will not either. Consequently, the 12-volt system should never be ignored when working on an HV system.

Most hybrid vehicles have a continuous drain on the 12-volt battery system. In addition, the HV battery pack will drain if it is not being used. When servicing the batteries on a HEV, remember that if the battery's voltage drops below a specific level, the emissions malfunction indicator lamp (MIL) and/or hybrid warning lights may illuminate. Due to the battery drain when the vehicle is not in use, most manufacturers recommend that HEVs be started and run for at least 10 minutes every month. This keeps the HV battery charged enough to operate the vehicle. However, it may not be enough to keep the 12-volt battery properly charged.

Special Tools
DMM
Scan tool

Before working on or around the HV battery and energy systems, be sure to review all safety warnings, precautions, and procedures outlined in Chapter 1. Remember that these are guidelines; specific instructions and safety warnings will be provided in the manufacturer's service literature. Following these safety precautions is not limited to servicing of the electrical system; all services including air-conditioning, engine, transmission, and bodywork may require services completed around and/or with HV systems. If there is any doubt as to whether something has high voltage or not, or if the circuit is sufficiently isolated, *test it!!*

A battery electronic control unit (ECU) monitors the condition of the HV battery assembly. The **battery ECU** determines the state of charge (SOC) of the HV battery by monitoring voltage, current, and temperature. The battery ECU collects data and transmits it to the HV ECU to be used for proper charge and discharge control.

The battery ECU also controls the operation of the battery blower motor to maintain proper HV battery temperature.

The HV battery stores power generated by MG1 and recovered by MG2 during regenerative braking (**Figure 5-29**). The HV battery must also supply power to the electric motor when the vehicle is first started from a stop or when additional power is needed. A typical HV battery uses several nickel-metal-hydride modules and can provide over 270 volts (**Figure 5-30**).

When the vehicle is moving, the HV battery is subjected to repetitive charge and discharge cycles. The HV battery is discharged by MG2 during acceleration mode and then is recharged by regenerative braking. An amperage sensor (**Figure 5-31**) is used so the battery ECU can transmit requests to the HV ECU to maintain the SOC of the HV battery. The battery ECU attempts to keep the SOC at 60%. The battery ECU also monitors delta SOC to determine if it is capable of maintaining acceptable levels of charge. The normal, low-to-high SOC delta is 20%.

If the battery ECU sends abnormal messages to the HV ECU, the HV ECU illuminates the warning light and enters fail-safe control. DTCs and informational codes are set along with freeze frame data. This will occur if the battery ECU determines the insulation resistance of the power cable to be 100k ohms or less.

MG1 functions as the control element for the planetary gear set. It recharges the HV battery and supplies electrical power to drive MG2. MG1 also functions as the starter for the engine. MG2 is used for power at low speeds and for supplemental power when needed at higher speeds.

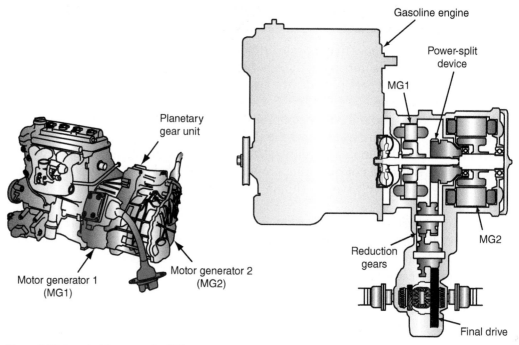

Figure 5-29 Layout of the generating HEV transaxle.

Figure 5-30 The modules of the HV battery.

SERVICE TIP If inspection and testing fail to locate the leak, then it is possible that water entered into the battery assembly or into the converter/inverter assembly.

Whenever an HV battery malfunction occurs, use the scan tool to view the "HV Battery Data List." This provides all HV battery system information. Since the HV battery is usually constructed of a series of modules, the scan tool may display the voltage for each module. Usually a module will have 7.2 volts so the scan tool reading should be between 7 and 8 volts for each module. Some manufacturers will combine two modules to make

Figure 5-31 The amperage sensor.

a block with a normal voltage of 14.4 volts. The scan tool reading for each block should be between 14 and 16 volts (**Figure 5-32**). Regardless of how the scan tool displays the voltage readings, compare each reading with the others. They should be within 0.3 volt of each other when the contactors are open. If any of the voltage readings are higher or lower than the others, that module or block may be damaged.

The scan tool may also display the HV battery SOC. Based on the SOC of the HV battery, the hybrid controller will perform the following:

- 6% SOC all vehicle operation shut down
- 35% SOC e-motors are not used (drive is disabled)
- 40% SOC e-motors are de-rated
- 50% SOC engine restarts to charge HV battery
- Approximately, 52% to 68% SOC normal operating range
- 80% SOC regenerative braking is disabled
- 94% SOC all vehicle operation shut down

If the scan tool does not display the HV battery voltages, then it will be necessary to test the battery with a digital multimeter (DMM). This may require removal of the HV battery pack from the vehicle and the removal of the battery's case.

Data Display - 8PCM			
Name	Value	Unit	
State of Change	45.9	X	Move Row Up
Battery Voltage (cell sum)	303.5	Volts	
Min Block Voltage	16.2	Volts	
Max Block Voltage	16.2	Volts	
Block 1 voltage	15.2	Volts	Move to Down
Block 2 voltage	15.2	Volts	
Block 3 voltage	15.2	Volts	
Block 4 voltage	16.2	Volts	
Block 5 voltage	15.2	Volts	
Block 6 voltage	15.2	Volts	
Block 7 voltage	15.2	Volts	Tech tips
Block 8 voltage	15.2	Volts	
Block 9 voltage	16.2	Volts	
Block 10 voltage	16.2	Volts	
Block 11 voltage	15.2	Volts	Toggle run Height
Block 12 voltage	15.2	Volts	
Block 14 voltage	15.2	Volts	
Block 15 voltage	15.2	Volts	
Block 16 voltage	15.2	Volts	
Block 17 voltage	15.2	Volte	
Block 19 voltage	15.2	Volts	
Block 29 voltage	15.2	Volts	Previous options
High Voltage Sensor	303.0	Volts	

Figure 5-32 Scan tool reading of HV battery voltages.

> **SERVICE TIP** Some battery blocks may read 0 volt when the service plug is
> removed. Be sure to confirm proper voltages with the service information.

High-Voltage Battery Charging

Special Tools

High-voltage battery
charger
Warning tape

⚠ **Caution**

Do not activate the
jump assist mode for
an extended period.
The 12-volt battery
will be damaged
because of excessive
current transfer.

Jump assist is also
referred to as "Charge
assist."

The HV battery is charged when the vehicle is driven, or when the engine is running. This
is the best method of charging the HV battery pack. If the SOC of the HV battery is too
low to allow the engine to run, the HV battery will need to be recharged. This may require
the use of a special high-voltage battery charger. In addition, some manufacturers will
allow only specially trained people to recharge the battery. Some manufacturers will not
even supply the charging equipment to the dealer; a representative of the company performs the task of recharging the HV battery.

To recharge the HV battery, a special Hybrid/EV battery service and de-power tool is
required. The de-power function discharges the battery pack or battery sections to make
them safe for transportation. Section balance equalizes the sections of the battery pack to
ensure optimal battery pack operation. Sections are charged or discharged individually to
bring the pack into alignment. The view pack function generates a detailed status report
of the HV battery, including temperature sensors and individual cell voltages.

HV battery recharging must be performed outside. The correct cable is connected
between the vehicle and the charger (**Figure 5-33**). When using the charger, the immediate area must be secured and marked with warning tape. It will require about 3 hours to
recharge the battery to an SOC of about 50%.

Some manufacturers provide a **jump assist** feature so the HV battery can be charged
without special charging equipment. The process can be initiated by pushing a button or
using a scan tool. The purpose of jump assist is to transfer enough energy from the 12-volt
battery to the HV battery to allow the engine to be started. Jump assist does not fully charge
the HV battery; it works to start the engine so the system can recharge the HV battery.
The HV system needs to recharge the HV battery after a jump assist is performed.

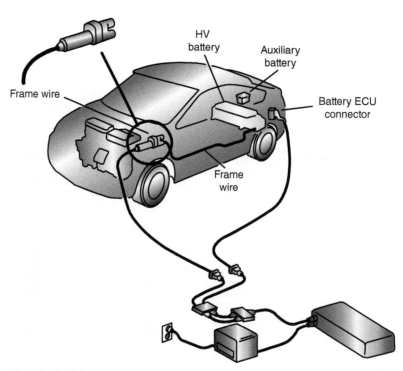

Figure 5-33 HV battery charger connection.

When jump assist is initiated, one of the hybrid system modules (depending on manufacturers and system) converts the 12 volts into high-voltage DC. The HV current is routed to the HV battery pack until a calibrated amount of charge (amp hrs) has been delivered and the HV battery is above a specified voltage (e.g., 300 volts). At this time, the charging process is complete and the electric motors should start the engine.

Depending on the manufacturer, HV battery condition, and temperature, the jump assist function may take up to an hour to complete.

CUSTOMER CARE It is unusual for the HV battery to lose enough charge to not start the engine; anytime the HV battery requires recharging or jump assist, inspect the HV system for failures. The HV battery will drain to a low SOC if the vehicle is not driven for an extended amount of time. Most manufacturers require the vehicle to be used at least 10 minutes every month to keep the HV battery charged.

On some vehicles, jump assist is automatically initiated after a failed crank has occurred. This will happen if an engine crank is requested from the driver and the HV system does not have enough energy to start the engine. For automatic jump assist to occur, the HV battery SOC must be at least 40% and the HV battery voltage must be between 200 and 358 volts. The driver is notified that the jump assist mode has been activated by a message on the HEV monitor screen (**Figure 5-34**).

Some manufacturers provide scan tool initiation of the jump assist. Scan tool initiation of jump assist mode bypasses any voltage requirements that occur with a failed crank. This mode usually requires that a 30-amp battery charge be applied to the 12-volt battery during the process. To assure a successful jump assist, note the following tips:

- The charger must have an output of at least 30 amps.
- Do not charge the battery with the charger set on "High" or "Jump."
- Make sure to leave the ignition key in the OFF position to provide the HV battery with all available current.
- To prevent damage to the 12-volt battery, do not charge the HV battery through the 12-volt battery for an extended period.

Figure 5-34 The driver is notified if an automatic jump assist is being performed after a failed crank.

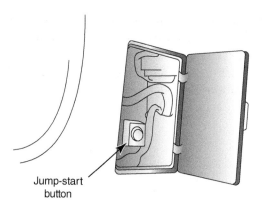

Jump-start
button

Figure 5-35 Jump-start button located in kick panel.

- Be sure to disconnect the battery charger as soon as the jump assist is completed.
- A low 12-volt system does not require a jump assist procedure. If the 12-volt battery is low, a traditional jump start is required.

Following a successful jump assist, the HEV monitor screen and the scan tool will confirm that the procedure is completed.

The jump assist feature can be aborted if any of the following occurs:

- The key is turned to the START position when the system is being charged.
- The 12-volt battery charger has been installed incorrectly (reversed cables).
- The 12-volt battery charger is removed.
- The control module determines conditions are not suitable for charging.

If a Ford Escape or Mariner Hybrid fails to start and the HV battery needs to be charged, a button is provided to initiate the process. First, make sure the ignition is in the OFF position. Next, open the access panel on the driver's side foot well. Located behind this access is the jump-start button (**Figure 5-35**). Press the button and then wait at least 8 minutes before continuing. If you attempt to continue before the 8 minutes have elapsed, the energy from the 12-volt battery will not be able to supply enough power to start the engine by the battery pack.

Pressing the jump-start button sends a request that the system send energy from the 12-volt battery to the HV battery pack. If the 12-volt battery has ample energy, it will be enough to start the engine. However, if the auxiliary battery is weak, it should be jump-started rather than the high-voltage battery pack.

After the 8-minute wait time, the HEV system warning lamp will start to blink. After the warning lamp stops blinking (this may take up to 2 minutes), attempt to start the engine. If the engine still does not start, wait a couple of minutes and repeat the procedure again. If the engine still does not start, the low-voltage battery must be recharged or the vehicle jump-started through the 12-volt battery.

RECYCLING BATTERIES

The materials used to make lead-acid batteries can be recycled. This is the best practice for removal of used batteries. Batteries should not be discarded with regular trash, as they contain metals and chemicals that are hazardous to the environment. The Rechargeable Battery Recycling Corporation (RBRC) was established to promote recycling of rechargeable batteries in North America. RBRC collects batteries from consumers and businesses and sends them to recycling companies. Ninety-eight percent of all lead-acid batteries are recycled. During the recycling process, the lead, plastic, and acid are separated. All of these compounds may be used in manufacturing new batteries.

 Caution

A battery should never be incinerated. Doing this can cause an explosion.

High-voltage NiMH batteries contain several components, including nickel, copper, and steel, which have value as recycled materials. Properly sealed NiMH batteries that are not leaking are considered dry cell batteries and are not hazardous waste. However, always refer to federal, state, local, and provincial laws and regulations governing the recycling of these batteries. If NiMH batteries are found to be leaking, they are regulated as hazardous waste under federal and state regulations.

AUTHOR'S NOTE In California, NiMH batteries must be managed under California Universal Waste Rules.

CASE STUDY

A customer states that his/her vehicle does not start without having to jump the battery. The technician learns that this happens every time the customer attempts to start the vehicle. The customer also says the voltmeter in the dash has been reading higher than normal.

The technician verifies the concern. The engine turns over very slowly for a few seconds and then does not turn. After jumping the battery to get the vehicle into the shop, the technician makes a visual inspection of the battery and cables. The open circuit voltage test shows a voltage of 12.5 volts across the terminals. When the battery is subjected to the capacity test, the voltage drops to 7.8 volts at 80°. After 10 minutes, the open circuit voltage is back up to 12.5 volts. The technician determines that the battery is sulfated. The battery can't handle the load of cranking the engine, which is why it always needs to be jumped. The higher-than-normal voltmeter readings indicate the charging system is trying to keep the battery charged. The technician calls the customer with a price quote. The customer agrees to have the battery replaced. While replacing the battery, the technician cleans the battery tray, the cable clamps, and sprays the clamps with a corrosion protector.

ASE-STYLE REVIEW QUESTIONS

1. Battery terminal connections are being discussed. *Technician A* says when disconnecting battery cables, always disconnect the negative cable first. *Technician B* says when connecting battery cables, always connect the negative cable first. Who is correct?

   A. A only
   B. B only
   C. Both A and B
   D. Neither A nor B

2. A customer's battery is always dead when she attempts to start her car in the morning. After jumping the battery one time in the morning, the car will start throughout the day with no problems. All of the following can be the cause **EXCEPT:**

   A. The starter motor is drawing too much current.
   B. The glove box light is staying on.
   C. A computer is not powering down.
   D. A relay contact is stuck closed.

3. The specific gravity of a battery has been tested. All cells have a corrected reading of about 1.200. *Technician A* says the battery needs to be recharged before further testing. *Technician B* says the battery is sulfated and needs to be replaced. Who is correct?

   A. A only
   B. B only
   C. Both A and B
   D. Neither A nor B

4. When charging a battery:

   A. Connect a voltmeter across the battery terminals while the battery is charging and keep the charge rate at fewer than 10 volts.
   B. Disconnect the negative battery cable before charging to prevent damage to the alternator and computers.
   C. Charge at an amperage rate of 50% of the CCA rating.
   D. All of the above.

5. Which statement concerning the battery leakage test is correct?

   A. The test is used to determine if the battery can provide current and voltage when loaded.

   B. A voltmeter reading of 0.05 when performing the test is acceptable.

   C. Both A and B.

   D. Neither A nor B.

6. The battery open circuit test is being discussed.
Technician A says the battery must be stabilized before the open circuit voltage test is performed.
Technician B says a test result of 12.4 volts is acceptable.
Who is correct?

   A. A only   C. Both A and B

   B. B only   D. Neither A nor B

7. A maintenance-free battery has failed the capacity test.
Technician A says if the voltage recovers to 12.45 volts, the battery is still good.
Technician B says if the voltage level does not return to 12.4 volts, recharge the battery and repeat the capacity test.
Who is correct?

   A. A only   C. Both A and B

   B. B only   D. Neither A nor B

8. *Technician A* says the best way to charge the HV battery is by allowing the engine to run.
Technician B says an on-board jump assist method may be available to start the engine if the HV battery SOC is too low.
Who is correct?

   A. A only   C. Both A and B

   B. B only   D. Neither A nor B

9. While jump-starting a vehicle with a second vehicle, a puff of smoke is observed and the engine ground cable is burned.
Technician A says this happened because the two vehicles were touching.
Technician B says this was caused by connecting the negative jumper cable to the disabled vehicle's engine ground.
Who is correct?

   A. A only   C. Both A and B

   B. B only   D. Neither A nor B

10. Ice crystals are found in the electrolyte. This can be caused by:

   A. A discharged battery.

   B. Use of tap water.

   C. Reversed battery connections.

   D. Improper hold-downs.

ASE CHALLENGE QUESTIONS

1. Charging a battery with a battery charger is being discussed.
Technician A says that battery-charging voltage should never exceed 15 volts.
Technician B says that a battery can be considered fully charged when its open circuit voltage exceeds 12.1 volts after it has been stabilized.
Who is correct?

   A. Technician A   C. Both A and B

   B. Technician B   D. Neither A nor B

2. *Technician A* says that a battery terminal test is performed by placing voltmeter leads between the positive battery post and the negative battery terminal.
Technician B says that the total amount of voltage drop between the negative battery post and the negative battery terminal should not exceed 300 mV.
Who is correct?

   A. Technician A   C. Both A and B

   B. Technician B   D. Neither A nor B

3. *Technician A* says that an ammeter is used when performing a battery leakage test.

   Technician B says that a fully charged battery will have a specific gravity of at least 1.265.

   Who is correct?

   A. Technician A   C. Both A and B

   B. Technician B   D. Neither A nor B

4. *Technician A* says that the 3-minute charge test is performed after a battery has failed a capacity test.

   Technician B says that if battery voltage is below 15.5 volts at the end of the 3-minute charge test, the battery is probably sulfated.

   Who is correct?

   A. Technician A   C. Both A and B

   B. Technician B   D. Neither A nor B

5. The battery current drain test is being discussed.

   Technician A says that an ammeter that can accurately read less than 1 amp should be used.

   Technician B says that a current drain caused by an internally shorted battery should not be measured with an ammeter.

   Who is correct?

   A. Technician A   C. Both A and B

   B. Technician B   D. Neither A nor B

Name _____   Date _____

INSPECTING AND CLEANING A BATTERY

Upon completion of this job sheet, you should be able to visually inspect a battery.

NATEF Correlation

This job sheet addresses the following **MLR/AST/MAST** task:

B.4.   Inspect and clean battery; fill battery cells; check battery cables, connectors, clamps, and hold-downs.

NATEF

Tools and Materials

- A vehicle with a 12-volt battery
- A DMM
- Safety glasses
- Basic tool kit
- Baking soda
- Terminal cleaning brush
- Battery cable puller
- Wire brush

Describe the vehicle being worked on:

Year _____ Make _____ Model _____

VIN _____ Engine type and size _____

Procedure   **Task Completed**

1. Describe the general appearance of the battery.

   _____

   _____

   _____

2. Describe the general appearance of the cables and terminals.

   _____

   _____

   _____

3. Check the tightness of the cables at both ends. Describe their condition.

   _____

   _____

   _____

4. Connect the positive lead of the DMM (set on DC volts) to the positive terminal of the   ☐
   battery.

5. Put the negative lead of the DMM on the battery case, and move it all around the top
   and sides of the case. What readings do you get on the voltmeter?

   _____

   _____

   _____

6. What is indicated by the readings?

_____

_____

_____

7. Measure the voltage of the battery. Your reading was: _____ volts.

8. Perform a battery terminal test for each cable and record your results:

   Positive Cable _____

   Negative Cable _____

9. What do you know about the condition of the battery based on the visual inspection
   and the tests that you did?

   _____

   _____

   _____

   _____

Instructor's Response

_____

_____

_____

Name _____   Date _____

REMOVING, CLEANING, AND INSTALLING A BATTERY

Upon completion of this job sheet, you should be able to remove, clean, and reinstall a battery. Also, you should be able to identify any modules that require memory functions or have reinitialization procedures.

NATEF Correlation

This job sheet addresses the following **MLR/AST/MAST** tasks:

B.3.   Maintain or restore electronic memory functions.

B.4.   Inspect and clean battery; fill battery cells; check battery cables, connectors, clamps, and hold-downs.

B.8.   Identify electrical/electronic modules, security systems, radios, and other accessories that require reinitialization or code entry after reconnecting vehicle battery.

NATEF

Tools and Materials

- A vehicle with a 12-volt battery
- A wash pan
- A box of clean baking soda
- Water
- Cleaning brush
- Hand tools
- Battery cable puller
- Battery terminal cleaner and brushes
- Cable end spreader
- Safety glasses
- Rubber gloves
- Memory keeper
- Service information

Describe the vehicle being worked on:

Year _____ Make _____ Model _____

VIN _____ Engine type and size _____

Procedure

Task Completed

NOTE: Before disconnecting the battery, record the stations of all presets on the radio. Also, use the service information and note any computer modules that require memory restore procedures, reinitialization, or code entry after reconnecting the battery.

_____

_____

_____

_____

_____

_____

1. Identify the battery type: ☐ Flooded   ☐ Gel cell   ☐ AGM

2. What is the rating of the battery?

_____

3. Is this the correct battery for the vehicle?   ☐ Yes   ☐ No

   If *no*, consult with your instructor.

4. Connect the memory keeper.   ☐

5. Disconnect the negative cable of the battery. Use the cable puller if the cable is difficult to remove. ☐

6. Carefully remove the battery hold-down strap. Describe the condition of the hold-down.

_____

_____

_____

7. Disconnect the positive cable of the battery. Use the cable puller if the cable is difficult to remove. Move anything that may interfere with the removal of the battery. ☐

8. With a terminal cleaner, clean off the terminals of the battery. ☐

9. With a battery-carrying strap, remove the battery from the vehicle and place it on a workbench or on the floor close to a drain. ☐

10. Inspect the battery tray. Describe its condition and your recommendations.

_____

_____

_____

11. Make a mixture of baking soda and water. The mixture should be like a paste. ☐

12. With the brush, scrub the top, sides, and bottom of the battery with the baking soda paste. Be careful not to allow the paste to get inside the battery. ☐

13. After all of the battery has been scrubbed, wash the paste off with clean water. ☐

14. Allow the battery to drip dry; then wipe the water off the battery with a clean rag. ☐

15. Use the leftover paste to clean the battery tray and the hold-down assembly. ☐

16. Rinse these off with water. ☐

17. Use a clean, dry cloth to wipe the tray dry. Then reinstall the battery. ☐

18. Clean the battery cable terminals and spread them slightly with the spreader tool. ☐

19. Install the battery hold-down assembly; make sure it is tight and that the battery is positioned properly. ☐

20. Connect the positive cable to the battery. ☐

21. Connect the negative cable to the battery. ☐

22. Reset all preset stations on the radio and other accessories with memory.

_____

_____

_____

Instructor's Response

_____

_____

_____

Name _____   Date _____

TESTING THE BATTERY'S CAPACITY

Upon completion of this job sheet, you should be able to test a battery's capacity and determine needed service actions. Also, you should be able to identify the procedures for testing a hybrid vehicle auxiliary battery.

NATEF Correlation

This job sheet addresses the following **MLR/AST/MAST** tasks:

B.1.   Perform battery state-of-charge test; determine necessary action.

B.2.   Confirm proper battery capacity for vehicle application; perform battery capacity test and load test; determine needed action.

B.9.   Identify hybrid vehicle auxiliary (12v) battery service, repair, and test procedures.

Tools and Materials

- A vehicle with a 12-volt battery
- Service information
- Starting charging system tester (VAT-60 or similar)
- Safety glasses
- Fender covers

Describe the vehicle being worked on:

Year _____ Make _____ Model _____

VIN _____ Engine type and size _____

Procedure   **Task Completed**

1. Perform a battery state of charge test:

   a. Record the specific gravity readings for each cell:

   (1) _____ (2) _____ (3) _____ (4) _____ (5) _____ (6) _____

   b. If the battery is a maintenance-free–type battery, what is the open circuit voltage? _____ volts

2. Summarize the battery's state of charge from the above.

   _____

   _____

   _____

3. Connect the starting charging system tester to the battery.   ☐

4. Locate the rating of the battery. What is the rating? _____

5. Based on the rating, how much load should be put on the battery during the capacity test? _____ amps

6. Conduct a battery load test.

   Battery voltage decreased to _____ volts after _____ seconds.

7. What is the specification for the vehicle being serviced?

   _____

8. Describe the results of the battery load (capacity) test. Include in the results your service recommendations and the reasons for them.

_____

_____

_____

_____

_____

_____

For this section, you will use the service information to identify service procedures to test the hybrid vehicle auxiliary battery.

Assigned vehicle:

Year _____ Make _____ Model _____

VIN _____ Engine type and size _____

1. What is the specified rating for the battery?

_____

2. Record any warnings or cautions posted in the service information:

_____

_____

_____

_____

3. Record any procedures that must be performed prior to testing the battery:

_____

_____

_____

_____

4. Describe the procedure for testing the battery:

_____

_____

_____

_____

Instructor's Response

_____

_____

_____

Name _____   Date _____

USING A BATTERY CONDUCTANCE TESTER

Upon completion of this job sheet, you should be able to test a battery using a conductance tester and determine needed service action.

NATEF Correlation

This job sheet addresses the following **MLR/AST/MAST** task:

B.2.   Confirm proper battery capacity for vehicle application; perform battery capacity test and load test; determine needed action.

ASE Correlation

This job sheet is related to the ASE Electrical/Electronic Systems *Battery Diagnosis and Service*; tasks: Confirm proper battery capacity for vehicle application; perform battery capacity test, and load test; determine needed action.

Tools and Materials

- A vehicle with a 12-volt battery
- Service information
- Conductance tester
- Safety glasses
- Fender covers

Describe the vehicle being worked on:

Year _____ Make _____ Model _____

VIN _____ Engine type and size _____

Procedure   Task Completed

1. The preferred test position is at the battery posts. Where is the battery located?

    _____

    _____

2. Inspect the battery and terminals per Job Sheets 23 and 24.   ☐

3. Check the electrolyte level. If low, add distilled water to fill to 1/2 inch (12 mm) above   ☐
    the top of the plates.

4. What is the CCA rating of the battery you are working on?

    _____

5. What is the battery type? Flooded _____ Gel Cell _____

6. If you must test at a remote-post location, it should have both a positive and a negative   ☐
    post. Otherwise, you must remove the battery.

7. If needed, plug the tester into the correct electrical supply.

    Connect the test leads to the battery terminals observing polarity. What is the voltage
    displayed by the tester?

    _____

8. Follow the on-screen instructions to select "Battery Test" and select if the battery is in   ☐
    the vehicle or removed.

9. Which cables need to be connected to the battery or jump posts?

_____

10. Continue to follow the on-screen instruction; then press ENTER to start the test.   ☐

11. Record the battery test results.

_____

_____

Instructor's Response

_____

_____

_____

Name _____ Date _____

PERFORMING A BATTERY DRAIN TEST

Upon completion of this job sheet, you should be able to test parasitic loads against battery and determine needed service action.

NATEF Correlation

This job sheet addresses the following **MLR** task:

A.8.   Measure key-off battery drain (parasitic draw).

This job sheet addresses the following **AST/MAST** task:

A.8.   Diagnose the cause(s) of excessive key-off battery drain (parasitic draw); determine necessary action.

ASE NATEF

Tools and Materials

- A vehicle with a 12-volt battery
- Service information
- DVOM capable of accurately reading less than 1 amp
- Test light
- Battery terminal pullers
- Safety glasses
- Fender covers

Describe the vehicle being worked on:

Year _____ Make _____ Model _____

VIN _____ Engine type and size _____

Procedure

1. According to the service information, what is the maximum amount of key-off current draw?

   _____

   _____

2. According to the service information, how long after key off does it take for the systems to shut down before accurate test results can be obtained?

   _____

   _____

3. Make sure the ignition is in the OFF position, all accessories are turned off, and all doors are closed. Are there any obvious causes of battery drain such as lights that did not turn off?   ☐ Yes ☐ No

4. If an amp clamp is not available, use the test light and connect it between the battery terminal and clamp. Being careful not to lose the connection of the test light, separate the cable clamp from the battery terminal. Is the test light illuminated?

   ☐ Yes   ☐ No

5. With the test light still connected between the battery terminal and the clamp, connect the DMM set to read amps in series between the clamp and battery terminal. Once the ammeter is connected, disconnect the test light. What is the indicated key-off draw?

_____

_____

6. Does the ammeter reading indicate a possible problem?

☐ Yes   ☐ No

7. If the answer to question 6 is *Yes*, reconnect the test light.   ☐

8. Making sure the test light connections stay intact, reconnect the cable to the battery   ☐
terminal.

9. Using the DVOM set to the DC mV scale, measure the voltage drop across each fuse. Did a fuse indicate that current was flowing through it?   ☐ Yes   ☐ No

10. If yes, use the wiring diagram for the vehicle and determine the affected circuit.

_____

_____

Trace the circuit to locate the cause for the excessive draw. Record your results:

_____

_____

Instructor's Response

_____

_____

_____

Name _____   Date _____

HV BATTERY SERVICE AND DIAGNOSIS

Upon completion of this job sheet, you should be able to diagnose the condition of the HV battery.

Tools and Materials

- Scan tool
- Safety glasses
- Service information

Describe the vehicle being worked on:

Year _____ Make _____ Model _____

VIN _____ Engine type and size _____

Procedure

⚡ WARNING **You will be working around high voltage while doing this job sheet. Follow all warnings and cautions in the manufacturer's service procedures. Even though the service disconnect is removed, the HV battery is still "live." Scan tool diagnostics**

1. Use the scan to check for DTCs related to the HV battery and record your results:

   _____

   _____

2. Navigate to the battery control module and record the following information, if available (Note: you may need to modify the chart for the vehicle you are working on.):

PID	READING	UNIT
Hybrid Battery State of Charge		%
Hybrid Battery Current Sensor		Amps
Module Voltage (1)		Volts
Module Voltage (2)		Volts
Module Voltage (3)		Volts
Module Voltage (4)		Volts
Module Voltage (5)		Volts
Module Voltage (6)		Volts
Module Voltage (7)		Volts
Module Voltage (8)		Volts
Module Voltage (9)		Volts
Module Voltage (10)		Volts
Module Voltage (11)		Volts
Module Voltage (12)		Volts
Module Voltage (13)		Volts
Module Voltage (14)		Volts
Module Voltage (15)		Volts
Module Voltage (16)		Volts

PID	READING	UNIT
Module Voltage (17)		Volts
Module Voltage (18)		Volts
Module Voltage (19)		Volts
Module Voltage (20)		Volts
HV Battery Impedance		Ohms
BPCM Package Voltage		Volts
BPCM HV Bus Voltage		Volts

3. Are all module voltages displaying approximately the same voltage values?

☐ Yes   ☐ No

4. Remove the HV battery service disconnect plug and note if the voltage in any of the modules or blocks changes.

_____

If there was a change, why did it occur?

_____

_____

_____

Instructor's Response

_____

_____

_____

Name _____   Date _____

BATTERY CHARGING

Upon completion of this job sheet, you should be able to properly recharge a 12-volt battery.

NATEF Correlation ────────────────────────────────

This job sheet addresses the following **MLR/AST/MAST** task:

B.5.   Perform slow/fast battery charge according to manufacturer's recommendations.

ASE NATEF

Tools and Materials

- Battery charging station
- DMM
- Memory keeper
- Safety glasses
- Service information

Describe the vehicle being worked on:

Year _____ Make _____ Model _____

VIN _____ Engine type and size _____

Procedure

⚠ WARNING **Be sure to have the proper charging equipment for the battery type. Also, the battery may release gasses that are flammable and harmful to breath. Perform the battery charging in an area that is well ventilated.**

1. Use the DMM to measure the current voltage of the battery:

   _____

2. Connect the memory keeper.

3. Disconnect the negative battery cable at the battery.

4. Use the service information to determine the charge rate and time based on your findings in step 1:

   _____

5. Follow the User's Guide for the charging station and recharge the battery. At the completion of the charging procedure, reconnect the negative cable. Reset or reinitialize any modules that require this procedure.

Instructor's Response

_____

_____

_____

Name _____   Date _____

JUMP-STARTING A VEHICLE

Upon completion of this job sheet, you should be able to properly jump start a vehicle using a booster battery or an auxiliary power supply.

NATEF Correlation

This job sheet addresses the following **MLR/AST/MAST** task:

B.6.   Jump-start vehicle using jumper cables and a booster battery or an auxiliary power supply.

NATEF

Tools and Materials

- Battery jumper
- Jumper cables
- Booster battery
- DMM
- Safety glasses

Describe the vehicle being worked on:

Year _____ Make _____ Model _____

VIN _____ Engine type and size _____

Procedure

⚡ **WARNING** Be sure to follow all battery safety cautions and warnings.

1. What method of jump-starting the vehicle has been assigned to you?

   Booster battery and jumper cables _____ (Go to step 2)

   Battery jumper _____ (Go to step 13)

Procedure for Using Booster Battery and Jumper Cables

2. Check the state of charge of the booster battery.

   _____

3. Engage the parking brake and put the transmission in neutral or park.

4. Turn off the ignition switch and all accessories.

5. Attach one end of the positive jumper cable to the vehicle's battery's positive terminal.

6. Connect the other end of the positive jumper cable to the booster battery's positive terminal.

7. Attach one end of the negative jumper cable to the booster battery's negative terminal.

8. Attach the other end of the negative jumper cable to an engine ground on the disabled vehicle.

⚡ **WARNING** **Do not connect this cable end to the battery negative terminal. Doing so may create a spark that will cause the battery to explode.**

9. Attempt to start the disabled vehicle. Did the vehicle start? Yes _____
   No _____

   If No, contact your instructor.

10. Once the disabled vehicle starts, disconnect the ground-connected negative jumper cable from its engine block.

11. Disconnect the negative jumper cable from the booster battery.

12. Disconnect the positive jumper cable from the booster battery and then from the other battery.

Procedure For Using Booster Battery and Jumper Cables

13. Check the state of charge of the battery jumper. _____

14. Connect the positive cable from the battery jumper to the battery's positive terminal.

15. Connect the negative cable from the battery jumper to an engine ground on the disabled vehicle.

⚠ **WARNING** **Do not connect this cable end to the battery negative terminal. Doing so may create a spark that will cause the battery to explode.**

16. Attempt to start the disabled vehicle. Did the vehicle start? Yes _____
    No _____

    If No, contact your instructor.

17. Once the disabled vehicle starts, disconnect the ground-connected negative jumper cable from its engine block.

18. Disconnect the positive jumper cable from the battery jumper.

Instructor's Response

_____

_____

_____

DIAGNOSTIC CHART 5-1

PROBLEM AREA:	Battery
SYMPTOMS:	No-start condition after vehicle is parked overnight.
POSSIBLE CAUSES:	**1.** Glove box, trunk, interior illumination lights not turning off due to defective switches. **2.** Circuit short to ground. **3.** Circuit copper-to-copper shorts. **4.** Contaminated battery case.

DIAGNOSTIC CHART 5-2

PROBLEM AREA:	Battery
SYMPTOMS:	Battery requires jump-starting after engine shutdown.
POSSIBLE CAUSES:	**1.** Low specific gravity. **2.** Defective battery cell(s).

DIAGNOSTIC CHART 5-3

PROBLEM AREA:	Battery
SYMPTOMS:	Inadequate current to start engine under heavy load conditions and battery will not accept a charge.
POSSIBLE CAUSES:	**1.** Contaminated battery case resulting in constant current draw. **2.** Contaminated battery terminals. **3.** Undercharged battery. **4.** Defective battery. **5.** Sulfated battery. **6.** Damaged battery.

DIAGNOSTIC CHART 5-4

PROBLEM AREA:	Battery
SYMPTOMS:	Starter cranks engine slowly or fails to turn engine; low state of charge.
POSSIBLE CAUSES:	**1.** Discharged battery. **2.** Contaminated terminal clamps. **3.** Defective battery cables.

DIAGNOSTIC CHART 5-5

PROBLEM AREA:	HV battery
SYMPTOMS:	Low SOC.
	Warning lamp illumination.
POSSIBLE CAUSES:	**1.** Poor electrical connections. **2.** Faulty motor generator or circuits. **3.** Faulty HV ECU or circuits. **4.** Faulty battery ECU or circuits. **5.** Faulty amperage sensor or circuits. **6.** Faulty HV battery.

CHAPTER 6

STARTING SYSTEM DIAGNOSIS AND SERVICE

Upon completion and review of this chapter, you should be able to:

- Perform a systematic diagnosis of the starting system.
- Determine what can cause slow- and no-crank conditions.
- Perform a quick check test series to determine the problem areas in the starting system.
- Perform and accurately interpret the results of a current-draw test.
- Perform and accurately interpret the results of an insulated circuit resistance test and a ground circuit test.

- Perform the solenoid test series and accurately diagnose the solenoid.
- Perform the no-crank test and recommend needed repairs as indicated.
- Diagnose the starter motor condition by use of the free speed test.
- Remove and reinstall a starter motor.
- Disassemble, clean, inspect, repair, and reassemble a starter motor.
- Identify stop/start system inhibits.

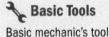

Basic Tools

Basic mechanic's tool set

Service information

Terms To Know

Crocus cloth
Current-draw test
Free speed test
Hydrostatic lock

No crank
Quick test
Relative compression testing

Slow cranking
Solenoid circuit resistance test

INTRODUCTION

Perhaps one of the most aggravating experiences to a car owner is to have an engine that will not start. However, not all starting problems are caused by the starting system. The ignition and fuel systems must also be in proper condition to perform their functions. In addition, the internal condition of the engine must be such that compression, correct valve timing, and free rotation are all obtained.

Classroom Manual
Chapter 6, page 140

AUTHOR'S NOTE It must be remembered that the starting system operates in conjunction with the battery and charging systems. You must keep the interdependency of these three systems in mind and that all three systems must perform properly. The starting, battery, and charging systems need to be diagnosed as a complete system.

In this chapter, you will perform the tests required to make an intelligent decision concerning the condition and operation of the components of the starting system. You will also remove, disassemble, reassemble, and reinstall the starter motor.

STARTING SYSTEM SERVICE CAUTIONS

Before beginning any service on the starter system, some precautions must be observed. Along with the precautions outlined in Chapter 1, when servicing the starter system several other precautions should be followed:

1. Refer to the manufacturer's manuals for correct procedures for disconnecting a battery. Some vehicles with on-board computers must be supplied with an auxiliary power source.
2. Disconnect the battery ground cable before disconnecting any of the starter circuit's wires or removing the starter motor.
3. Be sure the vehicle is properly positioned on the hoist or on safety jack stands.
4. Before performing any cranking test, be sure the vehicle is in PARK or NEUTRAL and the parking brakes are applied.
5. Follow the manufacturer's directions for disabling the ignition system.
6. Be sure the test leads are clear of any moving engine components.
7. Never clean any electrical components in solvent or gasoline. Clean with compressed air, denatured alcohol, or wipe with clean rags only.

STARTING SYSTEM TROUBLESHOOTING

No crank means that when the ignition switch is placed in the START position, the starter does not turn the engine. This may be accompanied by a buzzing noise, which indicates the starter motor drive has engaged the ring gear, but the engine does not rotate. There may also be no clicking sounds from the starter motor or solenoid.

Customer complaints concerning the starting system generally fall into four categories: **no crank, slow cranking**, starter spins but does not turn engine, and excessive noise. As with any electrical system complaint, a systematic approach to diagnosing the starting system will make the task easier. First, the battery must be in good condition and fully charged. Perform a complete battery test series to confirm the battery's condition. Many starting system complaints are actually attributable to battery problems. If the starting system tests are performed with a weak battery, the results can be misleading. The conclusions may be erroneous and costly.

Before performing any tests on the starting system, first begin with a visual inspection of the circuit. Repair or replace any corroded or loose connections, frayed wires, or any other trouble sources. The battery terminals must be clean and the starter motor must be properly grounded.

The diagnostic chart shows a logical sequence to follow whenever a starting system complaint is made (**Figure 6-1**). What tests are performed is determined by whether the starter will crank the engine.

Slow cranking means the starter drive engages the ring gear, but because the engine turns slowly, it cannot start. Some manufacturers provide specifications for engine cranking speed.

If the customer complains of a no-crank situation, attempt to rotate the engine by the crankshaft pulley bolt. Rotate the crankshaft two full rotations in a clockwise direction, using a large socket wrench. If the engine does not rotate, it may be seized due to its being operated with no oil, broken engine components, or **hydrostatic lock**. Since liquid cannot be compressed, if there is a leak that allows antifreeze from the cooling system to enter the cylinder, the cylinder can fill to such a level that the piston is unable to move upward. This condition is referred to as hydrostatic lock.

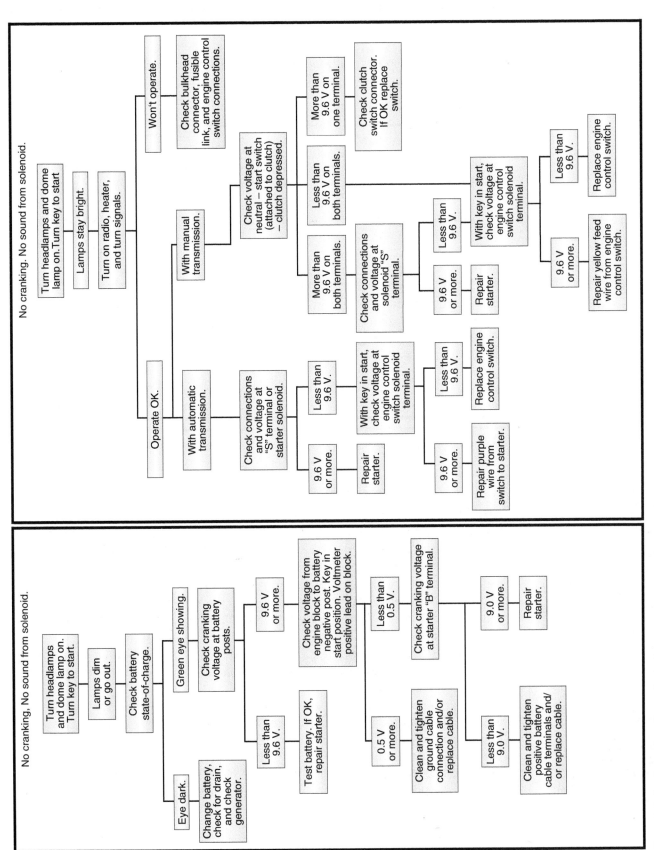

Figure 6-1 Starting system diagnostic chart.

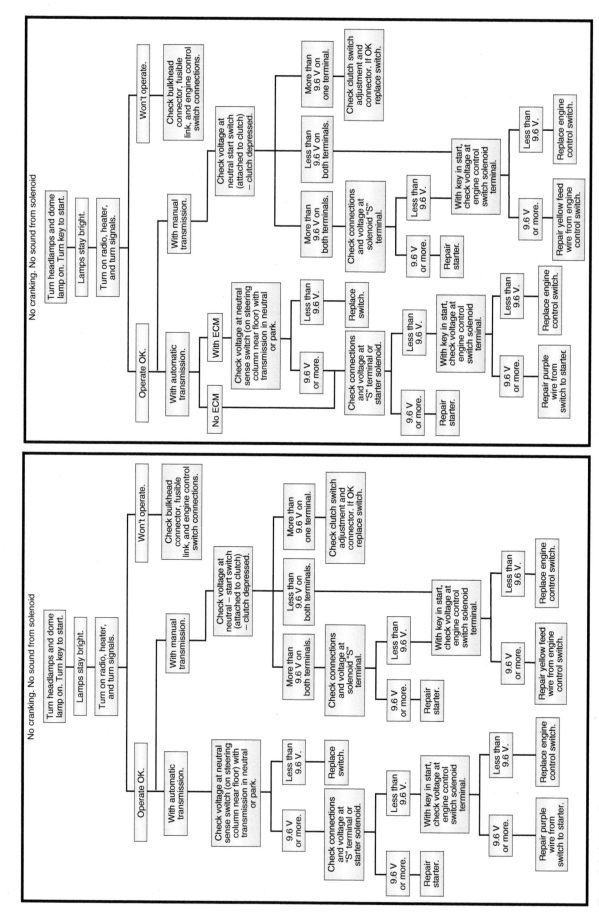

Figure 6-1 (Continued)

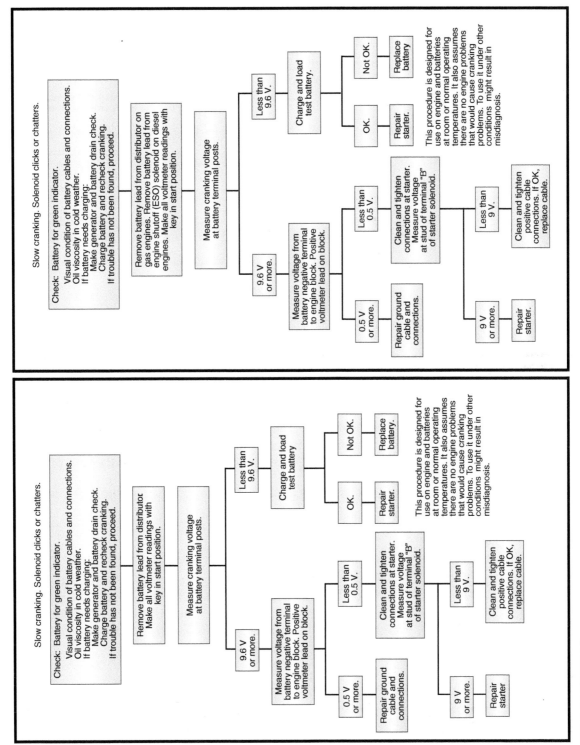

Figure 6-1 (Continued)

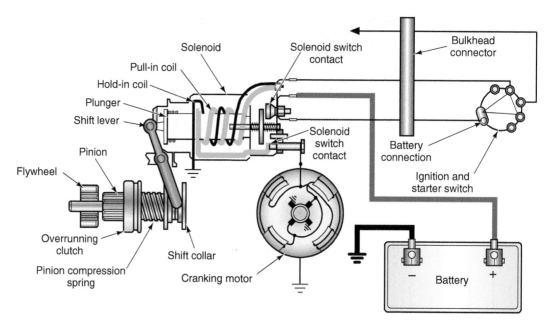

Figure 6-2 Excessive wear, loose electrical connections, or excessive voltage drop in any of these areas can cause a slow- or no-crank condition.

Several potential trouble spots in the circuit can cause a slow-crank or no-crank complaint (**Figure 6-2**). Excessive voltage drops in these areas will cause the starter motor to operate slower than required to start the engine. The speed that the starter motor rotates the engine is important to engine starting. If the speed is too slow, compression is lost and the air–fuel mixture draw is impeded. Most manufacturers require a speed of approximately 250 rpm during engine cranking.

If the starter spins but the engine does not rotate, the most likely cause is a faulty starter drive. If the starter drive is at fault, the starter motor will have to be removed to install a new drive mechanism. Before faulting the starter drive, also check the starter ring gear teeth for wear or breakage, and for incorrect gear mesh of the ring gear and starter motor pinion gear.

Most noises can be traced to the starter drive mechanism. The starter drive can often be replaced as a separate component of the starter.

CUSTOMER CARE Always treat the customer's car with respect. Place fender covers over the fenders when performing tasks under the hood. Do not lay tools on the vehicle's finish. Clean your hands before entering the vehicle. Place a seat protector over the seats and paper mats on the floor boards. Give the car back to the customer at least as clean as when you received it.

TESTING THE STARTING SYSTEM

As with the battery testing series, the tests for the starting system are performed with a starting/charging system tester (**Figure 6-3**). The starter and battery performances are so closely related that it is important for a full battery test series to be done before trying to test the starter system. If the battery fails the load test and is fully charged, it must be replaced before doing any other tests.

Figure 6-3 Starter tests can be performed with any tester capable of measuring high current, such as the GR8.

Quick Testing

The **quick test** will isolate the problem area and determine whether the starter motor, solenoid, or control circuit is at fault. If the starter does not turn the engine at all, and the engine is in good mechanical condition, the quick test can be performed to locate the problem area. To perform this test, make sure the transmission is in neutral and set the parking brake. Turn on the headlights. Next, turn the ignition switch to the START position while observing the headlights.

Three things can happen to the headlights during this test:

1. They will go out.
2. They will dim.
3. They will remain at the same brightness level.

If the lights go out completely, the most likely cause is a poor connection at one of the battery terminals. Check the battery cables for tight and clean connections. It will be necessary to remove the cable from the terminal and clean the cable clamp and battery terminals of all corrosion.

> **SERVICE TIP** Check the fusible link if the engine does not crank and the headlights do not come on.

If the headlights dim when the ignition switch is turned to the START position, the battery may be discharged. Check the battery condition. If it is good, then there may be a mechanical condition in the engine that is preventing it from rotating. If the engine rotates when turning it with a socket wrench on the pulley nut, the starter motor may have internal damage. A bent starter armature, worn bearings, thrown armature windings, loose

Figure 6-4 Bypassing the solenoid to determine if the solenoid or the control circuit is faulty.

pole shoe screws, or any other worn component in the starter motor that will allow the armature to drag can cause a high-current demand.

CUSTOMER CARE If the starter windings are thrown, this indicates several different problems. The most common is that the driver is keeping the ignition switch in the START position too long after the engine has started. Other causes include the driver opening the throttle plates too wide while starting the engine, which results in excessive armature speeds when the engine does start. Also, the windings can be thrown because of excessive heat buildup in the motor. The motor is designed to operate for very short periods of time. If it is operated for longer than 15 seconds, heat begins to build up at a very fast rate. If the engine does not start after a 15-second crank, the starter motor needs to cool for up to 2 minutes before the next attempt to start the engine.

If the lights stay brightly lit and the starter makes no sound (listen for a deep clicking noise), there is an open in the circuit. The fault is in either the solenoid or the control circuit. To test the solenoid, bypass the solenoid by bridging the BAT and S terminals (**Figure 6-4**).

⚠ WARNING **A starter can draw up to 400 amps (A). The tool used to jump the terminals must be able to carry this high current and must have an insulated handle.**

If the starter rotates with the solenoid bypassed, the control circuit is at fault. If the starter does not rotate and the lights do not dim, the solenoid is at fault. (Also, listen for the starter drive engaging.) If the starter rotates slowly and the headlights dim, there is excessive current draw and the system will have to be tested further.

DMM Testing

Another quick test of the starting system (and battery) is to use the Min/Max function of the DMM. First, measure the static voltage across the battery terminals. Activate the Min/Max feature on the DMM and start the engine. The DMM will record the minimum and maximum voltage values. The minimum voltage represents cranking battery voltage. The maximum voltage represents the output voltage from the charging system after the engine starts. If the cranking voltage is below specifications, the battery needs to be tested before continuing diagnostics of the starting system.

Current-Draw Test

The **current-draw test** measures the amount of current that the starter draws when actuated. It determines the electrical and mechanical condition of the starting system. If the starter motor cranks the engine, the technician should perform the current-draw test. The following procedure uses a typical starting/charging system tester and is similar to the procedure for other starting and charging system testers:

1. Connect the large red and black test leads across the battery, observing polarity.
2. Follow the manufacturer's instructions to zero the ammeter.
3. Connect the inductive amp probe around the battery ground cable. If more than one ground cable is used, clamp the probe around all of them (**Figure 6-5**).
4. Make sure all loads are turned off (lights, radio, etc.). Zero the ammeter.
5. Select STARTING TEST.
6. Follow the service information procedure to disable the ignition system to prevent the vehicle from starting. Some systems may require that the fuel pump or electronic fuel injection relay be removed or the ignition module be disconnected in order to prevent the engine from starting.
7. Crank the engine for 10–15 seconds and note the voltmeter and ammeter readings.

After recording the readings from the current-draw test, compare them with the manufacturer's specifications. If specifications are not available, correctly functioning systems, as a rule, will crank at 9.6 volts or higher. Current draw is dependent on engine size. Most V-8 engines will have a current draw of about 200 amps, six-cylinder engines about 150 amps, and four-cylinder engines about 125 amps.

If the readings obtained from the current-draw test are out of specifications, then additional testing will be required to isolate the problem. If the readings were on the borderline of the specifications or there is an intermittent problem, then detailed testing for bad components will pinpoint potential failures.

 Special Tools

Fender covers
Starting/charging
   system tester
Jumper wires

 Caution

Do not operate the starter motor for longer than 15 seconds. Allow the motor to cool between cranking attempts.

The specification for current draw is the maximum allowable; the specification for cranking voltage is the minimum allowable.

Figure 6-5 Test lead connections to perform the starter current-draw test.

The following provides an interpretation of the current-draw test:

- Voltage is 9.6 volts or more and amperage is higher than specified—indicates impedance to rotation of the starter motor. This includes worn bushings, a mechanical blockage, and excessive advanced ignition timing. Shorted starter motor windings can cause high-current draw.
- Voltage is 9.6 volts or more and amperage is lower than specified—indicates high electrical circuit resistance.
- Voltage is less than 9.6 volts and current is higher than specified—indicates the amperage is draining the battery. Perform the battery test series to confirm the battery is good. If the battery passes this test series, then see the first point above.
- Voltage is less than 9.6 volts and current is lower than specified—indicates a faulty battery.

Because the voltage reading obtained from the current-draw test was taken at the battery, this reading may not be an exact representation of the actual voltage delivered to the starter. Voltage losses due to bad cables, connections, and relays (or solenoids) may diminish the amount of voltage to the starter. These should be tested before removing the starter from the vehicle.

Photo Sequence 11 illustrates the use of the GR8 tester as an example of the use of computer testing equipment. Different models of GR8 are available with different options. Always refer to the instruction manual for the type of tester you are using.

Special Tools

Fender covers
GR8 system tester

Special Tools

Lab scope
Current clamp

Lab Scope Testing

A lab scope or graphing multimeter is a good method of testing a starting system. Prior to disabling the ignition system to prevent the engine from starting, use the ammeter probe to view the starter signal. Most current probes convert amps to millivolts to be read on a lab scope. The amp probe is connected over either the positive or negative battery cable. Be sure to orient the arrow on the current probe with the direction of current flow. Set the scope settings of 0.5 volt (V)/division and 100 ms/division to start with and adjust later as needed. Using the meter's record function will make interpretation of the pattern results easier.

Initial engine rotation requires high starter current. Once the engine is rotating, lower current is required. A typical pattern will indicate this higher current draw (rush-in current) at initial cranking (**Figure 6-6**). During this time, both the pull-in and hold-in windings of

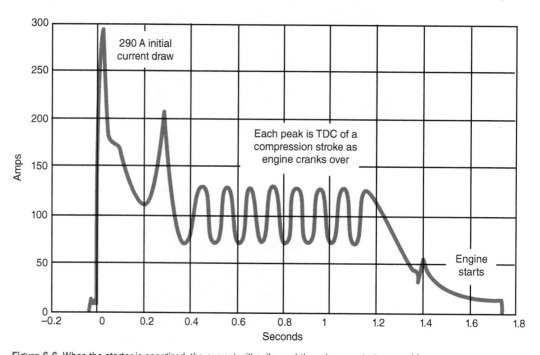

Figure 6-6 When the starter is energized, the current will spike and then drop as starter speed increases.

the starter solenoid are energized. Shortly after the high-current spike, the pull-in windings are de-energized and current will drop to a lower value. After the engine starts, the current draw drops to zero as the ignition switch is placed in the RUN position. Excessively high current indicates a shorted winding or high mechanical resistance within the starter motor or engine. Low current indicates high electrical resistance in the circuit.

Relative Compression Testing

Although usually a test procedure for determining internal engine condition, compression testing is also used to identify hard starting conditions. Good compression and a properly functioning starting system are critical for fast starts and good driveability. Due to the increased difficulty of accessing spark plugs, normal compression testing by removing the spark plugs and installing a compression gauge is being replaced with a nonintrusive diagnostic routine referred to as **relative compression testing**. Relative compression testing uses current draw during cranking to determine if a cylinder has lower compression relative to the other cylinders. This test also provides a good indication of the starting system.

PHOTO SEQUENCE 11
Using the Gr8 Starting System Test

P11-1 In the Main Menu select the System Test icon.

P11-2 Follow any on-screen instruction for connecting the test leads to the vehicle.

P11-3 When prompted, start the vehicle's engine.

P11-4 The complete results of the test will be displayed in a series of screens. Use the UP or DOWN ARROW keys to scroll to each screen.

P11-5 Screen 1 of 2 shows the average cranking voltage, average cranking current (if amp clamp is used), and cranking time in seconds.

P11-6 Screen 2 of 2 shows the system performance. The Y-axis represents cranking voltage and the X-axis represents time.

P11-7 The result can be printed by using the PRINT soft key.

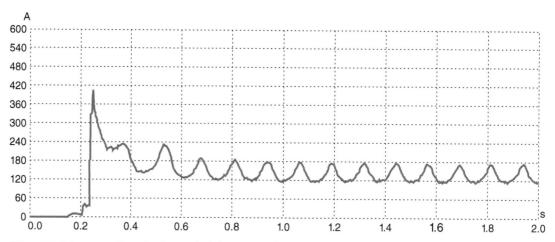

Figure 6-7 Lab scope pattern showing good relative compression between cylinders.

Every time a piston is forced up during the compression stroke, it takes work. Since current represents work, the lab scope provides a method of observing current on a small timeline scale; with the lab scope you can visualize the effect that each cylinder's compression stroke has on starter current while the engine is cranking.

To perform this test, clamp the lab scope's amp probe around the starter cable between the battery and the starter motor. With the ignition system disabled, crank the engine while monitoring the lab scope pattern. As discussed earlier, the initial current should go high and peak, followed by a ripple pattern (**Figure 6-7**). As the pistons approach top dead center compression stroke, the resistance increases and causes the current trace to also increase. This should result in a trace with even peaks across the screen. If compression of a cylinder is lower than that of other cylinders, then the current peak of that cylinder will also be lower (**Figure 6-8**).

By using an external trigger placed around a spark plug, it is possible to identify which cylinder has the low compression. Once the trigger is installed onto the spark plug, adjust the scope to trigger off this signal. For example, if the lab scope is set to trigger off of the rpm pickup that was placed on the number one plug wire, then every time the screen is updated, the number one cylinder is the first one on the screen. By following the engine's firing order, each cylinder number can be determined.

Most engine analyzers have a variation of this test that will test the relative current comparison of each cylinder during cranking. These analyzers provide a readout of the results and may suggest the faulty cylinder.

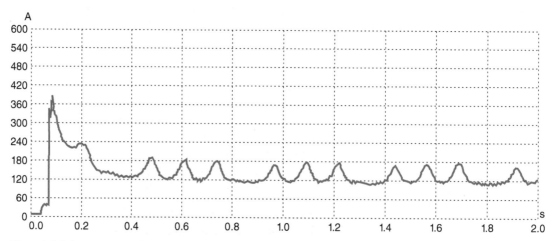

Figure 6-8 Lab scope pattern indicating a cylinder with lower relative compression.

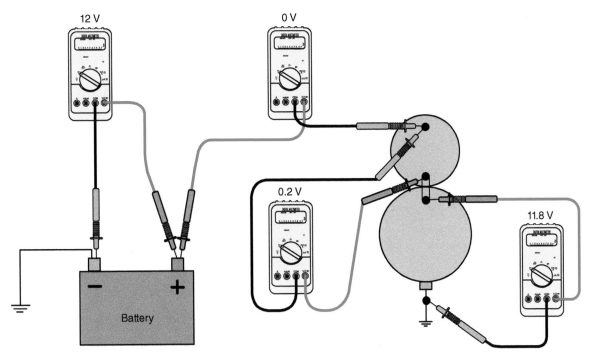

Figure 6-9 Voltage drop testing to identify sources of excessive resistance.

Insulated Circuit Resistance Test

An electrical resistance will have a different pressure or voltage on each side of the resistance. Voltage drops when current flows through resistance. Most manufacturers design their starting systems to have very little resistance to the flow of current to the starter motor. Most have less than 0.2 volt dropped on each side of the circuit. This means the voltage across the starter input terminal to the starter ground should be within 0.4 volt of battery voltage (**Figure 6-9**).

Voltage drops are measured by connecting a digital multimeter (DMM) set to the lowest voltmeter scale in parallel with the circuit section being tested. In order to obtain a voltage drop reading, a load on the circuit must be applied. The following is a typical test procedure:

>  **SERVICE TIP** As a general rule, allow up to 0.2 volt per cable and 0.1 volt per connection to be dropped. Switches can be as high as 0.3 volt. Use the wiring diagram for the vehicle to determine the number of conductors and connections used in the circuit. This will provide a specification for you if no other specifications are available.

1. Connect the test leads as shown in **Figure 6-10**, depending on the type of system being tested. Usually the positive lead of the voltmeter is connected to the positive battery post and the negative lead is connected to the starter battery (BATT) terminal.
2. Disable the ignition system as discussed in current-draw testing.
3. Crank the engine for 15 seconds and observe the voltmeter scale.

This tests for voltage drop in the entire circuit, so if voltage drop is excessive, the cause of the drop must be located. To locate the cause of the excessive voltage drop, move the voltmeter lead on the starter toward the battery. Check each connection while moving

Classroom Manual
Chapter 6, page 151

Special Tools
DMM fender covers
Starting/charging
system tester

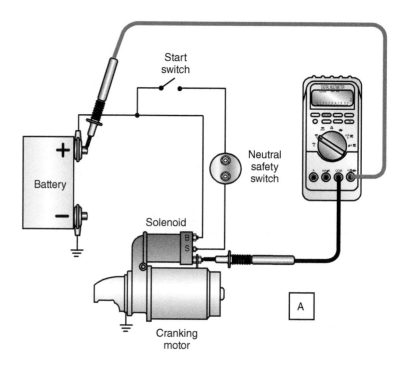

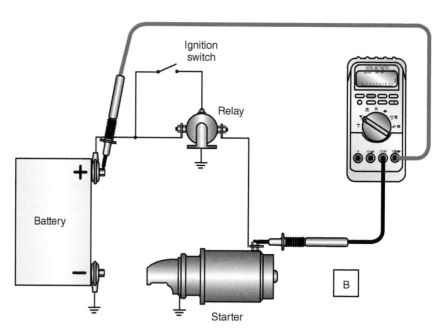

Figure 6-10 (A) Test lead connections for a starter-mounted solenoid. (B) Test lead connections for relay-controlled systems.

toward the battery. With each move of the test lead, crank the engine while observing the voltmeter reading. Continue to test each connection until a noticeable decrease in voltage drop is detected. The cause of the excessive voltage drop will be located between that point and the preceding point.

Ground Circuit Test

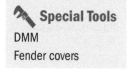

Special Tools

DMM
Fender covers

A ground circuit test is performed to measure the voltage drop in the ground side of the circuit (**Figure 6-11**). If the starter motor connection to ground is broken or loose, the circuit would be opened. This could cause an intermediate starter system problem or a

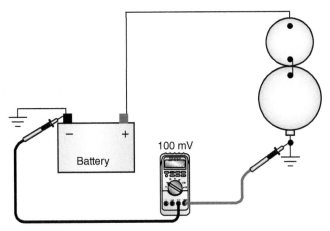

Figure 6-11 Voltage drop testing of the ground-side circuit of the starting system.

starter motor that will not crank the engine. To perform the ground circuit test, connect the voltmeter positive lead to the starter motor case and the negative test lead to the ground battery terminal. Make sure any paint is removed from the area where the lead is connected to the case. Crank the engine while observing the voltmeter.

Less than 0.2 volt indicates the ground circuit is good. If more than 0.2 volt is observed, then there is a poor ground circuit connection. A poor ground circuit connection could be the result of loose starter mounting bolts, paint on the starter motor case, or a bad battery ground terminal post connection. Also, check the ground cable for high resistance or for being undersized.

Solenoid Circuit Resistance Testing

High resistance in the solenoid will reduce the current flow through the solenoid windings and cause the solenoid to function improperly. If the solenoid has high resistance in the windings, it may result in the contacts burning and causing excessive resistance to the starter motor.

The **solenoid circuit resistance test** determines the electrical condition of the solenoid and the control circuit. To perform the solenoid circuit resistance test, first disable the ignition system. Using a voltmeter or starting/charging system tester, connect the positive voltmeter lead to the BAT terminal of the solenoid. Connect the negative test lead to the field coil terminal (M terminal). Crank the engine while observing the voltmeter reading. If the voltmeter reading indicates a voltage drop of greater than 0.2 volt, then the solenoid is defective. If this test proves the starter solenoid is good, then the solenoid switch circuit should be tested. Follow these steps:

1. Disable the ignition system.
2. Connect the voltmeter leads to both solenoid switch terminals, observing polarity.
3. Crank the engine and observe the voltmeter reading.

The total voltage drop should be less than 0.5 volt. If the indicated voltage drop is in excess of 0.5 volt, move the voltmeter leads up the circuit and test each component. The voltage drop across each component should be less than 0.1 volt.

Continue to move the voltmeter leads to test for voltage drop through the wires, starter relay, neutral safety switch, and ignition switch.

Open Circuit Test

In order to perform voltage drop tests, a load must be placed on the circuit. If there is a no-crank complaint, a voltage drop test may not be able to be performed. This is because

Classroom Manual
Chapter 6, page 151

Special Tools

DMM fender covers
Starting/charging
  system tester

Classroom Manual
Chapter 6,
pages 156, 158

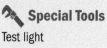
most no-crank problems are the result of opens in the circuit. The easiest way to diagnose this problem is with the use of a test light. On a system that uses a starter motor-mounted solenoid, the M terminal is the end of the circuit (**Figure 6-12**). Connecting the positive lead of the test light to the M terminal and the negative lead to a good ground, the light should be on when the ignition switch is located in the START position (**Figure 6-13**). If the light comes on, then the complete insulated circuit (including the ignition switch, wires, neutral safety switch, solenoid, and all connections) operates properly. The open is in either the starter motor or the ground circuit.

If the test light comes on very dim, then there is very high resistance in the circuit. By working the test light through the circuit and back to the battery, the reason for the high resistance should be found. If the test light does not come on, follow the same procedure of backtracking the circuit toward the battery until the open is found. Also, check for voltage at the B+ terminal (**Figure 6-14**). Connecting the test light as shown should light the

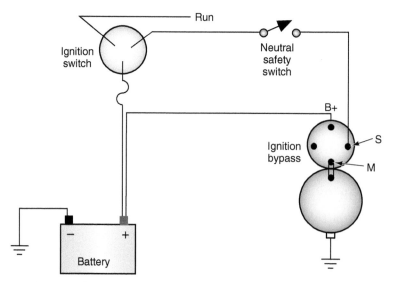

Figure 6-12 Starting system using a solenoid shift.

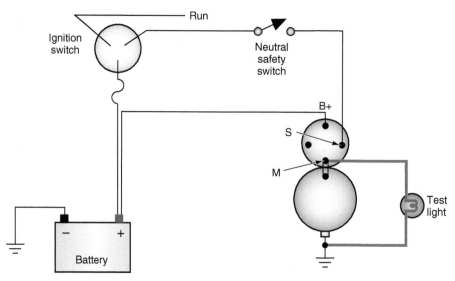

Figure 6-13 Test light connections for testing the solenoid and control circuit.

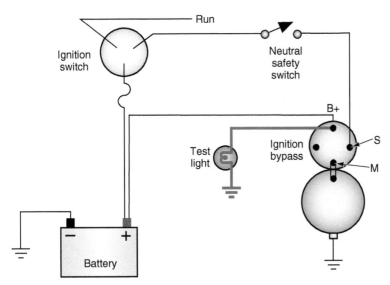

Figure 6-14 Test light connections for checking voltage at the BAT terminal.

bulb with the key off. The light should stay on when the ignition switch is turned to the START position. If the light goes out in the START position, repair the cable or end connections.

> ⚙ SERVICE TIP A voltmeter can be used in place of the test light. At the M terminal there should be more than 9.6 volts present when the ignition switch is in the START position.

Once the open has been found, it can be verified by using jumper wires to jump across the defective component or connection. If jumping across the solenoid, for example, and the starter spins, then there is an open in the solenoid. The same procedure can be used to jump across the ignition switch, neutral safety switch, or open wires.

If the test light did come on when connected to the M terminal, then make a simple test of the ground circuit. This is done by connecting the ground lead of the test light onto the starter body and the positive lead to the M terminal of the solenoid. The light should come on bright with the ignition switch in the START position. If the ground circuit is good, then the starter is suspect and should be bench tested.

The Ford starting system is tested in the same manner, except there is an additional battery cable to test.

> ⚙ SERVICE TIP Most starter safety switches are adjustable. Sometimes a no-start problem can be corrected by checking and adjusting (or replacing) the starter safety switch.

Free Speed Test

In the event that the starter has failed the previous tests, or a new starter is going to be installed, a **free speed test** should be performed. The free speed test determines the free rotational speed of the armature. Some manufacturers recommend this test procedure

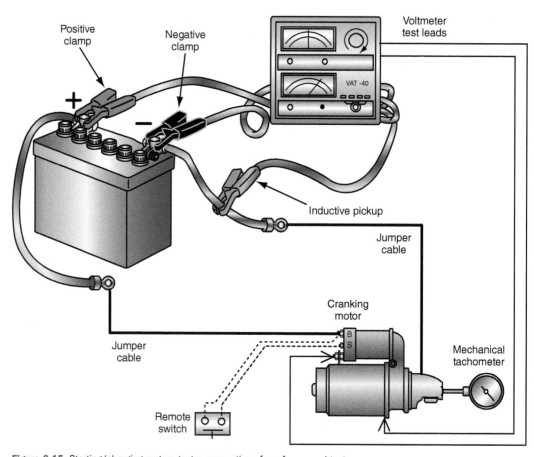

Figure 6-15 Starting/charging system tester connections for a free speed test.

Special Tools

Special starter tachometer

Jumper cables

Starting/charging system tester

Remote starter switch

Soft jaw vise

The free speed test is also referred to as the no-load test and the free spin test.

⚡ Caution

Failure to load the battery to 10 volts as instructed in step 9 can result in the armature windings being thrown. Because there is no load on the starter, the rpm's will be excessive if more than 10 volts are used.

over the current-draw test. The starter must be removed from the vehicle, as described in the next section. With the starter removed from the vehicle, perform the test as follows (**Figure 6-15**):

1. Place the starter motor into a secure vise. Be careful not to overtighten the vise against the frame assembly. It is possible to crack the frame and/or the pole shoes.
2. Attach an rpm indicator to the armature shaft at the drive housing end.
3. Connect a remote starter switch between the BAT and S terminals of the solenoid.
4. Connect the jumper cables, as shown in Figure 6-15.
5. Connect the large red and black test leads of the tester across the battery, observing polarity.
6. Follow the manufacturer's instructions for zeroing the ammeter.
7. Connect the amp inductive probe around the jumper cable from the battery negative terminal to the starter frame.
8. Place the test selector to the STARTING position.
9. Load the battery by rotating the load control knob until a voltage reading of 10 volts is obtained.
10. Close the remote starter switch while reading the ammeter, voltmeter, and tachometer scales.

Compare the test results with the manufacturer's specifications. General specifications will be about 6,000 to 12,000 rpm with a current draw of 60 to 85 amps. Voltage should remain at 10 volts. If the test results are within specifications, the starter motor is ready to be installed into the vehicle.

If the current draw is excessive and rpm slower than specifications, there is excessive resistance to rotation. This could be caused by:

1. Worn bushings or bearings.
2. Shorted armature.
3. Grounded armature.
4. Shorted field windings.
5. Bent armature.

If there is no current draw, and the starter does not rotate, this could be caused by one of the following:

1. Open field windings.
2. Open armature coils.
3. Broken brush or brush spring.

Low armature speed with low-current draw indicates excessive resistance. There may be a poor connection between the commutator and the brushes. Also, any connection in the starter and to the starter may be faulty.

If the armature speed and current draw readings are high, check for a shorted field winding.

STARTER MOTOR REMOVAL

If the tests indicate the starter motor must be removed, the first step is to disconnect the battery from the system. Remove the negative battery cable. It is a good practice to wrap the cable clamp with tape or enclose it in a rubber hose to prevent accidental contact with the battery terminal.

It may be necessary to place the vehicle on a lift to gain access to the starter motor. Before lifting the vehicle, disconnect all wires, fasteners, and so forth, which can be reached from the top of the engine compartment.

>  **Special Tools**
> Fender covers
> Battery cable puller

⚠ WARNING **Check for proper pad-to-frame contact after the vehicle is a few inches above the ground. Shake the vehicle. If there are any unusual noises or movement of the vehicle, lower it and reset the pads.**

Disconnect the wires leading to the solenoid terminals. To prevent confusion, it is a good practice to use a piece of tape to identify the different wires.

On some vehicles, it may be necessary to disconnect the exhaust system to be able to remove the starter motor. Spray the exhaust system fasteners with a penetrating oil to assist in removal. Loosen the starter mounting bolts and remove all but one. Support the starter motor; remove the remaining bolt. Then remove the starter motor.

⚠ WARNING **The starter motor is heavy; make sure it is secured before removing the last bolt.**

> ⚠ **Caution**
> Special care must be taken when handling the permanent magnet gear reduction starter. The permanent magnets are brittle and are easily destroyed if the starter is dropped or struck by another object.

Reverse the procedure to install the starter motor. Be sure all electrical connections are tight. If you are installing a new or remanufactured starter, remove any paint that may prevent a good ground connection. Be careful not to drop the starter. Make sure it is properly supported.

Some General Motors starters use shims between the starter motor and the mounting pad (**Figure 6-16**). To check this clearance, insert a flat-blade screwdriver into the access slot on the side of the drive housing. Pry the drive pinion gear into the engaged position.

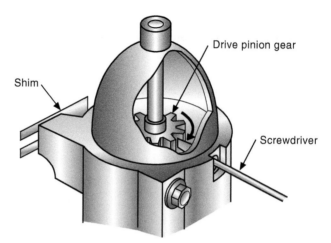

One shim will increase clearance by approximately 0.005".
More than one shim may be required.

Figure 6-16 Shimming the starter to obtain proper pinion-to-ring gear clearance.

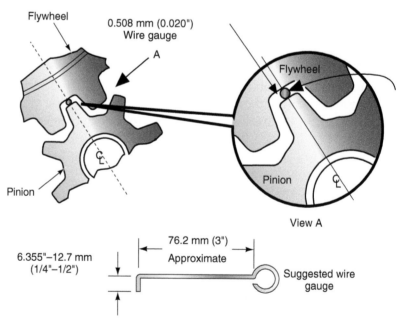

Figure 6-17 Checking the clearance between the pinion gear and ring gear.

The shims are used to provide proper pinion-turning clearance.

Use a piece of wire that is 0.020 inch (0.5 mm) in diameter to check the clearance between the gears (**Figure 6-17**).

If the clearance between the two gears is excessive, the starter will produce a high-pitched whine while the engine is being cranked. If the clearance is too small, the starter will make a high-pitched whine after the engine starts and the ignition switch is returned to the RUN position.

STARTER MOTOR DISASSEMBLY

If it is determined that the starter is the defective part, it can be disassembled, bench tested, and rebuilt. To reduce vehicle downtime to a minimum, many repair facilities do not rebuild starters. They replace them instead. However, many shops will replace the starter drive mechanism, which may require several of the following disassembling steps. The decision to rebuild or replace the starter motor is based on several factors:

Classroom Manual
Chapter 6,
pages 160, 163

⚙ SERVICE TIP The major cause of drive housing breakage is due to too small of a clearance between the pinion and ring gears. It is always better to have a little more clearance than too small of a clearance.

1. What is best for the customer.
2. Shop policies.
3. Cost.
4. Time.
5. Type of starter.

If the starter is to be rebuilt, the technician should study the manufacturer's service information to become familiar with the disassembly procedures for the particular starter. Always refer to the specific manufacturer's service information for the starter motor you are working on. The disassembled view of a Delco Remy starter is shown in **Figure 6-18**.

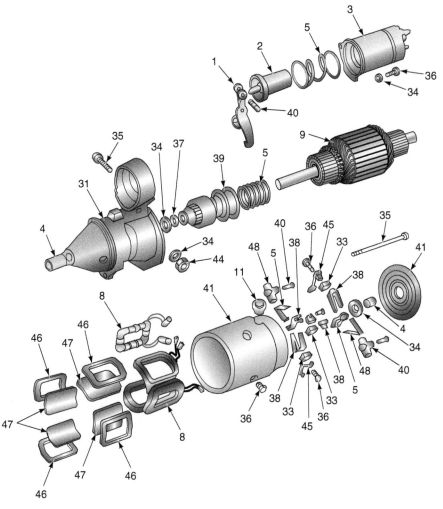

1. Lever
2. Plunger
3. Solenoid
4. Bushing
5. Spring
8. Coil
9. Armature
11. Grommet
31. Housing
32. Drive
33. Brushes
34. Washer
35. Bolt
36. Screw
37. Ring
38. Holder
39. Collar
40. Pin
41. Frame
44. Nut
45. Lead
46. Insulator
47. Shoe
48. Plate

Figure 6-18 Delco Remy 10MT starter.

⚠ **Caution**

Do not clean the starter motor components in solvent or gasoline. The residue left can ignite and destroy the starter. Wipe with clean rags, or use denatured alcohol to clean the starter components.

Photo Sequence 12 illustrates a typical procedure for disassembling a permanent magnet, gear reduction starter. Again, be sure to refer to the specific manufacturer's service information for the starter motor you are working on.

The starter motor can be cleaned and inspected when it is disassembled. Inspect the end frame and drive housing for cracks or broken ends. Check the frame assembly for loose pole shoes and broken or frayed wires. Inspect the drive gear for worn teeth and proper overrunning clutch operation. The commutator should be free of flat spots and should not be excessively burned. Check the brushes for wear. Replace them if worn past manufacturer's specifications.

STARTER MOTOR COMPONENT TESTS

🔧 **Special Tools**

Growler
Hacksaw blade
Continuity tester
Ohmmeter
Crocus cloth

Classroom Manual
Chapter 6, page 145

With the starter motor disassembled and the components cleaned, you are ready to perform tests that will isolate the reason for the failure. The armature and field coils are checked for shorts and opens. In most cases, the whole starter motor assembly is replaced if the armature or field coils are bad.

Field Coil Testing

The field coil and frame assembly should be tested for opens and shorts to ground. In most cases, if one of these conditions is found, the starter is considered unrebuildable in the field and will need to be replaced with a new unit.

Field coils can be wired in a number of different ways. The most effective testing of the coils for opens and shorts is determined by how the coils are wired. There are two things to do to determine the best way to check the field coils: refer to the service information for specific instructions and/or refer to the wiring diagram for the starting circuit. By looking at the wiring diagram, you will be able to tell where the coils get their power and where they ground. Knowing these things is critical to testing the coils. The following procedure is valid for many, but not all, vehicles.

Using an ohmmeter, place one lead on the starter motor input terminal. Connect the other lead to the insulated brushes (**Figure 6-19**). The ohmmeter should indicate zero resistance. If there is resistance in the field coil, replace the coil and/or the frame assembly.

Place one lead of the ohmmeter on the starter motor input terminal and the other lead on the starter frame (**Figure 6-20**). An infinite reading should be obtained. If the ohmmeter indicates continuity, there is a short to ground in the field coil.

Armature Short Test

Classroom Manual
Chapter 6, page 143

A growler produces a very strong magnetic field that is capable of inducing a current flow and magnetism in a conductor. It is used to test the armature for shorts and grounds (**Figure 6-21**).

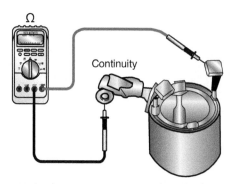

6-19 Testing the field coils for opens.

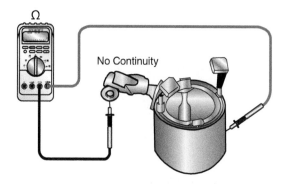

6-20 Testing the field coils for shorts to ground.

6-21 A growler is used to test the armature for shorts.

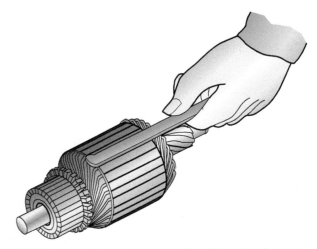

6-22 The growler generates a magnetic field. If there is a short, the hacksaw blade will vibrate over the area of the short.

To test the armature for shorts, place the armature in the growler and hold a thin steel blade parallel to the core (**Figure 6-22**). Slowly rotate the armature and observe the steel blade. If the blade begins to vibrate or pull toward the core, the armature is shorted and in need of replacement.

PHOTO SEQUENCE 12
Typical Procedure for Disassembly of Gear Reduction Starter

P12-1 Disconnect the lead to the field coil and remove the solenoid retainer screws.

P12-2 Remove the solenoid housing while working the plunger off of the drive lever.

P12-3 Remove the frame through bolts.

P12-4 Separate the drive end frame from the body. Remove the seal also.

P12-5 Remove the O-ring from the end of the drive gear and then remove the retainer ring and C-clip.

P12-6 Remove the drive from the output shaft.

P12-7 Separate the output shaft and stationary gear assembly from the armature. Be sure to locate and retain the thrust ball located in a seat in the output shaft.

P12-8 Remove the lock ring from the output shaft and remove the stationary gear from the shaft.

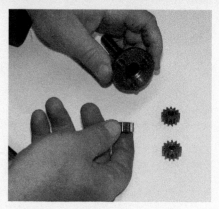

P12-9 Remove the planetary gears from the output shaft.

P12-10 Remove the fasteners that attach the end plate to the brush plate.

P12-11 Remove the armature and brush assembly from the body.

P12-12 Separate the brush assembly from the armature.

Armature Ground Test

With the armature placed in the growler, use a continuity tester or ohmmeter to check for continuity between the armature core and any bar of the commutator (**Figure 6-23**). If there is continuity, then the armature is grounded and in need of replacement.

Commutator Tests

If a growler is not available, the armature commutator can be tested for opens and grounds using an ohmmeter. The commutator should be cleaned with **crocus cloth**. To check for continuity, place the ohmmeter on the lowest scale. Connect the test leads to any two commutator sectors (**Figure 6-24**). There should be 0 ohm (Ω) of resistance. The armature will have to be replaced if there is resistance.

To test the armature for short to ground, use an ohmmeter and connect one of the test leads to the armature shaft. Connect the other lead to the commutator segments (**Figure 6-25**). Check each sector. There should be no continuity to ground. The armature will have to be replaced if there is continuity.

Classroom Manual
Chapter 6, page 143

Crocus cloth is used to polish metals. While polishing, it removes very little metal.

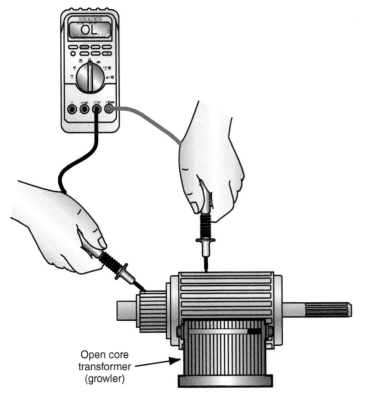

6-23 Checking an armature for a short.

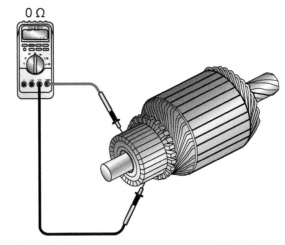

Figure 6-24 Testing the commutator for opens. There should be zero resistance between the segments.

Brush Inspection

Use an ohmmeter to test continuity through the brush holder (**Figure 6-26**). Connect one of the test leads to the positive brush and the other test lead to the negative brush. There should be no continuity between the brushes. If the ohmmeter indicated continuity, replace the brush holder.

Classroom Manual
Chapter 6, page 143

Another check of the brush assembly requires checking spring tension. To do this, install the brushes into the brush holder and slide the assembly over the commutator. Use a spring scale to measure the spring tension of the holders at the point where the spring

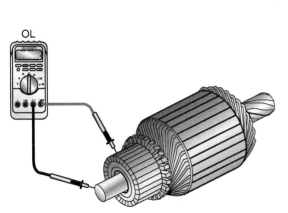

Figure 6-25 Testing the armature for short to ground. The meter should read infinite when placed on the shaft and the different segments of the commutator.

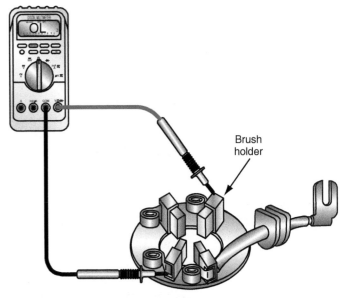

Figure 6-26 Typical brush holder.

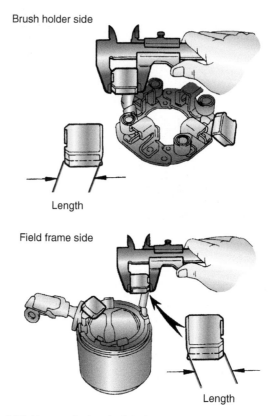

Brush holder side

Length

Field frame side

Length

Figure 6-27 Measure the length of the brushes to determine if they are worn.

lifts off the brush. If the spring tension is below specifications, replace the springs or the brush holder assembly.

Finally, measure the length of each brush (**Figure 6-27**). If they are shorter than specifications, replace the brushes. Some starters may not have serviceable brushes; thus, the brush holder assembly will need to be replaced.

Overrunning Clutch Inspection

The overrunning clutch may be inspected by sliding it onto the armature shaft. Attempt to rotate the clutch in both directions. If it is working properly, the clutch should rotate smoothly in one direction and lock in the other. If it fails to lock, or locks in both directions, replace the overrunning clutch.

Classroom Manual
Chapter 6, page 149

STARTER REASSEMBLY

If the brushes are worn beyond specifications, they must be replaced. Manufacturers use two methods of connecting the brushes; they are either soldered to the coil leads or screwed to terminals.

Special Tools

Feeler gauge set

100-watt soldering iron

Rosin core solder

Heat shrink tube

High-temperature grease

Two blocks of wood

Jumper cables

Jumper wire

Starting/charging system tester

Remote starter switch

SERVICE TIP To seat the new brushes to the commutator, slide the brushes into their holders and then place the assembly onto the commutator. Slide a piece of fine sandpaper between the commutator and the brushes with the grain facing the brushes. Rotate the armature to sand down the face of the brushes so their contour matches the commutator.

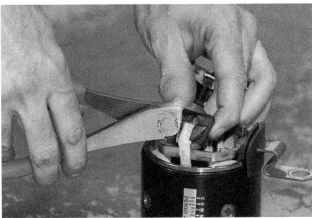

Figure 6-28 Removing the worn brushes.

Figure 6-29 Soldering a new brush to the field coil lead.

If the brushes are soldered to the coil leads, cut the old leads (**Figure 6-28**). Place a piece of heat shrink tube over the brush connector. Crimp the new brush lead connector to the coil leads. Solder the brush connector to the coil lead with rosin core solder (**Figure 6-29**). Slide the heat shrink tube over the soldered connection and use a heating gun to shrink the tube.

To reassemble the starter motor, basically reverse the disassembly procedures. Additional steps are listed here:

1. With a high-temperature grease, lubricate the splines on the armature shaft that the drive gear rides on.
2. To install the snap ring onto the armature shaft, stand the commutator end of the armature on a block of wood. Position the snap ring onto the shaft and hold in place with a block of wood. Hit the block of wood with a hammer to drive the snap ring onto the shaft (**Figure 6-30**).
3. Lubricate the bearings with high-temperature grease.
4. Apply sealing compound to the solenoid flange before installing the solenoid to the frame.
5. Use the scribe marks to locate the correct position of the frame-to-frame end and drive housing.

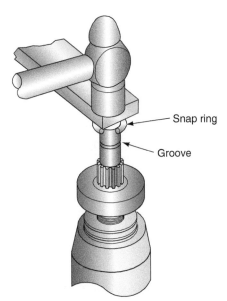

Figure 6-30 Once the snap ring is centered on the shaft, a hammer and block of wood can be used to install the ring onto the shaft.

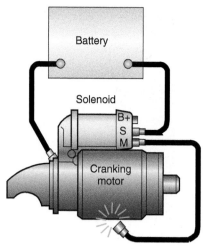

Figure 6-31 Jumper cable connections for checking the pinion gear clearance.

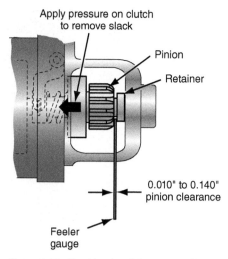

Figure 6-32 Checking the pinion gear to drive housing clearance.

6. Check the pinion gear clearance. Disconnect the M terminal to the starter motor's field coils. Connect a jumper cable from the battery positive terminal to the S terminal of the solenoid. Connect the other jumper cable from the battery negative terminal to the starter frame (**Figure 6-31**). Connect a jumper wire from the M terminal and momentarily touch the other end of the jumper wire to the starter motor frame. This will shift the pinion gear into the cranking position and hold it there until the battery is disconnected. Once the solenoid is energized, push the pinion back toward the armature; this removes any slack. Check the clearance with a feeler gauge (**Figure 6-32**). Compare clearance with specifications; normally, specifications call for a clearance of 0.010 to 0.140 inch (0.25 to 3.6 mm).
7. Perform the free spin test before installing the starter into the vehicle.

> ⚙ **SERVICE TIP** There is no provision on most starters to adjust the pinion clearance. However, if the clearance is excessive, it may indicate excessive wear of the solenoid linkage or shift lever.

STOP/START SYSTEM DIAGNOSTICS

Classroom Manual
Chapter 6, page 171

Diagnostics of the starter motor and the starting system circuits of the stop/start system is very much like that discussed in this chapter. However, performing some of the tests will require disabling the stop/start feature. Most manufacturers provide a means of disabling the system by push button activation. In addition, it may be possible to disable stop/start by opening the hood since the system uses hood ajar inputs to allow the feature to operate only when the hood is closed. Opening the hood requires the engine to be started by the ignition start switch.

> **AUTHOR'S NOTE** For the stop/start function to be operative, the vehicle must be initially started using the ignition system.

Figure 6-33 Use the scan tool to monitor all stop/start system inhibits.

If there is a malfunction in the stop/start system, the system will not shut off the engine (auto stop). A "SERVICE STOP/START SYSTEM" message or some other means will be used to alert the driver of a malfunction. If the stop/start system does not auto stop (engine does not shut off at stop), use a scan tool to monitor system inhibits (**Figure 6-33**). Typical auto stop inhibits include:

- The accelerator pedal is not in the idle position as reported by the throttle position sensor (TPS).
- The clutch is released (manual transmissions).
- The hood is open.
- The engine has not reached normal operating temperature.
- The driver's seat belt is not buckled.
- The outside temperature is below or above threshold.
- Cabin temperature is significantly different than temperature set on the temperature control.
- The temperature control system is set in the full defrost mode.
- The battery SOC is low.
- The transmission is in REVERSE.
- Transfer case is in 4LO mode.

There may be several other inhibitors that will prevent the engine from stopping. Reference the service information for a complete list. Confirm that all inhibitors are in their proper state. It may be necessary to diagnose the circuit of the offending inhibitor to locate the root cause of the issue.

CASE STUDY

A customer has a vehicle towed to the service center because it does not start. The technician verifies the complaint and observes that the starter drive engages the flywheel, but the engine does not rotate. By turning the crankshaft, the technician checks that the engine is able to rotate. The engine turns freely through two complete revolutions.

Next, the technician performs a visual inspection of the starting system. All connections are cleaned and tightened. A complete battery service is performed to confirm good battery condition. The battery passes all tests. The technician performs the solenoid circuit resistance test, and it indicates an excessive voltage drop as a result of burned contacts. The solenoid is replaced and the repair verified before returning the vehicle to the owner. The extra resistance caused by the burned contacts prevented sufficient current flow to the starter motor. This did not allow enough torque to be generated to rotate the engine.

ASE-STYLE REVIEW QUESTIONS

1. The starter circuit shown in **Figure 6-34** has a fully charged battery. The starter relay and solenoid do not operate when the ignition switch is placed in the START position.

   Technician A says this could be caused by a grounded circuit at terminal 85 of the starter relay.

   Technician B says this could be caused by an open to terminal 86 of the starter relay.

   Who is correct?

   A. A only   C. Both A and B

   B. B only   D. Neither A nor B

2. During a starter current-draw test the following results were noted:

   A. Voltage—9.4 volts.

   B. Current draw—363 amps.

   The specifications for the maximum current draw are 180 amps.

   Technician A says the test results indicate that the battery may be defective and should be tested.

   Technician B says the test results indicate a faulty natural safety start switch circuit.

   Who is correct?

   C. A only   E. Both A and B

   D. B only   F. Neither A nor B

3. A 600-mV drop is measured across the starter motor solenoid while the engine is being cranked. What is the repair?

   A. Replace the battery.

   B. Replace the starter motor.

   C. Confirm cable connections.

   D. All of the above.

4. In the starter system shown in **Figure 6-34**, a voltmeter indicates battery voltage to the input of the neutral safety switch and battery voltage on the output terminal with the transmission in neutral and the ignition switch in the START position. This indicates:

   A. Normal operation.

   B. The neutral safety switch is working properly.

   C. A faulty starter relay coil winding.

   D. An open neutral safety switch ground.

5. During a starter current-draw test, the voltage is 10.2 volts and amperage is above specifications. This could be caused by all of the following **EXCEPT**:

   A. Excessive circuit resistance.

   B. Shorted starter windings.

   C. Worn starter motor bushings.

   D. Internal engine failure.

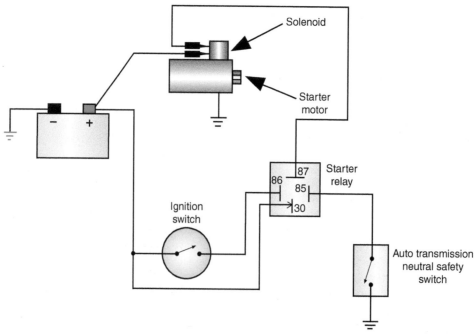

Figure 6-34 Starter circuit.

6. Armature testing is being discussed.

   Technician A says to test for shorts, place the armature in the growler and hold a thin steel blade parallel to the core and watch for blade vibrations that would indicate a short.

   Technician B says there should be zero resistance between the commutator sectors and the armature shaft.

   Who is correct?

   A. A only

   B. B only

   C. Both A and B

   D. Neither A nor B

7. *Technician A* says it is important that a full battery test series be done before trying to test the starter system.

   Technician B says the internal condition of the engine has little effect on the operation of the starting system.

   Who is correct?

   A. A only

   B. B only

   C. Both A and B

   D. Neither A nor B

8. The stop/start feature does not shut the engine off when the vehicle is stopped.

   Technician A says the feature may be inhibited due to a faulty seat belt buckle switch.

   Technician B says the defrost mode may be selected on the HVAC system.

   Who is correct?

   A. A only

   B. B only

   C. Both A and B

   D. Neither A nor B

9. Voltage drop testing of the solenoid control circuit is being discussed.

   Technician A says the maximum amount of voltage drop allowed is 0.9 volt.

   Technician B says the voltage drop across each wire should be less than 0.1 volt.

   Who is correct?

   A. A only

   B. B only

   C. Both A and B

   D. Neither A nor B

10. *Technician A* says most starter noises come from the armature.

    Technician B says if the engine does not rotate but the starter drive spins, the most likely cause is a shorted armature.

    Who is correct?

    A. A only

    B. B only

    C. Both A and B

    D. Neither A nor B

ASE CHALLENGE QUESTIONS

1. A vehicle is towed into the shop due to a no-start (no-crank) condition. The headlights are turned on, and when the ignition key is placed in the START position, the lights remain bright but the starter does not crank the engine.

   Technician A says that the starter may be binding internally.

   Technician B says that the battery may be discharged.

   Who is correct?

   A. A only

   B. B only

   C. Both A and B

   D. Neither A nor B

2. A vehicle is being tested for a slow-crank condition. The starter current draw is 525 amps.

   Technician A says that the field windings in the starter may be shorted.

   Technician B says that the starter drive gear may be slipping.

   Who is correct?

   A. A only

   B. B only

   C. Both A and B

   D. Neither A nor B

3. The starter draw of a vehicle with a slow-crank condition is 75 amps. The battery has a good state of charge and it passed a capacity test.

   Technician A says that a starting circuit voltage drop test should be performed.

   Technician B says that the neutral safety switch may have excessive resistance.

   Who is correct?

   A. A only   C. Both A and B

   B. B only   D. Neither A nor B

4. A voltmeter that is connected across a starter solenoid's battery and motor terminals indicates 12 volts when the ignition key is turned to the START position. A distinct click is heard from the solenoid when this occurs, but the engine does not crank.

   Technician A says that there is excessive voltage drop in the circuit between the battery and the starter solenoid.

   Technician B says that the starter solenoid probably needs to be replaced.

   Who is correct?

   A. A only   C. Both A and B

   B. B only   D. Neither A nor B

5. *Technician A* says incorrect pinion gear to ring gear clearance can result in noisy cranking and high starter draw.

   Technician B says starting circuit voltage drop should be checked before performing a starter draw test.

   Who is correct?

   A. A only   C. Both A and B

   B. B only   D. Neither A nor B

Name _____   Date _____

CURRENT-DRAW TEST

Upon completion of this job sheet, you should be able to measure the current draw of a starter motor and interpret the results of the test.

NATEF Correlation —————————————————————————————————

This job sheet addresses the following **MLR/AST/MAST** task:

C.1.   Perform starter current-draw test; determine necessary action.

This job sheet addresses the following **AST/MAST** task:

C.6.   Differentiate between electrical and engine mechanical problems that cause a slow- or no-crank condition.

ASE **NATEF**

Tools and Materials

- A vehicle with a 12-volt battery
- Starting/charging system tester
- Service information

Describe the vehicle being worked on:

Year _____ Make _____ Model _____

VIN _____ Engine type and size _____

Procedure   **Task Completed**

1. Perform a battery test series to confirm the battery is good. If necessary, refer to the job sheets in Chapter 5. What was the result of the battery test?

   PASS _____ FAIL _____

   WHY _____

   _____

2. Disable the ignition or fuel injection systems to prevent the engine from starting.

   a. How was this accomplished? _____

   _____

3. Expected starter current draw is: _____ amp(s).

   Voltage should not drop below: _____ volt(s).

4. Connect the starting/charging system tester cables to the vehicle.   ☐

5. Zero the ammeter on the tester.   ☐

6. Be prepared to observe the amperage when the engine begins to crank and while it is cranking. Also, note the voltage when you stop cranking the engine.

   The initial current draw was _____ amp(s)

   After _____ seconds, the current draw was _____ amps and the voltage dropped to _____ volt(s).

7. What is indicated by the test results? Compare your measurements to the specifications.

_____

_____

_____

_____

_____

8. If the current draw was high, how would you differentiate between electrical and engine mechanical problems?

_____

_____

_____

_____

9. Reconnect the ignition or fuel injection system and start the engine. ☐

Instructor's Response

_____

_____

_____

Name _____   Date _____

STARTER SYSTEM TEST USING THE GR8

Upon completion of this job sheet, you should be able to measure the current draw of a starter motor and interpret the results of the test.

NATEF Correlation

This job sheet addresses the following **MLR/AST/MAST** task:

C.1.   Perform starter current-draw test; determine necessary action.

AS **NATEF**

Tools and Materials

- A vehicle with a 12-volt battery
- GR8 or similar starting system tester
- Service information

Describe the vehicle being worked on:

Year _____ Make _____ Model _____

VIN _____ Engine type and size _____

Procedure

Task Completed

1. Perform a battery test series to confirm the battery is good. If necessary, refer to the job sheets in Chapter 5. What was the result of the battery test?

    PASS _____ FAIL _____

    WHY _____

    _____

2. Follow the on-screen prompts to select the Starter System Test.   ☐

3. Follow any on-screen instruction for connecting the tester to the vehicle.   ☐

4. When prompted, start the vehicle's engine.   ☐

5. What is the displayed test decision?   ☐

    _____

    _____

    _____

6. Are the complete results of the test able to be displayed?

    ☐ Yes   ☐ No

7. If yes, scroll to each screen and record the results here:

    Average cranking voltage: _____

    Average cranking current: _____

    Cranking time: _____

8. If possible, print off the test results and attach to this job sheet.   ☐

9. What is indicated by the test results? Compare your measurements to the specifications.

_____

_____

_____

_____

_____

Instructor's Response

_____

_____

_____

Name _____   Date _____

TESTING THE STARTING SYSTEM CIRCUIT

Upon completion of this job sheet, you should be able to visually inspect the starting circuit and perform voltage drop testing for excessive resistance.

NATEF Correlation

This job sheet addresses the following **MLR/AST/MAST** tasks:

C.2.   Perform starter circuit voltage drop tests; determine necessary action.

C.3.   Inspect and test starter relays and solenoids; determine necessary action.

C.5.   Inspect and test switches, connectors, and wires of the starter control circuits; determine necessary action.

Tools and Materials

- A vehicle with a 12-volt battery
- Wiring diagram for the vehicle assigned
- A DMM
- Highlighter markers

Describe the vehicle being worked on:

Year _____ Make _____ Model _____

VIN _____ Engine type and size _____

Procedure

Task Completed

1. Using the service information, retrieve the wiring diagram for the starting system of the assigned vehicle and print it off. Highlight the positive side of the system in yellow. Highlight the negative side of the system in green. Identify the control circuit with orange. Attach to Job Sheet.   ☐

2. Disable the ignition or fuel injection systems to prevent the engine from starting.   ☐

3. Connect the voltmeter across the battery's negative cable.

   Crank the engine with the starter and observe the readings on the meter.

   Your reading was _____ volt(s).

4. What does this indicate?

   _____

   _____

   _____

5. Connect the voltmeter across the battery's positive cable (from battery to starter motor). Crank the engine with the starter and observe the readings on the meter. Your reading was _____ volt(s).

6. This test measured the voltage drop across everything in the positive side of the starter motor circuit, including all relays and solenoids. How would you inspect a suspected faulty relay or solenoid?

   _____

   _____

   _____

7. Use the wiring diagram and perform a voltage drop test on the control circuit. ☐

8. What do the test results suggest? What are your recommendations?

_____

_____

_____

9. Reconnect the ignition or fuel injection systems to allow the engine to start. ☐

Instructor's Response

_____

_____

_____

Name _____   Date _____

REMOVE AND INSTALL STARTER

Upon completion of this job sheet, you should be able to remove and install the starter.

NATEF Correlation ——————————————————————————————————————

This job sheet addresses the following **MLR/AST/MAST** task:

C.4.   Remove and install starter in a vehicle.

NATEF

Tools and Materials

- A vehicle with a 12-volt battery
- Fender covers
- Battery cable puller
- Service information

Describe the vehicle being worked on:

Year _____ Make _____ Model _____

VIN _____ Engine type and size _____

Procedure

Task Completed

1. Position the vehicle over a hoist. Do not lift the vehicle at this time. ☐

2. Disconnect the negative battery cable from the battery terminal. Isolate the cable clamp so it cannot touch the negative battery post. ☐

3. Disconnect all wires, fasteners, and so on, which can be reached from the top of the engine compartment. ☐

4. Disconnect the wires leading to the solenoid terminals. To prevent confusion, it is a good practice to use a piece of tape to identify the different wires. ☐

5. On some vehicles, it may be necessary to disconnect the exhaust system to be able to remove the starter motor. Spray the exhaust system fasteners with a penetrating oil to assist in removal. ☐

6. Loosen the starter mounting bolts and remove all but one. ☐

7. Support the starter motor; remove the remaining bolt. Then remove the starter motor. ☐

8. Identify the type of starter motor.

_____

9. Does the motor require clearance shims?

Yes ☐   No ☐

If yes, describe the procedure for selecting the proper shim:

_____

_____

_____

_____

10. Reverse the procedure to install the starter motor. Be sure all electrical connections are tight. If you are installing a new or remanufactured starter, remove any paint that may prevent a good ground connection.   ☐

Instructor's Response

_____

_____

_____

Name _____   Date _____

STOP/START FAMILIARIZATION

Upon completion of this job sheet, you should be able to identify stop/start features and inhibits.

NATEF Correlation

This job sheet addresses the following **MLR** task:

C.6.   Demonstrate knowledge of an automatic idle stop/start-stop system.

This job sheet addresses the following **AST/MAST** task:

C.7.   Demonstrate knowledge of an automatic idle stop/start-stop system.

NATEF

Tools and Materials

- A vehicle stop/start with a 12-volt battery
- Scan tool
- Service information

Describe the vehicle being worked on:

Year _____ Make _____ Model _____

VIN _____ Engine type and size _____

Procedure

Task Completed

1. Based on the service information, describe how the stop/start feature is disabled for service:

   _____

   _____

   _____

2. Disable the stop/start feature.

   Was the process successful?   Yes ☐   No ☐

   If no, consult your instructor.

3. Enable the stop/start feature again.

   Was the process successful?   Yes ☐   No ☐

   If no, consult your instructor.

4. With the engine off, open the hood. Attempt to start the engine. What happened?

   _____

5. Shut the hood and assure the stop/start feature is enabled.   ☐

6. With the engine running, open the hood. What happened?

   _____

7. Use the scan tool to monitor the state of all inhibits listed in the service information. Are any input states causing the stop/start system from functioning?

_____

_____

_____

_____

Instructor's Response

_____

_____

_____

DIAGNOSTIC CHART 6-1

PROBLEM AREA:	Starting system operation
SYMPTOMS:	Starter fails to turn engine or operates at reduced efficiency. Excessive current draw.
POSSIBLE CAUSES:	**1.** Shorted armature. **2.** Worn starter bushings. **3.** Bent armature. **4.** Thrown armature windings. **5.** Loose pole shoes. **6.** Grounded armature. **7.** Shorted field windings.

DIAGNOSTIC CHART 6-2

PROBLEM AREA:	Starting system operation
SYMPTOMS:	No or reduced starter operation. Current draw too low.
POSSIBLE CAUSES:	**1.** Worn brushes. **2.** Excessive circuit voltage drop.

DIAGNOSTIC CHART 6-3

PROBLEM AREA:	Starting system operation
SYMPTOMS:	Starter fails to rotate the engine or operates at reduced efficiency. Excessive starter circuit resistance.
POSSIBLE CAUSES:	**1.** Excessive starter circuit voltage drop. **2.** Improper connections. **3.** Corroded ground connections. **4.** High resistance in solenoid or relay.

DIAGNOSTIC CHART 6-4

PROBLEM AREA:	Starting system operation
SYMPTOMS:	Starter fails to rotate engine or rotates too slowly. Failed starter test series.
POSSIBLE CAUSES:	**1.** Defective or worn starter. **2.** Worn brushes. **3.** Shorted field coils. **4.** Open field coils. **5.** Shorted armature. **6.** Open armature.

DIAGNOSTIC CHART 6-5

PROBLEM AREA:	Starter control circuit
SYMPTOMS:	Starter does not operate when ignition switch is located in the START position. No sounds.
POSSIBLE CAUSES:	**1.** Open circuit. **2.** Faulty ignition switch. **3.** Park/neutral switch faulty or misadjusted. **4.** Faulty starter relay/solenoid. **5.** Faulty starter.

DIAGNOSTIC CHART 6-6

PROBLEM AREA:	Starter control circuit
SYMPTOMS:	Starter does not operate when ignition switch is located in the START position. Relay/solenoid clicks.
POSSIBLE CAUSES:	**1.** High resistance in starter control circuit. **2.** High resistance in started circuit. **3.** Faulty starter relay/solenoid. **4.** Faulty starter.

DIAGNOSTIC CHART 6-7

PROBLEM AREA:	Starter drive
SYMPTOMS:	Starter spins but does not rotate the engine.
POSSIBLE CAUSES:	**1.** Defective one-way clutch. **2.** Broken teeth on ring gear. **3.** Faulty starter motor.

DIAGNOSTIC CHART 6-8

PROBLEM AREA:	Starter drive fails to disengage after engine starts.
SYMPTOMS:	Excessive noise after engine starts.
POSSIBLE CAUSES:	**1.** Faulty ignition switch. **2.** Faulty relay/solenoid. **3.** Faulty starter motor. **4.** Improper starter mounting.

DIAGNOSTIC CHART 6-9

PROBLEM AREA:	Stop/start
SYMPTOMS:	Engine will not shut off when vehicle is stationary.
POSSIBLE CAUSES:	**1.** The accelerator pedal is not in the idle position as reported by the throttle position sensor (TPS). **2.** The clutch is released (manual transmissions) or input is missing. **3.** The hood is open or faulty input switch. **4.** The engine has not reached normal operating temperature or faulty engine temperature sensors. **5.** The driver's seat belt is not buckled or faulty input. **6.** The outside temperature is below or above threshold or faulty ambient temperature sensor. **7.** Cabin temperature is significantly different than temperature set on the temperature control or faulty temperature sensors. **8.** The temperature control system is set in the full defrost mode or faulty input. **9.** The battery SOC is low. **10.** The transmission is in REVERSE or faulty range sensor. **11.** Transfer case is in 4LO mode or faulty input sensor.

CHAPTER 7

CHARGING SYSTEM TESTING AND SERVICE

Upon completion and review of this chapter, you should be able to:

- Diagnose charging system problems that cause an undercharge or no-charge condition.
- Diagnose charging system problems that cause an overcharge condition.
- Inspect, adjust, and replace generator drive belts, pulleys, and fans.
- Perform charging system output tests and determine needed repairs.
- Perform charging system circuit voltage drop tests and determine needed repairs.
- Perform voltage regulator tests and determine needed repairs.

- Test and replace AC generator diodes and/or rectifier bridge.
- Remove and replace the AC generator.
- Disassemble, clean, and inspect AC generator components.
- Inspect and replace AC generator brushes and brush holders.
- Test and diagnose the rotor.
- Test and diagnose the stator.
- Inspect the HV and LV circuits of the HEV inverter/converter module.

 Basic Tools

Basic mechanic's tool set
Service information

Terms To Know

Current output test	Full fielding	Voltage output test
Field current-draw test	Multiplying coil	
Full field test	Voltage drop test	

INTRODUCTION

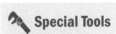 **WARNING** Always wear safety glasses when performing charging system tests.

Whenever there is a charging system problem, make sure a complete battery test series is conducted. The battery supplies the electrical power for the charging system. If the battery is faulty, the charging system cannot be expected to work its best. AC (alternating current) generators are designed to maintain the charge of a battery, not to charge a dead battery. It is important that the battery be in good condition in order to obtain accurate charging system test results. In addition, the battery must be fully charged before proceeding with the diagnosis of the charging system. It is also important to perform a preliminary

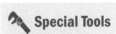 **Special Tools**

Battery/starting/ charging system tester
Belt tension gauge
Belt wear gauge

inspection of the charging system. Many problems can be detected during this simple step. Check the following items:

Classroom Manual
Chapter 7, page 181

1. Condition of the drive belt (**Figure 7-1**). If the drive belt is worn or glazed, it will not allow enough rotor rpms to produce sufficient current. A belt wear gauge can be used to determine wear on ribbed serpentine belts (**Figure 7-2**). If the gauge pin sits higher than the belt rib, the belt is good (**Figure 7-3**). A worn belt is indicated if the gauge pin is lower than the belt rib.
2. Drive belt tension (**Figure 7-4**).
3. Drive belt tensioner condition.

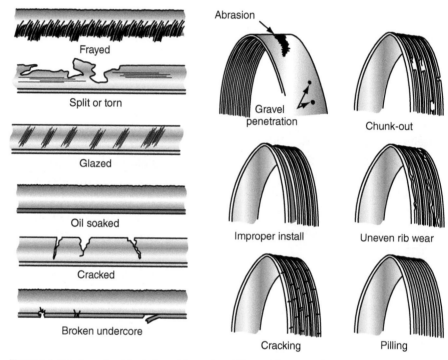

Figure 7-1 The generator drive belt must be replaced if any of these conditions exist.

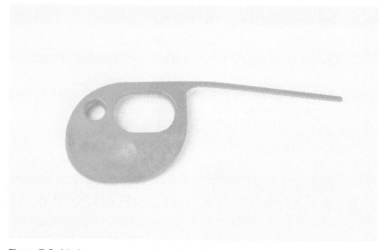

Figure 7-2 A belt wear gauge used to determine serpentine belt wear.

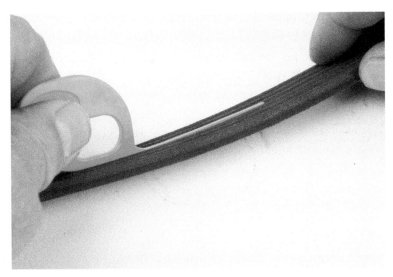

Figure 7-3 Using the belt wear gauge. If the gauge pin is above the rib, the belt is good.

Figure 7-4 Checking belt tension using a tension gauge.

4. Drive belt tensioner bearing failure.
5. Electrical connections to the AC generator.
6. Electrical connections to the regulator.
7. Ground connections at the engine and chassis.
8. Battery cables and terminals.
9. Fuses and fusible links.
10. Excessive current drain caused by a light or other electrical component remaining on after the ignition switch is turned off.
11. Symptoms of undercharging. These include slow-cranking, discharged battery, low instrument panel ammeter or voltmeter readings, and charge indicator lamp on.
12. Symptoms of overcharging. These include high ammeter and voltmeter readings, battery boiling, and charge indicator lamp on.

The manufacturer of the vehicle you are working on may have several additional tests to perform. It is important to always follow the procedures outlined by the manufacturer for the vehicle being tested.

> ⚙ SERVICE TIP To check the fusible link to the AC generator, use a voltmeter and test for voltage at the BAT terminal. Battery voltage should be present. No voltage indicates the possibility that the fusible link is burned out. A better test would be to measure the voltage drop across the link. This will identify any high resistance in the circuit.

Different types of testers can be used to test the charging system and AC generators. Some handheld multimeters have the ability to perform many tests. However, the best testers to use are those designed to test the entire system. These testers are commonly referred to as battery/starting/charging system testers. Always follow the operating procedures for the specific tester being used.

When performing the tests, be sure of the connections you are making. Refer to the service information for identification of the various terminals for the AC generator and regulator. Connecting a test lead to the wrong terminal can result in AC generator damage, as well as damage to other electrical and electronic components.

⚡ WARNING **Many charging system tests require that the vehicle be operated in the shop area. Always place wheel blocks against the drive wheels. Be sure there is proper ventilation of the vehicle's exhaust. Also, be aware of the drive belts and cooling fan. Be sure of where your hands and tools are at all times.**

CHARGING SYSTEM SERVICE CAUTIONS

The following are some of the general rules when servicing the charging system:

1. Do not run the vehicle with the battery disconnected. The battery acts as a buffer and stabilizes any voltage spikes that may cause damage to the vehicle's electronics.
2. Do not allow output voltage to increase over 16 volts (V) when performing charging system tests.
3. If the battery needs to be recharged, disconnect the cables while charging.
4. Do not attempt to remove electrical components from the vehicle with the battery connected.
5. Before connecting or disconnecting any electrical connections, the ignition switch must be in the OFF position.
6. Avoid contact with the BAT terminal of the AC generator while the battery is connected. Battery voltage is always present at this terminal.

AC GENERATOR NOISES

Classroom Manual
Chapter 7, page 181

Noises that come from the AC generator can be from three sources. The causes of the noises are identifiable by the types of noises they make. A loose belt will make a squealing noise. Check the belt condition and tension. Replace the belt if necessary.

A squealing noise can also be caused by faulty bearings. The bearings are used to support the rotor in the housing halves. To test for bearing noises, use a length of hose, a long screwdriver, or a technician's stethoscope. By placing the end of the probe tool close to the bearings and listening on the other end, any bearing noise will be transmitted so you will be able to hear it. Faulty bearings will require either AC generator replacement or disassembly of the AC generator to replace the bearings.

> ⚙ **SERVICE TIP** With the engine off, rub a piece of bar soap on the pulley surface of the drive belts. Do this one belt at a time until the noise stops. This way you will know which belt is the cause of the noise.

⚡ WARNING **This test is performed with the engine running. Use caution around the drive belts, fan, and other moving components.**

A whining noise can be caused by shorted diodes, a shorted stator, or by a worn rotor bearings. A quick way to test for the cause of a whining sound is to disconnect the wiring to the generator, and then start and run the engine. If there is no noise, the cause of the noise is a magnetic whine due to shorted diodes or stator windings. Use a lab scope to verify the condition of the diodes and stator. If the noise remains, the cause is mechanical and probably due to worn bearings.

CHARGING SYSTEM TROUBLESHOOTING

Troubleshooting charts assist the technician in diagnosing the charging system. They give several possible causes for the customer complaint. They also instruct the technician in what tests to perform or what service is required.

VOLTAGE OUTPUT TESTING

Once the visual inspection and preliminary checks are completed, the next step is to perform a **voltage output test**. The voltage output test is used to make a quick determination if the charging system is working properly. If the charging system is operating correctly, then check for battery drain. The following procedure is for performing the test:

1. Connect the voltmeter across the battery terminals, observing polarity.
2. Connect the tachometer, following the manufacturer's procedure. Alternately, a scan tool can be used to monitor engine speed.
3. With the engine off, record the base voltage value across the battery.
4. Start the engine and bring the rpm up to between 1,500 and 2,000.
5. Observe the voltmeter reading. It should read between 13.5 and 14.5 volts.

If the charging voltage is too high, there may be a problem in the following areas:

1. Poor voltage regulator ground connection.
2. High resistance in the "sense" circuit between the battery and the PCM or voltage regulator.

Classroom Manual
Chapter 7,
pages 179, 193

 **Special Tools**

Voltmeter
Tachometer
Scan tool

3. Short to ground in the field coil control circuit, causing the AC generator to full field.
4. Loose or corroded battery cable terminals.
5. Defective voltage regulator or PCM.

If the charging voltage is too low, the fault might be:

1. Loose or glazed drive belt.
2. Discharged battery.
3. Loose or corroded battery cable terminals.
4. Defective AC generator.
5. Defective voltage regulator or PCM.

If the voltage reading is correct, perform a load test to check the voltage output under a load condition:

1. With the engine running at idle, turn on the headlights and the heater fan motor to high speed.
2. Increase the engine speed to approximately 2,000 rpm.
3. Check the voltmeter reading. It should increase a minimum of 0.5 volt over the base voltage reading taken previously. Some vehicle manufacturer's specifications require a rise in voltage of 2.5 volts over the base voltage value.
4. If the voltage increases, the charging system is operating properly. If the voltage did not increase, perform the following test series to locate the fault.

> ⚙ SERVICE TIP If the charging system passes the no-load test but fails the load test, check the condition and tension of the drive belt closely.

It is also possible to use the scan tool to diagnose the charging system's output voltage on most late-model vehicles. Most PCM-controlled charging systems will use a battery voltage sense circuit to determine the state of charge for the battery (**Figure 7-5**). This circuit is also used to determine the output voltage of the AC generator. If the PCM is unable to operate the charging system correctly, a diagnostic trouble code (DTC) is set. In this case follow the appropriate diagnostic routine for the set DTC to isolate the cause. However, not all faults result in a DTC. For example, if the sense circuit had excessive resistance due to a poor connection, the PCM would increase the generator output until it achieved the desired voltage level. Since the PCM is performing what is believed the correct function, no DTCs are set. However, under this condition the battery would be overcharging.

The PCM will use a voltage sense input and a temperature sensor to determine the battery temperature. Based on this information it will determine a target voltage for the generator (**Figure 7-6**). In most cases, the PCM will energize the field circuit until it is about 0.50 volt above the target voltage and then turn off the field current until the voltage drops to a value 0.50 volt below the target voltage (**Figure 7-7**).

With a voltmeter connected across the battery terminals, compare the scan tool voltage display to the voltmeter reading. If there is a discrepancy between the two readings, the voltage sense circuit needs to be diagnosed.

> ⚙ SERVICE TIP Keep in mind that a faulty temperature sensor, or a fault in its circuit, can cause the generator to undercharge or overcharge the battery.

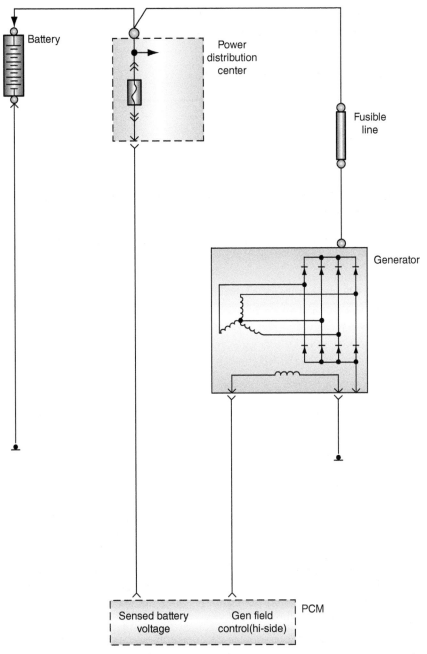

Figure 7-5 Computer-controlled charging system will use a sense circuit to monitor the state of charge (SOC) of the battery and the output of the generator.

CURRENT OUTPUT TESTING

In order to obtain maximum AC generator current output, the demands on the charging system must be increased. By connecting a carbon pile to maintain system voltage at 12.0 volts, the signal voltage to the regulator will be reduced. When this occurs, the regulator attempts to recharge the battery by full fielding. This will produce the maximum current output to the battery. The **current output test** will determine

Classroom Manual
Chapter 7,
pages 179, 184

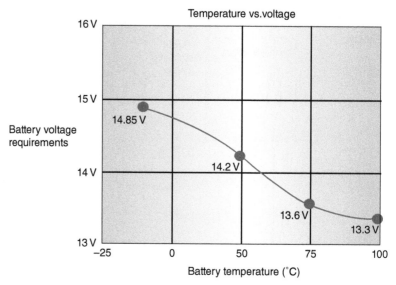

Figure 7-6 The temperature of the battery determines the charging system output voltage.

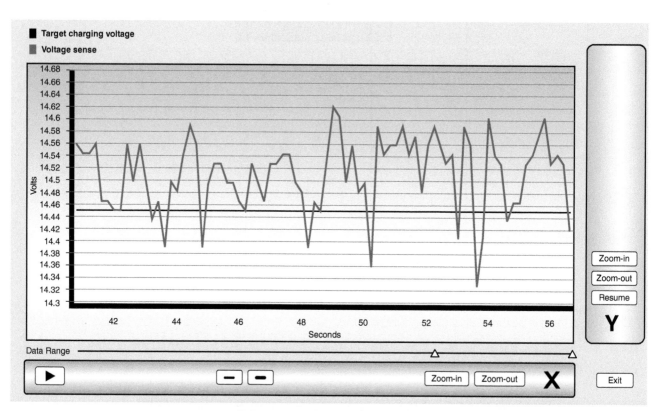

Figure 7-7 Scan tool graphing display of target voltage and sensed voltage.

the maximum output of the alternator. Follow the steps in **Photo Sequence 13** to perform the current output test using a battery/starting/charging system tester with a carbon pile. Most computer-based test equipment will automatically perform the current output test as part of a test series.

PHOTO SEQUENCE 13
Performing the Current Output Test

P13-1 Connect the large red and black cables across the battery, observing polarity.

P13-2 Select CHARGING.

P13-3 Adjust the ammeter reading to zero.

P13-4 Connect the inductive pickup around all battery ground cables.

P13-5 With the ignition switch in the RUN position, engine not running, observe the ammeter reading. This reading indicates how much current is required to operate any full-time accessories.

P13-6 Start the engine and hold engine speed between 1,500 and 2,000 rpm.

P13-7 Turn the load knob for the carbon pile slowly, until the highest ammeter reading possible is obtained. Do not reduce battery voltage below 12 volts.

P13-8 Return the load control knob to the OFF position.

P13-9 The highest reading indicates maximum current output.

Once the maximum current output is known, add the maximum output reading to the reading obtained in step 5 of the photo sequence. This total should be within 10% of the rated output of the AC generator.

If the ammeter reading indicates that output is 2 to 8 amps (A) below the specification, then an open diode or slipping belt may be the problem. An output reading that indicates 10 to 15 amps below specifications indicates that there is a possible shorted diode or slipping belt. If the AC generator output is below specifications, perform the full field test.

When testing the General Motors CS-130 or 144 AC generator, first use a voltmeter to test for voltage at the L and I terminals. Battery voltage should be indicated at both terminals with the ignition switch in the RUN position.

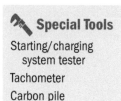

Special Tools

Starting/charging
  system tester
Tachometer
Carbon pile

VOLTAGE DROP TESTING

Excessive voltage drop is a primary cause of charging system problems. Testing of all wires and connections should be a part of the diagnostic approach. Any particular wire or connection should not exceed 0.2 volt drop while the total system drops should be less than 0.7 volt. The ground side voltage drop should be less than 0.2 volt. The **voltage drop test** determines if the battery, regulator, and AC generator are all operating at the same potential.

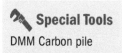

Special Tools

DMM Carbon pile

When performing the voltage drop tests, load the system using a carbon pile or by turning on the headlights and the heater blower motor onto high. Try to load the system to between 10 and 20 amps with the engine running at about 2,000 rpm. Voltage drop test the circuit between the output terminal of the AC generator and the battery positive terminal (**Figure 7-8A**). Also, perform the voltage drop test between the battery negative terminal and the AC generator housing (**Figure 7-8B**).

Some manufacturers recommend measuring voltage drop when the generator is putting out its maximum. Always follow the recommendations of the manufacturer. Remember, if the AC generator is not putting out any current, there will be no voltage drop even if the circuit is very corroded.

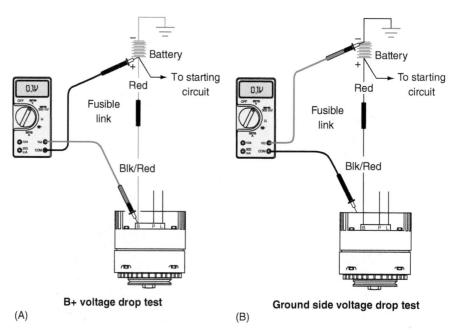

B+ voltage drop test

(A)

Ground side voltage drop test

(B)

Figure 7-8 Voltage drop testing of the generator output circuit (A) and the generator ground circuit (B).

SERVICE TIP Turning on the headlights may be substituted for the carbon pile.

The general specifications for voltage drop testing are:

Insulated circuit: less than 0.7 volt.
Ground circuit: less than 0.2 volt.

If a higher voltage drop is observed, work up the circuit to find the fault. Check every wire and connection.

FIELD CURRENT-DRAW TEST

Because field current is required for the production of a magnetic field, it is necessary to determine if current is flowing to the field coil. To perform the **field current-draw test** using the ammeter function of the digital multimeter (DMM), follow these steps:

1. Confirm that the battery is fully charged.
2. Connect a carbon pile across the battery. Make sure the knob is turned to the OFF position prior to making the connection.
3. Place the DMM in the 10-amp mode and locate the induction pickup clamp around the field circuit wire going to the AC generator (**Figure 7-9**). If a pickup clamp is not available, the DMM test leads will need to be connected in series to the field circuit.
4. Start the engine and run it at 2,000 rpm.
5. Using the carbon pile, load the alternator to the rated output current.
6. Measure the amperage in the field circuit. Check results with manufacturer's specifications. Readings typically range between 3 and 7 amps.

For GM, A circuit systems with internal regulators, the test will require removing the field circuit connector from the AC generator and connecting a Y-type connector between terminals 1 and 2. Connect the common terminal of the Y connector to battery positive. With the amp pickup around the wire (or the test leads in series), run the engine at 2,000 rpm. The field current should be between 2 and 5 amps. Use the carbine pile to decrease

Classroom Manual
Chapter 7, page 196

 **Special Tools**
DMM
Ammeter with inductive pickup
Multiplying coil
Carbon pile
Tachometer

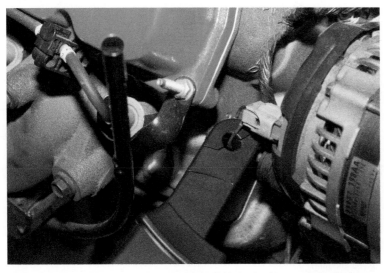

Figure 7-9 Connect the amp probe around the field circuit wire to measure field current.

battery voltage while observing the ammeter. As battery voltage decreases, the field current should increase. If the readings are within the specification limits, then the field circuit is good. If the readings are over specifications, a shorted field circuit or bad regulator may be the problem. If the readings are too low, then there is high electrical resistance that may be caused by worn brushes.

> ⚙ SERVICE TIP If the ammeter is not able to read low current use a **multiplying coil** connected in series with the field circuit and clamp the amp pickup to the coil (**Figure 7-10**). A multiplying coil is made of 10 wraps of wire. This multiplies the ammeter reading so that a starting/charging system tester's scale can be used to read lower current. For example, if the needle is pointing to 25 amps, when using the multiplying coil, the actual reading is 2.50 amps.

To test Ford's integral alternator/regulator (IAR) system, use a voltmeter as follows:

1. With the ignition switch in the OFF position, connect the negative voltmeter lead to the generator housing.
2. Connect the positive voltmeter lead to the F terminal screw of the regulator (**Figure 7-11**).
3. Check the voltmeter reading. It should indicate battery voltage. If it reads battery voltage, the field circuit is normal.
4. If the voltmeter reading is less than battery voltage, disconnect the wiring plug from the regulator.
5. Connect the positive voltmeter lead to the I terminal of the plug.
6. Check the voltmeter reading. It should indicate 0 volt. If there is voltage present, repair the I lead from the ignition switch. The I lead is receiving voltage from another source.
7. If there was no voltage present in step 6, connect the positive voltmeter lead to the S terminal of the regulator wiring plug.

Figure 7-10 Connecting the multiplying coil and amp pickup to an AC generator.

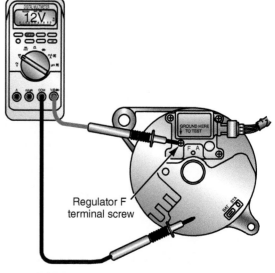

Figure 7-11 Testing the IAR generator field circuit.

8. Check the voltmeter reading. If there is 0 volt, replace or service the regulator.
9. If voltage is indicated, disconnect the wiring plug from the AC generator.
10. Check for voltage to the regulator wiring plug STA terminal.
11. If voltage is still present, repair the STA terminal wire lead to the AC generator. The STA terminal wire is receiving voltage from another source.
12. If no voltage is present, replace the rectifier bridge.

FULL FIELD TEST

The **full field test** will determine if the detected problem lies in the AC generator or the regulator. The full field test needs to be performed only if the charging system failed the output test. This test is performed by manually **full fielding** the AC generator with the regulator bypassed. Full fielding means the field windings are constantly energized with full battery voltage. Full fielding will produce maximum AC generator output. If this test still produces lower-than-specified output, the AC generator is the cause of the problem. If the output is within specifications with the regulator bypassed, then the problem is within the regulator.

When full fielding the system, the battery should be loaded to protect vehicle electronics and computers. With the voltage regulator bypassed, there is no control of voltage output. The AC generator is capable of producing well over 30 volts. This increased voltage will damage the circuits not designed to handle that high of a voltage. A 20% increase in voltage output results in a 20% increase in current flow that can damage sensitive electronic circuitry.

To perform the full field test on a General Motors' A circuit SI-type AC generator with an internal voltage regulator, insert a screwdriver into the D-shaped test hole (**Figure 7-12**). This test hole lines up with a small tab that is attached to the negative brush. By inserting a screwdriver into the "D" hole about 1/2 inch (12.7 mm) and grounding it to the housing, the regulator is bypassed. Perform the output test again with the regulator bypassed. If the output is within specifications, the regulator is at fault.

A variation of the test calls for shorting the negative brush in the "D" hole while the ignition switch is in the RUN position. If the brushes and rotor are good, then the rear bearing should be magnetized and attract a metal screwdriver (**Figure 7-13**).

Classroom Manual
Chapter 7,
pages 193, 195

Special Tools

Various jumper wires
Battery/starting/
  charging system
  tester
DMM
Scan tool
Tachometer

⚠ Caution

Not all AC generators can be full fielded. Check the manufacturer's procedures before attempting to full field an AC generator.

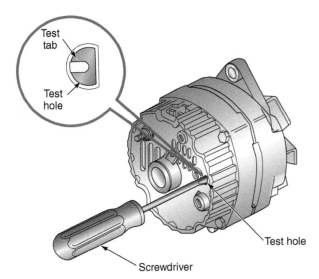

Figure 7-12 Full fielding the GM 10SI AC generator by grounding the tab in the "D" test hole.

Figure 7-13 Quick check of the rotor and brushes.

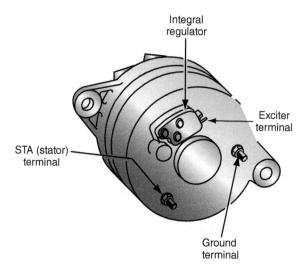

Figure 7-14 Integral regulator with exciter terminal.

If the vehicle is equipped with a CS or AD series generator, General Motors does not recommend a manual full field test. Instead, use the current output test to confirm proper operation of the generator.

Ford Motor Company has utilized different designs of the integral regulator. The early design had one terminal, called the exciter, which was connected to the outside of the regulator (**Figure 7-14**). The wiring schematic for this type of design is illustrated in **Figure 7-15**. By removing the protective cover from the field terminal (closest to the rear bearing), the field circuit can be grounded and the regulator bypassed.

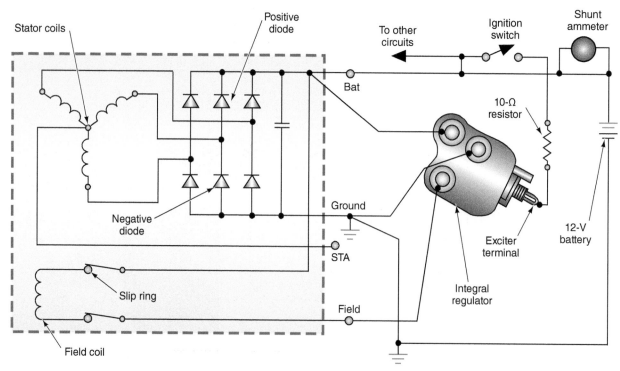

Figure 7-15 Wiring schematic of integral regulator with exciter terminal.

SERVICE TIP If the means of loading the battery is not available, do the full field test for a very short time. Do not allow voltage output to increase over 16 volts. Use the vehicle accessories to put a load on the battery.

Before full fielding the IAR AC generator, check the rotor and field circuit resistance by disconnecting the wiring plug to the regulator and connecting an ohmmeter between the regulator A and F terminals. Typically, the resistance should not be below 2.4 ohms (Ω).

If the resistance is less than 2.4 ohms, there is a short to ground somewhere in the circuit. Check for a failed regulator, a shorted rotor circuit, or a shorted field circuit.

Figure 7-16 shows the wiring of the IAR system. To full field this system, disconnect the wiring connector to the AC generator and install a 12-gauge wire jumper between the B+ terminal blades (**Figure 7-17**). Connect another jumper wire from the regulator F terminal screw to ground. Connect a voltmeter with the positive lead connected to one

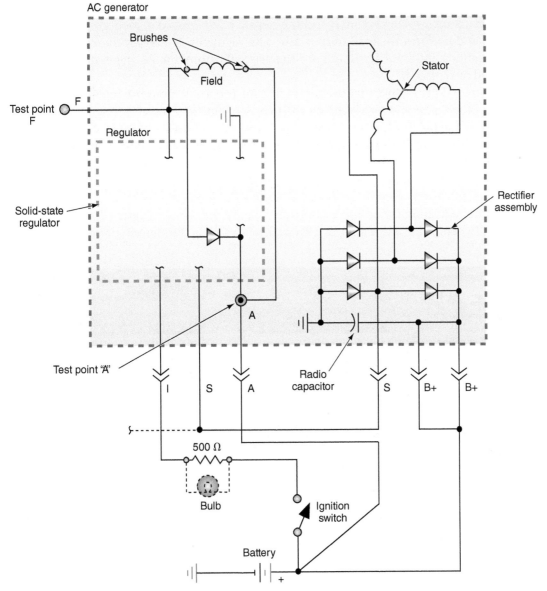

Figure 7-16 Wiring diagram of Ford's IAR charging system.

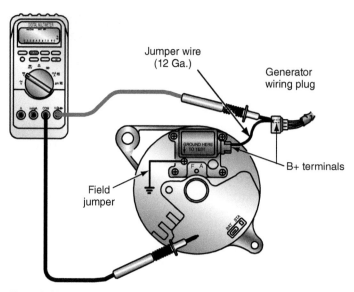

Figure 7-17 Jumper wire connections between B+ terminals.

of the B+ jumper wire terminals and the negative test lead to a good ground. Start the engine and perform the load output test. The regulator is faulty if the voltage rises to specifications. If the voltage does not rise to specifications, the AC generator needs to be serviced or replaced.

Classroom Manual
Chapter 7, page 201

Many AC generators that use internal regulators or computer-controlled regulators do not provide for a means of doing manual full fielding. Typically, they may provide a method of full fielding the generator using a scan tool (**Figures 7-18** and **7-19**). The scan tool will direct the PCM to full field the generator and display the sensed battery voltage. While full fielding, the voltage should increase. However, if the voltage does not increase, you still must determine if the fault is in the generator, the insulated side of the field circuit, the field circuit between the generator and the PCM, or the PCM itself.

If the regulation is computer controlled, study the wiring diagram for the charging system to determine if it is an A or a B circuit. By making this determination you can use

Figure 7-18 Some scan tools support an on-board generator full field test.

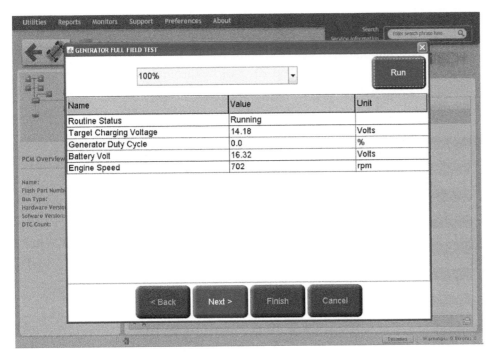

Figure 7-19 Running the full field test using a scan tool.

a jumper wire to connect the field circuit to either ground or voltage in place of the computer control. With a load against the battery, monitor the voltage output of the AC generator. If the voltage increases, the problem is not the AC generator but either the circuit between the AC generator and the computer or the computer itself. Be sure to remove the jumper wire as soon as the output voltage begins to increase. Never allow the voltage to increase over 16 volts.

⚙️ **SERVICE TIP** To test for proper PCM control of the field circuit, connect a battery charger to the battery and allow it to charge at a rate of about 15 volts with the engine running to simulate an overcharged condition. Monitor the PCM duty cycle of the field control circuit to see if the duty cycle is dropped.

AUTHOR'S NOTE When using the scan tool to perform the full field test, the PCM is not really bypassed; thus, it will still control generator output. Because of this, the output voltage will not be allowed to increase over a preset limit.

REGULATOR TEST

The regulator test is used to determine if the regulator is maintaining the correct voltage output under different load demands. To perform the regulator voltage test using a DMM, follow these steps:

1. Set the ammeter to the highest range (must be at least 10 amps).
2. Make sure the ammeter is reading zero.
3. Clamp the inductive pickup around the AC generator output wire.
4. Start the engine and hold between 1,500 and 2,000 rpm.

⚠️ **Caution**
Do not replace the regulator without first repairing any shorts in the rotor or field circuits. Doing so may damage the new regulator.

Classroom Manual
Chapter 7,
pages 195, 197, 201

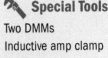 **Special Tools**
Two DMMs
Inductive amp clamp
Tachometer
Scan tool
Carbon pile

5. Allow the engine to run until the ammeter reads 10 amps or less. This indicates the battery is fully charged.
6. The voltmeter should read regulated voltage (13.5 to 14.5 volts).
7. Load the system to between 10 and 20 amps by turning on the headlights or using a carbon pile.
8. The voltmeter should still read the specified regulated voltage.

DIODE PATTERN TESTING

CUSTOMER CARE It is good practice to check the diode pattern of the AC generator anytime an electronic component fails. Because the electronics of the vehicle cannot accept AC current, the damage to the replaced component could have been the result of a bad diode. By performing this check, it is possible to find the cause of the problem.

Classroom Manual
Chapter 7, page 186

An AC generator may have an open diode, yet test close to manufacturer's specifications. If there is an open diode that is not determined in testing, a newly installed regulator may fail. In addition, an open diode can lead to the failure of other diodes.

A lab scope is an excellent diagnostic tool for testing the diodes. Set the lab scope on the lowest AC scale available. Connect the primary test leads onto the AC generator output terminal and ground. Start the engine and place a moderate load on the charging system (15 to 20 amps). Adjust the time scale so you can see at least 8 diode waveforms. Different patterns may appear. What is considered normal depends on the load placed on the system.

The diode pattern (**Figure 7-20**) illustrates a good pattern if the AC generator is under a full load. The pattern shown in **Figure 7-21** is a good pattern for some AC generators

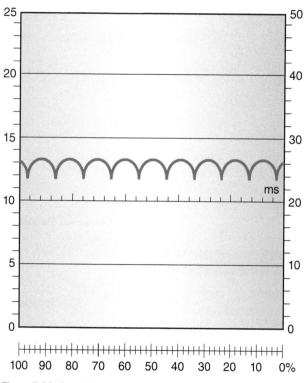

Figure 7-20 Good AC level from the generator.

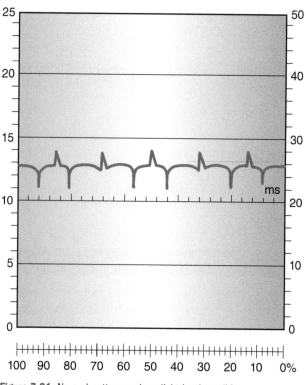

Figure 7-21 Normal pattern under a light-load condition.

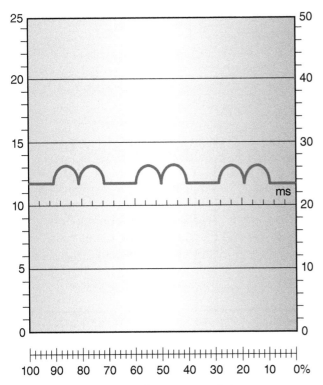

Figure 7-22 This waveform indicates an open diode.

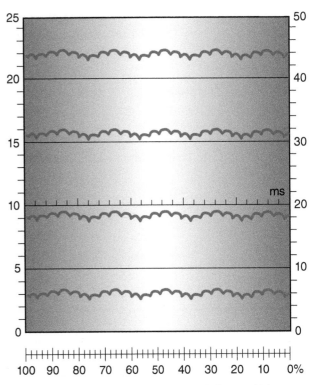

Figure 7-23 The uneven ripple of this pattern indicates a high resistance at the connection of the rectifier bridge and stator.

under a light-load condition. The spikes are the result of ignition and voltage regulator operation. The fast cycling of the regulator results in the inductive voltages that are indicated on the scope. This is one reason some manufacturers have started to use avalanche diodes in the output circuit of the generator.

Patterns that have high-resistance, open, and shorted diodes are illustrated in **Figures 7-22** through **7-25**. Remember to check the waveforms for electrical noise. If the diodes don't rectify all of the AC voltage, some will ride on the DC output.

> 🔧 **Special Tools**
> Lab scope
> DMM
> Carbon pile

> ⚙️ SERVICE TIP Instead of using a carbon pile, it is possible to place a moderate load on the charging system by turning on the headlights and a few other electrical accessories.

A DMM can be used as an alternate method to test the action and functionality of the diodes by checking for AC voltage at the output terminal of the AC generator. With the DMM set on a low AC voltage scale, connect the positive test lead to the AC generator's output terminal (not at the battery) and the negative test lead to the case of the generator. With the engine running, turn on enough accessories to cause a 10- to 20-amp flow from the generator. Ideally there should be 0 volt AC displayed on the meter. A voltage reading greater than 0.250 VAC indicates the diodes are not rectifying the AC output of the generator.

The DMM testing just described looks at the average of the three stator windings. Because of this, if one leg of the stator is beginning to have a problem, it may not be indicated by the average reading on the DMM. To see what is occurring with each leg, a lab scope is a better choice. A good pattern should not have a voltage over the peak-to-peak voltage limit of 0.250 VAC (**Figure 7-26**). Since each voltage pulse is an AC pattern from the generator, excessive AC voltage is easy to detect (**Figure 7-27**).

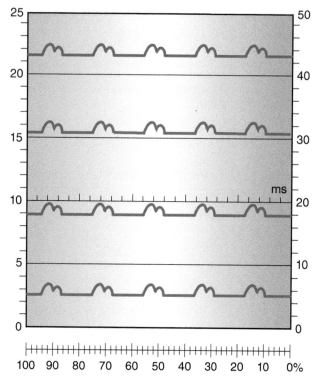

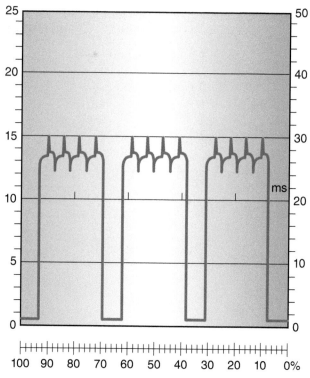

Figure 7-24 This uneven pattern indicates an open and a shorted diode.

Figure 7-25 A waveform indicating high resistance.

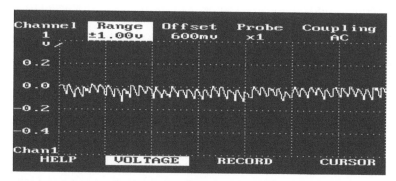

Figure 7-26 The waveform should not indicate a peak-to-peak voltage of over 0.250 volt AC.

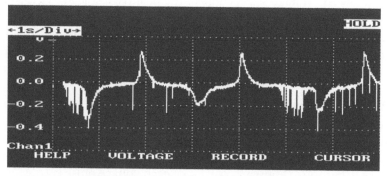

Figure 7-27 Excessive AC voltage indicates an open diode in the diode trio.

Figure 7-28 Using a DMM to test for faulty diodes.

> ⚙ SERVICE TIP In order to get accurate readings using a DMM, it must be capable of a root mean square reading.

The DMM can also be used to check for leakage from the diodes. Set DMM to the milli-amp scale and connect the test leads in series with the AC generator output terminal and the wire (**Figure 7-28**). With the engine off, read the current flow. Current flow is typically 0.5 mA, but may be as high as 2 mA. If the current is above specifications, a diode is leaking.

> ⚙ SERVICE TIP A leaking diode can result in a dead battery. Consider this when diagnosing a battery drain issue.

CHARGING SYSTEM REQUIREMENT TEST

It is possible to have a charging system that is working properly, yet not meet the requirements of the vehicle's electrical system. If an AC generator is installed on the vehicle that does not supply sufficient current output to meet the demands of the vehicle, the customer may have complaints that are identical to those of a charging system that is not functioning at all. The actual AC generator output should be at least 10% to 20% greater than the load demand. The charging system requirement test is used to determine the total electrical demand of the vehicle's electrical system.

Classroom Manual
Chapter 7, page 206

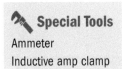 **Special Tools**
Ammeter
Inductive amp clamp

To determine the vehicle's electrical requirement:

1. Make sure the ammeter reads zero.
2. Clamp the inductive pickup around all of the negative battery cables.
3. Turn the ignition switch to the RUN position. Do not start the engine.
4. Turn on all accessories to their highest positions.
5. Read the ammeter. The indicated amperage is the total load demand of the vehicle.

USING COMPUTERIZED CHARGING SYSTEM TESTERS

Due to all of the different methods used by manufacturers for charging system operation and regulation, diagnostic equipment suppliers have developed several different computerized tools to assist the technician in properly diagnosing charging system issues. These testers typically will perform a series of tests of the system and then make a determination of the results. Once set up, the tester will guide the technician through the required procedures. The tester may require periodic updating of its database so current procedures and specifications are available.

To use the GR8 tester, follow **Photo Sequence 14**. Be sure to always reference the tester's instruction manual for the model of tester being used.

PHOTO SEQUENCE 14
Using a Computerized Charging System Analyzer

P14-1 Power on the tester and wait for the selection screen to appear.

P14-2 Proceed to the Charging System Test.

P14-3 As the test is being performed, the screen will notify the technician of what test leads need to be connected.

P14-4 During the "CHECKING FOR ALTERNATOR OUTPUT" test, the analyzer is testing the output voltage of the AC generator.

P14-5 Directions are provided in order to complete the test. After a directive is completed, press the NEXT key to continue.

P14-6 Continue to follow the instructions. In this step gradually increase the rpm until the tester screen indicates to maintain the engine speed.

P14-7 As the rpm is being held, the tester is recording measurements.

P14-8 If the test was successful, the tester will have you move into the next phase by pressing the NEXT key.

P14-9 The next test analyzes the AC generator functions at idle with no load. Data gathered during this time will be compared to data gathered in the following tests. Also, during this test, the tester looks for diode ripple to determine the condition of the diodes.

P14-10 You will be instructed to put a load on the system by turning on the headlights and heater blower motor to high speed while the engine is still idling. During this test it is determined if the AC generator can meet the electrical demands.

P14-11 Next, the tester will test the AC generator output at higher engine speeds while still having the same loads as in the previous step. Gradually increase the engine speed until the screen displays that the rpm should be held steady.

P14-12 Once again the tester will gather data. The engine speed must be held until this data is completely compiled.

P14-13 Finally the tester will indicate it is analyzing the retrieved data.

P14-14 Once the analyzing process is complete, you will be instructed to turn off all electrical loads and the shut off the engine.

P14-15 The test result will be displayed on three different screens. The first screen shows load and no-load test results.

P14-16 The second screen shows a graph of the voltage range under load and no-load test conditions.

P14-17 The third screen will show the results of the diode test.

P14-18 The test results can be printed off using the PRINT button.

If the test results for the charging system indicate that there may be a problem, the tester can perform a voltage drop test. The cable drop test uses conductance to send current through the circuit. This allows the test to be performed without having to start the engine or load the circuit. The voltage drop in the circuit is then calculated on the positive (+) and negative (−) sides of the circuit, along with total voltage drop. The amperage range for performing the cable drop test is 0 to 1,000 amps. The test requires two test lead connections. The tester's battery test leads are connected to the AC generator's output terminal and housing. The DMM test leads are connected across the battery.

To perform the cable drop test:

1. Select the Cable Drop Test icon in the Main Menu and follow the instructions on the display.
2. Select "ALT Circuit" function and press the NEXT key to continue.
3. Use the arrow keys to select the rated amperage of the AC generator and press the NEXT key to continue.
4. Connect the positive (+) clamp of the battery test leads to the AC generator output terminal.
5. Connect the negative (−) clamp to the AC generator housing.
6. Connect the positive (+) DMM clamp to the battery's positive (+) post.
7. Connect the negative clamp (−) to the battery's (−) negative post.

When the test starts, the GR8 screen will indicate that it is testing the circuit. The result screen indicates either PASS, CLEAN AND RETEST, or REPLACE.

AC GENERATOR REMOVAL AND REPLACEMENT

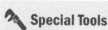

Special Tools

Fender covers
Pry bar
Belt tension gauge
Battery terminal puller

AC generator removal varies according to the engine size, engine placement, and vehicle accessories (such as power steering and air-conditioning). The following is a typical procedure for removal and replacement of the AC generator:

1. Place fender covers over the fenders.
2. Disconnect the battery ground cable.
3. Disconnect the wiring harness connections to the AC generator.
4. Loosen the drive belt tensioner.
5. Remove the drive belt.
6. Remove the upper bolt that attaches the AC generator to the mounting bracket.
7. Remove the generator lower bolt while supporting the generator.
8. Remove the generator from the vehicle.

⚠ WARNING **Never attempt to remove the AC generator or disconnect any wires to the generator without first disconnecting the battery's negative cable. Always wear safety glasses when working around the battery.**

Reverse the removal procedure to install the AC generator. Slow charge the battery to assure it is completely charged before starting the engine.

Belt Tension Adjustment

Classroom Manual
Chapter 7, page 181

There are various methods used by manufacturers to adjust the belt tension. However, regardless of the procedure, it is important that the use of the right belt is confirmed and that it is installed over the pulleys properly. The belt must be of proper length, width, depth, and type. Two types of belts are commonly used. The first is the "V" belt (**Figure 7-29**). This belt fits into a v-shaped groove in the pulleys. For proper operation, the belt must fit into the pulley so it makes full contact with the pulley (**Figure 7-30**).

Special Tools

Belt tension gauge
Belt tension tool kit

Most manufacturers now use a serpentine belt (**Figure 7-31**). This belt has a series of ribs around the inside that matches grooves on the pulleys. Make sure the bibs are properly fitted to the grooves of the pulleys (**Figure 7-32**).

Figure 7-29 "V" belt.

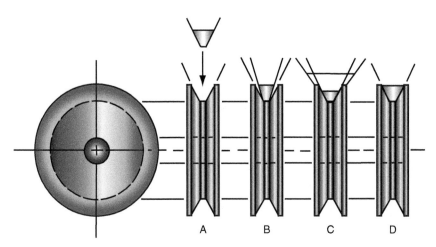

Figure 7-30 Proper installation and fit of the V belt is shown in D.

Figure 7-31 Serpentine belt.

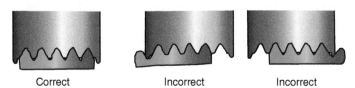

Correct   Incorrect   Incorrect

Figure 7-32 Proper installation of the serpentine belt.

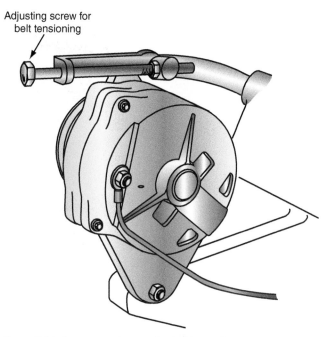

Adjusting screw for belt tensioning

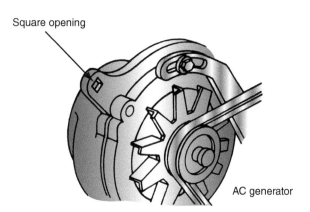

Square opening

AC generator

Figure 7-33 A typical method of adjusting belt tension at the brackets of the generator.

Figure 7-34 Some generators are fitted with an adjustment screw to adjust the belt tension.

 Caution

Do not pry on the rear housing of the generator. It may crack the housing.

 Caution

Do not remove or loosen the fastener that holds the tensioner to the engine block prior to removing the belt. Use a square drive ratchet to rotate the pulley out of the belt and remove the belt. Loosening or removing the tensioner mounting bolts before the belt is removed may damage the engine block or the tensioner.

Belt tension on some engine packages is adjusted at the generator mounting bracket (**Figure 7-33**). To adjust the belt tension, leave the pivot and adjusting arm bolts loose. Look up the correct belt tension specification for the vehicle you are working on. Install a belt tension gauge (refer to Figure 7-4) on the belt and use a square drive ratchet fitted into the square hole to rotate the generator until the proper belt tension is read on the gauge. If there is not a square hole or other method of moving the generator, then apply pressure with a pry bar to the front housing of the AC generator only. Once the correct tension reading is obtained, tighten the bolts to specified torque value. Manufacturers may provide different specifications for new and used belts. A belt is considered used if it has more than 15 minutes of run time.

Another method of adjusting the belt tension at the generator mounting brackets incorporates the use of an adjusting screw (**Figure 7-34**). This does away with the need to pry against the generator housing and provides an easier method of making the adjustment. The procedure is the same as just described, but turn the adjusting screw until proper belt tension is obtained, and then tighten the pivot bolts to the correct torque.

On many newer engines, the generator is rigidly mounted to the bracket or engine block and the belt tension is done at an idler pulley. Many engines will have a pivot bolt and an adjusting slot (**Figure 7-35**). Some provide a fitting hole for a ratchet handle to fit into to provide an easy method of applying tension to the belt.

Most current engines use an automatic belt tensioner (**Figure 7-36**). This system has a tensioner that uses a spring or hydraulic pressure to automatically maintain the proper belt tension. The auto tensioner maintains this belt tension throughout the life of the belt. Using an automatic belt tensioner system increases belt life and prevents bearing damage to driven components. A tab or boss on the back of the canister fits into a hole in the bracket or engine block. This provides an anchor point. The spring works to maintain a constant pressure that rotates the pulley assembly into the belt. The square hole is used to rotate the tensioner away from the belt during belt removal and installation. Special tools are available for use on different types of tensioners (**Figure 7-37**).

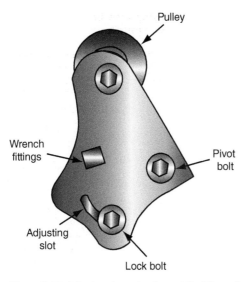

Figure 7-35 Adjustment may be done at the idler pulley.

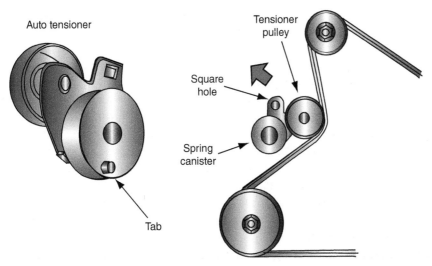

Figure 7-36 Most vehicles use an automatic belt tensioner.

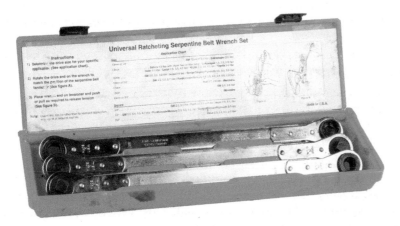

Figure 7-37 A belt tensioner tool is used to remove and install the belt on engines that use an automatic tensioner.

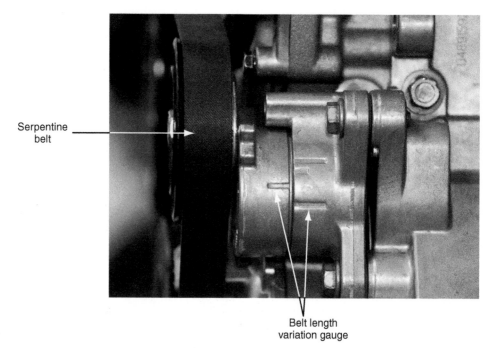

Figure 7-38 Some automatic tension adjusters have a wear indicator.

Some automatic tensioners have a built-in wear gauge (**Figure 7-38**). As the belt wears and stretches, the tensioner will continue to attempt to maintain the proper tension. To do this, the spring or hydraulics will force the tensioner to move into the belt. When the belt is worn beyond acceptable limits, the marks will align, indicating by a quick inspection that the belt requires replacement.

Some manufacturers give belt tension specifications in frequency. A frequency-measuring tool equipped with a special microphone probe is used to measure the frequency of the belt to determine proper tension. The end of the microphone probe is placed about 1 inch (25 mm) from the belt in the center of the span between the pulleys. Using your finger, pluck the belt a minimum of three times. The tool will display the frequency in hertz (Hz). Adjust the belt until the proper frequency is obtained.

Photo Sequence 15 illustrates a typical procedure for inspecting and replacing a serpentine belt.

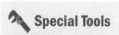 **Special Tools**

Ohmmeter
100-watt soldering iron
Resin core solder
Arbor press
Soft jaw vise
Scribe
High-temperature bearing grease
400-grain emery cloth
Pulley puller
Clean rags
Heat sink grease

AC GENERATOR DISASSEMBLY

If the AC generator fails the previous tests, the technician must decide whether to rebuild or replace the AC generator. This decision is based on several factors:

1. What is best for the customer?
2. Shop policies.
3. Cost.
4. Time.
5. Type of AC generator.
6. Availability of parts

Once the decision is made to disassemble the AC generator, the technician should study the manufacturer's service information and become familiar with the procedure for the particular AC generator being rebuilt. As an example, **Photo Sequence 16** shows the procedure for disassembling the Ford IAR AC generator. A disassembled view of this AC generator is shown (**Figure 7-39**).

Once the AC generator is disassembled, the components must be cleaned and inspected. Using a clean cloth, wipe the stator, rotor, and front bearing. Do not use solvent cleaners on these components. Inspect the front and rear bearings by rotating them on the rotor shaft. Check for noises, looseness, or roughness. Replace the defective bearing if any of these conditions are present.

PHOTO SEQUENCE 15
Typical Procedure for Inspecting and Replacing a Serpentine Belt

P15-1 Visually inspect both sides of the belt.

P15-2 The belt must be replaced if it is glazed.

P15-3 Check for cracking and tearing of the belt ribs.

P15-4 Identify the location of the belt tension adjustment.

P15-5 If needed, loosen the mounting bracket bolt for idler or generator.

P15-6 Pry the idler pulley or generator to release the tension on the belt and remove the belt.

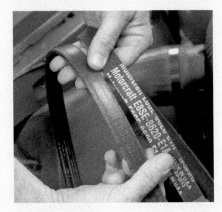

P15-7 Confirm that the new belt is the same size as the old one.

P15-8 Refer to the belt routing diagram that is located on a placard in the engine compartment.

P15-9 Install the new belt over each of the pulleys following the recommended sequence.

PHOTO SEQUENCE 15 (CONTINUED)

P15-10 Pry out the idler or generator pulley to put tension against the belt.

P15-11 Confirm that the belt is properly located on each pulley.

P15-12 Measure the deflection of the belt or use a gauge as recommended by the manufacturer.

P15-13 Pry the idler pulley or generator to tighten the belt tension until proper tension is obtained.

P15-14 Tighten the fasteners to the proper torque specifications.

P15-15 Start the engine and confirm proper belt operation and tracking.

Check the rotor shaft rear bearing surface. If the surface is not smooth, the rotor will have to be replaced. Visual inspection of the rotor includes checking the slip rings for smoothness and roundness. If the rings are discolored, dirty, scratched, nicked, or have burrs, they may be cleaned with fine-grit emery cloth. Caution must be observed to prevent creating flat spots while polishing the slip rings.

Inspect the terminals and wire leads of the rotor and stator. Also, check both the rotor and the stator for signs of burnt insulation of the windings. If there is damage to the insulation, replace the component.

Inspect the housing halves for cracks. Also, check the fan and pulley for looseness on the rotor shaft and for cracks. Replace any part that does not pass inspection. Remove the heat transfer grease that is in the rectifier mounting area with a clean cloth.

The AC generator's brushes should be inspected and tested anytime the unit is disassembled. Brushes should be replaced whenever they are worn shorter than 1/4 inch (6.35 mm) in length. Also, the brush springs must be checked for sufficient strength to keep constant contact of brushes with slip rings. Brush continuity may be checked using an ohmmeter. There should be zero resistance through the brush path. Replace the brushes if there is any resistance indicated.

Classroom Manual
Chapter 7, page 183

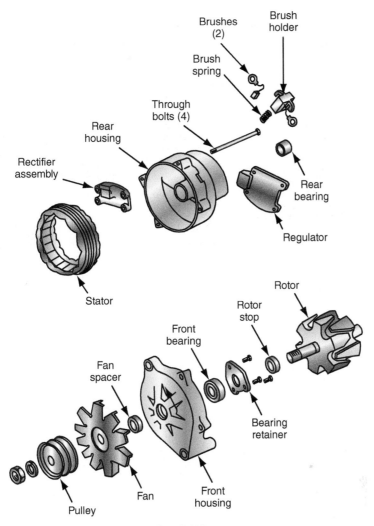

Figure 7-39 Disassembled view of Ford's IAR generator.

PHOTO SEQUENCE 16
Typical Procedure for IAR Generator Disassembly

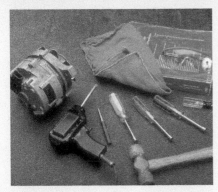

P16-1 Always have a clean and organized work area. Tools required to disassemble the Ford IAR AC generator: rags, T20 TORX wrench, plastic hammer, arbor press, 100-watt soldering iron, soft jaw vise, safety glasses, and assorted nut drivers.

P16-2 Using a T20 TORX, remove the four attaching screws that hold the regulator to the AC generator rear housing.

P16-3 Remove the regulator and brush assembly as one unit.

P16-4 Using a T20 TORX, remove the two screws that attach the regulator to the brush holder. Separate the regulator from the brush holder. Remove the A terminal insulator from the regulator.

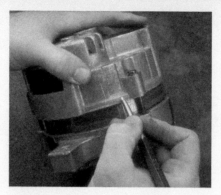

P16-5 Scribe or mark the two housing ends and the stator core for reference during assembly.

P16-6 Remove the three through bolts that attach the two housings.

P16-7 Separate the front housing from the rear housing. The rotor will come out with the front housing, while the stator will stay with the rear housing. NOTE: It may be necessary to tap the front housing with a plastic or dead weight hammer to get the halves to separate.

P16-8 Separate the three stator lead terminals from the rectifier bridge.

P16-9 Remove the stator coil from the housing.

P16-10 Using a T20 TORX, remove the four attaching bolts that hold the rectifier bridge.

P16-11 Remove the rectifier bridge from the housing.

P16-12 Use a socket to tap out the bearing from the housing.

P16-13 Clamp the rotor in a soft jaw vise.

P16-14 Remove the pulley-attaching nut, flat washer, drive pulley, fan, and fan spacer from the rotor shaft.

P16-15 Separate the front housing from the rotor. If the stop ring is damaged, remove it from the rotor; if not, leave it on the rotor shaft.

P16-16 Remove the three screws that hold the bearing retainer to the front housing.

P16-17 Remove the bearing retainer.

P16-18 Remove the front bearing from the housing. NOTE: It may be necessary to use an arbor press to remove the bearing if it does not slide out.

COMPONENT TESTING

If the AC generator is disassembled and cleaned, the individual components can be tested. The chart in **Figure 7-40** illustrates the test connections and results for the major components.

The most important test of a rotor is a complete visual inspection. Carefully check the rotor windings for signs of discoloration or overheating. If these signs are present, the rotor cannot be reused. Also, carefully inspect the slip rings; they should be flat, smooth, and free of damage. If the rotor passes the visual inspection, proceed to test it with an ohmmeter.

An ohmmeter can be used to measure the resistance between the slip rings (**Figure 7-41**). Always check the manufacturer's service information for the correct specification for the unit you are working on.

Connecting the ohmmeter from each of the slip rings to the rotor shaft should show infinite resistance (**Figure 7-42**). If any of the ohmmeter test results are not within specifications, replace the rotor.

When inspecting a stator, look for discoloration or other damage to the windings. Often the assembly will look fine but will actually be damaged due to excessive heat. One quick way of checking for this is to take the blade of a knife and scrape the windings.

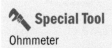 **Special Tool**
Ohmmeter

Classroom Manual
Chapter 7, page 182

Classroom Manual
Chapter 7, page 184

Component	Test Connection	Normal Reading	If Reading Was:	Trouble Is:
Rotor	Ohmmeter from slip ring to rotor shaft	Infinite resistance	Very low	Grounded
	Test lamp from slip ring to shaft	No light	Lamp light	Grounded
	Test lamp across slip rings	Lamp lights	No light	Open
Stator	Ohmmeter from any stator lead to frame	Infinite resistance	Very low	Grounded
	Test lamp from lead to frame	No light	Lamp lights	Grounded
	Ohmmeter across any pair of leads	Less than $1/2\,\Omega$	Any very high reading	Open
Diodes	Ohmmeter across diode, then reverse leads	Low reading one way; high reading other way	Both readings low / Both readings high	Shorted / Open
	12-V test lamp across diode, then reverse leads	Lamp lights one way, but not other way	No light either way / Lamp lights both ways	Open / Shorted

Figure 7-40 Guidelines for bench testing a generator.

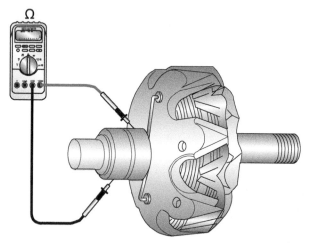

Figure 7-41 Test connections for checking for opens in the rotor windings.

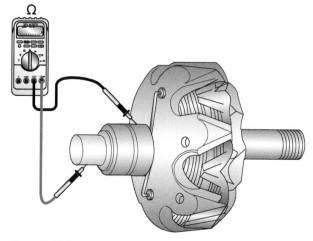

Figure 7-42 Testing the rotor for shorts to ground.

Remove the stator leads from the diodes before testing the stator windings.

If the coating or varnish flakes off, the windings have been overheated and the varnish is baked. Pay special attention to the connectors. Any signs of damage or breakage indicate the stator should not be reused. If the stator passes the visual inspection, it should be checked for opens and shorts to ground with an ohmmeter. To test for an open, connect the ohmmeter test leads to any pair of stator leads (**Figure 7-43**). Continue to test the stator until all combinations of pair connections are completed. If the ohmmeter reads infinity between any two leads, the stator has an open and it must be replaced.

To test the stator for a short to ground, connect the ohmmeter to the stator leads and stator frame (**Figure 7-44**). The ohmmeter should read infinity on all three stator leads. If the reading is less than infinity, the stator is shorted to ground and it must be replaced.

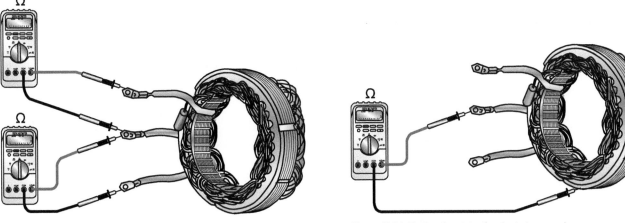

Figure 7-43 Testing the stator for opens.

Figure 7-44 Testing the stator for shorts to ground.

Because a diode should allow current to flow in only one direction, it must be tested for continuity in both directions. Test the diodes of the rectifier bridge using the diode test function of the DMM. Connect the test leads to the diode lead and the case (**Figure 7-45**). If the diode is good, it will show high resistance in one direction and about 600 mV in the opposite direction. Test all six diodes and replace any that are defective.

Classroom Manual
Chapter 7, page 186

> **SERVICE TIP** A shorted stator is difficult to test for because the resistance is very low for a normal stator. If all other components test okay, but output was low, a shorted stator is the probable cause.

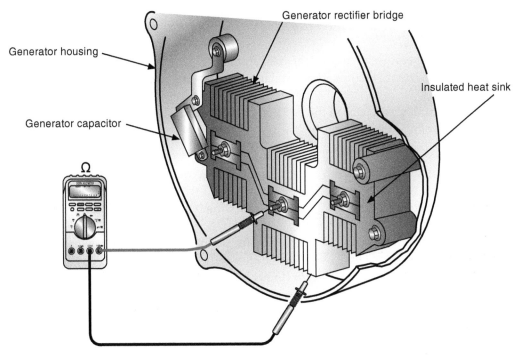

Generator rectifier bridge

Generator housing

Insulated heat sink

Generator capacitor

Figure 7-45 Testing the rectifier bridge.

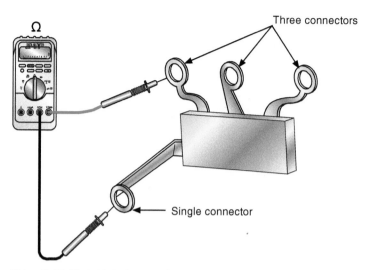

Three connectors

Single connector

Figure 7-46 Diode trio test connections.

Classroom Manual
Chapter 7, page 192

If the AC generator is equipped with a diode trio, it must be tested for opens and shorts. The procedure is much the same as with the rectifier bridge test. Connect one of the DMM test leads to the signal connector. Connect the other test lead to one of the three connectors (**Figure 7-46**). Test each of the three connectors in diode test mode. Reverse the test leads and retest. The DMM should read "OL" in one direction and above 600 mV in the other direction. These results should be obtained on all three connections. If not, replace the diode trio.

AC GENERATOR REASSEMBLY

The reassembly procedure is basically the reverse order of the disassembly. The following are suggestions to assist in the assembly process:

1. Always check the manufacturer's specifications for proper torque values of the attaching screws.
2. Use high-temperature grease on the bearings.
3. Check for free rotor rotation after installing and torquing the pulley to the rotor shaft.
4. Apply heat sink grease across the rectifier bridge base plate.
5. Protect the diodes from excess heat while soldering the connections. Use a pair of needle nose pliers as a heat sink.

When assembling a generator, always follow the recommendations of the manufacturer. It is critical that all screws and bolts be installed with the insulating washers that were present before disassembly. These insulators maintain proper circuit polarity. If a washer is left out, a short circuit will exist.

When installing the rotor into the brushes, most generators are equipped with a hole that allows a pin or paper clip to be inserted into the brush holder to keep the brushes back and allow the rotor to fit into them (**Figure 7-47**). After the rotor is in place, the pin can be removed and the brushes will snap into place on the slip rings. Before installing the pin to hold the brushes, make sure the brush springs are properly positioned behind the brushes.

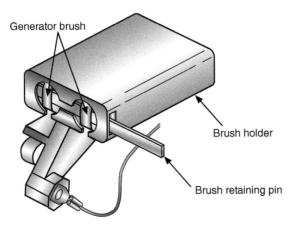

Generator brush

Brush holder

Brush retaining pin

Figure 7-47 When assembling a generator, use a pin to hold the brushes back so the rotor can be inserted into the brush holder.

HEV INVERTER/CONVERTER

Classroom Manual
Chapter 7, page 213

⚡ WARNING **Follow all safety precautions associated with working on or around the high-voltage system. Failure to follow all procedures may result in death.**

Special Tools
Isolation meter
DMM
Isolation gloves

The HEV uses an inverter to invert the DC voltage from the HV battery into three-phase AC voltage for use by the motors. It is also used to rectify the AC voltage from the motors when they are functioning as generators to charge the DC HV battery. The converter is a bidirectional DC/DC converter that changes the HV DC to LV DC to recharge the 12-volt battery.

Failures of these components to charge the batteries usually results in the setting of a DTC. This means that a majority of the diagnostics will be based on the trouble code retrieved and following the steps outlined by the manufacturer. Keep in mind that the condition of the HV cables, HV connections, LV connections, and circuits must be confirmed as in any system. In most cases, if there is no fault found in the cable or connections, the inverter or converter is replaced.

CASE STUDY

A customer states that his vehicle's engine dies and requires a jump-start. The engine runs for a few minutes and then dies again. If the headlights are turned on while the engine is running, the engine dies immediately.

The technician boost starts the engine to confirm the customer's complaint. The engine dies just as reported. The technician slow charges the battery to full capacity. Next, a full battery test series is performed. The battery passes all tests.

The technician then performs a visual inspection of the charging system. During this check, a worn and glazed AC generator drive belt is discovered. The technician does not know for certain yet if the belt is the only problem. However, the charging system tests will not be accurate if the belt is worn. The customer is informed of what the technician has found thus far, and that other tests will still need to be performed to confirm any other problems. The technician gives the customer an estimate for the belt and receives permission to replace it.

After the new belt is installed, the technician performs the voltage output test and the current output test. The requirement test is also performed to confirm that the correct-size AC generator is installed. The technician also checks the diode pattern. The charging system passes all tests.

The customer is notified that the car is ready to be picked up. When the customer arrives, the technician tactfully reminds the customer that all belts should be checked every six months and replaced per the manufacturer's maintenance schedule.

CASE STUDY

A customer states that her speedometer will read about 20 mph (32.2 kph) when the vehicle is stationary. Another shop has attempted to repair this condition by replacing the transmission output speed sensor, the PCM, and the underhood wiring harness. The technician attempts to duplicate the issue and notices that the problem only occurs if the transmission is in any gear other than PARK or NUETRAL.

Starting with the basics, the technician tests the battery and charging system. While monitoring voltage across the battery terminals with his DMM, while the charging system is loaded, he notices good AC generator voltage output. He then switches his DMM to read AC voltage and uses the min/max function to capture the voltage readings. The low voltage was −2.87 volts and the high voltage was 2.11 volts. Realizing that the voltage swing was too great, he determined that a diode was faulty in the AC generator.

With the customer's approval, the AC generator was replaced and the belt set to the proper tension. During the retest the voltage swing was eliminated. Also, the speedometer works properly now. The vehicle was returned to a very happy customer.

The AC voltage was "tricking" the PCM into thinking the vehicle was going 20 mph (32.2 kph). By performing the basic tests first, the technician was able to quickly identify the root cause of the issue.

ASE-STYLE REVIEW QUESTIONS

1. Charging system testing is being discussed.
   Technician A says before attempting to test the charging system, the battery must be checked.
   Technician B says the state of charge of the battery is not a concern to charging system testing.
   Who is correct?
   A. A only
   B. B only
   C. Both A and B
   D. Neither A nor B

2. AC generator noise complaints are being discussed.
   Technician A says a loose belt will make a grumbling noise.
   Technician B says a whining noise can be caused by a shorted diode.
   Who is correct?
   A. A only
   B. B only
   C. Both A and B
   D. Neither A nor B

3. *Technician A* says the voltage output test is used to make a quick determination concerning whether the charging system is working properly.
   Technician B says when testing a charging system, the first step is to perform a visual inspection and preliminary checks of the charging system.
   Who is correct?
   A. A only
   B. B only
   C. Both A and B
   D. Neither A nor B

4. Test results of the voltage output test are being discussed.
   Technician A says if the charging voltage is too high, there may be a loose or glazed drive belt.
   Technician B says if the charging voltage is too low, the fault might be a grounded field wire from the regulator (full fielding).
   Who is correct?
   A. A only
   B. B only
   C. Both A and B
   D. Neither A nor B

5. Voltage drop testing is being discussed.
   Technician A says the total system drops should be less than 0.7 volt.
   Technician B says the ground side voltage drop should be less than 0.2 volt.
   Who is correct?
   A. A only
   B. B only
   C. Both A and B
   D. Neither A nor B

6. *Technician A* says the field current-draw test determines if there is current available to the field windings.
   Technician B says a slipping belt can cause a low reading when performing the field current-draw test.
   Who is correct?
   A. A only
   B. B only
   C. Both A and B
   D. Neither A nor B

7. An ammeter is connected in series between the AC generator output terminal and the output circuit wire. The ammeter reads about 4 mA with the engine off.

   Technician A says the field winding is shorted.

   Technician B says this indicates a leaking diode.

   Who is correct?

   A. A only
   B. B only
   C. Both A and B
   D. Neither A nor B

8. *Technician A* says full fielding means the field windings are constantly energized with full battery voltage.

   Technician B says the full field test will isolate whether the detected problem lies in the AC generator or the regulator.

   Who is correct?

   A. A only
   B. B only
   C. Both A and B
   D. Neither A nor B

9. *Technician A* says if full fielding with the regulator bypassed produces lower-than-specified output, the regulator is the cause of the problem.

   Technician B says a DMM set to read AC voltage at the output terminal of the AC generator should read less than 0.250 VAC.

   Who is correct?

   A. A only
   B. B only
   C. Both A and B
   D. Neither A nor B

10. The results of current output testing are being discussed.

    Technician A says an open diode can result in an output that is 2 to 8 amps below specifications.

    Technician B says a current output reading of 10 to 15 amps below specifications indicates a possible slipping belt.

    Who is correct?

    A. A only
    B. B only
    C. Both A and B
    D. Neither A nor B

ASE CHALLENGE QUESTIONS

1. A vehicle's battery discharges in a very short period of time due to a shorted cell. Before replacing the battery, a charging system test is performed. The engine is started and all accessories are turned off.

   Technician A says that the charging system's amperage output will be lower than normal.

   Technician B says that the alternator's field current will be high when the engine is running.

   Who is correct?

   A. A only
   B. B only
   C. Both A and B
   D. Neither A nor B

2. The charging system voltage of a vehicle equipped with an external voltage regulator is 15.5 volts at 1,500 rpm.

   Technician A says that the voltage regulator sensing circuit may have excessive resistance.

   Technician B says that the alternator's field circuit may have excessive resistance.

   Who is correct?

   A. A only
   B. B only
   C. Both A and B
   D. Neither A nor B

3. Charging system voltage is being measured at two places at the same time with the engine running. Connecting the voltmeter's leads across the battery terminals results in a 12.8-volt reading, while connecting the voltmeter to the AC generator output terminal and the AC generator case results in a 14.2-volt reading.

   Technician A says that there may be excessive resistance on the ground side of the charging system.

   Technician B says that there may be excessive resistance on the positive side of the charging system.

   Who is correct?

   A. A only
   B. B only
   C. Both A and B
   D. Neither A nor B

4. A customer says that his battery completely discharges every few days even though he drives his car about 50 miles every day. A test of his charging system is performed and the following measurements are recorded with all accessories turned on.

At 1,300 rpm the charging system voltage and amperage are 12.4 volts and 60 amps. A replacement alternator produces the same measurements.

Technician A says that there may be excessive electrical demand on the alternator.

Technician B says that the charging system output circuit may have excessive voltage drop.

Who is correct?

A. A only

B. B only

C. Both A and B

D. Neither A nor B

5. Which of the following could result in battery overheating and eventual premature failure?

A. Open field circuit.

B. Shorted stator windings.

C. Open diode.

D. High resistance in the voltage regulator sensing circuit.

Name _____ Date _____

INSPECTING DRIVE BELTS

Upon completion of this job sheet, you should be able to visually inspect drive belts and check their tightness.

NATEF Correlation

This job sheet addresses the following **MLR** task:

D.2. Inspect, adjust, and/or replace generator (alternator) drive belts; check pulleys and tensioners for wear; check pulley and belt alignment.

This job sheet addresses the following **AST/MAST** task:

D.3. Inspect, adjust, and/or replace generator (alternator) drive belts; check pulleys and tensioners for wear; check pulley and belt alignment.

Tools and Materials

- Two vehicles, one with a serpentine belt and the other with V belts
- Service information for the vehicles assigned
- Belt wear gauge
- Belt tension gauge
- Belt tensioner tools

Describe the vehicles being worked on:

Year _____ Make _____ Model _____

VIN _____ Engine type and size _____

Procedure

1. On the vehicle with a serpentine belt, carefully inspect the belt and describe the general condition of the belt.

   _____

   _____

   _____

2. Use the belt wear gauge to determine if the belt is excessively worn. What is your determination?

   _____

   _____

   _____

3. Check the alignment of all pulleys that the belt uses. What are your recommendations?

   _____

   _____

   _____

4. Check the condition of all pulleys. What are your recommendations?

_____

_____

_____

5. With the proper belt tension gauge, check the tension of the belt. Belt tension should be _____. You found _____

6. Based on the above, what are your recommendations?

_____

_____

_____

7. Describe the procedure for adjusting the tension of the belt.

_____

_____

_____

8. Set the belt tension to the proper specifications.

9. On the vehicle with V belts, you will find more than one drive belt. List the different belts by their purpose.

_____

_____

_____

10. Carefully inspect the belts and describe the general condition of each one.

_____

_____

_____

11. Check the alignment of all pulleys that the belt(s) uses. What are your recommendations?

_____

_____

_____

12. Check the condition of all pulleys. What are your recommendations?

_____

_____

_____

13. Check the tension of the AC generator drive belt. Belt tensions should be _____. You found _____

14. Based on the above, what are your recommendations?

_____

_____

_____

15. Describe the procedure for adjusting the tension of the belt.

_____

_____

_____

16. Set the belt tension to the proper specifications.

_____

_____

_____

Instructor's Response

_____

_____

_____

Name _____   Date _____

TESTING CHARGING SYSTEM OUTPUT

Upon completion of this job sheet, you should be able to measure the output of the charging system.

NATEF Correlation

This job sheet addresses the following **MLR/AST/MAST** tasks:

D.1.   Perform charging system output test; determine necessary action.

This job sheet addresses the following **AST/MAST** task:

D.2.   Diagnose (troubleshoot) charging system for causes of undercharge, no-charge, or overcharge conditions.

NATEF

Tools and Materials

- A vehicle
- Service information for the vehicle assigned
- Starting/charging system tester
- Safety glasses

Describe the vehicle being worked on:

Year _____ Make _____ Model _____

VIN _____ Engine type and size _____

Procedure   **Task Completed**

1. Identify the type and model of AC generator. What type is it?

   _____

   What are the output specifications for this AC generator?

   _____ amps _____ and _____

   volts at _____ rpm.

2. Connect the starting/charging system tester to the vehicle.   ☐

3. Start the engine and run it at the specified engine speed.   ☐

4. Observe the output to the battery. The meter readings are

   _____ amps and _____ volts.

5. Compare readings to specifications and give recommendations.

   _____

   _____

   _____

   _____

6. Are the readings below or above the specifications? _____

   If below, go to step 7. If above, go to step 11.

7. Refer to the service information and list possible causes for the below specification results.

_____

_____

_____

8. If readings are below specifications, refer to the service information to find the proper way to full field the AC generator. Describe the method.

_____

_____

_____

_____

9. If possible, full field the generator and observe the output to the battery. The meter readings are _____ amps and _____ volts.

10. Compare readings to specifications and give recommendations.

_____

_____

_____

11. If the readings were above specifications, refer to the service information and list what could be the causes:

_____

_____

_____

12. Describe what test procedure you would do to isolate the cause of the overspecification test results.

_____

_____

_____

Instructor's Response

_____

_____

_____

Name _____   Date _____

TESTING THE CHARGING SYSTEM CIRCUIT

Upon completion of this job sheet, you should be able to visually inspect and test the insulated and ground side of the charging system circuit.

NATEF Correlation

This job sheet addresses the following **MLR** task:

D.4.   Perform charging circuit voltage drop tests; determine necessary action.

This job sheet addresses the following **AST/MAST** task:

D.5.   Perform charging circuit voltage drop tests; determine necessary action.

NATEF

Tools and Materials

• A vehicle
• Wiring diagram for the vehicle assigned
• A DMM

Describe the vehicle being worked on:

Year _____ Make _____ Model _____

VIN _____ Engine type and size _____

Procedure

Task Completed

1. Describe the general appearance of the AC generator and the wires that are attached to it.

   _____

   _____

   _____

2. Measure the open circuit voltage of the battery. Your measurement was
   _____ volts.

3. From the wiring diagram, identify the output, input, and ground wires for the AC generator. Describe these wires, by color and location.

   _____

   _____

   _____

   _____

4. Start the engine and allow it to run. Then turn on the headlights in the high-beam mode.   ☐

5. Connect the DMM across the charging system's output wire. Measure the voltage drop and record your readings.

   _____

   _____

6. Connect the DMM across the charging system's input wire. Measure the voltage drop and record your readings.

_____

7. Connect the DMM across the charging system's ground wire. Measure the voltage drop and record your readings.

_____

8. What do the test results indicate?

_____

_____

_____

Instructor's Response

_____

_____

_____

Name _____   **Date** _____

REMOVING AND INSTALLING THE AC GENERATOR

Upon completion of this job sheet, you should be able to properly remove, inspect and replace the AC generator.

NATEF Correlation ─────────────────────────────

This job sheet addresses the following **MLR** task:

D.3.   Remove, inspect, and/or replace the generator (alternator).

This job sheet addresses the following **AST/MAST** task:

D.4.   Remove, inspect, and/or replace the generator (alternator).

NATEF

Tools and Materials

- A vehicle
- Service information
- Safety glasses
- Fender covers
- Belt wear gauge
- Belt tension gauge
- Belt tensioner tools

Describe the vehicle being worked on:

Year _____ Make _____ Model _____

VIN _____ Engine type and size _____

Procedure   **Task Completed**

Note: It may be necessary to lift the vehicle on the hoist in order to access the ACV generator. Prior to raising the vehicle, disconnect/remove all components that can be accessed from the engine compartment.

_____

_____

_____

1. Place fender covers over the fenders.   ☐

2. Disconnect the battery ground cable.   ☐

⚡ **WARNING Never attempt to remove the AC generator or disconnect any wires to the generator without first disconnecting the battery's negative cable. Always wear safety glasses when working around the battery.**

3. Disconnect the wiring harness connections to the AC generator.   ☐

4. Describe the procedure for loosing the drive belt tension.   ☐

_____

_____

_____

Task Completed

5. Remove the drive belt and inspect it as outlined in Job Sheet 36. Describe the condition of the belt:

    _____

    _____

    _____

6. Remove the upper bolt that attaches the AC generator to the mounting bracket.   ☐

7. Remove the generator lower bolt while supporting the generator.   ☐

8. Remove the generator from the vehicle.   ☐

    Inspect the AC generator for indications of overheating, damaged/loose terminals, worn bearings, damaged cooling fan, and damaged pulley. What are your findings?

    _____

    _____

    _____

9. Check the condition of the belt drive, idler, and tension pulleys. Record your findings:

    _____

    _____

    _____

10. Reverse the removal procedure to install the AC generator.   ☐

11. Locate the drive belt over the pulleys. Are the pulleys and belt properly aligned?   Yes ☐   No ☐

    If no, consult with your instructor.

12. Adjust the belt tension as outlined in Job Sheet 36.   ☐

13. Connect all wires that attach to the AC generator.   ☐

14. Slow charge the battery to assure it is completely charged.   ☐

15. Connect the negative battery cable.   ☐

16. Check the output of the AC generator as outline in Job Sheet 37. What are your results?

    _____

    _____

    _____

    _____

Instructor's Response

_____

_____

DIAGNOSTIC CHART 7-1

PROBLEM AREA:	Charging system operation.
SYMPTOMS:	1. Battery is too low to start engine. 2. Headlight illumination dim. 3. Electrical accessories do not operate properly. 4. Charging indicator warning light illuminated.
POSSIBLE CAUSES:	1. Slipping or worn drive belt. 2. Poor battery cable connections. 3. Faulty voltage regulator. 4. Shorted stator. 5. Open stator. 6. Open diode. 7. Shorted diode. 8. Slipping or worn drive belt. 9. Worn brushes. 10. Open field coil circuit. 11. Shorted field coil. 12. Worn or slipping drive belt. 13. Worn brushes. 14. Faulty rectifier bridge. 15. Faulty diode trio. 16. Excessive resistance in system circuit. 17. Open in the circuit. 18. Improper ground connection. 19. Loose or corroded connection.

DIAGNOSTIC CHART 7-2

PROBLEM AREA:	Charging system operation (excessive charging).
SYMPTOMS:	1. Battery electrolyte level constantly low. 2. Bulbs burn out. 3. Brighter than normal bulb illumination. 4. Electrical system component failures.
POSSIBLE CAUSES:	1. Faulty voltage regulator. 2. Grounded field coil circuit. 3. Excessive resistance in sensing circuit. 4. Faulty generator.

DIAGNOSTIC CHART 7-3

PROBLEM AREA:	Voltage regulation.
SYMPTOMS:	1. Charging system overcharging the battery. 2. Charging system output below specifications.
POSSIBLE CAUSES:	1. Defective voltage regulator. 2. Open or short in sense circuit.

DIAGNOSTIC CHART 7-4

PROBLEM AREA:	Diodes.
SYMPTOMS:	**1.** Excessive noises. **2.** Low or no charging system output. **3.** AC generator fails output test.
POSSIBLE CAUSES:	**1.** Open diodes. **2.** Shorted diodes.

DIAGNOSTIC CHART 7-5

PROBLEM AREA:	AC generator pulleys.
SYMPTOMS:	**1.** Belts wear prematurely. **2.** Noises.
POSSIBLE CAUSES:	**1.** Pulley bent.

DIAGNOSTIC CHART 7-6

PROBLEM AREA:	AC generator.
SYMPTOMS:	**1.** Noises.
POSSIBLE CAUSES:	**1.** Bent fan blades. **2.** Worn or dry bearings.

DIAGNOSTIC CHART 7-7

PROBLEM AREA:	HEV 12-volt battery low SOC.
SYMPTOMS:	**1.** LV battery fails to charge. **2.** No HEV operation. **3.** MIL or HEV warning lamps.
POSSIBLE CAUSES:	**1.** 12-volt battery failure. **2.** Low SOC of 12-volt battery. **3.** High resistance in 12-volt battery cables or connections. **4.** Improper inverter/converter module ground connection. **5.** Poor or improper inverter/converter module LV connections. **6.** Excessive LV system loads. **7.** Faulty inverter/converter module.

DIAGNOSTIC CHART 7-8

PROBLEM AREA:	HEV 12-volt battery high SOC.
SYMPTOMS:	**1.** LV battery overcharged. **2.** No HEV operation. **3.** MIL or HEV warning lamps.
POSSIBLE CAUSES:	**1.** 12-volt battery failure. **2.** Improper jump-start procedures. **3.** Overcharging with a battery charger. **4.** High resistance in 12-volt battery cables or connections. **5.** Improper inverter/converter module ground connection. **6.** Poor or improper inverter/converter module LV connections. **7.** Faulty inverter/converter module.

CHAPTER 8
BODY COMPUTER SYSTEM DIAGNOSIS

Upon completion and review of this chapter, you should be able to:

- Describe the service precautions associated with servicing the BCM.
- Diagnose computer voltage supply and ground circuits.
- Distinguish between hard and intermittent codes.
- Perform flash code retrieval on various vehicles.
- Erase fault codes.
- Perform a complete visual inspection of the problem system.
- Enter BCM diagnostics by use of a scan tool.
- Enter BCM diagnostics through the ECC panel.
- Perform basic actuator tests.
- Perform basic sensor tests.
- Properly flash the BCM.

Basic Tools

Basic mechanic's tool set

Service information

Terms To Know

Activation test
Breakout box
Diagnostic trouble code (DTC)
Electrostatic discharge (ESD) strap
Fail-safe
Flash
Flash codes
Hard codes
Impedance
Intermittent codes

INTRODUCTION

Because the body control module (BCM) operates many functions of the vehicle's electrical systems, it is important for today's technician to be able to properly diagnose problems with this system. The use of body computers has expanded to include the functions of climate control, lighting circuits, cruise control, antilock braking, electronic suspension systems, electronic shift transmissions, and alternator regulation. In some systems, the direction lights, the rear window defogger, the illuminated entry, the intermittent wiper, and the antitheft systems are included in the body controller function (**Figure 8-1**).

> **AUTHOR'S NOTE** Many manufacturers are incorporating the functions of the typical BCM into other modules. For example, on many newer Chrysler vehicles the cabin compartment node (CCN) now performs the functions that were once the responsibility of the BCM. These vehicles no longer have a separate BCM. The term *BCM* will be used throughout this text in reference to the stand-alone module and the functions that may be contained within any other module.

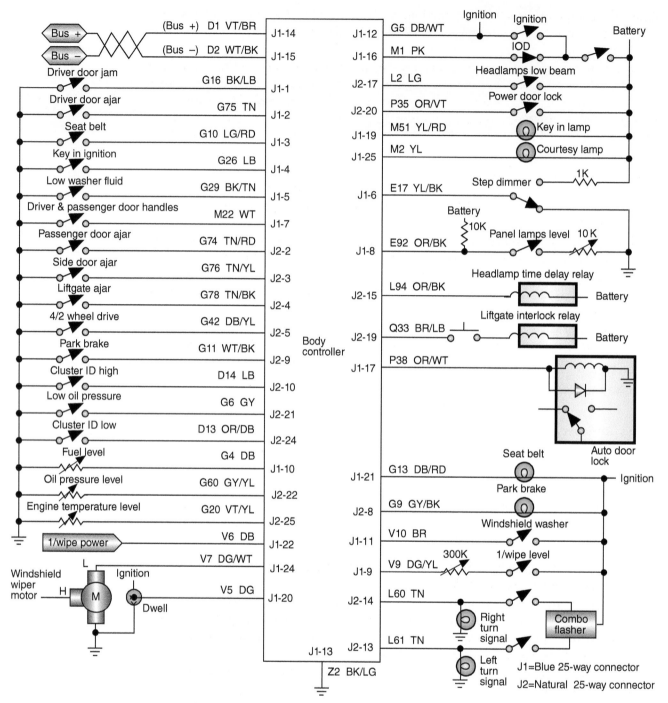

Figure 8-1 The body controller operates many of the vehicle's electrical systems.

As discussed in the Classroom Manual, a computer processes the physical conditions that represent information (data). The operation of the computer is divided into four basic functions: input, processing, storage, and output. Understanding these four computer functions will help you organize the troubleshooting process. When a system is tested, you attempt to isolate a problem with one of these functions.

In the process of controlling the various electrical systems, the BCM continuously monitors operating conditions for possible system malfunctions. The computer compares system conditions against programmed parameters. If the conditions fall outside of these limits, the computer detects a malfunction. A **diagnostic trouble code (DTC)** is set to indicate the portion of the system at fault. The technician can access this code for aid in troubleshooting.

If a malfunction results in improper system operation, the computer may minimize the effects by using **fail-safe** action. This provides limited system operation by substituting a fixed input value if a sensor circuit should fail. For example, if the automatic temperature control (ATC) system has a malfunction from the ambient temperature sensor, instead of shutting down the whole system, the computer will provide a fixed value as its own input. This fixed value can be programmed into the computer's memory, or it can be the last received signal from the sensor prior to failure. This allows the system to operate on a limited basis instead of shutting down completely. Some other faults may result in the ATC system switching to high fan speed, full heat, or defrost mode.

Some systems may substitute an invalid input value by referencing other inputs. The BCM uses these other inputs to calculate what the missing input value probably would be. For example, if the ambient air temperature sensor input is invalid, the BCM may look at the battery temperature sensor and the sun load sensor, and make a calculation of what the ambient temperature would be.

You need to know several things before learning how to access the computer's memory to gain information concerning system operation. You need to become familiar with what you're looking at and you must follow proper precautions when servicing these systems.

> **Fail-safe** action is commonly known as "limp-in mode" or "fail soft."

ELECTRONIC SERVICE PRECAUTIONS

The technician must take some precautions before servicing the body computer or any of its controlled systems. The BCM is designed to withstand normal current draws associated with normal operation. However, overloading any of the system circuits will result in damage to the BCM. To prevent BCM and circuit damage, follow these service precautions:

1. Do not ground or apply voltage to any controlled circuits unless the service information instructs you to do so.
2. Use only a high-**impedance** multimeter (10 megaohms or greater) to test the circuits. Impedance is the combined opposition to current created by the resistance, capacitance, and inductance of the meter. Never use a test light unless specifically instructed to do so in the service information.
3. Make sure the ignition switch is turned off before making or breaking electrical connections to the BCM.
4. Unless instructed otherwise in the service information, turn off the ignition switch before making or breaking any electrical connections to sensors or actuators.
5. Turn the ignition switch off whenever disconnecting or connecting the battery terminals. Also, turn it off when pulling and replacing the fuse.
6. Do not connect any other electrical accessories to the insulated or ground circuits of the computer-controlled systems.
7. Use only manufacturer's specific test and replacement procedures for the year and model of vehicle being serviced.
8. When handling a computer, wear an **electrostatic discharge (ESD) strap** to ground your body to prevent static discharges that may damage electronic components.

> Static electricity can be 25,000 volts or higher.

Figure 8-2 One type of Electrostatic Discharge (ESD) symbol used to warn the technician that the component or circuit is sensitive to static electricity.

By following these precautions, plus those listed in the service information, you can avoid having to replace expensive components.

Electrostatic Discharge

Some manufacturers mark certain components and circuits with a code or symbol to warn technicians that the units are sensitive to electrostatic discharge (**Figure 8-2**). Static electricity can destroy or render a component useless.

When handling any electronic part, especially those that are static sensitive, follow the listed guidelines to reduce the possibility of electrostatic buildup on your body and the inadvertent discharge to the electronic part. If you are not sure if a part is sensitive to static, treat it as if it is.

1. Always touch a known good ground before handling the part. This should be repeated while handling the part and more frequently after sliding across a seat, sitting down from a standing position, or walking a distance.
2. Avoid touching the electrical terminals of the part unless you are instructed to do so in the written service procedures. It is good practice to keep your fingers off all electrical terminals as the oil from your skin can cause corrosion.
3. When you are using a voltmeter, always connect the negative meter lead first.
4. Do not remove a part from its protective package until it is time to install the part.
5. Before removing the part from its package, ground yourself and the package to a known good ground on the vehicle.

Electromagnetic Interference

Electromagnetic interference (EMI) or radio frequency interference (RFI) can cause problems with the vehicle's on-board computers. Unfortunately, an automobile's spark plug wires, ignition coil, and generator possess the ability to generate these interferences. Under the right conditions, EMI and RFI can trigger sensors or actuators. The result may be an intermittent problem with system operation.

To minimize the effects of EMI and RFI, make sure your visual inspection is thorough. Also, check to make sure that sensor wires running to the computer are routed away from potential EMI and RFI sources. Rerouting a wire by no more than an inch or two (25 to 50 mm) may keep EMI and RFI from falsely triggering or interfering with computer operation. EMI and RFI can be present on a voltage signal or on a ground.

Most manufacturers shield their BCM from EMI and RFI. However, this shielding will work only if the BCM is properly grounded. Always confirm a good ground before condemning the BCM. Some BCMs may have up to five grounds.

In addition, confirm that all grounding straps are properly connected. A loose or missing grounding strap can cause erratic voltages to be seen by the computer and cause multiple issues.

DIAGNOSIS OF COMPUTER VOLTAGE SUPPLY AND GROUND CIRCUITS

Like any other electrical or electronic component, a computer cannot operate properly unless it has satisfactory voltage supply at the required terminals and proper ground connections. A computer wiring diagram for the vehicle being tested must be available for these tests. First, measure the voltage across the battery terminals as a reference. Now, backprobe the battery feed terminal to the computer (if allowed by manufacturer) and connect the digital volt/ohmmeter (DVOM) leads to this terminal and to ground. The voltage at the battery feed terminal should be 0.5 volt (V) of battery voltage with the ignition switch off. If the proper voltage is not present at this terminal, check the computer fuse and related circuit. Turn the ignition switch to the RUN position and check applied voltage to *all* battery and ignition feed terminals of the computer. The voltage measured at these terminals should be within 0.5 volt of battery voltage. If the specified voltage is not available, test the voltage supply wires to these terminals. These terminals may be connected through fuses, fuse links, or relays. Always refer to the vehicle manufacturer's wiring diagram for the vehicle being tested.

Classroom Manual
Chapter 8, page 225

Special Tools
DVOM
Lab scope

SERVICE TIP Never replace a computer unless the ground and voltage supply circuits are proven to be in satisfactory condition.

Computer ground wires usually extend from the computer to a ground connection on the engine or battery. Often there is more than one ground wire. In addition, the fasteners used to attach the computer to the vehicle chassis may be used as a ground. With the ignition switch in the RUN position, perform a voltage drop test across the ground wires. Compare your results with specifications. A good circuit should not drop more than 0.2 volt on the ground circuit.

If the manufacturer does not allow for backprobing of the connector, then use a **breakout box** or carefully forward probe the connector. In this case the voltage test is an open circuit. To achieve accurate results the circuit must be loaded. Use a test light that draws about 250 mA to load the circuit and then measure the voltage at the test lamp (**Figure 8-3**). When testing the battery and ignition feed circuits, the voltmeter should read within 500 mV of battery voltage. When reading the voltage of the ground circuit, the voltmeter should read less than 200 mV.

It is imperative to test all power and ground circuits. There may be multiple direct battery feeds and multiple grounds. Although these may be referred to as "redundant" powers and grounds, they do not actually become substitute circuits in the event that the other drops off. For example, a module may have two direct power feeds and two ground circuits. If one of the power feeds should open, the other does not make up for it. In this case, it is possible that each power feed supplies a microprocessor (µP), so if one is lost then one of the microprocessors fails to power up. The same may be true for the ground circuits. In fact, if a ground circuit is open, electrons will attempt to find alternate paths to grounds and the resulting feedback through the computer can result in a very strange system operation.

A **breakout box** is a special tool that connects to the computer connectors to allow the technician to test circuits, sensors, and actuators by providing test points.

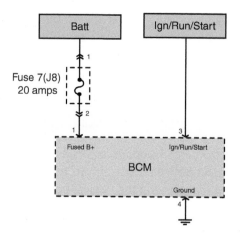

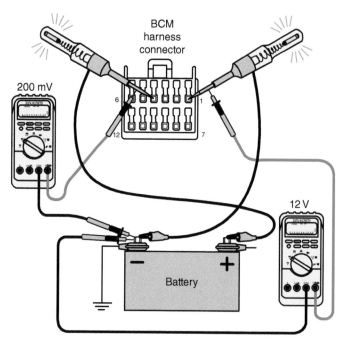

Figure 8-3 Using a test light to load the circuit when the connector is unplugged.

Sensor Return Circuit

Many manufacturers use a single sensor return circuit for all of the BCM sensors. This makes a good ground critical since a sensor return circuit that has high resistance will cause *all* of the sensors to indicate an erroneous voltage reading to the BCM. Consider a temperature sensing circuit as an example (**Figure 8-4**). (A) indicates the normal voltage sensed by the BCM with the NTC thermistor at 10 K ohms (O). (B) indicates the voltage sensed by the BCM with the NTC thermistor still at 10 K ohms but with an additional 200 ohms of resistance in the sensor's return circuit. The voltage is now off by 25 mV. The 200 ohms of resistance also affect all of the other sensors that share this return circuit and all of them will have a higher input voltage as a result. This may not seem like enough voltage to cause any problems, but remember that most systems are very sensitive to small voltage changes.

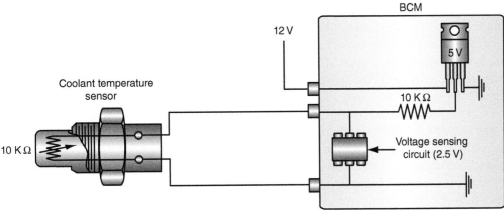

(A) NORMAL SENSOR RETURN CIRCUIT

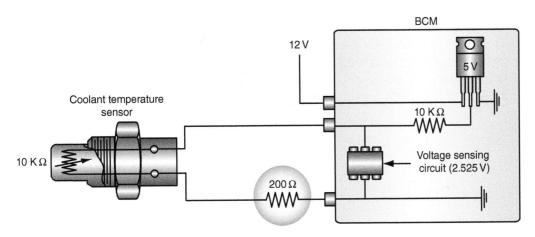

(B) RESISTANCE IN SENSOR RETURN CIRCUIT

Figure 8-4 The effect of resistance on the sensor return circuit. (A) Normal sensor voltage. (B) Voltage reading with NTC thermistor at the same resistance, but sensor return circuit has additional resistance.

⚙ SERVICE TIP When diagnosing computer problems, it is usually helpful to ask the customer about service work that has been performed lately on the vehicle. If service work has been performed, it is possible that a computer ground wire may be loose or disconnected.

In addition to causing some system operation problems, this type of fault may not set a fault code. The voltage values are within the normal operating parameters of the sensor, so continuity faults will not set. Also, since all sensors are affected by the resistance, rationality faults are not detected. System operation problems may occur as a result of the computer making bad decisions based on erroneous information.

Poor grounds can also allow EMI or noise to be present on the reference voltage signal. This noise causes minute changes in the voltage going to the sensor; therefore, the output signal from the sensor will also have these voltage changes. The computer will try to respond to these changes, which can cause a system operation problem. The best way to check for noise is to use a lab scope.

Connect the lab scope between the 5-volt reference signal to the sensor and ground. The trace on the scope should be flat (**Figure 8-5**). If noise is present, move the scope's negative probe to a known good ground. If the noise disappears, the sensor's ground circuit is faulty or has resistance. If the noise is still present, the voltage feed circuit is faulty or there is EMI in the circuit from another source, such as the AC generator. Find and repair the cause of the noise.

Circuit noise may be present in either the positive side or the negative side of a circuit. It may also be evident by a popping noise from the radio. However, noise can cause a variety of problems in any electrical circuit. The most common sources of noise are electric motors, relays, solenoids, AC generators, ignition systems, switches, and air-conditioning (A/C) compressor clutches. Typically, noise is the result of an electrical device being turned on and off. Sometimes the source of the noise is a defective suppression device. Some of the commonly used noise suppression devices are resistor-type secondary cables and spark plugs, shielded cables, capacitors, diodes, and resistors. If the source of the noise is not a poor ground or a defective component, check the suppression devices.

Resistors that are used for noise suppression do not eliminate the spikes, but they do limit their intensity. When testing a circuit that uses a resistor to limit noise, if the lab scope trace indicates large voltage spikes, the resistor may be bad. Clamping diodes are used to suppress noise and induced voltages in circuits such as A/C compressor clutches and relays. If the diode is bad, a negative spike will result (**Figure 8-6**). Capacitors or

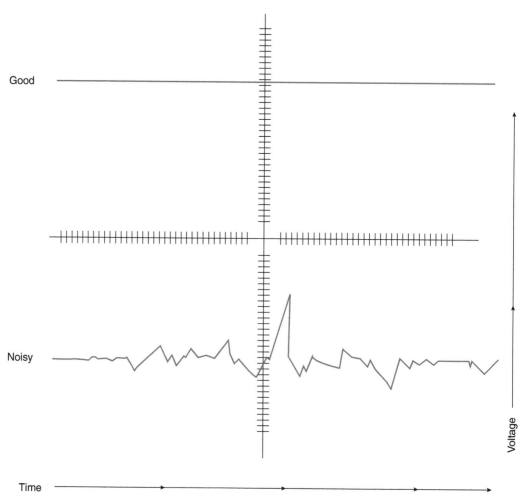

Figure 8-5 (Top) Voltage signal trace of a good circuit. (Bottom) Trace of a circuit with excess noise.

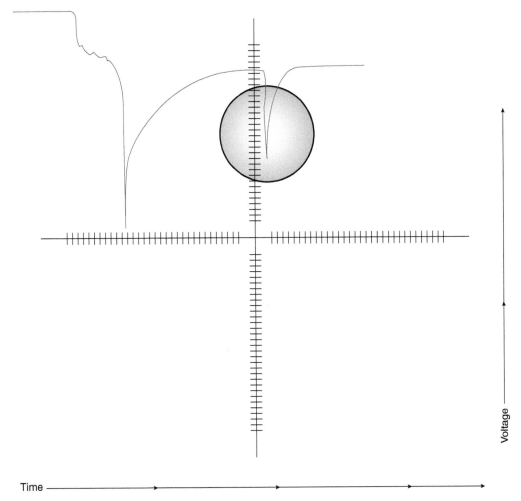

Figure 8-6 A voltage trace of an A/C circuit with a bad diode.

chokes are used to control noise from motors and generators (**Figure 8-7**). To avoid much frustration during diagnosis of a computer system (especially the inputs), check the integrity of the ground as one of the first steps.

> **SERVICE TIP** Considering the total number of vehicles being produced that use computer control systems, very few customer complaints are actually due to the fault of the computer. The computer should be replaced only if *all* other possible causes have been checked and confirmed as operating properly. Always check the condition of the battery feed and ground circuits of the computer before condemning the computer.

DIAGNOSTIC TROUBLE CODES

Once the computer performs a diagnostic test, or during the operation of the system, the microprocessor software determines whether the system and its circuits performed properly. If the system or circuit fails, a DTC is set.

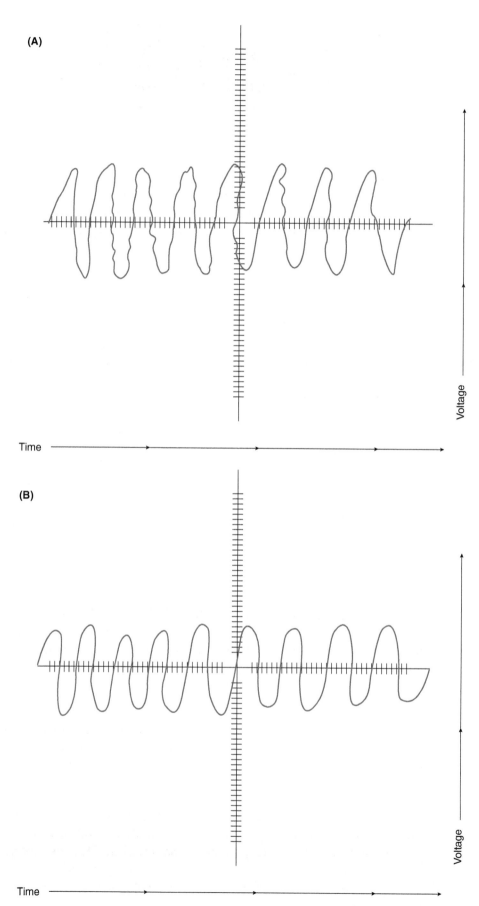

Figure 8-7 (A) Voltage trace of motor without a choke. (B) Voltage trace of a motor with a choke.

B2102

Figure 8-8 Example of a body function fault code.

Beginning in the 1996 model year, the Society of Automotive Engineers (SAE) published J2012 to describe industry-wide standards for a uniform DTC format. The trouble codes contain four characters, usually consisting of one letter and four numbers (**Figure 8-8**). The letter identifies the function of the device that has generated the fault code:

- P = Powertrain
- C = Chassis
- B = Body
- U = Network or DLC

The characters that follow identify the fault. These characters are usually numbers, but may also contain letters. Figure 8-8 is an example of a fault code set when the computer detects that the ignition switch input voltage is above a calibrated value. Using this code, the technician would then follow a diagnostic routine in the service information to locate the cause.

Most BCMs are capable of displaying the stored faults in memory. Diagnostic trouble codes can be displayed by the scan tool, and in early systems, by a method referred to as **flash codes**. Flash codes are DTCs displayed by flashing a lamp or light-emitting diode (LED). The method used to retrieve the codes varies greatly, and the technician must refer to the correct service information for the procedure. Some manufacturers provide a means of retrieving DTCs without a scan tool. They cannot use flash codes, but can display the code digitally. For example, Chrysler will display DTCs in the odometer if the ignition switch is cycled between OFF and RUN three times within 10 seconds.

AUTHOR'S NOTE Since the introduction of OBD II in 1996, the SAE no longer allows manufacturers to display emission-related DTCs by means of flash codes. Most manufactures have stopped using flash codes for DTC retrieval on most vehicle modules.

Depending on system design, the computer may store codes for long periods of time; some lose the code when the ignition switch is turned off. Systems that do not retain the code when the ignition is turned off require that the technician test drive the vehicle and attempt to duplicate the fault. Once the fault is detected by the computer, the code must be retrieved before the ignition switch is turned off again.

Diagnostic trouble codes are also called "fault codes."

The trouble code does not necessarily indicate the faulty component; it only indicates that circuit of the system is not operating properly. For example, the code displayed may be B1312, indicating an A/C high-side temperature sensor problem. This does not mean the sensor is bad; the fault is in that circuit, which includes the wiring, connections, sensor, and BCM. To locate the problem, follow the diagnostic procedure in the service information for the code received.

> **AUTHOR'S NOTE** Usually "B" codes will not illuminate the "Check Engine" (MIL) lamp. Since 1996, the MIL is only illuminated when there is an emissions related fault.

Differences between Hard and Intermittent Codes

Some BCMs will store trouble codes in their memory until they are erased by the technician or until a set amount of engine starts have passed. Some computers will display two sets of fault codes. Depending on the scan tool or method of DTC retrieval, the **hard codes** and intermittent codes can be displayed together or separately. Hard codes are failures that were detected the last time the BCM tested the circuit. **Intermittent codes** are those that have occurred in the past but were not present during the last BCM test of the circuit.

Most diagnostic charts cannot be used to locate causes of intermittent codes. This is because the testing at various points of the chart depends on the fault being present to locate the problem. If the fault is not present, the technician may be erroneously instructed to replace the BCM module, even though it is not defective.

Many intermittent problems are the result of poor electrical connections. Diagnosis should start with a good visual inspection of the connectors involved with the code. Even on hard codes, visually inspect the circuit before conducting any other tests.

Hard codes are also known as active codes. Intermittent codes are also called stored codes.

VISUAL INSPECTION

Perhaps the most important check to be made before diagnosing a BCM-controlled system is a complete visual inspection. The visual inspection can identify faults that could cause the technician to spend wasted time in diagnostics. In addition, the problem can be pinpointed without any further steps.

Inspect the following:

1. All sensors and actuators for physical damage.
2. Electrical connections into sensors, actuators, and control modules.
3. All ground connections.
4. Wiring for signs of burned or chafed spots, pinched wires, or contact with sharp edges or hot exhaust manifolds.
5. All vacuum hoses for pinches, cuts, or disconnects.

The time spent performing a visual inspection is worthwhile. Put forth the effort to check wires and hoses that are hidden under other components.

ENTERING DIAGNOSTICS

There are as many methods of entering BCM diagnostics as there are vehicle manufacturers. Most require a scan tool (**Figure 8-9**). A scan tool is a microprocessor designed to communicate with the BCM. It will access trouble codes and run system operation, actuator, and sensor tests. The scan tool is plugged into the diagnostic link connector (DLC) for the system being tested. Some manufacturers provide a single DLC, and the technician

Figure 8-9 A scan tool is used to enter the computer's diagnostic capabilities.

chooses the system to be tested through the scan tool. Always refer to the correct service information for the vehicle being serviced. Use only the methods identified in the service information for retrieving trouble codes. Once the trouble codes are retrieved, consult the appropriate diagnostic chart for instructions on isolating the fault. It is also important to check the codes in the order the manufacturer requires.

Using a Scan Tool

Connecting the scan tool into the DLC will allow the technician to access information concerning the operation of most vehicle systems. Some scan tools require the use of adapters, cartridges, or electronic keys. Follow **Photo Sequence 17** as a typical method used to enter body controller diagnostics.

When the technician has programmed the scan tool by performing the initial entries, some entry options appear on the screen. These entry options vary depending on the scan tool and the vehicle being tested. The technician makes the desired selection to proceed with the test procedure. In the first four selections, the tool is asking the technician to select the computer system to be tested. If "data line" is selected, the scan tool provides a voltage reading from each input sensor in the system.

When the technician makes a selection from the initial test selection menu, the scan tool moves on to the actual test selections. These selections vary depending on the scan tool and the vehicle being tested.

Since the scan tool provides the voltage value that the BCM is sensing from an input circuit, diagnosis of a circuit fault is made simpler. For example, if you are monitoring the input voltage of a two-wire sensor, such as the ambient temperature sensor, and the scan tool indicates 5.0 volts on the circuit, this indicates the circuit has an open. The open could be anywhere in the circuit between the BCM and sensor ground (including the sensor). If the monitored voltage shows 0 volt, this usually indicates a short to ground. The short could be the sensor or in the circuit between the BCM and the sensor. A reading of 0 volt may also be caused by a faulty BCM not sending the reference voltage to the circuit.

⚠ **Caution**

This section on entering BCM diagnostics is given as a guide and is intended to complement the service information. Improper methods of code retrieval may result in damage to the computer.

 **Special Tool**

Scan tool

⚠ **Caution**

The procedures may change between models, years, and the type of instrument cluster installed. Refer to the correct service manual for the vehicle being diagnosed.

PHOTO SEQUENCE 17
Typical Procedure for Scan Tool Diagnosis

P17-1 Turn the ignition switch to the OFF position.

P17-2 Select the proper scan tool data cable and adapters for the vehicle being tested. If required, connect the scan tool power cables to the vehicle's battery terminals (observing polarity) or to the accessory power outlet.

P17-3 Connect the scan tool cable to the DLC.

P17-4 Turn the ignition switch to the RUN (ON) position.

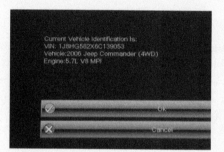

P17-5 Confirm the vehicle model year and engine size information.

P17-6 Select the system to be diagnosed.

P17-7 7 Operate the system being diagnosed and obtain the input sensor and output actuator data with the scan tool. It may be helpful to record and print the data. Compare the input sensor and output actuator data to specifications in the appropriate service information. Identify any data that is not within specifications.

Figure 8-10 Some manufactures will display DTCs in the odometer.

If the scan tool cannot make a connection to the BCM, the following may be the cause:

- Loss of battery supply voltage to the BCM.
- Loss of BCM ground circuit.
- Defective communication circuit.
- Problem with the scan tool.
- Defective DLC.
- Defective BCM.

Alternate BCM Trouble Code Retrieval

Many manufacturers will provide a method of retrieving diagnostic fault codes without using a scan tool. The methods used to display DTCs vary greatly between manufacturers and often between different lines of the same manufacturer. Always refer to the proper service information to determine the correct procedure to retrieve DTCs. Also, many manufacturers have discontinued the practice of displaying DTCs without a scan tool. Make sure that the vehicle you are working on supports this option.

Some manufacturers provide a method of retrieving fault codes by displaying them in the odometer (**Figure 8-10**). This may be accomplished by cycling the ignition switch from OFF to RUN three times within 10 seconds. Do not start the engine. The odometer will display all of the codes and then return to the odometer reading.

TESTING ACTUATORS

Testing of actuators is included here to orient you to the basic procedures. Specific procedures will be presented throughout this manual for individual systems and types of actuators.

Most computer-controlled actuators are electromechanical devices that convert the output commands from the computer into mechanical action. These actuators are used to open and close switches, control vacuum flow to other components, and operate doors or valves, depending on the requirements of the system.

Classroom Manual
Chapter 8, page 233

 **Special Tools**
Scan tool DMM

Most systems allow for testing of the actuator through the scan tool or while in the correct mode. Actuators that are duty cycled by the computer are more accurately diagnosed through this method. This will allow the technician to activate selected actuators to test their operation.

⚡ **WARNING** **When performing actuation tests on systems such as power windows and power seats, be sure that you are not in a position that will allow you to be pinched or caught by moving components.**

If the actuator needs to be tested by means other than the scan tool, follow the manufacturer's procedures very carefully. Because some of the actuators used by the BCM operate with 5 to 7 volts, do not connect a jumper wire from a 12-volt source unless the service information directs this. Some actuators are easily tested with a voltmeter by testing for input voltage to the actuator. If there is input voltage of the correct level, with the circuit under an electrical load, check for a good ground connection. If both of these circuits are good, then the actuator is faulty. If an ohmmeter needs to be used to measure the resistance of an actuator, disconnect the actuator from the circuit first.

An ammeter can also be used to measure the current of the actuator circuit. Simple on-off actuators should have the specified amperage anytime the actuator is energized. Higher than specified amperage reading would indicate that the actuator winding may be shorted, mechanical binding of the actuator, or a short in the circuit. Lower than specified amperage indicates high circuit resistance or the mechanical portion of the actuator is broken and thus not providing a mechanical resistance to movement. A reading of 0 amp would indicate an open in the circuit. If the actuator is controlled by pulse width modulation, the ammeter reading should increase and decrease as the actuator pulse width is increased and decreased, respectively.

Scan Tool Testing of Actuators

Often, the scan tool can be used to monitor actuator operation. This is done by watching the scan tool display of the computer controlling the actuator. As the computer turns the actuator on and off, the scan tool screen will indicate this command (**Figure 8-11**). Keep

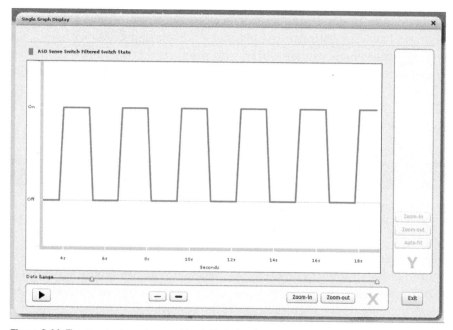

Figure 8-11 The scan tool may be capable of displaying the command of the computer to the actuator.

in mind this is indicating the *attempt* of the computer to turn the actuator on and off, not necessarily that the actuator is working. While the actuator is turned on and off, a voltmeter will indicate if the voltage levels are actually changing as the state of the actuator changes.

Some actuators will provide a feedback signal to the computer of correct operation. This feedback can be in the form of a potentiometer, commutator pulses, or a hard wire input back to the computer. Sometimes the feedback is accomplished by monitoring the voltage or the amperage of the control circuit. These values may be displayed on the scan tool. Consider the circuit of a BCM-controlled relay (**Figure 8-12**). With the relay de-energized the computer should see battery voltage on the control circuit. With the relay energized, the computer should see close to 0 volt. This may also be displayed as "High" or "Low" or as "True" or "False" on the scan tool (**Figure 8-13**).

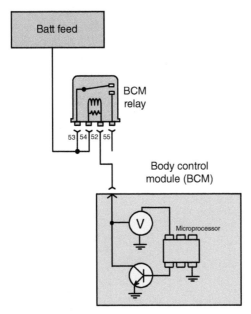

Figure 8-12 BCM relay driver control circuit.

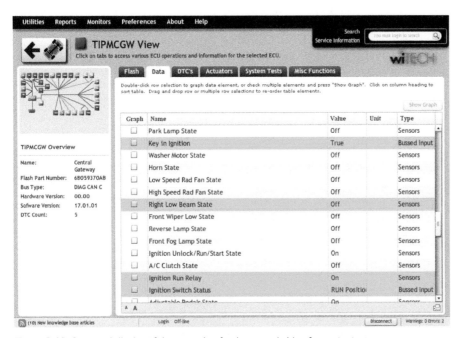

Figure 8-13 Scan tool display of the operation for the control side of an actuator.

Figure 8-14 illustrates an activator circuit that monitors the current flow. In this case the scan tool may display the interrupted current (**Figure 8-15**). This value should match the reading of a connected ammeter. Keep in mind that the scan tool may display 0 amp anytime the computer detects a fault in the circuit. The fault may be an open, short to ground, or any high current draw condition that will cause the microprocessor to shut down the circuit driver to prevent damage.

An additional function of the scan tool is the capacity to request an **activation test** (**Figure 8-16**). The scan tool activation tests places the scan tool into the function of the input to request the computer to follow through with the operation of the actuator. While the activation is being performed, a voltmeter or ammeter can be used to confirm whether the request is being performed. Often the activation can be observed by watching the actuator move. If the activation test works, but normal operation fails, then the problem is likely in the inputs to the computer.

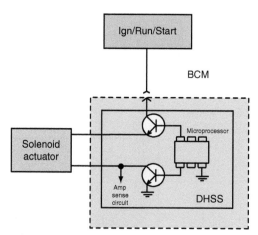

Figure 8-14 Dual high-side switch function that measures circuit current.

Figure 8-15 Scan tool display of the current required for operating the actuator.

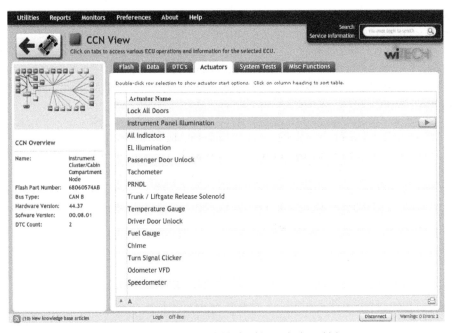

Figure 8-16 List of actuation tests that are available for this particular vehicle.

Testing Actuators with a Lab Scope

Since most actuators are electromechanical devices, when they fail it is because they are electrically faulty or mechanically faulty. By observing the action of an actuator on a lab scope, you will be able to watch an actuator's electrical activity. Normally, if there is a mechanical fault, it will affect the activator's electrical activity as well. Therefore, you get a good sense of the actuator's condition by watching it on a lab scope.

To test an actuator, you need to know what type it is. Most actuators are solenoids. The computer controls the action of the solenoid by controlling the pulse width of the control signal. By watching the control signal, you can see the turning on and off of the solenoid (**Figure 8-17**). The voltage spikes are caused by the discharge of the coil in the solenoid.

Some actuators are controlled by pulse-width modulated signals (**Figure 8-18**). These signals show a changing pulse width. These devices are controlled by varying the pulse width, signal frequency, and voltage levels.

Special Tool
Lab scope

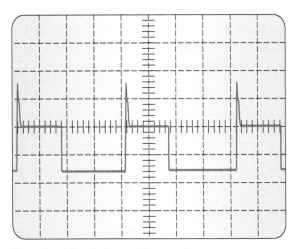

Figure 8-17 A typical solenoid control signal.

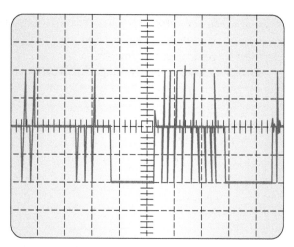

Figure 8-18 A typical pulse-width modulated solenoid control signal.

Both waveforms should be checked for amplitude, time, and shape. You should also observe changes to the pulse width as operating conditions change. A bad waveform will have noise, glitches, or rounded corners. You should be able to see evidence that the actuator immediately turns on and off according to the commands of the computer.

FLASHING THE BCM

Classroom Manual
Chapter 8, page 225

Most of the modules installed on today's vehicles are produced with EEPROM chips that allow the basic programming of the computer to be altered or rewritten. This enables manufacturers to reduce costs by having the technician **flash** the computer instead of replacing it. To flash a computer means to remove the existing programming and over-write it with new programming. The new program is downloaded into the scan tool and then downloaded into the BCM over a dedicated circuit.

When manufacturers first started developing ECUs that were capable of being reprogrammed, it required factory level equipment and proprietary access to the updated software. As a result, independent repair facilities were unable to perform this function for their customers. To remedy this, the Society of Automotive Engineers (SAE) developed J2534 as a communication standard for ECU reprogramming. Its purpose is to create an application programming interface (API) for all manufacturers, allowing the aftermarket the ability to reprogram ECUs using a compliant pass-through device.

The standard J2534 applies to 2004 MY and later vehicles. The J2534 mandate requires vehicle manufactures to support aftermarket repair shops with flash reprogramming for any emissions-related computers installed on a vehicle that can be reprogrammed by a manufacturer's dealership. Although this mandate effects 2004 and newer vehicles, some manufacturers provide J2534 support for older vehicles.

There are two subsets to the J2534 standard: J2534-1 and J2534-2. J2534-1 is primarily associated with emission modules. The J2534-1 reprogramming system consists of two components: subscription software and a J2534-1 compliant pass-thru device. The subscription software is available directly from the manufacturer and operates on the shop's PC or laptop.

SAE J2534-2 defines optional features of the reprogramming device on non-emission ECUs. This standard allows the original equipment manufacturer (OEM) to reprogram all controllers on their vehicles if they release and publish the necessary protocol information about their non-emission ECUs.

Manufacturers will specify minimum equipment requirements to perform a flash. The shop must have a high-speed internet connection that is reliable (DSL or Cable Modem).

A reliable connection is critical for any application that requires a live internet connection. If the connection is interrupted, it may render the PCM useless.

Prior to performing the reprogramming function, always check the condition of the vehicle's battery. A battery failure during the reprogramming process can result in a failure of the files to write correctly. Sometimes a failed operation can be recovered; however, there is a chance that failed reprogramming could permanently damage the control module. All manufacturers recommend that a constant voltage at the battery be maintained during the reprogramming procedure. A constant 12 volts must be supplied to the module at all times.

The shop computer used for reprogramming should be dedicated to this purpose only. The computer must meet the OEM requirements and have the supported operating system. Internet options such as enabling pop-ups and cookies, disabling firewalls, turning off all power saving modes, turning off screen savers, and so forth must also comply with stated requirements.

A subscription to the OEM websites is required to download the reprogramming files and programs. Some sites are free; others require the purchase of a subscription. Some manufacturers allow the purchase of temporary subscriptions that last for a few days.

Many manufacturers require an API be downloaded to the shop computer. The API is a library that includes specifications for routines, data structures, object classes, and variables.

The J2534-1 compliant pass-thru device (**Figure 8-19**) is a gateway between the vehicle's on-board computers and the shop's PC or laptop. The pass-thru device translates messages from the PC into the protocols used by the on-board computer and vice versa. The pass-thru device connects to the PC using either USB, serial cable, Ethernet cable, or wireless connections.

An important step in the reprogramming service process is to determine if the control module actually requires a calibration update. When a computer is flashed, the part number or version number is updated. Usually a TSB is issued if a new calibration is required to address issues. The TSB will typically list the part number or version number the flash will apply for and may specify the new part number. In addition, the subscription software typically informs the technician if an update is needed. The scan tool may provide ECU data that will list the current part/version number. If the current number does not match the latest version, then the update may apply. However, it is not a good practice to arbitrarily update controllers just because one is available. It is best to only update the controller if it addresses a customer concern. This is suggested since reprogramming is a procedure that typically cannot be reversed. Once the calibration has been updated, there is no method to reload the old calibration. There is a chance the new calibration will have noticeable changes to the operating characteristics of the vehicle. There may be no way to revert back if the customer does not like one of the new settings.

Figure 8-19 The J2534-1 compliant pass-thru device is gateway between the on-board computer and the shop's computer.

SERVICE TIP: If the reprogramming operation should fail, immediately retry the operation. Do not disconnect the interface tool.

Special Tools

Scan tool
Computer with internet access
Battery charger

Photo Sequence 18 shows a typical procedure for flashing a BCM. It is important to follow the service information procedures for the vehicle you are working on. Also, connecting a battery charger to keep the battery at 13.5 volts will prevent problems during the download resulting from low battery voltage. Be sure to check and clear any DTCs that may have been set during the flash procedure.

PHOTO SEQUENCE 18
Flashing a BCM

P18-1 Use the scan tool to obtain the part number of the BCM and record the number.

P18-2 Connect a battery charger to the vehicle's battery. Maintain about 13.5 volts on the battery.

P18-3 Connect the scan tool to a PC that links the tool to the flash software. Some scan tools will connect directly to an internet site through a LAN or wireless connection.

P18-4 Follow the scan tool instructions to enter the computer's part number or the VIN in the field and select the NEXT button.

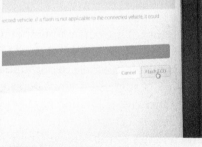

P18-5 Access the ECU requiring the flash update and select the desired flash file and select the "Flash ECU" button to begin the download to the scan tool.

P18-6 Monitor the progress of the download to the scan tool.

P18-7 Follow any instructions concerning cycling of the ignition.

P18-8 Once the flash is complete, a confirmation screen will show if the file transfer was successful.

P18-9 DTCs may have been set during the flash process. Erase any DTCs.

CASE STUDY

A customer is experiencing intermittent problems with several electrical options on his or her vehicle. The customer states that the systems will either operate erratically or stop operating completely. Retrieval of DTCs indicates several codes in many of the vehicle modules. Most of the DTCs relate to improper input values or feedback signals. The technician asks the customer if any services have been performed on the vehicle lately and is informed that some body work was done a couple of weeks ago. Realizing that all of the DTCs are related to a possible single cause, the technician performs a visual inspection of all ground straps. He locates a ground strap that is disconnected between the bulkhead and the engine hood. He repairs this connection and continues to verify that all grounds are properly connected. The disconnected ground strap was used to suppress EMI that can result from the sheet metal behaving as a capacitor and the air space between the sheet metal forming an electrostatic field. Without proper grounding of the sheet metal, interference with computer-controlled circuits that are routed near the sheet metal can result. Upon verifying the repair, the vehicle was returned to the customer.

ASE-STYLE REVIEW QUESTIONS

1. All of the following statements about servicing the BCM are true EXCEPT:
   A. Always check voltage supply circuits before replacing the BCM.
   B. Analog voltmeters must be used to test circuit voltage.
   C. Always check grounds before replacing the BCM.
   D. Turn the ignition switch to the OFF position before disconnecting or connecting components.

2. Diagnostic trouble codes are being discussed.
   Technician A says hard code failures are those that have occurred in the past but were not present during the last BCM test of the circuit.
   Technician B says intermittent codes are those that were detected the last time the BCM tested the circuit.
   Who is correct?
   A. A only
   C. Both A and B
   B. B only
   D. Neither A nor B

3. *Technician A* says DTCs will indicate the exact failure in the circuit.
   Technician B says DTCs will direct the technician to the circuit with a fault in it.
   Who is correct?
   A. A only
   C. Both A and B
   B. B only
   D. Neither A nor B

4. Accessing trouble codes is being discussed.
   Technician A says a scan tool is required to access DTCs on all vehicles since 1996.
   Technician B says the body codes are prefixed with a "U."
   Who is correct?
   A. A only
   C. Both A and B
   B. B only
   D. Neither A nor B

5. When flashing a computer, all of the following are true EXCEPT:
   A. A battery charger should be connected to the vehicle's battery.
   B. Special instructions may be displayed on the scan tool screen during the flash.
   C. The flash process loads new software programming into the computer.
   D. The BCM must be removed from the vehicle before it is flashed.

6. *Technician A* says open circuit testing of the BCM power supplies will indicate if there is high resistance in the circuit.
   Technician B says that open circuit testing has the potential to lead to improper diagnostics.
   Who is correct?
   A. A only
   C. Both A and B
   B. B only
   D. Neither A nor B

7. If the scan tool fails to make connection with the BCM, all of the following could be the cause EXCEPT:

   A. Loss of BCM power ground.

   B. Open actuator circuit.

   C. Open battery feed circuit to the BCM.

   D. Faulty scan tool.

8. *Technician A* says a PWM actuator cannot be tested with a DMM.

   Technician B says the PWM actuator may be tested with a lab scope.

   Who is correct?

   A. A only   C. Both A and B

   B. B only   D. Neither A nor B

9. A high-side controlled actuator fails to operate. The scan tool displays 0 amp on the control circuit when the system is attempted to be turned on.

   Technician A says this may indicate an open in the circuit.

   Technician B says this may indicate a short to ground in the circuit.

Who is correct?

A. A only   C. Both A and B

B. B only   D. Neither A nor B

10. The relay coil circuit is controlled by a low-side driver. Direct battery feed is connected to the other terminal of the coil. If the BCM monitors voltage on the control circuit to determine proper operation, which of the following statements is **not** true?

    A. The voltage reading by the BCM before the low-side driver should be battery voltage with the coil turned off.

    B. The voltage reading by the BCM before the low-side driver should be close to 0 volt with the coil turned on.

    C. A constant low-voltage reading by the BCM may be the result of an open in the circuit between the battery and the coil.

    D. The BCM will not be capable of setting control circuit faults.

ASE CHALLENGE QUESTIONS

1. The BCM-controlled ATC system is operating at full heat, high fan speed, with output from the defrost grids. This occurs whenever the vehicle is started and cannot be changed by the vehicle operator.

   Technician A says a sensor circuit may have failed.

   Technician B says for some reason the BCM may have initiated a fail-safe action.

   Who is correct?

   A. A only   C. Both A and B

   B. B only   D. Neither A nor B

2. BCM system troubleshooting is being discussed.

   Technician A says if the BCM "sees" a problem with any of its input sensors or output devices it will always command the "Check Engine" lamp to illuminate.

   Technician B says most diagnostic charts cannot be used to diagnose intermittent problems.

   Who is correct?

   A. A only   C. Both A and B

   B. B only   D. Neither A nor B

3. The in-car temperature sensor of a vehicle equipped with a BCM and ATC occasionally shorts internally, resulting in erratic operation of the ATC system. However, when the vehicle is brought into the shop for testing, the ATC system performs correctly.

   Technician A says an intermittent trouble code may be stored by the BCM.

   Technician B says an intermittent and a hard trouble code will be stored by the BCM.

   Who is correct?

   A. A only   C. Both A and B

   B. B only   D. Neither A nor B

4. The diagnosis of BCM-controlled output actuators are being discussed.

   Technician A says if the actuator works during the activation test, the problem is probably associated with a faulty BCM.

   Technician B says some scan tools can control the operation of output actuators.

   Who is correct?

   A. A only   C. Both A and B

   B. B only   D. Neither A nor B

5. An actuator that receives direct battery feed and is controlled by a low-side driver has a short to ground in the control circuit.

   Technician A says this may cause the battery to drain overnight.

   Technician B says the BCM will no longer have control of the system and may set a DTC.

   Who is correct?

   A. A only   C. Both A and B

   B. B only   D. Neither A nor B

6. The flashing of a computer is done to perform which task?

   A. Repair hardware problems within the computer-controlled system.

   B. To update the software within the computer to correct or enhance operation.

   C. Determine if driver habits contribute to EEPROM issues.

   D. Correct for faulty input signals.

7. Which method may be used to check a computer-controlled actuator?

   A. Use a lab scope to monitor voltage levels.

   B. Use an ammeter to monitor current flow.

   C. Use a scan tool to activate the component.

   D. All of the above.

8. When measuring the voltage on the power feed circuit to the computer, the voltmeter reads 12 volts with the connector unplugged. When load tested with a test light, the voltmeter reads 12 volts. When the connector is plugged back in and backprobed, the voltmeter reads 9.5 volts.

   Technician A says this may be due to an internal problem in the computer.

   Technician B says this may be due to an open computer ground circuit.

   Who is correct?

   A. A only   C. Both A and B

   B. B only   D. Neither A nor B

9. The scan tool indicates that an actuator is being turned on and off; however, the actuator does not operate. What is the *least* likely cause?

   A. Faulty computer.

   B. Open in the control circuit of the actuator.

   C. Open in the ground circuit of the actuator.

   D. Defective actuator device.

10. Using **Figure 8-20** of a BCM-controlled headlight system, what is the most likely cause that the headlights operate dimly on both sides?

    A. Poor battery feed connection to the BCM.

    B. Worn or corroded relay contacts.

    C. Faulty BCM.

    D. High resistance in the headlamp ground circuits.

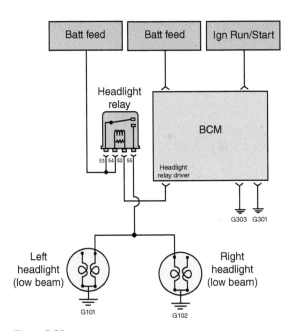

Figure 8-20

Name _____   **Date** _____

TESTING COMPUTER POWER AND GROUND CIRCUITS

Upon completion of this job sheet, you should be able to test the power and ground circuits of the BCM.

NATEF Correlation

This job sheet addresses the following **MLR** tasks:

A.4.   Demonstrate proper use of a digital multimeter (DMM) when measuring source voltage, voltage drop (including grounds), current flow, and resistance.

A.5.   Demonstrate knowledge of the causes and effects from shorts, grounds, opens, and resistance problems in electrical/electronic circuits.

This job sheet addresses the following **AST/MAST** tasks:

A.3.   Demonstrate proper use of a digital multimeter (DMM) when measuring source voltage, voltage drop (including grounds), current flow, and resistance.

A.4.   Demonstrate knowledge of the causes and effects from shorts, grounds, opens, and resistance problems in electrical/electronic circuits.

Describe the vehicle being worked on:

Year _____ Make _____ Model _____

VIN _____ Engine type and size _____

Tools and Materials

- DMM
- Test light
- Breakout box
- Backprobe tools
- Service information

Procedure

1. Using the proper service information, identify the following information for each direct battery feed circuit:

    BCM connector number and cavity: _____

    Circuit identification: _____

    Wire color code: _____

2. Identify the following information for each ignition feed or switched battery feed circuit:

    BCM connector number and cavity: _____

    Circuit identification: _____

    Wire color code: _____

3. Identify the following information for each ground circuit:

    BCM connector number and cavity: _____

    Circuit identification: _____

    Wire color code: _____

4. Identify the following information for each sensor return circuit:

   BCM connector number and cavity: _____

   Circuit identification: _____

   Wire color code: _____

5. Disconnect the connector(s) that contain the direct battery feed circuits and measure the voltage at each of the battery feed terminals.

   _____

   _____

   _____

6. Use a test light to load the circuit and retest the battery voltage at each terminal identified in step 1. Record your results.

   _____

   _____

   _____

7. Were the readings different from those taken in step 5 and step 6? ☐ Yes ☐ No Explain why or why not.

   _____

   _____

8. Reconnect the connector(s) and backprobe the terminals and measure the direct battery feed circuits identified in step 1. Record your results.

   _____

   _____

   _____

9. If any of the direct battery feed circuits voltage values were below specifications, what would your next step be?

   _____

   _____

   _____

10. Test each of the ignition or switched battery feed circuits identified in step 2 using open circuit, test light load, and backprobing methods and record your conclusions.

    _____

    _____

    _____

    _____

11. Disconnect the connector(s) that contain the circuits for the BCM grounds and measure the voltage at each of the ground terminals. Record your results.

    _____

    _____

12. Use a test light connected to battery positive to load the ground circuits and re-measure the voltage.

_____

_____

13. Reconnect the BCM connector and backprobe the ground circuits to measure voltage with the ignition in the RUN position. Record your results.

_____

_____

14. Disconnect the negative battery terminal and test the BCM ground circuits using an ohmmeter to test continuity from the disconnected BCM connector(s) and chassis ground. Record your results.

_____

_____

15. Comparing the tests performed on the ground circuit in steps 11 through 14, which do you believe to be the most accurate method of locating circuit faults and why?

_____

_____

_____

16. Describe how you would proceed to test the sensor return circuit.

_____

_____

_____

17. After reviewing step 16 with your instructor, perform the test you outlined and record your results.

_____

_____

_____

Instructor's Response

_____

_____

_____

Name _____   Date _____

USING A SCAN TOOL

Upon completion of this job sheet, you should be able to hook up a scan tool and retrieve the codes from a computer.

NATEF Correlation

This job sheet addresses the following **AST/MAST** task:

G.5.   Diagnose body electronic system circuits using a scan tool; check for module communication errors (data bus system); determine needed repairs.

ASE CERTIFIED NATEF

Tools and Materials

• Vehicle equipped with a BCM   • Scan tool

Describe the vehicle being worked on:

Year _____ Make _____ Model _____

VIN _____ Engine type and size _____

Procedure   **Task Completed**

1. Locate the data link connector for the BCM. Where is the connector located?

2. Connect the scan tool to the DLC and turn the ignition switch to the RUN position.   ☐

   What version is the scan tool software?

3. Select the body computer function. Record the following information about the BCM.

   Part number: _____

   Software version: _____

   Hardware version: _____

4. How is this information useful? _____

5. Retrieve any fault codes that may be in the BCM's memory. Record the codes below.

   _____

   _____

   _____

   _____

6. Are the fault codes hard codes or intermittent? _____

7. For the system with fault code retrieved in step 5, use the scan tool to monitor the desired state of the actuator while you attempt to operate the system inputs. Describe your results.

   _____

   _____

   _____

8. Identify the control and ground circuit for the actuator.

   _____

   _____

   _____

9. Following recommended practices of the vehicle manufacturer, either backprobe the actuator connector, use a jumper harness, or use a breakout box to measure the voltage on each circuit with the ignition key off.

_____

_____

_____

10. Repeat step 9 with the ignition switch in the RUN position.

_____

_____

_____

11. Observe the voltmeter reading while attempting to operate the system and record your results.

_____

_____

_____

12. Based on your testing thus far, can you make any conclusions?

_____

_____

_____

13. Connect an ammeter to the control circuit of the actuator and attempt to operate the system again. Record your results.

_____

_____

_____

14. Use the scan tool's activation test function to command the operation of the actuator while observing the ammeter. Explain your results.

_____

_____

_____

15. Remove the ammeter from the circuit and replace it with a lab scope. Repeat step 14 and record your results.

_____

_____

_____

16. What are your conclusions?

_____

_____

_____

Instructor's Response

_____

_____

_____

Name _____   Date _____

FLASHING A BCM

Upon completion of this job sheet, you should be able to flash a BCM.

NATEF Correlation

This job sheet addresses the following **AST/MAST** task:

G.6.   Describe the process for software transfer, software updates, or reprogramming of electronic modules.

Tools and Materials

- Scan tool
- Access to flash software
- Battery charger
- Pass-thru device
- Shop computer
- Internet access
- Subscription

Describe the vehicle being worked on:

Year _____ Make _____ Model _____

VIN _____ Engine type and size _____

Procedure:

1. Is there a TSB that addresses flashing the BCM for your assigned vehicle?

   ☐ Yes   ☐ No

   If yes, what is the document number? _____

   If no, go to step 3.

2. According to the TSB, what is the purpose of the flash?

   _____

   _____

   _____

3. Connect the scan tool to the DLC and access the BCM. Obtain the BCM part number and record.

   _____

4. Describe the procedure used to obtain the flash download into your shop PC or scan tool.

   _____

   _____

5. Follow the procedure outlined in step 4 and download the flash file. Describe the navigation required to download the file to the scan tool.

   _____

   _____

6. How do you know the download was successful?

   _____

   _____

7. Return to the vehicle and connect a battery charger to the battery. Why is this step important?

   _____

   _____

8. Perform the flash of the BCM. Record any steps the scan tool directs you to perform during the flash.

   _____

   _____

   _____

9. How do you know if the flash update was successful?

   _____

10. After the flash is completed, check for DTCs. Were any DTCs set?   ☐ Yes   ☐ No

    If yes, erase DTCs.

Instructor's Response

_____

_____

_____

DIAGNOSTIC CHART 8-1

PROBLEM AREA:	Scan tool does not connect to the BCM.
SYMPTOMS:	No response from the BCM.
POSSIBLE CAUSES:	**1.** Poor connection at the DLC. **2.** Loss of battery feed to the DLC. **3.** Loss of ground connection at the DLC. **4.** Loss of BCM ground connection(s). **5.** Loss of direct battery feed circuits to the BCM. **6.** Loss of ignition switch or switched battery feed circuits to the BCM. **7.** Faulty scan tool. **8.** Defective BCM.

DIAGNOSTIC CHART 8-2

PROBLEM AREA:	BCM functions.
SYMPTOMS:	No BCM functions and safety systems are in default mode.
POSSIBLE CAUSES:	**1.** Loss of BCM ground connection(s). **2.** Loss of direct battery feed circuits to the BCM. **3.** Loss of ignition switch or switched battery feed circuits to the BCM. **4.** Defective BCM.

DIAGNOSTIC CHART 8-3

PROBLEM AREA:	BCM software.
SYMPTOMS:	TSB update and/or improper system operation.
POSSIBLE CAUSES:	**1.** BCM requires flashing. **2.** Defective BCM.

DIAGNOSTIC CHART 8-4

PROBLEM AREA:	Activators.
SYMPTOMS:	System fails to function.
POSSIBLE CAUSES:	**1.** Blown circuit fuse. **2.** Open in the control circuit. **3.** Open in the ground circuit. **4.** Faulty actuator. **5.** Open or poor BCM power and ground circuits. **6.** Defective BCM.

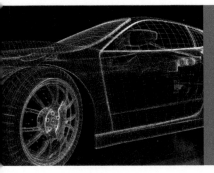

CHAPTER 9
SENSOR DIAGNOSTIC ROUTINES

Upon completion and review of this chapter, you should be able to:

- Comprehend the complexities of diagnosing intermittent faults.
- Properly utilize all tools and tool features to isolate the cause of an intermittent fault.
- Test thermistors used for temperature sensing for proper operation.
- Diagnose the thermistor circuit for proper operation.
- Perform proper diagnostics of pressure switch inputs.
- Properly diagnose Wheatstone bridge pressure sensors and their circuits.
- Diagnose frequency-generating capacitance discharge sensors.
- Determine proper operation of a pressure sensor by monitoring its signal with a lab scope.
- Use a digital multimeter (DMM) or lab scope to test and determine needed repairs of the potentiometer and its circuits.
- Use a DMM to diagnose the magnetic induction sensor.

- Use a DMM to diagnose the magnetic induction sensor output.
- Use a lab scope to determine magnetic induction sensor output functionality.
- Determine causes for magnetic induction sensor dropout.
- Properly diagnose the magnetic induction sensor circuits for opens, shorts, and short to ground conditions.
- Use a scan tool to diagnose the magnetic induction sensor.
- Use a DMM, scan tool, and lab scope to diagnose magnetoresistive (MR) sensors.
- Use a DMM, scan tool, and lab scope to diagnose the MR sensor circuits for opens, shorts, and short to ground conditions.
- Use a lab scope and DMM to determine proper operation of the Hall-effect sensor and its circuits.

🔧 **Basic Tools**

Basic hand tools
Service information

Terms To Know

Air gap
Capacitance discharge sensor
Cold soak
Dual ramping
Environmental data
Freeze frame
Knock
Knock sensor

Magnetic induction sensors
Magnetoresistive (MR) sensors
Negative temperature coefficient (NTC)
Piezoelectric sensors
Piezoelectric transducer
Piezoresistive

Positive temperature coefficient (PTC)
Potentiometer
Sensor dropout
Thermistor
Throttle position sensor (TPS)
Wheatstone bridge

AUTHOR'S NOTE Although introduced in the Classroom Manual, photo cell service and diagnostics are covered in Chapter 11 where operation with advanced lighting systems is discussed.

INTRODUCTION

Inputs are used by the automotive computer as a means of gathering information. Missing or corrupted input data results in poor system performance. This chapter explores the methods used to diagnose temperature sensors, different types of pressure sensors, position sensors, motion detection sensors, and the related circuits for each.

All sensor diagnostics begin with a thorough inspection of the sensor and its related circuits. This includes looking for physical damage and proper electrical connection. If a sensor has a hose or other mechanical connection, it must also be inspected. In addition, carefully examine the wiring harness and connectors between the sensor and the control module, including any dedicated ground circuits. All connectors and connector terminals must be tested for looseness, pullout, or damage. Any other conditions that can cause the sensor to not perform properly must be investigated.

Examples of diagnostic routines will be provided. Although the diagnostics discussed may pertain to a specific function, the diagnostics of the circuits of most sensor types will be compatible.

INTERMITTENT FAULTS

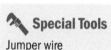

Special Tools

Jumper wire
Scan tool
DMM
Backprobing tools

Probably one of the most aggravating repair situations for the vehicle owners is to have their vehicle returned to them with "No Problem Found" noted on the repair order. It is also very aggravating to the technicians when they are unable to duplicate the problem the customer is experiencing. Intermittent problems challenge the technician to really examine the system to determine possible causes for the fault. The answer is not to throw parts at it; instead a methodical diagnostic routine must be utilized.

Intermittent faults with sensors usually set a diagnostic trouble code (DTC). However, the DTC is stored instead of being active when the technician retrieves them. This means the fault occurred at some point, but is not active right now and that the conditions necessary to set this DTC are not currently present.

Begin by reviewing any technical bulletins that are related to the problem the customer is experiencing. Perform any repairs directed by the bulletin. If there is not a bulletin that addresses the concern, continue with the diagnostics. This includes a thorough visual inspection of the sensor, sensor circuit wires, connectors, and terminals. Most intermittent conditions are the result of a poor connection somewhere in the circuit. It is a good practice to disconnect and clean all grounds involved in the system. If there are multiple stored DTCs, use the wiring information to determine common connections or grounds for all of the affected circuits. If the visual inspection fails to isolate the cause, further testing will need to be performed.

If the scan tool provides **environmental data** information that is associated with the stored DTC, this may be helpful in duplicating the problem. Environmental data is a snap shot of conditions when the fault occurs. It also indicates how many ignition cycles since the fault was set and for how long it was an active fault (**Figure 9-1**). In addition, OBD II faults may have an associated **freeze frame** with the DTC (**Figure 9-2**). Freeze frame is similar to environmental data, but contains OBD II–mandated information for when the fault first occurred. Both of these assist the technician in determining the operating conditions that set the code so they can attempt to duplicate the conditions.

EV Data-PCM		
Names	**Value**	**Units**
Manifold Absolute Pressure Sensor Circuit High	P0108	
DTC Readiness Flag	Not Complete	
Odometer	1232.8	miles
Starts Since Set Counter	3	
Warning Indicator Request State	On	
Good Trip Counter	0	
DTC	01 08	
Number DTC	1	
Ignition Key Cycles	3	
Accumulation Timer	3	minutes
Warm-Up Cycles	0	
Key Cycles Since DTC Last Set Counter	4	
DTC Storage State	Stored	

Figure 9-1 Scan tool screen showing the details associated with the setting of a fault code.

CARB			
	Name	**Value**	**Units**
	PCM Mileage Since MIL On	0.0	miles
▲	Freeze Frame Caused by DTC	01 08	
	Freeze Frame DTC Priority	03	
	PCM Odometer	1236.9921	miles
	Open Loop due to Driving Conditions - Bank 2	False	
	Open Loop - Bank 1	False	
	Closed Loop - Bank 1	True	
	Open Loop with DTC - Bank 1	False	
	Closed Loop with DTC - Bank 1	False	
	Open Loop due to Driving Conditions - Bank 1	False	
	Open Loop - Bank 2	False	
	Closed Loop - Bank 2	True	
	Open Loop with DTC - Bank 2	False	
	Closed Loop with DTC - Bank 2	False	
	MAP	15	psi
	MAP Voltage	4.94	volts
	Calculated MAP	8.0	psi
	Vacuum	16.17	in. Hg
	Barometric Pressure	15	psi
	Engine Load	58.8	%
	TPS 2 Voltage	4.22	volts
	APP Pedal Percent	0.0	%
▼	APP 1 Voltage	0.4497	volts
	APP 2 Voltage	0.22	volts

Figure 9-2 Partial screen capture of scan tool display for freeze frame.

> ⚙ **SERVICE TIP** Do not erase the fault code until environmental or freeze frame data has been reviewed. Clearing the DTC may also clear this data.

Next, use the scan tool to clear the DTC. With the ignition in the RUN position, monitor the scan tool for at least 2 minutes while wiggling the harness and connectors for the sensor circuits. Monitor the scan tool while looking for the sensor data to change or for the DTC to reset. If this does not cause a change in the sensor data or the DTC to return, cycle the ignition key off and on several times, leaving the ignition on for at least 10 seconds at a time. If the DTC does not return after this test, start the engine and allow it to reach normal operating temperature.

If the code appears during any of the tests, then perform the diagnostic routines for an active code. If the code does not return, duplicate the conditions indicated by the environmental or freeze frame data. Using the data recording function of the scan tool may assist the technician in locating the fault. Also, a lab scope is a very useful tool to help diagnose intermittent conditions.

> **CUSTOMER CARE** If you are unable to duplicate an intermittent fault, do not attempt the repair by guessing. Some scan tool and OEM manufacturers provide a data recording tool that can be connected to the vehicle's bus network. When the problem the owner is experiencing occurs, they push a button and the data recording is saved. Upon returning to the shop, the technician is then able to retrieve the information and pinpoint the root cause.

DIAGNOSING TEMPERATURE SENSORS AND THEIR CIRCUITS

Temperature sensors of various types are used in several different systems on today's vehicles. Due to their prominence, it is imperative that today's technician be able to competently diagnose these sensors and their associated circuits. This section details diagnostic routines for the most common forms of temperature-sensing systems.

Diagnosing Thermistor Circuits

Classroom Manual
Chapter 9, page 240

Special Tools
DMM
Scan tool
Lab scope
Jumper wires
Backprobing tools

A **thermistor** is a solid-state variable resistor made from a semiconductor material that changes resistance as a function of temperature. Two types of thermistors are used to measure temperature changes: negative temperature coefficient and positive temperature coefficient. **Negative temperature coefficient (NTC)** thermistors reduce their resistance as the temperature increases. **Positive temperature coefficient (PTC)** thermistors increase their resistance as the temperature increases.

Regardless of the purpose or type, the thermistor circuits operate under the same principles (**Figure 9-3**). In most cases, a problem in the circuit will be detected by the control module and a DTC will be recorded. The DTC can relate to a circuit voltage too high, circuit voltage too low, or a rationality fault.

The temperature sensor can be tested while they are installed in the system by backprobing the terminals to connect a digital voltmeter or lab scope to the sensor terminals. The sensor should provide the specified voltage drop at any temperature. The scope trace of an NTC thermistor should indicate a smooth transition from a high voltage (about 4 volts) to a low voltage as the sensor warms.

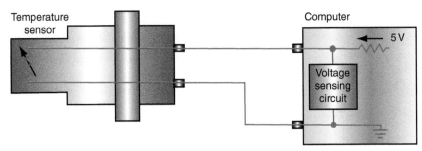

Figure 9-3 Typical temperature sensor circuit.

COLD		HOT	
10 KΩ resistor		909 Ω resistor	
−20°F	4.7 V	110°F	4.2 V
0°F	4.4 V	130°F	3.7 V
20°F	4.1 V	150°F	3.4 V
40°F	3.6 V	170°F	3.0 V
60°F	3.0 V	180°F	2.8 V
80°F	2.4 V	200°F	2.4 V
100°F	1.8 V	220°F	2.0 V
120°F	1.2 V	240°F	1.6 V

Figure 9-4 Example of voltage specifications for a dual ramping temperature sensor circuit.

Dual ramping is also referred to as dual resolution or dual range.

Classroom Manual
Chapter 9, page 241

⚠ **Caution**

Before disconnecting any computer system component, be sure the ignition switch is turned off. Disconnecting components may cause high induced voltages and computer damage.

If the system uses **dual ramping**, there should be a change in voltage to a higher reading at the specified voltage or temperature reading (**Figure 9-4**). If this switch does not occur, the fault is within the control module.

Consider Figure 9-3 as an example for circuit diagnosis. Simply disconnect the sensor and observe the voltage on the scan tool. The displayed voltage should equal that used by the control module signal circuit (**Figure 9-5**). If the voltage displayed by the scan tool is 0, the problem is that the signal circuit is shorted either to chassis ground or to the sensor ground circuit, or a faulty control module.

If the specified voltage is present with the sensor unplugged, use a jumper wire to connect the two terminals at the sensor harness connector. The scan tool should display 0 volt (V).

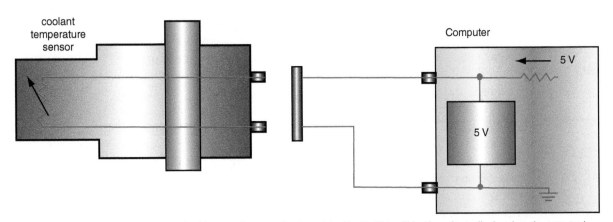

Figure 9-5 Disconnecting the sensor should cause the controller to register 5 volts. This will be the voltage displayed on the scan tool.

Data Display - PCM		
Name	Value	Unit
Fuel Level Percent	49.0	%
Engine Coolant Temp	120.2	F
Engine Coolant Temp Volt	0.0587	Volts
Intake Air Temp Deg	93.2	F
Intake Air Temp Volt	3.1821	Volts
Ambient Temp	44.6	F
Ambient Temp Voltage	2.69	Volts
CAT Modeled Temp	−83.2	F

Figure 9-6 This voltage displayed on the scan tool indicates that there is resistance in the circuit.

⚠ Caution

Never apply an open flame to a coolant temperature for test purposes. This action will damage the sensor.

If the voltage is still 5 volts, connect the signal circuit wire to chassis ground. If the reading is now 0 volt, the sensor ground circuit is open. While doing these steps if the voltage is above 0 volt (but less than 5 volts), this indicates resistance in the circuit (**Figure 9-6**). If the voltage is high when the signal circuit is connected to chassis ground, the resistance is in the signal circuit. If the voltage reading is above 0 volt when the jumper wire is connected across the two terminals, the resistance is in the sensor ground circuit.

The circuits can also be tested using an ohmmeter. Disconnect the sensor and the control module connectors. Connect an ohmmeter from each sensor terminal to the

⚙ SERVICE TIP Because current flow through the sensor will affect the readings, do not leave the ohmmeter connected for longer than 15 seconds.

Cold soak means to allow the vehicle to sit without the engine running long enough that the coolant temperature equalizes with ambient temperature.

control module terminal to which the wire is connected. Both sensor wires should indicate less resistance than specified by the vehicle manufacturer. If the wires have higher resistance than specified, the wires or wiring connectors must be repaired. Also, test circuits for being shorted together.

If the control module sets a rationality-type DTC for the thermistor circuit, this means the sensed value does not agree with other inputs. To test for this type of fault, allow the vehicle to **cold soak** long enough to allow all the temperature sensors to be at room temperature. Use the scan tool to compare the different temperature values for such sensors such as engine coolant temperature (ECT), intake air temperature (IAT), ambient temperature, and battery temperature. If a sensor value is different than the others, diagnose the sensor and circuit as described earlier.

Intermittent faults may be located by using a lab scope or the data-recording function of the scan tool. Since these instruments indicate a change in voltage over time, any change that occurs on the trace that cannot actually happen in the time frame indicates a problem within the circuit. For example, engine temperature cannot increase 80 degrees in 2 seconds. Since the ECT sensor is used by some body systems (such as air conditioning), a fault with this sensor may result in many different customer concerns being expressed. **Figure 9-7** is a data recording of a normally operating ECT and **Figure 9-8** is a data recording of an intermittent fault.

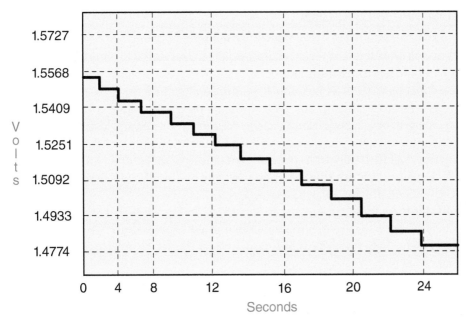

Figure 9-7 Normal NTC thermistor operation as temperature changes from cold to warm.

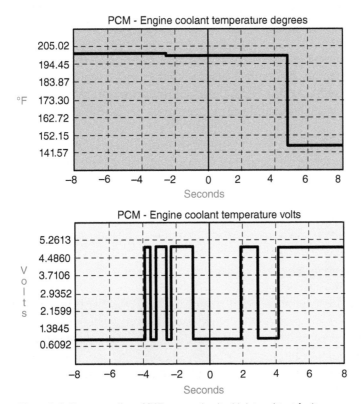

Figure 9-8 Data recording of ECT sensor circuit with intermittent fault.

DIAGNOSING PRESSURE SENSORS

Inputs used to determine pressures may be something as simple as a switch or can be a pressure transducer. This section discusses the routines used to diagnose pressure monitoring input circuits.

Classroom Manual
Chapter 9,
pages 243, 261

Special Tools

Scan tool
DMM
Jumper wires
Backprobing tools

Pressure Switch Diagnosis

Pressure switch circuits are simple on-off type switches. They can be used to control the operation of a warning lamp—for example, an oil pressure warning lamp. The pressure switch can be used to turn on and off a system. An example of this would be a high-pressure cut-off switch used in some air-conditioning systems. Also, the pressure switch may be an input to a control module to indicate proper operation of the system. This is the case of a pressure switch used in the hydraulic circuits of an electronically controlled automatic transmission.

Regardless of their function, pressure switches are set to open and close at a certain pressure. Usually this is done by a calibrated spring or disc. We will first look at the process to diagnose an oil pressure warning lamp circuit.

Consider the simple circuit that is illustrated in **Figure 9-9**. This circuit uses a normally closed oil pressure switch that will open at a preset pressure. To test faulty warning lamp operation, first turn the ignition switch to the RUN position. The lamp should illuminate.

If the light does not come on during the prove-out, disconnect the oil pressure switch connector (**Figure 9-10**). Use a jumper wire to connect the switch circuit from the ignition switch to ground. With the ignition switch in the RUN position, the warning lamp should light. If the light does not come on, either the bulb is burned out or the wiring is faulty. If the lamp comes on, the problem is a faulty oil pressure switch or its ground.

If voltage is not present to the oil pressure switch, the bulb may be burned out. At this point, the instrument cluster will need to be removed. With the cluster removed, check for battery voltage to the panel connector. If voltage is present, substitute a known good bulb and test again.

If the customer states that the warning light stays on after the engine is started, test in the following manner: disconnect the lead to the sender switch. The light should go out with the ignition switch in the RUN position. If it does not, there is a short to ground in the wiring between the lamp and the oil pressure switch. If the light goes out, replace the oil pressure switch.

Computer-driven instrument cluster warning lamps may use normally open pressure switches that close when oil pressure increases. These systems use the switch as an input

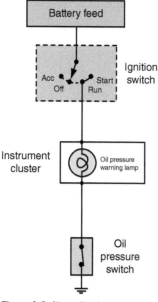

Figure 9-9 Normally closed oil pressure switch circuit.

Figure 9-10 Oil pressure switch and connector.

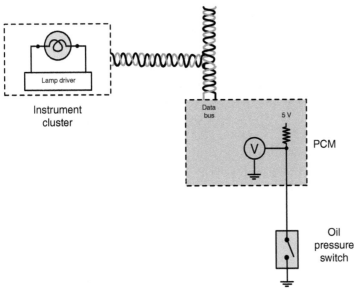

Figure 9-11 Computer-controlled warning lamp systems can use the pressure switch as an input.

that is used to turn on or off the warning lamp; they do not directly operate the warning lamp. Consider the system illustrated in **Figure 9-11**. The switch is a direct input to the powertrain control module (PCM). The PCM will send a sense voltage to the switch through a pull-up resistor. When oil pressure is below the threshold of the switch, the switch is open and the voltage sense will be high. A data bus message will be sent to the instrument cluster requesting a light on activation. When the switch closes, the sense circuit is pulled low and the light-off request is sent to the instrument cluster.

If the customer states that the oil pressure light does not go out after the engine is started, first use a shop oil pressure gauge to confirm adequate oil pressure.

> ⚙ **SERVICE TIP** If there was a circuit problem in the data bus between the PCM and the instrument cluster, there would be multiple symptoms such as speedometer, tachometer, and engine temperature gauges not operating.

Preliminary testing of this circuit can be performed by simply unplugging the oil pressure switch. With the ignition switch in the RUN position, the lamp should be on. Unplugging the oil pressure switch should turn off the lamp. If these results are not accomplished, then the cause will need to be pinpointed.

A scan tool accessing the instrument cluster module can be used to command activation of the lamp. If the lamp does not come on when commanded, the problem is a faulty lamp, circuit board, or instrument cluster module. If the lamp does light when commanded, the fault is in the signal to the instrument cluster. Use the scan tool to access the data stream to the instrument cluster module. Confirm that the proper message is being received.

If unplugging the oil pressure switch does not turn the warning lamp off and the data bus message seen on the scan tool confirms that the requested state is to turn the lamp on, test the circuit from the PCM to the oil pressure switch for a short to ground. Use a voltmeter to test for the proper level of sense voltage from the PCM. If the voltmeter reads 0 volt, either the circuit is shorted to ground or the PCM is faulty. The circuit is not open because the symptom in the lamp is always off. Use an ohmmeter to test for continuity between the oil pressure switch connector and engine ground. The meter should have a very high-resistance reading. If the resistance is low, the circuit is shorted to ground.

Unplug the PCM connector and test again between the circuit and engine ground. If it is now reading infinite, the PCM has an internal short.

If the lamp is always on (and the data bus indicates this is the requested state of the lamp), test the sense circuit between the PCM and the oil pressure switch for an open circuit. Unplug the oil pressure switch and use a jumper wire to connect the circuit to ground. The requested lamp state displayed on the scan tool should indicate a lamp-off command. If it does, replace the oil pressure switch. If the request does not change from on to off, backprobe the PCM connector for the sense circuit and use a jumper wire to connect the circuit to ground. If the lamp request changes to off, repair the open in the circuit between the PCM and the oil pressure switch. If the requested state does not change, the PCM is faulty.

Diagnosing Wheatstone Bridge Pressure Sensors

Pressure sensors are generally a form of strain gauge that determines the amount of applied pressure by measuring the strain a material experiences when subjected to the pressure. Most strain gauge pressure sensors are a form of **piezoresistive** construction. A piezoresistive sensor changes in resistance value as the pressure applied to the sensing material changes. A common piezoresistive pressure sensor is the **Wheatstone bridge**.

The Wheatstone bridge is used to measure small changes in resistance in a strain gauge. The manifold absolute pressure (MAP) sensor commonly uses the Wheatstone bridge as a sensing element. The MAP sensor is a primary sensor for determining the quantity of fuel to be injected into the combustion chamber. It is used in the process of determining the amount of air that is entering the combustion chamber at that instant. In most systems, it also serves as the means of determining the barometric pressure prior to starting the engine. Since the MAP sensor input indicates engine load, this information is also used by other body and chassis systems.

A defective MAP sensor may cause a rich or lean air-fuel ratio, excessive fuel consumption, and engine surging. It can also result in improper automatic transmission operation and/or improper air-conditioning system operation.

Diagnosis of MAP sensors differs between types. If the MAP sensor signal voltage is within the programmed value range of the PCM, a DTC may not set. For example, an open signal circuit will set a DTC, but resistance in the circuit that causes the voltage to be off by a few millivolts may not set a fault code. Also, a vacuum leak will cause the voltage values from the signal circuit to be different from specifications.

A very simple test of the Wheatstone bridge MAP sensor is to monitor the MAP vacuum and voltage values displayed on the scan tool (**Figure 9-12**). With the ignition switch in the RUN position (engine off), record the MAP reading. At this time, the reading should be actual atmospheric pressure and the voltage should confirm the reading. Compare this reading with the actual barometric pressure and the two should be equal. A MAP sensor that does not read correct barometric pressure cannot read correct vacuum. If the reading is not correct, verify the condition of all the circuits to the MAP sensor, as discussed later. If the circuits are good, replace the sensor.

If the readings are within the normal ranges, start the engine and allow it to idle. The MAP vacuum and MAP voltage should indicate correct vacuum readings. This may be confirmed by a mechanical vacuum gauge connected to a port in the intake manifold.

Classroom Manual
Chapter 9, page 245

Special Tools

Scan tool
Vacuum gauge
Jumper wire
DMM

> **SERVICE TIP** Anytime you are using a scan tool to monitor a pressure reading, do not accept the displayed value until you confirm the voltage. If the system is substituting a pressure value you may be fooled into thinking the sensor is operating properly. By correlating the signal voltage to the pressure reading, a fault in the circuit can be determined.

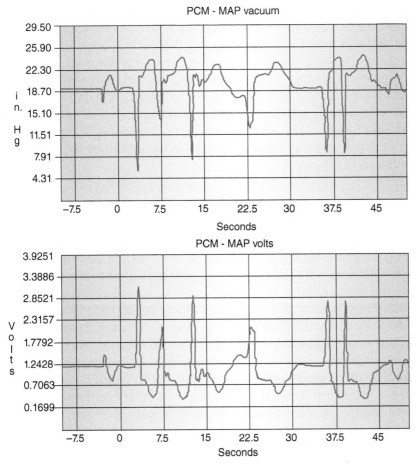

Figure 9-12 Correlation between MAP vacuum and MAP volts as different loads on engine are applied.

To perform a quick test of the circuits, consider the schematic of a Wheatstone bridge sensor circuit (**Figure 9-13**). If the scan tool indicates a voltage reading of 5 volts, this would be caused by an open or short to voltage in the signal circuit, an open in the return circuit, or a faulty sensor. Unplug the sensor and use a jumper wire to jump the signal circuit to ground. The scan tool should now read 0 volt. If it still reads 5 volts, the circuit is open between the control module and the sensor connector. If the voltage is above 0 volt, there is excessive resistance in the signal circuit between the control module and the sensor connector. If the reading is 0 volt, move the jumper wire to short the signal circuit to the return circuit. The scan tool should now read 0 volt in a proper

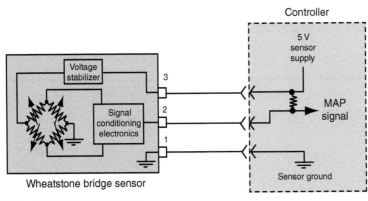

Figure 9-13 MAP sensor circuit.

operating circuit. If the scan tool displays 5 volts, the return circuit is open. If the reading is above 0 volt, there is excessive resistance in the return circuit. If all of the readings are normal, the sensor is faulty.

If the scan tool reading with the ignition in the RUN position and sensor connected is 0 volt, the problem is either an open 5-volt supply circuit or a short to ground in the signal circuit. If the voltage changes to 5 volts when the sensor is unplugged, the sensor is faulty. If the voltage remains 0 when the sensor is unplugged, use a voltmeter to check the supply circuit for 5 volts. If the voltage is low, there is either high resistance, an open, or a fault in the control module. These problems can be isolated by using an ohmmeter to test the continuity of the circuit.

DMM Testing. To test the Wheatstone bridge sensor with a DMM, backprobe the sensor signal wire and turn the ignition switch to the RUN position (engine off). Connect the DMM between the signal wire and ground. The voltage reading indicates the barometric pressure signal from the sensor to the control module. Usually this voltage is about 4 volts, depending on altitude. If the signal circuit voltage is not within the vehicle manufacturer's specifications compared with actual barometric pressure, confirm that the sensor's voltage supply and return circuits are good. If these circuits are good, replace the sensor.

If the pressure sensor is installed into a system that is not exposed to atmospheric pressures (e.g., air-conditioning system high-side pressure sensor), mechanical gauges may need to be used to confirm proper voltage readings for the measured pressure. In some instances, it may be possible to remove the sensor from the system so it is exposed to atmospheric pressure.

If the voltage is acceptable with the ignition in the RUN position and the engine off, the next step is to confirm a change in voltage as condition that the sensor is exposed to changes. For example, a MAP sensor is exposed to engine vacuum once the engine is started, and air-conditioning (A/C) pressure transducer is exposed to high-side pressures as the A/C system cycles on and off, a boost pressure sensor is exposed to pressures greater than atmospheric as engine speed is increased. In the example of a MAP sensor, start the engine and observe the signal voltage. The voltage should have changed to a low value. If the voltage does not change, the sensor pressure chamber is obstructed or damaged and the MAP sensor requires replacement. If the voltage changes, allow the engine to warm up to normal operating temperature while watching the voltage. Compare your voltage reading to specifications. If they are within specifications and the MAP sensor tests good, further testing of other components will be necessary.

For the Wheatstone bridge sensor circuit shown in Figure 9-13, if the voltage reading obtained in the initial test was 0 volt, unplug the MAP sensor connector and measure the voltage on each terminal with the ignition in the RUN position. On a three-wire connector, the voltage will usually be 5 volts on the supply circuit, 5 volts on the signal circuit, and 0 volt on the ground circuit. If these readings are obtained, the MAP sensor is suspect. If these voltages are not obtained, perform the same tests discussed earlier. In this case, confirm the changes in voltages as read on the DMM. If needed, use the ohmmeter function to test for continuity between the PCM connector and the MAP connector on the suspect circuit with the ignition in the OFF position. If there is continuity in the circuit, confirm PCM powers and grounds. If these are good, replace the PCM.

> **AUTHOR'S NOTE** Some sensor signal circuits may have 0 volt when the connector is separated. Always confirm the expected voltage values with the proper service information.

If the voltage recorded in the first step is inaccurate, backprobe the 5-volt supply circuit wire (connector plugged to the sensor) with a voltmeter with the ignition in the RUN position. If the reference wire is not supplying the specified voltage, check the voltage on this wire at the PCM connector. If the voltage is within specifications at the PCM, but low at the sensor, repair the 5-volt supply circuit wire for high resistance or an open. If this voltage is low at the control module, disconnect the control module's connector and use an ohmmeter to test for a short to ground. If there is no short to ground indicated, check the voltage supply and ground circuits for the control module. If these circuits are satisfactory, replace the control module.

If the voltage supply circuit is good, then test the sensor return (ground) circuit. With the ignition switch in the RUN position, backprobe the ground circuit terminal at the connector (with the connector plugged into the MAP sensor). Connect the voltmeter from the sensor ground wire to the battery ground. If the voltage drop across this circuit exceeds specifications, test the voltage drop on the PCM's ground circuits. If this is good, repair the ground wire from the MAP sensor to the control module.

Capacitance Discharge Sensor Diagnostics

Another variation of the piezo sensor uses capacitance discharge. Instead of using a silicon diaphragm, the **capacitance discharge sensor** uses a variable capacitor. Some of these sensors will use the discharged voltage as the signal and will be diagnosed in the same manner as the Wheatstone bridge sensor. Other types of capacitance discharge sensors produce a digital voltage signal of varying frequency. On these types of sensors a voltmeter can be used to check the 5-volt reference and the ground circuits. However, to test the signal circuit of this type of sensor requires the use of a sensor tester that changes the sensor varying frequency to a voltage reading or a scan tool that performs this function. **Photo Sequence 19** illustrates the use of the special sensor tester to diagnose the frequency-generating MAP sensor.

If the special sensor tester is not available, diagnosis can be performed using a DMM that has a frequency measurement function. For example, to test a sensor used to measure vacuum, connect the positive meter lead to the sensor signal wire and the negative lead to ground. Set the meter to read hertz with a + trigger. With the ignition in the RUN position, measure the frequency on the sensor signal wire with no vacuum applied. Using a hand vacuum pump, increase the vacuum against the sensor in 5-in. Hg increments. Record the hertz reading at each step. **Figure 9-14** is an example of expected test results for a good sensor. Be sure to refer to the correct service information for the sensor you are testing. In this example, the hertz readings decrease as vacuum is applied.

Lab Scope Testing of the Pressure Sensor

A good method of testing the pressure sensor is to use a lab scope. As the exposed pressures change, the sensor voltage signal should increase and decrease (**Figure 9-15**). Lack of change, a slow response, or an erratic signal may be the result of the sensor or connecting wires being defective.

Classroom Manual
Chapter 9, page 247

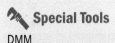
Special Tools
DMM
MAP sensor tester
Scan tool
Hand vacuum pump
Backprobing tools

Caution

Connecting any type of voltmeter directly to the voltage signal wire on a Ford MAP sensor may cause damage to the sensor.

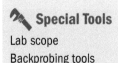
Special Tools
Lab scope
Backprobing tools

Vacuum Applied	Output Frequency
0 in. Hg	152–155 Hz
5 in. Hg	138–140 Hz
10 in. Hg	124–127 Hz
15 in. Hg	111–114 Hz
20 in. Hg	93–98 Hz

Figure 9-14 Examples of MAP sensor frequency with various vacuum values.

PHOTO SEQUENCE 19
Typical Procedure for Testing Frequency-Generating Sensors

P19-1 Remove the MAP sensor connector and vacuum hose.

P19-2 Connect the MAP sensor tester leads to the MAP sensor.

P19-3 Connect the MAP sensor connector to the harness connector.

P19-4 Connect the MAP sensor tester to the DMM.

P19-5 With the DMM set to DC voltage, observe the MAP sensor barometric pressure (BARO) reading on the voltmeter and compare this reading to specifications.

P19-6 Connect a hand vacuum pump to the MAP sensor and supply 5-in. Hg. Observe the MAP sensor reading and compare to specifications.

P19-7 Increase the vacuum applied to the MAP sensor to 10-in. Hg and observe the reading. Compare this reading to specifications.

P19-8 Apply 15-in. Hg to the MAP sensor and compare the reading to specifications.

P19-9 Increase the vacuum 20-in. Hg to the MAP sensor and compare the reading to specifications. If any of the MAP sensor readings do not meet specifications, replace the MAP sensor.

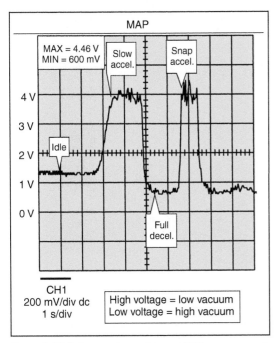

Figure 9-15 Lab scope trace of normal MAP sensor operation.

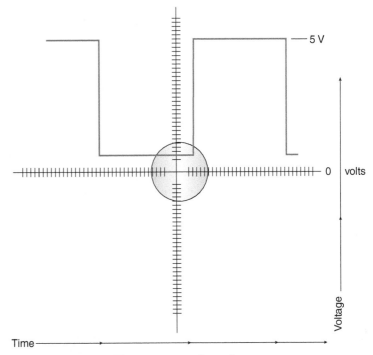

Figure 9-16 Example of frequency output of capacitance-type sensor.

A frequency-generating sensor produces a frequency that should change as the exposed pressures change. The upper voltage value of the trace should be 5 volts, while the lower voltage should be close to 0 volt (**Figure 9-16**). If the frequency is erratic or does not change or the voltage values are incorrect, the sensor is defective.

Diagnosing Piezoelectric Transducers

Piezoelectric sensors produce a proportional voltage output resulting from deformation of the element as pressure is applied. Since the **piezoelectric transducer** is capable of measuring the pressures associated with ultrasonic waves, it can be used for many functions, including that of a **knock sensor** (detonation sensor). The knock sensor measures engine knock, or vibration, and converts the vibration into a voltage signal. Piezoelectric sensors are also used in passive restraint systems.

We will consider the knock sensor in this example since it is one of the easiest piezoelectric-type sensor to understand and test. Diagnosing of other piezoelectric sensors will be covered during the discussions on system diagnosis in later chapters.

Knock is caused by excessive spark advance for the given engine-operating conditions. The output voltage from the knock sensor circuit represents the strength of the engine knock and is read by the PCM. The knock sensor is constantly producing an output voltage due to engine background noise, even when knock is not present (**Figure 9-17**). If knock occurs, the voltage output will increase (**Figure 9-18**). At this threshold, knock has occurred and the PCM calculates the necessary amount of short-term spark retard required to be subtracted from the spark advance based on the severity of the knock event. Severity is determined by the amount of knock sensor voltage that is greater than the knock threshold voltage level.

The procedure for checking a knock sensor varies depending upon the vehicle make and year. Always follow the vehicle manufacturer's recommended test procedure and specifications. Some knock sensor circuits use a single wire and the sensor case to engine

Classroom Manual
Chapter 9, page 238

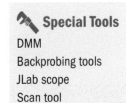 **Special Tools**

DMM
Backprobing tools
JLab scope
Scan tool

Knock is the spontaneous auto-ignition of the remaining fuel-air mixture in the engine combustion chamber that occurs after normal combustion has started, causing the formation of standing ultrasonic waves.

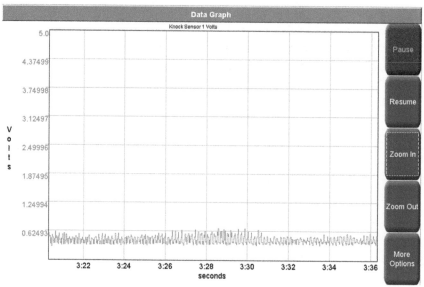

Figure 9-17 Normal knock sensor activity at idle.

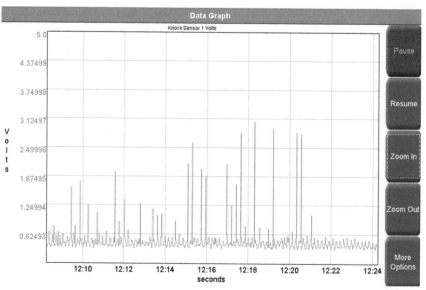

Figure 9-18 Knock sensor activity when knock is occurring.

block connection as ground. If there are two wires in the circuit, one is the signal circuit and the other is ground. Follow these steps for a typical knock sensor diagnosis:

1. Disconnect the knock sensor wiring connector, and turn on the ignition switch.
2. Connect a voltmeter from the disconnected knock sensor signal circuit to ground. The voltage should be 4 to 6 volts. If the voltage is within specifications, the ground circuit should be tested as in step 3. If the voltage is about 12 volts, the circuit is shorted to a power circuit. If the specified voltage is not available at this signal circuit, backprobe the knock sensor wire at the PCM (or use a breakout box) and read the voltage at this terminal. If the voltage is satisfactory at this terminal, repair the knock sensor signal circuit for an open. If the voltage is not within specifications at the PCM terminal, replace the PCM.
3. If the voltage at the signal circuit connection to the sensor is within specifications, move the negative voltmeter lead to the ground circuit terminal of the sensor connector. If there is no voltage reading, the ground circuit is open. If the voltage reading is lower now than in the initial test, check for high resistance in the circuit.

4. If the above tests indicate there is not a problem with the signal or the ground circuits, then test the knock sensor. Use an ohmmeter connected between the knock sensor signal terminal and the ground terminal. If the knock sensor uses only one wire, the ground terminal is the case. Compare the ohmmeter reading with specifications, usually between 3,300 and 4,500 ohms (Ω). If the knock sensor does not have the specified resistance, replace the sensor.

Operation of the knock sensor can be observed with a lab scope. Some scan tools may also provide a graphed data display or recorded event display. As discussed, the sensor is active at idle but when knock is occurring the sensor should show increased voltage output.

DIAGNOSING POSITION AND MOTION DETECTION SENSORS

Input data concerning position, motion, and speed are needed for many automotive systems such as electronic vehicle stability control, antilock brakes, air bags, and roll-over mitigation. Today's technician will be called upon to diagnose these sensors. A common type of sensor to determine position and motion is the potentiometer. Motion and speed sensors that use magnetism include magnetoresistive (MR), inductive, variable reluctance (VR), and Hall-effect sensors. In addition, photoelectric sensors, solid-state accelerometers, axis rotation sensors, yaw sensors, and roll sensors are common components on many systems.

Testing the Potentiometer

A **potentiometer** sensor can be used to measure linear or rotary movement. These sensors are tested by measuring the input voltage to the sensor and the feedback voltage to the computer. The feedback voltage to the computer should change smoothly as the resistance value of the sensor changes. To test these voltage signals, a series of jumper wires may be required (**Figure 9-19**). The jumper wires provide a method of gaining access to the terminals of weather-pack connectors without breaking the wire insulation.

Classroom Manual
Chapter 9, page 249

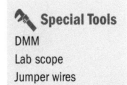 **Special Tools**
DMM
Lab scope
Jumper wires

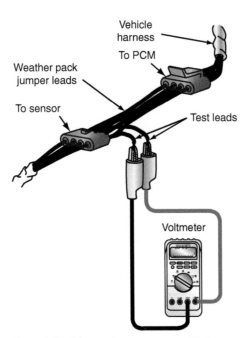

Figure 9-19 A jumper harness connected between the sensor and the wiring harness allows the technician to probe for voltage or test resistance without damaging the wiring.

The order of diagnostic testing is not set in stone. The presentation given is an example of testing that can be done on the **throttle position sensor (TPS)** circuit to locate the fault. Not all of the tests will need to be done to isolate the problem.

An ohmmeter can be used to measure changes in resistor values as the wiper is moved across the internal fixed resistor. Disconnect the sensor from the system. Connect the ohmmeter leads to the reference and ground terminals (**Figure 9-20**). This measures the value of the fixed resistor. Check the results against specifications. If good, connect the ohmmeter test leads between the reference terminal and the feedback terminal (**Figure 9-21**). Move the sensor's measurement arm or lever and observe the ohmmeter. The resistance should change smoothly and consistently as the wiper position is changed.

To illustrate a typical test procedure for a three-wire potentiometer circuit (**Figure 9-22**), begin by disconnecting the sensor connector and measure the voltage on the three terminals with the ignition in the RUN position. Compare the results with specifications since some systems will have 5 volts, 5 volts, 0 volt on the terminals (terminal A to terminal C), while others will have 5 volts, 0 volt, 0 volt. If these voltages are within specifications, reconnect the sensor and backprobe the signal wire from the sensor.

With the ignition switch in the RUN position, connect a voltmeter between the signal wire and ground. Typical voltage readings are 0.5 volt to 1 volt, with the potentiometer in the "at home position." Always refer to the vehicle manufacturer's specifications. As the

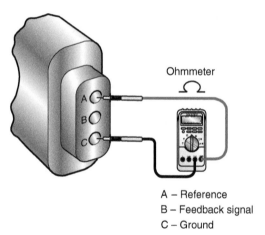

A – Reference
B – Feedback signal
C – Ground

Figure 9-20 Connecting an ohmmeter to test the potentiometer. This will give the fixed resistance value.

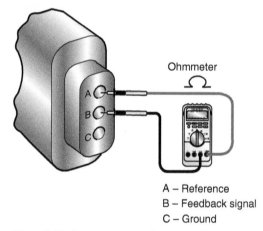

A – Reference
B – Feedback signal
C – Ground

Figure 9-21 Ohmmeter connection to test the wiper movement. As the wiper is moved from one end to the other, the resistance should change smoothly.

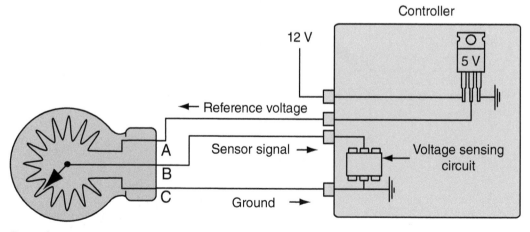

Potentiometer sensor

Figure 9-22 Potentiometer sensor circuit.

potentiometer wiper is slowly moved, observe the voltmeter reading. It should climb smoothly to the maximum specified voltage. Typical maximum voltage is between 3.5 and 4.5 volts. If the potentiometer does not have the specified voltage or if the voltage signal is erratic, replace the sensor.

To test the sensor ground (return) circuit, do a voltage drop test. With the ignition switch in the RUN position, connect the voltmeter between the sensor ground wire at the sensor's harness connector and the battery ground. If the voltage drop across this circuit exceeds specifications (and the control module ground circuit is good), repair the ground wire from the sensor to the control module.

SERVICE TIP While the potentiometer is being moved through its travel and while observing the sensor voltage signal, tap the sensor lightly and watch for fluctuations on the voltmeter reading. Fluctuations indicate a defective sensor.

The lab scope is an excellent tool for testing the potentiometer since it displays every voltage value it sees. This means there is less chance of missing something while trying to use a voltmeter or an ohmmeter. Each time the wiper is moved across the fixed resistor, the sensor should provide a smooth analog voltage signal (**Figure 9-23**). If the sensor is defective, glitches will appear in the sensor signal as the wiper is moved (**Figure 9-24**). Depending on circuit design, the glitch for an open can go either upward or downward. A glitch for a short will always go downward.

When looking at the scope trace, remember you are looking at voltage over time. Evaluate the trace, asking yourself if what you see could actually occur with a good sensor. For example, when the potentiometer is moved from one extreme position to the other, there should still be a ramping of the voltages involved. As the voltage moves from low to high and back again there should also be a "rounded corner" indicated on the trace. If the trace spikes straight up or down, this would indicate a problem with the sensor or circuit.

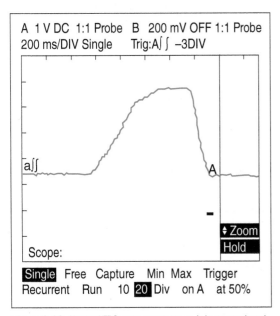

Figure 9-23 Normal TPS scope pattern as it is opened and closed again.

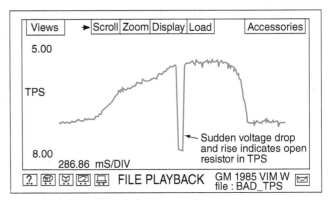

Figure 9-24 Faulty TPS waveform.

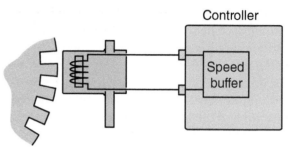

Figure 9-25 Magnetic inductive speed sensor circuit.

Diagnosing Magnetic Induction Sensors

Classroom Manual
Chapter 9, page 249

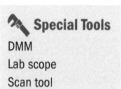

Special Tools
DMM
Lab scope
Scan tool

Magnetic induction sensors use the principle of inducing a voltage into a winding by use of a moving magnetic field (**Figure 9-25**). These sensors are commonly used to send data to the control module about the speed or position of the monitored component. For example, vehicle speed or wheel speed sensors (WSSs) can be of this design.

Improper timing pickup or rotational speed signals can be the result of circuit resistance. The zero cross characteristics of the magnetic induction sensor will accurately provide a timing reference provided that the target tooth width is close to the diameter of the sensor pole piece. However, the ideal timing signal occurs at the zero cross only if there is no electrical load on the sensor. A resistance load in the circuit will cause the inductance of the sensor coil to have a current that lags the open circuit generator voltage. This causes a phase shift in the output voltage.

> **SERVICE TIP** In the case of the wheel speed sensor used on many antilock brake systems (ABS), loose wheel bearings or worn parts can alter the air gap.

Another factor that may result in improper operation of the sensor is improper tooth gap. The amplitude of the signal is directly related to the distance between the sensor coil and the toothed ring. The distance is referred to as the **air gap**. The space of the air gap is more critical at lower tone wheel speeds. Improper air gap can cause **sensor dropout**. This can occur when the sensor will not produce an output voltage at slower speeds.

> **AUTHOR'S NOTE** On some vehicles that use the ABS wheel speed sensor signal as an input to the speedometer, a customer may complain that the speedometer drops off if they are moving at about 10 mph. This can be caused by sensor dropout due to an excessive air gap.

To test the magnetic induction sensor, first check the resistance value of the coil. This will indicate if the coil is intact, open, or shorted. Disconnect the sensor from the system and use an ohmmeter to test the resistance value of the coil. Connect the ohmmeter across the coil terminals and record the reading. Compare the results with specifications. A lower than specified reading indicates a shorted winding. A higher than specified reading indicates resistance or an open.

The voltage generation of the sensor can be tested by connecting a voltmeter across the sensor terminals. The voltmeter must be in the alternating current (AC) position and on the lowest scale. Rotate the shaft while observing the voltage signal. It should increase and decrease with changes in shaft speed. Also, the frequency function of the DMM can be used to monitor the sensor output.

Magnetic induction sensors can also be tested with a lab scope. Connect the lab scope leads across the sensor's terminals and rotate the shaft. The expected pattern is an AC signal that should be a perfect sine wave when the speed is constant (**Figure 9-26**). When the speed is changing, the AC signal should change in amplitude and frequency.

Regardless of the instrument used, you need to look for intermittent drop out, erratic frequency changes, and unstable readings. If the signal output indicates a potential problem, be sure to inspect the tone wheel (**Figure 9-27**), and check and adjust the air gap before condemning the sensor (**Figure 9-28**).

If the sensor performs properly, test the circuits between the computer and the sensor. This can be done by disconnecting both harness connectors and using an ohmmeter to check the integrity of the circuit. Check for opens and shorts to ground in both circuits. Also, test for the two circuits being shorted together. Another method is to reconnect the sensor and backprobe the harness connector at the computer and retest at this location. If there is a problem with the signal output now, the circuit wires are at fault.

Magnetic induction sensors are also called magnetic pulse generators.

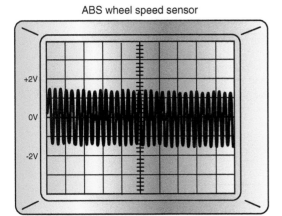

Figure 9-26 WSS waveform pattern.

Figure 9-27 Inspect the tone wheel for damage.

Figure 9-28 Measuring the air gap.

A scan tool can also be used to monitor the sensor output. Usually the scan tool will display the data as rpm or miles per hour. To accurately determine if a sensor is putting out the wrong signal, you must have a base to measure from. For example, on an ABS system you can compare all four of the wheel sensor inputs to see if they all read the same value. On a transmission input speed sensor, it should read the same speed as the engine crankshaft speed sensor when the torque converter is locked. The transmission output speed will need to be calculated using the input speed and the current gear ratio.

Some magnetic induction sensor circuits use a bias voltage supplied from the system's computer for use in fault detection (**Figure 9-29**). In addition, the bias voltage also elevates the sensor signal off the common ground plane of the vehicle electrical system to reduce signal interference. The bias voltage varies from manufacturer to manufacturer. Typically, it is 5 volts; however, some manufacturers use a bias voltage of 1.5, 1.8, or 2.29 volts. **Figure 9-30** illustrates a signal voltage from a system that biases the voltage to 2.29 volts. Notice that the sensor voltage signal is shifted to the positive. Always refer to the manufacturer's specifications to determine the required bias voltage when troubleshooting a magnetic induction sensor circuit.

The computer monitors the sensor signal at a point between the fixed pull-up resistor and the pickup coil. When power is applied to the circuit, current flows through the pull-up resistor and through the pickup coil to ground. The voltage drop at the signal monitor point is a predetermined portion of the reference voltage and a known value that is part of the computer program.

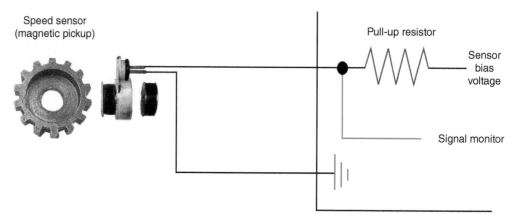

Figure 9-29 A speed sensor circuit that biases the signal circuit.

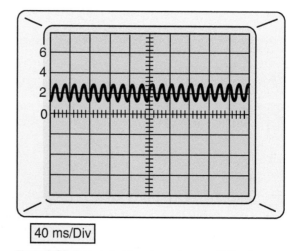

40 ms/Div

Figure 9-30 Output of a biased speed sensor. Notice the voltage is above 0 volt.

If an open circuit exists, no current flows through the circuit and no voltage is dropped across the pull-up resistor. The signal monitor voltage will be high. In this case, the computer will immediately set a trouble code for an open circuit fault.

If a shorted circuit exists, all or nearly all of the bias voltage is dropped across the pull-up resistor. The signal monitor voltage will be lower than the programmed signal monitor voltage.

The simple voltage divider circuit shown in Figure 9-29 allows the computer to detect an electrical fault as soon as the ignition is turned on. The shaft does not need to be rotating to detect these types of circuit defects. You can verify an open or short circuit fault by connecting a voltmeter between the high-voltage side of the pickup coil circuit and ground. Depending on the circuit fault, the meter should read close to full bias voltage or close to 0 volt with the ignition on.

Diagnosing the Magnetoresistive Sensor

Like the magnetic induction sensor, the air gap of the **magnetoresistive (MR) sensors** is an important factor in sensor operation. If the air gap is too wide, the current change in the circuit will cease to vary between 7 mA and 14 mA (**Figure 9-31**) and will remain constant at one of these values. The sensor signal decreases as the air gap increases and will become too small to be recognized by the signal conditioning electronics in the computer.

The following describes the procedure for diagnosing the MR sensor that uses active target tone wheels that have alternating magnetic poles (**Figure 9-32**). This type of sensor is common in ABS.

If there is an active ABS fault code for a WSS, visually inspect the wheel speed sensors, related wiring and electrical connections, and the ABS controller for obvious problems. If no problem is found, carefully backprobe the WSS harness connector and use the DMM to test for voltages on the supply circuit. With the ignition switch in the RUN position, the DMM should read above 10 volts. If the voltage is below 10 volts, check the circuit for a short to ground, high resistance, or an open.

To test for a short to ground, disconnect the WSS harness connector and the control module connector. Be sure the ignition is in the OFF position prior to disconnecting the components. Use a 12-volt test light that is connected to the positive post of the battery and probe the supply circuit at the WSS connector with the light. If the light illuminates, the circuit is shorted to ground.

Use the ohmmeter function of the DMM to test the supply circuit for an open or high resistance. Another method to test the supply circuit for an open is to connect a jumper

Classroom Manual
Chapter 9, page 251

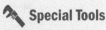 **Special Tools**
Scan tool
DMM
Lab scope
12-volt test light
Jumper wires

The alternating magnetic poles allow the computer to determine the direction of rotation.

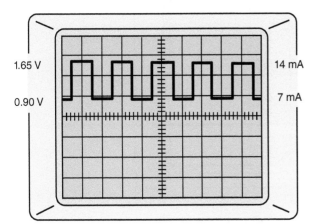

Figure 9-31 Output waveform of the magnetoresistive sensor.

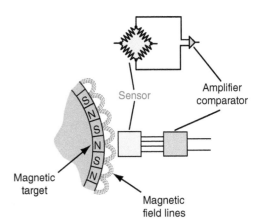

Figure 9-32 Active target with alternating magnetic poles.

wire between ground and the WSS supply circuit at the control module harness connector. Use a test light that is connected to a 12-volt source and probe the supply circuit at the WSS harness connector. If the light illuminates, test the power and ground circuit of the controller. If these are good, repair or replace the controller. If the light does not illuminate, the circuit is open.

If the supply circuit has greater than 10 volts, move the voltmeter test lead to measure the voltage on the signal circuit. The voltage here should be approximately 0.8 or 1.6 volts, depending on the position of the tone wheel. If the voltage reading is too high, test the circuit for a short to voltage condition. It is possible that the supply and the signal circuits are shorted together.

If 0 volt is read during this test, test the circuit for a short to ground in the same manner as discussed for testing for short to ground in the supply circuit.

The signal circuit can be tested for an open condition by disconnecting the harness connector at the control module and connecting a jumper wire between ground and the signal circuit. Using a 12-volt test light that is connected to a 12-volt source, probe the signal circuit at the WSS harness connector. If the light does not illuminate, the circuit is open.

WSS operation can be verified by slowly rotating the wheel by hand while monitoring the voltage displayed on the DMM. This is done with the test lead still connected to the signal circuit. The sensor signal voltage should alternate between about 0.8 and 1.6 volts.

To test the circuit with a lab scope, connect the leads as you would for a voltmeter. Adjust the scope settings to read 0.5-volt divisions at a rate of about 20 ms. A good WSS scope waveform should have sharp square corners on the DC signal circuit to the control module (**Figure 9-33**).

> **SERVICE TIP** Erratic signal outputs can be caused by damage, missing teeth, cracks, corrosion, or looseness of the tone wheel. Also, wheel bearing failure can cause the WSS signal to be erratic.

A scan tool can be used to monitor the WSS inputs for comparative reasons (**Figure 9-34**). While the vehicle is being accelerated in a straight line, monitor the sensors while looking for an indication of one dropping out or reading different than the others. Although this identifies that there is a problem, you will still need to use the DMM or lab scope to isolate the root cause.

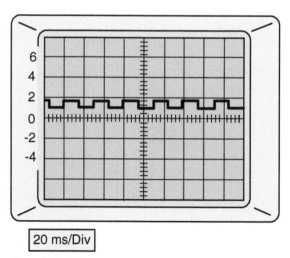

20 ms/Div

Figure 9-33 Active target speed sensor output.

Data Display - ABS		
Name	Value	Unit
LF Wheel Speed	31.73	mph
RF Wheel Speed	31.89	mph
LR Wheel Speed	31.89	mph
RR Wheel Speed	31.89	mph
ABS Pump Feed	SNA/Not Programmed	Volts
Valve Feed	10.0	Volts
Brake Switch Status	Not Pressed	
Steering Angle Sensor Position	−92.96875	Degrees
Yaw Sensor	0.0	Degrees/sec
Pressure Sensor	16.24	psi
Lateral Acceleration	0.02	G
Rolls Complete	True	

Figure 9-34 The wheel speed sensor output can be read by a scan tool.

CUSTOMER CARE Uneven tire air pressures can cause the WSSs to read different speeds from each wheel position. This may cause the vehicle to enter ABS mode when not needed or cause the control module to set a DTC and inhibit ABS operation. Let your customers know that proper tire pressure maintenance is important for proper ABS operation.

Alternate Magnetoresistive Sensor Diagnosis. Another method for testing MR sensors is to design a tool that will visually display if the sensor is operating (**Figure 9-35**). The tool is assembled using a 9-volt battery, a 220-ohm resistor, and a light-emitting diode

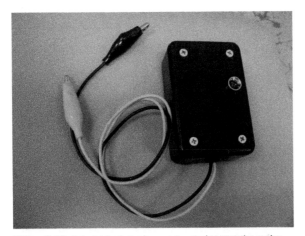

Figure 9-35 Special tool can be constructed to test the active speed sensor.

(LED). Construction of this tool is performed in Job Sheet 48. The tool is connected to the sensor harness connector so that the positive side of the 9-volt battery is connected to the voltage input wire of the WSS. The signal side of the sensor is connected to the 220-ohm resistor, an LED, and the negative post of the 9-volt battery. As the tone wheel is rotated, the changes in the magnetic poles cause the LED to blink if the WSS is operating properly.

Testing Hall-Effect Sensors

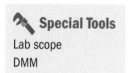

Special Tools
Lab scope
DMM

Classroom Manual
Chapter 9, page 253

The best way to test the performance of Hall-effect inputs is to use a lab scope. A DMM can be used to confirm that the proper voltages and grounds are supplied to the sensor, but it will not be able to indicate the quality of the signal. With a lab scope, the unit can be checked while the monitored component is operating.

Connect the positive lead of the scope to the signal wire by backprobing the connector. The connector must be plugged into the Hall-effect sensor. With the component operating, or being rotated, the trace should show a clean 5-volt or 12-volt square wave pattern (based on design) that increases in frequency as shaft rpm increases (**Figure 9-36**). The voltage value should go to a full 5 or 12 volts. If it does not, then there is resistance on the signal circuit between the Hall-effect sensor and the computer. The voltage should also return to 0 volt. If it does not, then there is resistance on the ground circuit. In addition, a sloping rising and falling line may indicate that the tone wheel is too far away from the magnet of the Hall-effect sensor or that the transistor is faulty (**Figure 9-37**). Check the trace for glitches and noise that may be the result of radio frequency interference (RFI) or electromagnetic interference (EMI) (**Figure 9-38**).

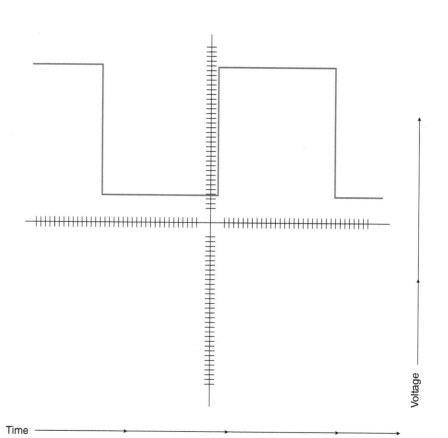

Figure 9-36 A good Hall-effect waveform should be a clean square wave pattern.

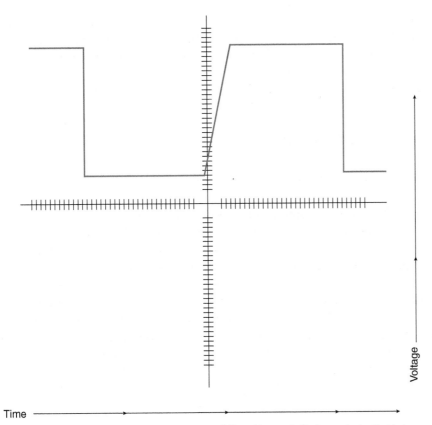

Figure 9-37 A sloping edge on either the rising or falling side may indicate an air gap that is too wide or the transistor in the Hall-effect sensor is faulty.

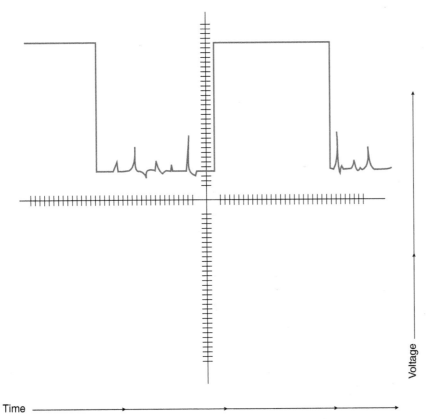

Figure 9-38 The Hall-effect trace pattern should be free of glitches and noise.

SERVICE TIP In many systems that use a Hall-effect sensor, it is possible to test the sensor by performing a scratch test. For example, if the fuel ignition system uses a Hall-effect crankshaft sensor, disconnect the sensor connector and use a terminal probe and a jumper wire to momentarily ground the signal wire to the PCM. If the circuit is good, the PCM should respond by activating the fuel pump and/or the shutdown relays. If the PCM responds now, the problem is the Hall-effect sensor.

CASE STUDY

A customer brings his or her minivan to the shop with a concern that the power lift gate does not open when the key FOB or inside button is pressed. The technician confirms the problem. While testing the operation of the power lift gate, he notices that the latch releases when the outside handle is used. Using his scan tool he accesses the lift gate module, but no DTCs are set. While looking through the sensor data display on the scan tool, he notices that the door temperature sensor is reading 20°F (−6.7°C). The temperature in the shop is 77°F (25°C). Knowing that the lift gate will not open if the temperature is too cold, he removes the sensor and tests its resistance. The sensor tested good. He then inspected the sensor connector, but found no problems. He then inspected the connector at the lift gate module, but could not find a problem. After consulting the wiring diagram, he located an in-line connector in the D-pillar area. After de-trimming the pillar, he noticed that the connector was corroded. Once the connector was cleaned, the sensor reading matched the temperature in the shop, and the lift gate operated properly. Prior to installing the D-pillar trim, he inspected for the cause of the corrosion and found signs of a small water leak from the taillight lens area. He also repairs the leak to assure the customer would not experience a repeat failure.

ASE-STYLE REVIEW QUESTIONS

1. Using the circuit illustrated in Figure 9-3, the computer has set a fault code for the sense circuit being high.

   Technician A says this can be caused by a short to chassis ground in the return circuit.

   Technician B says the signal circuit may be shorted to the return circuit.

   Who is correct?

   A. A only   C. Both A and B

   B. B only   D. Neither A nor B

2. What is being tested in **Figure 9-39**?

   A. Open in sensor signal circuit

   B. Open in sensor return circuit

   C. Short to ground in the signal circuit

   D. Both A and B

3. An oil pressure switch circuit that uses a normally open oil pressure switch is being tested with the engine running. When the switch is unplugged the warning light is on. Unplugging the oil pressure switch does not turn the warning lamp off.

   Technician A says if the data bus message confirms that the requested state is to turn the lamp on, test the circuit from the PCM to the oil pressure switch for a short to ground.

   Technician B says to test the circuit between the switch and the PCM for an open.

   Who is correct?

   A. A only   C. Both A and B

   B. B only   D. Neither A nor B

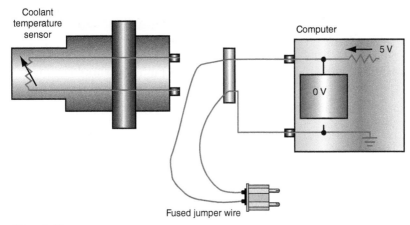

Coolant temperature sensor

Computer

5 V

0 V

Fused jumper wire

Figure 9-39

4. Using a scan tool, a low voltage on the pressure sensor's supply circuit is observed. This could be caused by any of the following EXCEPT:

A. An open in the supply circuit wire

B. A short to ground in the supply circuit wire

C. High internal resistance in the sensor

D. Excessive voltage drop in the supply circuit

5. Using a scan tool, a high voltage on the pressure sensor signal circuit is observed. The most likely cause would be:

A. An open in the supply circuit

B. An open in the signal circuit

C. A short to ground in the supply circuit

D. A short to ground in the signal circuit

6. *Technician A* says the knock sensor output voltage should increase as it detects an engine knock.
Technician B says the knock sensor voltage should be a square wave pattern.
Who is correct?

A. A only   C. Both A and B

B. B only   D. Neither A nor B

7. *Technician A* says a Hall-effect sensor that shows a trapezoidal-shaped waveform on the lab scope indicates that the tone wheel air gap is too wide.
Technician B says the Hall-effect waveform should have round edges.
Who is correct?

A. A only   C. Both A and B

B. B only   D. Neither A nor B

8. A potentiometer is suspected of being faulty. Voltmeter testing indicates there are 5 volts in the signal circuit when backprobed at the sensor, regardless of the wiper position.
Technician A says this can be caused by an open sensor return circuit.
Technician B says this can be caused by a short to ground in the supply circuit.
Who is correct?

A. A only   C. Both A and B

B. B only   D. Neither A nor B

9. *Technician A* says a bent or damaged tone wheel could cause sensor dropout.
Technician B says a damaged tone wheel could cause the output voltage to increase.
Who is correct?

A. A only   C. Both A and B

B. B only   D. Neither A nor B

10. When inspecting magnetic reluctant type sensors, all of the following must be checked EXCEPT:

A. Bias voltage

B. Proper contact between the sensor and the tone wheel

C. Proper mounting of the sensor

D. Tone wheel teeth condition

ASE CHALLENGE QUESTIONS

1. *Technician A* says when diagnosing intermittent faults, it is good practice to substitute control modules to see if the problem goes away.

   Technician B says a circuit performance fault indicates that the continuity of the circuit is suspect.

   Who is correct?

   A. A only
   B. B only
   C. Both A and B
   D. Neither A nor B

2. The scan tool displays 5 volts for the ambient temperature sensor. This indicates:

   A. An open in the sensor return circuit
   B. An open in the signal circuit
   C. An open in the sensor
   D. All of the above

3. A customer states that their air-conditioning system does not work. It is observed that the A/C clutch does not turn on when the A/C button is activated.

   Technician A says this can be caused by an open in the high-pressure cutout switch circuit.

   Technician B says this can be caused by a faulty A/C pressure transducer.

   Who is correct?

   A. A only
   B. B only
   C. Both A and B
   D. Neither A nor B

4. *Technician A* says the MAP sensor reading with the key on, engine off should equal barometric pressure.

   Technician B says when the engine is started, the MAP sensor signal voltage should increase.

   Who is correct?

   A. A only
   B. B only
   C. Both A and B
   D. Neither A nor B

5. A vehicle with four-wheel ABS has a problem with the right rear wheel locking during heavy braking.

   Technician A says this could be caused by a bad speed sensor mounted at the wheel.

   Technician B says the speed sensor mounted at the differential could cause this problem.

   Who is correct?

   A. A only
   B. B only
   C. Both A and B
   D. Neither A nor B

Name _____   Date _____

TESTING AN ENGINE COOLANT TEMPERATURE SENSOR

Upon completion of this job sheet, you should be able to check the operation of an ECT sensor and its associated circuits.

NATEF Correlation ─────────────────────────────

This job sheet addresses the following **MLR** tasks:

A.3. Use wiring diagrams to trace electrical/electronic circuits.

A.4. Demonstrate proper use of a digital multimeter (DMM) when measuring source voltage, voltage drop (including grounds), current flow, and resistance.

A.7. Use fused jumper wires to check operation of electrical circuits.

This job sheet addresses the following **AST/MAST** tasks:

A.3. Demonstrate proper use of a digital multimeter (DMM) when measuring source voltage, voltage drop (including grounds), current flow, and resistance.

A.6. Use fused jumper wires to check operation of electrical circuits.

A.7. Use wiring diagrams during the diagnosis (troubleshooting) of electrical/electronic circuit problems.

G.5. Diagnose body electronic system circuits using a scan tool; check for module communication errors (data bus systems); determine needed action.

ASE NATEF

Tools and Materials

- Service information
- DVOM
- Lab scope
- Scan tool
- Fused jumper wires
- Backprobing tools

Describe the vehicle being worked on:

Year _____ Make _____ Model _____

VIN _____ Engine type and size _____

TASK ONE—TESTING THE SENSOR

Procedure

Task Completed

1. Let the engine cool down completely.

☐

2. Describe the location of the ECT sensor.

_____

3. What color of wires is connected to the sensor?

_____

4. Record the resistance specifications for a normal ETC sensor for this vehicle at shop temperature.

_____

5. Disconnect the electrical connector to the sensor.   ☐

6. Measure the resistance of the sensor. Record your results: _____ ohms at approximately _____ degrees.

7. If not within specifications, what is your next step?

    _____

    _____

8. If the sensor resistance in step 6 is good, reconnect the sensor wires and backprobe the signal circuit with a voltmeter or lab scope.   ☐

9. Start the engine and allow it to warm to normal operating temperature while recording your voltage readings at different temperatures.

    _____

    _____

10. Is the voltage change smooth or erratic?

    _____

11. Is the ETC a negative temperature coefficient (NTC) or positive temperature coefficient (PTC) thermistor?

    _____

12. Conclusions:

    _____

    _____

TASK TWO—CIRCUIT TESTING

Procedure

1. With the scan tool connected to the DLC, disconnect the sensor and observe the voltage on the scan tool with the ignition switch in the RUN position. Record your results.

    _____

    _____

2. If the voltage displayed is not correct, what would your next step be?

    _____

    _____

3. Use a fused jumper wire to short the two ECT sensor terminals together at the connector and observe the voltage displayed on the scan tool. Record your results.

    _____

4. If the voltage did not change, what would be your next step?

    _____

    _____

5. Describe what steps 1 and 3 have proven.

_____

_____

6. With the ignition switch in the OFF position, disconnect the PCM connector that ☐
houses the ECT sensor circuits.

7. Connect an ohmmeter test lead to each end of the ECT sensor sense circuit and
record your results.

_____

8. If the ohmmeter reading displayed is not correct, what would your next step be?

_____

_____

9. Connect an ohmmeter test lead to each end of the ECT sensor ground (return) circuit
and record your results. _____

10. If the ohmmeter reading displayed is not correct, what would your next step be?

_____

_____

11. Connect one ohmmeter test lead to the ECT sensor ground (return) circuit and the
other lead to the sense circuit at either connector. Record your results.

_____

12. If the ohmmeter reading displayed is not correct, what would your next step be?

_____

_____

13. Conclusions:

_____

_____

Instructor's Response

_____

_____

_____

Name _____   Date _____

PRESSURE SWITCH ANALYSIS

Upon completion of this job sheet, you will be able to test and determine the operation of a pressure switch.

NATEF Correlation —————————————————————————

This job sheet addresses the following **MLR** tasks:

A.3.   Use wiring diagrams to trace electrical/electronic circuits.

A.4.   Demonstrate proper use of a digital multimeter (DMM) when measuring source voltage, voltage drop (including grounds), current flow, and resistance.

A.7.   Use fused jumper wires to check operation of electrical circuits.

This job sheet addresses the following **AST/MAST** tasks:

A.3.   Demonstrate proper use of a digital multimeter (DMM) when measuring source voltage, voltage drop (including grounds), current flow, and resistance.

A.6.   Use fused jumper wires to check operation of electrical circuits.

A.7.   Use wiring diagrams during the diagnosis (troubleshooting) of electrical/electronic circuit problems.

G.5.   Diagnose body electronic system circuits using a scan tool; check for module communication errors (data bus systems); determine needed action.

ASE NATEF

Tools and Materials

- Service information
- DMM
- Fused jumper wires
- Any vehicle with a pressure switch included in system operation (oil pressure switch, A/C high-pressure cutout switch, transmission pressure switch, etc.)

Describe the vehicle being worked on:

Year _____ Make _____ Model _____

VIN _____ Engine type and size _____

Procedure

1. What is the function of the pressure switch for the vehicle and system assigned to you?

   _____

   _____

2. Identify the circuit(s) to the pressure switch and draw a simple schematic of the circuit.

   _____

   _____

3. Disconnect the switch connector and measure the voltage on the signal circuit with the ignition in the RUN position. Record your reading.

   _____

4. Is this within specifications?   ☐ Yes   ☐ No

   If no, list possible causes for the reading obtained.

   _____

   _____

5. Operate the system while the switch is disconnected. Record your results.

   _____

   _____

6. With the system still operating, use a fused jumper wire and short the signal circuit to the ground circuit. Record your results.

   _____

   _____

7. Based on your observations, is the switch a normally open or normally closed switch?

   _____

   _____

8. With the system turned off, measure the resistance of the switch. Record your readings.

   _____

9. With the system operating, measure the resistance of the switch. Record your readings.

   _____

10. Are the results those expected from the operation you determined in step 7?

    ☐ Yes   ☐ No

    If no, what could be the cause?

    _____

    _____

11. Based on your observations, describe the results of your evaluation of the knock sensor circuit.

    _____

    _____

Instructor's Response

_____

_____

_____

Name _____   Date _____

TESTING THE POTENTIOMETER

Upon completion of this job sheet, you should be able to inspect and test a potentiometer.

NATEF Correlation

This job sheet addresses the following **MLR** tasks:

A.3.   Use wiring diagrams to trace electrical/electronic circuits.

A.4.   Demonstrate proper use of a digital multimeter (DMM) when measuring source voltage, voltage drop (including grounds), current flow, and resistance.

This job sheet addresses the following **AST/MAST** tasks:

A.3.   Demonstrate proper use of a digital multimeter (DMM) when measuring source voltage, voltage drop (including grounds), current flow, and resistance.

A.7.   Use wiring diagrams during the diagnosis (troubleshooting) of electrical/ electronic circuit problems.

G.5.   Diagnose body electronic system circuits using a scan tool; check for module communication errors (data bus systems); determine needed action.

Check electrical/electronic circuit waveforms; interpret readings and determine needed repairs.

This job sheet addresses the following **AST/MAST** tasks:

A.11.   Check electrical/electronic circuit waveforms; interpret readings and determine needed repairs.

Tools and Materials
- Service information
- Scan tool
- DMM
- Lab scope

Describe the vehicle being worked on:

Year _____ Make _____ Model _____

VIN _____ Engine type and size _____

Procedure

1. Identify the purpose of the potentiometer for your assigned task.

   _____

   _____

2. Identify the color code and purpose of the circuits to the potentiometer.

   _____

   _____

3. Unplug the harness connector at the potentiometer and measure the voltage at each terminal with the ignition switch in the RUN position.

   _____

   _____

4. Do the voltage readings match specifications? ☐ Yes ☐ No

   If no, what problem is indicated?

   _____

   _____

5. Measure the resistance across the potentiometer between the supply and ground terminals.

   _____

   Specification: _____

6. Measure the resistance between the signal and ground circuits as you move the wiper of the potentiometer. Describe your results.

   _____

   _____

7. Reconnect the harness connector to the potentiometer and backprobe the signal circuit terminal. With the ignition switch in the RUN position, what is the voltage on the signal circuit?

   _____

8. Move the potentiometer through its entire sweep while observing the voltmeter. Describe your observations.

   _____

   _____

9. What was the highest voltage observed? _____

   Specification: _____

10. Repeat steps 7 through 9 with a lab scope and record your observations.

    _____

    _____

Instructor's Response

_____

_____

_____

Name _____   Date _____

DMM TESTING OF THE MAGNETIC INDUCTION SENSOR

Upon completion of this job sheet, you will have inspected and tested the magnetic induction speed sensor used to monitor wheel speed for the ABS system using a DMM.

NATEF Correlation

This job sheet addresses the following **MLR** tasks:

A.3.   Use wiring diagrams to trace electrical/electronic circuits.

A.4.   Demonstrate proper use of a digital multimeter (DMM) when measuring source voltage, voltage drop (including grounds), current flow, and resistance.

This job sheet addresses the following **AST/MAST** tasks:

A.3.   Demonstrate proper use of a digital multimeter (DMM) when measuring source voltage, voltage drop (including grounds), current flow, and resistance.

A.7.   Use wiring diagrams during the diagnosis (troubleshooting) of electrical/ electronic circuit problems.

G.5.   Diagnose body electronic system circuits using a scan tool; check for module communication errors (data bus systems); determine needed action.

Check electrical/electronic circuit waveforms; interpret readings and determine needed repairs.

ASE **NATEF**

Tools and Materials

- Service information
- DMM
- Lift or jacks with stands

Describe the vehicle being worked on:

Year _____ Make _____ Model _____

VIN _____ Engine type and size _____

Procedure

Task Completed

1. Safely lift the wheels of the vehicle from the ground.

2. Referring to the service information, identify the location of the RF wheel speed sensor connector. Where is it located?

_____

_____

3. Other than ABS, what other vehicle systems require information from the wheel speed sensors?

_____

_____

_____

_____

4. Locate and disconnect the speed sensor two-way connector harness for the assigned wheel location.   ☐

5. Connect your DMM to the sensor side of the connector and measure the resistance across the two terminals of the sensor.

_____

Specifications: _____

6. Set the DMM to read DC voltage and measure the voltage across the harness side of the connector with the ignition switch in the RUN position. _____

Specifications: _____

7. What is the purpose of this voltage?

_____

_____

8. Reconnect the speed sensor and backprobe the (+) side of the connector. Read the voltage at this terminal _____

Did the voltage drop? _____

9. Backprobe the (+) and (−) terminals of the speed sensor and connect the voltmeter across the two terminals. With the voltmeter set to read AC voltage, have an assistant start the vehicle and accelerate to 10 mph and maintain that speed. Record the voltmeter reading.

_____

10. Accelerate to 20 mph and record the voltmeter reading.

_____

11. Explain your observations.

_____

_____

12. Switch the DMM to read frequency and perform steps 8 and 9 again. Record your observations.

_____

_____

Instructor's Response

_____

_____

_____

Name _____   Date _____

LAB SCOPE TESTING A MAGNETIC INDUCTION SENSOR

Upon completion and review of this job sheet, you should be able to inspect and test an ABS wheel speed sensor with a lab scope.

NATEF Correlation

This job sheet addresses the following **MAST** tasks:

A.7. Use wiring diagrams during the diagnosis (troubleshooting) of electrical/electronic circuit problems.

A.11. Check electrical/electronic circuit waveforms; interpret readings and determine needed repairs.

NATEF

Tools and Materials

- Wiring diagram
- Lab scope
- Lift or jacks with stands

Describe the vehicle being worked on:

Year _____ Make _____ Model _____

VIN _____ Engine type and size _____

Procedure

Task Completed

1. Safely lift the wheels of the vehicle from the ground.   ☐

2. Referring to the service information, identify the location of the assigned wheel speed sensor connector. Where is it located?

   _____

   _____

3. Other than ABS, what other vehicle systems require information from the wheel speed sensors?

   _____

   _____

   _____

   _____

4. Backprobe the speed sensor two-way connector harness for the RF wheel.   ☐

5. Connect the lab scope and observe the trace while the wheel is rotated at a constant speed.

   General results:

   _____

   _____

6. Increase the speed of the wheel while observing the graph. General results:

   _____

   _____

7. Record the operational action of the speed sensor and make any recommendation.

_____

_____

Instructor's Response

_____

_____

_____

Name _____   Date _____

BUILDING A MAGNETORESISTIVE SENSOR TESTER

Tools and Materials

- 9-volt battery
- Project enclosure (Radio Shack #270-1801) 9-volt battery
- Snap connector (Radio Shack #270-324)
- Mini alligator clips (Radio Shack #270-1540)
- 2 feet of red wire
- 2 feet of black wire
- Green LED (Radio Shack #276-022)
- LED holder (Radio Shack #276-080)
- 220-ohm resistor (Radio Shack #271-1111)

Procedure

1. Use the schematic shown in **Figure 9-40** to construct the sensor tester.

2. Backprobe the terminals of an active wheel speed sensor and turn the ignition switch to the RUN position. Connect the tester leads to the sensor terminals, observing polarity, and rotate the wheel. Describe your results.

_____

_____

Instructor's Response

_____

_____

_____

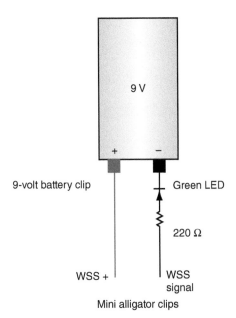

Figure 9-40 Magnetoresistive sensor tester schematic.

Name _____   Date _____

ACTIVE SPEED SENSOR CIRCUIT DIAGNOSIS

Upon completion of this job sheet, you will have inspected and tested the speed sensor circuit.

NATEF Correlation ————————————————————————

This job sheet addresses the following **MLR** tasks:

A.3.   Use wiring diagrams to trace electrical/electronic circuits.

A.4.   Demonstrate proper use of a digital multimeter (DMM) when measuring source voltage, voltage drop (including grounds), current flow, and resistance.

This job sheet addresses the following **AST/MAST** tasks:

A.3.   Demonstrate proper use of a digital multimeter (DMM) when measuring source voltage, voltage drop (including grounds), current flow, and resistance.

A.7.   Use wiring diagrams during the diagnosis (troubleshooting) of electrical/electronic circuit problems.

G.5.   Diagnose body electronic system circuits using a scan tool; check for module communication errors (data bus systems); determine needed action.

Check electrical/electronic circuit waveforms; interpret readings and determine needed repairs.

This job sheet addresses the following **MAST** task:

A.11.   Check electrical/electronic circuit waveforms; interpret readings and determine needed repairs.

ASE NATEF

Tools and Materials
- Service information
- Scan tool
- DMM

Describe the vehicle being worked on:

Year _____   Make _____   Model _____

VIN _____   Engine type and size _____

Procedure   **Task Completed**

1. Using the service information, determine what other vehicle systems require information from the wheel speed sensors (other than the ABS system)?

   _____

   _____

   _____

   _____

2. Establish communication with the assigned vehicle using the scan tool and check for DTCs in the ABS system. Are any DTCs present?   ☐ Yes   ☐ No

   If so, record them.

   _____

   _____

3. If applicable, list the possible causes for the DTC(s).

_____

_____

4. Visually inspect the sensor wiring. Check for any chafed, broken, or pierced wires. Also, visually inspect related harness connectors for broken, bent, spread, or pushed out terminals. Record your findings.

_____

_____

5. Inspect the tone wheel teeth for missing teeth, cracks, or looseness. Teeth should be perfectly square, not bent or nicked. Record your results.

_____

_____

6. Disconnect the WSS harness connector and use a DMM to measure the voltage of the supply circuit at the harness connector. What was the measured voltage?

_____

7. Using a DMM, measure the voltage between the supply circuit and the signal circuit at the WSS harness connector. What were the results?

_____

8. What do the results indicate?

_____

_____

9. Turn the ignition off and disconnect the ABM harness connector and the WSS connector.   ☐

10. Using a DMM, measure the resistance between the WSS signal circuit and the WSS supply circuit. What were the results?

_____

_____

11. What do the results indicate?

_____

Instructor's Response

_____

_____

_____

Name _____   Date _____

TESTING THE HALL-EFFECT SENSORS

Upon completion of this job sheet, you will be able to test a camshaft and crankshaft position sensor.

NATEF Correlation

This job sheet addresses the following **MLR** tasks:

A.3.   Use wiring diagrams to trace electrical/electronic circuits.

A.4.   Demonstrate proper use of a digital multimeter (DMM) when measuring source voltage, voltage drop (including grounds), current flow, and resistance.

This job sheet addresses the following **AST/MAST** tasks:

A.3.   Demonstrate proper use of a digital multimeter (DMM) when measuring source voltage, voltage drop (including grounds), current flow, and resistance.

A.7.   Use wiring diagrams during the diagnosis (troubleshooting) of electrical/ electronic circuit problems.

G.5.   Diagnose body electronic system circuits using a scan tool; check for module communication errors (data bus systems); determine needed action.

Check electrical/electronic circuit waveforms; interpret readings and determine needed repairs.

This job sheet addresses the following **MAST** task:

A.11.   Check electrical/electronic circuit waveforms; interpret readings and determine needed repairs.

ASE NATEF

Tools and Materials

- A vehicle equipped with Hall-effect sensor
- Service information for the selected vehicle
- Lab scope
- DMM

Describe the vehicle being worked on:

Year _____ Make _____ Model _____

VIN _____ Engine type and size _____

Procedure

1. Describe the location on the vehicle for the Hall-effect sensor assigned to you.

   _____

   _____

2. What is the function of this sensor in the system?

   _____

   _____

3. According to the service information, what should be the voltages at each of the terminals of the sensor with the ignition switch in the RUN position?

   _____

   _____

4. Disconnect the sensor connector and measure the voltage to each terminal (harness side) with the ignition switch in the RUN position. Do the voltages agree with those found in step 3?   ☐ Yes ☐ No

5. If the voltages do not agree, what is the likely cause?

_____

_____

6. Reconnect all disconnected connectors. Connect the lab scope to read the signal from the sensors to the control module. Select the appropriate voltage level for the sensor.

7. Operate the system while observing the sensor signals. Describe the signals received.

_____

_____

8. Do any of the patterns indicate a problem? ☐ Yes ☐ No

   If so, what is the likely cause?

_____

_____

Instructor's Response

_____

_____

_____

DIAGNOSTIC CHART 9-1

PROBLEM AREA:	Temperature sensor circuit performance.
SYMPTOMS:	Implausible voltage on the temperature sensor signal circuit; Improper system operation; Limp-in mode initiated.
POSSIBLE CAUSES:	1. Sensor signal circuit shorted to voltage. 2. Sensor signal circuit shorted to ground. 3. Sensor signal circuit shorted to sensor ground circuit. 4. Signal circuit open. 5. High resistance in signal circuit. 6. Sensor ground circuit open. 7. High resistance in sensor ground circuit. 8. Faulty sensor. 9. Internal controller fault.

DIAGNOSTIC CHART 9-2

PROBLEM AREA:	Temperature sensor circuit voltage low.
SYMPTOMS:	Improper system operation; Limp-in mode initiated.
POSSIBLE CAUSES:	1. Sensor signal circuit shorted to ground. 2. Signal circuit open. 3. High resistance in signal circuit. 4. Sensor ground circuit open. 5. High resistance in sensor ground circuit. 6. Faulty sensor. 7. Internal controller fault.

DIAGNOSTIC CHART 9-3

PROBLEM AREA:	Temperature sensor circuit voltage high.
SYMPTOMS:	Improper system operation; Limp-in mode initiated.
POSSIBLE CAUSES:	**1.** Sensor signal circuit shorted to voltage.
	2. Signal circuit open.
	3. High resistance in signal circuit.
	4. Sensor ground circuit open.
	5. High resistance in sensor ground circuit.
	6. Faulty sensor.
	7. Internal controller fault.

DIAGNOSTIC CHART 9-4

PROBLEM AREA:	Malfunction pressure switch or circuit.
SYMPTOMS:	Improper warning lamp operation; Improper system operation; Limp in mode initiated.
POSSIBLE CAUSES:	**1.** Open sense circuit.
	2. Open ground circuit.
	3. Sense circuit shorted to ground.
	4. Sense circuit shorted to voltage.
	5. Faulty pressure switch.

DIAGNOSTIC CHART 9-5

PROBLEM AREA:	Pressure sensor signal voltage high.
SYMPTOMS:	Improper system operation; Limp-in mode initiated.
POSSIBLE CAUSES:	**1.** Open signal circuit.
	2. Signal circuit shorted to voltage.
	3. Supply circuit shorted to battery voltage.
	4. Sensor ground circuit open.
	5. Faulty sensor.
	6. Internal controller fault.

DIAGNOSTIC CHART 9-6

PROBLEM AREA:	Pressure sensor signal voltage low.
SYMPTOMS:	Improper system operation; Limp-in mode initiated.
POSSIBLE CAUSES:	**1.** Signal circuit shorted to ground.
	2. Supply circuit shorted to ground.
	3. Open supply circuit.
	4. Faulty sensor.
	5. Internal controller fault.

DIAGNOSTIC CHART 9-7

PROBLEM AREA:	Piezoelectric sensor high-voltage fault.
SYMPTOMS:	System malfunction.
POSSIBLE CAUSES:	**1.** Sensor signal circuit shorted to battery voltage.
	2. Signal circuit open.
	3. Sensor ground circuit open.
	4. High resistance sensor ground circuit.
	5. Faulty sensor.
	6. Internal controller fault.

DIAGNOSTIC CHART 9-8

PROBLEM AREA:	Incorrect reading from potentiometer.
SYMPTOMS:	Sensor voltage at the controller is less than specifications.
POSSIBLE CAUSES:	**1.** Supply circuit open. **2.** Supply circuit shorted to ground. **3.** Signal circuit shorted to ground. **4.** Signal circuit shorted to the sensor ground circuit. **5.** Faulty sensor. **6.** Internal controller fault.

DIAGNOSTIC CHART 9-9

PROBLEM AREA:	Incorrect reading from potentiometer.
SYMPTOMS:	Sensor voltage at the controller is higher than specifications.
POSSIBLE CAUSES:	**1.** Signal circuit shorted to battery voltage. **2.** Signal circuit open. **3.** Signal circuit shorted to the supply circuit. **4.** Sensor ground circuit open. **5.** Faulty sensor. **6.** Internal controller fault.

DIAGNOSTIC CHART 9-10

PROBLEM AREA:	Incorrect reading from potentiometer.
SYMPTOMS:	Sensor voltage at the controller is not correct for position indicated by other inputs.
POSSIBLE CAUSES:	**1.** High resistance in the SIGNAL CIRCUIT. **2.** High resistance in the sensor ground circuit. **3.** High resistance in the voltage SUPPLY CIRCUIT. **4.** Faulty sensor. **5.** Internal controller fault.

DIAGNOSTIC CHART 9-11

PROBLEM AREA:	No signal from magnetic induction sensor.
SYMPTOMS:	No signal input detected by controller.
POSSIBLE CAUSES:	**1.** Sensor (+) circuit short to ground. **2.** Sensor (−) circuit short to ground. **3.** Sensor (+) circuit open. **4.** Sensor (−) circuit open. **5.** Sensor (+) circuit shorted to voltage. **6.** Sensor (−) circuit shorted to voltage. **7.** Sensor (−) circuit shorted to the sensor (+) circuit. **8.** Excessive sensor air gap. **9.** Faulty sensor. **10.** Internal controller fault.

DIAGNOSTIC CHART 9-12

PROBLEM AREA:	Intermittent signal from magnetic induction sensor.
SYMPTOMS:	Sudden change of output signal from the sensor.
POSSIBLE CAUSES:	**1.** Sensor (+) circuit short to ground. **2.** Sensor (−) circuit short to ground. **3.** Sensor (+) circuit open. **4.** Sensor (−) circuit open. **5.** Sensor (+) circuit shorted to voltage. **6.** Sensor (−) circuit shorted to voltage. **7.** Sensor (−) circuit shorted to the sensor (+) circuit. **8.** Excessive sensor air gap. **9.** Faulty sensor. **10.** Internal controller fault.

DIAGNOSTIC CHART 9-13

PROBLEM AREA:	Incorrect signal from magnetic induction sensor.
SYMPTOMS:	Incorrect reading from sensor as compared to other inputs.
POSSIBLE CAUSES:	**1.** High resistance in the sensor (+) circuit. **2.** High resistance in the sensor (−) circuit. **3.** Excessive sensor air gap. **4.** Faulty sensor. **5.** Internal controller fault.

DIAGNOSTIC CHART 9-14

PROBLEM AREA:	Incorrect speed input from active magnetoresistive sensor.
SYMPTOMS:	Incorrect reading from sensor as compared to other inputs.
POSSIBLE CAUSES:	**1.** Wiring harness, terminal, connector damage. **2.** Loose sensor mounting. **3.** Damaged tone wheel. **4.** Faulty sensor. **5.** Internal controller fault.

DIAGNOSTIC CHART 9-15

PROBLEM AREA:	Erratic performance from active magnetoresistive sensor.
SYMPTOMS:	Signal intermittently missing; Periodic drop off of signal.
POSSIBLE CAUSES:	**1.** Wiring harness, terminal, connector damage. **2.** Loose sensor mounting. **3.** Damaged tone wheel. **4.** Faulty sensor. **5.** Internal controller fault.

DIAGNOSTIC CHART 9-16

PROBLEM AREA:	Active magnetoresistive sensor circuit failure
SYMPTOMS:	**1.** No sensor output. **2.** Sensor circuit fails the diagnostic test. **3.** Sensor circuit low. **4.** Sensor circuit high.
POSSIBLE CAUSES:	**1.** Wiring harness, terminal, connector damage. **2.** 12-volt supply circuit shorted to ground. **3.** Signal circuit shorted to ground. **4.** 12-volt supply circuit shorted to voltage. **5.** 12-volt supply circuit open. **6.** Signal circuit shorted to voltage. **7.** Signal circuit open. **8.** Signal circuit shorted to 12-volt supply circuit. **9.** Faulty sensor. **10.** Internal controller fault.

DIAGNOSTIC CHART 9-17

PROBLEM AREA:	Hall-effect sensor input missing.
SYMPTOMS:	No signal is present during motion of monitored component.
POSSIBLE CAUSES:	**1.** Sensor voltage supply circuit shorted to battery voltage. **2.** Sensor voltage supply circuit open. **3.** Sensor voltage supply circuit shorted to ground. **4.** Sensor signal circuit open. **5.** Sensor signal circuit shorted to battery voltage. **6.** Sensor signal circuit shorted ground. **7.** Sensor signal circuit shorted to the voltage supply circuit. **8.** Sensor ground circuit open. **9.** Excessive sensor air gap. **10.** Faulty sensor. **11.** Internal controller fault.

DIAGNOSTIC CHART 9-18

PROBLEM AREA:	Intermittent Hall-effect sensor input.
SYMPTOMS:	Intermittent signal during motion of monitored component.
POSSIBLE CAUSES:	**1.** Sensor voltage supply circuit open. **2.** Sensor voltage supply circuit shorted to ground. **3.** Sensor signal circuit shorted to battery voltage. **4.** Sensor signal circuit open. **5.** Sensor signal circuit shorted ground. **6.** Sensor signal circuit shorted to the voltage supply circuit. **7.** Sensor ground circuit open. **8.** Excessive sensor air gap. **9.** Faulty sensor. **10.** Damaged tone wheel. **11.** Internal controller fault.

CHAPTER 10
VEHICLE MULTIPLEXING DIAGNOSTICS

Upon completion and review of this chapter, you should be able to:

- Describe the purpose of U- and B-codes.
- Properly diagnose an ISO 9141-2 bus system and determine needed repairs.
- Properly diagnose an ISO-K bus system and determine needed repairs.
- Properly diagnose a class A bus system and determine needed repairs.

- Properly diagnose a J1850 bus system and determine needed repairs.
- Properly diagnose a controller area network (CAN) bus system and determine needed repairs.
- Properly diagnose a local interconnect network (LIN) bus system and determine needed repairs.

Basic Tools

Hand tools
Fender covers
Service information

Terms To Know

B-codes
J1962 breakout box (BOB)

U-codes
Vehicle module scan

INTRODUCTION

If a vehicle's multiplexing system should fail, symptoms can range from a single function (such as instrument gauges) not operating to multiple function failures, including engine no-start. Diagnosing the bus system is not much different from diagnosing any other electrical system. Begin by verifying the customer's complaint; determine if there are any related symptoms, and then analyze the symptoms to develop a logical troubleshooting plan. It is important to understand how the bus system you are diagnosing should operate and the normal voltages and resistances on the system. Bus system failures include circuit opens, shorts, high resistance, and component failures. In addition, do not be quick to condemn the bus system if a module is not communicating. It is possible that the module is not powering up due to loss of battery voltage feed or loss of ground. This chapter discusses those items that the technician must be aware of while diagnosing different bus networks. The most common bus systems are discussed here.

Since most bus systems communicate with the scan tool, they will have a point of connection at the data link connector (DLC). To assist in testing of the data bus, a **J1962 breakout box (BOB)** is available (**Figure 10-1**). Since J1962 is the mandated DLC configuration for on-board diagnostics, second generation (OBD II), this tool will work on any OBD II–compliant vehicle. The J1962 BOB provides a pass-through test point that connects in series between the DLC and the scan tool. This provides easy testing of voltage and resistance of any of the DLC circuits without damaging the DLC.

Figure 10-1 The J1962 BOB makes pin out testing of the DLC easier.

COMMUNICATION FAULT CODES

Diagnostic trouble codes (DTCs) assigned to the vehicle communication network are called **U-codes**. These codes follow the same Society of Automotive Engineers (SAE) guideline as the P-codes used for powertrain faults. The prefix *U* indicates the fault is associated with network communications.

In addition, DTCs that are assigned to the vehicle's body systems and control modules are called **B-codes**. Typically, these codes refer to a failure of the system the module operates (such as a sensor failure). If a module relies on a bus message from another module but does not receive it, the first module may set both a U-code and a B-code. The U-code would be due to the loss of communication with the second module, and the B-code due to the system not able to perform a function. B-codes also follow the same SAE guideline as the P-codes.

Most modules on the bus network are capable of setting U-codes if they detect abnormal conditions. Most bus modules can detect loss of communication conditions with one or more modules or a bus failure. Some modules may also be able to monitor the actual voltage on the bus circuits and set additional trouble codes for conditions such as voltage high, low, shorted, or open bus circuits.

In addition, most modules report the status of the DTCs. If the conditions currently exist, then the DTC is reported as being active. If the conditions no longer exist, then the DTC is reported as being stored.

> **AUTHOR'S NOTE** It is important to remember that one fault can set multiple DTCs in many different modules. Just because a module recorded a fault does not mean the problem is in that module; it could be anywhere in the bus circuits.

ISO 9141-2 BUS SYSTEM DIAGNOSTICS

Classroom Manual
Chapter 10, page 269

The ISO 9141-2 standard bus system provides for communication links between the scan tool and the module. Some OBD II vehicles use this protocol for communication with the powertrain control module (PCM). In addition, some manufacturers use the system for communication between the scan tool and other modules on the vehicle (**Figure 10-2**).

The ISO 9141-2 bus uses a K-line to transmit data from the module to the scan tool and an L-line for the module to receive data from the scan tool (**Figure 10-3**). The scan

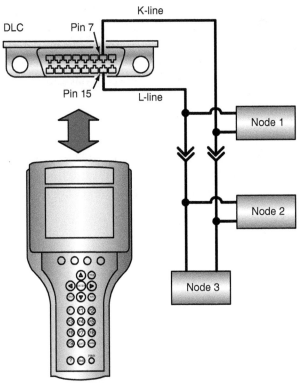

Figure 10-2 ISO 9142-2 bus system used to communicate between the scan tool and modules on the vehicle.

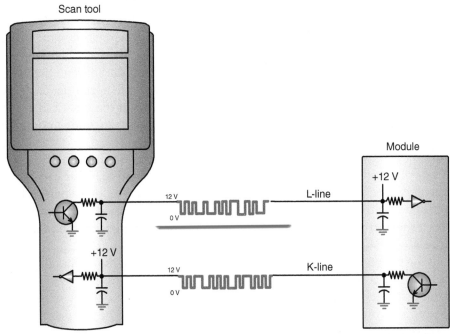

Figure 10-3 The K-line transmits data from the module to the scan tool and an L-line receives data from the scan tool.

tool supplies bias to the module on the K-line, while the module supplies bias to the scan tool on the L-line. Communication occurs when the transmitting node pulls the voltage low. If a failure occurs in this bus system, then communications between the scan tool and the module will not be possible. Since this bus system is not used for communications between modules on the vehicle, the customer may not have any noticeable problems with vehicle or accessory operation.

> **SERVICE TIP** Some scan tools provide a **vehicle module scan** function that will query all of the modules on the bus to respond and then list those that did reply. This makes it simple to see if any other modules are responding and which ones are not.

When diagnosing a failure due to the scan tool not being able to communicate with the PCM, it is important to analyze the symptoms. Review the wiring diagram and bus system information in the proper service information to determine if other modules on the vehicle are diagnosed using the ISO 9141-2 bus system. If other modules are on the bus, then use the scan tool to attempt to connect with each of these modules. If the scan tool connects to any module using the ISO 9141-2 bus, then it is not a total bus failure and the technician will need to diagnose for a partial bus failure. If no modules respond, then the technician will need to diagnose for a total bus failure.

Referring to Figure 10-2, if node 1 responded, but nodes 2 and 3 did not, then the first location to check would be the in-line connector. If only nodes 1 and 3 responded, then the problem is in the bus circuit to node 2. Since the circuit is wired in parallel, the problem cannot be a short to ground or voltage. Either the bus has an open between the splice and the node, or the node is not powering up due to faulty battery feed or ground circuits.

If no module responds, then check voltages on the K-line and the L-line. Connect the J1962 BOB to the DLC and test for voltages at the proper terminals. Without the scan tool connected, there should be voltage only on the L-line. Voltage on the L-line is supplied by the module. Zero voltage here can indicate an open circuit or short to ground. Disconnect the battery and use an ohmmeter to determine the type of fault. If an open is indicated, the most likely location is between the DLC and the splice to the first node. A short to ground can be anywhere in the circuit. To locate the short, refer to the service information and determine if there is an in-line connector in the circuit. If the in-line connector is accessible, disconnect it and see if the resistance changes. If it does, then the short is downstream of the in-line connector. If the ohmmeter reading does not change, then the fault is between the DLC and the in-line connector. Next disconnect the modules on the side of the in-line connector that the fault was isolated to. After each module is disconnected, check the resistance reading. If disconnecting a module changes the reading, then that module has an internal fault and needs to be replaced. If the ohmmeter still indicates a short to ground after all of the modules are disconnected, then the fault is in the harness.

With the scan tool connected, voltage should be present on the K-line. Since the voltage on the K-line is supplied from the scan tool, 0 volt (V) on this line means that either the circuit is shorted to ground or the scan tool is faulty. Use an ohmmeter to determine if the K-line is shorted to ground. If so, follow the same procedure just described for the L-line to locate the short.

ISO-K BUS SYSTEM DIAGNOSTICS

Classroom Manual
Chapter 10, page 269

Vehicles that use ISO-K for the communication connection between the module and the scan tool may have their own dedicated lines from individual terminals in the DLC to the module (**Figure 10-4**). The K-line from the DLC to the PCM will be from terminal 7 of the DLC. The ISO-K bus is used only for communications between the scan tool and the module on a signal wire; it is not used for intermodule communications. The scan tool supplies the voltage onto the K-line.

If communications between the scan tool and module are not possible, begin diagnosis by connecting the J1962 BOB to the DLC. With the ignition key in the OFF position and the scan tool disconnected, there should be 0 volt on the K-line. If voltage is present on this circuit at this time, there is a short to voltage between the DLC and the module, or internal to the module.

If 0 volt is indicated, test for short to ground with an ohmmeter. If a short to ground is indicated, it may be in the wiring between the DLC and the module, or internal to the module.

Next, connect the scan tool with the ignition switch in the OFF position. It would be best to use a lab scope to observe the voltages. Set the scope to read 10 ms per division on the time frame. At this time, up to 12 volts should be seen on the scope trace. While observing the scope trace, turn the ignition key to the RUN position and attempt to establish communication between the scan tool and the module. The scope trace should indicate a digital signal as communication is established and as data is transmitted (**Figure 10-5**). Notice in the scope trace that the biased voltage from the scan tool is about 8.0 volts. When the scan tool is attempting to request data from the module, the voltage is pulled low, to −0.5 volt. Finally, when the module is transmitting data to the scan tool, the voltage is pulled to +0.5 volt.

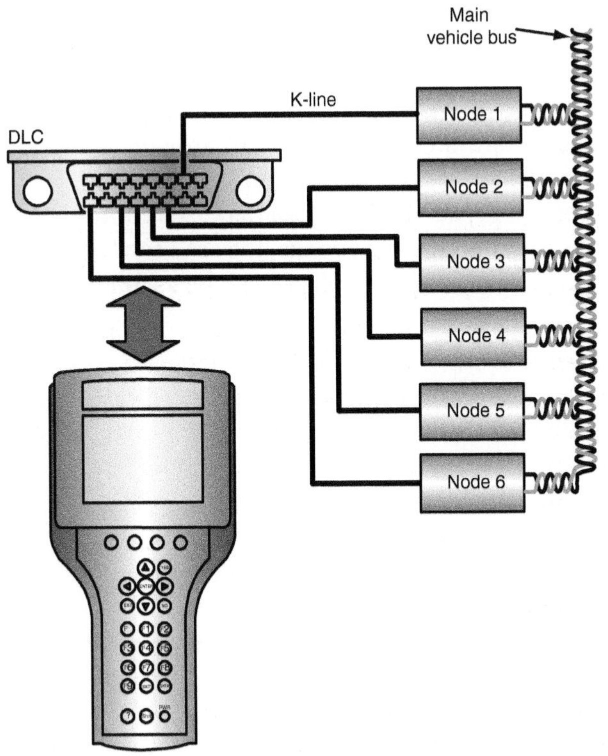

Special Tools
J1962 BOB
Scan tool
DMM
Lab scope

Figure 10-4 An ISO-K bus circuit used to connect several modules to the scan tool. Each module uses its own dedicated circuit from the DLC.

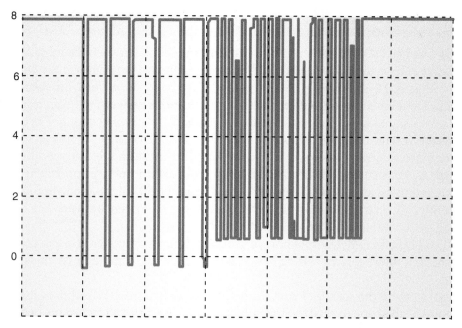

Figure 10-5 ISO-K transmission trace. The negative voltage is from the scan tool attempting to request data from the module. When the module is transmitting data to the scan tool, the voltage is pulled down to +0.5 volt.

CLASS A BUS SYSTEM DIAGNOSTICS

Classroom Manual
Chapter 10, page 271

Special Tools
J1962 BOB
Scan tool
DMM
Lab scope

As discussed in the Classroom Manual, the class A bus system is a slow-speed bus. This system will have a master module and several slave modules. The master module will supply the bias voltage onto the bus system. If this module should fail, then no bus communications will be possible. In the Classroom Manual, the Chrysler Collision Detection (CCD) multiplexing system was presented as an example of how a class A bus operates. In this chapter, the CCD bus system will be used to provide an example of failure modes and how to diagnose a typical class A bus system.

Depending on scan tool type, the scan tool may attempt to diagnose the bus system if communication fails to occur. If this happens, the scan tool will display the cause of the problem on the screen. The messages displayed can be as follows:

- Bus (−) open.
- Bus (+) and bus (−) open.
- Bus (+) open.
- Bus (+) shorted to bus (−).
- Bus bias level too high.
- Bus bias level too low.
- Bus shorted to 5 volts.
- Bus shorted to battery voltage.
- Bus shorted to ground.
- No bus bias.
- No bus termination.
- Not receiving bus messages correctly.

The CCD bus network uses the voltage difference between the two wires to transmit data (**Figure 10-6**). The bus bias on the two wires is about 2.50 volts. The bus can operate between 1.5 and 3.5 volts. Voltages that are outside of this range will not transmit data.

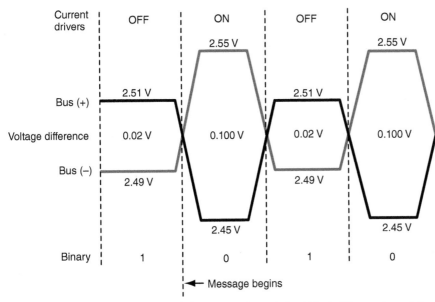

Figure 10-6 Normal CCD bus voltages during message transmission. Voltage difference is used to transmit bit 1 or bit 0.

Since the CCD bus is a vehicle-wide bus system, failure of this bus will result in electrical systems not operating properly. First, determine if it is a total bus failure or a partial bus failure by attempting to communicate with all of the modules on the vehicle. If it is possible to communicate with a module on the bus system, then approach the diagnostics as a partial bus failure. If no modules can communicate with the scan tool, then a total bus failure is indicated.

If a total bus failure is indicated, then possible causes include the following:

- A faulty master module.
- Faulty power or ground circuits to the master module.
- An open in one of the bus circuits from the master module.
- A short to ground in one of the bus circuits.
- A short to voltage in one of the bus circuits.
- The two bus wires are shorted together.

> ⚙️ **SERVICE TIP** The PCM does not communicate with the scan tool on the CCD bus. It uses the K-line for this purpose. Do not be fooled into thinking that the problem is a partial CCD bus failure based on the response of the PCM.

To determine the type of fault that is causing total bus failure, use the J1962 BOB and connect it to the DLC. Measure the voltage on each bus circuit. The CCD system uses pin 3 of the DLC for CCD (+) and pin 11 for CCD (−). CCD (+) should have 2.49 volts with the ignition key in the RUN position. CCD (−) should have 2.51 volts.

If the voltmeter reads 12 volts on either circuit, that circuit is shorted to battery voltage. To isolate the location of the short, turn the ignition switch to the OFF position and watch the voltmeter. If the voltage drops, then the short is to ignition-switched voltage. If the voltage is still reading battery voltage, the circuit is shorted to a direct battery feed. Remove fuses and relays one at a time while watching the voltmeter. If the voltage drops after a fuse or relay is pulled out, note the identification of the fuse or relay and use the service information to determine what circuits and components are protected by that fuse or controlled by that relay. Once this is known, return to the vehicle, reinstall the fuse, and begin to unplug each component one at a time while observing the voltmeter. Once a component is unplugged and the voltage drops to normal, you have located the faulty component. If all components are unplugged and the voltage is still high, the fault is in the wiring of the circuit for those components.

If the voltmeter indicates 0 volt on either or both of the bus circuits, then the cause can be an open or a short to ground. To determine the type of fault, use an ohmmeter and measure the termination resistance of the bus system by disconnecting the battery and placing the test leads into pins 3 and 11. Most CCD bus systems will have 60 ohms (Ω) of termination resistance. However, some vehicles will have 120 ohms of termination resistance, so it is important to check the service information.

If the system should have 60 ohms and the ohmmeter reads 120 ohms, then one of the termination resistors is open or there is an open in the bus circuit to a module that has one of the termination resistors. Usually the termination resistors will reside in the PCM and the body control module (BCM). Each of these modules will have a 120-ohm termination resistor (**Figure 10-7**). To locate the open resistor, disconnect either the PCM or BCM while observing the ohmmeter. If the resistance does not change, then the termination resistor that is open is in the module that was just unplugged. If the resistance changes to infinite, the remaining termination resistor in the circuit is just removed, so the fault is within the termination of the other module. Before replacing the module, check for an open in the bus circuits to that module. If the bus circuits are good, and module power and ground circuits are confirmed, replace the module.

If termination resistance is normal and there is no indication of a short to ground on any of the bus circuits, the fault is probably that the master module is not powering up and thus not able to supply the bias for the bus. Usually the master module is the BCM in the CCD bus system. Confirm if the power and ground circuits are good before replacing the BCM.

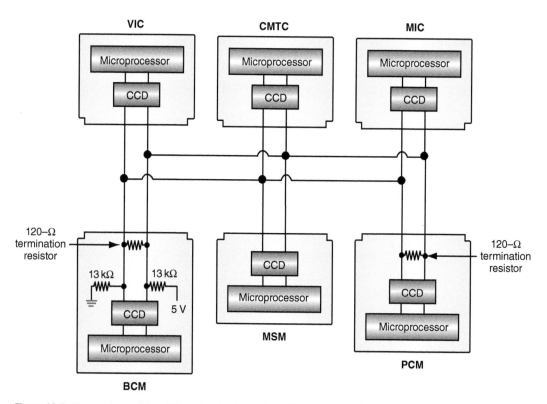

Figure 10-7 Usually two modules will have termination resistors. Total circuit resistance will be 60 ohms.

If ohmmeter testing indicates a short to ground in one of the bus circuits, isolation of the fault must still be performed. The fault may be a faulty module that has an internal short, or it could be the bus wire that is shorted to ground within the wiring harness. The system circuit needs to be broken down into smaller sections to isolate where the fault is. Examine the wiring diagram for the vehicle and determine if there is a connector that would separate the bus network. Identify any connectors that can be used and their location. With the ohmmeter measuring resistance between the affected circuit and chassis ground, unplug the connector. If the ohmmeter still reads low resistance, the fault is between the DLC and the connector. If the ohmmeter reads infinite, the fault is downstream of the connector. Now that this is known, move the ohmmeter test leads (if necessary) to observe resistance readings, as modules on the appropriate side of the connector are unplugged. If the readings change after a module is unplugged, then the fault is in the module. If the reading is still low after all modules are unplugged, the fault is in the wiring harness.

> ⚙ **SERVICE TIP** If the scan tool cannot communicate with the PCM, and other modules on the CCD bus system have logged a loss of communication fault code against the PCM, check the common voltage supply circuits to the sensors. If these circuits short to ground, the PCM will turn off. This will also result in a no-start, and the PCM may not store any DTCs.

If the initial tests indicate a partial bus failure, try to determine which modules are not communicating. If there is more than one, use the wiring diagrams for the vehicle to determine if there is anything common between the affected modules. For example, look for a common connector, common splice, common voltage supply to the modules, or common ground for the modules. Locate this common component on the vehicle and confirm proper function and repair as needed.

When diagnosing the CCD bus system, remember the following:

1. Normal bus voltages with the ignition switch in the RUN position are approximately 2.5 volts.
2. Normal bus voltages with the ignition switch in the OFF position are 0 volt on both circuits.
3. Normal termination resistance is typically 60 ohms.
4. The wires must be twisted at the rate of one twist every 1¾ inches (44 mm).

J1850 BUS SYSTEM DIAGNOSTICS

The J1850 bus system can be either a 10.4 kb/s variable pulse width–modulated (VPWM) system or a 41.6 kb/s pulse width–modulated (PWM) system. In the Classroom Manual, the Programmable Communication Interface (PCI) bus system that Chrysler uses was discussed. This system is similar to the class 2 bus system that General Motors uses as well. This system will be discussed here as an example of failure modes and how to diagnose a typical J1850 bus system.

Since the PCI bus is a vehicle bus system, a failure in it will cause electric systems to not operate properly. Like the other bus system diagnostics just discussed, begin by determining if the problem is a total bus failure or a partial bus failure.

If a total bus failure is indicated, then the system needs to be tested for proper voltage and termination resistance. Use the J1962 BOB to provide a diagnostic test point to measure these values. Total bus failure can occur only if there is a short to voltage or a short

Classroom Manual
Chapter 10,
page 270, 273

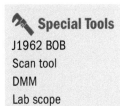
Special Tools
J1962 BOB
Scan tool
DMM
Lab scope

to ground in the circuit. Termination resistance will be different on every vehicle since each module provides its own termination. Since vehicles will have different options on them, the resistance will be different between vehicles. However, by measuring the resistance between pin 2 of the DLC (at the J1962 BOB) and chassis ground, a low reading will confirm a short to ground in the system. The procedure to locate this type of fault is presented later.

The normal voltage on the PCI bus is 0 volt when the bus is at rest, and the voltage pulls up to near 7.5 volts when the bus is active (**Figure 10-8**). The best method of determining proper voltage on this system is with a lab scope. A DMM set to the voltmeter will not be able to give actual voltage values since the activity is too fast, so the meter will display the average. The MIN/MAX feature of the voltmeter will freeze the readings, but they still will not be actual voltages. However, the voltmeter will indicate if the bus is shorted to battery voltage (12 volts) or is a short to ground (0 volt). A reading of 0 volt may also indicate an open, but this type of fault will not result in total bus failure.

If a short to battery voltage is indicated, then the location of the fault needs to be isolated. In order to do this, the system needs to be broken down into manageable pieces or smaller circuits. With the voltmeter still measuring PCI voltage, turn the ignition key to the OFF position. Wait for 20 seconds and read the voltmeter again. If the voltmeter now reads 0 volt, then the bus network is shorted to ignition-switched voltage. If the voltmeter still reads battery voltage, then the bus is shorted to direct battery voltage. In either case, the circuit that is shorted to the bus network can be further isolated by pulling fuses and relays one at a time while watching the voltmeter. If the voltage drops after a fuse or relay is pulled, note the identification of the fuse or relay and use the service information to determine what circuits and components are protected by that fuse or controlled by that relay. Once this is known, return to the vehicle, reinstall the fuse, and begin to unplug each component one at a time while observing the voltmeter. Once a component is unplugged and the voltage drops to normal, that component is the fault. If all components are unplugged and the voltage is still high, the fault is in the wiring of the circuit for those components.

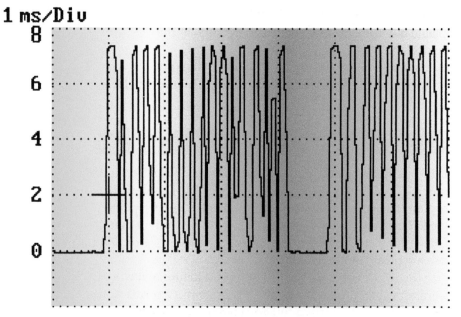

Figure 10-8 The normal voltage on the PCI bus is 0 volt when the bus is at rest. To transmit data, the voltage is pulled up to near 7.5 volts.

If a short to ground is indicated while testing the termination resistance as discussed earlier, isolation of the fault must still be performed. The fault may be a faulty module that has an internal short, or it could be the PCI bus wire that is shorted to ground within the wiring harness. Again, the trick is to break the system down into smaller circuits. Depending on how the vehicle is wired, some may use a common hub for all of the bus circuits. If this is the case, this location provides an excellent place to begin diagnostics. For example, **Figure 10-9** shows a PCI bus system that uses the BCM as a common hub for the bus. The BCM has three connectors attached to it that contain the different bus circuits from all of the modules. By unplugging these connectors (one at a time) and observing system operation, the system can be broken down into a few circuits. Once a connector is unplugged and some of the bus systems become active, the fault is within the bus circuits of that connector. It is possible that the connector has several bus wires in it. Use the ohmmeter to determine which bus wire is shorted to ground. Use the service information to determine what module is on that circuit and unplug the module. Again, measure resistance between the BCM connector and chassis ground. If the reading is infinite, the fault is in the module. If the reading is still low, the wire is shorted to ground.

If the bus system does not have a common hub, examine the wiring diagram for the vehicle and determine if there is a connector that would separate the bus network. Identify any connectors that can be used and their location. With the ohmmeter measuring resistance between pin 2 of the DLC (using the J1962 BOB) and chassis ground, unplug the connector. If the ohmmeter still reads low resistance, the fault is between the DLC and the connector. If the ohmmeter reads any other value, the fault is downstream of the connector. Now that this is known, move the ohmmeter test leads (if necessary) to observe resistance readings, as modules on the appropriate side of the connector are unplugged. As before, if the readings change after a module is unplugged, then the fault is in the module. If the reading is still low after all modules are unplugged, the fault is in the wiring harness.

If the initial tests indicate a partial bus failure, try to determine which modules are not communicating. If there is more than one, use the wiring diagrams for the vehicle to determine if there is anything common between the affected modules. For example, look for a common connector, common splice, common voltage supply to the modules, or common ground for the modules. Locate this common component on the vehicle and confirm proper function and repair as needed.

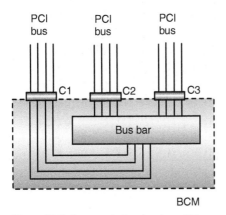

Figure 10-9 Common hub or bus bar within a controller helps make isolating the cause of bus failures easier for the technician.

CONTROLLER AREA NETWORK BUS DIAGNOSTICS

Classroom Manual
Chapter 10,
page 270, 276

Special Tools

J1962 BOB
Scan tool
DMM
Lab scope

Vehicles that utilize controller area network (CAN) bus systems will generally have two or three different CAN systems. The vehicle body systems will communicate over the medium-speed CAN B network. Those systems such as engine controls and antilock brakes that require data at a faster rate will communicate on the CAN C bus. Beginning in the 2005 model year, manufacturers have started migrating to the new requirement for using a CAN bus network for diagnostics (often referred to as diagnostic CAN C).

The first step after verifying the customer's complaint is to determine which bus network is at fault. Usually a scan tool can easily determine this. Use the scan tool to access the central gateway (CGW) module for the bus networks. Since this is the common component for all three CAN bus networks, it will usually provide a DTC that will indicate which bus network has the fault. If the scan tool cannot communicate to the central gateway module, then the diagnostic CAN C circuit is faulty.

Once the faulty network is identified, diagnostics of either CAN network can be performed using a DMM or a lab scope. However, methods are different based on the network and the type of fault. The following diagnostic procedure should be followed to isolate the cause:

1. If the fault code indicates an open circuit fault with either the CAN B (+) or the CAN B (−) circuit (**Figure 10-10**), then determine and locate an easily accessed module on the CAN B bus system. At the connector for this component, create a short to ground on the opposite CAN B circuit. For example, if the fault code indicates that the CAN B (−) is open, then short the CAN B (+) circuit to ground. Since CAN B buses that operate at speeds up to 125 kb/s are fault tolerant and will operate in single-wire mode if an electrical potential exists between one of the circuits and ground, the open may not cause any noticeable symptoms. However, when the other circuit is shorted to ground, that module can no longer communicate. All other modules will communicate using the single-wire mode (**Figure 10-11**). With the wire shorted to ground, use the scan tool to determine which module is not communicating on the bus. Locate this module on the vehicle. The fault will be between the splice from the affected bus circuit to the module.

2. If the fault code indicates that one of the CAN B circuits is high or low, then use a DMM to determine if the fault is a short to voltage (12 volts) or shorted to ground (0 volt). If the voltage on both circuits is between 500 mV and 1.0 volt, the two circuits are shorted together. Locating a shorted circuit will be similar to the procedure described earlier for the J1850 VPWM bus. Use the wiring diagram for the vehicle and determine if connectors that can break the circuits down into smaller sections are used. Return to the vehicle and, with a voltmeter connected to the shorted circuit, observe the readings while opening the connectors. Once it is determined which side of the connector has the short, modules or other connectors on that branch of the circuit can then be disconnected to locate the root cause of the failure.

3. If the fault code indicates both CAN B (+) and CAN B (−) circuits are low or high, this indicates that a total bus failure has occurred. Symptoms may include headlamps on, warning indicator lamps on in the instrument panel, instrument panel backlighting at full intensity, gauges inoperative, and possible no-start. To locate the cause of this failure, use the diagnostic procedure for locating a short circuit, as previously described in step 2.

4. Some modules do not set specific codes that isolate the type of fault. They will set a generic "CAN B BUS" fault. First determine if the single-wire failure is an open

or short by measuring the voltages at an easily accessed module connector. If this fault is set due to an open, short to ground each CAN B circuit one at a time while monitoring the scan tool and observing symptoms. If shorting one of the circuits to ground does not affect operation or set additional DTCs, remove the short and repeat the procedure on the other CAN B circuit to determine where the open exists. The open will be on the CAN B circuit that was *not* shorted to ground. If the tests indicate that the circuit(s) is shorted to ground, shorted to voltage, or shorted to each other, then follow the diagnostic procedure outlined previously in step 2.

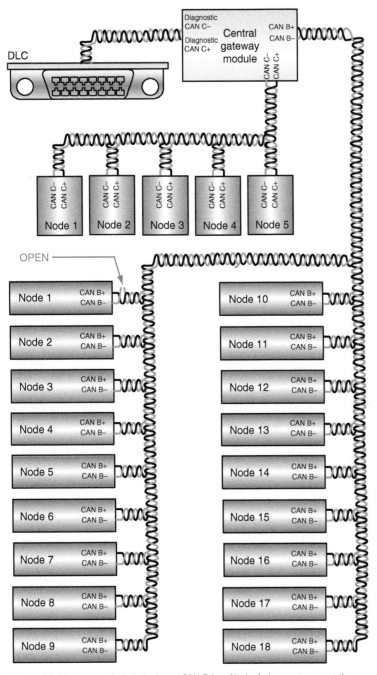

Figure 10-10 An open bus (−) wire at CAN B bus Node 1 does not prevent the other nodes from being active on the bus.

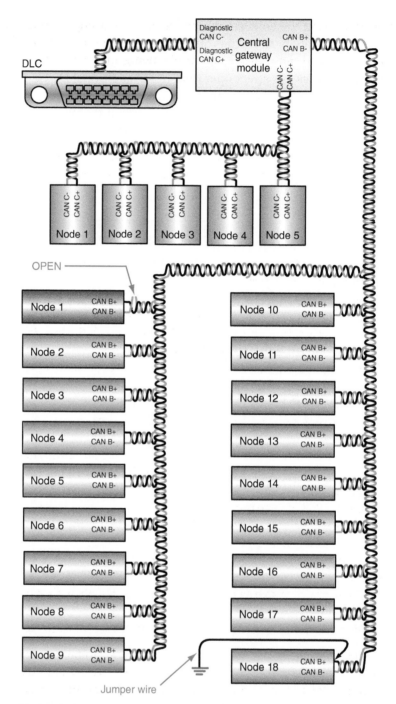

Figure 10-11 When the bus (+) wire at CAN B bus Node 18 is shorted to ground, all modules continue to communicate except Node 1. Node 1 cannot communicate since it no longer has a circuit.

5. If the fault code indicated that a CAN C circuit is shorted, this would cause a total failure of the CAN C bus. The short can be to voltage, ground, or the two circuits together. Symptoms may include illuminated warning indicator lamps in the instrument panel, inoperative gauges, and possible no-start. To locate the short circuit, determine if it is caused by a defective module by disconnecting the modules one at a time and observing symptoms and ohmmeter readings on the CAN circuit. If disconnecting all of the modules fails to indicate a faulty module, the problem is in the wiring harness.

6. If the fault code indicates a loss of communication with a single module (on either the CAN B or the CAN C network), check for proper power and ground circuit at the affected module. If these circuits are confirmed good, then test the CAN circuits for opens between the splice to the network and the module. If the circuits do not have an open, replace the module.

> **CUSTOMER CARE** Modules can be very expensive and should be replaced only after all power feeds and ground circuits are tested and confirmed good.

When diagnosing the CAN bus networks, remember these facts:

1. A normal voltmeter reading on the CAN B (+) circuit with the ignition key in the RUN position is between 280 and 920 mV.
2. A normal voltmeter reading on the CAN B (−) circuit with the ignition key in the RUN position is between 4.08 and 4.72 volts.
3. A normal voltmeter reading on the CAN B (+) circuit with the ignition key in the OFF position and the bus asleep is 0 volt.
4. A normal voltmeter reading on the CAN B (−) circuit with the ignition key in the OFF position and the bus asleep is battery voltage.
5. CAN B termination resistance cannot be measured. However, it should be infinite with the battery disconnected. A low reading indicates a short.
6. The CAN B bus can become active from inputs or from the ignition switch.
7. Normal CAN C bus termination is 60 ohms. However, if other modules on the CAN C bus network have termination resistance, then the actual reading on an ohmmeter will be less than 60 ohms.
8. Normal diagnostic CAN C termination is 60 ohms.
9. A normal voltmeter reading on the CAN C (+) circuit with the ignition key in the RUN position is about 2.60 volts.
10. A normal voltmeter reading on the CAN C (−) circuit with the ignition key in the RUN position is about 2.4 volts.
11. The CAN C bus may be either event driven or be active only with the ignition in the RUN position.
12. The wires must be twisted at a rate of one twist every inch (25.4 mm).

Classroom Manual
Chapter 10, page 282

If the bus network is designed as a star system, diagnostics can be done at the DLC and the common point connectors (star connectors). The star network architecture will have all bus circuits connected to this common point (**Figure 10-12**). Since these systems typically use CAN buses, voltage and termination resistance values are the same as for the stub architecture. In Figure 10-12, the vehicle does not use a diagnostic CAN C bus; instead, the CAN C and interior bus networks are both in the DLC. This means the J1962 BOB can be used to measure voltage and resistance of each bus. Each terminal of the star connector goes to a module, the DLC, or to another star connector (**Figure 10-13**). The star connector provides an ideal location for isolating the circuits during diagnosis. For example, if the voltage and resistance readings at the DLC indicate that one of the CAN C bus circuits is shorted to ground, this would be the reason for the CAN C to be totally down. While leaving the ohmmeter connected to the terminal of the J1962 BOB that houses the faulty circuit, disconnect the star connector terminals one at a time until the ohmmeter no longer indicates a short to ground. The fault lies in the circuit that was just disconnected. The short will be between the star connector and the module, or there may be an internal fault within the module.

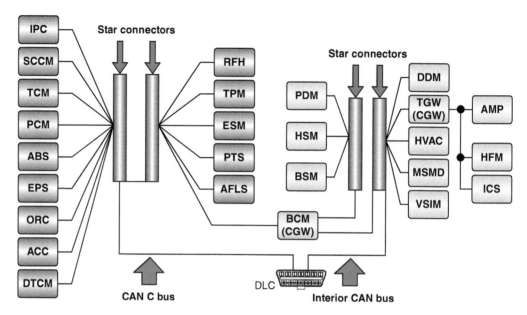

Figure 10-12 Star architecture typically does not use the diagnostic CAN C bus since the vehicle's bus networks are all located in the DLC.

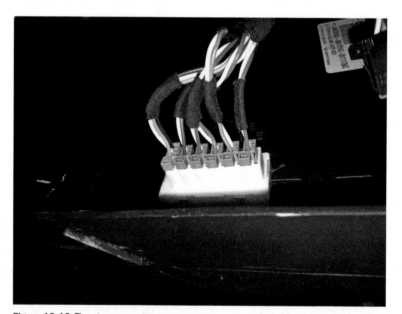

Figure 10-13 The star connectors are a common connection point that makes an ideal location for diagnosing bus network faults. Each terminal of the connector goes to one module, the DLC, or to the other connector.

SERVICE TIP If there is more than one star connector in the bus network, voltage measurements at the DLC may look close to normal if one of the termination resistors was open. Use an ohmmeter to confirm that the proper termination resistance is on the circuit.

AUTHOR'S NOTE The CAN C and the interior bus may each have more than one star connector. Since termination resistance can be located in the star connector, if one is used, it will have 60 ohms; if two are used, then each will have 120 ohms.

LOCAL INTERCONNECT NETWORK BUS DIAGNOSTICS

The local interconnect network (LIN) bus is a supplemental bus network that is used along with the vehicle's main multiplexing system. The master module communicates data from the slave modules onto the main bus network. The scan tool will not have access to the slave module. All diagnostics with the scan tool are performed through the master module.

Figure 10-14 is an illustration of a LIN bus system used for steering wheel–mounted radio controls. The steering column module (SCM) is the master module and is also connected to the CAN B bus network. The right steering wheel switch assembly is actually the slave module and is capable of sending data to the SCM and of receiving data from the SCM. The left steering wheel switch is a multiplex switch system that uses different resistance values for each switch position. The different resistances result in a unique voltage drop that is sensed by the slave module. The request action of the driver that is indicated by pressing a switch is received by the slave module and sent to the master module. The master module then puts the data onto the CAN B bus network, and the message is sent to the radio, which performs the requested action.

In this example, the scan tool will communicate with the SCM. By observing the data display information on the scan tool, each switch position input should be indicated. In addition, the master module (the SCM in this case) will store a DTC if a system indicates a fault. For example, if a switch in the left steering wheel switch assembly indicates that it

Classroom Manual
Chapter 10, page 283

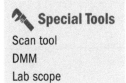 **Special Tools**
Scan tool
DMM
Lab scope

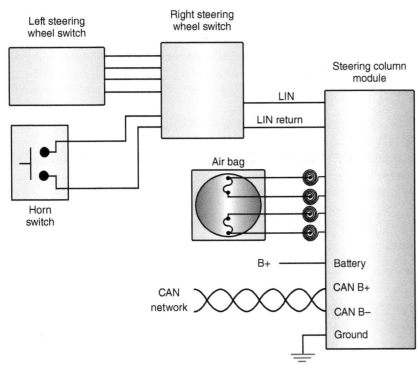

Figure 10-14 LIN bus system used for remote radio controls mounted on the steering wheel.

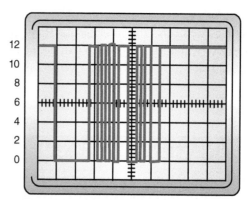

Figure 10-15 Normal LIN bus trace.

is stuck in the pressed position for an excessive amount of time, the SCM will set a DTC for this condition.

If a customer states that the steering wheel radio controls fail to operate, use the scan tool to see if any of the button pushes are indicated by the master module. If any are indicated, then the LIN bus circuit is working. If none of the switch inputs are indicated on the scan tool, then the LIN bus circuit, the slave module, or the master module has failed.

Normal voltages on the LIN bus is up to 12 volts, which is pulled low during communications (**Figure 10-15**). However, if the slave module is disconnected from the master module and the LIN bus circuit is tested with a voltmeter with the ignition switch in the RUN position, the reading will be about 9.5 volts. This is due to the master module pulse width modulating the signal in an attempt to communicate with the slave module. The voltmeter is indicating an average reading. If the slave module is plugged in and the voltage is tested, the reading will now be about 8.3 volts since both modules are pulling the circuit low in an attempt to communicate. The most accurate way of determining LIN bus voltages is with a lab scope.

If 0 volt is indicated by the voltmeter or lab scope when the slave module is disconnected, the cause could be a faulty master module, an open circuit between the master and slave module connector, or a short to ground in the LIN circuit. Use the voltmeter and ohmmeter to diagnose and locate the fault.

If normal voltage is indicated with the slave module disconnected, but then reads 0 volt when it is reconnected, the fault is in the slave module. In this case the slave module has an internal short to ground.

CUSTOMER CARE: Many people do not realize the subjectivity of the communication networks to RFI and EMI. These signals may interfere with module communications. If the customer says he is experiencing intermittent problems with the vehicle, it is possible the problem is due to an add-on feature such as a radio, cell phone, entertainment system, and so on. If this is found to be the problem, inform the customer that these systems may require special shielding and location alterations in order to prevent future instances.

ENHANCED SCAN TOOL DIAGNOSTICS

Some scan tools provide enhanced diagnostics of the CAN bus networks. In addition to performing the module scan, these units may provide advanced loss of communication screens, network topology screens, and so on. All of these additional features are designed to assist the technician in diagnosing the CAN bus network. **Photo Sequence 20** provides an example of these features and how they are accessed. Since scan tool manufacturers differ in their approach to diagnostics and display of information, this photo sequence is provided only as a guide. Always refer to the instructions for the scan tool you are using.

PHOTO SEQUENCE 20
Advanced Scan Tool Functions

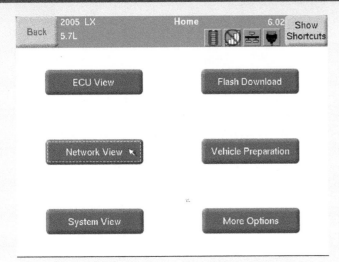

P20-1 The "Home" screen of the scan tool. Select "Network View."

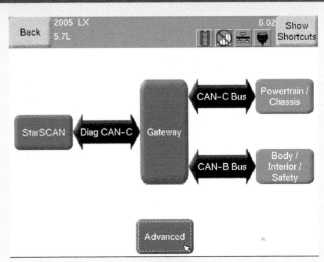

P20-2 The network view provides an icon look of the CAN bus network from the scan tool to the different bus systems. Red-colored icons indicate a fault has been recorded in that system. From this screen select the "Advanced" button.

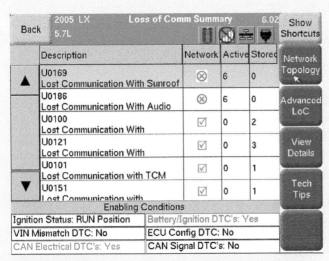

P20-3 This is the Loss of Communications Summary screen. A green check mark indicates that the module is communicating with the scan tool. The red Xs indicate that those modules are not communicating on the network bus. Under the "Active" column, the "6" indicates that six other modules have logged loss of communication fault codes against the module and the faults are currently present. If any logged-against faults were stored, they would be under the "Stored" column. Press the "Network Topology" button.

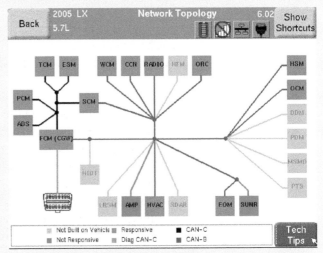

P20-4 This is a live layout of the vehicle's communication system. The color codes indicate if the module is active on the bus, if the vehicle is not built with that module, and the different bus networks. This is an actual representation of the vehicle's communication wiring. The red icons indicate modules that are not communicating. In this case, a common connector or splice is not the cause of the problem. The next logical diagnostic step would be to see if the two affected modules share a common power circuit or ground.

PHOTO SEQUENCE 20 (CONTINUED)

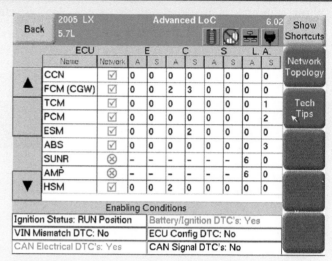

P20-5 Another option from the "Loss of Communication" screen is to press the "Advanced LoC" button. This takes the technician to another loss of communications screen.

P20-6 This advanced loss of communications (LoC) screen indicates the type of fault code as "E" for electrical, "C" for communications, "S" for implausible signal, and "L.A." for logged against. Also, columns indicate the number of active or stored codes and which modules have logged the fault.

DATA BUS CIRCUIT REPAIR

Repairing the data bus circuit wires must be done in such a manner that the integrity of the system is not compromised. Always repair the circuit by soldering, never use a crimp connector. If the circuit is shielded, be sure to follow the repair procedure outlined in Chapter 4. When repairing the data bus circuit(s), consider the following:

1. Refer to the service information for the specification for the number of twists per meter or inch.
2. Any new wire(s) added must be the same length as that removed. The data bus + and – circuits must be the same length. Since the bus operated as a high speed, deviations in circuit length can disrupt the timing and result in improper data transmission.
3. If both wires are to be replaced, twist the wires together prior to soldering them in place.
4. Use the same size wire as that removed.
5. If repairing both wires, stagger the splices.
6. Use heat shrink tube to provide a good insulating seal. Do not use electrical tape.

CASE STUDY

A customer has had his vehicle towed to the repair facility. Other shops have attempted to repair the problem of an intermittent no-start condition, but the problem persists. The technician attempts to duplicate the problem, but the engine will now start and run fine. The technician connects the scan tool and accesses the PCM to read fault codes. There are no active or stored DTCs. Realizing that the vehicle he is working on is equipped with the CAN bus network, he accesses the central gateway module to see if it has any DTCs stored in it. Here he finds a stored "Loss of communication with PCM" fault. There are no active faults. Realizing that this is an intermittent problem, he checks the connectors at the PCM. Here he finds a loose terminal and makes the necessary repairs. The intermittent no-start was due to the vehicle's security system not being able to communicate with the PCM; thus the PCM would not allow the engine to start. No fault codes were set in the PCM since it was doing what it was supposed to. Other modules did log the loss of communication fault against the PCM, since they could not communicate with the PCM when the terminal connection came loose.

ASE-STYLE REVIEW QUESTIONS

1. In a J1850 VPWM bus system, how can a short to ground be isolated?
   A. Use a jumper wire to supply battery voltage to pin 2 of the DLC while observing the DMM.
   B. Disconnect in-line connectors one at a time while observing the DMM.
   C. Use a jumper wire to short pin 15 of the DLC to ground while observing the DMM.
   D. All of the above.

2. If one of the CAN C bus circuits were shorted to ground, all of the following could occur EXCEPT:
   A. No-start.
   B. No communication between scan tool and any CAN C modules.
   C. CGW module logs a loss of communication fault code against CAN C modules.
   D. The bus will operate in single-wire mode.

3. All of the following can cause total bus failure of the J1850 VPWM bus system EXCEPT:
   A. Open bus wire to the BCM.
   B. Bus wire shorted to ground.
   C. Bus wire shorted to battery voltage.
   D. Internal short in a module.

4. The lab scope trace of an ISO-K bus system shows the voltage going to − 0.5 volt. This would indicate:
   A. The scan tool bias voltage.
   B. The module attempting to transmit data to the scan tool.
   C. The scan tool attempting to request data from the module.
   D. Excessive electrical noise on the circuit.

5. A class A bus that has 0 volt on both bus circuits may indicate:
   A. Both circuits are shorted to ground.
   B. The master module's power feed circuit is open.
   C. An open bus connector to the master module.
   D. All of the above.

6. To isolate a bus circuit that is shorted to battery voltage, all of the following methods can be used EXCEPT:
   A. Use a jumper wire and connect the faulty circuit to ground.
   B. Turn the ignition switch off to see if it is switched voltage.
   C. Pull fuses one at a time to determine the shorted circuit.
   D. Pull relays one at a time to determine shorted circuits or components.

7. On a LIN bus, 0 volt is indicated on the bus circuit with the slave module disconnected. What is the **LEAST LIKELY** cause?

   A. An open circuit between the master module and the slave module connector.

   B. A short circuit between the master module and the slave module connector.

   C. Faulty master module.

   D. Faulty slave module.

8. *Technician A* says on the LIN bus system the scan tool accesses the slave modules to display their data.

   Technician B says the LIN slave module can send data only to the master module.

   Who is correct?

   A. A only

   B. B only

   C. Both A and B

   D. Neither A nor B

9. When the scan tool is connected, voltage on the K-line reads zero.

   Technician A says this means the circuit may be shorted to ground.

   Technician B says the scan tool may be faulty.

   Who is correct?

   A. A only

   B. B only

   C. Both A and B

   D. Neither A nor B

10. *Technician A* says normal voltages on the LIN bus is up to 12 volts.

    Technician B says 12 volts on the LIN bus indicates data transmission.

    Who is correct?

    A. A only

    B. B only

    C. Both A and B

    D. Neither A nor B

ASE CHALLENGE QUESTIONS

1. *Technician A* says on a CAN B system that if one of the bus circuits is shorted to ground, it may be possible to locate the fault by jumping battery voltage to the faulty circuit.

   Technician B says to locate the short to ground, use a jumper wire and ground the opposite circuit.

   Who is correct?

   A. A only

   B. B only

   C. Both A and B

   D. Neither A nor B

2. *Technician A* says on an ISO 9141-2 bus system, without the scan tool connected, there should be voltage only on the L-line.

   Technician B says the module supplies the voltage on the L-line.

   Who is correct?

   A. A only

   B. B only

   C. Both A and B

   D. Neither A nor B

3. *Technician A* says on CAN B bus systems it may be possible for the bus to operate if there is an electrical potential between one of the circuits and ground.

   Technician B says if the two CAN B circuits are shorted together, total bus failure will result.

   Who is correct?

   A. A only

   B. B only

   C. Both A and B

   D. Neither A nor B

4. One module fails to communicate on the bus.

   Technician A says this can be due to an internal short of one of the bus circuits.

   Technician B says this could be due to the module ground circuit being faulty.

   Who is correct?

   A. A only

   B. B only

   C. Both A and B

   D. Neither A nor B

5. *Technician A* says U-codes identify a failure with a body system function.

   Technician B says B-codes are set to indicate bus network communication failures.

   Who is correct?

   A. A only

   B. B only

   C. Both A and B

   D. Neither A nor B

Name _____   Date _____

ISO 9141-2 BUS SYSTEM DIAGNOSIS

Upon completion of this job sheet, you will be able to determine normal ISO 9141-2 bus operation. You will also observe faulty bus system operation.

NATEF Correlation ────────────────────────────────────

This job sheet addresses the following **AST/MAST** tasks:

A.7.   Use wiring diagrams during the diagnosis (troubleshooting) of electrical/electronic circuit problems.

This job sheet addresses the following **AST** tasks:

A.11.   Check electrical/electronic circuit waveforms; interpret readings and determine needed repairs.

G.5.   Describe body electronic systems circuits using a scan tool; check for module communication errors (data bus systems); determine needed action.

This job sheet addresses the following **MAST** task:

G.5.   Diagnose body electronic systems circuits using a scan tool; check for module communication errors (data bus systems); determine needed action.

NATEF

Tools and Materials

- J1962 BOB
- Scan tool
- DMM
- Lab scope
- Fused jumper wires
- Service information
- Vehicle with an ISO 9141-2 bus system

Describe the vehicle being worked on:

Year _____ Make _____ Model _____

VIN _____ Engine type and size _____

Procedure

Task 1

1. Use the service information to identify DLC cavities assigned to the ISO 9141-2 circuits. These circuits are usually referred to as the K-line and the L-line.

   a. K-line

      Cavity number _____

      Circuit identification _____

      Color code _____

   b. L-line

      Cavity number _____

      Circuit identification _____

      Color code _____

2. Connect the J1962 BOB to the DLC. Use the voltmeter function of the DMM to measure the voltage on each bus circuit without a scan tool connected to the DLC.

K-line _____

L-line _____

3. Connect the scan tool to the DLC and observe and record the voltages for each bus circuit.

K-line _____

L-line _____

4. Establish communications between the scan tool and the PCM and record the voltages on each bus circuit.

K-line _____

L-line _____

5. Remove the scan tool from the DLC and use a lab scope to trace the voltage on each bus circuit with the ignition switch in the OFF position. If available, use the print function to print the trace for each circuit. If not available, describe what was observed.

K-line _____

L-line _____

6. Connect the scan tool, and with the ignition switch in the OFF position, use a lab scope to trace the voltage on each bus circuit. If available, use the print function to print the trace for each circuit. If not available, describe what was observed.

K-line _____

L-line _____

7. Turn the ignition switch to the RUN position, and use a lab scope to trace the voltage on each bus circuit. If available, use the print function to print the trace for each circuit. If not available, describe what was observed.

K-line _____

L-line _____

8. Establish communication between the scan tool and the PCM while observing the lab scope trace. If available, use the print function to print the trace for each circuit. If not available, describe what was observed.

K-line _____

L-line _____

9. Based on your observations of normal bus operation, what can you conclude concerning normal voltage?

_____

_____

_____

Task 2

1. Use a fused jumper wire and short the K-line from the J1962 BOB to chassis ground.

2. Connect the scan tool, and turn the ignition switch to the RUN position; attempt to establish communication with the PCM. Record the results.

_____

_____

3. Use a lab scope to trace the voltages on both circuits. If available, use the print function to print the trace for each circuit. If not available, describe what was observed.

K-line _____

L-line _____

4. Based on your observations, what is the effect of the short to ground?

_____

_____

Instructor's Response

_____

_____

_____

Name _____   Date _____

ISO-K BUS SYSTEM DIAGNOSIS

Upon completion of this job sheet, you will identify normal operating characteristics of the ISO-K bus using a lab scope and a scan tool.

NATEF Correlation

This job sheet addresses the following **AST/MAST** tasks:

A.7.   Use wiring diagrams during the diagnosis (troubleshooting) of electrical/electronic circuit problems.

This job sheet addresses the following **AST** tasks:

G.5.   Discribe body electronic systems circuits using a scan tool; check for module communication errors (data bus systems); determine needed action.

This job sheet addresses the following **MAST** task:

A.11.   Check electrical/electronic circuit waveforms; interpret readings and determine needed repairs.

G.5.   Diagnose body electronic systems circuits using a scan tool; check for module communication errors (data bus systems); determine needed action.

ASE CERTIFIED NATEF

Tools and Materials

- J1962 BOB
- Scan tool
- DMM
- Lab scope
- Service information
- Vehicle with an ISO-K bus system

Describe the vehicle being worked on:

Year _____ Make _____ Model _____

VIN _____ Engine type and size _____

Procedure

1. Use the service information to identify DLC cavity assigned to the ISO-K circuit to the PCM.

    a. K-line

    Cavity number _____

    Circuit identification _____

    Color code _____

2. Connect the J1962 BOB to the DLC. Use a lab scope to trace the voltage on the K-line with the ignition switch in the OFF position. If available, use the print function to print the trace for the K-line circuit. If not available, describe what was observed.

    _____

3. Connect the scan tool, and with the ignition switch in the OFF position, use a lab scope to trace the voltage on the K-line. If available, use the print function to print the trace for the K-line circuit. If not available, describe what was observed.

    _____

4. With the ignition switch still in the OFF position, attempt to communicate with the PCM. Use a lab scope to trace the voltage on the K-line. If available, use the print function to print the trace for the K-line circuit. If not available, describe what was observed.

_____

5. Turn the ignition switch to the RUN position. Establish communication between the scan tool and the PCM while observing the lab scope trace. If available, use the print function to print the trace for the K-line circuit. If not available, describe what was observed.

_____

6. What is indicated by each of the voltage values seen on the trace?

_____

_____

_____

Instructor's Response

_____

_____

_____

Name _____   Date _____

J1850 BUS SYSTEM DIAGNOSIS

Upon completion of this job sheet, you will be able to use the scan tool and a DMM to monitor normal J1850 bus activity. You will also use the lab scope to observe normal bus operation. Finally, you will use the DMM to diagnose full and partial failures of the J1850 bus.

NATEF Correlation

This job sheet addresses the following **AST/MAST** tasks:

A.7.   Use wiring diagrams during the diagnosis (troubleshooting) of electrical/electronic circuit problems.

This job sheet addresses the following **AST/MAST** task:

G.5.   Describe body electronic systems circuits using a scan tool; check for module communication errors (data bus systems); determine needed action.

This job sheet addresses the following **MAST** task:

A.11.   Check electrical/electronic circuit waveforms; interpret readings and determine needed repairs.

G.5.   Diagnose body electronic systems circuits using a scan tool; check for module communication errors (data bus systems); determine needed action.

Tools and Materials

- J1962 BOB
- Scan tool
- DMM
- Lab scope
- Fused jumper wires
- Service information
- Vehicle with J1850 bus
- Instructor-installed bugs

Describe the vehicle being worked on:

Year _____ Make _____ Model _____

VIN _____ Engine type and size_____

Procedure   Task Completed

Task 1

1. Use the service information to identify the DLC cavity number(s), circuit identification, and color code(s). Record your findings.

   a. Cavity number(s) _____

   b. Circuit identification _____

   c. Color code(s) _____

2. Connect the J1962 BOB to the DLC, and connect the scan tool to the J1962 BOB. Turn the ignition switch to the RUN position.   ☐

3. If available, use the scan tool to perform an 1850 module scan. List all active modules on the J1850 bus.

   _____

   _____

   _____

4. Does the J1850 bus appear to be operating normally?   ☐ Yes   ☐ No

   Explain your answer.

   _____

   _____

5. Connect the DMM positive lead to the J1850 bus circuit of the J1962 BOB (a fused jumper wire may be required). Connect the negative lead to a good chassis ground. With the DMM set to read DC volts, observe and record the voltage range:

   MIN _____ MAX _____

6. Remove the DMM, and use a lab scope to observe the J1850 bus activity. If available, use the print function to print off the trace. If not available, describe the lab scope display.

   _____

   _____

   _____

7. Turn the ignition switch to the OFF position, and disconnect the vehicle battery.   ☐

8. Use the ohmmeter function of the DMM to measure the resistance between the J1962 BOB cavity for the J1850 bus and chassis ground. Observe and record the J1850 bus termination resistance. _____

9. Disconnect a regular node from the J1850 bus and record the bus termination resistance. _____

10. Disconnect the scan tool from the J1850 bus and record the bus termination resistance. _____

11. Disconnect the PCM from the J1850 bus and record the bus termination resistance.

    _____

12. Based on your observations, are the nodes wired in series or parallel? _____

13. Based on your observations, which module had the greatest impact on bus terminal resistance and why?

    _____

    _____

14. Reconnect all modules and then reconnect the vehicle battery.   ☐

Task 2

1. Start the engine, and use a fused jumper wire to short the J1850 bus circuit of the J1962 BOB to chassis ground. What are the observable conditions?

   _____

   _____

   _____

2. Using the scan tool, perform a J1850 module scan and record the results.

_____

3. Use the voltmeter function of the DMM to observe and record the J1850 bus voltage.

_____

4. Does the measured voltage indicate normal operation?   ☐ Yes ☐ No

5. Turn the ignition switch to the OFF position, and disconnect the vehicle battery.   ☐

6. Use the ohmmeter function of the DMM; observe and record the J1850 bus termination resistance. _____

7. Explain the difference in this reading compared to Task 1, step 11.

_____

_____

_____

8. Remove the fused jumper wire used to short the bus to ground and reconnect the vehicle battery.   ☐

9. With the engine running, use a fused jumper wire to short the J1850 bus circuit of the J1962 BOB to battery voltage. What are the observable conditions?

_____

_____

_____

10. Use the scan tool to perform a J1850 module scan and record the results.

_____

_____

11. Use the voltmeter function of the DMM; observe and record the J1850 bus voltage.

_____

12. Turn the ignition switch to the OFF position. Observe and record the J1850 bus voltage. _____

13. If the voltage changed between step 11 and step 12, what does this indicate?

_____

_____

14. Based on your observations, can the J1850 bus be diagnosed with a DMM?
☐ Yes   ☐ No Explain your answer.

_____

_____

Task 3

1. A vehicle is brought to the shop with inoperative gauges and several electrical accessories not functioning properly. Can the condition be verified?   ☐ Yes   ☐ No

2. Use the scan tool to perform a J1850 module scan. Are there any modules present?   ☐ Yes   ☐ No

3. Disconnect the scan tool. With the ignition switch in the RUN position, use the J1962 BOB and the voltmeter function of the DMM to measure and record the J1850 bus voltage. _____

   Is the voltage within normal limits?   ☐ Yes   ☐ No

4. What does the voltmeter reading indicate?

   _____

   _____

5. Turn the ignition switch to the OFF position, and measure and record the J1850 bus voltage at the BOB.

   _____

6. What conclusions, if any, can be made at this time?

   _____

   _____

   _____

7. While the voltmeter is reading the indicated voltage of step 3, monitor the voltage while removing fuses from the power distribution center and junction box one at a time. Which fuses, when removed, caused a change in bus voltage?

   _____

8. Use the service information to determine which circuits or systems are implicated by the fuses removed in step 7.

   _____

9. At the vehicle, isolate the implicated components one at a time by disconnecting them and observing the voltmeter readings. Does isolating the component correct the problem?   ☐ Yes   ☐ No

   If yes, identify the faulty component.

   _____

   If no, what is the next step?

   _____

10. If the answer to step 9 was no, perform the tests you indicated that still need to be done. Was the root cause of the fault able to be determined? ☐ Yes ☐ No
    If yes, describe the fault.

    _____

    If no, consult with your instructor.

    _____

Task 4

1. Your instructor will inform you of the customer complaint. Record the problem.

   _____

   _____

2. Can the problem be verified?   ☐ Yes   ☐ No

3. Use the scan tool to communicate with the module(s) that operates the faulty system. Are communications with the module possible?   ☐ Yes   ☐ No

   If yes, record any DTCs that are present.

   _____

   _____

   If no, perform a J1850 module scan and record your results.

   _____

   _____

4. What problems are indicated?

   _____

   _____

5. Based on these observations and your understanding of J1850 bus operation, what is the next logical step?

   _____

   _____

6. Based on your answer to step 5, assemble the necessary information needed to diagnose the fault. Record this information.

   _____

   _____

   _____

7. Based on the information gathered, return to the vehicle and determine the cause of the fault. What is the necessary repair action?

   _____

   _____

   _____

Instructor's Response

_____

_____

_____

Name _____   Date _____

STUB NETWORK CAN BUS SYSTEM DIAGNOSIS

Upon completion of this job sheet, you will be able to identify the circuits that make up the stub-type network CAN bus and determine its operating characteristics. You will also see how circuit defects affect bus operation to aid in determining communication faults.

NATEF Correlation

A.7.   Use wiring diagrams during the diagnosis (troubleshooting) of electrical/electronic circuit problems.

This job sheet addresses the following **AST** task:

G.5.   Describe body electronic systems circuits using a scan tool; check for module communication errors (data bus systems); determine needed action.

This job sheet addresses the following **MAST** task:

A.11.   Check electrical/electronic circuit waveforms; interpret readings and determine needed repairs.

G.5.   Diuagnose body electronic systems circuits using a scan tool; check for module communication errors (data bus systems); determine needed action.

Tools and Materials
- J1962 BOB
- Scan tool
- DMM
- Lab scope
- Fused jumper wires
- Back probing tools
- Service information
- Vehicle with full vehicle CAN bus system

Describe the vehicle being worked on:

Year _____ Make _____ Model _____

VIN _____ Engine type and size _____

Procedure   Task Completed

Task 1

1. Use the service information to identify the following CAN bus circuit numbers and color codes:

   NOTE: The circuits may be called by other names than those listed here.

   a. CAN B (+) _____

   b. CAN B (−) _____

   c. CAN C (+) _____

   d. CAN C (−) _____

   e. Diagnostic CAN C (+)_____

   f. Diagnostic CAN C (−)_____

2. Identify the cavity numbers the diagnostic CAN bus uses at the DLC.

   a. Diagnostic CAN C (+) _____

   b. Diagnostic CAN C (–) _____

3. Identify the cavity numbers of a connector to a module on the CAN B bus network.

   a. Module _____

   i. CAN B (+) cavity _____

   ii. CAN B (–) cavity _____

4. Identify the cavity numbers for the CAN C bus network at the PCM connector.

   a. CAN B (+) _____

   b. CAN B (–) _____

5. Using the service information, determine the baud rate of each of the bus networks.

   a. CAN B (+) _____

   b. CAN B (–) _____

   c. CAN C (+) _____

   d. CAN C (–) _____

   e. Diagnostic CAN C (+) _____

   f. Diagnostic CAN C (–) _____

Task 2

1. Locate the CAN B module from Task 1, step 3, on the vehicle. With the ignition switch in the RUN position, use back probing tools to connect the voltmeter leads to the CAN B (+) circuit and chassis ground. Record your reading._____

2. Read the voltage on the CAN B (–) circuit with the ignition switch in the RUN position. _____

3. Turn the ignition switch to the OFF position, and close all doors so the bus can power down. While the bus is powering down, record the voltages on each circuit.

   a. CAN B (+) _____

   b. CAN B (–) _____

4. Leave the ignition switch in the OFF position, and open a door. Record the bus voltages now.

   a. CAN B (+) _____

   b. CAN B (–) _____

5. Based on your observations, what is required to activate the CAN B bus network?

   _____

6. Turn the ignition switch to the OFF position and disconnect the vehicle battery. ☐

7. Measure the resistance between the two CAN B bus circuits. _____

   Explain your results. _____

   _____

Task 3

1. Connect the J1962 BOB to the DLC, and turn the ignition switch to the RUN position. Use the scan tool to determine which modules on the vehicle are on the CAN B bus and record them.

_____

_____

_____

2. Access the CAN B module connector from Task 1, step 3, and use a fused jumper wire to short one of the CAN B bus circuits to ground. Use the scan tool to determine which modules are active on the CAN B bus network.

_____

_____

_____

3. Using a fused jumper wire and a back probing tool, short both CAN B bus circuits together. Use the scan tool to determine which modules are active on the CAN B bus network.

_____

_____

_____

4. Using a fused jumper wire and a back probing tool, short both CAN B bus circuits to ground. Use the scan tool to determine which modules are active on the CAN B bus network.

_____

_____

_____

5. Use the scan tool and access the central gateway module. Record any fault codes listed.

_____

_____

_____

6. Remove the short from the bus circuit. ☐

7. Based on your observations, is this bus network fault tolerant?   ☐ Yes   ☐ No
Explain your results.

_____

_____

Task 4

1. Use the scan tool to identify the modules that are on the CAN C bus network and record them.

_____

_____

_____

2. Locate the PCM connector identified in Task 1, and disconnect it from the PCM. Use
   a fused jumper wire and a back probing tool to short one of the CAN C bus circuits to
   chassis ground.   ☐

3. Use the scan tool and attempt to establish communication with any of the modules
   identified in step 1 (except the PCM). Is communication possible?

   ☐ Yes   ☐ No

   Explain your results. _____

   _____

4. Use a fused jumper wire and a back probing tool to short both of the CAN C bus
   circuits together. Use the scan tool and attempt to establish communication with any
   of the modules identified in step 1 (except the PCM). Is comunication possible?

   ☐ Yes   ☐ No

   Explain your results. _____

   _____

5. Based on your observations, is this bus network fault tolerant?   ☐ Yes   ☐ No
   Explain your results.

   _____

   _____

6. Remove all faults, but leave the PCM connector unplugged.

Task 5

1. Remove the scan tool from the DLC. Use a probing tool and the voltmeter function
   of the DMM to obtain and record the CAN C bus voltages during the following
   conditions.

   a. Ignition switch in RUN position.

      CAN C (+) _____ CAN C (−) _____

   b. Ignition switch in OFF position.

      CAN C (+) _____ CAN C (−) _____

   c. Ignition switch in OFF position and CAN B bus powered down.

      CAN C (+) _____ CAN C (−) _____

   d. Ignition switch in OFF position; open a door.

      CAN C (+) _____ CAN C (−) _____

2. Based on your observations, what is required to activate the CAN C bus network?

   _____

3. Turn the ignition switch to the OFF position and disconnect the vehicle battery.   ☐

4. Measure the resistance between the two CAN C bus circuits. _____
   Explain your results. _____

   _____

5. Reconnect all connectors.   ☐

Task 6

1. Connect the scan tool to the J1962 BOB and the DLC. Use a fused jumper wire to short one of the diagnostic CAN C bus circuits at the BOB to ground. Attempt to communicate with the PCM. Was communication possible?   ☐ Yes ☐ No

2. Remove the short, and move the fused jumper wire to short the other diagnostic CAN C bus circuit to ground. Attempt to communicate with the PCM. Was communication possible?   ☐ Yes ☐ No

3. Use a fused jumper wire to short the two diagnostic CAN C bus circuits together. Attempt to communicate with the PCM. Was communication possible?   ☐ Yes ☐ No

4. Based on your observations, what is your conclusion?

   _____

   _____

   _____

5. Turn the ignition switch to the OFF position, and disconnect the vehicle battery.   ☐

6. Measure the resistance between the two diagnostic CAN C bus circuits. _____

   Explain your results._____

   _____

Instructor's Response

_____

_____

_____

Name _____   Date _____

STAR NETWORK CAN BUS SYSTEM DIAGNOSIS

Upon completion of this job sheet, you will be able to identify the circuits that make up the star-type network CAN bus and determine their operating characteristics. You will also see how circuit defects affect bus operation to aid in determining communication faults.

NATEF Correlation

This job sheet addresses the following **AST/MAST** tasks:

A.7.   Use wiring diagrams during the diagnosis (troubleshooting) of electrical/electronic circuit problems.

This job sheet addresses the following **AST** task:

G.5.   Describe body electronic systems circuits using a scan tool; check for module communication errors (data bus systems); determine needed action.

This job sheet addresses the following **MAST** task:

A.11.   Check electrical/electronic circuit waveforms; interpret readings and determine needed repairs.

G.5.   Diagnose body electronic systems circuits using a scan tool; check for module communication errors (data bus systems); determine needed action.

Tools and Materials

- Scan tool
- J1962 BOB
- DMM (2)
- Lab scope
- Service information
- Vehicle with star CAN bus system

Describe the vehicle being worked on:

Year _____ Make _____ Model _____

VIN _____ Engine type and size _____

Task 1

1. Using the service information, locate the following components on the vehicle and record their location:

   CAN C star connector(s) _____

   _____

   Interior bus star connector(s) _____

   _____

2. What components must be removed to access the star connectors?

   _____

   _____

   _____

3. List the circuit identification and wire colors for the CAN C and interior CAN circuits.

   NOTE: The circuits may be called by other names than those listed here.

Circuit Name	Circuit ID	Wire Color
CAN C (+)		
CAN C (−)		
Interior CAN (+)		
Interior CAN (−)		

4. Record the connector cavities where the bus circuits are located in the DLC:

Circuit Name	Cavity Number
CAN C (+)	
CAN C (−)	
Interior CAN (+)	
Interior CAN (−)	

Task 2

Voltage Measurements

This task uses the voltmeter function of the DMM to monitor normal bus activity.

1. Connect the J1962 BOB to the DLC.   ☐

2. Using two voltmeters, connect the red lead of one voltmeter to the CAN C bus   ☐
   (+) terminal on the BOB and the red lead of the other voltmeter to CAN C bus (−)
   terminal on the BOB. Connect both black leads to chassis ground.

3. Connect a scan tool to the J1962 BOB and start a diagnostic session. Measure the
   voltage levels on the CAN C bus with the ignition ON and ignition OFF.

   CAN C (+) Ignition ON _____ Ignition OFF _____

   CAN C (−) Ignition ON _____ Ignition OFF _____

4. Perform the same measurements for the interior CAN bus and record all voltage
   measurements:

   Scan tool disconnected

   Interior CAN (+) Ignition ON _____ Ignition OFF _____

   Interior CAN (−) Ignition ON _____ Ignition OFF _____

   Scan tool connected

   Interior CAN (+) Ignition ON _____ Ignition OFF _____

   Interior CAN (−) Ignition ON _____ Ignition OFF _____

5. Disconnect the scan tool and allow the bus networks to enter sleep mode. How do you know that the bus networks entered sleep mode?

_____

_____

_____

6. With the bus networks in sleep mode, perform each of the following while observing the voltmeters. Record which bus network becomes active after each event.

Cycle the ignition key _____

Open the passenger door _____

Press the power door unlock button _____

Task 3

Resistance Measurements

This task uses the ohmmeter function of the DMM to measure the termination resistance of the CAN C and interior CAN buses.

1. Disconnect the battery.   ☐

2. Connect the J1962 BOB to the DLC.   ☐

3. Measure the resistance between CAN C (+) and CAN C (−): _____
Are these readings within specifications?   ☐ Yes   ☐ No

4. Measure the resistance between interior CAN (+) and interior CAN (−) _____
Are these readings within specifications?   ☐ Yes   ☐ No

Task 4

Testing CAN C for a Short to Ground

In this task you will diagnose a simulated communication network fault that was installed by your instructor.

1. Connect the J1962 BOB to the DLC and connect the scan tool to the BOB. Start a diagnostic session. Is communication with the vehicle possible?   ☐ Yes   ☐ No

2. List any communication network(s) not communicating. _____

_____

3. Were any active DTCs set? If yes, list them:

_____

_____

4. Using the service information, locate all communication network wiring schematics   ☐
for the vehicle.

5. Using two voltmeters, measure the voltage levels on the CAN C bus with the ignition ON:

CAN C (+) _____

CAN C (−) _____

6. Using two voltmeters, measure the voltage levels on the interior CAN bus with the ignition ON:

   Interior CAN (+) _____

   Interior CAN (−) _____

7. Based on the voltage readings, what fault is present?

   _____

   _____

8. Refer to the communication network wiring schematics; where is the best place to disconnect and isolate the communication bus for diagnosis?

   _____

   _____

9. Disconnect the battery.   ☐

10. Connect the ohmmeter between the terminal at the J1962 BOB for the faulty circuit and chassis ground.   ☐

11. While observing the ohmmeter, disconnect each connector one at a time. Were you able to isolate the fault? ☐ Yes ☐ No

    If no, consult with your instructor.

    Where is the fault isolated to?

    _____

    _____

12. How would this fault be repaired?

    _____

    _____

    _____

Instructor's Response

_____

_____

_____

Name _____   Date _____

LIN BUS SYSTEM DIAGNOSIS

Upon completion of this job sheet, you will be able to determine normal LIN bus operation.
You will also observe faulty bus system operation.

NATEF Correlation

This job sheet addresses the following **AST/MAST** tasks:

A.7. Use wiring diagrams during the diagnosis (troubleshooting) of electrical/
electronic circuit problems.

This job sheet addresses the following **AST** task:

G.5. Describe body electronic systems circuits using a scan tool; check for module
communication errors (data bus systems); determine needed action.

This job sheet addresses the following **MAST** task:

A.11. Check electrical/electronic circuit waveforms; interpret readings and
determine needed repairs.

G.5. Diagnose body electronic systems circuits using a scan tool; check for module
communication errors (data bus systems); determine needed action.

Tools and Materials

- Scan tool
- DMM
- Lab scope
- Fused jumper wires
- Back probing tool
- Service information
- Vehicle with LIN bus system

Describe the vehicle being worked on:

Year _____ Make _____ Model _____

VIN _____ Engine type and size_____

Procedure

1. Use the service information to identify the function(s) of the LIN bus(es) on the
assigned vehicle.

_____

_____

_____

2. Identify the LIN bus cavities of the master module.

_____

_____

_____

3. Identify the LIN bus cavities of the slave module.

_____

_____

_____

4. Does the slave module have any inputs to it from other sources?   ☐ Yes   ☐ No

   If yes, list them.

   _____

   _____

5. With the ignition switch in the RUN position, connect the scan tool and navigate to the master module. Are any items available in the data display to assist in diagnosing a system failure?   ☐ Yes   ☐ No

   If yes, list them.

   _____

   _____

6. Press and release any switch inputs that were identified while observing the scan tool. Do the switch states change on the scan tool?   ☐ Yes   ☐ No

7. Press and hold one of the switch buttons for a minimum of 60 seconds and then release. Check and record any DTCs.

   _____

   _____

8. Is the DTC active or stored? _____

   a. If active, press and release the same button again. Did it change to stored now?

   ☐ Yes   ☐ No

9. Disconnect the electrical connector to a switch input to the slave module. Check and record any DTCs.

   _____

   _____

10. If no DTCs were set, why not?

    _____

11. Press and release another switch input to the slave module while observing the scan tool. Does the scan tool indicate the switch presses?   ☐ Yes   ☐ No

12. Explain your results.

    _____

    _____

13. With the slave module still disconnected, use the voltmeter function of the DMM to measure the voltage on the LIN bus. _____

14. Reconnect the slave module and record the voltage on the LIN bus. _____

15. Are the voltages the same? ☐ Yes ☐ No

    If no, explain.

    _____

16. Use a lab scope to observe the trace of the LIN bus voltages. Describe what is observed.

    _____

17. Short the LIN bus circuit to ground by back probing the slave module connector and using a fused jumper wire connected to a chassis ground. Check and record any DTCs.

_____

_____

18. **Caution: Verify circuit identification before performing this step, or else vehicle damage could occur.**

    Short the LIN bus circuit to power by back probing the right steering wheel switch connector; use a fused jumper wire connected to vehicle power. Check and record any DTCs.

    _____

    _____

19. Record your conclusions.

    _____

    _____

    _____

Instructor's Response

_____

_____

_____

Name _____   Date _____

DATA BUS CIRCUIT WIRE REPAIR

Upon completion of this job sheet, you will be able to properly repair the wires of a data bus network circuit.

NATEF Correlation ———————————————————————

This job sheet addresses the following **MAST** task:

A.12.   Repair data bus wiring harness.

NATEF

Tools and Materials

- 100-watt soldering iron
- 60/40 rosin core solder
- Crimping tool
- Splice clip
- Heat shrink tube
- Heating gun
- Safety glasses
- Service information
- Vehicle with a data bus network system

Describe the vehicle being worked on:

Year _____ Make _____ Model _____

VIN _____ Engine type and size _____

Procedure

Task Completed

1. Refer to the service information and record the specification for the number of twists per meter or inch.

   _____

2. Cut out the portion of the damaged wiring being sure that the wire slices will be staggered.   ☐

3. What is the length of each wire that was removed?

   (+) _____

   (-) _____

4. Why is this measurement important?

   _____

   _____

   _____

5. Using the correct size stripper, remove about 1/2 inch (12 mm) of the insulation from the ends of all wires. Be careful not to nick or cut any of the wires.   ☐

6. Select the proper size splice clip to hold the splice.   ☐

7. Place the correct length and size of heat shrink tube over the wire. Slide the tube far enough away so the wires are not exposed to the heat of the soldering iron.   ☐

Task Completed

8. Overlap the two splice ends and hold in place your thumb and forefinger. While the wire ends are overlapped, center the splice clip around the wires and crimp into place. ☐

9. Heat the splice clip with the soldering iron while applying solder to the opening in the back of the clip. Apply only enough solder to make a good connection. The solder should travel through the wire. ☐

10. After the solder cools, slide the heat shrink tube over the splice and heat the tube with the hot air gun until it shrinks around the splice. Do not overheat the tube. ☐

11. Repeat the soldering process for all wires. ☐

12. Use the service information to identify DLC cavity assigned to the ISO-K circuit to the PCM.

   a. K-line

     Cavity number _____

     Circuit identification _____

     Color code _____

13. Connect the J1962 BOB to the DLC. Use a lab scope to trace the voltage on the K-line with the ignition switch in the OFF position. If available, use the print function to print the trace for the K-line circuit. If not available, describe what was observed.

_____

14. Connect the scan tool, and with the ignition switch in the OFF position, use a lab scope to trace the voltage on the K-line. If available, use the print function to print the trace for the K-line circuit. If not available, describe what was observed.

_____

15. With the ignition switch still in the OFF position, attempt to communicate with the PCM. Use a lab scope to trace the voltage on the K-line. If available, use the print function to print the trace for the K-line circuit. If not available, describe what was observed.

_____

16. Turn the ignition switch to the RUN position. Establish communication between the scan tool and the PCM while observing the lab scope trace. If available, use the print function to print the trace for the K-line circuit. If not available, describe what was observed.

_____

17. What is indicated by each of the voltage values seen on the trace?

_____

_____

_____

Instructor's Response

_____

_____

_____

DIAGNOSTIC CHART 10-1

PROBLEM AREA:	Vehicle communication systems.
SYMPTOMS:	Scan tool cannot communicate with a module.
POSSIBLE CAUSES:	**1.** Faulty battery feed circuit to module. **2.** Faulty ignition feed circuit to module. **3.** Poor module ground. **4.** Open data bus circuit to module.

DIAGNOSTIC CHART 10-2

PROBLEM AREA:	Vehicle communication systems.
SYMPTOMS:	Scan tool cannot communicate with any module.
POSSIBLE CAUSES:	**1.** Open bus circuit at DLC. **2.** Bus circuit shorted to ground. **3.** Bus circuit shorted to voltage. **4.** Internal failure in a module causing short to power. **5.** Internal failure in a module causing short to ground. **6.** Open termination resistors. **7.** Faulty scan tool.

DIAGNOSTIC CHART 10-3

PROBLEM AREA:	Vehicle communication systems.
SYMPTOMS:	Total bus failure indicated by default conditions and possible no-start.
POSSIBLE CAUSES:	**1.** Open bus circuit at DLC. **2.** Bus circuit shorted to ground. **3.** Bus circuit shorted to voltage. **4.** Internal failure in a module causing short to power. **5.** Internal failure in a module causing short to ground. **6.** Open termination resistors.

DIAGNOSTIC CHART 10-4

PROBLEM AREA:	Vehicle communication systems.
SYMPTOMS:	System fails to operate and/or warning lamp illumination.
POSSIBLE CAUSES:	**1.** Bus circuit shorted to ground. **2.** Bus circuit shorted to voltage. **3.** Open in bus circuit. **4.** Internal failure in a module causing short to power. **5.** Internal failure in a module causing short to ground. **6.** Open termination resistors.

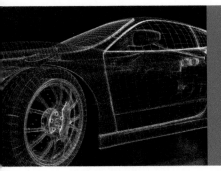

CHAPTER 11

LIGHTING CIRCUITS DIAGNOSTICS AND REPAIR

Upon completion and review of this chapter, you should be able to:

- Correctly replace sealed-beam and composite headlights.
- Correctly service the high-intensity discharge (HID) lamp and ballast.
- Diagnose the computer-controlled headlight system.
- Diagnose the automatic headlight system.
- Perform a functional test of the automatic headlight system.
- Test and replace the automatic headlight system's photo cell.
- Test the automatic headlight system's amplifier.
- Perform SmartBeam camera calibration and adjustment.
- Diagnose the automatic headlight leveling system.
- Diagnose the adaptive headlight system.
- Correctly aim sealed-beam and composite headlights.
- Diagnose the cause of brighter-than-normal lights.

- Diagnose the cause of dimmer-than-normal lights.
- Remove and replace dash-mounted and steering column–mounted headlight switches.
- Replace multifunction switches.
- Test and determine needed repairs of the dimmer switch and related circuits.
- Replace the dimmer switch.
- Diagnose incorrect taillight assembly operation.
- Diagnose the turn signal system for improper operation.
- Replace the turn signal switch.
- Diagnose the interior lights, including courtesy, instrument, and panel lights.
- Diagnose control module–operated illuminated entry systems.
- Diagnose BCM-controlled illuminated entry systems.
- Diagnose fiber-optic systems.

Basic Tools

Basic mechanic's tool set

Service information

Terms To Know

Active headlight system (AHS)

Curb height

Demonstration mode

Dimmer switch

Feedback

Illuminated entry actuator

Multifunction switch

Optics test

Photo cell resistance assembly

INTRODUCTION

Vehicle lighting systems are becoming very complex. Any failure of the lighting system requires a systematic approach to diagnose, locate, and correct the fault in the minimum amount of time.

501

The importance of a proper operating lighting system cannot be overemphasized. The lighting system should be checked whenever the vehicle is brought into the shop for repairs. Often a customer may not be aware of a light failure. If a lighting circuit is not operating properly, there is a potential danger to the driver and other people. When today's technician performs repairs on the lighting systems, the repairs must assure vehicle safety and meet all applicable laws. Be sure to use the correct lamp type and size for the application (**Figure 11-1**).

Before performing any lighting system tests, check the battery for state of charge. Also, be sure all cable connections are clean and tight. Visually check the wires for damaged insulation, loose connections, and improper routing. A troubleshooting chart for the lighting system is located at the end of this chapter.

When troubleshooting the lighting system, if only one bulb does not operate, it is usually faster to replace it with a known good unit first. Check the connector or socket for signs of corrosion. When testing the circuit with a voltmeter, ohmmeter, or test light, check the most easily reached components first.

AUTHOR'S NOTE Some headlight bulbs are very expensive. Before replacing these bulbs be sure to perform complete circuit diagnostics.

Although lighting circuits are largely regulated by federal laws, and the systems are similar between the various manufacturers, there are variations in operation. Before attempting to do any repairs on an unfamiliar circuit, the technician should always refer to the manufacturer's service information.

Diagnosis of computer-controlled lighting systems is designed to be as easy as possible. The controller may provide trouble codes to assist the technician in diagnosis. Most manufacturers provide a detailed diagnostic chart for the most common symptoms. The most important thing to remember when diagnosing these systems is to follow the diagnostic procedures in order. Do not attempt to get ahead of the chart by assuming the outcome of a test. This will lead to replacement of good parts and lost time.

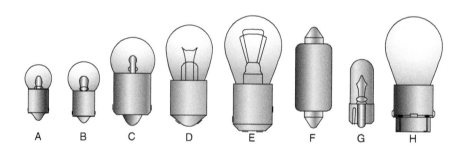

Common automotive bulbs:

A, B – Miniature bayonet for indicator and instrument lights
C – Single-contact bayonet for license and courtesy lights
D – Double-contact bayonet for trunk and underhood lights
E – Double-contact bayonet with staggered indexing lugs for stop, turn, and brake lights
F – Cartridge type for dome lights
G – Wedge base for instrument lights
H – Blade double-contact for stop, turn, and brake lights

Figure 11-1 Correct selection of the lamp is important for proper operation of the system.

In this chapter, you will perform selected service samples on the computer-controlled concealed headlight, automatic headlight, and illuminated entry systems. It is out of the scope of this manual to provide service procedures for the different manufacturers that use these systems. Technicians must follow the procedure in the service information for the vehicle they are diagnosing.

HEADLIGHT REPLACEMENT

One of the most common lighting system repairs is replacing the headlight. After a period of time, the filament may burn through or the lens may be broken. Before the headlight is replaced, however, a voltmeter or test light should be used to confirm that voltage is present. Next, check the ground for proper connections. Confirm that the connector or socket does not have any corrosion and is in good condition. If these test good, the headlight is probably faulty and needs to be replaced. If there is no voltage present at the connector, work back toward the switch and battery until the fault is located.

The procedure for replacing the headlight differs depending on the type of bulb used. Most conventional sealed-beam and halogen sealed-beam headlights are replaced in the same manner. Composite headlights require different procedures. Always refer to the service information for the vehicle you are working on.

Sealed-Beam Headlights

Photo Sequence 21 illustrates a common procedure for replacing a sealed-beam headlight. After confirming proper headlight operation, check headlight aiming as described later in this chapter.

> **CUSTOMER CARE** Because the filament of the halogen lamp is contained in its own bulb, cracking or breaking of the lens does not prevent headlight operation. The filament will continue to operate as long as the filament envelope is not broken. However, a broken lens will result in poor light quality and should be replaced for the safety of the customer.

🔦 Special Tools
Fender covers
Safety glasses
Torx drivers

⚠ Caution
Because of the construction and placement of the prisms in the lens, it is important that the headlight is installed in its proper position. The lens is usually marked "TOP" to indicate the proper installed position.

Classroom Manual
Chapter 11, page 294

PHOTO SEQUENCE 21
Replacing a Sealed-Beam Headlight

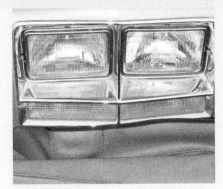

P21-1 Place fender covers around the work area.

P21-2 Sealed-beam headlight replacement usually requires the removal of the bezel. Some vehicles may require that the turn signal light assembly be removed before the headlight is accessible.

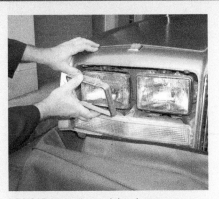

P21-3 Remove the retaining ring screws and the retaining trim. Do not turn the two headlight-aiming adjustment screws.

PHOTO SEQUENCE 21 (CONTINUED)

P21-4 Remove the headlight from the shell assembly.

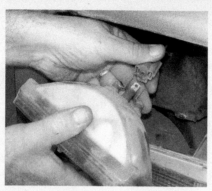

P21-5 Disconnect the wire connector from the back of the lamp.

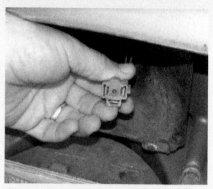

P21-6 Check the wire connector for corrosion or other foreign material. Clean as necessary.

P21-7 Coat the connector terminals and the prongs of the new headlight with dielectric grease to prevent corrosion.

P21-8 Install the wire connector onto the new headlight's connector prongs.

P21-9 Place the headlight into the shell assembly. When positioning the headlight, be sure that the embossed number is at the top. Many headlights are marked "TOP."

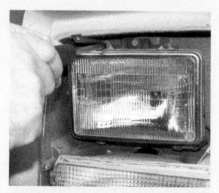

P21-10 Install the retainer trim and fasteners.

P21-11 Install the headlight bezel.

P21-12 Check operation of the headlight.

Composite Headlights

Photo Sequence 22 illustrates a typical procedure for replacing a composite headlight. Whenever replacing a halogen bulb, care must be taken not to touch the envelope with your fingers. Staining the bulb with normal skin oil can substantially shorten the life of the bulb. Handle the lamp only by its base. Also, dispose of the old lamp properly. Wear safety glasses when handling halogen bulbs. In addition, be careful not to get any of the dielectric grease onto the bulb. The bulb's life will be shortened. After the bulb is replaced, the headlight aiming should be checked and corrected if needed.

Classroom Manual
Chapter 11, page 298

HID Lamp and Ballast Service

⚡ WARNING **Before testing the HID system, remove the headlight assembly from the vehicle. The HID system operates on high voltage and current. If the system is accidentally shorted, personal injury or death may occur.**

Classroom Manual
Chapter 11, page 296

The high-intensity discharge (HID) lamps require battery voltage and ground to operate, as any other lamp. The ballast has internal circuit protection to prevent damage resulting from an open or shorted circuit. In addition, the circuit protection will come into effect if an overcharge or undercharge condition exists. If any of these faults are detected, the headlights will not turn on for that key cycle, or until the fault is corrected. If the customer states that both headlights do not operate, recycle the ignition or the headlight switch to see if the condition recurs. If the lights still do not come on, then check battery condition and the charging system. It is unlikely that both lamps failed at the same time, so check the switch input, HID relay, and common power feeds and grounds.

A burned-out lamp will appear black or smoky. If a visual inspection does not indicate a faulty lamp, then check for battery voltage to the ballast. If battery voltage is within specifications, check the ground circuit for the ballast. If neither of these circuits has an open, short, or excessive resistance, then substitute a known good ballast. If the headlight on the opposite side of the vehicle is operating properly, substitute its ballast with the side that does not operate. If the headlight illuminates, replace the ballast. If the light still does not come on, replace the bulb/ignitor element.

Depending on design, it may be necessary to remove the headlight assembly prior to replacing the bulb and ignitor. Some vehicles provide access that allows the bulb and ignitor to be serviced without removing the headlight assembly. Regardless of how the bulb element is accessed, replacement procedures are very similar.

PHOTO SEQUENCE 22
Replacing a Composite Headlight Bulb

P22-1 Place fender covers around the work area.

P22-2 On some vehicles it may be necessary to remove the lamp unit.

P22-3 Disconnect the wiring harness from the socket.

PHOTO SEQUENCE 22 (CONTINUED)

P22-4 Unlock the socket by rotating the retaining ring 1/8 of a turn.

P22-5 Gently pull the bulb straight back out of the lamp unit. To prevent breaking the bulb and locating tabs, do not rotate it while pulling.

P22-6 Check the connector terminals for corrosion or other foreign material. Clean as needed.

P22-7 Coat the connector terminals and the prongs of the new headlight with dielectric grease to prevent corrosion.

P22-8 Align the socket with the keyed opening of the lamp unit. When properly aligned, the socket will seat completely onto the headlight unit.

P22-9 Insert the retaining ring into the lamp unit and rotate it to lock it in place.

P22-10 Reconnect the wiring harness to the socket.

P22-11 If removed, install the lamp unit.

P22-12 Check headlight operation.

Because the housings of a composite headlight and HID system are vented, condensation may develop inside the lens assembly. This condensation is not harmful to the bulb and does not affect headlight operation.

Prior to servicing the headlight element, disconnect and isolate the battery negative cable or remove the HID relay. This will prevent accidental operation of the headlights.

⚡ WARNING **The HID system produces a high voltage and current. Do not attempt to operate the headlights without all connections properly made.**

⚡ WARNING **Never attempt to open the ballast or ignitor assembly. The ignitor is a component of the bulb and cannot be separated.**

To gain access to the bulb assembly, remove the rubber boot that protects the element (**Figure 11-2**). At the ignitor, disconnect the cable coming from the ballast (**Figure 11-3**).

Figure 11-2 The HID element is protected by a rubber boot on the back of the headlight assembly.

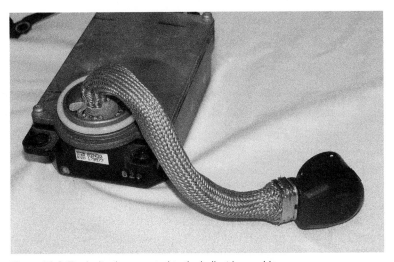

Figure 11-3 The ignitor is connected to the ballast by a cable.

A wire clip may retain the element to the headlight assembly. Release the clip and remove the element by grabbing the ignitor and pulling the unit rearward. It may be necessary to rock the ignitor slightly as it is being pulled.

To install the new element, push the bulb element assembly into the headlight unit. Once the bulb element assembly is properly seated, secure it with the retainer clip. Connect the cable from the ballast to the ignitor's connector. Install the rubber protective cover, making sure that the cover is properly seated. If the cover is not seated, water may enter the element and damage the unit. If necessary, install the headlight assembly. Reconnect the battery or connect the relay and test for proper operation.

The ballast can be attached to the bottom of the headlight assembly, to the vehicle's frame, or to the inner fender well. A cable connects the ballast to the ignitor. In addition, another connector attaches the ballast to the vehicle's wiring system (**Figure 11-4**). It may be necessary to remove the headlight assembly to gain access to the ballast. Prior to removing the ballast, disconnect and isolate the battery negative cable or remove the HID relay.

Remove the screws that retain the ballast to the headlight assembly or frame. Next, disconnect the main electrical connector and the cable to the ignitor from the ballast. The cable may be secured by a wire lever that must be rotated down. Installation of the new ballast is reverse order of the removal procedure. Be sure to properly secure the wire retainer around the cable connector.

Figure 11-4 HID ballast connectors. Notice the warning sticker on the ballast.

TESTING COMPUTER-CONTROLLED HEADLIGHT SYSTEMS

Classroom Manual
Chapter 11, page 303

Special Tools

Scan tool
DMM
Fused jumper wires
Test light

A basic computer-controlled headlight system will use the headlight switch as an input to the BCM, which then activates the required relays to perform the requested function. The system may have separate relays for the park lights, fog lights, and headlights. The circuit operation for each of these relays is usually performed by low-side drivers to complete the path to ground for the relay coil. The following will focus on testing the headlight system, although diagnostics for the park and fog light systems are similar.

Using **Figure 11-5** as a schematic of the system, we will go through the steps to diagnose a customer concern that the headlights do not come on. Always refer to the proper wiring diagram for the vehicle you are working on. As with any electrical system problem, begin by confirming that the system is malfunctioning. Also, note any other systems that are not working properly. In the system shown, use the multifunction switch to activate the flash-to-pass feature. This will test the operation of the high-beam lights and relay.

Check any associated fuses and repair as needed. Next, conduct a visual inspection and repair any faults found. If the cause of the problem is not located, substitute the low-beam relay with a known good relay of the same type and test operation. If the system works now, the relay was at fault and replacing the relay will complete the repair. If the system still does not work, begin testing the rest of the system.

If a scan tool is available, check to see if any diagnostic trouble codes (DTCs) have been set. If so, use the diagnostic chart for the DTC to locate the problem. If there are no DTCs (not all systems will set codes), use the scan tool to perform an activation test of the head lights. If the headlights work during the activation test, the problem is in the input side of the system. Use the scan tool to monitor the headlight switch while it is placed in all switch positions. Record the voltages for each position and compare to specifications. In this example, if 5 volts is displayed in all positions, there is an open in the headlight switch MUX circuit. The open could be in the wire between the BCM and the switch, in the switch itself, or in the ground circuit. Test each of these components with an ohmmeter. The switch is a resistive multiplex switch that should have a different resistance value for each switch position. With the switch disconnected, attach the ohmmeter across terminals 1 and 2 of the switch. Place the switch in each position and record the results. Check the results with the service information. If the values are out of range, replace the switch.

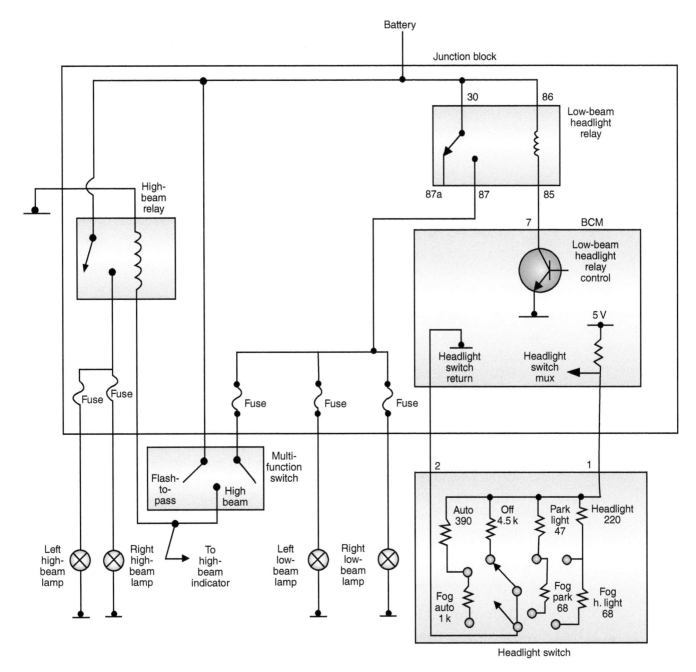

Figure 11-5 Schematic of a computer-controlled headlight system.

When testing the input signal with the scan tool, if the voltage displayed is 0 volt, this would indicate that the circuit is shorted to ground or a faulty BCM. Test the circuit for a short. If there are no shorts, check the voltage input and ground connections of the BCM. If these are good, replace the BCM.

If the headlights do not come on when the scan tool activates the system, the problem is in the relay control circuit, the BCM, the high-current circuit from the battery through the relay to the headlights, or in the headlight ground. To determine which circuit is at fault, remove the low-beam headlight relay. Use a voltmeter or test light to confirm whether battery voltage is present at pins 30 and 86 of the junction block terminals. If voltage is not present, there is an open or a short between the junction block and the battery. If voltage is present at both pins, use a fused jumper wire to connect pin 30 to pin 87. If the lights come on now, the problem is in the control side circuit of the relay (the relay itself was tested earlier when it was substituted).

> ⚙️ SERVICE TIP If a scan tool is not available, the input can be tested along with the control side of the relay by connecting a test light across terminals 85 and 86. If the test light comes on when the switch is placed in the headlamp position, the BCM receives the input and carries out the command.

To test the control side circuit, connect a test light between pins 86 and 85 of the junction block terminals (relay removed). Place the headlight switch into the headlamp position and observe the test light. Since the BCM should now complete the path to ground, the test light should illuminate. If the test light fails to turn on, go to pin 7 of the BCM and back probe a fused jumper wire into the connector. Touch the other end of the fused jumper wire to a good ground. If the test light comes on, check the ground connections of the BCM. If the grounds are good, replace the BCM. If the test light still does not come on, there is an open in the circuit between pin 85 of the junction block and the BCM. In this case, the BCM attaches to the back of the junction block and the circuit wire is internal to the junction block. If this circuit is bad, the junction block will need to be replaced.

Special Tools
Scan tool
DMM

If the lights do not come on when the fused jumper wire is connected across pins 30 and 87 of the junction block, the problem is between the relay and the headlights. Connect a fused jumper wire from the battery positive post to the feed into one of the headlight connectors. If the headlight illuminates, check the circuit from the junction block to the headlight. If the circuit is good, the problem is probably at the splice in the junction block and the junction block will need to be replaced. If the headlight still does not turn on, check the ground connections. If the grounds are good, the bulb is burned out. It is unlikely that both headlight bulbs will burn out at the same time unless there are other problems. If both bulbs are burned out, check the charging system for over voltage output. Also, use a lab scope connected to the battery positive post and ground to look for voltage spikes above the charging system's target voltage value. Voltage spikes may indicate that the battery connections are not clean (remember, the battery acts as a buffer), the battery is bad, or there is resistance on the charging system's sense circuit.

HSD-Controlled Headlights

Classroom Manual
Chapter 11, page 305

Systems that use high-side drivers (HSDs) to operate the headlights typically will use pulse width modulation (PWM) to keep the voltage to the bulbs at a consistent value. Based on configuration, the HSD may provide short and open circuit diagnostics. The HSD can also act as the fuse and turn off if a short is determined. Some HSDs can also detect a circuit fault even when the headlights are turned off. In HSD-controlled lighting systems, the driver uses a diagnostic voltage that is typically between 8 and 12 volts. When the driver is off, the diagnostic voltage is sent through the bulb to ground. The current in the circuit is too low to illuminate the bulb. If the circuit is intact, the voltage will be pulled low. If there is an open, then the voltage will be high. If the voltage is high, the module will set a fault and (in some systems) send a message to the vehicle information module or cabin compartment node (CCN) to turn on the lamp outage warning. This process of fault detection can be applied to all of the exterior lighting systems.

Classroom Manual
Chapter 11,
page 343, 346

Diagnostics of the HSD headlight system is DTC driven. If a fault is detected, the module records the DTC. Once the scan tool retrieves the DTC, follow the troubleshooting tree to isolate the fault. Circuit testing is the same as with any other circuit. Confirm that the power and ground circuits are intact using voltage drop and resistance testing methods.

⚙ **SERVICE TIP** Keep in mind that many HSD-controlled systems will disable the HSD when a fault is detected. The HSD may remain off for the remainder of the key cycle. If a connector is unplugged to measure for voltage, the system will detect the open and turn off the HSD. This will result in misdiagnosis if you are not aware of this function. It may be necessary to apply an electrical load to the circuit in order to get accurate test results.

AUTOMATIC HEADLIGHT SYSTEM DIAGNOSIS

The automatic on-off with time delay has two functions: to turn on the headlights automatically when ambient light decreases to a predetermined level and to allow the headlights to remain on for a certain amount of time after the vehicle has been turned off (providing light for the passengers exiting the vehicle). This system can be used in combination with automatic dimming systems.

Classroom Manual
Chapter 11, page 307

The common components of the automatic on-off with time delay include the following:

1. Photo cell and amplifier.
2. Power relay.
3. Timer control. Control of time delay can be a potentiometer incorporated into the headlight switch. It can also be a programmable feature through the CCN or overhead console. The timer control unit controls the automatic operation of the system and the length of time the headlights stay on after the ignition switch is turned off.

🔧 **Special Tool**
Bright flashlight

If the headlights do not turn on, check the regular headlight system first before condemning the automatic headlight system. If the headlights do not illuminate when the automatic system is turned off, the problem is in the basic circuit.

To perform a functional test of the automatic headlight system, turn the headlight switch into the appropriate position to activate the automatic headlight function. Cover the photo cell and start the engine. Some systems may be activated with the ignition switch in the RUN position; however, many systems are designed to delay turning on the headlights until an engine run signal is received. The headlights should turn on within 30 seconds. Remove the cover you placed over the photo cell and shine a bright light onto it. The headlights should turn off after 10 seconds but within 60 seconds. Cover the photo cell again. When the headlights turn on, wait for 15 seconds, then turn off the ignition switch. The headlights should turn off after the selected amount of time delay.

Once it is determined that the fault is within the automatic headlight system, make a few quick checks of the system. Inspect the photo cell lens for obstructions. Check all connections from the headlight switch, as well as all fuses used in the system.

The following are some of the most common concerns that result from problems in the automatic headlight:

1. Lights turning on and off at wrong ambient light levels.
2. Lights that do not turn on in darkness.
3. Lights not turning off in bright light.
4. Lights not staying on for an adjustable time after the ignition switch is turned off.
5. Lights that do not turn off after the ignition switch is turned off.

These problems can be caused by faults in the headlight switch, ignition switch, amplifier, or photo cell. To locate the problem, perform the following test series. As a service sample of automatic headlight system diagnosis, refer to **Figure 11-6**. The following test

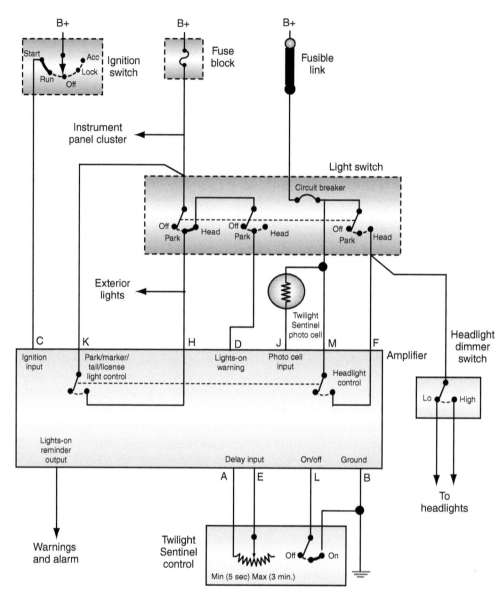

Figure 11-6 Twilight Sentinel schematic.

connections will refer to those this system uses. Use the correct service information for the system you are diagnosing.

Photo Cell Test

Classroom Manual
Chapter 11, page 307, and Chapter 9, page 238

To test the photo cell, use a scan tool to monitor photo cell voltages or a DMM to measure the voltage of the photo cell signal circuit to the BCM. With the engine running, observe the voltage readings. First, cover the photo cell's lens with a piece of cardboard. The voltage should read 0 volt. Next, shine a flashlight onto the photo cell's lens. The voltage should go up to 5 volts. If the voltages do not change, check the circuit, and if there are no problems, replace the photo cell.

Another method of testing the photo cell is to unplug the photo cell connector and start the engine. Place the headlight switch in the correct position to turn on the automatic headlight feature. If the headlights turn on within 60 seconds, replace the photo cell. If the lights do not turn on, test the amplifier (if equipped).

Photo Cell Replacement. Photo Sequence 23 illustrates the procedure for replacing the photo cell.

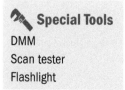

Special Tools
DMM
Scan tester
Flashlight

PHOTO SEQUENCE 23
Typical Procedure for Replacing a Photo Cell Assembly

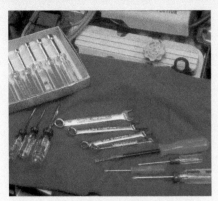

P23-1 Tools required to replace the photo cell assembly include nut driver set, thin flat-blade screw driver, Phillips screwdriver set, battery terminal pullers, battery pliers, box-end wrenches, safety glasses, seat covers, and fender covers.

P23-2 Place the fender covers over the vehicle's fenders and disconnect the battery negative terminal.

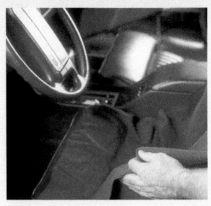

P23-3 Protect the seats by placing the seat covers over them.

P23-4 Using a clean shop rag to protect the dash pad, insert a long flat-blade screwdriver into the defroster and pry it up and out.

P23-5 Carefully pry out the vent outlets from the panel. Carefully work your way down the length of the grill, pulling it free from its seat.

P23-6 Gently work the grill up and out to gain access to the photo cell socket.

P23-7 Twist the socket to free it from the grill.

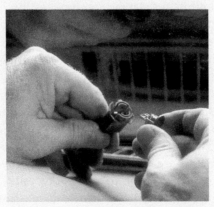

P23-8 Gently work the photo cell free from its socket.

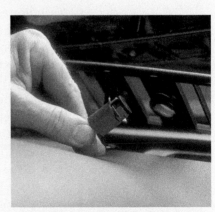

P23-9 Replace the photo cell and reinstall its socket into the grill.

PHOTO SEQUENCE 23 (CONTINUED)

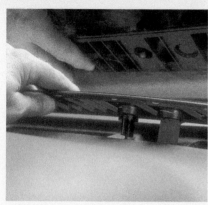

P23-10 Position the backside of the grill into place.

P23-11 Work the leading edge of the grill back into place using your thumb.

Classroom Manual
Chapter 11, page 307

Special Tools
Ohmmeter
Fused jumper wire
Test light
Fender covers
Safety glasses

AMPLIFIER TEST

Refer to Figure 11-6 as an example of how to test the amplifier.

1. Disconnect the wire connector to the amplifier.
2. Turn the headlight switch to the OFF position.
3. Disconnect the negative battery terminal.
4. Turn the control switch to the ON position.
5. Use an ohmmeter to measure the resistance between the wire connector terminal L and ground. There should be 0 ohm of resistance. If there is more than 0 ohm of resistance, check the wire from amplifier terminal L to control switch terminal C for opens. Also, check the circuit from amplifier terminal B for opens. If the ohmmeter indicated 0 ohm of resistance, continue testing.
6. Turn the control switch to the OFF position.
7. The ohmmeter should read infinite when connected between the L terminal and ground. If there is low resistance, check the circuit for a short. If the ohmmeter indicates infinite resistance, continue testing.
8. Place the headlight switch in the PARK position and connect the ohmmeter between terminals H and D. There should be 0 ohm of resistance. If not, check the two circuits for an open. If the ohmmeter indicates 0 ohm of resistance, continue testing.
9. Place the control switch in the middle of MIN and MAX settings.
10. Connect the ohmmeter between terminals A and E. There should be between 500 and 250,000 ohms of resistance. If the resistance value is not within this range, check the circuits between the amplifier and the control switch for opens.
    - If the circuits are good, observe the ohmmeter as the control switch is moved from the MIN position to the MAX position. The resistance value should change smoothly and consistently from one position to the next. If the ohmmeter indicates a resistance value out of limits, or the reading is erratic, replace the control switch.
11. If all of the resistance values tested are correct, continue testing.

12. Turn off all switches and reconnect the amplifier wire connector. Reconnect the battery negative terminal.
13. Turn the ignition switch to the RUN position. Turn the headlight switch to the OFF position, and turn the control switch to ON.
14. Connect a fused jumper wire between amplifier terminals J and M. The headlights should go off within 60 seconds. If the headlights go off within 60 seconds, but do not go off normally in bright light, check the circuit for opens. If the circuit is good, replace the photo cell.
15. Turn off all switches.
16. Disconnect the wire connector to the amplifier.
17. Turn the ignition switch to RUN and turn the headlight switch to OFF.
18. Connect a test light between terminal M and ground. The test light should light. If not, there is a short or an open in the circuit between the battery and the amplifier or in the circuit between the amplifier and the photo cell. If the test light illuminates, continue testing.
19. Connect the test light between terminals M and B. The test light should light. If not, check the ground circuit for an open. If the test light illuminates, continue testing.
20. Connect the test light between terminal K and ground. The test light should light. If not, check the fuse and the circuit for an open. If the test light illuminates, continue testing.
21. Connect the test light between terminal C and ground. The test light should light. If not, check the circuit for an open. If this circuit is good, check the ignition switch. If the test light illuminates, continue testing.
22. Turn the ignition switch to OFF.
23. With the test light connected as in step 21, the test light should turn off. If not, replace the ignition switch. If the test light turns off, continue testing.
24. Connect the test light between terminal H and ground. If the test light illuminates, replace the light switch. If the test light is off, continue testing.
25. Connect the test light between terminal F and ground. If the test light illuminates, replace the light switch. If the test light remains off, continue testing.
26. With the test light connected as in step 25, place the headlight switch in the HEAD position. If the test light does not turn on, check the circuit and the headlight switch for opens. If the test light lights, and all other tests have the correct results, replace the amplifier.

Use a 20-amp (A) fuse in the jumper.

⚠ **Caution**

Skipping any of the tests will result in replacement of the amplifier, even if it is good.

Resistance Assembly Test

The resistance assembly test is performed when the lights turn on and off at the wrong ambient light levels. Use **Figure 11-7** to construct a **photo cell resistance assembly**, which replaces the photo cell to produce predictable results.

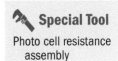
Special Tool
Photo cell resistance assembly

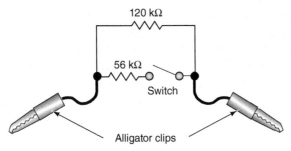

Figure 11-7 Photo cell resistance assembly.

Replace the photo cell with the resistance assembly. Turn the resistance assembly switch to the OFF (open) position and start the engine. If the lights do not come on within 60 seconds, replace the amplifier.

If the lights turn on, wait 30 seconds and turn the resistance assembly switch on. If the lights turn off within 60 seconds, replace the photo cell. If the lights do not turn off within 60 seconds, replace the amplifier.

SmartBeam Diagnostics

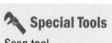
Classroom Manual
Chapter 11, page 313

SmartBeam is one of the systems that a manufacturer may use to control automatic high-beam operation. This system is presented as an example of the diagnostic procedures used to determine and repair faults.

Customer concerns that can be related to the operation of SmartBeam will usually be in one of two categories: either the system is totally inoperative or it is not performing properly. When diagnosing a concern that the SmartBeam system is inoperative, it is important that the following checks are performed:

Special Tools

Scan tool
Calibration target
Grease pencil
Tape measure

- Verify that the headlights work properly on both high and low beams when operated manually.
- Verify the power and ground circuits of the automatic high-beam module (AHBM).
- Verify that the system status indicator light-emitting diode (LED) in the mirror (**Figure 11-8**) is on steady. If the LED flashes, then a fault has been detected.
- Verify the automatic high-beam/low-beam function has been enabled (**Figure 11-9**).
- Verify that the headlight switch is set to the AUTO position.
- Verify that the headlight beam select switch is in the low-beam position.

Concerns related to poor system performance are usually associated with sensitivity of the system. If the system is oversensitive, the customer may state that the high beams come on too late and go off too early. If the system is under-sensitive, then the high beams will come on too early and go off too late. These types of sensitivity problems can be caused by:

- Camera not properly aimed.
- Loose camera mounting or improperly positioned mirror button.
- Obstruction in front of the camera.
- Improper headlight aiming.
- Vehicle overloading resulting in the rear of the vehicle sitting lower than the front.

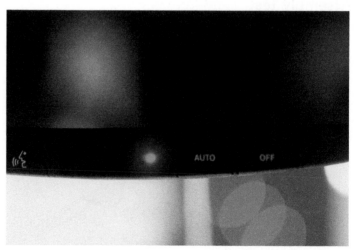

Figure 11-8 The LED can be used to indicate auto high-beam faults.

Figure 11-9 For the auto high-beam function to operate, it must be activated.

SERVICE TIP Because the SmartBeam requires the function of several vehicle modules, diagnosis of the system may require verifying the correct operation of modules other than the AHBM itself.

Whenever possible, system operation can be verified on stationary vehicles by using the scan tool to manually activate the system.

CUSTOMER CARE Inform the customer that the camera must have an unobstructed view from the front of the windshield for the system to perform properly. Hanging items from the mirror or placing toll road transponders in front of the camera will result in poor performance of the system.

The SmartBeam automatic high-beam system may have a **demonstration mode** that can be used to assist in diagnostics. The demonstration mode allows the function of the automatic high beams and high-beam indicator to be demonstrated while the vehicle is stationary and under any ambient lighting conditions. To initiate the demonstration mode function:

SERVICE TIP Remember that the SmartBeam system controls the high-beam portion of the headlamp system. The SmartBeam system varies the high-beam headlamp illumination intensity from low-beam level headlamp illumination intensity to full high-beam headlamp illumination intensity, and any level of headlamp illumination intensity in between.

1. Begin with the ignition switch in the OFF position.
2. Depress and hold the AUTO button on the inside mirror (refer to Figure 11-8).
3. While continuing to depress the AUTO button, turn the ignition switch to the RUN position.

4. Continue to hold the AUTO button depressed until the demonstration mode begins as indicated by the high beams ramping up in intensity.
5. Release the button.

The system will complete three cycles of ramping up the headlamp high beams to full intensity and then ramping them down. During this time, the high-beam indicator should also come on and go off with the high beams. The high beams will cycle three times; then the system will return to normal operation.

The LED in the rearview mirror is also used to assist in diagnostics. Usually this LED is on steady to indicate that the auto dimming function of the electrochromic mirror is on. A flashing LED indicates the system has detected a problem. The LED can flash at different frequencies and at different sequences to indicate a problem. For example, if the LED is continuously flashing at a rate of 1 flash per second, this indicates that the camera needs calibrating. To correct this condition, the camera calibration procedure must be performed.

If the LED continuously flashes at a rate of 2 flashes per second, this indicates that the system failed its last attempt to calibrate. To correct this condition, the camera calibration procedure must be performed.

The last possible LED indication is a series of flashes when the ignition switch is first placed in the RUN position, followed by the LED staying on steady. The flashing LED can indicate a hardware failure that may require AHBM replacement.

Camera Calibration Test and Adjustment. It is critical that the camera's field of view is maintained to specifications. If the camera's field of view is no longer within specifications, the performance of the system is seriously degraded. For proper operation, the camera's field of view must be maintained within 2° of the vehicle centerline and within 10° horizontally. If the camera is aligned within these specifications, the AHBM can adjust and fine-tune the alignment based on sensed lighting inputs while driving.

The camera calibration procedure must be performed any time the inside rearview mirror is replaced with a new unit, the rearview mirror mounting button has been replaced, or the windshield has been replaced. The calibration procedure ensures that the field of vision for the camera is aimed at the proper path ahead of the vehicle. If a new camera is installed, it is shipped with the calibration mode initiated. Once the camera is installed and connected, the LED will flash once every second while the ignition switch is in the RUN position. Before attempting to calibrate the camera, the following should be performed:

- Clean the windshield glass in front of the camera lens.
- Check for proper mounting of the mirror assembly and that the set screw which secures the assembly to the button is properly torqued at 15 inch pounds (1.7 Nm).
- Repair or replace any faulty, worn, or damaged suspension components.
- Verify proper tire inflation pressures.
- Verify that there is no load in the vehicle, except for the driver.
- The fuel tank should be full. Add 6.5 pounds (2.94 kg) of weight over the fuel tank for each gallon of missing fuel.

To calibrate the camera, the centerline of the vehicle must be established and marked. A special alignment target is also required. The target consists of a black, square field containing three red LEDs positioned in a diagonal pattern (**Figure 11-10**). The target is placed a specified distance in front of the vehicle. Once the proper height of the target is established to match the height of the camera lens, the target must be positioned within 1/2° of centerline of the vehicle. When in calibration mode, the AHBM is programmed to look for the red LED light pattern of the target from the camera. When the target is properly placed, the AHBM will identify it and the center of the target will become the new center of the field of view. The flashing green LED changes to a steady ON state when calibration is successful. **Photo Sequence 24** illustrates the typical procedure for setting up and calibrating the camera.

Figure 11-10 Calibration target.

If the camera cannot locate the target's pattern, the calibration fails. The LED will blink at a 2-hertz rate, indicating that the calibration procedure has failed and that a fault has been set. This may be caused by the target being incorrectly placed or the camera being improperly aligned.

PHOTO SEQUENCE 24
Automatic High-Beam Camera Calibration

P24-1 Locate the "ASI" mark on the windshield.

P24-2 Using the "ASI" marks on both sides of the windshield, measure across the windshield using the shaft of the tint band arrow as the reference points.

P24-3 Divide the measurement in half and mark that dimension near the lower edge of the tint band using a grease pencil.

P24-4 Measure across the upper edge of the blackout area at the bottom of the windshield using the inside corner of the intersection between the side blackout area and the lower blackout area as the reference points.

P24-5 Divide the lower measurement in half and mark that dimension near the upper edge of the blackout area using a grease pencil.

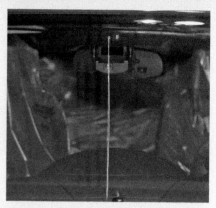

P24-6 To locate the centerline of the camera lens, measure and mark the glass 7/8 inch (21 mm) toward the passenger side of the windshield from the upper and lower glass centerline marks and draw a vertical line connecting these two marks.

P24-7 Remove the grease pencil mark in front of the lens.

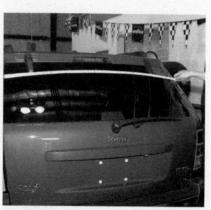

P24-8 To locate the camera centerline on the rear glass, measure across the upper portion of the rear glass using the vertical edges of the body opening as the reference points.

P24-9 Divide the upper measurement in half and mark that dimension on the glass using a grease pencil.

P24-10 Measure across the rear glass at the bottom using the vertical edges of the body opening as the reference points.

P24-11 Divide the lower measurement in half and mark that dimension on the glass using a grease pencil.

P24-12 Measure and mark 7/8 inch (21 mm) toward the passenger side of the rear glass on both the upper and lower references and draw a vertical line connecting these two marks.

PHOTO SEQUENCE 24 (CONTINUED)

P24-13 Locate the calibration target 50 inches (127 cm) in front of the vehicle measuring from the foremost center of the front fascia.

P24-14 While sighting through the V-notch in the upper edge of the calibration target, move the target left or right to align the target to the camera centerline marks on both the windshield and the rear glass.

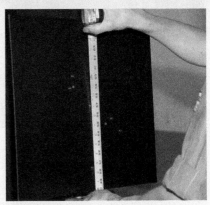

P24-15 Adjust the tripod so that the center LED is on the target 57 inches (145 cm) from the floor.

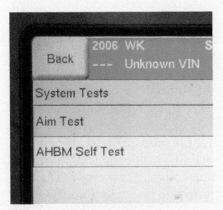

P24-16 Use the scan tool to enter the SmartBeam unit into calibration mode. When the calibration mode is entered, the LED in the mirror assembly will flash once per second.

P24-17 Turn on the LEDs in the calibration target.

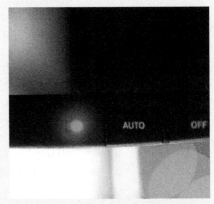

P24-18 The LED in the mirror assembly should continue to flash for 5 to 10 seconds, and then will stop flashing to indicate that it has completed calibration.

If the camera fails to calibrate, a DTC will be set, indicating that the performance is off in any of the four directions. To correct this condition, you will have to verify that the mounting of the camera and the position of the target are correct. If they are correct, then adjustment of the camera will be necessary.

> **SERVICE TIP** If more than one high-beam camera alignment performance fault DTC were retrieved, then one screw and one shim will be common to both faults. Make the correction at this location.

Each DTC for alignment has the description of the direction the camera should be moved to make a correction. For example, if the DTC is "High-Beam Camera Alignment Performance Bottom," the camera locates the target but it is at the bottom of the field of view. To correct this, the camera needs to be tilted downward.

Figure 11-11 An adjustment shim is used to align the camera.

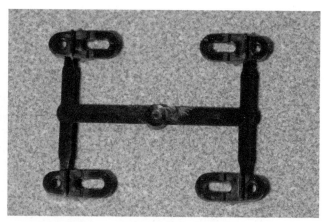

Figure 11-12 The adjustment shim.

Camera alignment is done by the placement of a spacer between the camera mount and the camera housing (**Figure 11-11**). The adjustment spacer consists of four stepped shims connected by an integral plastic tree (**Figure 11-12**).

Determine what shim movement will be required to correct the alignment based on the DTCs retrieved (**Figure 11-13**). To adjust the camera alignment, move the inside rear view mirror head downward so access to the two screws securing the rear cover can be obtained. Remove the two screws, and then rotate the mirror head up toward the headliner to its uppermost position. Unsnap and pull the upper edge of the rear cover away from the housing far enough to disengage the tabs at the lower edge of the cover; then remove the rear cover.

With the mounting bracket now accessible, carefully cut the plastic tree for the shims at the appropriate location(s) to allow movement of the shims determined earlier. With the shims separated from the tree, loosen the attaching screws one-half turn. The screw holes in the shims are slotted to allow sliding of the shims. Slide the shim beneath each of the loosened screws to its most outboard position. Once the shims are relocated into the desired position, the attaching screws are then tightened to 7 inch pounds (0.8 Nm). As the screws are tightened, the camera is pulled into the new alignment position. The total amount of correction is 2° to 2.5°. Only two shims can be moved for correction. The shims moved must be either of the following:

- Both top shims, to tilt the camera upward.
- Both bottom shims, to tilt the camera downward.
- Both left shims, to tilt the camera to the left.
- Both right shims, to tilt the camera to the right.

Reinstall the mirror assembly rear cover. Use the scan tool to erase any DTCs, and perform the calibration procedure again. If the correct shims were moved and the necessary correction was within the range of adjustment, calibration should be successful.

Camera Optics Test. If the automatic high-beam system is performing poorly, and the LED does not flash, the camera's optics may be dirty or obstructed. The camera's optics can be tested by entering the demonstration mode. The **optics test** will confirm that the camera can recognize ambient light through the lens and the windshield. To perform the optics test, first initiate the automatic high-beam demonstration mode. While observing the LED, obstruct the view of the imager by placing a piece of cardboard between the camera lens and the windshield. The LED should turn off each time the optics is obstructed by the cardboard. Once the cardboard is removed, the LED should illuminate again.

Failure of the LED to respond to these inputs indicates that the imager optics is obstructed. Clean the lens and the windshield glass and/or remove any obstructions from the windshield in front of the imager. To clean the imager lens, spray a small amount of glass cleaner onto a soft cloth and gently wipe the lens clean. After cleaning the lens and the windshield, repeat the test to confirm proper automatic high-beam system operation.

 Caution

Do not spray glass cleaners directly onto the imager lens. Doing so may damage the imager optics and electronics.

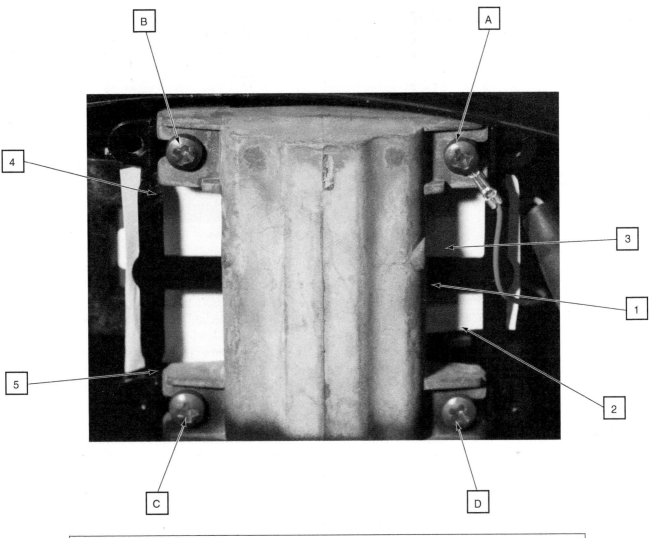

Camera adjustment table		
Fault	Tree cut location	Shim + screw location
Right	1	A + D
Left	1	B + C
Top	3 + 4	A + B
Bottom	2 + 5	C + D

Figure 11-13 Adjustment chart and locations.

HEADLIGHT LEVELING SYSTEMS

The headlight leveling system is designed to adjust the beam projection of the headlights based on vehicle load and other considerations. Two types of systems are commonly used: driver-initiated level by use of a switch or thumb wheel, and automatic leveling.

The first requires the driver to adjust the headlights using a switch or thumb wheel to select the level. The switch is located in the instrument panel bezel. Generally, there will be four positions that will lower the headlight beam as the vehicle load increases. The switch is a MUX switch that sends a voltage signal to the controller board and logic circuitry of the headlamp leveling motor. The motor is located in the headlight housing

Classroom Manual
Chapter 11, page 316

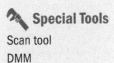

Special Tools

Scan tool
DMM

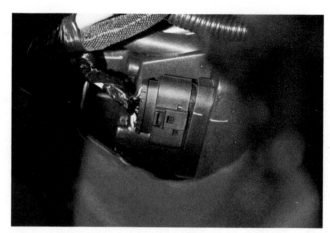

Figure 11-14 Headlight leveling motor and linkage.

(**Figure 11-14**). When the motor is energized, it will extend or retract the motor pushrod through the integral screw-drive transmission. The other end of the pushrod is snapped into a socket on the back of the reflector, which causes the reflector to move as the pushrod is extended or retracted. The reflector position changes the angle at which the light is projected from the headlamp.

The leveling motors and switch have a direct connection to chassis ground. Although there are differences between manufacturers on power supply to the switch and motors, a common method is to supply power only when the headlights are turned on. This can be accomplished by supplying voltage to the switch and motors through the fused park lamp relay output circuit.

The headlamp leveling switch as well as the hard-wired inputs and outputs of the switch may be diagnosed using a DMM. Use the appropriate wiring information when testing the circuits. The following is an example of testing the switch function using a DMM. The switch terminals are identified in **Figure 11-15**.

Using the ohmmeter function of the DMM, perform the resistance tests at the terminal pins in the switch connector receptacle as shown in **Table 11-1**. If the switch fails any of the resistance tests, replace the switch.

Because of active electronic elements located in the leveling motor, it cannot be tested using conventional diagnostic tools and procedures. If the headlamp leveling motor is believed to be defective, the hard-wired headlamp leveling motor circuits and the leveling switch must be tested before considering motor replacement.

The second system automatically levels the headlamps, regardless of cargo or passenger loads or road conditions. When the vehicle is loaded with passengers and/or cargo, the system will use inputs from the front and rear sensors and make an adjustment based on the feedback from the sensor to see if it has approximately 27.5° from its zero or neutral

> **⚠ Caution**
>
> To avoid serious or fatal injury, disable the supplemental restraint system before attempting any component diagnosis or service. Disconnect and isolate the battery negative cable and wait at least 2 minutes to allow the system capacitor to discharge before performing further diagnosis or service.

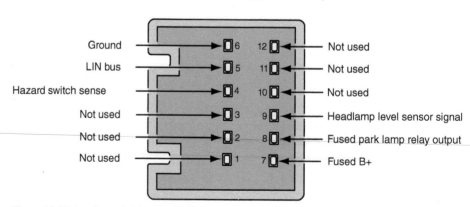

Figure 11-15 Leveling switch terminal callouts.

Table 11-1   Table of Resistance Test for Leveling Switch

	Headlamp Leveling Switch Tests	
Switch Position	Resistance (ohms) ± 1% between Pins 6 and 9	Resistance (ohms) ± 1% between Pins 6 and 8
0	1,518	1,750
1	1,971	2,203
2	3,661	3,893
3	9,851	10,183

position. The sensors determine if the vehicle has had its ride height or angle changed, due to passenger or cargo additions and road conditions.

There is one sensor mounted on the spring link on the right side of the rear suspension (**Figure 11-16**). The other sensor is mounted on the upper control arm on the right side, in the front of the vehicle (**Figure 11-17**).

This system will use a headlamp stepper motor that adjusts the headlamp reflector up or down for proper headlamp aiming. The headlamp level will be adjusted while driving if required based on sensor inputs. A headlight leveling module uses the input from the sensors to control the operation of the stepper motor. The module is also capable of monitoring the system for faults. If a system fault occurs, the module will store the DTC and enable it to be read using the scan tool.

Usually, the sensor linkage is not adjustable. However, anytime the sensors are removed or replaced, the system needs to be calibrated using a scan tool. This is also true if the headlamp leveling module is replaced.

When performing the calibration function, the vehicle must be level. Place the ignition in the RUN position and confirm that the vehicle doors, trunk or liftgate, and hood are all closed. Make sure no one is seated in the vehicle and that the vehicle is not disrupted (bounced or bumped) until the calibration procedure is complete. Also, the headlamps must be ON to power the stepper motors. Calibration will take approximately 12 to 15 seconds to complete. The scan tool will display a "status of calibration" message. The message will be either "still in progress," "passed," or "failed."

The headlight aiming module is also referred to as the high-intensity discharge translator module.

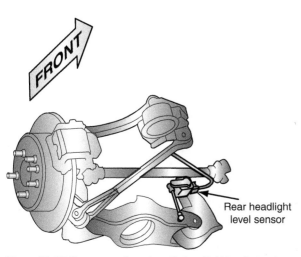

Figure 11-16 Rear sensor for automatic headlight leveling system.

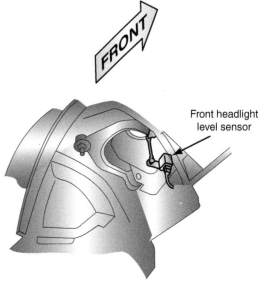

Figure 11-17 Front sensor for automatic headlight leveling system.

If the calibration failed, use the wiring diagram to confirm that all circuits are functioning properly. If there is no problem with the circuits, the module is probably the fault. Confirm power and grounds to the module before replacing it.

ACTIVE HEADLIGHT SYSTEM SERVICE

Classroom Manual
Chapter 11, page 317

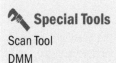

Special Tools

Scan Tool
DMM

The **active headlight system (AHS)** is designed to enhance night time safety by providing drivers additional time to act or react to approaching road hazards by predicting upcoming corners or turns in the road. If the system detects a fault, it reverts to a conventional system, sets a DTC, and notifies the driver by illumination of a warning light or by displaying a message.

Like most computer-driven systems, active headlight system diagnostics rely heavily on DTCs. Remember that DTCs do not always identify the component that has failed. Instead, they indicate a circuit that has failed, an invalid input received, no expected change after an activation, or a rational test failure.

The actuator motors can be controlled by the scan tool. Most systems will not activate the active headlight system unless vehicle speed is greater than a programmed threshold. Thus, the scan tool will be necessary to activate the motors while the vehicle is stationary. If the motor fails to operate when commanded, back probe the motor connector's power and ground circuits and connect a voltmeter across the terminals. Activate the motor with the scan tool and record the voltage. If the voltage is equal to battery voltage, the motor has failed. If no voltage is recorded, move the negative test lead to a chassis ground location and activate the motor again. Voltage at this time indicates an open ground circuit; no voltage indicates an open power circuit.

If the motors operate when using scan tool bidirectional controls, check the input sensors. Typical sensors required for operation include the vehicle speed sensor, steering angle sensor, and the yaw sensor. If all of the inputs are correct, the active headlight module is faulty.

Replacement of any parts of the system may require initialization or calibration. Typically, these procedures will require the use of a scan tool.

DAYTIME RUNNING LAMPS

Classroom Manual
Chapter 11, page 320

Special Tools

Test light
DMM
Fused jumper wire

Some manufacturers use a daytime running lamp (DRL) module that receives inputs from either the headlight switch or the BCM to determine switch position. If the switch is in the OFF or AUTO position (and ambient light is high enough that the headlights do not need to be on), the DRL module will send a PWM signal to the headlight beam using HSDs. This turns the headlight on, but at reduced illumination levels.

Some vehicles use parking lamps for DRL function. In this case, the headlight position is used to determine if the headlights are on. If they are not, the BCM or DRL module will turn on the park lamp relay.

Diagnostics of either type of system is performed by using a scan tool or DMM. As with any system, determine proper power and ground circuit operation. Also, remember that the parking brake switch is used as an input. If the switch should fail and indicates to the DRL module that the parking brakes are applied, the DRL function is disabled.

HEADLIGHT AIMING

Classroom Manual
Chapter 11, page 294

The headlights should be checked for proper aiming whenever the lamps are replaced. Proper aiming is important for good light projection onto the road and to prevent discomfort and dangerous conditions for oncoming drivers.

CUSTOMER CARE The Department of Transportation reports that about 50% of all vehicles on the road have at least one headlight that is not properly aimed. This is a very dangerous condition and your customers must be educated of the importance of proper headlight aiming. Suggest that their headlight aiming be checked at least once a year.

Correct headlight beam position is so critical that government regulations control limits for headlight aiming. For example, a headlight that is misaimed by 1° downward will reduce the vision distance by 156 feet (47.5 meters). The following are maximum allowable limits that have been established by all states:

Special Tools

Headlight-aiming unit
Torx driver
Fender covers
Safety glasses

1. Low beam: In the horizontal plane, the left edge of the headlight high-intensity area should be within 4 inches (102 mm) to the right or left of the vertical centerline of the lamp. In the vertical plane, the top edges of the headlight high-intensity area should be within 4 inches (102 mm) above or below the horizontal centerline of the lamp (**Figure 11-18**).
2. High beam: In the horizontal plane, the center of the headlight high-intensity area should be within 4 inches (102 mm) to the right or left of the vertical centerline of the lamp. In the vertical plane, the center of the headlight high-intensity area should be within 4 inches (102 mm) above or below the horizontal centerline of the lamp (**Figure 11-19**).

Before the headlights are aimed, the vehicle must be checked for proper **curb height**. Curb height is the height of the vehicle when it has no passengers or loads and has normal fluid levels and tire pressure. This includes checking the suspension, tire inflation, removing any additional load in the vehicle, a half-filled fuel tank, and removing dirt, ice, snow, and so on from the vehicle.

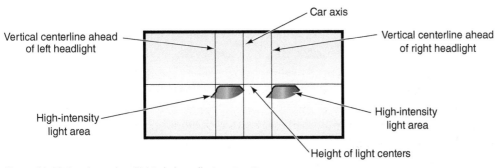

Figure 11-18 Low-beam headlight aiming adjustment pattern.

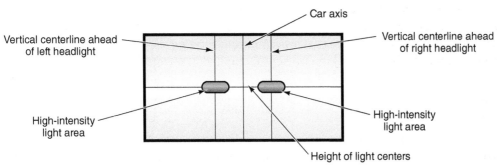

Figure 11-19 High-beam headlight aiming adjustment pattern.

Classroom Manual
Chapter 11, page 294

Sealed Beam

Portable mechanical aiming units can be used to aim most sealed-beam-type headlights (**Figure 11-20**). These are secured to the headlight lens by suction cups (**Figure 11-21**). The aiming unit should have a variety of adapters to attach to the various styles of headlights. Before using the aiming equipment, be sure to follow the manufacturer's procedure for calibration. Park the vehicle on a level floor area and place fender covers around the work area. It may be necessary to remove the trim and bezel from around the headlight. Using the correct adapter, connect the calibrated aimer units to the headlights. Be sure the adapters fit the headlight-aiming pads on the lens (**Figure 11-22**). Zero the horizontal adjustment dial. Confirm that the split-image target lines are visible in the view port (**Figure 11-23**). If the target lines are not seen, rotate the aimer unit. Turn the headlight horizontal adjusting screw until the split-image target lines are aligned. Repeat for the headlight on the other side.

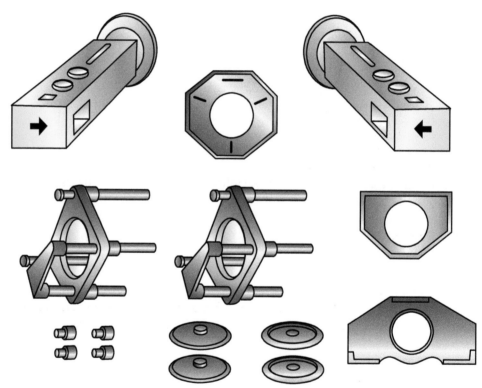

Figure 11-20 Typical portable mechanical headlight-aiming equipment and adapters.

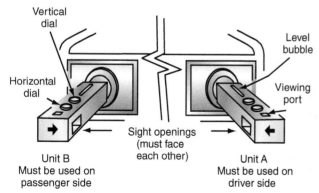

Figure 11-21 The aiming units attach to the headlight lens with suction cups.

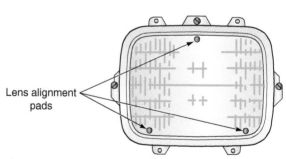

Lens alignment pads

Figure 11-22 Headlight-aiming pads.

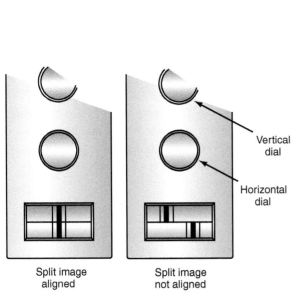

Figure 11-23 Split-image target.

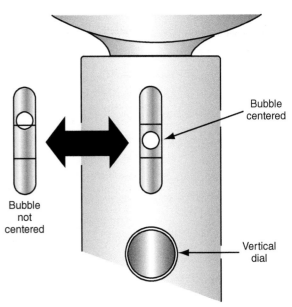

Figure 11-24 Center the spirit level by turning the vertical aiming screw.

To set the vertical aim of the headlight, turn the vertical adjustment dial on the aiming unit to zero. Turn the vertical adjustment screw until the spirit-level bubble is centered (**Figure 11-24**). Recheck the horizontal aiming on each headlight. The vertical adjustment may have altered the original adjustments.

If the vehicle is equipped with a four-headlight system, repeat the procedures for the other pair of lamps.

Composite and HID

To adjust some composite and HID headlight designs, special adapters are required (**Figure 11-25**). Also, the lens must have headlight-aiming pads to be able to use a mechanical aiming unit. The headlight assembly will have a number molded on it. The adjustment rod setting must be set to that number and locked in place. The aiming unit is attached to the headlight lens in the same manner as with sealed-beam headlights (**Figure 11-26**).

Classroom Manual
Chapter 11, page 296

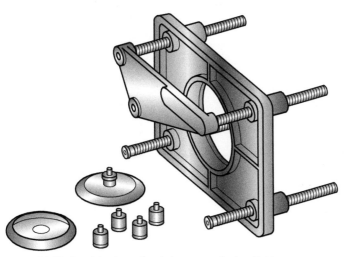

Figure 11-25 Special adapter for aiming composite headlights.

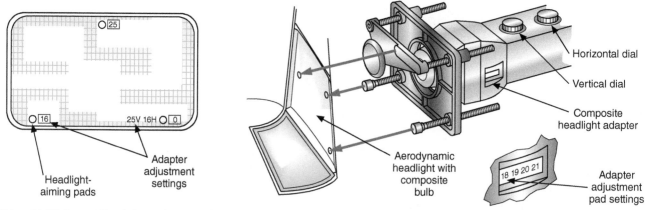

Figure 11-26 Connect the aiming equipment to the headlight lens. The lens must have aiming pads.

The adjustment procedure of composite and HID headlights is identical to that of the sealed-beam headlights. **Figure 11-27** shows the typical location of the headlight adjusting screws.

Many composite and HID headlight designs do not have alignment pads on the lens. These systems usually adjust the beam location by moving the reflector position. Since the lens does not move, conventional headlight aimers are not used. One method of aiming these systems is by locating the vehicle 25 feet (7.6 meters) away from a blank wall with the vehicle on a level surface. The wall is marked, based on the centerline of the vehicle; then the location of the beam on the wall is adjusted to meet manufacturer's specifications. Some manufacturers may require that the headlights be adjusted with the high beams.

With complex lens designs and multiple bulb configurations, many service facilities use an optical aiming system (**Figure 11-28**). Optical headlight aimers are often required

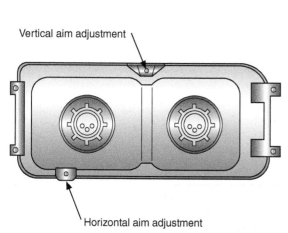

Figure 11-27 Composite headlight aiming screw locations.

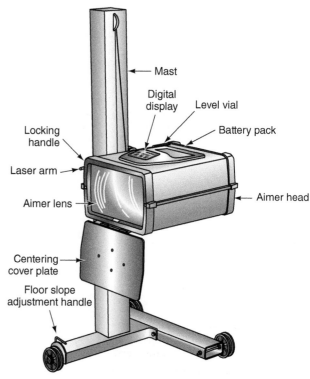

Figure 11-28 Typical optical headlight aimer.

Figure 11-29 Optical aimer screen.

if the state has mandatory safety inspections that include checking the headlight aiming. These units have an optical-grade lens that reproduces the headlight beam image. The image is then transmitted to a screen to show the technician where the high intensity of the light is being projected (**Figure 11-29**).

To accurately use the optical sensor, any slope in the bay floor must be compensated for. Typically, this is done by using a laser light beam projected from the optical aimer unit to a measuring device at the rear wheel of the vehicle. An adjustment wheel is used to set the optical aimer unit so front to rear vehicle measurements are the same.

Once the initial set-up is complete, the center of the headlight bulb is determined and the aimer is aligned to it. Depending on the equipment, this may involve locating the grill's exact center and then the machine will automatically detect the beam center. Some equipment requires the beam center to be manually located and then the machine is aligned to the vehicle by using two symmetrical points on the vehicle (such as radiator support bolts, top of the headlight lens, or strut bolts). The headlights are turned on and set to low beam and the screen displays the high-intensity pattern. The low beams are aimed so the top edge of the high-intensity pattern is along the horizontal axis, and the left side of the zone pattern aligns with the vertical centerline (**Figure 11-30**). Typically, the headlights are aimed using only the low beam. High beam can be checked and the high-intensity zone should be center on the screen (**Figure 11-31**). Some aimers use a digital display to indicate if the beam is aligned. An "X" in the vertical and horizontal positions indicates that the beam is properly aligned. If the beam is not properly aligned, the screen will indicate the direction of the misalignment.

⚙ SERVICE TIP If the headlight lens is clear, the center of the bulb can be determined visually. Some lens will have a dot to indicate the location of the bulb's center. Another method to locate the center of the bulb is to cover the lens with a cloth shop rag and turn the headlights on. Since the center of the bulb will be dark, the light through the rag will identify the center of the bulb (**Figure 11-32**).

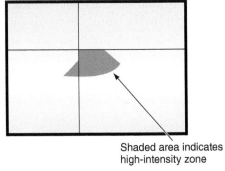

Shaded area indicates
high-intensity zone

Figure 11-30 Correct low-beam high-intensity location.

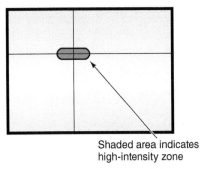

Shaded area indicates
high-intensity zone

Figure 11-31 Correct high-beam high-intensity location.

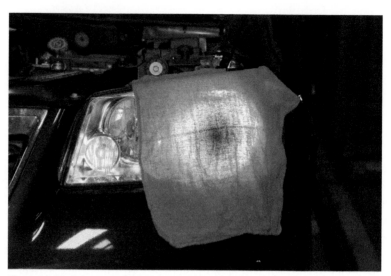

Figure 11-32 Using a shop rag to locate the center of the headlight bulb.

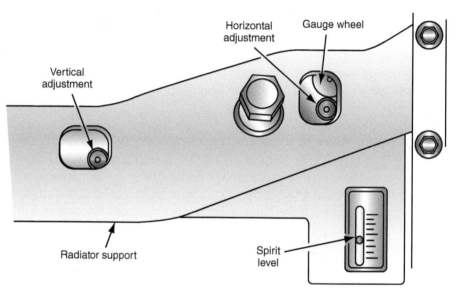

Figure 11-33 Some headlight assemblies are equipped with spirit levels to provide easy adjustments to offset vehicle loads.

Many late-model vehicles have spirit levels built into the headlamp assembly (**Figure 11-33**). These are not always to be used for initial headlamp adjustment. They are supplied for the driver to adjust his headlights based on the load in the vehicle. For example, if the trunk is loaded, the front of the vehicle is lifted and the light beam is too high. By turning the adjuster wheels until the bubble is in the middle of the level, the beam is returned to its original position. After the load is removed from the trunk, the headlights are adjusted until the bubble is returned to the middle. Whenever the headlights are adjusted, the technician should adjust the spirit level also.

DIAGNOSING DIMMER- OR BRIGHTER-THAN-NORMAL LIGHTS

Classroom Manual
Chapter 11, page 302

The complete headlight circuit consists of the headlight switch, dimmer switch, high-beam indicator, and the headlights (**Figure 11-34**). Excessive resistance in these units, or at their connections, can result in lower illumination levels of the headlights.

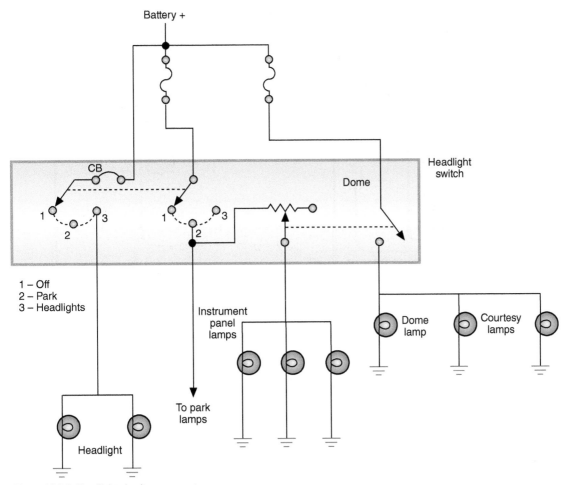

Figure 11-34 Headlight circuit components.

The extra resistance can be on the insulated side or the ground side of the circuit. To locate the excessive resistance, perform a voltage drop test (**Figure 11-35**). Consult the wiring diagram to determine the number of connectors and switches. This will provide you with the specification for maximum voltage drop. Start at the light and work toward the battery.

All headlight systems are wired in parallel. If both headlights are dim, then the excessive resistance is in the common portions of the circuit. Dim headlights can also be the result of low-charging system output.

Other causes of dim lights can be the use of the wrong lamps, improper circuit routing, addition of extra electrical loads to the circuit, and wrong-size conductors.

Brighter-than-normal lights can be the result of higher-than-specified charging system output or improper lamp application. It is also possible that the dimmer switch contacts are stuck in the high-beam position.

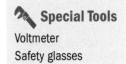 **Special Tools**

Voltmeter
Safety glasses

Although this procedure is being shown for the headlights, it is identical for all lighting systems. The only difference in the test results will be if the circuit uses insulated bulbs or grounded bulbs. If testing the turn signal circuit, bypass the flasher with a fused jumper wire.

⚙ **SERVICE TIP** Headlights do not wear out and get dimmer with age. If one of the headlights is dimmer than the other, there is excessive voltage drop in that circuit. If a new headlight is installed, the breaking and making of the socket connection may clean the terminals enough to make a good contact. Once the new headlight is installed, it may operate properly. Do not be fooled. It was not the headlight that was at fault. It was the connection.

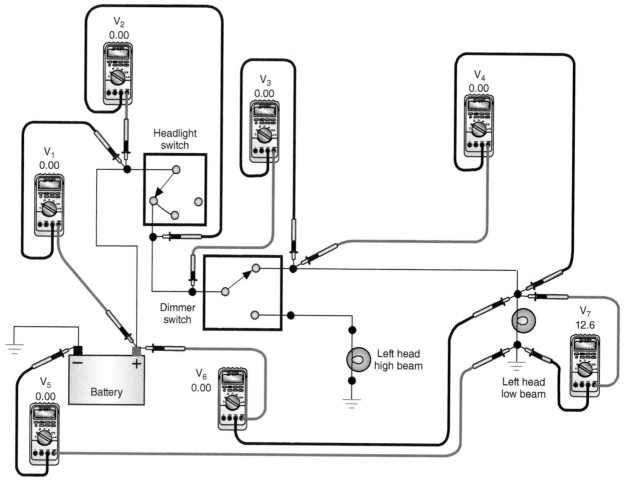

Figure 11-35 Voltage drop testing the headlight system.

Classroom Manual
Chapter 11, page 301

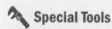 **Special Tools**

12-volt test light
Ohmmeter or self-
    powered test light
Fused jumper wire
Safety glasses
Battery terminal pliers
Terminal pullers

HEADLIGHT SWITCH TESTING AND REPLACEMENT

The headlight switch controls most of the vehicle's lighting systems. The headlight switch will generally receive direct battery voltage to two of its terminals. Disconnect the battery before removing the headlight switch.

In the headlight switch circuit, a rheostat is used to control the instrument cluster illumination lamp brightness. Most dash-mounted headlight switches incorporate the rheostat into the switch assembly. Steering column–mounted switches may have the rheostat located on the dash.

Many customer concerns associated with the lighting systems can be the result of a faulty headlight switch. For example, dim or no instrument panel lights, dim or no headlights, dim or no parking lights, and improperly operating dome lights can all be caused by the headlight switch.

Dash-Mounted Switches

Many methods are used to retain the headlight switch to the dash. Consult the service information of the vehicle you are working on. The following is a common method of removing the headlight switch:

⚡ WARNING **If the vehicle is equipped with air bags, disable the supplemental restraint system before attempting any component diagnosis or service. Failure to take proper precautions could result in accidental air bag deployment.**

1. Place fender covers on the fenders.
2. Install a memory keeper and disconnect and isolate the battery negative cable.
3. Remove the lower cluster bezel from the instrument panel.
4. Disconnect the wire harness connector from the back of the headlamp switch.
5. Remove the fasteners that secure the headlamp switch to the back of the cluster bezel.
6. Remove the headlamp switch from the cluster bezel.

> **SERVICE TIP** Headlights that flash on and off as the vehicle goes over road irregularities indicate a loose connection. Headlights that flash on and off at a constant rate indicate that the circuit breaker is being tripped. There is an overload in the circuit that must be traced and repaired.

With the switch removed, it can be tested for continuity and the connector plug will serve as a test point for the lighting circuits. First, test at the connector.

The following is a typical procedure for testing the headlight switch connector on the harness side. A test light and fused jumper wires are used to test the circuits. This procedure would be very similar for any non-computer-controlled headlight system—just use the service information to determine the function of each terminal. In this procedure, the terminals are identified as follows (**Figure 11-36**):

- Terminal B—battery.
- Terminal A—fuse.
- Terminal H—headlights.
- Terminal R—rear park and side marker lights.
- Terminal I—instrument panel lights.
- Terminal D1—dome light feed.
- Terminal D2—dome light.

Consult the service information for terminal identification for the vehicle you are working on. If this is not listed in a separate chart, you should be able to identify the circuits from the wiring diagrams.

Some headlight switches are retained by spring clips. Remove them by compressing the springs.

1. Connect the 12-volt (V) test light across terminal B and ground. The test light should light. If not, there is an open in the circuit back to the battery.

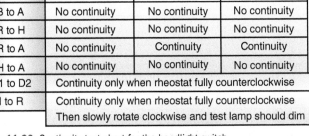

Switch terminals	Switch positions		
	Off	Park	Headlamp
B to H	No continuity	No continuity	Continuity
B to R	No continuity	No continuity	No continuity
B to A	No continuity	No continuity	No continuity
R to H	No continuity	No continuity	No continuity
R to A	No continuity	Continuity	Continuity
H to A	No continuity	No continuity	No continuity
D1 to D2	Continuity only when rheostat fully counterclockwise		
I to R	Continuity only when rheostat fully counterclockwise		
	Then slowly rotate clockwise and test lamp should dim		

Figure 11-36 Continuity test chart for the headlight switch.

2. Connect the test light across terminal A and ground. The test light should come on. If not, repair the circuit back to the fuse panel.
3. Connect a fused jumper wire between terminals B and H. The headlights should come on. If the headlights fail to turn on, trace the H circuit to the headlights. Also, check the ground circuit side from the headlights.
4. Connect a fused jumper wire between terminals A and R. The rear lamps should illuminate. If not, trace the circuit to the rear lights. Also, check the ground return path.
5. Connect a fused jumper wire between terminals A and I. The instrument panel lights should come on. If not, trace the circuit to the panel lights.

If all the tests performed at the switch connector pass, then the problem is in the headlight switch. Figure 11-36 indicates the test results that should be obtained when testing the headlight switch for continuity for the system just discussed. Use an ohmmeter or a self-powered test light to test the switch. Most service information will provide a table similar to that in **Figure 11-37**. If a chart is not available, use the wiring diagram to determine which terminals should or should not have continuity in the different switch positions.

Steering Column–Mounted Switches

On some vehicles, it is possible to test the steering column–mounted switch without removing it. The test is conducted at the connector at the base of the column (Figure 11-37). However, on some models, it is necessary to remove the column cover and/or the steering wheel to gain access.

A common procedure for removing and testing the **multifunction switch** is shown in **Photo Sequence 25**. The switch is called a multifunction switch because it can have a combination of any of the following switches in a single unit: headlights, turn signal, hazard, dimmer switch, horn, and flash-to-pass (**Figure 11-38**).

Dimmer Switch Testing and Replacement

The **dimmer switch** is connected in series within the headlight circuit and controls the current path for high and low headlight beams. Most dimmer switches are located on the steering column or incorporated within the multifunction switch. Testing of the switch is done by using a set of fused jumper wires to bypass the switch. If the headlights operate with the switch bypassed, the switch is faulty.

The steering column–mounted dimmer switch can be operated by an actuator control rod from the lever to a remotely mounted switch (**Figure 11-39**). Another style incorporates the dimmer switch into the multifunction switch (**Figure 11-40**).

To remove the remote switch, first place fender covers on the fenders of the vehicle and disconnect the battery negative cable. Disconnect the wire connector at the switch. Remove the two switch mounting screws and disengage the switch from the actuator rod.

When installing the switch, make sure the actuator rod is firmly seated into the switch. During the installation, adjust the position of the switch so that all actuator rod slack is taken up. If the switch has alignment holes, compress the switch until two appropriately sized dowels can be inserted into the alignment holes (**Figure 11-41**). While applying a slight rearward pressure, install and tighten the mounting bolts.

When the switch is adjusted properly, it will click when the lever is lifted and again when it is returned to its downward position. The second click should occur just before the stop.

If the dimmer switch is a part of the multifunction switch, follow the general procedure shown in Photo Sequence 25.

To reinstall the multifunction switch, reverse the procedure. Torque the steering column–attaching nuts to the amount specified in the service information.

Classroom Manual
Chapter 11, page 303

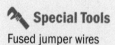 **Special Tools**
Fused jumper wires
Safety glasses
Battery terminal pliers
Terminal pullers

Use a memory keeper before disconnecting the battery.

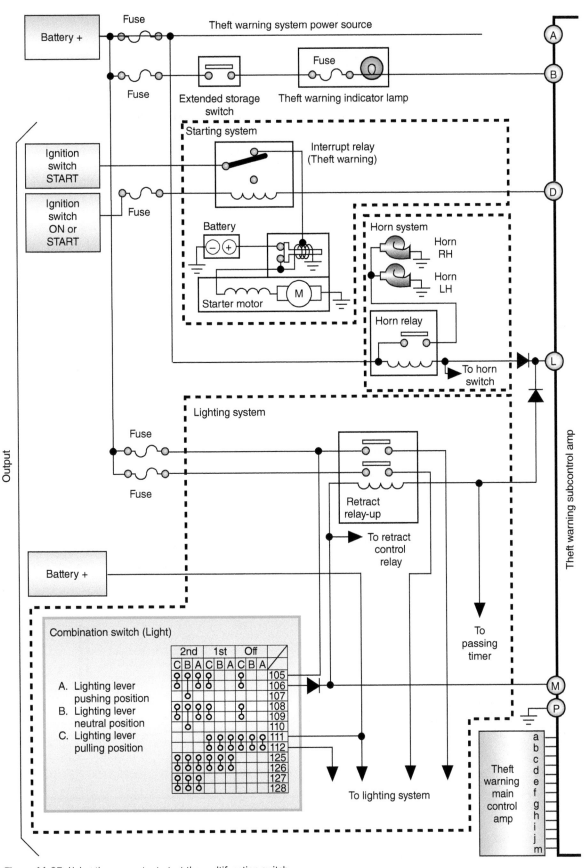

Figure 11-37 Using the connector to test the multifunction switch.

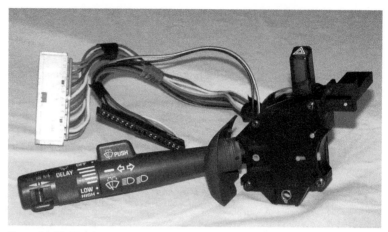

Figure 11-38 Multifunction switch.

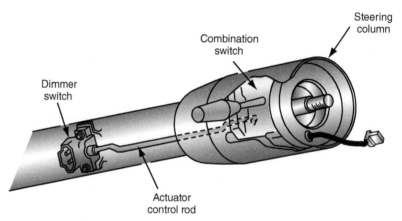

Figure 11-39 Steering column–mounted dimmer switch.

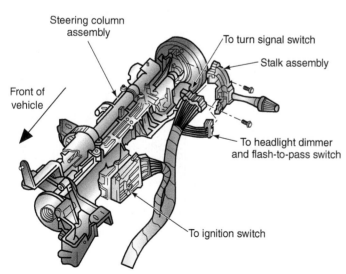

Figure 11-40 Dimmer switch incorporated into the multifunction switch.

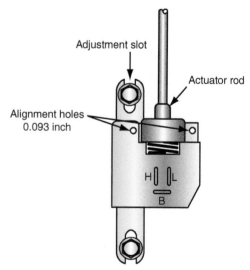

Figure 11-41 Insert a dowel into the alignment holes to adjust the dimmer switch.

PHOTO SEQUENCE 25
Removal and Testing of the Multifunction Switch

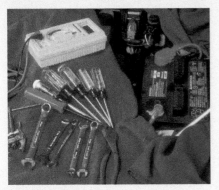

P25-1 Tools required to remove and test the multifunction switch: fender covers, battery terminal pliers, terminal pullers, assorted combination wrenches, torx drivers, and ohmmeter.

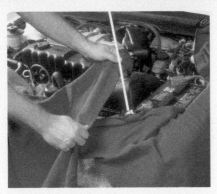

P25-2 Place the fender covers around the battery work area.

P25-3 Loosen the negative battery clamp bolt and remove the clamp using terminal pullers. Place the battery cable where it cannot contact the battery.

P25-4 Remove the shroud-retaining screws and remove the lower shroud from the column.

P25-5 If needed to gain access to the upper shroud, loosen the steering column–attaching nuts. Do not remove the nuts.

P25-6 Lower the steering column enough to remove the upper shroud.

P25-7 Remove the turn signal switch lever by slightly rotating the outer end of the lever and then pulling straight out on the lever. Some levers may be attached with fasteners.

P25-8 Peel back the foam shield from the turn signal switch.

P25-9 Disconnect the turn signal switch electrical connectors.

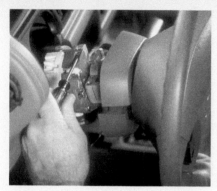

P25-10 Remove the screws attaching the turn signal switch to the lock cylinder assembly.

P25-11 Disengage the switch from the lock assembly.

P25-12 Use the ohmmeter to test the switch. Check for continuity from terminal 15 to 13 when the dimmer switch is in the low-beam position.

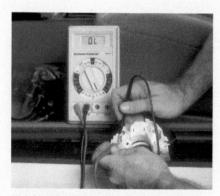

P25-13 With the switch in the low-beam position, the circuit should be open between terminals 15 and 12.

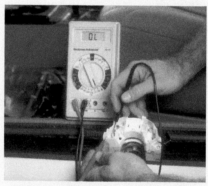

P25-14 The circuits between terminals 196 and 13 and 196 and 12 should be open in the low-beam position.

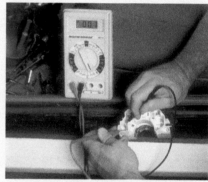

P25-15 With the switch in the high-beam position, continuity should be between terminals 15 and 12. Circuits 15 to 13, 196 to 13, and 196 to 12 should be open.

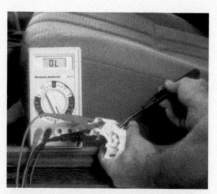

P25-16 When the dimmer switch is placed in the flash-to-pass position, there should be a closed circuit between terminals 196 and 12 and an open circuit between 196 and 13.

TAILLIGHT ASSEMBLIES

In a three-bulb taillight system, the brake lights are controlled directly by the brake light switch (**Figure 11-42**). The brake lights on both sides of the vehicle are wired in parallel. Most brake light systems use dual-filament bulbs that perform multiple functions. In this type of circuit, the brake lights are wired through the turn signal and hazard switches (**Figure 11-43**).

If all of the taillights do not operate, check the condition of the fuse. If it is good, use a voltmeter to test the circuit. With the headlight switch in the PARK position (first detent), check for voltage at the last common connection between the switch and the lamps. If battery voltage is present, then the problem is in the individual circuits from that connector to the lamps. If no battery voltage is present, test for voltage from the switch terminal. If no voltage is present at this terminal yet the headlights operate when in the ON position, replace the switch. If battery voltage is present, the problem is between the switch and the last common connection. If there was no voltage present at the switch terminal, check for battery voltage into the switch.

Most taillight bulbs can be replaced without removal of the lens assembly. The bulb and socket are removed by twisting the socket slightly and pulling it out of the lens assembly (**Figure 11-44**). To remove the bulb from the socket, push in on the bulb slightly while turning it. When the lugs align with the channels of the socket, pull the bulb from the socket (**Figure 11-45**).

If the bulb is a blade base-type bulb, pull it straight out of the socket without twisting. On all bulb types, it is a good practice to use an oil-free rag or wear nylon gloves to grasp the bulb. This will keep oil off the bulb but also will protect you if the bulb should break.

Figure 11-46 shows how the lens assembly is fastened to the vehicle body. Remove the attachment nuts from the back of the assembly to remove it.

Turn Signals and Brake Lights

To test the turn signal switches, use a 12-volt test light to probe for voltage into and out of the switch. The ignition switch must be in the RUN position for the circuit to operate. If voltage is present on the input side of the switch but not on the output side, the switch is faulty.

Classroom Manual
Chapter 11, page 322

 Caution
Using the wrong type of lamp for the socket and application can result in "crazy lights." This is a result of feedback caused by the incorrect bulb (**Figure 11-47**).

Classroom Manual
Chapter 11,
pages 323, 327

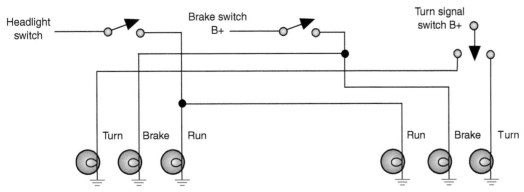

Figure 11-42 Three-bulb taillight circuit has individual control for each bulb.

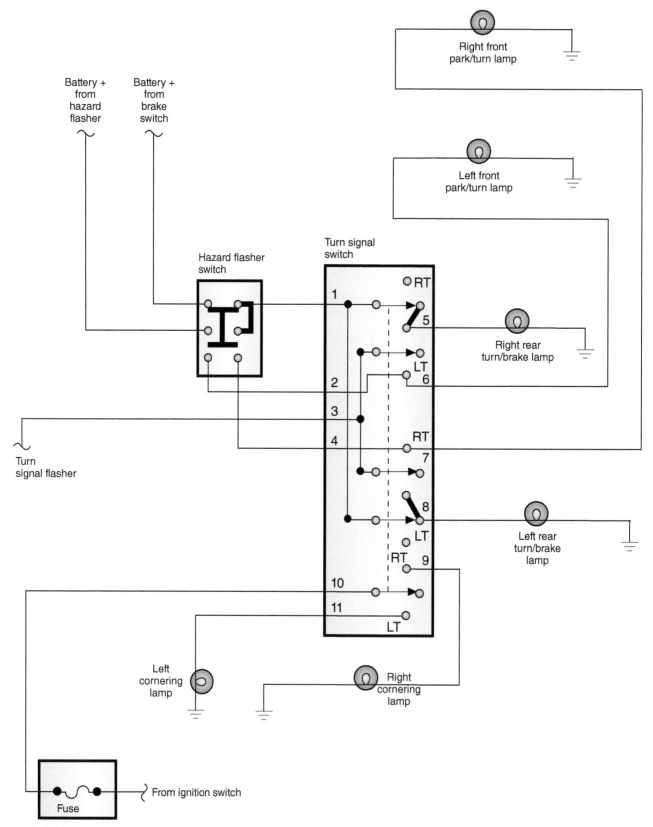

Figure 11-43 Turn signal switch used in a two-bulb taillight circuit.

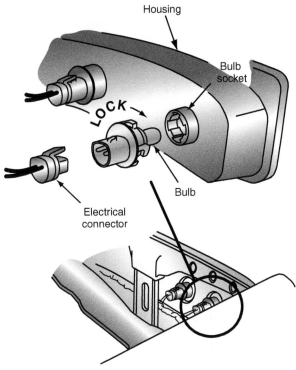

Figure 11-44 Bulb and socket removal from the taillight lens assembly.

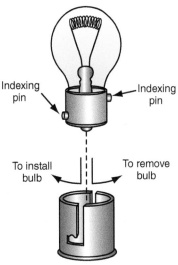

Figure 11-45 Removing the bulb from the socket.

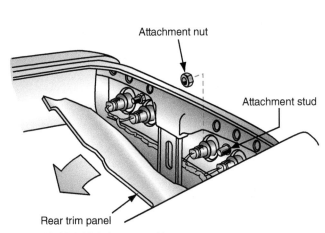

Figure 11-46 Taillight lens assembly.

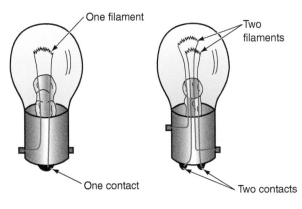

Figure 11-47 If the single-filament bulb is mistakenly installed into a socket designed for dual filament, the single contact of the bulb will short across the two contacts of the socket. This will result in lighting circuits operating when they are not supposed to.

Check brake light operation through the turn signal switch in the same manner. Also, check the brake light switch for proper adjustment and operations (**Figure 11-48**).

Many early vehicles use a turn signal switch that is separate from the multifunction switch. The steering wheel will have to be removed to gain access to the turn signal switch on these vehicles. The following procedure is a common method of turn signal switch replacement:

1. Place the fender covers on the fenders.
2. Install a memory keeper and disconnect the battery negative cable.

 **Special Tools**

12-volt test light
Steering wheel puller
Lock ring compressor
Safety glasses

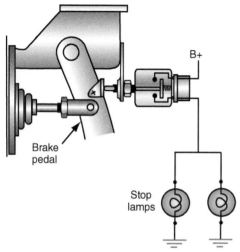

Figure 11-48 Typical brake light switch operation. Check for continuity in both positions. Should be open when at rest and closed when the pedal is depressed.

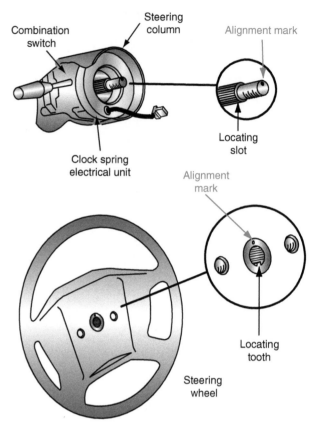

Figure 11-49 Steering wheel attachment.

> ⚡ WARNING **If the vehicle is equipped with an air-bag system, wait for the recommended amount of time before removing any other components. Fifteen minutes or more may be required to discharge the capacitors that are used to fire the air bags.**

Most turn signal switches receive their voltage from the ignition switch when it is in the RUN position only. This prevents the turn signals from operating while the ignition switch is in the OFF position.

3. Remove the steering column trim.
4. Remove the horn pad from the steering wheel and make alignment marks on the steering wheel and shaft (**Figure 11-49**).
5. If equipped, remove the steering shaft nut and horn collar.
6. Use a suitable puller to remove the steering wheel.
7. If needed, use a suitable compressor to compress the preload spring to the lock plate (**Figure 11-50**). Compress the spring only enough to remove the snap ring.
8. Use a pick and a small flat-blade screwdriver to remove the snap ring.
9. Remove the lock plate, horn contact carrier, and spring.
10. Remove the bolts at the upper steering column support and the upper mounting bracket from the column.
11. Disconnect the turn signal wiring connector.
12. Wrap tape around the wire and connector.
13. Remove the hazard warning knob from the column.
14. Remove the switch-retaining screws and remove the switch.

> ⚙ SERVICE TIP Two-bulb system switches also control some brake light functions through a complex system of contacts. Many brake light problems are caused by worn contacts in the turn signal switch.

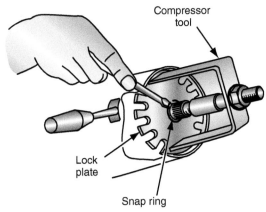

Figure 11-50 To remove the snap ring, use the compressing tool to relieve the pressure against the snap ring.

Flashers. The flasher uses a bimetallic strip and a heating coil to flash the turn signals. A common customer concern that is attributable to the flasher is related to the flashing speed. If the flasher is of the wrong type and rating, the amount of time required to heat the coil will differ from what the manufacturer designed into the circuit. Also, newer flashers that use electronic circuits will flash at an increased speed if one of the turn signal bulbs is burned out or the circuit is defective. If the flasher is rated higher than required, the flashing rate is reduced because it takes longer for the current to heat the coil.

Check the size and type of light bulbs in the circuit. Use only the lamp size recommended by the manufacturer. If these checks do not correct the problem, test the charging system output. Voltage output that is higher or lower than specified may cause the flasher rate to be incorrect. If the charging system output is within specifications, check for excessive resistance in the turn signal circuit. Check both sides of the circuit.

If none of the turn signals operate, check the fuse. Next, check the flasher. Remove the flasher from the fuse box (**Figure 11-51**). Connect a fused jumper wire across the fuse box terminals (**Figure 11-52**). If the turn signal lamps come on with the lever in either indicator position, the flasher is faulty. If the lights still do not illuminate, test the turn signal switch.

Classroom Manual
Chapter 11, page 328

 **Special Tools**
Fused jumper wires
Safety glasses

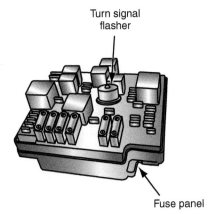

Figure 11-51 Many manufacturers locate the flasher unit in the fuse panel.

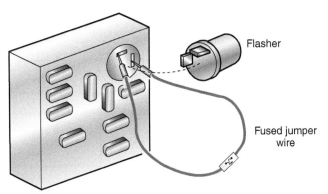

Figure 11-52 Connecting a fused jumper wire across the terminals to by pass the flasher unit.

SERVICE TIP If the turn signals operate properly in one direction but do not flash in the other, the problem is not in the flasher unit. A burned-out lamp filament will not cause enough current to flow to heat the bimetallic strip sufficiently to cause it to open. Thus, the lights do not flash. Locate the faulty bulb and replace it.

INTERIOR LIGHTS

Not all flashers are located in the fuse box. Use the component locator to find where the flasher is installed.

Classroom Manual
Chapter 11, page 335

Interior lighting includes courtesy lights, map lights, and instrument panel lights. These lights may also be incorporated into the illuminated entry feature.

Courtesy Lights

Courtesy lights operate from the headlight and door switches. They receive their power source directly from a fused battery connection. The switches can control either the ground circuit (**Figure 11-53**) or the insulated circuit (**Figure 11-54**). The courtesy lights may also be activated from the headlight switch by turning the switch knob to the extreme counterclockwise position. The contacts in the switch close and complete the circuit.

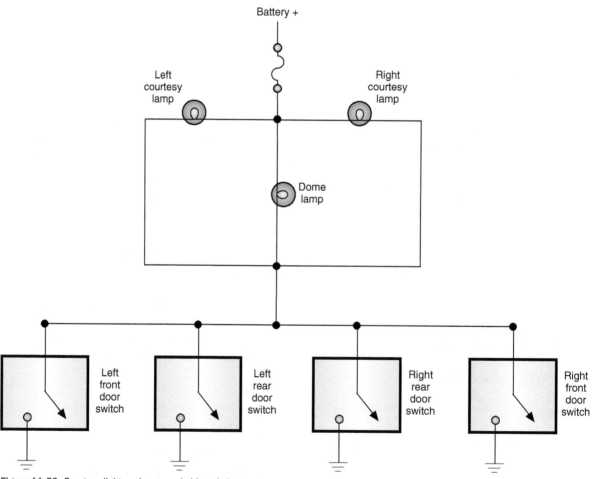

Figure 11-53 Courtesy lights using ground side switches.

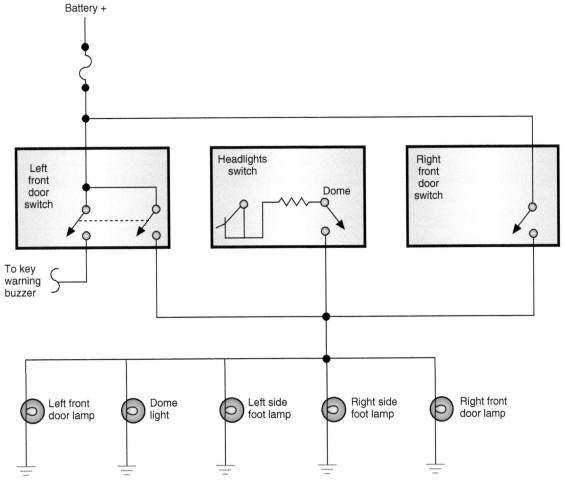

Figure 11-54 Courtesy lights using insulated side switches.

Figure 11-55 provides a systematic approach to troubleshooting courtesy lights. Follow the steps in proper order to locate the fault.

If all the lights of the circuit do not light, begin by checking the fuses. If the fuse is good, then use a voltmeter to check for battery voltage at the last common connection. If voltage is present at the fuse box but not at the common connection, the problem is

Any time a blown fuse is found, the cause of the circuit overload must be traced.

COURTESY LAMPS DO NOT TURN ON WHEN ONE DOOR IS OPENED
OK WHEN OTHER DOORS ARE OPENED

Test step	Result	Action to take
A0 **Verify condition**		Go to A1
A1 **Check power** Check for power at door switch	(OK̶)	Check the power circuit back to fuse
	(OK)	Go to A2
A2 **Check the door switch** Check the door switch for proper operation	(OK̶)	Replace the switch
	(OK)	Check circuit from switch to lamp

Figure 11-55 A diagnostic chart for the courtesy light system (continued on next page).

COURTESY LAMPS DO NOT COME ON WHEN HEADLAMP SWITCH IS TURNED COUNTERCLOCKWISE TO STOP

Test step	Result	Action to take
B0 **Verify condition**		Go to B1
B1 **Check operation of door switches** Check to see if courtesy lamps operate from door switches	NOT OK	Go to chart C
	OK	Go to B2
B2 **Check for power** Check for power at headlamp switch	NOT OK	Check circuit back to fuse
	OK	Go to B3
B3 **Check for continuity** Check continuity of headlamp switch	NOT OK	Replace headlamp switch
	OK	Check circuit from switch to lamp

COURTESY LAMP DOES NOT COME ON WHEN ALL DOORS ARE OPENED

Test step	Result	Action to take
C0 **Verify condition**	Vehicle with only one courtesy lamp	Go to C1
	Vehicle with more than one courtesy lamp	Go to C2
C1 **Check operation of fuse circuit** Check operation of other circuits that share the same fuse	NOT OK	Go to C4
	OK	Go to C2
C2 **Check for power** Check for power at the bulb	NOT OK	Replace bulb
	OK	Go to C3
C3 **Check for continuity** Check continuity of bulb	NOT OK	Replace bulb
	OK	Check bulb ground
C4 **Check fuse** Check continuity of fuse	NOT OK	Replace fuse
	OK	Go to C5
C2 **Check for power** Check for power through the fuse	NOT OK	Check power feed circuit
	OK	Check for open circuit between fuse and common point in lamp circuit

Figure 11-55 Continued

between these two points. If battery voltage is present at the common connection, trace the individual circuits until the cause(s) for the open is located.

If the courtesy lights do not come on when only one of the doors is opened but do come on when any of the other doors are opened, the problem is in the affected door's switch or circuit. In order to check the switch and circuit, bypass the switch with a fused jumper wire. If the lights come on with the switch bypassed, it is a faulty switch. If the lights do not come on with the switch bypassed, there is a problem in the circuit.

Illuminated Entry System Diagnostics

The diagnostic procedures for testing the illuminated entry system depend upon the manufacturer. Always refer to the service information for the vehicle you are working on. Always perform a visual inspection before performing any tests. First check the fuse. Then check to make sure all connections are tight and clean. Inspect the ground wires for good connections. Check all visible wires for fraying or damaged insulation, especially where they go through body parts. Make sure all doors are closed properly and the headlight switch is in the detent position.

The following are typical procedures for diagnosing the illuminated entry system. Systems that are activated by lifting the outside-door handle use a contact switch that momentarily closes to complete the ground circuit of the **illuminated entry actuator** module. The module activates the interior lights for 25 seconds or until the ignition switch is placed in the RUN or ACC position.

A logic circuit is included in the module to prevent battery drain if the door handle is held up for longer than 25 seconds. The system will operate as normal until the 25 seconds has elapsed, and then the module will turn off the lights. The lights will remain off and cannot be reactivated until the handle is returned to the released position.

This type of system has four main components: fuse, ignition switch, actuator module, and door handle switch assemblies (**Figure 11-56**). The door lock cylinder uses an LED to provide the illumination of the cylinder. The lens of the LED is built into the cylinder.

To test the system, disconnect the actuator harness from the actuator and connect the test light between terminal 8 and ground. The test light should illuminate. If the test light fails to come on, trace the circuit back to the battery to locate the problem.

Connect the test light between terminal 7 and ground. The test light should not glow with the ignition switch in the OFF position. When the ignition switch is turned to the RUN or ACC position, the test light should come on. If the test light does not turn on and off as the ignition switch is turned, trace the circuit to the fuse box and the ignition switch to locate the problem.

Connect a fused jumper wire between connector terminals 6 and 8. Make sure all doors are closed. The courtesy lights and door lock cylinders should be illuminated. If the lights do not operate, trace the circuit from terminal 6 to the LEDs and lights to locate the problem.

Connect an ohmmeter between connector terminal 2 and chassis ground. The ohmmeter should indicate an infinite reading. However, a minimum of 10,000 ohms is acceptable. Lift up on each of the outside-door handles to close the latch switch. Hold the handle up while observing the ohmmeter. The ohmmeter should indicate a resistance value of 50 ohms maximum. If either of the ohmmeter readings is out of specifications, trace the circuit to the latch switches. Also, test the latch switches for correct operation.

With the test light connected between connector terminals 1 and 8, the light should be on. If the test light fails to come on, trace the circuit from terminal 1 to ground.

If the preceding tests do not indicate any problems, the actuator module assembly is faulty. The module must be replaced.

Classroom Manual
Chapter 11, page 337

Special Tools
DMM
Test light
Fused jumper wire
Safety glasses

Courtesy lights and illuminated entry lights usually use the same bulbs. Courtesy lights are lights that come on when the door is open or the switch is turned on. *Illuminated entry* is a term used for turning on lamps prior to entering the vehicle, such as when the handle is lifted or remote keyless entry is used.

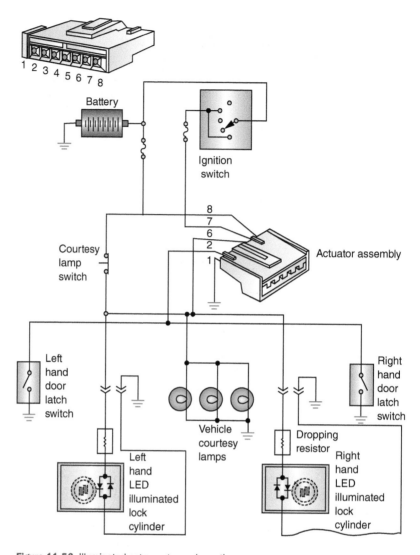

Figure 11-56 Illuminated entry system schematic.

Follow these steps to diagnose the system illustrated in **Figure 11-57**:

1. Move the dimmer control to the center position.
2. Open the driver-side door to activate the courtesy lights. If none of the courtesy lights turn on, continue testing. If only one bulb is inoperative, check its circuit and the bulb.
3. Lower the driver-side window and close all doors. Manually lock the driver's door. Wait 30 seconds with the ignition switch off.
4. Activate the illuminated system by lifting the driver's door handle. If the lights come on, repeat the test for the right-side door. If the system does not operate when the right-side door handle switch is closed, refer to the service information for the circuits to be tested. The procedure will be the same as when the left door is inoperative. However, the circuit designations are different.
5. If the lights do not turn on when the door handles are lifted, connect the scan tool to the diagnostic connector. Maneuver through the menu screens to locate the door handle switch state.
6. With the ignition switch on, observe the scan tool display while lifting the door handle. The display should indicate the switch closed when the handle is lifted. If the switch operates correctly, go to step 7. If the display does not indicate proper

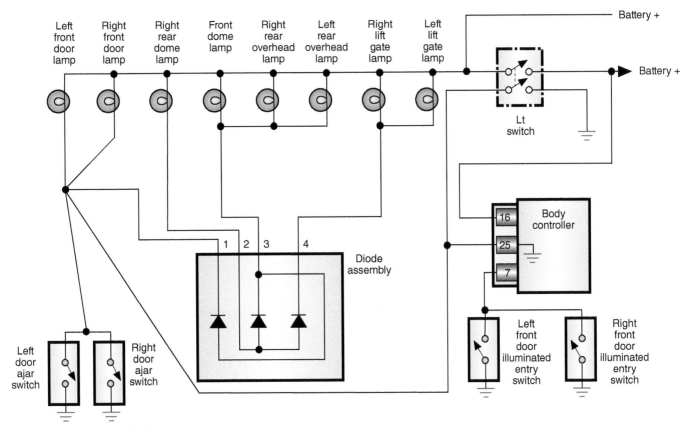

Figure 11-57 Schematic of an illuminated entry system.

switch operation, connect a fused jumper wire from the controller terminal identified in the service information to ground. This would be terminal 7 at the BCM in Figure 11-57. Observe the scan tool. If it indicates that the circuit is closed, follow the service information procedure for testing the door switches. If the display indicates that the circuit is open, replace the BCM.

7. Open the driver-side door. If the courtesy lights do not turn on, go to step 8. Close all doors and jumper the terminal of the BCM identified in the service information as lamp control (terminal 25 in Figure 11-57) to ground. If the lights do not turn on, there is an open circuit between the BCM and the lamps. If the lights turn on when the BCM is jumped, replace the BCM.

8. If the lights do not come on in step 7 when the door is opened, gain access to the driver-side door ajar switch harness and disconnect it. Use a fused jumper wire to jump across the two terminals of the connector on the harness side. This should complete the circuit to ground and the lights come on. If the lights do not turn on, check both circuits for an open. If the lights come on when the connector is jumped, replace the switch. Test the door ajar switch for the passenger side in the same manner. Use a fused jumper wire to jump across the two switch connector terminals on the harness.

Some BCM-controlled illuminated entry systems do not use a door handle switch. The system is activated when the doors are opened, or the remote keyless entry system is used to unlock the doors. The system uses the same lamp driver as the courtesy lamps. Check for operation of the system by opening each door, one at a time. Usually, the three-passenger door ajar switches are connected in parallel with the driver's door on its own circuit (**Figure 11-58**). If the system does not activate when the doors are opened, use the

On some systems, terminal 4 is used. On these systems, jumping between terminals 6 and 8 will illuminate the courtesy lights. Jumping between terminals 4 and 8 will light only the door lock cylinders. If they do not, trace circuit 464.

⚠ **Caution**

The terminal to jump from the controller is different between years and models. Be sure to refer to the service information for the correct terminals.

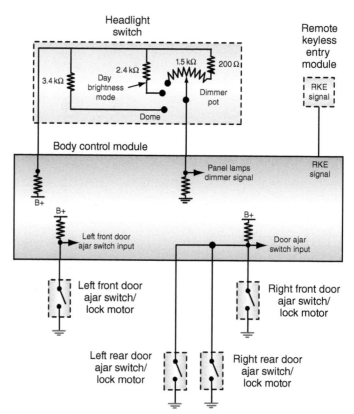

Figure 11-58 The inputs used for the illuminated entry system. Note that the passenger door ajar switches are in parallel.

remote keyless entry fob and press the UNLOCK button. If the system works now, the problem is in the door ajar input circuits or the BCM. If using the fob does not activate the system, the fault is probably in the BCM or the control circuits to the lamps. However, it is possible that all inputs are missing. To confirm this, use the scan tool to activate the system. If the lamps illuminate now, each input will need to be tested.

To test the inputs to the BCM, use a scan tool to monitor the door ajar switches. The switches should change state as the doors are opened and closed. The scan tool will display the state of the switches, not the door. The switches are open when the door is closed, and the switches are closed when the door is open. If no change of state is seen, test the door ajar signal circuit by disconnecting the door ajar switch. Connect a fused jumper wire across the connector on the BCM side. If the illuminated entry system activates now (or the scan tool displays the switch as closed), there is a faulty switch. If this still does not activate the system, test for battery voltage to the switch connector. If battery voltage is not present, test the circuit back to the BCM by back probing for voltage at the BCM connector. If voltage is present now, there is an open in the wire between the BCM and the switch. If voltage is not present, check for good battery feed to the BCM and proper ground connections. If these are good, replace the BCM. If battery voltage is present at the switch connector when tested above, the fault is in the ground circuit for the switch.

> **⚙ SERVICE TIP** You can also use the headlight switch dimmer function to test the lamps. Turn the switch to the DOME position. If the light comes on now, the output circuit from the BCM is working fine and the problem is in the inputs.

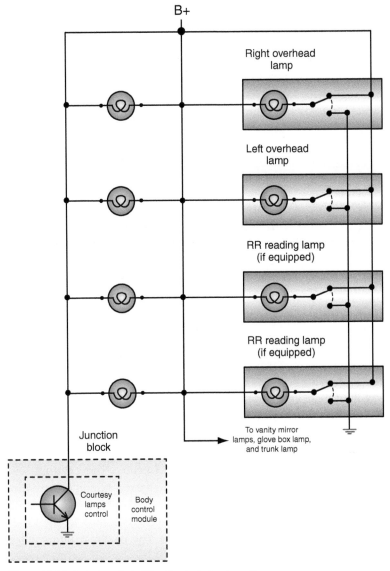

B+

Right overhead lamp

Left overhead lamp

RR reading lamp (if equipped)

RR reading lamp (if equipped)

Junction block

To vanity mirror lamps, glove box lamp, and trunk lamp

Courtesy lamps control

Body control module

Figure 11-59 Output control of the illuminated entry system.

If the scan tool indicates proper input from the door ajar switches and the activation test fails to illuminate the lamps, then the output side of the system must be tested (**Figure 11-59**). Locate the proper wire into the BCM connector for the courtesy lamp control. Using a back-probing tool, jump this terminal to a good ground. If the lamps turn on now, check for proper battery feed and grounds to the BCM. If these are good, replace the BCM. If the lamps do not turn on, use the manual switches to turn on each light. If they do not turn on, trace the circuit from the battery to the common connection or splice.

Instrument Cluster and Panel Lights

The power source for the instrument panel lights is provided through the headlight switch. The contacts are closed when the headlight switch is located in the PARK or HEADLIGHT position. The current must flow through a variable resistor (rheostat) that is either a part of the headlight switch or a separate dial on the dash. The resistance of the rheostat is varied by turning the knob. By varying the resistance, changes in the current flow to the lamps control the brightness of the lights.

Classroom Manual
Chapter 11, page 339

Special Tools

DMM
Safety glasses

Test for voltage output from the headlight switch to determine if the switch is operating properly. Vary the amount of resistance in the rheostat while observing the voltmeter. The voltage reading from the rheostat should vary as the knob is turned. If voltage is present to the printed circuit, check the ground.

If all connections are good, remove the dash and test the printed circuit board. Use an ohmmeter to check for opens and shorts in the printed circuit board from the connector plug to the lamp sockets. If the printed circuit is bad, it must be replaced. There are no repairs to the board.

LIGHTING SYSTEM COMPLEXITY

Today's vehicles have a sophisticated lighting system and electrical interconnections. It is possible to have problems with lights and accessories that cause them to operate when they are not supposed to. This is through a condition called **feedback**. Feedback occurs when electricity seeks a path of lower resistance. This alternate path operates another component than the one intended. If there is an open in the circuit, electricity will seek another path to follow. This may cause any lights or accessories in that path to turn on.

Examples of feedback and how it may cause undesired operation are illustrated in **Figures 11-60** through **11-67**. Figure 11-60 shows a system that has the dome light, taillight, and brake light circuits on one fuse; the cigar lighter and courtesy light circuits share a second fuse. The two fuses are located in the main fuse block and share a common bus bar on the power side.

If the dome light fuse blew and the headlight switch was in the PARK or HEADLIGHT position, the courtesy lights, dome light, taillight, parking lights, and instrument lights would all be very dimly lit (**Figure 11-61**). Current would flow through the cigar lighter fuse to the courtesy light and on to the door light switch. Current will then continue through the dome light to the headlight switch. Because the headlight switch is now closed, the instrument panel lights are also in the circuit. The lights are dim because all the bulbs are now connected in series.

If the dome light fuse is blown and the headlight switch is in the OFF position, all lights will turn off. However, if the door is opened, the courtesy lights will come on but the dome light will not (**Figure 11-62**).

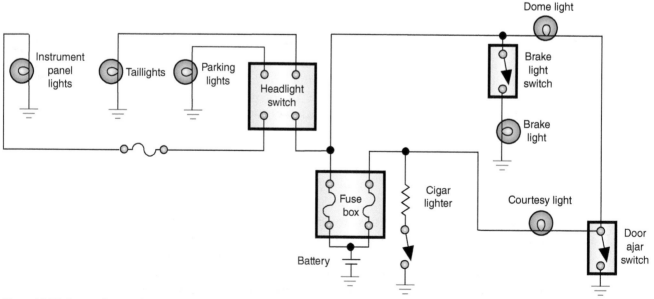

Figure 11-60 A normally operating light circuit.

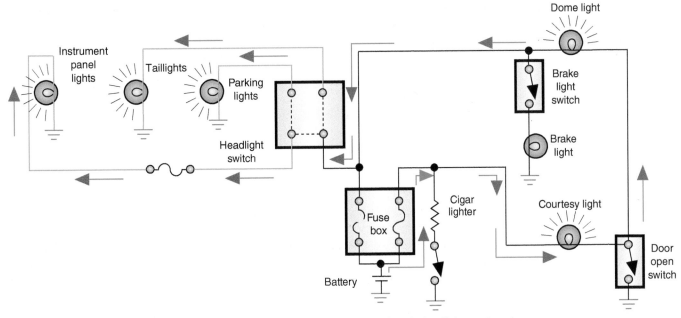

Figure 11-61 An open (blown) dome light fuse can cause feedback into other circuits when the headlights are turned on.

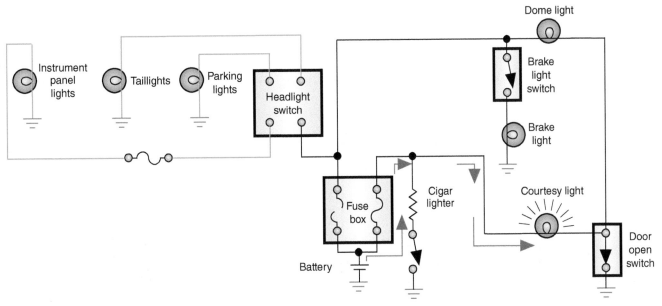

Figure 11-62 Circuit operation with a blown dome light fuse and the door switch closed.

With the same blown fuse and the brake light switch closed, the dome light, courtesy light, and brake light will all illuminate dimly because the loads are in series (**Figure 11-63**).

In this example, if the dome light and courtesy lights come on dimly when the cigar lighter is pushed in, the problem can be caused by a blown cigar lighter fuse (**Figure 11-64**). With the cigar lighter pushed in, a path to ground is completed. The lights and cigar lighter are now in series; thus, the lights are dim and there is not enough current to heat and release the cigar lighter. If the cigar lighter was left in this position, the battery would eventually drain down.

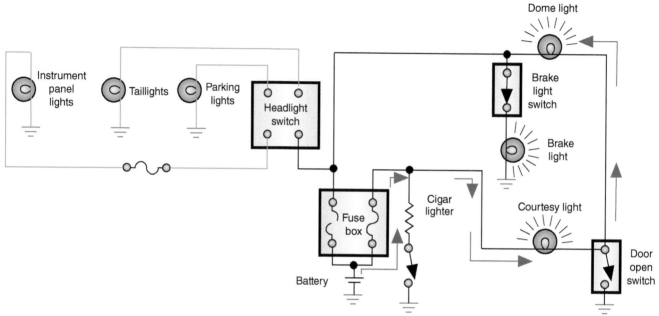

Figure 11-63 Feedback when brake light switch is closed.

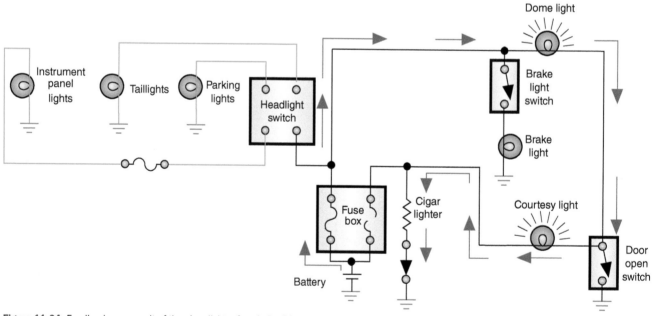

Figure 11-64 Feedback as a result of the cigar lighter fuse being blown.

A blown cigar lighter fuse will also cause the dome light to get brighter when the doors are open, and the courtesy lights will go out (**Figure 11-65**). Also, if the lighter is pushed in and the brake light switch is closed, the dome and courtesy lights will go out (**Figure 11-66**).

Feedback can also be the result of a conductive corrosion that is developed at a connection. If the corrosion allows for current flow from one conductor to an adjacent

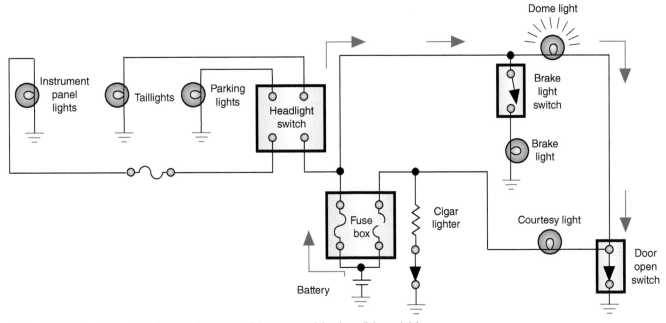

Figure 11-65 Opening the door will make the courtesy light go out and the dome light get brighter.

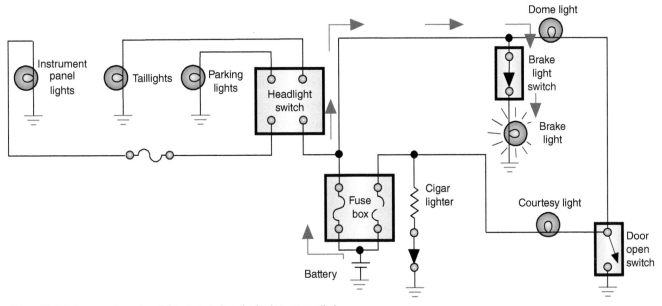

Figure 11-66 Dome and courtesy lights go out when the brakes are applied.

conductor in the connection, the other circuit will also be activated. **Figure 11-67** shows how corrosion in a common connector can cause the dome light to illuminate when the wiper motor is turned on. Because the wiper motor has a greater resistance than the light bulb, more voltage will flow through the bulb than through the motor. The bulb will light brightly, but the motor will turn very slowly or not at all. The same effect will result if the courtesy light switch is turned on with the motor switch off.

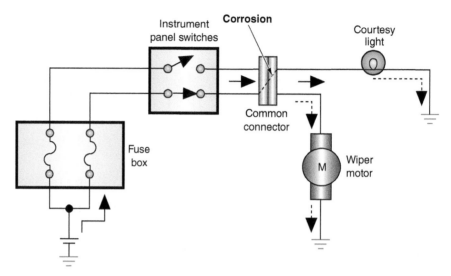

Figure 11-67 A corroded common connection can cause feedback.

AUTHOR'S NOTE Dual-filament light bulb sockets are subject to corrosion that can cause feedback. Also, dual-filament bulbs can have a filament burn out and attach to the other filament and result in feedback. The most common indicator of this problem is parking lights that illuminate when the brake pedal is applied.

FIBER OPTICS DIAGNOSIS

Classroom Manual
Chapter 11, page 341

The fiber-optic system uses plastic strands to transmit light from the source to the object to be illuminated. The strands of plastic are sheathed by a polymer that insulates the light rays as they travel within the strands. The light rays travel through the strands by means of internal reflections. Fiber optics can be used to provide light in areas where bulbs would be inaccessible for service.

If the fiber optics do not illuminate, most likely the light source has failed. Check whether the bulb illuminates. If the bulb turns on, check that the fiber-optic lead is connected to the light source and to the lens. If these are good, the only other cause is that the cable is cut. It will need to be replaced.

CASE STUDY

A customer brings his vehicle into the shop because of an intermittent problem with the headlight delay feature. This person has taken the vehicle to other shops and spent several dollars in repair bills, but the problem has not been corrected. Using a systematic diagnostic approach and following the tests outlined in the service information lead the technician to test the potentiometer in the control switch. The resistance value is within specifications. However, while moving the potentiometer from the MIN to the MAX position, the ohmmeter reading is erratic in one portion. This area is the usual setting selected by the driver. The technician calls the customer and receives approval to replace the control switch. The new switch cures the problem. The technician opens the old switch and finds that carbon from electrical arcing has built up in the problem area of the potentiometer.

ASE-STYLE REVIEW QUESTIONS

1. The results of a functional test on the automatic headlight system are being discussed.

   Technician A says when the photo cell is covered and the engine is running, the headlights should turn on within 30 seconds.

   Technician B says when a bright light is shone onto the photo cell, the headlights should turn off within the specified period of time.

   Who is correct?

   A. A only

   B. B only

   C. Both A and B

   D. Neither A nor B

2. The automatic high-beam function does not operate properly. The LED in the mirror assembly is flashing continuously at a 1-hertz frequency. This can indicate:

   A. The mirror assembly needs to be calibrated.

   B. The system failed its last attempt to calibrate.

   C. A hardware failure has occurred.

   D. None of the above.

3. An indicator that uses fiber optics is not functioning. This can be caused by:

   A. A bent fiber-optics cable.

   B. A faulty light source.

   C. Electromagnetic interference.

   D. None of the above.

4. The most likely cause of the automatic headlights failing to activate in low ambient light conditions is:

   A. A faulty ignition switch.

   B. Camera angle alignment out of specifications.

   C. Burned-out headlight elements.

   D. Faulty headlight switch.

5. *Technician A* says problems with the automatic headlight system can be the fault of the headlight switch.

   Technician B says a bad ignition switch may cause the lights to not come on.

   Who is correct?

   A. A only

   B. B only

   C. Both A and B

   D. Neither A nor B

6. The headlights work in the manual position but do not turn on in the AUTO position. What is the most likely cause?

   A. A bad headlight ground.

   B. A faulty photo cell assembly.

   C. A faulty headlight relay.

   D. A bad headlight relay ground connection.

7. *Technician A* says the photo cell signal voltage should be 5 volts, with the lens covered.

   Technician B says the signal voltage should be 0 volt, with the flashlight shining on the lens.

   Who is correct?

   A. A only

   B. B only

   C. Both A and B

   D. Neither A nor B

8. When the headlights are switched from low beam to high beam, all headlights go out. All of the following can cause this **EXCEPT**:

   A. Faulty dimmer switch.

   B. Defective high-beam relay.

   C. Defective ignition switch.

   D. Open in high-beam circuit.

9. None of the turn signals operate. The **LEAST LIKELY** cause of this is:

   A. Burned-out bulbs.

   B. Faulty turn signal switch.

   C. Faulty turn signal flasher.

   D. Open circuit from ignition feed to turn signal.

10. A customer states that the headlights are brighter than normal and that she has to replace the lamps regularly.

    Technician A says this can be caused by too high charging system output.

    Technician B says this can be caused by excessive voltage drop in the circuit.

    Who is correct?

    A. A only

    B. B only

    C. Both A and B

    D. Neither A nor B

ASE CHALLENGE QUESTIONS

1. The high-beam headlamps of a vehicle equipped with a Twilight Sentinel automatic headlamp system are inoperative; the low beams are working correctly.

   A voltmeter connected across the high-beam contacts of the dimmer switch indicates 0.15 volt when the high beams are "on."

   Technician A says the circuit from the Sentinel amplifier to the dimmer switch has excessive resistance.

   Technician B says the dimmer switch is faulty.

   Who is correct?

   A. A only
   B. B only
   C. Both A and B
   D. Neither A nor B

2. The headlights work normally, except they do not turn on in the automatic mode. What is the **LEAST LIKELY** cause?

   A. Faulty photo cell.
   B. Faulty amplifier.
   C. Faulty headlight switch.
   D. Faulty headlight relay.

3. The brake lights of a vehicle equipped with dual-function turn signal/brake lamps are inoperative. The turn signals are functioning properly. However, there is no power at the brake light terminals when the brake pedal is depressed.

   Technician A says the brake light switch may be faulty.

   Technician B says the turn signal switch may be faulty.

   Who is correct?

   A. A only
   B. B only
   C. Both A and B
   D. Neither A nor B

4. The left turn signals of a vehicle are flashing very slowly; the right turn signals are operating correctly.

   Technician A says this may be caused by excessive circuit resistance on the left turn signal circuit.

   Technician B says this may be occurring because someone may have installed bulbs with higher-than-normal wattage ratings on the left side of the vehicle.

   Who is correct?

   A. A only
   B. B only
   C. Both A and B
   D. Neither A nor B

5. All of the following could cause premature failure of a composite bulb **EXCEPT**:

   A. High charging system voltage.
   B. Excessive bulb circuit ground resistance.
   C. Improper bulb handling.
   D. Cracked lamp housing.

Name _____   Date _____

DIAGNOSING LIGHTING SYSTEMS

Upon completion of this job sheet, you should be able to diagnose the cause of a no light operation, brighter-than-normal, and dimmer-than-normal operation; and determine needed repairs.

NATEF Correlation

This job sheet addresses the following **MLR** tasks:

A.3.	Use wiring diagrams to trace electrical/electronic circuits.
A.4.	Demonstrate proper use of a digital multimeter (DMM) when measuring source voltage, voltage drop (including grounds), current flow, and resistance.
A.5.	Demonstrate knowledge of the causes and effects from shorts, grounds, opens, and resistance problems in electrical/electronic circuits.
A.6.	Check operation of electrical circuits using a test light.
A.11.	Identify electrical/electronic system components and configuration.
E.1.	Inspect interior and exterior lamps and sockets including headlights and auxiliary lights (fog lights/driving lights); replace as needed.
E.3.	Identify system voltage and safety precautions associated with high-intensity discharge headlights.

This job sheet addresses the following **AST/MAST** tasks:

A.3.	Demonstrate proper use of a digital multimeter (DMM) when measuring source voltage, voltage drop (including grounds), current flow, and resistance.
A.4.	Demonstrate knowledge of the causes and effects from shorts, grounds, opens, and resistance problems in electrical/electronic circuits.
A.5.	Demonstrate proper use of a test light on an electrical circuit.
A.7.	Use wiring diagrams during the diagnosis (troubleshooting) of electrical/electronic circuit problems.
E.1.	Diagnose (troubleshoot) the causes of brighter-than-normal, intermittent, dim, or no light operation; determine necessary action.
E.2.	Inspect interior and exterior lamps and sockets including headlights and auxiliary lights (fog lights/driving lights); replace as needed.
E.4.	Identify system voltage and safety precautions associated with high-intensity discharge headlights.

ASE **NATEF**

Tools and Materials

- Vehicle
- Wiring diagram for the vehicle
- Digital multimeter (DMM)
- Test light

Describe the vehicle being worked on:

Year _____ Make _____ Model _____

VIN _____ Engine type and size _____

Procedure   **Task Completed**

1. Record the concern the customer would have with the lighting system.

   _____

   _____

   _____

2. Confirm the customer's concern by performing a check of the system and record your results.

   _____

   _____

   _____

3. Determine if there are any other symptoms that may be related to the customer's concern by testing the operation of other systems. Record your results.

   _____

   _____

   _____

   _____

4. Analyze the symptoms and reference the wiring diagram to determine the most likely location of the fault. Record the component(s) and its (their) location(s).

   _____

   _____

   _____

5. If the vehicle is equipped with HID headlights, record all safety concerns and warnings associated with the system.

   _____

   _____

   _____

   _____

   _____

6. Locate a component from step 4 and return to the vehicle to test it. Describe the test   ☐
   procedure used and record the results.

7. Based on your diagnostic checks, was the cause of the problem located?

   ☐ Yes   ☐ No

   If yes, record your findings.

   _____

   _____

   _____

   If no, continue to test other components identified in step 4.

8. If the fault is determined, make the repair and verify proper operation.

   Was the repair successful?   ☐ Yes   ☐ No

   If no, continue testing the system.

Instructor's Response

_____

_____

_____

Name _____   **Date** _____

CHECKING A HEADLIGHT SWITCH

Upon completion of this job sheet, you should be able to check the operation of a headlight switch with an ohmmeter.

NATEF Correlation

This job sheet addresses the following **MLR** tasks:

A.3.   Use wiring diagrams to trace electrical/electronic circuits.

A.4.   Demonstrate proper use of a digital multimeter (DMM) when measuring source voltage, voltage drop (including grounds), current flow, and resistance.

A.5.   Demonstrate knowledge of the causes and effects from shorts, grounds, opens, and resistance problems in electrical/electronic circuits.

A.6.   Check operation of electrical circuits using a test light.

A.11.   Identify electrical/electronic system components and configuration.

E.1.   Inspect interior and exterior lamps and sockets including headlights and auxiliary lights (fog lights/driving lights); replace as needed.

This job sheet addresses the following **AST** task:

A.10.   Inspect and test switches, connectors, relays, solenoid solid state devices, and wires of electrical/electronic circuits; determine necessary action.

This job sheet addresses the following **AST/MAST** tasks:

A.3.   Demonstrate proper use of a digital multimeter (DMM) when measuring source voltage, voltage drop (including grounds), current flow, and resistance.

A.4.   Demonstrate knowledge of the causes and effects from shorts, grounds, opens, and resistance problems in electrical/electronic circuits.

A.7.   Use wiring diagrams during the diagnosis (troubleshooting) of electrical/electronic circuit problems.

E.1.   Diagnose (troubleshoot) the causes of brighter-than-normal, intermittent, dim, or no light operation; determine necessary action.

E.2.   Inspect interior and exterior lamps and sockets including headlights and auxiliary lights (fog lights/driving lights); replace as needed.

ASE **NATEF**

Tools and Materials

- A vehicle
- A wiring diagram for the vehicle
- A DMM

Describe the vehicle being worked on:

Year _____ Make _____ Model _____

VIN _____ Engine type and size _____

Procedure   **Task Completed**

1. Put the headlight switch in all possible positions and observe which lights are controlled by each position. List each position and the controlled lights here.

_____

_____

_____

_____

_____

2. Locate the headlight switch in the wiring diagram and print out or draw the switch with each possible connection and possible position. Label the lights controlled by each position of the switch. Highlight the path of each position from the power source to ground. Highlight different circuitry with different colors.   ☐

3. Remove the circuit protection device to the headlights or disconnect the battery's negative cable. Remove the headlight switch according to the procedures outlined in the service information. Describe the procedure to your instructor before removing it.

   Instructor's OK to move to the next step. _____.

4. Identify the various terminals of the switch and list the different terminals that should have continuity in the various switch positions.

_____

_____

_____

5. Connect the ohmmeter across these terminals, one switch position at a time, and record your readings.

_____

_____

_____

_____

6. Based on the test, what are your conclusions about the switch?

_____

_____

_____

_____

7. Reinstall the switch and connect the negative battery cable or reinstall the fuse. Then check the operation of the headlights.   ☐

8. Instructor's verification that the vehicle is properly reassembled and the lighting system is operating properly.

   Instructor's OK _____

Instructor's Response

_____

_____

_____

Name _____   **Date** _____

HEADLIGHT AIMING

Upon completion of this job sheet, you should be able to adjust the aim of headlights using portable headlight-aiming equipment.

NATEF Correlation

This job sheet addresses the following **MLR** task:

E.2.   Aim headlights.

This job sheet addresses the following **AST/MAST** task:

E.3.   Aim headlights.

ASE NATEF

Tools and Materials

- A vehicle with adjustable headlights
- Portable headlight-aiming kit
- Hand tools

Describe the vehicle being worked on:

Year _____ Make _____ Model _____

VIN _____ Engine type and size _____

Procedure

Task Completed

1. Describe the type of headlights used on the vehicle.

   _____

   _____

2. Park the vehicle on a level floor.   ☐

   Install the calibrated aiming units to the headlights. (Make sure that the adapters fit the headlight-aiming pads on the lens.)

3. Zero the horizontal adjustment dial. Are the split-image target lines visible in the view port? _____ If the lines cannot be seen, what should you do?

   _____

4. Turn the headlight horizontal adjusting screw until the split-image target lines are aligned. Then repeat this for the other headlight. List any problems you may have had doing this.

   _____

   _____

   _____

5. Turn the vertical adjustment dial on the aiming unit to zero. Turn the vertical   ☐
   adjustment screw until the spirit-level bubble is centered. Recheck your horizontal setting after adjusting the vertical.

6. List any problems you had making the vertical adjustment.

   _____

   _____

   _____

7. If the headlight assembly has four-lamp assemblies, repeat steps 4 and 5 on the other two lamps. List any problems you may have had doing this.

_____

_____

_____

Instructor's Response

_____

_____

_____

Name _____   Date _____

AIMING COMPOSITE OR HID HEADLIGHTS USING AN OPTICAL AIMER

Upon completion of this job sheet, you should be able to adjust the aim of headlights using portable headlight-aiming equipment.

NATEF Correlation

This job sheet addresses the following **MLR** task:

E.2.   Aim headlights.

This job sheet addresses the following **AST/MAST** task:

E.3.   Aim headlights.

Tools and Materials

- A vehicle with adjustable headlights
- Optical headlight aimer
- Hand tools

Describe the vehicle being worked on:

Year _____ Make _____ Model _____

VIN _____ Engine type and size _____

Procedure

1. Describe the type of headlights used on the vehicle.

   _____

   _____

2. Park the vehicle on a level floor. Perform the process of floor slope compensation used by your assigned equipment. Record the dial settings.

   _____

   _____

3. What is the recommended distance from the headlight should the aimer be located?

   _____

   _____

4. Does your assigned equipment require locating the exact center of the grill?

   ☐ Yes   ☐ No

   If yes, perform the procedure; then go to step 5.

   If no, go to step 5.

5. Describe the process for locating the center of the beam.

   _____

   _____

   _____

   _____

6. Set the aligner in the proper position for the center of the beam and have your instructor confirm. Instructor's initials _____

7. If needed, use the alignment lens and two symmetrical locations on the vehicle to complete set up procedure. ☐

8. Turn on the headlights and follow the equipment manufacturer's instructions for determining the current headlight beam aim. Record your results.

_____

_____

_____

9. If needed, adjust headlights.

Instructor's Response

_____

_____

_____

Name _____   Date _____

TESTING AN AUTOMATIC HEADLIGHT SYSTEM

Upon completion of this job sheet, you should be able to diagnose an automatic headlight system and test the individual components of the system.

NATEF Correlation

This job sheet addresses the following **MLR** tasks:

A.3. Use wiring diagrams to trace electrical/electronic circuits.

A.4. Demonstrate proper use of a digital multimeter (DMM) when measuring source voltage, voltage drop (including grounds), current flow, and resistance.

A.5. Demonstrate knowledge of the causes and effects from shorts, grounds, opens, and resistance problems in electrical/electronic circuits.

A.7. Using fused jumper wires, check operation of electrical circuits.

A.11. Identify electrical/electronic system components and configuration.

E.1. Inspect interior and exterior lamps and sockets including headlights and auxiliary lights (fog lights/driving lights); replace as needed.

This job sheet addresses the following **AST/MAST** tasks:

A.3. Demonstrate proper use of a digital multimeter (DMM) when measuring source voltage, voltage drop (including grounds), current flow, and resistance.

A.4. Demonstrate knowledge of the causes and effects from shorts, grounds, opens, and resistance problems in electrical/electronic circuits.

A.6. Use fused jumper wires to check operation of electrical circuits.

A.7. Use wiring diagrams during the diagnosis (troubleshooting) of electrical/electronic circuit problems.

E.1. Diagnose (troubleshoot) the causes of brighter-than-normal, intermittent, dim, or no light operation; determine necessary action.

E.2. Inspect interior and exterior lamps and sockets including headlights and auxiliary lights (fog lights/driving lights); replace as needed.

G.1. Diagnose operation of comfort and convenience accessories and related circuits (such as power windows, power seats, pedal height, power locks, trunk locks, remote start, moon roof, sun roof, sun shade, remote keyless entry, voice activation, steering wheel controls, back-up camera, park assist, cruise control, and auto dimming headlamps); determine needed repairs.

ASE NATEF

Tools and Materials

- A vehicle with an automatic headlight system
- Wiring diagram for the chosen vehicle
- Component locator for the chosen vehicle
- Service information for the chosen vehicle
- A fused jumper wire
- A DMM

Describe the vehicle being worked on:

Year _____ Make _____ Model _____

VIN _____ Engine type and size _____

Procedure

1. Locate the photo cell. Then disconnect the connector to the photo cell. Turn the ignition switch and the automatic headlamp control to ON.

   Turn the headlight switch off.

   If the lights come on within 60 seconds and the automatic headlights don't work as before, the photo cell must be bad. If the lights still don't come on, test the amplifier unit.

   Describe what happened.

   _____

   _____

   _____

   _____

   _____

   _____

2. Turn the ignition off. Disconnect the negative cable from the battery. Locate the amplifier assembly and disconnect the electrical connector to it. Turn the headlight switch off.

   Turn the automatic headlight control to its ON position.

   Locate the resistance checks of the amplifier circuit in the service information. Briefly outline those procedures.

   _____

   _____

   _____

   _____

3. Follow the previously described procedures and list the results.

   _____

   _____

   _____

   _____

4. Turn the automatic headlight control to its ON position.

   Locate the resistance checks of the amplifier circuit in the service information. Briefly outline those procedures.

   _____

   _____

   _____

   _____

5. Follow the procedures in step 4 and list the results.

_____

_____

_____

_____

6. List any additional diagnostic steps that the manufacturer recommends in the case where the previously described tests did not identify the problem.

_____

_____

_____

_____

_____

_____

_____

_____

Instructor's Response

_____

_____

_____

Name _____   Date _____

TESTING THE BCM-CONTROLLED HEADLIGHT SYSTEM

Upon completion of this job sheet, you should be able to test the computer-controlled headlight system and determine needed repairs.

NATEF Correlation

This job sheet addresses the following **MLR** tasks:

A.3.	Use wiring diagrams to trace electrical/electronic circuits.
A.4.	Demonstrate proper use of a digital multimeter (DMM) when measuring source voltage, voltage drop (including grounds), current flow, and resistance.
A.5.	Demonstrate knowledge of the causes and effects from shorts, grounds, opens, and resistance problems in electrical/electronic circuits.
A.6.	Check operation of electrical circuits using a test light.
A.7.	Using fused jumper wires, check operation of electrical circuits.
A.11.	Identify electrical/electronic system components and configuration.

This job sheet addresses the following **AST** task:

G.5.	Describe body electronic systems circuits using a scan tool; check for module communication errors (data bus systems); determine needed action.

This job sheet addresses the following **AST/MAST** tasks:

A.3.	Demonstrate proper use of a digital multimeter (DMM) when measuring source voltage, voltage drop (including grounds), current flow, and resistance.
A.4.	Demonstrate knowledge of the causes and effects from shorts, grounds, opens, and resistance problems in electrical/electronic circuits.
A.5.	Demonstrate proper use of a test light on an electrical circuit.
A.6.	Use fused jumper wires to check operation of electrical circuits
A.7.	Use wiring diagrams during the diagnosis (troubleshooting) of electrical/electronic circuit problems.
E.1.	Diagnose (troubleshoot) the causes of brighter-than-normal, intermittent, dim, or no light operation; determine necessary action.
G.1.	Diagnose operation of comfort and convenience accessories and related circuits (such as: power windows, power seats, pedal height, power locks, trunk locks, remote start, moon roof, sun roof, sun shade, remote keyless entry, voice activation, steering wheel controls, back-up camera, park assist, cruise control, and auto dimming headlamps); determine needed repairs.

This job sheet addresses the following **MAST** task:

G.5.	Diagnose body electronic systems circuits using a scan tool; check for module communication errors (data bus systems); determine needed action.

Tools and Materials

- A vehicle equipped with BCM-controlled headlights
- Scan tool
- DMM
- Test light

- Fused jumper wires
- Wiring diagram for the vehicle

Describe the vehicle being worked on:

Year _____ Make _____ Model _____

VIN _____ Engine type and size _____

Procedure

1. Test the operation of the headlights. Describe the symptoms.

   _____

   _____

2. Are there any other related symptoms? ☐ Yes ☐ No

   If yes, describe the symptom(s).

   _____

   _____

3. Describe the basic operation of the system you are working on.

   _____

   _____

   _____

4. Check any associated fuses and conduct a visual inspection of the system. List any problems found.

   _____

   _____

   Repair any faults found and test system operation. Did this fix the problem? ☐ Yes ☐ No

5. If used, substitute the headlamp relay with a known good relay of the same type and test operation. Test the system operation. Did this fix the problem? ☐ Yes ☐ No

6. Connect the scan tool and record any DTCs.

   _____

   _____

7. Use the scan tool to perform an activation test of the headlights. Did the headlamps come on during the activation? ☐ Yes ☐ No

   What do you know about the problem so far?

   _____

   _____

8. Use the scan tool to monitor the headlight switch as it is placed in all switch positions.

   Record the voltages for each position and compare to specifications.

   _____

   _____

9. What do you know about the system so far?

_____

_____

10. Remove the headlight relay and use a voltmeter or test light to confirm battery voltage is present at pins 30 and 86 of the junction block terminals. Is voltage present at these terminals? ☐ Yes ☐ No

   If *no*, what would you test next?

_____

_____

11. If voltage is present at both terminals in step 10, use a fused jumper wire to connect pin 30 to pin 87. Did the lights come on? ☐ Yes ☐ No

   What do you know about the circuit now?

_____

_____

12. Connect a test light between pins 86 and 85 of the junction block terminals (relay removed). Place the headlight switch into the headlamp position and observe the test light. What does this test check?

_____

_____

13. Based on your results for step 12, what do you know about the problem?

_____

_____

14. What other checks need to be performed to pinpoint the fault?

_____

_____

15. Perform the tests you listed and record your findings and recommendations.

_____

_____

Instructor's Response

_____

_____

_____

Name _____   **Date** _____

DIAGNOSING HSD-CONTROLLED LIGHTING SYSTEMS

Upon completion of this job sheet, you should be able to diagnose the cause or causes of brighter-than-normal, intermittent, dim, or no light operation and determine necessary action for HSD-controlled lighting systems.

NATEF Correlation

This job sheet addresses the following **MLR** tasks:

A.3.   Use wiring diagrams to trace electrical/electronic circuits.

A.4.   Demonstrate proper use of a digital multimeter (DMM) when measuring source voltage, voltage drop (including grounds), current flow, and resistance.

A.5.   Demonstrate knowledge of the causes and effects from shorts, grounds, opens, and resistance problems in electrical/electronic circuits.

A.11.   Identify electrical/electronic system components and configuration.

This job sheet addresses the following **AST** task:

G.5.   Describe body electronic systems circuits using a scan tool; check for module communication errors (data bus systems); determine needed action.

This job sheet addresses the following **AST/MAST** tasks:

A.3.   Demonstrate proper use of a digital multimeter (DMM) when measuring source voltage, voltage drop (including grounds), current flow, and resistance.

A.4.   Demonstrate knowledge of the causes and effects from shorts, grounds, opens, and resistance problems in electrical/electronic circuits.

A.7.   Use wiring diagrams during the diagnosis (troubleshooting) of electrical/electronic circuit problems.

E.1.   Diagnose (troubleshoot) the causes of brighter-than-normal, intermittent, dim, or no light operation; determine necessary action.

E.2.   Inspect interior and exterior lamps and sockets including headlights and auxiliary lights (fog lights/driving lights); replace as needed.

This job sheet addresses the following **MAST** task:

G.5.   Diagnose body electronic systems circuits using a scan tool; check for module communication errors (data bus systems); determine needed action.

Tools and Materials

- Vehicle with HSD-controlled headlights
- Scan tool
- DMM
- Back-probing tools
- Service information

Describe the vehicle being worked on:

Year _____ Make _____ Model _____

VIN _____ Engine type and size _____

Procedure

For this task you will identify the circuits of the front lighting system.

Task One

1. Use the service information to identify the right headlight and park/turn assembly connector and the circuit information. Record the information in the following table:

Connector	CAVITY	Circuit ID	Wire Color
Headlamp low beam			
Low-beam ground			
Headlamp high beam			
High-beam ground			
Park lamp			
Turn signal			
Park/turn ground			

Task Two

For this task, you will identify the operational characteristics of the HSD-controlled lighting system.

1. Connect the scan tool and navigate to the module responsible for headlight control. Which module is responsible? _____

2. Clear any DTCs.   ☐

3. List any actuator tests that are available for the exterior lighting and perform the tests. Record your results.

   _____

   _____

   _____

   _____

   _____

4. With the headlamps turned off, disconnect the right headlamp connector from the low-beam bulb socket. Are any DTCs set?   ☐ Yes   ☐ No

   If so, record them. _____

5. Are there any observable symptoms?

   _____

   _____

6. Turn on the headlamps. Are there any DTCs?   ☐ Yes   ☐ No

   If so, record them.

   _____

   _____

   _____

7. Is there a lamp-out indicator illuminated?   ☐ Yes   ☐ No

8. Turn off the headlights and reconnect the headlamp low-beam connector.

9. With the headlamps off, disconnect the right high-beam connector. Are there any DTCs?   ☐ Yes   ☐ No

If so, record them.

_____

_____

_____

10. Turn on the high beams. Are there any DTCs?   ☐ Yes   ☐ No

If so, record them.

_____

_____

_____

11. Is there a lamp-out indicator illuminated?   ☐ Yes   ☐ No

12. Turn off the headlights and connect the headlamp high beam. Are there any DTCs?

☐ Yes   ☐ No

If so, record them.

_____

_____

_____

13. Turn on the park lamps and right turn signal. Are there any DTCs?   ☐ Yes   ☐ No

If so, record them.

_____

_____

_____

14. Is there a lamp-out indicator illuminated?   ☐ Yes   ☐ No

15. Is there any indication that a lamp is out?   ☐ Yes   ☐ No

If so, what is it?

_____

_____

Task Three

For this task you will determine the voltages applied to the HSD-controlled lighting system.

1. Refer to the table in Task One. Use a DMM to measure the voltages at the headlight assembly connector(s). Measure with the connector disconnected and then with it reconnected. Measure voltage on the headlamps; turn lamp and park lamp circuits. Record the results in the following table:

	Lamp-Connected Volts	**Lamp-Disconnected Volts**
Low beam		
High beam		
Turn		
Park		

2. Explain your results.

_____

_____

_____

_____

_____

3. When are the HSD circuits monitored?

_____

_____

_____

4. When do the DTCs set?

_____

_____

_____

5. Clear all fault codes.   ☐

Instructor's Response

_____

_____

_____

Name _____   Date _____

DIAGNOSING AUTOMATIC HIGH-BEAM SYSTEMS

Upon completion of this job sheet, you should be able to diagnose the cause of poor automatic high-beam system operation and determine needed repairs.

NATEF Correlation

This job sheet addresses the following **MLR** tasks:

A.11.   Identify electrical/electronic system components and configuration.

This job sheet addresses the following **AST** tasks:

G.1.   Describe operation of comfort and convenience accessories and related circuits (e.g., power windows, power seats, pedal height, power locks, trunk locks, remote start, moon roof, sun roof, sun shade, remote keyless entry, voice activation, steering wheel controls, back-up camera, park assist, cruise control, and auto dimming headlamps); determine needed repairs.

G.5.   Describe body electronic systems circuits using a scan tool; check for module communication errors (data bus systems); determine needed action.

This job sheet addresses the following **AST/MAST** tasks:

A.7.   Use wiring diagrams during the diagnosis (troubleshooting) of electrical/ electronic circuit problems.

E.1.   Diagnose (troubleshoot) the causes of brighter-than-normal, intermittent, dim, or no light operation; determine necessary action.

This job sheet addresses the following **MAST** tasks:

G.1.   Diagnose operation of comfort and convenience accessories and related circuits (such as: power windows, power seats, pedal height, power locks, trunk locks, remote start, moon roof, sun roof, sun shade, remote keyless entry, voice activation, steering wheel controls, back-up camera, park assist, cruise control, and auto dimming headlamps); determine needed repairs.

G.5.   Diagnose body electronic systems circuits using a scan tool; check for module communication errors (data bus systems); determine needed action.

ASE NATEF

Tools and Materials

- Vehicle with SmartBeam
- Calibration target
- Grease pencil
- Scan tool
- Tape measure
- Piece of cardboard
- Service information

Describe the vehicle being worked on:

Year _____   Make _____   Model _____

VIN _____   Engine type and size _____

Procedure

Task Completed

For this job sheet task, you will identify the operational characteristics of the SmartBeam automatic high-beam system and calibrate the camera.

1. Turn on the ignition. Observe the SmartBeam camera status LED and determine if the camera is calibrated. Is the SmartBeam camera calibrated?   ☐ Yes   ☐ No

NOTE: If the camera is not calibrated, it must be calibrated prior to entering demonstration mode.

2. Turn off the ignition. Press and hold the AUTO button on the rearview mirror while turning the ignition switch to the RUN position. Continue to depress the AUTO button until the LED blinks and the high beams turn on; then release the button. Describe what happens.

_____

_____

_____

3. Repeat the steps to enter the demonstration mode again. While the lamps are illuminated during the demonstration mode, slide a piece of cardboard between the camera lens and the windshield. Describe the results.

_____

_____

4. Use the service information and determine the alignment specification of the camera. Record the specifications.

_____

_____

5. Connect the scan tool and navigate to the "automatic high-beam module (AHBM)." Locate the camera aim test function and activate the test. Describe the status of the LED.

_____

6. What does this indicate?

_____

7. Turn off the ignition, and use a tape measure and service information procedures to determine the centerline of the camera lens; mark the windshield and rear glass with a grease pencil.   ☐

8. Using a tape measure, determine a location in front of the vehicle which is the specified distance from the vehicle. What is the specified distance? _____

9. Locate the calibration target in front of the vehicle, and align it with the marks on the windshield.   ☐

10. Using a tape measure, determine the distance the target's center LED should be from the floor. Record this distance. _____

11. Adjust the calibration target to the required height.   ☐

12. Turn on the power supply to the calibration target, and then place the ignition switch in the RUN position. Describe the results.

_____

_____

13. Use the scan tool to determine and record the calibration status.   ☐

14. Check for DTCs and record.

_____

_____

_____

15. Clear any DTCs.

Instructor's Response

_____

_____

_____

DIAGNOSTIC CHART 11-1

PROBLEM AREA:	Headlights—park/taillights.
SYMPTOMS:	Bright illumination, early bulb failure.
POSSIBLE CAUSES:	**1.** Alternator output too high. **2.** Defective dimmer switch.

DIAGNOSTIC CHART 11-2

PROBLEM AREA:	Headlights—park/taillights.
SYMPTOMS:	Intermittent headlight or park/taillights operation, headlights flicker.
POSSIBLE CAUSES:	**1.** Defective circuit breaker. **2.** Overload in circuit. **3.** Improper connection. **4.** Defective switch. **5.** Poor ground. **6.** Excessive resistance.

DIAGNOSTIC CHART 11-3

PROBLEM AREA:	Headlights.
SYMPTOMS:	Dim headlight illumination.
POSSIBLE CAUSES:	**1.** Poor ground connection. **2.** Corroded headlight socket. **3.** Poor battery cable connections. **4.** Low generator output. **5.** Loose or broken generator drive belt.

DIAGNOSTIC CHART 11-4

PROBLEM AREA:	Headlights—park/taillights.
SYMPTOMS:	No or improper headlight or park/taillights operation.
POSSIBLE CAUSES:	1. Burned-out headlights. 2. Defective headlight switch. 3. Open circuit. 4. Defective circuit breaker. 5. Overload in circuit. 6. Improper or poor connection. 7. Poor ground. 8. Excessive resistance. 9. Defective relay. 10. Blown fuse. 11. Faulty dimmer switch. 12. Short in insulated circuit. 13. Improper bulb application. 14. Improper headlight aiming.

DIAGNOSTIC CHART 11-5

PROBLEM AREA:	Active Headlights.
SYMPTOMS:	Headlights fail to flow into turn.
POSSIBLE CAUSES:	1. Headlight swivel motor or circuits. 2. Yaw sensor input missing or incorrect. 3. Steering angle sensor input missing or incorrect. 4. Steering angle sensor requires calibration. 5. Active headlight ECU.

DIAGNOSTIC CHART 11-6

PROBLEM AREA:	Instrument cluster lighting.
SYMPTOMS:	Intermittent brightness control of instrument cluster light circuits. Dash lights flicker.
POSSIBLE CAUSES:	1. Improper connection. 2. Defective headlight switch rheostat. 3. Poor ground. 4. Excessive resistance. 5. Faulty printed circuit.

DIAGNOSTIC CHART 11-7

PROBLEM AREA:	Instrument cluster lighting.
SYMPTOMS:	Low-level light intensity from panel illumination lights.
POSSIBLE CAUSES:	1. Burned-out bulbs. 2. Defective headlight switch rheostat. 3. Improper bulb application. 4. Improper connection. 5. Poor ground. 6. Excessive resistance. 7. Defective or faulty printed circuit.

DIAGNOSTIC CHART 11-8

PROBLEM AREA:	Instrument cluster lighting.
SYMPTOMS:	No bulb illumination.
POSSIBLE CAUSES:	**1.** Blown circuit protection device. **2.** Burned-out bulbs. **3.** Defective headlight switch rheostat. **4.** Open circuit. **5.** Improper connection. **6.** Poor ground. **7.** Excessive resistance. **8.** Improper bulb application. **9.** Defective printed circuit.

DIAGNOSTIC CHART 11-9

PROBLEM AREA:	Instrument cluster lighting.
SYMPTOMS:	No dash light brightness control.
POSSIBLE CAUSES:	**1.** Defective headlight switch rheostat.

DIAGNOSTIC CHART 11-10

PROBLEM AREA:	Courtesy lights.
SYMPTOMS:	Intermittent courtesy light operation.
POSSIBLE CAUSES:	**1.** Improper connection. **2.** Defective headlight switch. **3.** Defective door jamb switch. **4.** Defective or sticking door switch. **5.** Poor ground. **6.** Excessive resistance.

DIAGNOSTIC CHART 11-11

PROBLEM AREA:	Courtesy lights.
SYMPTOMS:	Dimmer-than-normal courtesy lights. Battery condition good.
POSSIBLE CAUSES:	**1.** Improper bulb application. **2.** Improper connection. **3.** Poor ground. **4.** Excessive resistance.

DIAGNOSTIC CHART 11-12

PROBLEM AREA:	Courtesy light operation.
SYMPTOMS:	No courtesy light illumination.
POSSIBLE CAUSES:	**1.** Blown circuit protection device. **2.** Burned-out bulbs. **3.** Defective headlight switch. **4.** Defective door switches. **5.** Open circuit. **6.** Improper connection. **7.** Poor ground. **8.** Excessive resistance. **9.** Improper bulb application.

DIAGNOSTIC CHART 11-13

PROBLEM AREA:	Courtesy lights.
SYMPTOMS:	Courtesy lights stay on all of the time.
POSSIBLE CAUSES:	**1.** Defective door jamb switch. **2.** Defective headlight switch. **3.** Shorted circuit.

DIAGNOSTIC CHART 11-14

PROBLEM AREA:	Stop (brake) lamp operation.
SYMPTOMS:	Intermittent stop lamp operation.
POSSIBLE CAUSES:	**1.** Misadjusted brake light switch. **2.** Poor ground connection. **3.** Excessive resistance. **4.** Faulty sockets. **5.** Poor connections. **6.** Faulty turn signal switch contacts. **7.** Defective brake light switch.

DIAGNOSTIC CHART 11-15

PROBLEM AREA:	Stop (brake) lamp operation.
SYMPTOMS:	Dimmer-than-normal stop lights.
POSSIBLE CAUSES:	**1.** Excessive circuit resistance. **2.** Poor ground connection. **3.** Improper bulb application. **4.** Improper connections. **5.** Faulty turn signal switch contacts. **6.** Improper bulb application.

DIAGNOSTIC CHART 11-16

PROBLEM AREA:	Stop (brake) lamp operation.
SYMPTOMS:	No stop lamps illuminate. Stop lights fail to illuminate when the brakes are applied.
POSSIBLE CAUSES:	**1.** Faulty brake light switch. **2.** Open in the circuit. **3.** Improper bulb application. **4.** Faulty turn signal switch. **5.** Improper common ground connection. **6.** Burned-out light bulbs.

DIAGNOSTIC CHART 11-17

PROBLEM AREA:	Turn signal operation.
SYMPTOMS:	Turn signals do not operate in either direction.
POSSIBLE CAUSES:	**1.** Blown fuse. **2.** Defective or worn flasher unit. **3.** Defective or faulty turn signal switch. **4.** Open circuit.

DIAGNOSTIC CHART 11-18

PROBLEM AREA:	Turn signals.
SYMPTOMS:	Turn signal lamp does not illuminate.
	Turn signal indicator illuminates but does not flash.
POSSIBLE CAUSES:	**1.** Improper bulb. **2.** Burned-out bulb. **3.** Open circuit. **4.** Failed flasher unit.

DIAGNOSTIC CHART 11-19

PROBLEM AREA:	Hazard light operation.
SYMPTOMS:	Hazard lights fail to operate when activated.
POSSIBLE CAUSES:	**1.** Blown fuse. **2.** Defective or worn flasher unit. **3.** Defective or faulty hazard light switch. **4.** Open circuit. **5.** Defective turn signal switch.

DIAGNOSTIC CHART 11-20

PROBLEM AREA:	Back-up light operation.
SYMPTOMS:	Back-up lights fail to operate some of the time.
POSSIBLE CAUSES:	**1.** Misadjusted back-up light switch. **2.** Poor ground connection. **3.** Excessive resistance. **4.** Faulty sockets. **5.** Poor connections.

DIAGNOSTIC CHART 11-21

PROBLEM AREA:	Back-up light operation.
SYMPTOMS:	Dimmer-than-normal back-up lights
POSSIBLE CAUSES:	**1.** Excessive circuit resistance. **2.** Poor ground connection. **3.** Improper bulb application. **4.** Improper connections. **5.** Faulty back-up switch contacts.

DIAGNOSTIC CHART 11-22

PROBLEM AREA:	Back-up light operation.
SYMPTOMS:	Back-up lights fail to illuminate when the transmission is in reverse.
POSSIBLE CAUSES:	**1.** Faulty back-up light switch. **2.** Misadjusted back-up light switch. **3.** Blown fuse. **4.** Open in the circuit. **5.** Improper bulb application. **6.** Improper common ground connection. **7.** Burned-out light bulbs.

DIAGNOSTIC CHART 11-23

PROBLEM AREA:	Back-up light does not operate.
SYMPTOMS:	One back-up light fails to illuminate when the transmission is in reverse.
POSSIBLE CAUSES:	**1.** Burned-out lamp. **2.** Loose connection. **3.** Open circuit to lamp.

DIAGNOSTIC CHART 11-24

PROBLEM AREA:	Concealed headlight system.
SYMPTOMS:	Headlight doors fail to open or close.
POSSIBLE CAUSES:	**1.** Faulty headlight switch. **2.** Power feed to relay or relay module. **3.** Faulty relay or relay module. **4.** Faulty door motor. **5.** Open, short, or high resistance in motor control circuits. **6.** Controller power and ground circuits. **7.** Faulty controller.

DIAGNOSTIC CHART 11-25

PROBLEM AREA:	Computer-controlled headlight system.
SYMPTOMS:	Headlights fail to turn on.
POSSIBLE CAUSES:	**1.** Faulty headlight switch. **2.** Power feed to relays. **3.** Faulty relay(s). **4.** Open, short, or high resistance in relay control circuit. **5.** Headlamp circuit failure. **6.** Burned-out headlamp elements. **7.** Controller power and ground circuits. **8.** Faulty controller.

DIAGNOSTIC CHART 11-26

PROBLEM AREA:	Automatic headlight system.
SYMPTOMS:	Headlights fail to turn on in automatic mode; headlights work in manual mode.
POSSIBLE CAUSES:	**1.** Open input circuit from switch to control module. **2.** Faulty module. **3.** Faulty switch.

DIAGNOSTIC CHART 11-27

PROBLEM AREA:	Automatic headlight system.
SYMPTOMS:	Headlights turn on in daytime when switch is in AUTO mode.
POSSIBLE CAUSES:	**1.** Faulty photo cell. **2.** Open photo cell circuit. **3.** Shorted photo cell circuit. **4.** Faulty control module. **5.** Bus communications error. **6.** Immobilizer system inoperative.

DIAGNOSTIC CHART 11-28

PROBLEM AREA:	Automatic high/low beam.
SYMPTOMS:	Headlights fail to automatically switch to high beam.
POSSIBLE CAUSES:	**1.** Obstruction in front of camera. **2.** Improper headlight aiming. **3.** System not initialized. **4.** Camera alignment. **5.** System voltage to module. **6.** Module ground circuit. **7.** Bus network failure. **8.** Faulty controller.

DIAGNOSTIC CHART 11-29

PROBLEM AREA:	Headlight leveling system.
SYMPTOMS:	Headlights fail to level properly.
POSSIBLE CAUSES:	**1.** Improper headlight aiming. **2.** Faulty switch input or circuit. **3.** Faulty headlight level sensor or sensor circuits. **4.** Disconnected link to headlight level sensor(s). **5.** Circuits to motor(s). **6.** Defective motor(s). **7.** System voltage to module. **8.** Module ground circuit. **9.** Bus network failure. **10.** System not calibrated. **11.** Faulty controller.

DIAGNOSTIC CHART 11-30

PROBLEM AREA:	Daytime running lights.
SYMPTOMS:	DRL fail to turn on.
POSSIBLE CAUSES:	**1.** Faulty headlight switch input or circuit. **2.** Power circuit to DRL relay or relay module. **3.** Defective relay or relay module. **4.** Parking brake switch or circuit. **5.** System voltage to module. **6.** Module ground circuit. **7.** Bus network failure. **8.** Faulty controller.

DIAGNOSTIC CHART 11-31

PROBLEM AREA:	Illuminate entry system.
SYMPTOMS:	Courtesy lights fail to turn on.
POSSIBLE CAUSES:	**1.** Faulty headlight switch input or circuit. **2.** Faulty door ajar switches or circuits. **3.** Dead battery in remote FOB. **4.** Defective FOB. **5.** Open, short, or high resistance in the lamp driver circuit. **6.** Power circuit to the module. **7.** Defective relay or relay module. **8.** Module ground circuit. **9.** Bus network failure. **10.** Burned-out lamp elements. **11.** Faulty controller.

DIAGNOSTIC CHART 11-32

PROBLEM AREA:	Illuminate entry system.
SYMPTOMS:	Courtesy lights fail to turn off.
POSSIBLE CAUSES:	**1.** Faulty headlight switch input or circuit. **2.** Faulty door switches or circuits. **3.** Short to ground in the lamp driver circuit. **4.** Power circuit to the module. **5.** Defective relay or relay module. **6.** Module ground circuit. **7.** Bus network failure. **8.** Faulty controller.

CHAPTER 12

INSTRUMENTATION AND WARNING LAMP SYSTEM DIAGNOSIS AND REPAIR

Upon completion and review of this chapter, you should be able to:

- Remove and replace the instrument cluster.
- Remove and replace the printed circuit.
- Diagnose and repair causes for erratic and inaccurate speedometer readings.
- Diagnose and repair causes for tachometer malfunctions.
- Diagnose and repair faulty gauge circuits.
- Diagnose and repair the cause of multiple gauge failure.
- Diagnose sender units, including thermistors, piezoresistive, and mechanical variable styles.
- Diagnose and repair warning light circuits.
- Diagnose and repair the cause of multiple warning light failures.

- Enter diagnostic mode and retrieve trouble codes from BCM-controlled electronic instrument clusters.
- Perform self-diagnostic tests on the electronic instrumentation system.
- Determine faults as indicated by the self-test.
- Diagnose computer-driven speedometer and odometer instrumentation malfunctions.
- Test magnetic pickup speed sensors.
- Test optical-type speed sensors.
- Determine the cause of constant low gauge readings in computer-controlled instrumentation systems.
- Diagnose and locate the cause for constant high gauge readings in computer-controlled instrumentation systems.
- Diagnose head-up display operation.

Basic Tools

Basic mechanic's tool set
Service information

Terms To Know

Bimetallic gauges
Electromagnetic gauges
Gauge

Odometer
Prove-out circuit
Quartz swing needle

Resistive shorts
Tachometer

INTRODUCTION

The instrument panel gauges and warning lamps monitor the various vehicle operating systems and provide information to the driver about their operation. Most problems in the gauges or warning lamps are caused by an open circuit in the wiring, open circuit in the printed circuit board, improper gauge calibration, loose connections, or excessive circuit resistance. In addition, other causes include defective bulbs, gauges, or sending units.

In this chapter you will learn how to diagnose the gauge, lamp, and sending unit of the various styles of conventional instrument panels. You will learn how to remove the instrument panel to replace the printed circuit, and how to repair the speedometer cable core.

This chapter will also introduce you to common service procedures used to diagnose and repair computer-driven instrumentation systems. These systems include the speedometer, odometer, fuel, and engine instrumentation. The computer-driven instrument cluster uses a microprocessor (μP) to process information from various sensors and to control the gauge display. Depending on the manufacturer, the microprocessor can be a separate computer that receives direct information from the sensors and makes the calculations, or the body control module (BCM) is used to perform all functions. The computer may control a digital or a quartz swing needle instrument cluster.

It is out of the scope of any textbook to cover service procedures for every type of instrument cluster. To illustrate, in one model year alone, Chrysler offered two different electronic clusters (Huntsville and Motorola). The Huntsville cluster was available in four different variations, one with a message center, one with a trip computer, and two options that offered tachometers. In addition, the system could also have a 24 voice alert function. In the same year, Ford offered five different electronic clusters, and each division of General Motors (GM) offered its own electronic instrument cluster. Add to this the many different types of import vehicles that use their own systems. This chapter will familiarize you with general procedures; however, it is important to remember that each system uses its own diagnostic procedures.

Usually the technician is required to isolate the faulty component and replace it. Most instrument panel components are not repaired or serviced in the shop but sent back to the manufacturer or specialty shop for rebuilding.

INSTRUMENT PANEL AND PRINTED CIRCUIT REMOVAL

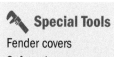

Special Tools

Fender covers
Safety glasses
Battery terminal
Pliers

Many times it may be necessary to remove the instrument panel to replace defective gauges, lamps, or printed circuits. Before removing the instrument panel, always disconnect the negative battery cable. Consult the service information for the procedure of the vehicle you are working on. The following is a common method of removing the instrument cluster and printed circuit:

1. Place fender covers on the fenders and disconnect the battery negative cable.
2. Remove the retaining screws to the steering column cover. Then remove the cover.
3. Remove the finish-panel retaining screws. On some models it may be necessary to remove the radio knobs.
4. Remove the finish panel.
5. Remove the retaining bolts that hold the cluster to the dash.
6. Reach behind the instrument panel and disconnect the speedometer cable.
7. Gently pull the cluster away from the dash.
8. Disconnect the cluster feed plug from the printed circuit receptacle. Be careful not to damage the printed circuit.
9. Remove the instrument voltage regulator (IVR) and all illumination and indicator lamp sockets (**Figure 12-1**).
10. Remove the charging system warning lamp resistor if applicable.
11. Remove all printed circuit attaching nuts and remove the printed circuit.

Many instrument clusters must be replaced as a unit and do not allow for the servicing of individual gauges, lamps, or the circuit board. Since the cluster may also be a module, it is important to check the service procedures before replacing the cluster. It may be necessary to retrieve data from the cluster that must be transferred to the new cluster.

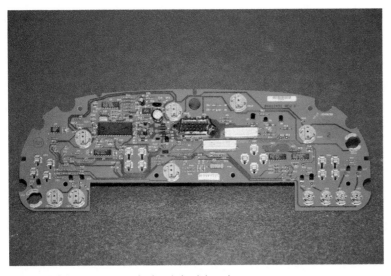

Figure 12-1 Instrument panel printed circuit board.

This data may include mileage, vehicle identification number (VIN), theft deterrent codes, vehicle build configurations, and so forth. If the data transfer procedure is not properly done, the vehicle may not start.

BCM DIAGNOSTICS

The BCM may be capable of running diagnostic checks of the electronic instrument cluster to determine if a fault is present. If the values received from monitored functions are outside of programmed parameters, a diagnostic trouble code (DTC) is set. This code can be retrieved by the technician to aid in troubleshooting. Depending upon the vehicle, code retrieval is done through the ECC, instrument panel cluster (IPC), jumping terminals in the data link connector (DLC), or by a scan tool.

To retrieve diagnostic codes in some GM vehicles with an electronic climate control (ECC) display, turn the ignition switch to the RUN position and simultaneously press the OFF and WARMER buttons on the climate control panel. Engine control module (ECM) codes will be displayed first, followed by BCM codes. The system will display codes twice. The first pass is all codes in memory. Codes that are in the first set but not in the second are history codes. All codes displayed during the second pass are current codes.

After the trouble codes have been retrieved from memory, refer to the proper diagnostic chart to isolate the fault. It is possible for a problem to exist that does not set a trouble code. In these instances, use the symptom or troubleshooting chart in the service information to locate the fault.

Some manufacturers provide a means of overriding the instrument cluster display and change the parameters to allow for testing. By changing the parameters in this test mode, the gauge changes its indicated reading. If the gauge changes its readings correctly, the fault is in the control module.

Classroom Manual
Chapter 12,
pages 357, 362

Special Tool

Scan tool

SELF-DIAGNOSTICS

Most instrument panel display modules have a diagnostic mode within their programming. The diagnostic mode allows the module to isolate any faults within the instrument panel cluster. In most systems, if the module is not able to complete its self-diagnostic test, the fault is within the module and it must be replaced. Successful completion of the self-diagnostic test indicates the problem is not in the module. The following are examples of self-diagnostic procedures.

Diagnosis of a Typical Electronic Instrument Cluster

All electronic instrument clusters (EICs) are sensitive to static electricity damage, and EIC cartons usually have a static electricity warning label. When servicing EICs:

1. Do not open the EIC carton until you are ready to install the component.
2. Ground the carton to a known good ground before opening the package.
3. Always touch a known good ground before handling the component.
4. Do not touch EIC terminals with your fingers.
5. Follow all service precautions and procedures in the vehicle manufacturer's service information.

Prove-Out Display. Most EICs have a prove-out display each time the ignition switch is turned on. During this display, all the EIC segments are illuminated and then turned off momentarily (**Figure 12-2**). The EIC returns to normal display after the prove-out. If the EIC is not illuminated during the prove-out display, check the power supply and grounds to the EIC. If these are good, replace the EIC.

If some of the segments do not illuminate during the prove-out display, the EIC is defective and must be replaced. During the prove-out mode, the turn signal and high-beam indicators are not illuminated. Other indicator lights remain on when the EIC display is turned off momentarily in the prove-out mode. After the prove-out mode is completed, the indicator lights go out shortly after the EIC returns to normal display.

Function Diagnostic Mode. The diagnostic procedure for EICs varies depending on the vehicle makes and model year. Always follow the diagnostic procedures in the vehicle manufacturer's service information. Some EICs have a function diagnostic mode that provides diagnostic information in the display readings if certain defects occur in the system. For example, if the coolant temperature sender has a shorted circuit, the two top and bottom bars are illuminated in the temperature gauge and the ISO symbol is extinguished (**Figure 12-3**). If the engine coolant never reaches normal operating temperature

Caution

VFD displays are easily damaged by physical shock. When handling EICs, do not drop or jar them.

Caution

When servicing EICs, follow all service precautions related to static discharge in the vehicle manufacturer's service information to avoid EIC damage.

Figure 12-2 All EIC segments are illuminated during the prove-out display.

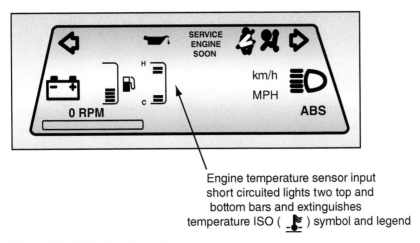

Engine temperature sensor input
short circuited lights two top and
bottom bars and extinguishes
temperature ISO ( ) symbol and legend

Figure 12-3 EIC function diagnostic mode.

or the coolant temperature sender circuit has an open circuit, the bottom bar in the temperature gauge is illuminated with the ISO symbol.

If the fuel gauge sender develops a short or open circuit, the two top and bottom bars in the fuel gauge are illuminated and the ISO symbol is not illuminated. A shorted fuel gauge sender causes CS to be displayed in the fuel remaining or distance to empty displays. If the fuel gauge sender has an open circuit, CO is displayed in the fuel remaining and distance to empty displays. When the function diagnostic mode indicates short or open circuits in the inputs, the cause of the problem must be located by performing voltmeter and ohmmeter tests in the circuit with the indicated problem. These voltmeter and ohmmeter tests are included in the vehicle manufacturer's service information.

When the word *ERROR* appears in the odometer display, the EIC computer cannot read valid odometer information from the nonvolatile memory chip.

Special Test Mode. Many EICs have a special test mode to determine if the display is working properly. Typically this mode is entered by pressing buttons on the EIC. For example, on some systems this mode is entered when the E/M and SELECT buttons are pressed simultaneously and the ignition switch is turned from the OFF to the RUN position. When this action is complete, a number appears in the speedometer display and two numbers are illuminated in the odometer display. The gauges and message center displays are not illuminated. If any of the numbers are flashing in the speedometer or odometer displays, the EIC is defective and must be replaced.

⚠ WARNING **If the odometer has been repaired or replaced and it cannot indicate the same mileage as before it was removed, in most areas the law requires that an odometer mileage label must be attached to the left front door frame. Failure to comply with this procedure could lead to court action.**

Electromechanical Cluster Self-Test

Like the EIC test, the electromechanical instrument cluster (MIC) self-test is usually initiated by button pushes and cycling of the ignition switch. **Photo Sequence 26** illustrates a procedure for entering the MIC self-test and the resulting displays as the test is run. During the actuator test, the circuitry positions each of the gauge needles at various calibration points, illuminates each of the segments in the VFD units, and turns all of the indicators ON and OFF again.

PHOTO SEQUENCE 26
Self-Test of a MIC

P26-1 With the ignition switch in OFF position, press and hold the trip odometer reset button.

P26-2 While continuing to hold the reset button, turn the ignition switch to the RUN position.

P26-3 Release the odometer/trip odometer switch button. The instrument cluster will perform a bulb check of each LED indicator, and each gauge needle is swept to its high calibration point.

P26-4 The VFD segments and LED indicators remain illuminated as the gauges move to their next calibration points.

P26-5 As the gauges move to additional calibration points, the VF displays perform a segment test.

P26-6 The gauges will continue to move to each calibration point.

P26-7 The gauges then move to their low calibration points.

P26-8 The VF display continues to perform its segment test.

P26-9 When the test is completed, the cluster will automatically exit the self-diagnostic mode and return to normal operation.

Confirmation of proper cluster operation indicates that there may be problems elsewhere in the system. Use the service information to determine where each message or hard wire input originates. Problems with incorrect or erratic gauge and indicator operation may be caused by faults in the data bus, the originating control module, or the inputs to one of the control modules. Use a diagnostic scan tool to diagnose these components.

Chrysler Electromechanical Cluster Self-Test

Many Chrysler MICs can be tested with or without a scan tool. Usually a self-test of the cluster can be performed. This test does not check any of the inputs, only the cluster.

To enter self-diagnostics, the ignition key must be in the LOCK position. Push and hold the TRIP and RESET buttons on the cluster at the same time. While continuing to hold these buttons, turn the ignition switch to the RUN position. Note that the cluster will illuminate in the UNLOCK position but will not activate self-diagnosis. Continue to hold the two buttons until CODE is displayed in the odometer. Release the buttons. If there are any fault codes, they will be displayed in the odometer. A code 999 means there are no faults. If fault codes are present, use the correct diagnostic manual to diagnose the system.

After the codes are displayed, the cluster goes through a series of tests as follows:

Check 0. Tests all of the vacuum fluorescent (VF) display segments in the odometer and PRND3L. All segments should be illuminated.
Check 1. Tests the operation of all gauges. The gauge swing needles will move to programmed values.
Check 2. Illuminates each odometer VF segment individually.
Check 3. Tests the PRND3L display.
Check 4. Illuminates all of the warning lamps that are controlled by the MIC.

Observe operation of the MIC during each test. If any of these tests fail proper operation, the MIC must be replaced.

An additional feature of this cluster is that the technician can calibrate the gauges. This requires the use of Chrysler's scan tool. The following steps provide a guide to recalibrating the speedometer. All gauges are calibrated in the same manner.

Special Tool

Scan tool

1. Plug the scan tool cable into the DLC. The ignition switch does not need to be in the RUN position. However, the bus must be active. Opening a door will awake the BCM.
2. Once the scan tool powers up the MAIN MENU is displayed.
3. Select BODY.
4. Select ELECTRO/MECH CLUSTER from the BODY menu.
5. Select MISCELLANEOUS.
6. Select CALIBRATE GAUGES.
7. The scan tool will ask if the cluster has a tachometer. Answer with the YES or NO key.
8. The scan tool will ask if the vehicle is a diesel. Answer with the YES or NO key.
9. The scan tool will ask if the cluster units are in MPH. Answer with the YES or NO key.
10. Place the ignition switch into the UNLOCK position.

This places the scan tool in the mode to run gauge calibration. The first gauge to be calibrated will be the speedometer. The scan tool screen displays that it is sending a signal to the MIC to set the mph at 0. If the needle is not aligned with the 0, then use the up or down arrow keys to move the needle until it is aligned. Once the gauge is calibrated to 0, press the enter key and the scan tool will move to the next calibration unit. The next unit is 20 mph. Follow the same procedure to align the needle with the 20-mph mark on the cluster. After each calibration, press the enter key to move to the next unit. The other calibration units are 55 and 75 mph.

The tachometer, fuel, and temperature gauges are calibrated in the same manner. Once all of the gauges are calibrated, the scan tool instructs the MIC to write the new values to memory.

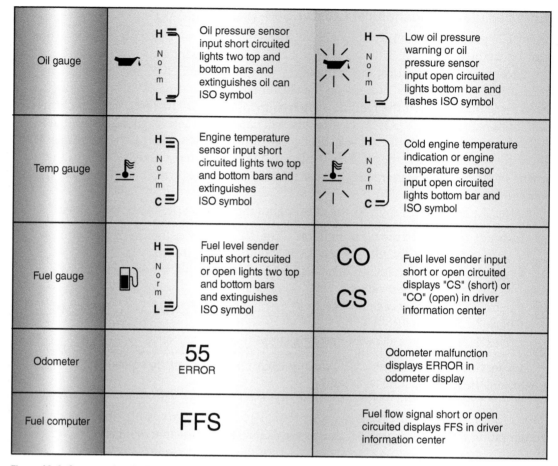

Figure 12-4 Gauge readout indicates the nature of the fault when the system is in the diagnostic mode.

Classroom Manual
Chapter 12, page 355

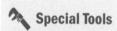 **Special Tools**

Fender covers
Safety glasses
Battery terminal pliers
Terminal pliers

 Caution

It is possible to short out wires while reaching behind the instrument panel. Disconnect the battery negative cable before removing the speedometer cable assembly.

Ford Electronic Cluster Self-Diagnostic Test

The electronic cluster is capable of indicating a fault and providing an explanation of the cause. Use **Figure 12-4** as a guide to the function of the gauges when in diagnostic mode. Use the gauge display to determine the nature of the fault. Then refer to the service information for diagnostic charts to locate the problem.

Speedometer Diagnosis and Repair

Speedometer complaints range from chattering noises when cold, to inaccurate readings, to not operating at all. Diagnosis and repair of the speedometer depends on the type: conventional or electronic.

Conventional Speedometer

In instrument clusters that use conventional speedometers, often the problem of noise, erratic, or inaccurate readings can be corrected by lubricating the cable with an approved lubricant. If the cable is dry, it will bind as it attempts to rotate in the housing. However, the cause of the noise must be isolated since just applying lubricant may only stop the noise temporarily. It is a good practice to remove the cable core to clean and inspect it before adding lubricant. If the cable is well lubricated and the problem is still present, check the condition of the speedometer drive gear. If the cable and the gears are not the problem, the speedometer head may be faulty and need to be replaced.

To remove the cable core, disconnect the speedometer cable assembly from the back of the speedometer head. For most vehicles, this is done by reaching behind the

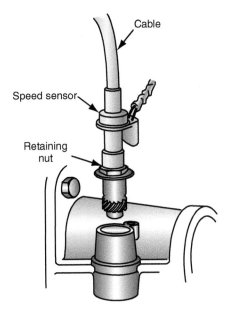

Figure 12-5 Speedometer connection at the transaxle.

instrument panel and pressing down on the flat surface of the plastic quick connect. On some vehicles, it may be necessary to remove the instrument panel to gain access to the speedometer cable.

CUSTOMER CARE The speedometer cable should be lubricated every 10,000 miles. This practice reduces speedometer cable problems that cause noisy, erratic, or inaccurate readings.

It may be possible to remove the core by pulling it out of the housing. If the core cannot be removed in this way, disconnect the speedometer retainer from the transaxle (**Figure 12-5**). Pull the core out of the speed sensor. With the cable core removed from the housing, clean it with solvent and wipe it dry. Place the core on a flat surface and stretch it out straight. Roll it back and forth while looking for signs of kinks or other damage. If the core is damaged, it must be replaced.

If the speedometer assembly needs to be replaced, usually a new **odometer** is included with the assembly. The odometer is a mechanical counter in the speedometer unit that indicates total miles driven by the vehicle. Be sure to follow the manufacturer's procedures for setting the odometer to the correct reading.

Classroom Manual
Chapter 12, page 362

Electronic Speedometers and Odometers

Diagnosis of the sending units and input circuits to the gauges of electromechanical instrument clusters typically follow normal testing procedures. The following is an example of diagnosing the **quartz swing needle** electronic speedometer and odometer gauges of an electromechanical cluster (**Figure 12-7**). Computer-driven quartz swing needle displays are similar in design to the air-core electromagnetic gauges used in conventional analog instrument panels.

Classroom Manual
Chapter 12,
pages 355, 360, 361

🔧 **SERVICE TIP** If the electronic speedometer uses a cable-driven sensor, the cable may be serviced as described in the previous section.

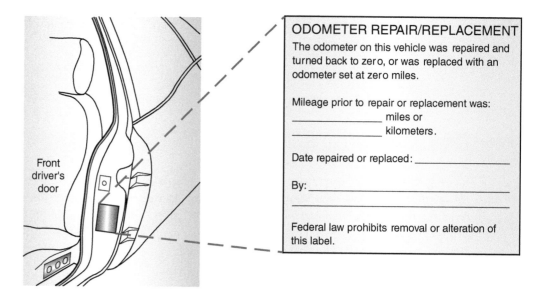

Figure 12-6 An odometer repair label.

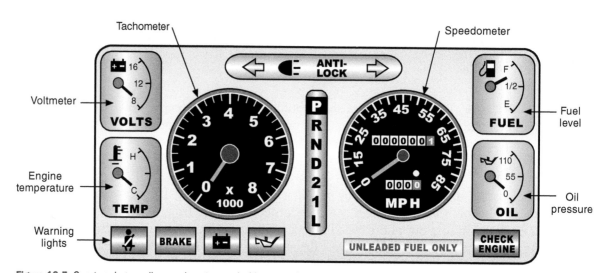

Figure 12-7 Quartz swing needle speedometer used with conventional gauges.

On most systems, the odometer and speedometer receive their input from the vehicle speed sensor. If there is a fault with the speed sensor, other systems (such as cruise control) will also be affected. When test driving the vehicle, attempt to activate the cruise control system to determine if it is operating properly. If the cruise control system fails, the problem can be the speed sensor, its circuit, BCM, or the ECM.

Some systems may provide for replacement of the stepper motor or odometer chip separate of the cluster. Always refer to the service information for the vehicle being diagnosed.

Generally, if the BCM and/or cluster module pass their self-diagnostic tests, the fault will be in the speed sensor circuit. Common test procedures for the speed sensor are presented later in this chapter.

If the speedometer is not operating, but the odometer works properly, the fault is in the instrument cluster. Likewise, if the speedometer operates but the odometer fails, the fault is in the cluster. In either case, the cluster must be replaced. If the speedometer and/ or odometer are inaccurate, or both do not operate, check the following items:

1. If the system uses an optical vehicle speed sensor, check the speedometer cable for kinks, twists, or other defects that will cause an inaccurate reading.
2. See if there are shorts in the wiring circuit of the speed sensor.
3. Make sure of proper gear ratio and tire size. Both of these items will affect correct speedometer and odometer operation. Changing tire size from that intended by the manufacturer has the same effect as changing gear ratios in the differential.

If the speedometer and/or odometer display illuminates but remains at 0 (or any other digit), or operates erratically or intermittently, check the connector at the speed sensor for proper installation and corrosion. Next, check the wiring circuit for any shorts or opens. If these tests do not isolate the problem, the speed sensor should be tested. Testing of the speed sensor will depend on the type of sensor used.

Magnetic Pickup Speed Sensor Testing. Disconnect the wire connector at the vehicle speed sensor. With the ignition switch placed in the RUN position, use a jumper wire to make and break the connection between the two wires (**Figure 12-8**). This should cause the speedometer display to change. Change the rate of speed at which you make and break the connection and the display should indicate the changes in speed. The faster you make and break the connection, the higher the speedometer reading.

If there is no change in the speedometer display, check for opens and shorts in the sensor circuit. If the cluster passes its self-test (and you did not skip any steps), this is the only area in which the fault can be located.

If the speedometer changes speed, the problem is in the speed sensor. To test the speed sensor, remove it from the transaxle. Connect an ohmmeter to the connector terminals of the sensor and select the lowest scale (**Figure 12-9**). Rotate the sensor gear while observing the ohmmeter. Distinct pulses should be detected on the ohmmeter. Compare the number of pulses per revolution with specifications. Also, compare the resistance value with specifications. If the number of pulses and resistance values are within specifications, the sensor is good.

Classroom Manual
Chapter 12, page 356

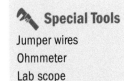

Special Tools

Jumper wires
Ohmmeter
Lab scope

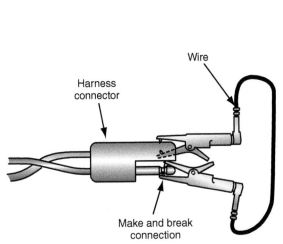

Figure 12-8 Testing the magnetic pickup speed sensor circuit. Making and breaking the connection should produce a reading in the speedometer window.

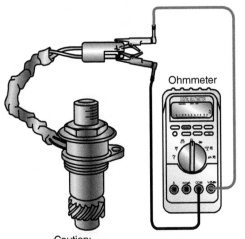

Caution:
Do not use a test light to check the distance sensor. Damage to the sensor may result.

Figure 12-9 Using an ohmmeter to test the speed sensor.

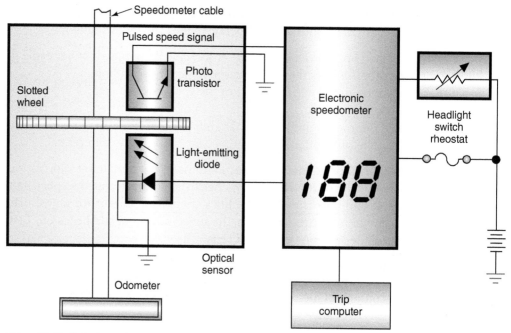

Figure 12-10 Optical speed sensor circuit.

Optical Speed Sensor Testing. Disconnect the speedometer cable at the transaxle and rotate the cable in its housing as fast as possible. If the speedometer display operates properly, check the speedometer pinion and drive gear for damage.

If there is no speedometer operation, check for a broken speedometer cable. This can usually be determined by feeling for resistance while turning the cable by hand. Little resistance indicates a broken cable. Excessive resistance indicates a damaged cable or sensor head.

Classroom Manual
Chapter 12, page 361

If the cable is good, the problem is in the sensor or in the wiring between the sensor and the speedometer. Follow the manufacturer's procedure for removing the instrument cluster. Connect a digital multimeter (DMM) to read the pulsed speed signal from the sensor (**Figure 12-10**). Rotate the speedometer cable while observing the voltmeter. Compare the pulses per revolution and pulse output values with specifications. Replace the sensor if the values are not within specifications.

Hall-Effect Sensors. In Chapter 9, the operation of the Hall-effect sensor was covered. As discussed, using a lab scope provides a fast and accurate test of the switch. Refer to Chapter 9 for the test procedures of this type of speed sensor.

> **SERVICE TIP** A reversible, variable speed drill can be used to rotate the speedometer cable if you are sure there are no twists or kinks in the cable.

> **CUSTOMER CARE** Often inaccurate speedometers are caused by improper tire sizes being used on the vehicle. The larger the tire diameter over original equipment, the slower the speedometer will read. Not all manufacturers provide for entering different tire sizes into the computer, and aftermarket "black boxes" may be illegal to install, depending on state and local ordinances. In most cases, it is in the customer's best interest to keep original equipment size tires on the vehicle.

TACHOMETER

The **tachometer** is a gauge instrument used to display the speed of the engine in revolutions per minute (rpm). Most electrically operated tachometers receive their reference pulses from the ignition system. **Figure 12-11** illustrates a troubleshooting chart to use as a guide to diagnosing the electrically operated tachometer. If the tachometer is faulty, it must be replaced; there is no servicing the meter itself.

Computer-driven tachometers receive their signals from the crankshaft position sensor (CKP). If the tachometer is not functioning, but the engine starts, check the circuit

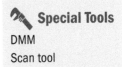

Classroom Manual
Chapter 12, page 364

Special Tools
DMM
Scan tool

TACHOMETER INOPERATIVE, ERRATIC, WRONG OPERATION

Test step	Result	Action to take
1. **Check operation** Check tachometer operation.	⊗ OK	Go to 2.
	OK	Test complete. Check for intermittent operation.
2. **Check fuse** Check tachometer fuse.	⊗ OK	Replace fuse and determine cause.
	OK	Go to 3.
3. **Check wiring** Check connections and wiring in engine compartment and at instrument cluster.	⊗ OK	Repair connections or wiring.
	OK	Go to 4.
4. **Check resistance and voltage** Disconnect battery. Remove instrument cluster and make resistance and voltage checks at lower wire harness connector as follows: (1) Check pin 5 resistance to ground. Should read 1 Ω or less. (2) Check pin 17 resistance to negative terminal of ignition coil. Should read 15 Ω or less. (3) Connect battery. Turn ignition switch to RUN position. Check for +12 V at pin 14. Turn ignition off and disconnect battery.	⊗ OK	Repair wiring for open or high resistance.
	OK	Go to 5.
5. **Check fasteners** Check for loose fasteners on rear of instrument cluster, or damaged printed circuit.	⊗ OK	Tighten fasteners/Replace printed circuit.
	OK	Replace tachometer.

Tach signal
+12 V to tach   Ign coil neg (−)

Tach ground

Figure 12-11 Troubleshooting chart for testing the tachometer.

from the sensor to the instrument cluster. If the instrument cluster receives the signal from the data bus, use the scan tool to determine if the signal is being sent correctly. Usually the CKP sensor is monitored by the powertrain control module (PCM), and then sent on the data bus. If the message is sent to the BCM, then to the instrument cluster, check the input message to both modules. If the bus message is being sent properly, the problem is the gauge or the instrument cluster. Perform the self-diagnostic test, or scan tool activation test, for the tachometer. If the tachometer fails to operate during this test, replace the gauge or cluster. Some manufacturers do not allow for individual gauge replacement; in such cases, the entire instrument cluster must be replaced.

CONVENTIONAL INSTRUMENT CLUSTER GAUGE DIAGNOSIS

Classroom Manual
Chapter 12, page 351

 Caution

These gauges are called analog because they use needle movement to indicate current levels. However, many modern instrument panels use computer-driven analog gauges that operate under different principles. It is important that the technician follow the manufacturer's procedures for testing the gauges or else gauge damage will result.

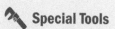 **Special Tools**

DMM
12-volt test light
Fender covers
Safety glasses
Shop oil pressure
Gauge

A **gauge** is a device that displays the measurement of a monitored system by the use of a needle or pointer that moves along a calibrated scale. The sender unit is the sensor for the gauge. It is a variable resistor that changes resistance values with changing monitored conditions.

The IVR provides a constant voltage to the gauge regardless of the voltage output of the charging system. The gauge is called an electromechanical device because it is operated electrically, but its movement is mechanical.

With the exception of the voltmeter and ammeter, all electromechanical gauges (whether **bimetallic** or **electromagnetic gauges**) use a variable resistance sending unit. Bimetallic gauges (or thermoelectric gauges) are simple dial and needle indicators that transform the heating effect of electricity into mechanical movement. Electromagnetic gauges produce needle movement by magnetic forces instead of heat. The types of tests performed depend on the nature of the problem and if the system uses an IVR.

> ⚙ SERVICE TIP   It is common for the fuel gauge sender unit to get corrosion on the ground wire connection. Before replacing the sending unit, clean the ground connections and test for proper operation.

Single Gauge Failure

If the gauge system does not use an IVR, check the gauge for proper operation as follows:

1. Check the fuse panel for any blown fuses. The gauge that is not operating may share a fuse with some other circuit that is separate from the other gauges.
2. Disconnect the wire connector from the sending unit of the malfunctioning gauge.
3. Check the terminal connectors for signs of corrosion or damage.
4. Use a test light to confirm that voltage is present to the connector with the ignition switch in the RUN position. If the test light does not illuminate, check the circuit back to the gauge and battery.
5. Connect a 10-ohm resistor in series with the lead from the gauge to the sending unit. Connect the lead to ground (**Figure 12-12**).
6. With the ignition switch in the RUN position, watch the gauge. Depending on the gauge design, the needle should indicate either high or low on the scale. Check the service information for correct results.
7. Remove the 10-ohm resistor and replace with a 73-ohm resistor between the sensor lead and ground. Repeat step 6.

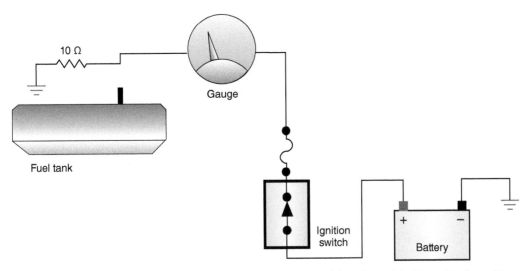

Figure 12-12 Testing gauge operation by putting the gauge lead to ground through a resistor. The resistor is used to protect the circuit.

8. If the test results are in the acceptable range, the sending unit is faulty.
9. If the gauge did not operate properly in steps 5 and 7, check the wiring to the gauge. If the wiring is good, replace the gauge.

If the gauge circuits use an IVR, follow steps 1 through 4 as described. The test light should flicker on and off. If it does not illuminate, reconnect the sending unit lead and check for voltage at the sender unit side of the gauge (**Figure 12-13**). If there is voltage at this point, repair the circuit between the gauge and the sending unit. If voltage is not shown, test for voltage at the battery side of the gauge. If voltage is present at this point, the gauge is defective and must be replaced. If no voltage is present at this terminal, continue to check the circuit between the battery and the gauge.

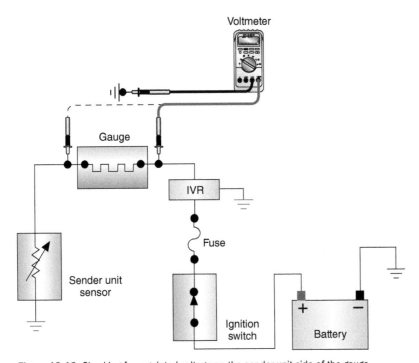

Figure 12-13 Checking for regulated voltage on the sender unit side of the gauge.

If the IVR is working properly and voltage was present to the sender unit, follow steps 5 through 9.

> ⚙ **SERVICE TIP** It is unlikely that all of the gauges would fail at the same time. If the diagnostic tests indicate that the gauges are defective, bench test the gauges before replacing them. Use an ohmmeter to check the resistance. Most electromagnetic gauges should read between 10 and 14 ohms. On systems that do not use an IVR and all of the gauges are defective, check the charging system for excessive output.

Special Tools

12-volt test light
10-ohm resistor
73-ohm resistor
Fender covers
Safety glasses

Multiple Gauge Failure

If all gauges fail to operate properly, begin by checking the circuit fuse. Test for voltage to the fuse. If voltage is not present at this point, the fault is between the fuse and the battery. Remember, most systems supply battery voltage to the instrument panel gauges through the ignition switch. If voltage is present at the fuse, then continue through the circuit by testing for voltage at the last common circuit point (**Figure 12-14**). If voltage is not present at this point, work toward the fuse to find the fault. Keep in mind that this common connection point may be on the printed circuit board. If this is the case, test for voltage at the instrument panel connector first.

If the system uses an IVR, use a voltmeter to test for regulated voltage at a common point to the gauges (**Figure 12-15**). If the voltage is out of specifications, check the ground circuit of the IVR. If that is good, replace the IVR. If there is no voltage present at the common point, check for voltage on the battery side of the IVR. If voltage is present at this point, then replace the IVR. If battery voltage is not present on the battery side of the IVR, the problem is in the circuit between the fuse and the IVR.

If regulated voltage is within specifications, test the printed circuit from the IVR to the gauges. If there is an open in the printed circuit, replace the board.

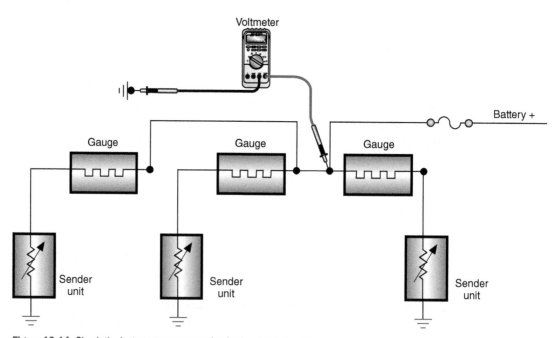

Figure 12-14 Check the last common connection in the circuit for voltage.

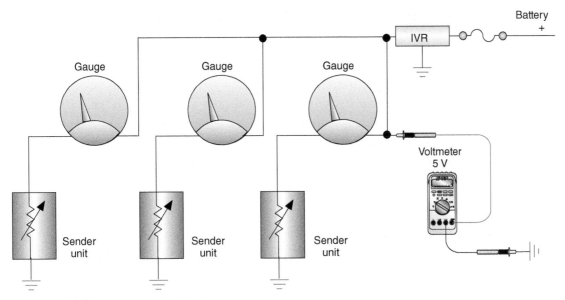

Figure 12-15 Testing for correct IVR operation.

ELECTRONIC GAUGE DIAGNOSIS

The following are guidelines for testing the individual gauges for proper operation. These tests are to be conducted after the self-diagnostic test indicates there is no problem with the cluster or the module. Test procedures will vary between manufacturers. To properly troubleshoot the gauges, you will need the manufacturer's diagnostic procedure, specifications, and circuit diagram.

Gauge Reads Low Constantly

A gauge that constantly reads low when the ignition switch is in the RUN position indicates an open in the gauge circuit. To locate the open, follow these steps:

1. Disconnect the wire harness from the sending unit.
2. Connect a jumper wire between the wire circuit from the gauge and ground.
3. Turn the ignition switch to the RUN position. The gauge should indicate maximum.

   If the gauge reads high, check the sending unit ground connection. If the ground is good, the sending unit is faulty and must be replaced.

   If the gauge continues to read low, follow the circuit diagram for the vehicle being serviced to test for opens in the wire from the sending unit. If the circuit is good, test the control module following recommended diagnostic procedures.

Gauge Reads High Constantly

A gauge that reads high when the ignition switch is placed in the RUN position indicates there is a short to ground in the circuit. To test the circuit, disconnect the wire harness at the gauge sending unit. Place the ignition switch in the RUN position while observing the gauge. If the gauge reads low, the sending unit is faulty and needs replacement.

   If the gauge continues to read high, use the circuit diagram to test for shorts to ground in the circuit from the sending unit. If the circuit is good, test the control module following recommended diagnostic procedures.

Classroom Manual
Chapter 12,
pages 355, 359

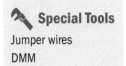 **Special Tools**
Jumper wires
DMM

> ⚙️ **SERVICE TIP** Although some service information procedures do not require it, it is a good practice to connect a 10-ohm resistor into the jumper wire when performing these tests. This prevents a nonresistive short to ground, yet does not noticeably affect gauge operation.

Inaccurate Gauge Readings

Inaccurate gauge readings are usually caused by faulty sending units. To test the operation of the gauge, you will need the manufacturer's specifications concerning resistance values as they relate to gauge readings. Gauge testers are available to test the units as different resistance values are changed.

Other reasons for inaccurate gauge readings include poor connections, **resistive shorts**, and poor grounds. Resistive shorts are shorts to ground that pass through a form of resistance first. Also, look for damage around the sending unit. For example, a damaged fuel tank can result in inaccurate gauge readings.

GAUGE SENDING UNITS

Classroom Manual
Chapter 12,
pages 357, 366

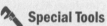

Special Tools

12-volt test light
10-ohm resistor
73-ohm resistor
Fender covers
Safety glasses
DMM

There are three types of sending units associated with electromechanical gauges: a thermistor, a piezoresistive sensor, and a mechanical variable resistor. Most of these can be tested before replacement to confirm the fault.

The fuel level sending unit can be tested in or out of the tank. If it is tested in the tank, add and remove fuel to change the level. The easiest method is to bench test the unit. **Photo Sequence 27** illustrates how to bench test a mechanical variable resistor sending unit.

To test the coolant temperature sensing unit, use an ohmmeter to measure the resistance between the terminal lead and ground (**Figure 12-16**). The resistance value of the variable resistor should change in proportion to the coolant temperature. Check the test results with the manufacturer's specifications.

To test a piezoresistive sensor sending unit used for oil pressure gauges, connect the ohmmeter to the sending unit terminal and ground (**Figure 12-17**). Check the resistance

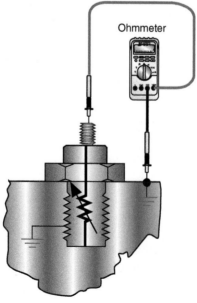

Figure 12-16 Testing a thermistor with an ohmmeter.

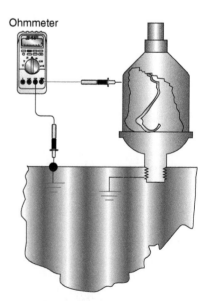

Figure 12-17 Using an ohmmeter to test a piezoresistive sensor.

with the engine off and compare to specifications. Start the engine and allow it to idle. Check the resistance value and compare to specifications. Before replacing the sending unit, connect a shop oil pressure gauge to confirm that the engine is producing adequate oil pressure.

Classroom Manual
Chapter 12, page 370

WARNING LAMPS

A warning light may be used to warn of low oil pressure, high coolant temperature, a defective charging system, or a brake failure. Unlike gauge sending units, the sending units for a warning light are nothing more than simple switches. The style of switch can be either normally open or normally closed, depending on the monitored system.

Special Tools
Jumper wires
Scan tool
DMM

PHOTO SEQUENCE 27
Bench Testing the Fuel Level Sender Unit

P27-1 Tools required to perform this task: DMM, jumper wires, and service information.

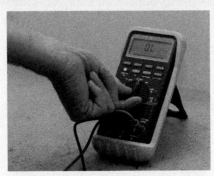

P27-2 Select the ohmmeter function of the DMM.

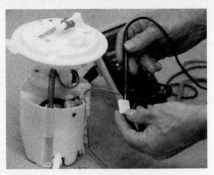

P27-3 Connect the negative test lead of DMM to the ground terminal of the sender unit.

P27-4 Connect the positive test lead to the variable resistor terminal.

P27-5 Holding the sender unit in its normal position, place the float rod against the empty stop.

P27-6 Read the ohmmeter and check the results with specifications. UNF p 581-7

P27-7 Slowly move the float toward the full stop while observing the ohmmeter. The resistance change should be smooth and consistent.

P27-8 Check the resistance value while holding the float against the full stop. Check the results with specifications.

P27-9 Check the float that it is not filled with fuel, distorted, or loose.

> ⚙️ **SERVICE TIP** With the bulbs used for most warning light circuits, it is hard to determine whether or not the filament is good. When a test procedure requires that a bulb be checked, it is usually easier to replace the bulb with a known good one.

It is not likely that all of the warning lights would fail at the same time. Check the fuse if all of the lights are not operating properly. Next, check for voltage at the last common connection. If voltage is not present, then trace the circuit back toward the battery. If voltage is present at the common connection, test each circuit branch in the same manner as described here for individual lamps.

To test a faulty warning lamp on a system with a normally open switch (sending unit), turn the ignition switch to the START position. The **prove-out circuit** should light the warning lamp. A prove-out circuit completes the warning light circuit to ground through the ignition switch when it is in the START position. The warning light is on during engine cranking to indicate to the driver that the bulb is working properly. If the light does not come on during the prove-out, disconnect the sender switch lead (**Figure 12-18**). Use a jumper wire to connect the sender switch lead to ground. With the ignition switch in the RUN position, the warning lamp should light. If the lamp is illuminated, test the prove-out circuit for an open. If the light does not come on, either the bulb is burned out or the wiring is damaged. Use a test light to confirm that voltage is present at the sensor terminal connector. If there is voltage, the bulb is probably bad. If voltage is not present to the sending unit, the bulb may be burned out. At this point, the instrument cluster will need to be removed. With the cluster removed, check for battery voltage to the panel connector. If voltage is present, substitute a known good bulb and test again.

If the customer states that the oil pressure light does not go out after the engine is started, use a shop oil pressure gauge to confirm adequate oil pressure.

If the system uses a normally closed switch, test in the same manner. However, there will not be a separate prove-out circuit.

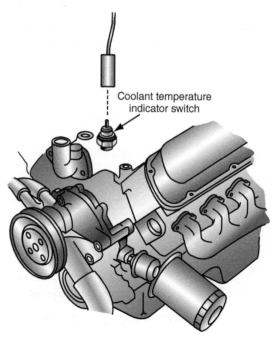

Coolant temperature
indicator switch

Figure 12-18 Coolant temperature sensor switch and lead.

If the customer states that the warning light stays on, test in the following manner: disconnect the lead to the sender switch. The light should go out with the ignition switch in the RUN position. If it does not, there is a short to ground in the wiring from the sender switch to the lamp. If the light goes out, replace the sender switch.

Computer-driven instrument cluster warning lamps are diagnosed using a scan tool. If the customer states that the lamp does not operate, the scan tool can be used to command activation of the lamp. If the lamp does not come on when commanded, the problem is a faulty lamp, circuit board, or instrument cluster module. If the lamp does light when commanded, the fault is in the signal to the instrument cluster.

If the customer states that the lamp is on all of the time, this indicates a problem in the monitored system. Use the scan tool to check for DTCs and use the service information to diagnose the fault. If there are no problems in the monitored circuit, then you will need to diagnose the module that receives the lamp's sensor input. If there is no problem found in this module, the instrument cluster module will need to be diagnosed. Follow the service information procedures for all diagnostic tests.

Classroom Manual
Chapter 12, page 373

TRIP COMPUTERS

Simple trip computers use inputs from the speed sensor and the fuel gauge to perform their functions. Like most electronic instrument clusters, trip computers usually have a self-diagnostic test procedure. Follow the manufacturer's procedure for initiating this test. If the trip computer passes the self-test, check the fuel gauge and speed sensor inputs.

Complex trip computers may receive inputs from several different areas of the vehicle (**Figure 12-19**). These systems require the use of specific manufacturer diagnostic charts, specifications, and procedures. Using the skills you have acquired, you should be able to perform these tests competently.

Classroom Manual
Chapter 12, page 369

Trip computer systems may have several different names. The most common are Traveler, Vehicle Information Center, and Drive Information Center.

MAINTENANCE MINDERS

Most manufacturers will display a warning lamp or message when oil change or other services are due (**Figure 12-20**). The determination of when the service due can be calculated strictly off of mileage driven; however, many systems take into account many

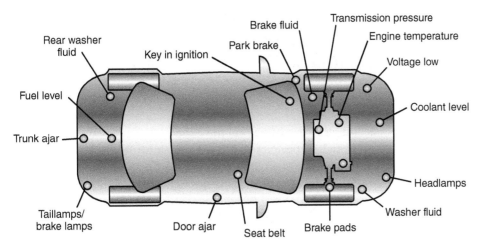

Figure 12-19 Complex information centers use many inputs to monitor the vehicle's subsystems.

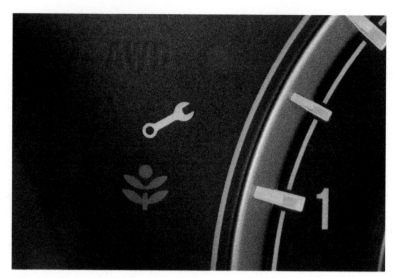

Figure 12-20 Maintenance minder indicator.

factors. Some systems will calculate the maintenance service requirement based on vehicle mileage, engine run hours, engine load, climatic conditions, and drive cycles.

Resetting the maintenance indicator may require the use of a scan tool. However, most manufacturers will provide an alternate method. Use the service information, or owner's manual, to determine the process for resetting the minder. There are various methods in use that range from a simple button press to complex activation of several inputs in a specific order. For example, one manufacturer requires the ignition switch to be placed in the RUN position, then the accelerator pedal pressed three times within 10 seconds, followed by the ignition switch being turned off.

Some maintenance minder systems may allow the vehicle operator to change the mileage intervals between services. Usually, this will not alter calculations that use factors other than just vehicle mileage. For example, the mileage interval may be set for 6,000 miles (9,656 Km) but the minder may come on at just 3,000 miles (4,828 Km) based on heavy engine loads in a cold climate.

HEAD-UP DISPLAY DIAGNOSTICS

Special Tools

Scan tool
DMM
Back probe tools
Test light

Classroom Manual
Chapter 12, page 368

We will look at the head-up display (HUD) used by General Motors as an example of typical diagnostic procedures for this system (**Figure 12-21**). The head-up display is a slave device to the instrument cluster and communicates directly with instrument cluster via the serial data bus. The image information is received from the instrument cluster through a video cable. Since the HUD module is a slave, the system has limited self-diagnosis capacity. Head-up display DTCs are reported through the instrument cluster.

Improper multiplexed switch inputs will set DTC B361B. Use the scan tool to monitor the switch data. With the ignition ON, hold each of the available switch positions. As the switch position is held, the scan tool should report that switch as being activated.

If the scan tool parameters does not equal the expected result with the switch in the appropriate state, test the switch circuits. With the ignition ON back probe the switch harness connector terminal 3. Use a test light with the clamp attached to battery positive and locate the probe onto the backprobe tool. The test light should illuminate. Use a DMM to measure the voltage at the tip of the test light probe. If the reading is greater than 0.02 volt, the circuit has excessive resistance. Move the backprobe tool to terminal 8 of the HUD module and retest. If the voltage is still high, test the HUD ground circuit. If the ground circuit is good, replace the HUD module.

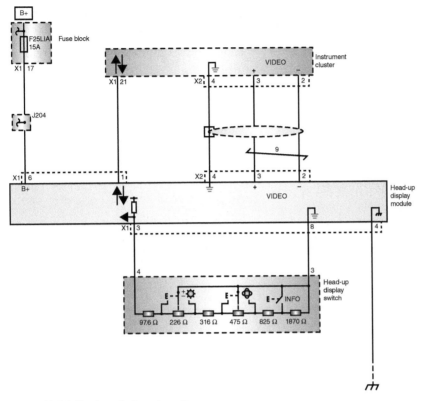

Figure 12-21 Head-up display schematic.

If the switch return circuit is good, turn the ignition OFF and test for less than 1 volt between the switch signal circuit terminal 4 and ground. If the voltage is greater than 1 volt, disconnect the HUD module connector and measure the voltage at terminal 4 of the harness connector. If voltage is present, the circuit is shorted to voltage.

If voltage was not present at terminal 4 of the switch connector, turn the ignition to the RUN position and measure for battery voltage. If less than battery voltage, test the circuit between the HUD module and the switch for a short to ground and/or an open. If battery voltage is present at switch connector terminal 4, remove the switch and test the resistance values between terminals 3 and 4 while each switch position is activated. If the resistance values do not match specifications, replace the switch. Diagnostics of the HUD module is within the instrument cluster. Other than testing for good voltage feed and ground circuits, no external diagnostics are done.

CASE STUDY

A customer brings his vehicle into the shop because the fuel level gauge does not operate. It remains on empty all the time, regardless of how full the tank is. The technician checks the fuse box for any blown fuses and finds that all are good. He disconnects the lead wire to the fuel tank sending unit and probes for voltage. The test light comes on when he places the ignition switch in the ON position. When the sending unit lead is connected to ground through a 10-ohm resistor, the gauge needle moves to the FULL position. Upon checking with the service information, the technician determines that this is normal operation. Before draining and removing the fuel tank, he uses a jumper wire to jump from the ground terminal of the sending unit to a known good ground on the frame. When he reconnects the lead wire and places the ignition switch in the RUN position, the gauge works properly. The technician cleans the ground connections for the sending unit and returns the vehicle to the customer.

ASE-STYLE REVIEW QUESTIONS

1. In a conventional instrument cluster, all of the gauges are inoperable. What is the **LEAST LIKELY** cause of this?
   A. Faulty IVR.
   B. Blown fuse.
   C. Shorted sending unit.
   D. Faulty printed circuit.

2. All of the following statements concerning gauge sending units are true **EXECPT**:
   A. A gauge can use a switch as a sensor.
   B. A gauge can use a thermistor as a sensor.
   C. A gauge can use a piezoresistive sensor.
   D. A gauge can use a variable resistor as a sensor.

3. The oil pressure warning light will not turn off with the engine running. A shop gauge confirms good oil pressure. What is the most likely cause?
   A. Faulty IVR.
   B. Damaged printed circuit board.
   C. Open in the wire to the oil pressure switch.
   D. Short to ground in the wire to the oil pressure switch.

4. When testing the IVR-regulated voltage, the indicated reading was 2.5 volts over specifications.
   Technician A says the IVR ground connection may be faulty.
   Technician B says the alternator output may be too high.
   Who is correct?
   A. A only   C. Both A and B
   B. B only   D. Neither A nor B

5. *Technician A* says to use a voltmeter to test a thermistor sensor unit.
   Technician B says the thermistor sensor measures pressure.
   Who is correct?
   A. A only   C. Both A and B
   B. B only   D. Neither A nor B

6. *Technician A* says when the ignition switch is in the START position the proving circuit should light the warning lamps.
   Technician B says if the sender switch lead is grounded, and the ignition switch is in the RUN position, the warning lamp should come on.
   Who is correct?
   A. A only   C. Both A and B
   B. B only   D. Neither A nor B

7. A digital speedometer constantly reads 0 mph.
   Technician A says the speed sensor may be faulty.
   Technician B says the throttle position sensor may have an open.
   Who is correct?
   A. A only   C. Both A and B
   B. B only   D. Neither A nor B

8. All gauges read low in a computer-controlled instrument cluster.
   Technician A says the connector to the cluster may be loose.
   Technician B says the cluster module may be at fault.
   Who is correct?
   A. A only   C. Both A and B
   B. B only   D. Neither A nor B

9. An electronically controlled instrument cluster is being diagnosed for no speedometer operation. Which of the following statements is most correct?
   A. If the BCM and/or cluster module passes its self-test, the fault is probably in the vehicle speed sensor circuit.
   B. If the speedometer does not work but the odometer does, then the problem is probably in the vehicle speed sensor circuit.
   C. If the speedometer fails to move during the self-test, the problem is probably in the vehicle speed sensor circuit.
   D. All of the above.

10. Ford diagnostic mode displays are being discussed.
    Technician A says a CO displayed in the fuel gauge window indicates that the fuel level is near empty.
    Technician B says a CS indicates an open in the sender circuit.
    Who is correct?
    A. A only
    B. B only
    C. Both A and B
    D. Neither A nor B

ASE CHALLENGE QUESTIONS

1. The odometer of a vehicle equipped with a cable-driven speedometer is inoperative; however, the speedometer is working correctly.

   Technician A says that the speedometer cable may be faulty.

   Technician B says that the speedometer drive gear at the transmission may be stripped.

   Who is correct?

   A. A only
   B. B only
   C. Both A and B
   D. Neither A nor B

2. The fuel gauge of a multiple bimetallic-type gauge instrument cluster is reading higher than normal.

   Technician A says that the wire leading to the gauge sender unit may be open.

   Technician B says that the resistance of the sender unit may be lower than normal.

   Who is correct?

   A. A only
   B. B only
   C. Both A and B
   D. Neither A nor B

3. The water temperature warning light of a vehicle is on whenever the ignition key is in the ON position; when the engine is started, the light remains illuminated.

   Technician A says that the sending unit wire may be shorted to ground.

   Technician B says that the water temperature switch may be electrically open.

   Who is correct?

   A. A only
   B. B only
   C. Both A and B
   D. Neither A nor B

4. The diagnosis of a digital instrument cluster that is missing two segments in its display is being discussed.

   Technician A says that the cluster has excessive ground circuit resistance.

   Technician B says that the cluster should be replaced.

   Who is correct?

   A. A only
   B. B only
   C. Both A and B
   D. Neither A nor B

5. A vehicle equipped with an electronic instrument cluster has a "service engine" lamp illuminated; a speed sensor code (indicating the computer does not "see" vehicle speed) has been generated by the computer. The speedometer is working correctly; however, a scan tool that is connected to the vehicle's on-board computer indicates 0 mph whenever the vehicle is in motion.

   Technician A says that the vehicle speed sensor may be faulty.

   Technician B says that the on-board computer may be faulty.

   Who is correct?

   A. A only
   B. B only
   C. Both A and B
   D. Neither A nor B

Name _____   Date _____

CHECKING A FUEL GAUGE

Upon completion of this job sheet, you should be able to diagnose an inaccurate or inoperative fuel gauge.

NATEF Correlation

This job sheet addresses the following **MLR** tasks:

A.3.   Use wiring diagrams to trace electrical/electronic circuits.

A.4.   Demonstrate proper use of a digital multimeter (DMM) when measuring source voltage, voltage drop (including grounds), current flow, and resistance.

A.5.   Demonstrate knowledge of the causes and effects from shorts, grounds, opens, and resistance problems in electrical/electronic circuits.

A.11.   Identify electrical/electronic system components and configuration.

E.7.   Verify operation of instrument panel gauges and warning/indicator lights; reset maintenance indicators.

This job sheet addresses the following **AST/MAST** tasks:

A.3.   Demonstrate proper use of a digital multimeter (DMM) when measuring source voltage, voltage drop (including grounds), current flow, and resistance.

A.4.   Demonstrate knowledge of the causes and effects from shorts, grounds, opens, and resistance problems in electrical/electronic circuits.

A.7.   Use wiring diagrams during the diagnosis (troubleshooting) of electrical/electronic circuit problems.

F.1.   Inspect and test gauges and gauge sending units for causes of abnormal gauge readings; determine necessary action.

Tools and Materials

- A vehicle
- DMM
- Service information for the chosen vehicle
- Wiring diagram for the chosen vehicle
- Vehicle hoist

Describe the vehicle being worked on:

Year _____ Make _____ Model _____

VIN _____ Engine type and size _____

Procedure

1. From the procedures listed in the service information, describe the procedures for testing the fuel gauge and fuel gauge sending unit on this vehicle.

_____

_____

_____

_____

_____

_____

2. Locate the fuel gauge circuit wiring diagram and print out the schematic map of the   ☐
   circuit from power to ground. Attach the print out to this job sheet.

3. Follow the procedure to simulate a full fuel tank. Describe what happened when you
   followed this procedure; include in your answer your conclusions from this test.

   _____

   _____

   _____

   _____

   _____

4. Follow the procedure to simulate an empty fuel tank. Describe what happened when
   you followed this procedure; include in your answer your conclusions from this test.

   _____

   _____

   _____

   _____

   _____

Instructor's Response

_____

_____

_____

Name _____   Date _____

TESTING AN OIL PRESSURE WARNING LIGHT CIRCUIT

Upon completion of this job sheet, you should be able to test the oil pressure warning circuit and determine needed repairs.

NATEF Correlation

This job sheet addresses the following **MLR** tasks:

A.3.   Use wiring diagrams to trace electrical/electronic circuits.

A.4.   Demonstrate proper use of a digital multimeter (DMM) when measuring source voltage, voltage drop (including grounds), current flow, and resistance.

A.5.   Demonstrate knowledge of the causes and effects from shorts, grounds, opens, and resistance problems in electrical/electronic circuits.

A.11.   Identify electrical/electronic system components and configuration.

E.7.   Verify operation of instrument panel gauges and warning/indicator lights; reset maintenance indicators.

This job sheet addresses the following **AST/MAST** tasks:

A.3.   Demonstrate proper use of a digital multimeter (DMM) when measuring source voltage, voltage drop (including grounds), current flow, and resistance.

A.4.   Demonstrate knowledge of the causes and effects from shorts, grounds, opens, and resistance problems in electrical/electronic circuits.

A.7.   Use wiring diagrams during the diagnosis (troubleshooting) of electrical/electronic circuit problems.

F.1.   Inspect and test gauges and gauge sending units for causes of abnormal gauge readings; determine necessary action.

F.2.   Diagnose (troubleshoot) the causes of incorrect operation of warning devices and other driver information systems; determine needed action.

ASE NATEF

Tools and Materials

- Vehicle with conventional analog instrument cluster with oil pressure warning lamp
- Wiring Diagram
- DMM

Describe the vehicle being worked on:

Year _____ Make _____ Model _____

VIN _____ Engine type and size _____

Procedure

1. Describe the normal conditions in which the oil pressure warning lamp should be on.

   _____

   _____

   _____

2. At what pressure should the oil pressure warning lamp turn off?

   _____

3. Referring to the service information and wiring diagrams, is the oil pressure warning lamp sending unit a normally open or normally closed switch?

   _____

4. Locate the oil pressure warning lamp circuit and print it out. Map the circuit from power to ground, identifying components of the circuit. Attach the printed copy to this job sheet.

5. List all possible causes for an oil pressure warning lamp that does not turn on.

_____

_____

_____   ☐

_____

_____

6. Unplug the sending unit connector and use the DMM to test for voltage with the ignition key in the RUN position with the engine off. Is voltage present?

_____

7. If voltage is not present, what will be your next step?

_____

_____

_____

8. If voltage is not present in step 6, test the circuit. What are your results and recommendations?

_____

_____

_____

_____

9. If voltage was present in step 6, describe the method to be used to test the sending unit.

_____

_____

_____

_____

_____

10. Perform the test described in step 9 and record your results and recommendations.

_____

_____

_____

_____

_____

Instructor's Response

_____

_____

_____

Name _____   Date _____

TESTING THE ELECTRONIC INSTRUMENT CLUSTER

Upon completion of this job sheet, you should be able to test the electronic instrument cluster using a scan tool or stand-alone diagnostic routines and determine needed repairs.

NATEF Correlation

This job sheet addresses the following **MLR** task:

E.7.   Verify operation of instrument panel gauges and warning/indicator lights; reset maintenance indicators.

This job sheet addresses the following **AST** task:

G.5.   Describe body electronic systems circuits using a scan tool; check for module communication errors (data bus systems); determine needed action.

This job sheet addresses the following **MAST** task:

G.5.   Diagnose body electronic systems circuits using a scan tool; check for module communication errors (data bus systems); determine needed action.

This job sheet addresses the following **AST/MAST** tasks:

F.1.   Inspect and test gauges and gauge sending units for causes of abnormal gauge readings; determine necessary action.

F.2.   Diagnose (troubleshoot) the causes of incorrect operation of warning devices and other driver information systems; determine needed action.

ASE CERTIFIED   NATEF

Tools and Materials

- A vehicle equipped with electronic instrument cluster
- Scan tool
- Service information

Describe the vehicle being worked on:

Year _____ Make _____ Model _____

VIN _____ Engine type and size _____

Procedure

1. Test the operation of the gauges. Describe the symptoms.

   _____

   _____

2. Are there any other related symptoms? ☐ Yes   ☐ No

   If yes, describe the symptom.

   _____

   _____

3. Refer to the proper service information to determine if a self-diagnostic routine can be performed. If so, describe how to enter the diagnostics.

   _____

   _____

   _____

   _____

4. Is the self-diagnostic routine capable of displaying fault codes?

   ☐ Yes ☐ No

5. Perform the procedure listed in step 3 and record the results and any DTCs.

   _____

   _____

6. Use a scan tool and access the electronic instrument cluster. Does the scan tool indicate that DTCs are present?

   ☐ Yes ☐ No

   If yes, record the DTCs.

   _____

   _____

7. Based on the tests so far, is the fault in the instrument cluster or in the sensor circuits?

   ☐ CLUSTER ☐ SENSOR

8. Based on the results so far, what tests need to be performed to find the cause of the fault?

   _____

   _____

9. Perform the tests listed in step 8. What is your determination and recommendation?

   _____

   _____

Instructor's Response

_____

_____

Name _____   Date _____

DIAGNOSING THE HEAD-UP DISPLAY

Upon completion of this job sheet, you should be able to diagnose the HUD system.

NATEF Correlation

This job sheet addresses the following **MLR** tasks:

A.3.   Use wiring diagrams to trace electrical/electronic circuits.

A.4.   Demonstrate proper use of a digital multimeter (DMM) when measuring source voltage, voltage drop (including grounds), current flow, and resistance.

A.5.   Demonstrate knowledge of the causes and effects from shorts, grounds, opens, and resistance problems in electrical/electronic circuits.

A.6.   Check operation of electrical circuits using a test light.

A.11. Identify electrical/electronic system components and configuration.

This job sheet addresses the following **AST** tasks:

G.4.   Describe operation of safety systems and related circuits (such as horn, airbags, seat belt pretensioners, occupancy classification, wipers, washers, speed control/collision avoidance, heads-up display, park assist, and back-up camera); determine needed repairs.

G.5.   Describe body electronic systems circuits using a scan tool; check for module communication errors (data bus systems); determine needed action.

This job sheet addresses the following **AST/MAST** tasks:

A.3.   Demonstrate proper use of a digital multimeter (DMM) when measuring source voltage, voltage drop (including grounds), current flow, and resistance.

A.4.   Demonstrate knowledge of the causes and effects from shorts, grounds, opens, and resistance problems in electrical/electronic circuits.

A.5.   Demonstrate proper use of a test light on an electrical circuit.

A.7.   Use wiring diagrams during the diagnosis (troubleshooting) of electrical/electronic circuit problems.

This job sheet addresses the following **MAST** tasks:

G.4.   Diagnose operation of safety systems and related circuits (such as: horn, airbags, seat belt pretensioners, occupancy classification, wipers, washers, speed control/collision avoidance, heads-up display, park assist, and back-up camera); determine needed repairs.

G.5.   Diagnose body electronic systems circuits using a scan tool; check for module communication errors (data bus systems); determine needed action.

Tools and Materials

- A vehicle equipped with HUD
- Scan tool
- DMM
- Test light
- Service information for the chosen vehicle
- Wiring diagram for the chosen vehicle

Describe the vehicle being worked on:

Year _____ Make _____ Model _____

VIN _____ Engine type and size _____

Procedure   **Task Completed**

1. Use the scan tool to monitor the switch data as each switch is pressed. While holding each switch listed below, verify the scan tool HUD switch state parameter is as follows:

DIM + switch: _____

DIM − switch: _____

UP switch: _____

DOWN switch: _____

PAGE switch: _____

2. Are all switch parameters the expected value?

   Yes ☐   No ☐

3. Identify the switch terminal for the switch return circuit.

   _____

4. Identify the switch terminal for the switch sense circuit.

   _____

   Ignition OFF/Vehicle OFF and all vehicle systems OFF, disconnect the harness connector at the S27 Head-up display switch. It may take up to 2 minutes for all vehicle systems to power down.

5. Use a test light and DMM to test the switch return circuit.

   Record your results.

   _____

6. If the voltage reading in step 5 was out of specifications, what would be your next step?

   _____

   _____

   _____

7. With the ignition in the OFF position, measure the voltage on the switch sense circuit. Record your results.

   _____

8. What do the test results in step 7 indicate?

   _____

9. Turn the ignition OFF and remove the HUD switch assembly.   ☐

10. Measure the resistance between the sense circuit terminal and the return terminal in all switch positions:

   No switch pressed: _____

   With the DIM + switch pressed _____

   With the DIM − switch pressed _____

   With the UP switch pressed _____

   With the DOWN switch pressed _____

   With the PAGE switch pressed _____

11. Based on the test results, what are your conclusions concerning the HUD switch function?

   _____

Instructor's Response

_____

_____

Name _____   Date _____

RESETTING MAINTENANCE INDICATORS

Upon completion of this job sheet, you should be able to properly reset the maintenance indicator.

NATEF Correlation

This job sheet addresses the following **AST/MAST** task:

F.3.   Reset maintenance indicators as required.

NATEF

Tools and Materials

- A vehicle equipped with maintenance indicator
- Scan tool
- Service information

Describe the vehicle being worked on:

Year _____ Make _____ Model _____

VIN _____ Engine type and size _____

Procedure

1. Reference the service information to determine the process for resetting the maintenance indicator. Record the procedure below:

   _____

   _____

   _____

   _____

2. Can the reset procedure be performed with a scan tool?

   ☐ Yes   ☐ No

   If yes, describe the procedure.

   _____

   _____

3. Can the service interval mileage be changed?

   ☐ Yes   ☐ No

   If yes, describe the procedure.

   _____

   _____

4. Perform the maintenance reset procedure. How do you know the procedure was successful?

   _____

   _____

   _____

   _____

Instructor's Response

_____

_____

_____

DIAGNOSTIC CHART 12-1

PROBLEM AREA:	Instrument cluster gauges.
SYMPTOMS:	One or all gauges fluctuate from low or high to normal readings.
POSSIBLE CAUSES:	**1.** Poor ground connection. **2.** Excessive resistance. **3.** Poor connections. **4.** Faulty sending unit. **5.** Defective printed circuit.

DIAGNOSTIC CHART 12-2

PROBLEM AREA:	Instrument cluster gauges.
SYMPTOMS:	One or all gauges read high.
POSSIBLE CAUSES:	**1.** Faulty instrument voltage regulator. **2.** Shorted printed circuit. **3.** Faulty sending unit. **4.** Short to ground in sending unit circuit. **5.** Faulty gauge. **6.** Poor sending unit ground connection.

DIAGNOSTIC CHART 12-3

PROBLEM AREA:	Instrument cluster gauges.
SYMPTOMS:	One or all gauges read low.
POSSIBLE CAUSES:	**1.** Faulty instrument voltage regulator. **2.** Poor sending unit ground connection. **3.** Improper bulb application. **4.** Improper connections. **5.** Faulty back-up switch contacts.

DIAGNOSTIC CHART 12-4

PROBLEM AREA:	Instrument cluster gauges.
SYMPTOMS:	One or all gauges fail to read.
POSSIBLE CAUSES:	**1.** Blown fuse. **2.** Open in the printed circuit. **3.** Faulty gauge. **4.** Poor common ground connection. **5.** Open in the sending unit circuit. **6.** Faulty sending unit. **7.** Faulty instrument voltage regulator. **8.** Poor electrical connection to cluster. **9.** Shorted sending unit circuit.

DIAGNOSTIC CHART 12-5

PROBLEM AREA:	Warning light operation.
SYMPTOMS:	Warning light remains on all the time.
POSSIBLE CAUSES:	**1.** Sending unit circuit grounded. **2.** Faulty sending unit switch.

DIAGNOSTIC CHART 12-6

PROBLEM AREA:	Warning light operation.
SYMPTOMS:	Warning light fails to operate on an intermittent basis.
POSSIBLE CAUSES:	**1.** Loose sending unit circuit connections. **2.** Faulty sending unit.

DIAGNOSTIC CHART 12-7

PROBLEM AREA:	Warning light operation.
SYMPTOMS:	One or all warning lights fail to operate.
POSSIBLE CAUSES:	**1.** Blown fuse. **2.** Burned-out bulb. **3.** Open in the circuit. **4.** Defective sending unit switches.

DIAGNOSTIC CHART 12-8

PROBLEM AREA:	Electronic instrument panel.
SYMPTOMS:	Digital display does not light.
POSSIBLE CAUSES:	**1.** Blown fuse. **2.** Inoperative power and ground circuit. **3.** Faulty instrument panel.

DIAGNOSTIC CHART 12-9

PROBLEM AREA:	Electronic instrument panel.
SYMPTOMS:	Speedometer reads wrong speed.
POSSIBLE CAUSES:	**1.** Faulty speedometer. **2.** Wrong gear on vehicle speed sensor (VSS). **3.** Wrong tire size.

DIAGNOSTIC CHART 12-10

PROBLEM AREA:	Electronic instrument panel.
SYMPTOMS:	Fuel gauge displays two top and bottom bars.
POSSIBLE CAUSES:	**1.** Open or short in circuit.

DIAGNOSTIC CHART 12-11

PROBLEM AREA:	Electronic instrument panel.
SYMPTOMS:	Fuel computer displays CS or CO.
POSSIBLE CAUSES:	**1.** Open or short in fuel gauge sender. **2.** Inoperative instrument panel.

DIAGNOSTIC CHART 12-12

PROBLEM AREA:	Electronic instrument panel.
SYMPTOMS:	Odometer displays error.
POSSIBLE CAUSES:	**1.** Inoperative odometer memory module in instrument panel.

DIAGNOSTIC CHART 12-13

PROBLEM AREA:	Electronic instrument panel.
SYMPTOMS:	Fuel gauge display is erratic.
POSSIBLE CAUSES:	**1.** Sticky or inoperative fuel gauge sender. **2.** Fault in circuit. **3.** Inoperative fuel gauge.

DIAGNOSTIC CHART 12-14

PROBLEM AREA:	Electronic instrument panel.
SYMPTOMS:	Fuel gauge will not display FULL or EMPTY.
POSSIBLE CAUSES:	**1.** Sticky or inoperative fuel gauge sender.

DIAGNOSTIC CHART 12-15

PROBLEM AREA:	Electronic instrument panel.
SYMPTOMS:	Fuel economy function of message center is erratic or inoperative.
POSSIBLE CAUSES:	**1.** Inoperative fuel flow signal. **2.** Faulty wiring. **3.** Inoperative instrument panel.

DIAGNOSTIC CHART 12-16

PROBLEM AREA:	Electronic instrument panel.
SYMPTOMS:	Extra or missing display segments.
POSSIBLE CAUSES:	**1.** Inoperative instrument panel.

DIAGNOSTIC CHART 12-17

PROBLEM AREA:	Electronic instrument panel.
SYMPTOMS:	Speedometer always reads zero.
POSSIBLE CAUSES:	**1.** Faulty wiring. **2.** Inoperative instrument panel.

DIAGNOSTIC CHART 12-18

PROBLEM AREA:	Electronic instrument panel.
SYMPTOMS:	Temperature gauge displays two top and bottom bars.
POSSIBLE CAUSES:	**1.** Short in circuit. **2.** Inoperative coolant temperature sender. **3.** Inoperative instrument panel.

CHAPTER 13

ELECTRICAL ACCESSORIES DIAGNOSIS AND REPAIR

Upon completion and review of this chapter, you should be able to:

- Identify the causes of no operation, intermittent operation, or constant horn operation.
- Diagnose the cause of poor sound quality from the horn system.
- Perform diagnosis and repair of no–windshield wiper operation at only one speed or at all speeds.
- Identify causes for slower-than-normal wiper operation.
- Determine the cause for improper park operation.
- Identify the causes of continuous wiper operation.
- Diagnose faulty intermittent wiper system operation.
- Remove and install wiper motors and wiper switches.
- Diagnose computer-controlled wiper systems.
- Determine the causes for improper operation of the windshield washer system and be able to replace the pump if required.
- Perform diagnosis of problems associated with blower motor circuits.
- Diagnose and repair electric rear window defoggers.

- Diagnose the power window system.
- Diagnose common problems associated with the power seat system.
- Diagnose the memory seat function.
- Diagnose climate-controlled seat systems.
- Perform diagnosis of the power door lock and automatic door lock systems.
- Diagnose vehicle alarm systems.
- Perform tests on the antitheft controller to determine proper operation.
- Use self-diagnostic tests on alarm systems that provide this feature.
- Perform self-diagnostic procedures on electronic cruise control systems.
- Diagnose causes for no, intermittent, and erratic cruise control operation.
- Test the cruise control servo assembly for proper operation.
- Replace the servo assembly and adjust actuator cable.
- Replace the cruise control switch assembly.
- Diagnose the adaptive cruise control system.
- Determine the circuit fault causing a malfunction in sunroof operation.

🔧 **Basic Tools**

Basic mechanic's tool set

Service information

Terms To Know

Adaptive cruise control (ACC)

Dump valve

False wipes

Grids

KOEO

KOER

Park switch

Pinpoint test

Resistor block

Servo

Tin

INTRODUCTION

The electrical accessories included in this chapter represent the most often performed electrical repairs. Most of the systems discussed do not provide for rebuilding of components. A technician must be capable of diagnosing faults and then replacing the defective part. As with any electrical system, always use a systematic diagnostic approach to finding the cause. Refer to the service information to obtain information concerning correct system operation. The fault is easier to locate once you understand how the system is supposed to operate.

Included in this chapter are diagnostic and repair procedures for horn systems, windshield wipers and washer systems, blower motors, electric defogger systems, power seat and window systems, and power door locks.

Also, you will learn how to service electronic accessories designed to increase passenger comfort, provide ease of operation, and increase passenger safety. These systems include electronic cruise control, memory seats, sunroof, automatic door locks, keyless entry, and antitheft systems.

Most of these systems are additions to existing systems. For example, the memory seat feature is an addition to the conventional power seat system. As vehicles and accessories become more sophisticated, these luxury features will become more commonplace. Today's technician is expected to accurately and quickly diagnose malfunctions in these systems.

Although the procedures presented here are typical, always refer to the proper service information for the vehicle you are servicing.

HORN DIAGNOSIS

Classroom Manual
Chapter 13,
pages 378, 380

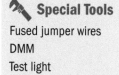 **Special Tools**
Fused jumper wires
DMM
Test light
Fender covers

Customer complaints associated with the horn system can include no operation, intermittent operation, continuous operation, or poor sound quality. Testing of the horn system varies between systems that do and do not use a relay.

No Horn Operation

Systems with Relay. When a customer complains of no horn operation, first confirm the complaint by depressing the horn button. If it is mounted in the steering wheel, rotate the steering wheel from stop to stop while depressing the horn button. If the horn sounds intermittently while the steering wheel is turned, the problem is probably in the sliding contact ring in the steering column, or the tension spring is worn or broken. Also, a faulty clockspring may cause intermittent horn operation as the steering wheel is turned. Check the air bag system for any diagnostic trouble codes (DTCs) associated with the clockspring circuit. If the horn does not sound during this test, continue to check the system as follows:

1. Check the fuse or fusible link. If defective, replace as needed. After replacing the fuse, operate all other circuits it protects. It is possible that another circuit is faulty but the customer has not noticed. Isolate the cause of the blown fuse or fusible link.
2. Connect a fused jumper wire from the battery positive terminal to the horn terminal. If the horn sounds, continue testing; if the horn does not sound, check the ground connection. Replace the horn if the ground is good.
3. Remove the relay from its connector and check for voltage at terminals 30 and 86 (**Figure 13-1**). If there is no voltage at these points, trace the wiring from the relay to the battery to locate the problem. Continue testing if voltage is present at the power feed terminals.
4. With the relay removed, connect a test light across terminals 85 and 86 of the relay connector (**Figure 13-2**). Press the horn switch and observe the test light. The test light should illuminate. If it does not, the horn switch or its circuit is faulty—go to step 7. If the light illuminates, continue testing.

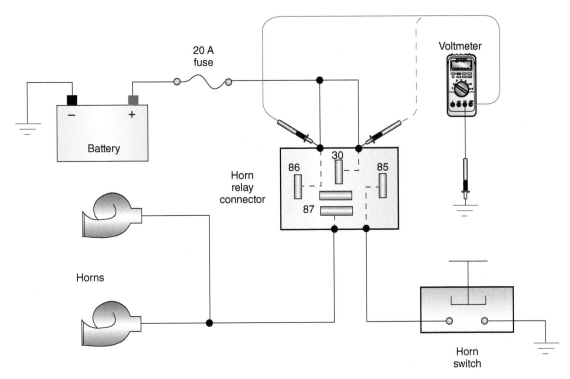

Figure 13-1 Testing for voltage to the relay connector.

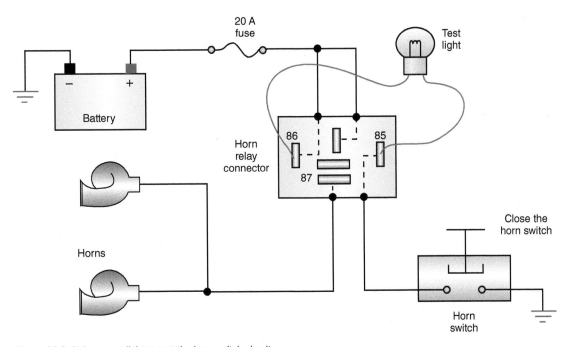

Figure 13-2 Using a test light to test the horn switch circuit.

5. Use a fused jumper wire and jump terminals 30 and 87 (**Figure 13-3**). If the horns sound, replace the relay. If the horns do not sound, continue to test.

6. In a multiple-horn system, test for voltage at the last common connection between the horn relay and the horns (**Figure 13-4**). On a single-horn system, test for voltage at the horn terminal. Voltage should be present at this connection only when the horn button is depressed. If there is no voltage at this connection, repair the open between the relay and the common connection. If voltage is present, check the individual circuits from the connection to the horns; repair as needed.

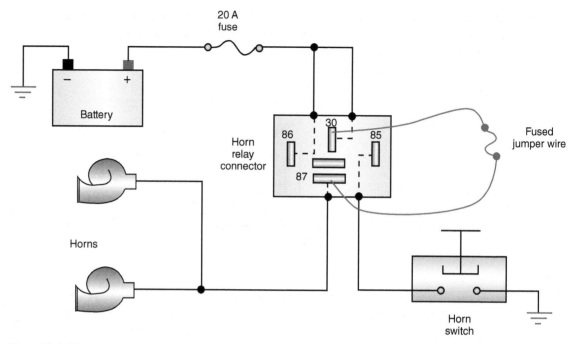

Figure 13-3 Using a fused jumper wire to test the horn circuit. If the horn sounds, the relay is faulty.

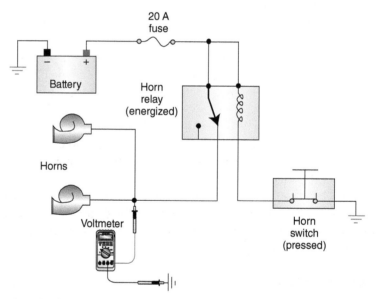

Figure 13-4 Testing for voltage at the last common connection.

7. Check for voltage on the battery side of the horn switch. If there is no voltage at this location, the fault is between the relay and the switch. Continue testing if voltage is present.
8. Check for continuity through the switch. If good, check the ground connection for the switch. Then recheck operation. Replace the horn switch if there is no continuity when the button is depressed.

Systems without Relay. Follow the steps described under "Systems with Relay" to confirm the complaint. If this step confirms that the horn is not operational, perform the following tests to locate the fault:

1. Check the fuse or fusible link. If defective, replace as needed. After replacing the fuse, operate all other circuits it protects. It is possible that another circuit is faulty but the customer has not noticed. Isolate the cause of the blown fuse or fusible link.

2. Connect a fused jumper wire from the battery positive terminal to the horn terminal. If the horn sounds, continue testing; if the horn does not sound, check the ground connection. Replace the horn if the ground is good.

3. In a multiple-horn system, test for voltage at the last common connection between the horn switch and the horns. On a single-horn system, test for voltage at the horn terminal. Voltage should be present at this connection only when the horn button is depressed. Continue testing if there is no voltage at this connection. If voltage is present, check the individual circuits from the connection to the horns; repair as needed.

4. Check for voltage at the horn side of the switch when the button is depressed. If voltage is present, the problem is in the circuit from the switch to the horn(s). Continue testing if there is no voltage at this connection.

5. Check for voltage at the battery side of the switch. If voltage is present, the switch is faulty and must be replaced. If there is no voltage at this terminal, the problem is in the circuit from the battery to the switch.

Classroom Manual
Chapter 13, page 380

Poor Sound Quality

Poor sound quality can be the result of several factors. In a multiple-horn system, if one of the horns does not operate, the horn sound quality may suffer. Other reasons include a damaged diaphragm, excessive circuit resistance, poor ground connections, or improperly adjusted horns.

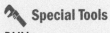

Special Tools
DMM
Fused jumper wires
Fender covers

If one horn of a multiple-horn system does not operate, use a fused jumper wire from the battery positive terminal to the horn terminal to determine whether the fault is in the horn or in the circuit. If the horn sounds, the problem is in the circuit between the last common connection and the affected horn.

If one or all horns produce poor-quality sound, use a voltmeter to measure the voltage at the horn terminal when the horn switch is closed. The voltage should be within 0.5 volt (V) of battery voltage. If the voltage measured is less than the battery voltage, there is excessive voltage drop in the circuit. Work back through the circuit measuring voltage drop across connectors, relays, and switches to find the source of the high resistance.

If the voltage to the horn is of proper value, test the ground circuit. Use the voltage drop method while the horn switch is closed. If the voltage drop is excessive, check the connections to ground.

If the circuits test good, connect a fused jumper wire between the battery positive terminal and the horn terminal to activate the horn. With the horn sounding, turn the adjusting screw counterclockwise one-quarter to three-eighths of a turn. Replace the horn if the sound quality cannot be improved.

Horn Sounds Continuously

A horn that does not shut off is usually caused by a sticking horn switch or sticking contact points in the relay. To find the fault, disconnect the horn relay from the circuit. Use an ohmmeter to check for continuity from the battery feed terminal of the relay to the horn circuit terminal (**Figure 13-5**). If there is continuity, the relay is defective. If there is no continuity through the relay, test the switch.

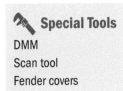

Special Tools
DMM
Scan tool
Fender covers

The easiest way to test the switch circuit is to measure for continuity between the relay connector terminal 85 and ground. With the horn switch released, there should not be continuity to ground. If there is continuity, then the circuit to the horn switch is shorted to ground or the switch contacts are stuck. Use an ohmmeter to test the switch in the normal manner.

Systems that control the horn function by computers have the microprocessor control the relay coil ground circuit or use high-side drivers (HSDs) to operate the horns. Often a scan tool can be used to actuate the horns. If the system works properly with the scan tool, the problem is in the inputs to the computer. If the system does not operate when actuated with the scan tool, the problem may be the driver circuits, faulty horn, faulty horn ground circuit, or defective module.

Classroom Manual
Chapter 13, page 380

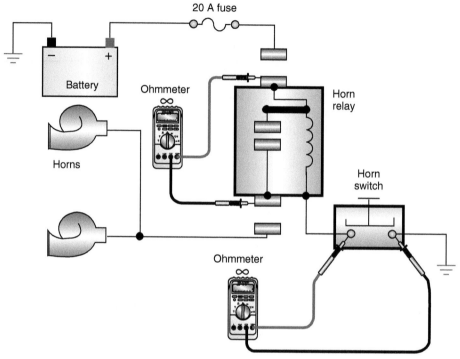

Figure 13-5 Ohmmeter tests to find cause of continuous horn operation.

> ⚙ **SERVICE TIP** Most modules that use high-side drivers are capable of detecting an open circuit. Often, once a fault in the circuit is determined, the HSD is turned off for the remainder of that ignition cycle. If you disconnect the circuit from the module to the horn, the HSD may have been shut down and you will need to cycle the ignition switch between each test.

Classroom Manual
Chapter 13, pages
382, 383, 384

WIPER SYSTEM SERVICE

Customer complaints concerning windshield wiper operation can include no operation, intermittent operation, continuous operation, and wipers that will not park. Other complaints have to do with blade adjustment (such as blades slapping the molding or one blade that parks lower than the other).

When a customer brings the vehicle into the shop because of faulty windshield wiper operation, the technician needs to determine if it is an electrical or a mechanical problem. To do this, disconnect the arms to the wiper blades from the motor (**Figure 13-6**). Turn on the wiper system. Observe operation of the motor. The problem is mechanical if the motor operates properly.

Classroom Manual
Chapter 13, page 384

No Operation in One Speed Only

Problems that cause the system to not operate in only one switch position are usually electrical. Use the wiring schematic to determine proper operation. For example, use the three-speed wiper system schematic illustrated in **Figure 13-7** to determine the cause of a motor that does not operate only in the MEDIUM-speed position. The problem is that the 7-ohm (Ω) resistor is open. The problem could not be the shunt field in the motor because LOW and HIGH speeds operate; nor could the problem be in the wiring to the motor because this is shared by all speeds.

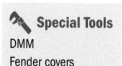

Special Tools

DMM
Fender covers

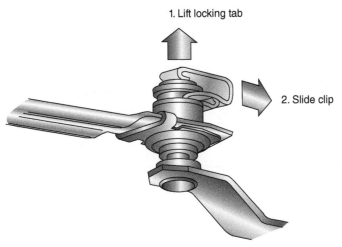

Figure 13-6 To remove the clip, lift up the locking tab and pull the clip.

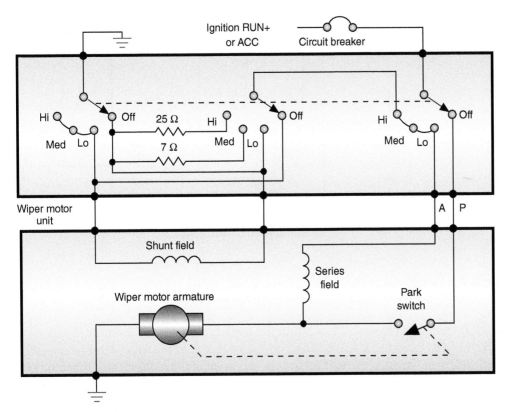

Figure 13-7 Three-speed wiper system schematic.

An opened resistor can be verified by using a voltmeter to measure voltage at the terminal leading to the shunt field. If it drops to 0 volt (V) in the MEDIUM position, the switch must be replaced. By proper use of the wiring schematic and by understanding the correct operation of the system, you are able to diagnose this problem quickly.

In two-speed systems, the motor operating in only one-speed position can be caused by several different faults. It will require the use of wiring schematics and test equipment to locate. Use **Figure 13-8** to step through a common test sequence to locate the reason why the motor does not operate in the HIGH position.

Turn the ignition switch to the ACC or RUN position and place the wiper switch in the HIGH position. Use a voltmeter to test for voltage at the high-speed connector of the

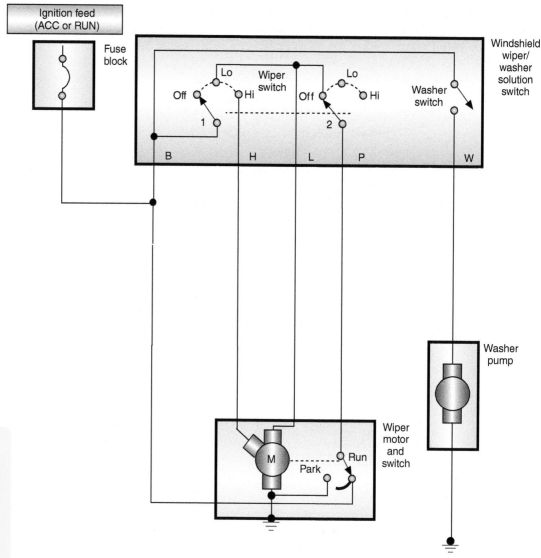

Figure 13-8 Two-speed wiper circuit.

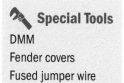

Caution

Do not leave the ignition switch in the RUN position without the engine running for extended periods of time. This may result in damage to the ignition system components.

This instance is used as an example of troubleshooting the wiper system. Be sure to use the correct schematic for the vehicle you are working on.

Classroom Manual
Chapter 13, page 382

Special Tools

DMM
Fender covers
Fused jumper wire

motor. If voltage is present at this point, the high-speed brush is worn or the wire from the terminal to the brush is open. Most shops do not rebuild the wiper motor; replacement is usually the preferred service. If no voltage is present at the high-speed connector, check for voltage at connector terminal H for the switch. If voltage is present at this point, the fault is in the circuit from the switch to the motor. If no voltage is present at this point, replace the switch.

To test for no LOW-speed operation only, use the same procedure used for testing the low-speed circuit.

No Wiper Operation

If the wiper motor does not operate in any speed position, check the fuse or circuit breaker. If the fuse is blown, replace it and test the operation of the motor. Also check for binding in the mechanical portion of the system. This can cause an overload and blow the fuse. The cause for the blown fuse or tripped circuit breaker must be identified.

If the fuse is good, check the motor ground by using a fused jumper wire from the motor body to a good chassis ground. If the motor operates when the ignition switch is in the RUN position and the wiper switch is placed in all speed positions, repair the ground connection. Continue testing if the motor does not operate.

Use a voltmeter to check for voltage at the low-speed terminal of the motor with the ignition switch in the RUN or ACC position and the wiper switch in LOW position. If there is no voltage at this point, test for voltage on the low-speed terminal of the wiper switch. If the voltmeter indicates battery voltage at this terminal, the fault is in the circuit between the switch and the motor. Look for indications of burned insulation or other damage that would affect both the high- and low-speed circuits.

No voltage at the low-speed terminal of the wiper switch indicates that the fault may be in the switch or the power feed circuit. Test for battery voltage at the battery supply terminal of the switch. If there is voltage at this point, the switch is faulty and needs to be replaced. If no voltage is at the supply terminal, trace the circuit back to the battery to locate the fault.

If battery voltage is present at the low-speed terminal of the motor, check for voltage at the high-speed terminal. Voltage at both of these terminals indicates that the motor is faulty and needs to be replaced. If there is no voltage at the high-speed terminal, use the procedure just described to trace the high-speed circuit.

> ⚙ **SERVICE TIP** No voltage at either terminal means the problem is probably at a shared location. In most systems, this would be the power supply portion of the wiper switch and the switch itself. If the switch is good the problem is in the wiring loom.

Slower-than-Normal Wiper Speeds

Slower-than-normal wiper speeds can be caused by electrical or mechanical faults. An ammeter can be used to determine the current draw of the motor with and without the mechanical portion connected. If the draw changes substantially when the mechanical portion is removed, the fault is in the arms and/or wiper blades.

Most electrical circuit faults that result in slow wiper operation are caused by excessive resistance. If the complaint is that all speeds are slow, use the voltage drop test procedure to check for resistance in the power feed supply circuit to the wiper switch. If the power supply circuit is good, then check the switch for excessive resistance.

If the insulated side voltage tests fail to locate the problem, check the voltage drop on the ground side of the wiper motor. Connect the voltmeter positive lead to the ground terminal of the motor (or motor body) and the negative lead to the vehicle chassis. The voltage drop should be no more than 0.1 volt. If excessive, repair the ground circuit connections. If voltage drop on both the insulated and the ground sides of the motor is within specifications, the fault is in the motor.

Wipers Will Not Park

The most common complaint associated with a faulty park switch is that the wipers stop in the position they are in when the switch is turned off. This may not be the direct fault of the **park switch**, however. The park switch is located inside the motor assembly. It supplies current to the motor after the wiper control switch has been turned to the PARK position. This allows the motor to continue operating until the wipers have reached their park position.

The operation of the park switch can usually be observed by removing the motor cover (**Figure 13-9**). Operate the wipers through three or four cycles while observing the latch arm. When the wiper switch is placed in the OFF position, the park switch latch must be in position to catch the drive pawl. Check to make sure the drive pawl is not bent. If good, replace the park switch.

A faulty wiper switch can also cause the park feature to not operate. Using **Figure 13-10**, if wiper 2 is bent or broken so that it does not make an electrical connection with the contacts, the wipers will not park even with the park switch in the PARK position, as shown.

Classroom Manual
Chapter 13, pages
383, 384

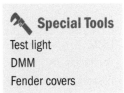
Special Tools
Test light
DMM
Fender covers

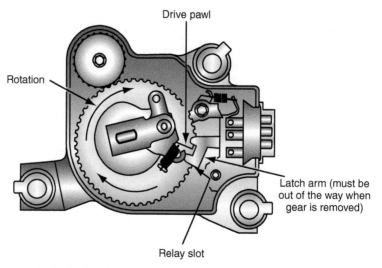

Figure 13-9 Checking the operation of the park switch while the motor is operating.

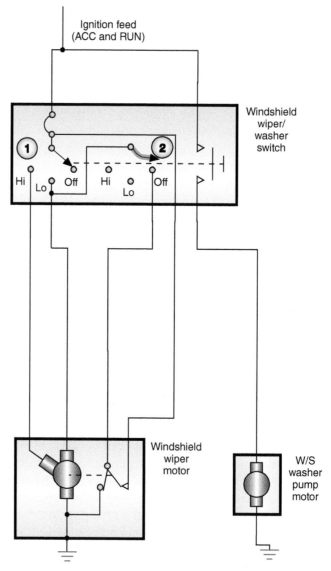

Figure 13-10 A faulty contact wiper in the switch can cause the park feature to not operate.

To test the switch, check for voltage at the low-speed circuit when the switch is moved from the LOW to the OFF position. If the switch operates properly, there should be voltage present for a few seconds after the switch is in the OFF position. No voltage at this circuit when the wiper switch is turned off indicates that the problem may be in the switch.

Some motors provide for replacement of the park switch. However, most shops replace the motor.

The park switch operation can also be checked by using a test light to probe for voltage on the park switch circuit when the wiper switch is turned off. Probing for voltage at this circuit should produce a pulsating light when the motor is running.

If the wiper blades continue to operate with the wiper switch in the OFF position, the most probable cause is welded contacts in the park switch (**Figure 13-11**). If the park switch does not open, current continues to flow to the wiper motor. The only way to turn off the wipers is to turn off the ignition switch, physically remove the wires to the motor, or pull the fuse.

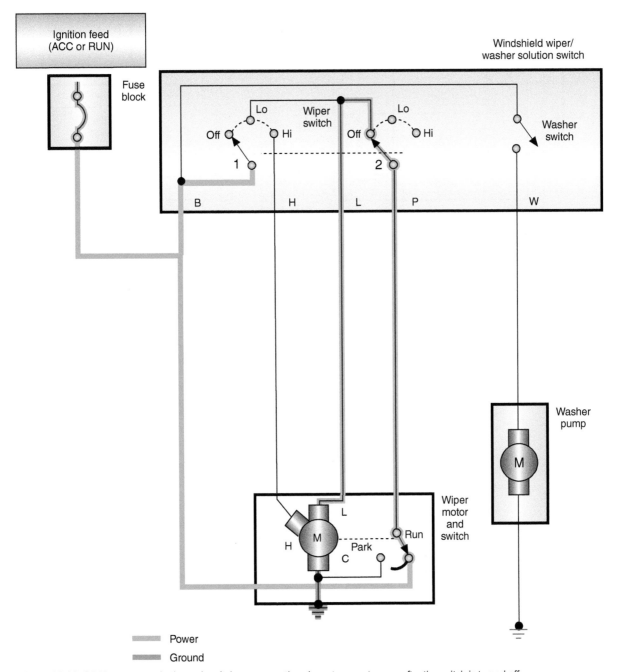

Figure 13-11 Sticking contacts in the park switch can cause the wipers to operate even after the switch is turned off.

Classroom Manual
Chapter 13, page 389

Intermittent Wiper System Diagnosis

The illustration shown in **Figure 13-12** is a schematic of the intermittent wiper system that Ford uses. If the intermittent function is the only portion of the system that fails to operate properly, begin by checking the ground connection for the timer module. If the ground is good, perform a continuity test of the switch using an ohmmeter (**Figure 13-13**). If the switch is good and all wires and connections are good, then replace the module.

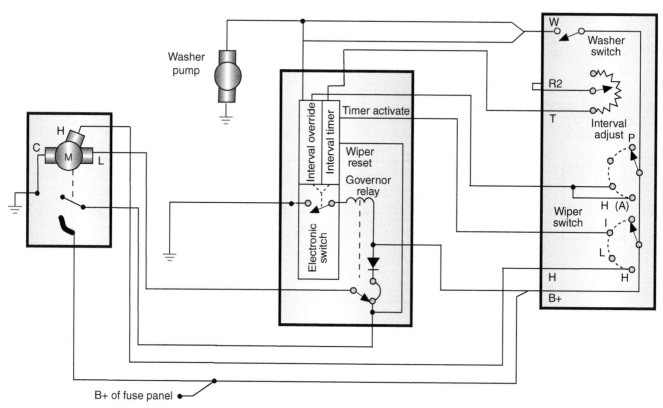

Figure 13-12 Intermittent wiper system schematic.

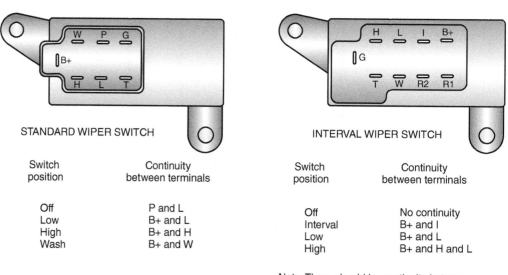

STANDARD WIPER SWITCH

Switch position	Continuity between terminals
Off	P and L
Low	B+ and L
High	B+ and H
Wash	B+ and W

INTERVAL WIPER SWITCH

Switch position	Continuity between terminals
Off	No continuity
Interval	B+ and I
Low	B+ and L
High	B+ and H and L

Note: There should be continuity between terminals R1 and R2 throughout variable resistance range.

Figure 13-13 Wiper switch continuity chart.

Wiper Motor Removal and Installation

Removal procedures differ among manufacturers. Some motors are situated in areas that may require removing several engine compartment components. Always refer to the correct service information for the recommended procedure. **Photo Sequence 28** provides a common method of removing a wiper motor.

PHOTO SEQUENCE 28
Wiper Motor Removal

P28-1 Place fender covers over the vehicle's fenders.

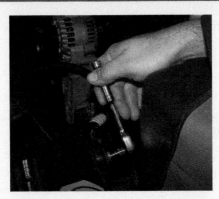

P28-2 Disconnect the battery negative cable.

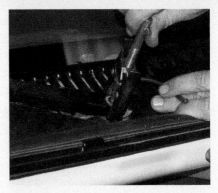

P28-3 Disconnect the wiper arms from the linkage.

P28-4 To gain access to the motor, remove the shield cover from the cowl.

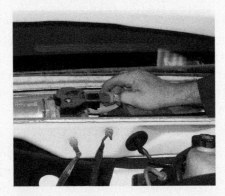

P28-5 Disconnect the wire connector at the motor.

P28-6 Remove the linkage from the motor.

P28-7 Remove the attaching bolts from the motor assembly.

P28-8 Remove the motor.

⚠ **Caution**

The internal permanent magnets of the motor are constructed of ceramic material. Use care while handling the motor to avoid damaging the magnets.

On some vehicles, the linkage is removed by lifting the locking tab and pulling the clip away from the pin (refer to Figure 13-6). Installation is basically the reverse of the removal procedure, but be sure to attach the ground wire to one of the mounting bolts during installation.

Wiper Switch Removal and Installation

The wiper switch removal procedure differs among manufacturers and depends on switch location. The procedures presented here are common. However, always refer to the manufacturer's service information for correct procedures. Always protect the customer's investment by using fender covers while disconnecting the battery negative cable.

Dash-Mounted Switches. Depending on the location of the switch control, it may be necessary to remove the finish panel. Usually the finish panel is held in place by a combination of fasteners and clips.

Remove the switch housing retaining screws. Then remove the housing. Pull off the wiper switch knob. Disconnect the wire connectors from the switch. Remove the switch from the dash.

Reverse the procedure to install the switch.

Steering Column–Mounted Switches. Remove the upper and lower steering column shrouds to expose the switch. Disconnect the wire connectors to the switch. It may be necessary to peel back the foam to gain access to the retaining screws. Remove the screws and the switch.

DIAGNOSING COMPUTER-CONTROLLED WIPER SYSTEMS

Classroom Manual
Chapter 13, page 391

Most computer-controlled wiper systems are capable of performing a diagnostic routine and setting DTCs if a problem is indicated. The following are some common DTCs that are associated with the wiper system and common diagnostic procedures for locating the fault. Use **Figure 13-14** as an example of the system.

Wiper Park Switch Input Circuit Low

Whenever the ignition switch is in the RUN position and the wipers are requested to be turned ON, the front control module (FCM) monitors the park switch circuit. As the wipers sweep across the windshield and return to their parked position, the switch should cycle from open to close. If the FCM detects a short to ground condition, then the "Wiper park switch input circuit low" fault is set. This DTC can indicate that the front wiper park switch sense circuit is shorted to ground, the wiper motor is faulty (internal park switch failure), or a fault in the FCM. The customer would probably state that the wipers turn off immediately after the switch is set to the OFF position, regardless of the position of the wipers on the windshield.

Special Tools
Scan tool
DMM

The BCM can perform the functions of the FCM described here.

Begin by inspecting the related wiring for any chafed, pierced, pinched, or partially broken wires. If no faults are found, erase the fault codes and disconnect the front wiper motor harness connector. Turn the ignition to the RUN position and activate the wipers for about 30 seconds, and then check for a "Wiper park switch input circuit high" DTC. If this fault code does set, then replace the wiper motor.

If the circuit low fault is set again, measure the resistance between ground and the wiper park switch sense circuit. If the resistance is below 5 ohms (Ω), the circuit is shorted to ground. If the resistance is more than 5 ohms, replace the FCM.

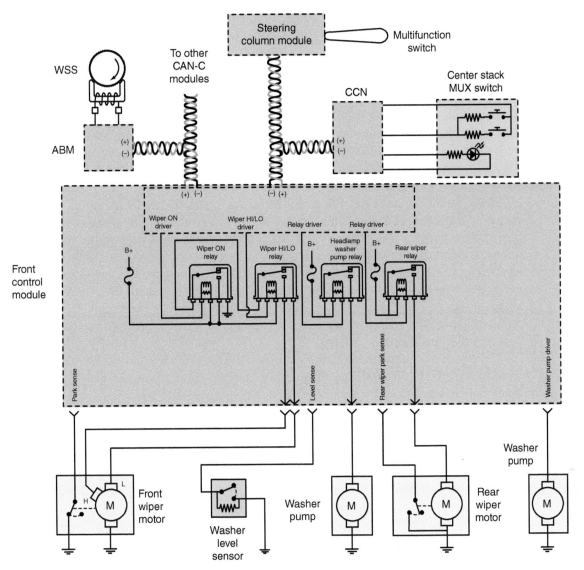

Figure 13-14 Computer-controlled wiper system.

Wiper Park Switch Input Circuit High

The "Wiper park switch input circuit high" fault is set when the FCM detects an open circuit. This DTC can also set if the park switch sense circuit is shorted to voltage. This DTC can indicate that the front wiper park switch sense circuit is shorted to ground, the wiper motor is faulty (internal park switch failure), or a fault in the FCM. The customer would probably state that the wipers turn off immediately after the switch is set to the OFF position, regardless of the position of the wipers on the windshield.

Begin by inspecting the related wiring for any chafed, pierced, pinched, or partially broken wires. If no faults are found, erase the fault codes and disconnect the front wiper motor harness connector and the FCM connector. Measure for voltage on the park switch sense circuit. Since both connectors are disconnected, there should be 0 volt on the circuit. If there is measured voltage, the circuit is short to a voltage supply circuit.

With the connectors still disconnected, measure the resistance in the park switch circuit. The resistance should be close to 0 ohm. If the ohmmeter indicates high resistance or open circuit, repair the sense circuit. If the problem still has not been located, test the park switch ground circuit.

If all tests indicate there is no problem, the fault is in the FCM. Be sure to test all connections, power, and ground feeds before condemning the FCM.

Wiper On-Off Control Circuit Low

The "Wiper On-Off control circuit low" DTC is set if the FCM detects a short to ground or low-voltage condition in the ON relay's control circuit. This could be caused by an open in the fused battery feed circuit to the relay, a faulty relay coil, an open in the control circuit between the relay and the FCM, or a short to ground in the control circuit. In addition, the FCM may be the faulty component. The customer would probably state that his or her wipers do not work in any switch position.

Begin by checking the condition of the fuse that protects the voltage supply circuit to the relay. If the fuse is blown, then there is a short to ground in the circuit between the fuse and the relay, or the relay coil is shorted. If the fuse is good, replace the relay with a known good unit and test operation. If the wipers still do not operate, continue testing by removing the relay and turning the ignition to the RUN position. Measure for voltage at cavity 86 of the relay. If the voltage is not the same as battery voltage, there is an open or a high resistance in the circuit. It is always a good practice to load the circuit when testing. Removing the relay means that the test is an open-circuit test. Load the circuit with a test light and measure available voltage.

If the fault has not yet been located, reinstall the relay and backprobe the relay control circuit at the FCM. Measure the voltage at this location. If the voltmeter reads 0 volt, there is an open between the relay and the FCM. If the voltage reads battery voltage, use a fused jumper wire to jump the backprobed connector to ground. If the wipers come on during this test, the FCM has a faulty driver.

AUTHOR'S NOTE The input from the multifunction switch is not being tested because of the fault code retrieved. In this instance, the DTC indicates the problem area as being the relay control circuit. If the wipers did not operate and there were not any DTCs, then the input side would be tested.

Wiper On-Off Control Circuit High

The "Wiper On-Off control circuit high" DTC is set if the FCM detects a short to voltage condition in the ON relay's control circuit. This could be caused by a shorted relay coil or an internal fault in the FCM. The customer would probably state that his or her wipers do not work in any switch position.

To test for this type, begin by erasing the fault codes and replace the relay with a known good unit. Operate the system for about 30 seconds and then turn it off. Use the scan tool to see if the DTC returns. If it does not return, the relay is the problem. If the DTC returns, the FCM is faulty. Be sure to test all connections, powers, and grounds before faulting the FCM.

Wiper HI/LOW Control Circuit Low

The "Wiper HI/LOW control circuit low" DTC is set when the FCM detects a short to ground or low-voltage condition on the HI/LOW relay control circuit. When the driver in the FCM is turned off, the FCM should sense 12 volts on the control circuit. When the driver is turned on, the voltage should go to about 700 mV. If the voltage is always low on this circuit, the FCM will set the code. The customer will probably state that their wipers work on low and intermittent speeds, but not on high speed. Causes for this fault to set include an open battery feed circuit, open relay coil, short to ground in the control circuit, and an internal problem within the FCM.

> **AUTHOR'S NOTE** If the control circuit is shorted to ground between the relay and FCM, the customer will probably state that the wipers appear to be working only on high speed.

Begin by testing all fuses associated with the relay control circuit. Referring back to Figure 13-14, if the wipers work on low speed but not on high speed, the problem is between the common splice in the 12-volt feed circuit and the FCM.

Remove the HI/LOW relay and measure the voltage at cavity 86 with the ignition in the RUN position. It would be best to load the circuit with a test light when measuring this voltage. If the voltage is about equal to battery voltage, the circuit to this point is good. If the voltage is 0 volt, there is an open in the circuit toward the common splice.

Replace the relay with a known good unit and clear all DTCs. Attempt to operate the wipers on high speed; then turn the wipers off again. Use the scan tool to read DTCs; if the fault code does not come back, the relay is the problem; if the fault comes back, then check the circuit between the relay and the FCM for an open or a short to ground.

Backprobe the FCM connector for the HI/LOW relay control circuit. With the wipers operating at low speed, jump the backprobed terminal to ground. If the wipers go to high speed, the fault is in the FCM. If the wipers still do not operate at high speed, check for an open in the relay control circuit between the relay and the FCM.

Wiper HI/LOW Control Circuit High

The "Wiper HI/LOW control circuit high" DTC is set when the FCM detects a short to voltage or high-voltage condition on the HI/LOW relay control circuit. When the driver in the FCM is turned off, the FCM should sense 12 volts on the control circuit. When the driver is turned on, the voltage should go to about 700 mV. If the voltage is always high on this circuit, the FCM will set the code. The customer will probably state that his or her wipers work on low and intermittent speeds, but not on high speed. Causes for this fault to set include a short to battery voltage on the circuit between the relay and the FCM, a faulty relay, or an internal problem within the FCM.

Begin erasing all DTCs and replacing the relay with a known good unit. Operate the wipers for about 30 seconds, and then turn them off. Check to see if the DTC returns; if not, the relay is the problem. If the DTC returns, test for voltage on the circuit between the relay and the FCM by disconnecting the connector on both ends and measuring for voltage. If there is a measured voltage, then isolate the shorted circuit and repair. If the circuit is not shorted to voltage, replace the FCM.

AUTOMATIC WIPER SYSTEM DIAGNOSIS

Classroom Manual
Chapter 13, page 393

A vehicle that is equipped with the automatic wiper system uses the same type of computer-controlled wipers, but adds a rain sensor. Also, when automatic wipers are enabled, the intermittent position on the multifunction switch control knob is used to set the desired system sensitivity. Using the schematic of an automatic wiper system (**Figure 13-15**), you can see that the right multifunction switch is a MUX switch to the left multifunction switch, which is a LIN slave module to the cabin compartment node (CCN). The electronic wiper switch sensitivity message is sent to the CCN over the LIN

Special Tools
Scan tool
DMM

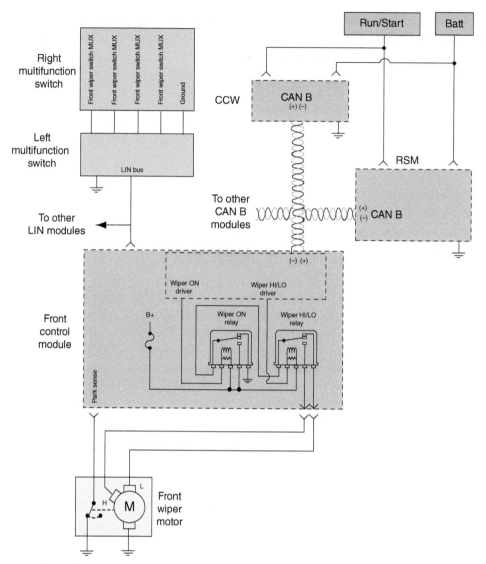

Figure 13-15 Automatic wiper system.

data bus, and then the CCN relays the message to the rain sensor module (RSM) over the CAN B data bus. The RSM monitors an area within the wipe pattern of the windshield glass for the accumulation of moisture. Based upon internal programming and the selected sensitivity level, when sufficient moisture has accumulated, the RSM sends the appropriate electronic wipe command messages to the FCM over the CAN B data bus. The FCM operates the front wiper system accordingly.

AUTHOR'S NOTE The CCN is another term used to describe the electronic instrument cluster when it is a dominant module on the CAN bus network.

If the wipers do not operate properly in either the manual mode or the automatic mode, the problem is in the basic wiper system. If the wipers operate properly in the manual mode, but not in the automatic mode, then the problem is in the automatic function components. This includes the RSM and the data bus network to it. It is also possible that the system is disabled.

⚠ WARNING **To avoid serious or fatal injury, disable the supplemental restraint system (SRS) before attempting to diagnose or remove parts on the steering wheel, steering column, or instrument panel. Disconnect and isolate the battery negative (ground) cable; then wait for 2 minutes for the system capacitor to discharge before performing further diagnosis or service. Failure to follow this warning could result in accidental air bag deployment.**

The hardwired wiper system circuits and components may be diagnosed using conventional diagnostic tools and procedures. However, conventional diagnostic methods will not prove conclusive in the diagnosis of the front wiper and washer system or the electronic controls or communication between other modules and devices that provide the automatic wiper feature. The most reliable, efficient, and accurate means to diagnose the front wiper and washer system or the electronic controls and communication related to front wiper and washer system operation require the use of a diagnostic scan tool.

A problem that the customer may experience with automatic wiper systems is called **false wipes**. False wipes are unnecessary wipes that occur when the automatic wiper system is enabled but there is no apparent rain or moisture on the windshield. When diagnosing this type of complaint, keep in mind that the system is designed to operate whenever it detects moisture. Any road spray, bug splatters, or mist from passing vehicles may cause the wipers to cycle. These are normal characteristics of this system and are not false wipes.

False wipes are usually the result of a foreign material on the windshield. In some instances, false wipes can result from flaws in the windshield that interfere with the system optics. Anything that distorts the intensity of the IR light beams or affects the ability of the photo diodes to accurately measure the returning beams can result in the RSM logic misinterpreting the input data as moisture on the windshield. The optics for this system include the lenses of the RSM, the lenses of the RSM bracket, the adhesive pad layer that bonds the bracket to the inside of the windshield, and the windshield glass.

Usually the system can detect faults in the electronics or bus communications that may cause the system to not function properly. Use a scan tool to check for DTCs associated with the front wiper system, the multifunction switch, the CCN, and the RSM. If there are any DTCs in the components that operate within the system, use the appropriate diagnostic procedures to isolate the problem. If no DTCs are present, then perform the routine illustrated in **Photo Sequence 29**. If a problem is found within the RSM mounting bracket, it is usually serviced as a unit with the windshield glass. If either the bracket or the windshield glass is the cause of the system problem, the entire RSM bracket and windshield glass unit must be replaced.

WINDSHIELD WASHER SYSTEM SERVICE

Many windshield washer problems are due to restrictions in the delivery system. To check for restrictions, remove the hose from the pump and operate the system. If the pump ejects a stream of fluid, then the fault is in the delivery system. If the pump does not deliver a spray of fluid, continue testing using the following procedure:

1. Make checks of obvious conditions such as low fluid level, blown fuses, or disconnected wires.
2. Activate the washer switch while observing the motor. If the motor operates but does not squirt fluid, check for blockage at the pump. Remove any foreign material. If there is no blockage, then replace the motor.

Classroom Manual
Chapter 13, page 394

Special Tools
Fender covers
DMM
Test light
Safety glasses

3. If the motor does not operate, use a voltmeter or test light to check for voltage at the washer pump motor with the switch closed. If there is voltage, then check the ground circuit with an ohmmeter. If the ground connection is good, then replace the pump motor.

4. If there is no voltage to the pump motor in step 3, trace the circuit back to the switch. Test the switch for proper operation. If there is power into the switch but not out of it to the motor, replace the switch.

If the motor is in need of replacement, follow this procedure for pumps installed in the reservoir (**Figure 13-16**). Disconnect the wire connector and hoses from the pump. Remove the reservoir assembly from the vehicle. Use a small blade screwdriver to pry out the retainer ring.

PHOTO SEQUENCE 29
Inspection of Rain Sensor Module

P29-1 Carefully inspect the outer surface of the windshield glass for physical damage, including scratches, cracks, or chips in the vicinity of the RSM mounting bracket lenses. If damage is present, replace the windshield.

P29-2 From the outside of the windshield glass, carefully inspect the adhesive layer between the windshield glass and the RSM bracket for any voids greater than 0.040 inch (1 mm). If an adhesive void is detected, replace the flawed RSM mounting bracket and windshield unit.

P29-3 Remove the inside rearview mirror.

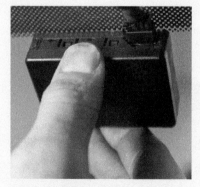

P29-4 Lightly pull the RSM away from the windshield bracket to confirm that both module sliding cam locks are fully engaged with all four pins of the mounting bracket on the inside of the windshield glass. If needed, reinstall the module onto the bracket.

P29-5 Remove the RSM from the mounting bracket.

P29-6 Inspect the RSM lenses and the mounting bracket lenses for contamination.

PHOTO SEQUENCE 29 (CONTINUED)

P29-7 Clean any foreign material from each of the lenses using rubbing alcohol and a lint-free cloth.

P29-8 Carefully inspect the RSM for any physical damage, including scratches on the RSM lenses.

P29-9 Carefully inspect the RSM mounting bracket for any physical damage, including scratches on the RSM bracket lenses. If any damage is found, replace the bracket and windshield as a unit.

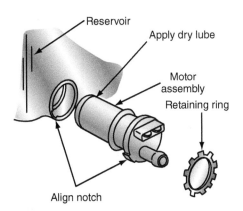

Figure 13-16 Reservoir-mounted washer pump and motor.

⚡ WARNING **Wear safety glasses to prevent the ring from striking your eyes. Also be careful to position the palm of your hand so that if the screwdriver slips it will not puncture your skin.**

Use a pair of pliers to grip one of the walls that surround the terminals. Pull out the motor, seal, and impeller.

Before installing the pump assembly, lubricate the seal with a dry lubricant. The lubricant is used to prevent the seal from sticking to the wall of the reservoir. Align the small projection on the motor with the slot in the reservoir and assemble. Make sure the seal seats against the bottom of the motor cavity. Use a 12-point, 1-inch socket to hand press the retaining ring into place.

⚠ Caution

Do not operate the washer pump without fluid. Doing so may damage the new pump motor.

Classroom Manual
Chapter 13, page 396

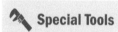 **Special Tools**

Fused jumper wires
Test light
DMM
Fender covers

Replace the reservoir assembly in the vehicle. Reconnect the hose and wires. When refilling the reservoir, do so slowly to prevent air from being trapped in the reservoir. Check system operation while checking for leaks.

BLOWER MOTOR SERVICE

Conventional blower motor speed is controlled by sending current through a **resistor block**. The resistor block is a series of resistors with different values. There is usually one less resistor than there are fan speed positions because the high-speed circuit bypasses the resistors. The higher the resistance value, the slower the fan speed. The position of the switch determines which resistor will be added to the circuit. Circuits can use either ground side switches or insulated side switches.

If the customer complaint is that the fan operates in only a couple of speed positions, the most likely cause is an open resistor in the resistor block. Using the illustration in **Figure 13-17**, if resistor 1 is open, the motor will not operate in any position except high speed. If resistor 2 is open, the motor would operate only in high and M2 speeds. If resistor 3 is open, the motor would operate in all speeds except low.

If the motor operates in any one of the speed select positions, the fault is not in the motor. If the motor fails to operate at all, begin by inspecting the fuse. If the fuse is good,

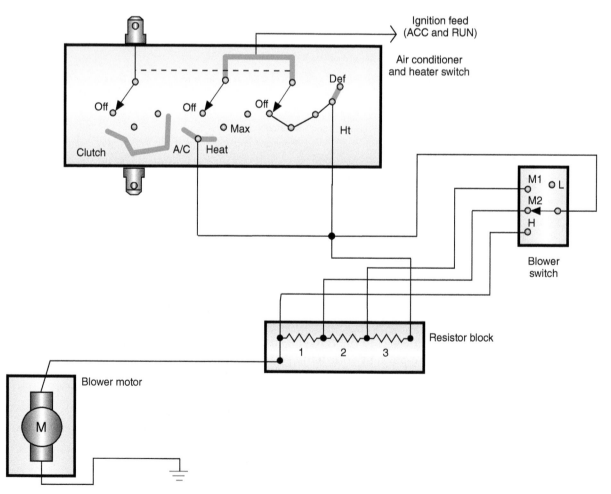

Figure 13-17 Typical wiring for a four-speed fan motor circuit.

use the correct wiring schematic to determine whether ground or insulated side switches are used. The diagnostic procedure used depends on the circuit design.

Inoperable Motor with Insulated Switches

Use a fused jumper wire to bypass the switch and resistor block to check motor operation. Connect the fused jumper wire from a battery positive supply to the motor terminal. If the motor does not operate, connect a second fused jumper wire from the motor body to a good ground. Replace the motor if it still does not operate.

If the motor operates when the switch and resistor block are bypassed, trace up the circuit toward the switch. Use a voltmeter or test light to check for voltage in and out of the blower speed control switch. The switch is faulty if voltage is at the input terminal but not at any of the output terminals. No voltage at the input terminal indicates an open in the circuit between the battery and the switch.

Inoperable Motor with Ground Side Switches

Using **Figure 13-18** as an example of a negative side switch blower motor circuit to test the motor, connect the fused jumper wire from the motor negative terminal to a good ground. This bypasses the switch and resistor block. If the motor does not operate, use a voltmeter or test light to check for voltage at the battery terminal of the motor. If voltage is present, then the motor is defective.

If there is no voltage at the battery terminal of the motor, the problem is in the circuit from the battery to the motor. Be sure to check the circuit breaker.

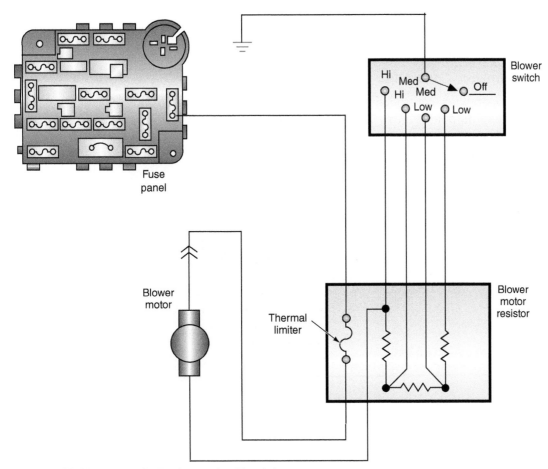

Figure 13-18 Blower motor circuit using negative side switch.

⚠ Caution

The resistor block is mounted in the heater/air-conditioning housing where it is cooled by airflow from the fan. Do not run the fan motor with the resistor block removed from the airflow because it may overheat and burn the coils.

⚙ **SERVICE TIP** Because the high-speed circuit bypasses the resistor block, it is doubtful that no motor operation would be the fault of open resistors. However, an open in the wire from the block to the motor can cause the problem. Most likely the switch is bad and in need of replacement. Always confirm your diagnosis by doing a continuity test on the switch.

Operation of the motor when the jumper wire is connected to the ground terminal indicates that the problem is in the switch side of the circuit. Use an ohmmeter to check the ground connection for the switch. A fused jumper wire or test light can also be used to test this connection.

If the ground connection is good, use a voltmeter or test light to probe for voltage at any of the circuits from the resistor block to the switch. Replace the switch if there is power to this point. Replace the resistor block if there is no power at these points.

Constantly Operating Blower Fans

General Motors designed many of its blower fans to constantly operate. Do not confuse this normal operation with a circuit defect.

Constantly running blower motors are more common in ground side switch systems. A short to ground at any point on the ground side of the circuit will cause the motor to run. Other areas to check include the switch and the circuit between the switch and the resistor block.

In insulated side switch circuits, check for copper-to-copper shorts in the power side of the system. If the motor receives power from another circuit, due to a copper-to-copper short, the motor will continue to run whenever current flows through that circuit. Some systems may incorporate a relay, and if the contact points fuse together, the motor will continue to operate.

Computer-Controlled Fans

Classroom Manual
Chapter 13, page 398

Most computer-controlled motor circuits control motor speed by duty cycling. Typically, a power module (**Figure 13-19**) controls the motor based on drive signals from the

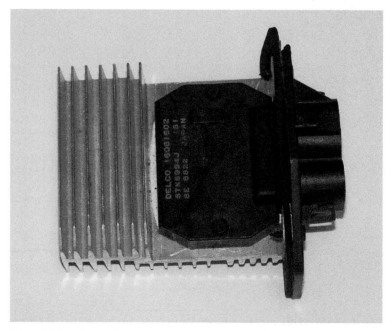

Figure 13-19 Power module.

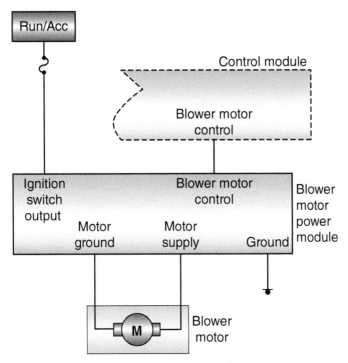

Figure 13-20 Example of a motor circuit using a power module.

control module. The control module sends a pulse-width modulated (PWM) signal to the power module, and the power module amplifies the signals to provide the variable motor speed.

A scan tool can be used to activate the motor at different speeds. However, if the blower motor does not operate properly, diagnostics with a DMM or lab scope will need to be performed to isolate the root cause. Consider the system illustrated by the wiring diagram in **Figure 13-20**. Backprobe the power module at the terminals for both the control and motor ground circuits. Using two DMMs set to read voltage, connect one between the control circuit terminal and chassis ground, and the second between the motor ground terminal and chassis ground. The same set-up can be done with a dual-channel lab scope.

For this example system, since the blower motor is controlled by low-side switching, with the blower motor off, both the input voltage and the voltage on the ground terminal should be 12 volts. As the blower motor switch is rotated from LOW speed to HIGH speed, the control circuit voltage reading should drop. If a lab scope is used, a duty cycle will be shown. The duty cycle controls the voltage level across the blower motor. The voltage on the motor ground circuit will also decrease as the blower motor speed increases.

Special Tools

Scan tool
DMM (2)
Back-probing tools
Fused jumper wires

ELECTRIC DEFOGGER DIAGNOSIS AND SERVICE

If the rear window defogger fails to operate when the switch is activated, use a test light to test the **grids**. The rear window defogger grids are a series of horizontal, ceramic, silver-compounded lines that are baked into the surface of the window. Under normal conditions, the test light should be bright on one side of the grid and off on the other side. If the test light has full brilliance on both sides of the grid, then the ground connection for the grid is broken.

If the test light does not illuminate at any position on the grid, use normal test procedures to check the switch and relay circuits. There may be several fuses involved in the system. Use the correct wiring diagram to determine the fuse identification.

Classroom Manual
Chapter 13, page 399

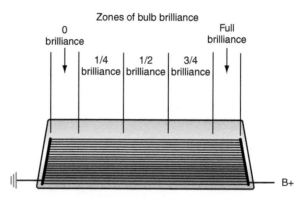

Figure 13-21 Zones of test light brilliance while probing a rear window defogger grid.

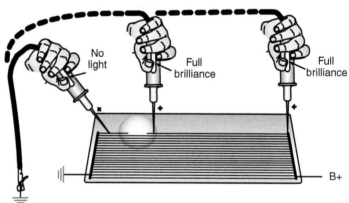

Figure 13-22 Test light brilliance when passed over a break.

Most rear window defogger complaints are associated with broken grids. These will generally be complaints that only a portion of the window is cleared while the rest remains foggy. Some grid wire breaks are easily detected by visual inspection. However, many are too small to see. To test the grid lines, start the engine and activate the system. (Remember that the system is controlled by a timer.) Use a test light to check each grid wire to locate the breaks. Test each grid in at least two places—one on each side of the centerline. The test results that should be obtained on each grid are illustrated in **Figure 13-21**.

If the test light does not indicate normal operation on a specific grid line, place the test light probe on the grid at the left bus bar and work toward the right until the light goes out. The point where the light goes out is the location of a break (**Figure 13-22**). Mark the location of the break with a grease pencil on the *outside* of the glass.

The rear window defogger should turn off about 10 minutes after activation. If the circuit fails to turn off, check the ground for the control module (**Figure 13-23**). If the ground is good, replace the module.

Grid Wire Repair

If the grid wire is broken, it is possible to repair the grid with a special repair kit. Follow the procedures in **Photo Sequence 30** to repair the grid.

Bus Bar Lead Repair

The bus bar lead wire can be resoldered using a solder containing 3% silver and a rosin flux paste. Clean the repair area using a steel wool pad. Apply the rosin flux paste in small quantities to the wire lead and bus bar. **Tin** the solder iron tip with the solder. Finish the repair by soldering the wires to the bus bar. Be careful not to overheat the wire.

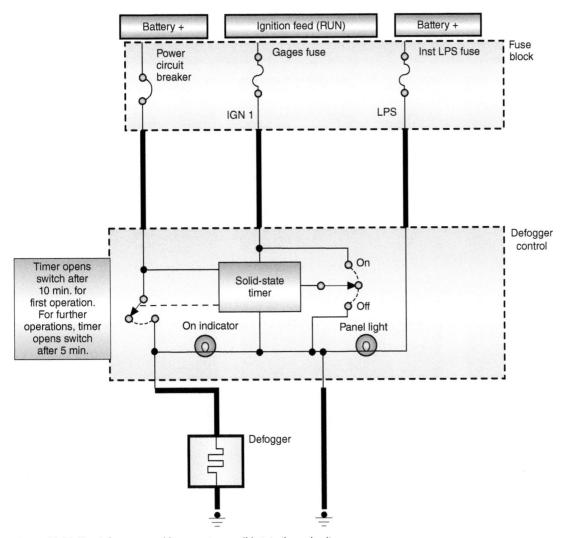

Figure 13-23 The defogger control incorporates a solid-state timer circuit.

PHOTO SEQUENCE 30
Typical Procedure for Grid Wire Repair

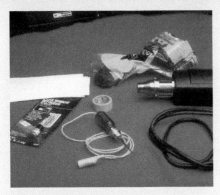

P30-1 Tools required to perform this task include masking tape, repair kit, 500°F heat gun, test light, steel wool, alcohol, and a clean cloth.

P30-2 Clean the grid line area to be repaired by buffing with steel wool and wiping clean with a cloth dampened with alcohol. Clean an area about 1/4 inch (6 mm) on each side of the break.

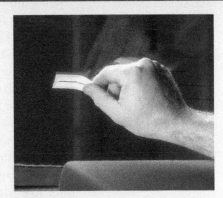

P30-3 Position a piece of tape above and below the grid. The tape is used to control the width of the repair.

PHOTO SEQUENCE 30 (CONTINUED)

P30-4 Mix the hardener and silver plastic thoroughly. If the hardener has crystallized, immerse the packet in hot water.

P30-5 Apply the grid repair material to the repair area using a small stick.

P30-6 Remove the tape.

P30-7 Apply heat to the repair area for 2 minutes. Hold heat gun 1 inch (25 mm) from the repair.

P30-8 Inspect the repair. If it is discolored, apply a coat of tincture of iodine to the repair. Allow to dry for 30 seconds, and then wipe off the excess with a cloth.

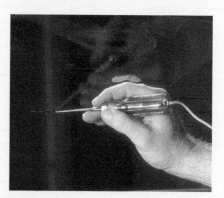

P30-9 Test the repair with a test light. Note: it takes 24 hours for the repair to fully cure.

POWER WINDOW DIAGNOSIS

Classroom Manual
Chapter 13, page 405

Usually the door panel needs to be removed to gain access to the window motor and regulator. Several methods are used to attach the trim panel to the door frame. Common methods include screws, push pins, clips, and "L" brackets.

Figure 13-24 illustrates a typical door panel removal procedure. Begin by locating and removing any screws that fasten the panel to the frame. This panel uses "L" hooks molded into the panel. This requires the panel to be lifted off the attachment hooks and the lock rod at the same time in order to remove the panel. Once the panel is no longer attached to the door frame, disconnect any electrical connectors and actuating rods to free the panel. Lay the panel on a clean work bench and protect it from damage.

Special Tools

Test light
DMM
Fused jumper wires

If the trim panel is attached with push pins, use a trim stick to gently pry the panel from the frame. Be sure to have the push pins properly aligned with their receiver holes prior to pushing the panel back into position on the frame.

Use the illustration in **Figure 13-25** as a guide to diagnosing the power window circuit. If the window does not operate, begin by testing the circuit breaker. Use a test light or voltmeter to test for voltage on both sides of the circuit breaker. If voltage is present on

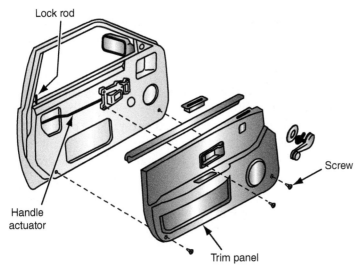

Figure 13-24 Door panel attachment.

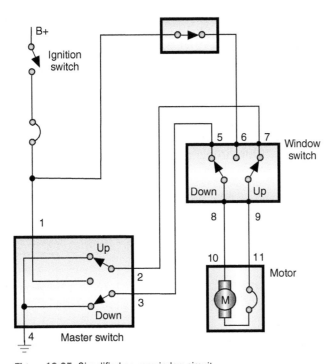

Figure 13-25 Simplified power window circuit.

both sides, then the circuit breaker is good. If there is voltage into the circuit breaker but not out of it, the circuit breaker is faulty or tripped due to a short. If there is no voltage into the circuit breaker, then there is an open in the feed from the battery.

If the circuit breaker is good, use fused jumper wires to test the motor. The motor is a reversible motor, so connections to the motor terminals are not polarity sensitive. Disconnect the wire connectors to the motor. Connect battery positive to one of the terminals and ground the other. If the motor does not operate, reverse the fused jumper wire connections. The motor should reverse directions when the polarity is reversed. If the motor does not operate in one or both directions, it is defective and needs to be replaced.

> ⚠ **Caution**
>
> Study the service information prior to attempting to remove the panel. Sometimes there will be hidden screws. Failure to remove these screws before attempting to pull the panel will damage the panel.

⚡ **WARNING** Do not place your hands into the window's operating area. Make the final test connections outside of the door where there is no danger of getting caught in the window track.

If the motor operates when the switches are bypassed, the problem is in the control circuit. To test the master switch, connect the test light between terminals 1 and 2 (**Figure 13-26**). When the master switch is in the rest (OFF) position, the test light should illuminate. If the light does not glow, there is an open in the circuit to the window switch or from the window switch to ground at terminal 4. Check the ground at terminal 4 for good connections. If the connections are good, continue testing.

If the test light illuminates when connected across terminals 1 and 2, place the switch in the UP position. The test light will go out if the switch is good. Repeat the test between terminals 1 and 3. Place the switch in the DOWN position.

If the master switch is good, test the window switch. Battery voltage should be present at terminal 6. If not, check to see if the lockout switch is closed. Check the circuit from terminal 6 to the circuit breaker. Connect the test light across terminals 5 and 6 (**Figure 13-27**). The light should come on. Move your test light to connect between terminals 6 and 7. Again, the test light should come on. Next, connect the test light between terminals 8 and 9 of the window switch. With the switch in the at-rest position, the test light should be off. Placing the window switch in either the UP or the DOWN position should illuminate the test light. If the light does not come on, you need to isolate the problem. It may be the switch, the circuits between the switch and the motor, or the motor itself. Use common test methods to determine the fault.

Slower-than-normal operating speeds are an indication of excessive resistance or of binding in the mechanical linkage. Use the voltage drop test method to locate the cause of excessive resistance. Excessive resistance can be in the switch circuits, the ground circuit, or in the motor. If the problem is mechanical, lubricate the track and check for binding or bent linkage.

The circuit used in this example is typical. However, use the service information for the vehicle you are working on to get the correct wiring schematic.

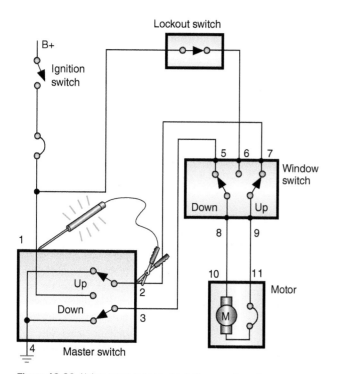

Figure 13-26 Using a test light to check the operation of a power window master switch.

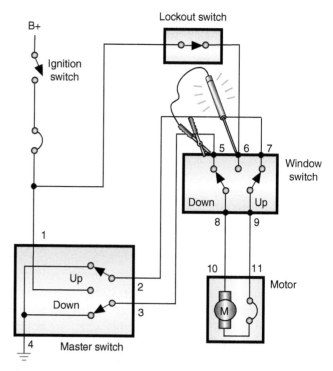

Figure 13-27 Test light connections for testing the window switch.

⚠ WARNING **Follow the manufacturer's recommended procedure when removing the power window motor. The springs used in window regulators can cause serious injury if removed improperly.**

CUSTOMER CARE Testing and repair of the power window system usually require the door panels to be removed. Several methods are used by manufacturers to secure the door panel. Always refer to proper service information to determine the correct methods of removal and installation of the panel to prevent damage. Also, use new clips (if applicable) to assure a tight connection and eliminate noise from the panel. Most doors will have a sound dampening material behind the panel. Therefore, you must remember to reinstall this material.

Computer-Controlled Window Systems

Computer-controlled window systems have the same type of motor and regulator as the standard power window system. Thus, these components are tested in the same manner. The main differences between the two systems are that on computer-controlled windows, the switches are used as inputs and motor control is by commands from the module. Like most computer-controlled systems, diagnostics will be enhanced by the setting of DTCs. The scan tool can also be used to perform actuation tests of the system.

Figure 13-28 illustrates a system that uses door modules to control the operation of windows and other accessories. Begin by accessing the door module with the scan tool and see if DTCs have been set. If there are DTCs, then follow the diagnostic procedure for that code. If there are no DTCs, then access the actuation mode of the scan tool and attempt to command the window up and down. If the window operates properly, the problem is likely within the input side of the system. Use the scan tool to see the data stream from the input switches. If the input is not being seen by the module, use a DMM to confirm proper voltage to and from the switch.

If the activation fails to operate the window motor, then use a DMM to monitor voltages while performing the activation. Be sure to test both sides of the circuit since the module will typically use H-gate circuits. Connect a DMM test lead to each terminal at the motor. When actuating the window motor, the DMM should show battery voltage. When the motor operates in the opposite direction, the DMM should read a reversed polarity with the same voltage value. If proper voltages are present, replace the motor. If the voltage is not correct, trace the circuit between the motor and the module for opens or shorts. If the circuits are good, replace the module.

⚠ WARNING **Be careful when using the scan tool to command window operation. Do not place your hands, arms, or head where they can become pinched or trapped when the window moves.**

POWER SEAT DIAGNOSIS

The power seat system is usually very simple to troubleshoot. Test for voltage to the input of the switch control. If voltage is available to the switch, remove it from the seat or arm rest. Using a continuity chart from the service information, test the switch for proper operation. If the switch operates properly, it may be necessary to remove the seat to test the motors and circuits to the motors.

The power seat motors are tested in the same manner as the power window motor. Be sure to test each armature of the trimotor. If any of the armatures fail to operate, the trimotor must be replaced as a unit.

Classroom Manual
Chapter 13, page 407

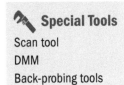

Special Tools
Scan tool
DMM
Back-probing tools

Classroom Manual
Chapter 13, page 409

Special Tools
Fused jumper wires
DMM
Test light

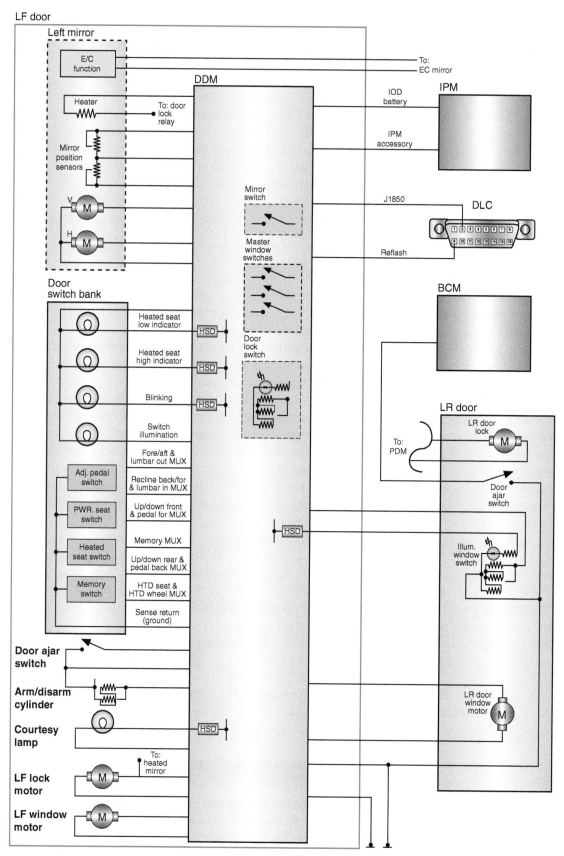

Figure 13-28 Computer-controlled power window system.

⚡ WARNING Be careful when making the fused jumper wire connection to test the motor. Do not place your hands in locations where they can become pinched or trapped when the seat moves.

⚡ WARNING If the trimotor needs to be replaced, follow the manufacturer's service procedures closely. Improper removal of the springs may result in personal injury.

Noisy operation of the seat can generate from the motor, transmission, or cable. If the motor or transmission is the cause of the noise, it must be replaced. A noisy cable can usually be cured with a dry lubricant, provided the cable is not damaged.

MEMORY SEAT DIAGNOSIS

If the seat motors fail to operate under any condition, test the motors and switches as outlined earlier. This section relates only to that portion of the system that operates the memory function.

Most modern memory seat systems provide diagnostics by use of a scan tool. Although not all systems store fault codes, the scan tool can usually be used to check inputs and perform actuator tests. If the seat operates properly when the power seat switches are used, but not the memory feature, the problem is either the inputs or the memory seat control module. Use the scan tool to check the inputs from all switches. Also, test the input from the keyless entry system if it is tied into the memory seats. If the scan tool indicates that the inputs are operating properly, check the connector from the control module to the motors. Also, check the powers and ground for the module. If these are good, replace the module.

If the memory seat system does not support scan tool diagnostics, the problem must be isolated using basic electrical diagnostics. The following is a typical example of testing the system. You need to reference proper service information for the vehicle being diagnosed in order to determine what voltage values are used and which circuits to test.

Using the illustration in **Figure 13-29** of a memory seat circuit, this system would be diagnosed as follows: All tests are performed at memory seat module connectors C1 and C3. The connectors are disconnected from the module to perform the tests. Place the ignition switch in the RUN position, with the gear selector in the PARK position.

With the test light connected between C1 connector terminal B and ground, the lamp should illuminate. If the light does not come on, check for a circuit fault in the battery feed circuit. Connect the test light between terminals A and B at the C1 connector. If the test light fails to illuminate, check the ground circuit for an open. Move the test light between circuit 39 and ground. The light should turn on. If not, there is an open in circuit 39.

Connect the test light between terminal D of connector C3 and ground. If the light does not turn on, there is a problem in the neutral safety switch circuit. Check the adjustment of the neutral safety switch and circuits 75 and 275 for opens. With the test light connected between terminal B of the C3 connector and ground, the test light should remain off. An illuminated light indicates that the left-seat switch assembly must be replaced.

Leave the test light connected between terminal B of the C3 connector and ground. Place the memory select switch in position 1. If the test light does not illuminate, check circuit 615 for an open. If the wire is good, use an ohmmeter to test the memory select switch for an open. Release the memory select switch and press the EXIT button. The test light should light. If the light fails to illuminate, there is a fault in the left-seat switch assembly. It therefore must be replaced.

Move the probe of the test light to terminal A of the C3 connector and press the exit button. If the light fails to illuminate, the problem is in circuit 616 or in the exit switch.

Classroom Manual
Chapter 13, page 411

Special Tools
Test light
DMM
Scan tool

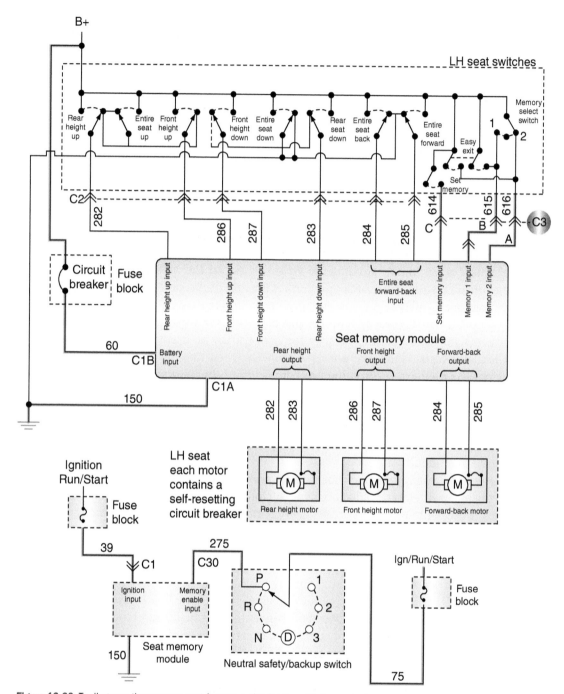

Figure 13-29 To diagnose the memory seat feature, a circuit schematic is required.

With the test light still connected to the A terminal of the C3 connector, release the EXIT button. The test light should turn off. If the test light remains illuminated, replace the left-seat switch assembly.

Continue to leave the test light connected to terminal A. Place the memory select switch in the number 2 position. The test light should light. If not, replace the left-seat switch assembly.

With the test light connected between C3 connector terminal C and ground and the memory select switch released, the test light should be off. If it remains on, the seat switch assembly is defective. Press the set memory switch. The test light should illuminate. If not, check the set memory switch and circuit 614 for an open.

If all the test results are correct, the fault is in the control module. The module must be replaced.

CLIMATE-CONTROLLED SEAT SERVICE

Changing the duty cycle to an output changes the average voltage applied to a device. This is the premise used to control the temperature of heated seats. In the example given here, we will discuss the diagnostic procedure for the positive temperature coefficient-type heater element.

The intent of the seat heater grid is to get the seat cushions up to operating temperature as fast as possible and then to maintain the heat level. Many heated seats use an HSD-controlled heating circuit. Most of these systems can be diagnosed using a scan tool since the system records DTCs for out-of-range voltage values that may indicate electrical shorts or opens. The system may also diagnose the input switches. Most climate-controlled seat systems rely on several modules to perform their functions. The data bus network is used to transfer the information between these modules. Be sure to include testing of the data bus when diagnosing the system. The following are typical diagnostic procedures for seats that are heated by a grid in the cushion and seat back (**Figure 13-30**). Note that this system uses a cushion and seat back grid that is wired in parallel. If the seat cushion warms but the seat back does not (or vice versa), the module and ground are not at fault. The fault would be in the shunt circuit to the inoperative grid or the affected grid itself. Some systems use a series connection between the seat cushion and seat back grids.

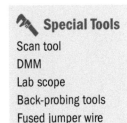

Classroom Manual
Chapter 13, page 413

Special Tools

Scan tool
DMM
Lab scope
Back-probing tools
Fused jumper wire

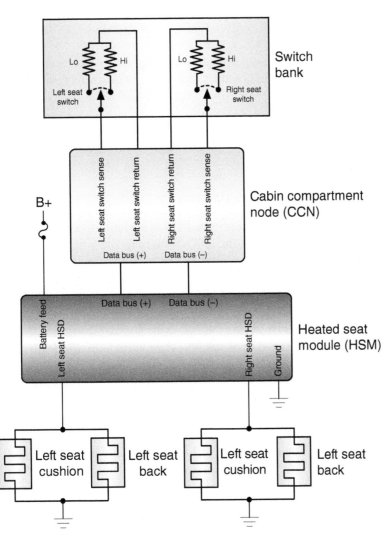

Figure 13-30 Heated seat circuit.

⚙️ **SERVICE TIP** Some manufacturers use a low voltage cut-off feature that turns off the heated seat system if battery voltage drops below a specified value. Be certain to check the vehicle's electrical system for proper voltage anytime the power seat system appears inoperative, especially for intermittent operation complaints.

⚙️ **SERVICE TIP** When measuring the resistance of the heating element, stress the grid by putting pressure against the cushion. This will help determine if an intermittent open or short in the element is present.

Since this system uses HSDs, they may deactivate the system if a fault is detected. For example, if an open is detected in the circuit prior to any of the elements or on the common ground so no current flows, HSDs will turn off the system. If the technician were to check for voltage from the heated seat module (HSM) at this time, the reading would be 0.0 volt. This may fool the technician into thinking the HSM is defective when in actuality it is not. For this reason, it is best to test for continuity through the heating element circuits using an ohmmeter. To test the grids, access the wire harness connectors for the heating elements. Usually this connection is under the seat. With the seat element connector disconnected, measure the resistance between the circuits leading into and out of the inoperative heated seat element. Compare the reading to specifications. The grid wires of each element are wired in parallel, so the measured resistance must fall within the specification range listed. A higher-than-specified reading indicates one or more grid wires are open, and a lower-than-specified reading indicates the grid wires are shorted.

⚙️ **SERVICE TIP** On some vehicles, the HSDs are supplied voltage from a relay. If all other circuits that receive voltage from this relay function properly, the relay is not at fault.

To test the heater circuit with a DMM or lab scope, the harness connector needs to be backprobed. If the circuit is disconnected, the HSD fault detection will determine that there is an open circuit and shut off the HSD. Be sure to allow the grid to cool completely. With the DMM or lab scope connected between the heater voltage supply wire and ground, and the ignition switch in the RUN position and the heated seats turned to the HIGH position, the specified voltage should be supplied on this wire. If the voltage is less than specification, repair the supply circuit. Some systems will supply a constant 12 volts, while others use PWM or duty cycle.

If PWM or duty cycle is used to control the circuit, the measured voltage on the supply circuit should begin to drop as the heater grid warms. This drop in voltage is due to the decrease in duty cycle since the seat is heating. If the voltage does not drop, the heater element or ground circuit is suspect. Use a DMM to perform a voltage drop test of the ground circuit. If excessive resistance is found, repair the cause and retest the heater.

A scan tool may also have the capabilities of monitoring the duty cycle of the heater circuit. The duty cycle should decrease as the heater grid warms. Some systems also display the temperature of the grid on the scan tool. Both of these indicate whether the heater element and circuit are operating properly.

If the heating element is faulty, it can usually be replaced without having to replace the entire cushion. Some systems require that the faulty grid be removed from underneath the seat trim cover and a new unit be attached. Other systems leave the old grid in place and apply the new element directly on top of the old element. In this case, the seat trim cover is removed and the wires that lead from the inoperative heating element are cut off flush with the edge of the old heating element. The new element is installed by peeling off the adhesive backing and sticking the element directly on top of the original heating element. It is important to take care not to fold or crease the element assembly while attaching it. If this occurs, premature failure of the new element may result.

If one or both seats fail to operate, diagnose the input switches for proper operation and transmission of data over the vehicle data bus network. Many systems do not wire the switches directly to the HSM. Instead, the switches are hardwired to a close module and bussed. Study the wiring diagram and service information for the system to determine what modules are involved. Diagnoses of the switches will be performed through the module that receives the hardwired input by use of the scan tool. A DMM can be used to check the integrity of the circuit between the switch and the module. Also, an ohmmeter can be used to test the switches.

> **SERVICE TIP** Not all HSD heaters are duty-cycle controlled. Be sure to check with proper service information.

If switch inputs are being received by the HSM and the elements are not faulty, the module is suspect. Most systems use one control module for both front seats. If one seat operates but the other does not, then the power and ground circuits to the HSM are not at fault. However, if both seats fail to operate, check powers and grounds before deciding that the module is faulty.

Seats can be heated or cooled using a Peltier element. When used with a climate-controlled seat, the Peltier element is integral to the climate controller (**Figure 13-31**). The controller cools or warms the airflow from the climate control fan motor based on the climate control ECU activation. A temperature sensor is used to monitor the surface temperature of the seat back and cushion. A fan motor provides airflow to the seat cushion and seat back (**Figure 13-32**).

Classroom Manual
Chapter 13, page 414

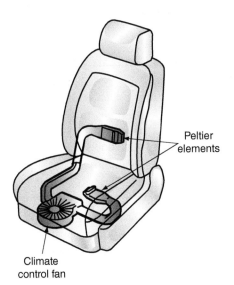

Figure 13-31 Climate-controlled seat using a Peltier element and a fan.

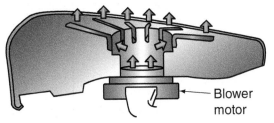

Figure 13-32 Airflow is directed through the seat cushion by the fan.

> ⚙️ **SERVICE TIP** Some systems do not turn on the heated seat function if the seat is not occupied. A faulty sensor or switch used to determine seat occupancy may prevent the heated seat system from operating. Also, during diagnostics, weight may need to be added to the seat for the heaters to turn on.

Diagnostics of the Peltier element–controlled device are done with a scan tool and using the DTCs. If the circuits are good, the Peltier element is suspect. The module/element is usually serviced as an assembly. Like any other diagnosis routine, this is a process of elimination. Eliminate all of the other possibilities such as switch inputs and circuits, and what is left is the faulty component.

The climate-controlled seats will also use a fan. Most systems use the fan only for cooling, but it may also be used for heating. Consider **Figure 13-33** of a seat fan system as an example for diagnosing the circuit. In this example, the HSM controls fan operation. The customer's concern is that the right front seat cooling operation does not function properly.

Use a scan tool to record any DTCs recorded by the HSM. If active codes are found, follow the diagnostic routine in the service information for the code(s). If the codes are stored, this may indicate that an intermittent problem exists. In this case, erase the DTCs with the scan tool, cycle the ignition switch to OFF, and then start the engine. Turn the seat cooling ON, and wait at least for 12 seconds before proceeding. Use the scan tool to see if any DTCs reset. If the DTCs do not reset, the conditions that caused the codes to set originally are not present at this time. Verify that the seat harness is routed correctly to the HSM. Check for any harness damage or pinching possibilities along the seat track.

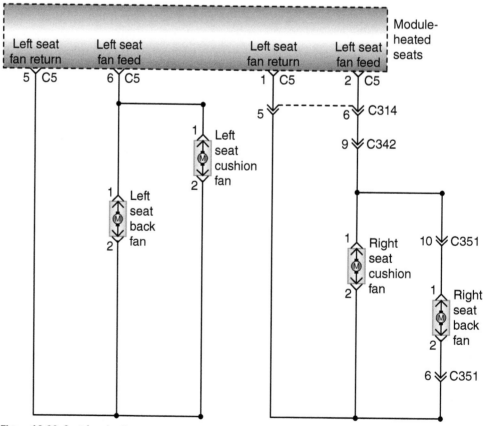

Figure 13-33 Seat fan circuit.

Confirm that the connector terminals are not damaged or pulled out slightly. Also, look for any corroded or contaminated terminals.

⚙ **SERVICE TIP** The battery must be fully charged and proper operation of the charging system verified before diagnosing the fan circuit.

⚙ **SERVICE TIP** Adjust the seat into several different positions to see if the DTC resets. This would indicate that the harness may be pulled.

If an active DTC returns, turn the ignition switch to the OFF position and disconnect the HSM C5 connector. Inspect the connector and repair if necessary. If no obvious problem is found, connect an ammeter between the return circuit (P922) and chassis ground at the HSM C5 connector. Connect a fused jumper wire from battery voltage to the motor feed circuit (P422) terminal at the HSM C5 connector. Once the jumper wire is connected, both the seat cushion and seat back fans should operate. In this system, the normal amperage reading should be between 720 and 880 mA. If 0 amp (A) is indicated, one of the circuits is open. Lower-than-normal amperage indicates high resistance in the circuit. Higher-than-normal amperage indicates that a motor may have a shorted winding. If the amperage is in accordance with specifications, the HSM is suspect. Confirm proper powers and grounds to the HSM before condemning the module.

If the amperage is not within specifications, isolate whether the problem is with the seat cushion fan or the seat back fan circuit. Use the fused jumper wires and the ammeter to measure the amperage of the seat back motor by itself at connector C351. The reading should be between 320 and 480 mA. If the reading is not within the specifications, the problem is in the circuit or fan of the seat back. If the reading is within specifications, the problem is in the circuit or fan of the seat cushion. Test the circuits using either voltage drop or an ohmmeter. If the circuit is good, replace the fan.

POWER DOOR LOCK DIAGNOSIS

To test the door lock motor, apply 12 volts directly to the motor terminals. The actuator rod should complete its travel in less than 1 second. Reverse polarity to test operation in both directions.

The switch is checked for continuity using an ohmmeter. There should be no continuity between any terminals when the switch is in its neutral position. Use the circuit schematic to determine when there should be continuity between terminals.

If the system uses a relay, use the schematic to determine relay circuit operation. In this example, battery voltage should be present at terminal 4 of the connector. Using an ohmmeter, check the ground connections of terminals 1 and 5 of the connector. To test the relay, connect a test light across terminal 3 and ground. Ground terminal 1 and apply power to terminals 2 and 4. The test light will light if the relay is good.

AUTOMATIC DOOR LOCK SYSTEM TROUBLESHOOTING

Some systems offer self-diagnostics through the body computer. The service information will provide the steps required to enter diagnostics on these vehicles. The following is an example of locating the fault in vehicles that do not provide this feature when the door locks work but they do not lock or unlock automatically. As with any electrical diagnosing,

Classroom Manual
Chapter 13, page 415

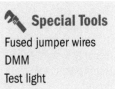

Special Tools
Fused jumper wires
DMM
Test light

Classroom Manual
Chapter 13, page 418

you will need the circuit diagram for the system you are working on. The following steps relate to the system shown (**Figure 13-34**):

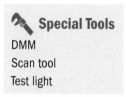

Special Tools

DMM
Scan tool
Test light

1. Locate the controller and backprobe for voltage at the power input terminal D with the ignition switch in the RUN position. If there is no voltage, there is an open in circuit 39.
2. Backprobe for voltage between terminals A and D. If there is no voltage, check for an open in circuit 150.
3. Make sure the courtesy lights are off and all doors are closed. With the gear selector in the PARK position, turn the ignition switch to the RUN position.
4. Connect a test light between controller terminal B and a good ground. If the neutral safety switch circuit operates properly, the test light illuminates.
5. With the test light connected as in step 4, move the gear selector to any other position. The test light should go out. If the light does not go out, check the neutral safety switch. It may be out of adjustment or faulty.
6. Leave the gear selector as in step 5 and connect the test light between terminals C and D. The test light should not illuminate. If it does, check circuit 156 and the light switch and doorjamb switches.
7. Return the gear selector to the PARK position. Connect the test light between terminal H and ground.
8. Observe the test light while the gear selector is moved from PARK to REVERSE. The test light should flash once. If not, replace the controller.

If the circuits pass all tests, check circuits 207 and 195, and the left switch assembly for opens.

ANTITHEFT SYSTEM TROUBLESHOOTING

Classroom Manual
Chapter 13, page 420

As with many electrical systems, manufacturers take many approaches in designing their antitheft system. Most of the testing of relays, switches, and circuits requires only basic electrical troubleshooting capabilities. Use the troubleshooting chart as a guide in locating the fault. Refer to the service information for the correct procedure of arming the system you are diagnosing.

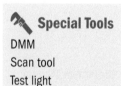

Special Tools

DMM
Scan tool
Test light

Self-Diagnostic Systems

Some antitheft systems offer self-diagnostic capabilities. Follow the service procedures for the proper method of entering diagnostics for the vehicle you are working on. The following is a typical example of entering diagnostics.

Some vehicle theft security systems enter the diagnostic mode when the ignition switch is cycled three times from the OFF to accessory position. When the vehicle theft security system is in the diagnostic mode, the horn should sound twice and the park lamps and tail lamps should flash. If the horn does not sound or the lights do not flash, voltmeter and ohmmeter tests are required to locate the cause of the problem.

This system is also referred to as vehicle security system and vehicle theft security system.

The scan tester may be used to diagnose many vehicle theft alarm systems. Follow the scan tester manufacturer's recommended procedure to enter the vehicle theft alarm system diagnostic mode. When this diagnostic mode is entered, the horn may sound twice to indicate that the trunk lock cylinder is in the proper position. When the key is placed in the ignition switch, the park lamps and tail lamps should begin to flash.

The following procedures should cause the horn to sound once if the system is operating normally:

1. Activating the power door locks to the LOCKED and UNLOCKED positions.
2. Using the key to lock and unlock each front door.
3. Turning on the ignition switch.

When the ignition switch is turned on in step 3, the diagnostic mode is exited.

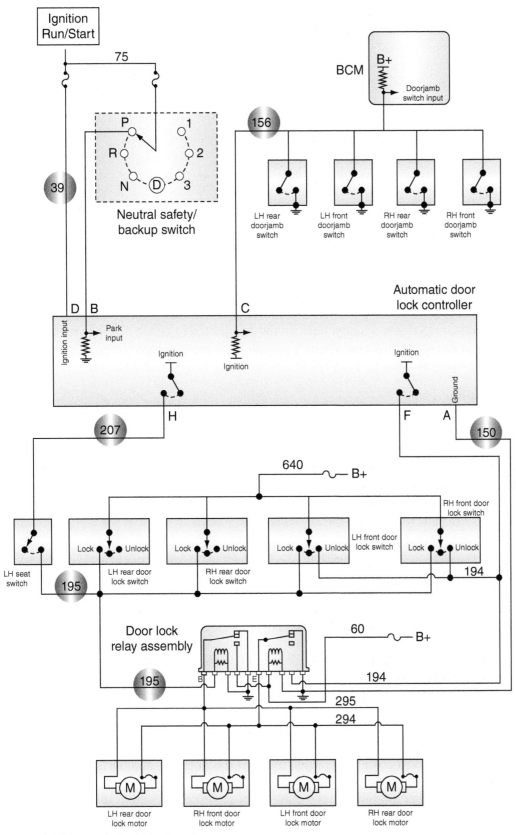

Figure 13-34 Automatic door lock schematic.

CUSTOMER CARE Always check the indicator lights in a customer's vehicle. These lights may be indicating a dangerous situation, but the customer may not have noticed them. For example, the vehicle theft security system set light may not be flashing when the normal system arming procedure is followed. This indicates an inoperative security system, and someone could break into the car without triggering the alarms. The customer has paid a considerable amount of money to have this system on the car. Therefore, it should be working. If this defect is brought to the customer's attention, he will probably have you repair the system and will appreciate your interest in the vehicle.

Alarm Sounds for No Apparent Reason

Mechanical and corrosion factors on the cylinder tamper switches can cause the system to activate for no apparent reason. If the customer complains of this condition, check the lock cylinder for looseness. Any looseness of the cylinder can cause the switch to activate the alarm.

Other causes of alarm system activation include loose, corroded, or improperly adjusted jamb switches. The switches should be adjusted to assure that they remain in the OFF position when the doors are fully closed. The switch is adjusted by a nut located at the base of the switch.

Controller Test

Caution

Failure to disconnect both harnesses will lead to false test indications.

To test the controller used in **Figure 13-35**, disconnect the harness from the controller and the harness to the relay. Connect the test light between the N terminal and ground. The test light should illuminate to indicate voltage to the horns and controller.

Move the test light probe to terminal M. The light should turn on only when the electrical door lock switch is moved to the UNLOCK position. Next, connect the test light between terminal B and a 12-volt source. The light should illuminate only if the doors are locked. The light should go out if any doors are unlocked.

Probe terminal K for voltage. The test light should illuminate only when the ignition switch is placed in the RUN position. Check for a blown fuse or an open circuit if the test light does not illuminate.

To test whether the cylinders are operating properly and have not been tampered with, connect the test light between terminal J and a 12-volt source. The test light should illuminate only if a door is open. If it glows with the doors closed, inspect the lock cylinders for damage.

With the test light connected between terminal H and a 12-volt source, the test light should be on only when the outside door key is turned to the unlock position. Move the test light between terminal G and a good ground. The light should illuminate when the electric door lock switch is operated. This indicates that there is electrical power to the switch. The test light should go on in the LOCK position and go out in the UNLOCK position.

Reconnect the relay harness. When the test light is connected between the F terminal of the controller connector and ground, the horn should sound and the lights should turn on. This indicates that the relay coil is functioning.

Next, turn the ignition switch to the RUN position with the test light connected between terminal E of the controller connector and ground. Use a voltmeter to measure voltage to the starter. There should be 0 volt. This indicates the starter interrupt relay is opening to prevent engine starting.

Connect a fused jumper wire between terminal D and a good ground while observing the security warning light. The warning light should be on. Connect a test light between terminal A of the controller connector and a 12-volt power supply. The test light should illuminate. If not, there is a problem in the ground circuit.

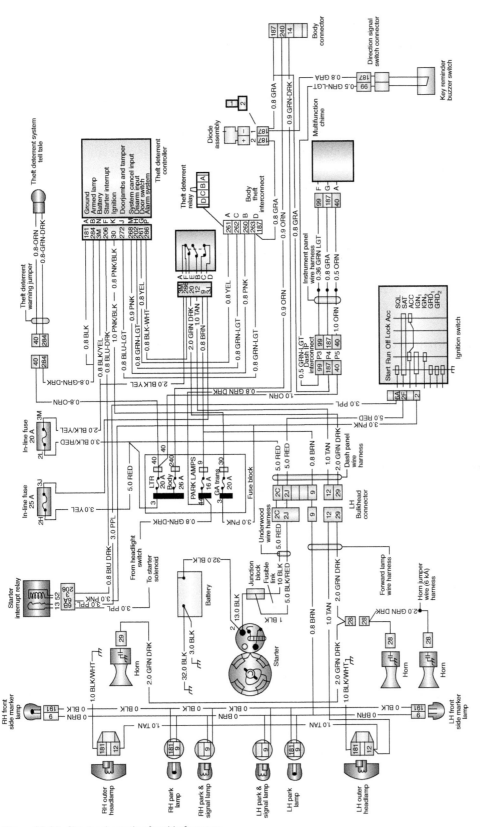

Figure 13-35 Circuit schematic of antitheft system.

⚙️ **SERVICE TIP** In some instances, it may be easier to attempt starting the engine than to check for voltage at the starter. If the relay works properly, the engine will not start.

DIAGNOSIS AND SERVICE OF ELECTRONIC CRUISE CONTROL SYSTEMS

Classroom Manual
Chapter 13, page 424

Cruise control systems are also called speed control.

The cruise control system is one of the most popular electronic accessories installed on today's vehicles. During open-road driving, it will maintain a constant vehicle speed without the continued effort of the driver. This reduces driver fatigue and increases fuel economy.

Problems with the system can vary from no operation, to intermittent operation, to not disengaging. To diagnose these system complaints, today's technicians must be able to rely on their knowledge and diagnostic capabilities. Most of the system is tested using familiar diagnostic procedures. Build on this knowledge and ability to diagnose cruise control problems. Use system schematics, troubleshooting diagnostics, and switch continuity charts to assist in isolating the cause of the fault.

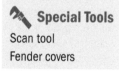 **WARNING** **When servicing and testing the cruise control system, you will be working close to the air bag and the antilock brake systems. The service manual will instruct you when to disarm and/or depressurize these systems. Failure to follow these procedures can result in injury and additional costly repairs to the vehicle.**

Self-Diagnostics

Special Tool

Scan tool

Most vehicle manufacturers have incorporated self-diagnostics into their cruise control systems. This allows some means of retrieving trouble codes to assist the technician in locating system faults. Modern vehicles usually require the use of a scan tool to retrieve the DTCs and to perform diagnostic tests of the cruise control system.

On any vehicle, perform a visual inspection of the system. Check the vacuum hoses for disconnects, pinches, loose connections, and so forth. Inspect all wiring for tight, clean connections. Also, look for good insulation and proper wire routing. Check the fuses for opens and replace as needed. Check and adjust linkage cables or chains, if needed. Some manufacturers will require additional preliminary checks before entering diagnostics. In addition, perform a road test (or simulated road test) in compliance with the service information to confirm the complaint.

⚠ Caution

Do not attempt to place the transmission back into PARK at any time during the test without first stopping the drive wheels with the brakes. Doing so may result in damage to the transmission.

Ford IVSC System Diagnostics. Ford's integrated vehicle speed control (IVSC) system has self-test capabilities that are contained within the **KOEO** and **KOER** routine of the electronic control assembly (ECA) (**Figure 13-36**). KOEO stands for Key On, Engine Off. It is a static test of the IVSC inputs and outputs. KOER stands for Key On, Engine Running. It is a dynamic check of the engine in operation. Testing of the IVSC system is broken down into two divisions: quick tests and pinpoint test.

The quick test will check the operation and function of all system components except the vehicle speed sensor. The quick test is performed first. Then, if any failure codes are displayed, the **pinpoint test** is performed. This is a specific component test service. If there is a complaint with the cruise control system, and the quick test does not indicate any faults, test the speed sensor.

Special Tools

Scan tool
Fender covers

After the system has been serviced, perform the quick test to verify proper operation.

The processor stores the self-test program within its memory. When this test is activated, the processor initiates a function test of the IVSC system to verify that the sensors and actuators are connected and operating properly. The quick test can be performed with a scan tool. The quick test will detect faults that are present at the time of the test. It will not store history codes. Follow the troubleshooting procedures in the service information for any DTCs retrieved. Also, perform the KOEO, KOER, and intermittent (wiggle) test procedures.

Diagnosing Systems without Trouble Codes

Classroom Manual
Chapter 13, page 426

Systems that do not provide for trouble code diagnostics require the technician to perform a series of diagnostic tests. The test performed will depend on the symptom. The following sections discuss areas of generic troubleshooting procedures for all types of systems.

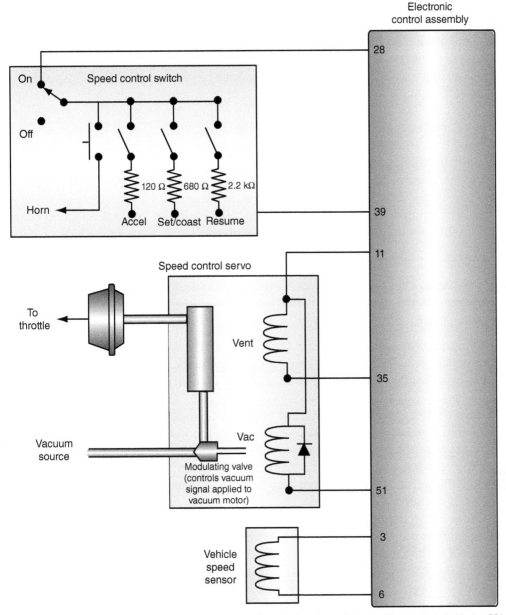

Figure 13-36 The IVSC system is an integrated system where the functions of the amplifier are included in the ECA.

Simulated Road Test. The simulated road test will allow the technician to perform a road test without leaving the shop. Before performing this test, connect the shop's ventilation system to the vehicle's exhaust pipe. Lift the drive wheels from the floor and place jack stands under the vehicle. If the vehicle is equipped with CV joint shafts, place the jack stands under the lower control arms so the shafts are in their normal drive position. If the vehicle is rear-wheel drive with a solid axle, place the jack stands under the axle.

⚡ **WARNING Block the wheels that are to remain on the ground. The wheels must remain blocked throughout the test.**

Start the engine and place the transmission into drive. Turn the speed control switch into the ON position. Accelerate and hold the speed at 35 mph (56 km/h). Press and release the SET ACCEL button. Maintain a slight foot pressure on the accelerator. The speed should be maintained at 35 mph (56 km/h) for a short period of time and then gradually start to surge. The engine surge is caused by operating the system while there is no load on the engine and is normal.

⚠ **Caution**

During this test, it is possible the engine will overspeed. If the system should appear to go out of control, the technician must be ready to turn it off. This can be done by turning off the ignition or turning off the speed control switch.

⚠ **Caution**

Do not exceed 50 mph (80.5 Km/h), or damage to the differential assembly may result.

Press the OFF button and the engine should decelerate to an idle speed. Stop the drive wheels by lightly applying the brakes. Press the ON button and accelerate to 35 mph (56 km/h). Press and hold the SET/ACCEL button and gradually remove your foot from the accelerator pedal. The engine rpm should begin to increase. Continue to hold the SET ACCEL button until the indicated speed reaches 50 mph (80.4 km/h); then release the button. Vehicle speed should remain at 50 mph (80.4 km/h) for a short period of time; then the engine will start to surge.

Press the COAST button and hold it. The engine rpm should return to idle speed. Allow the indicated speed to slow to 35 mph (56 km/h) without applying the brakes. When the speed is returned to 35 mph (56 km/h), release the COAST button. The speed should be held at 35 mph (56 km/h) for a short period of time; then the engine will begin to surge.

Tap the brake pedal, which will cause the speed control system to shut off and engine speed to return to idle. Set the indicated speed to 50 mph (80.4 km/h); then use the brakes to slow to 35 mph (56 km/h). Maintain 35 mph (56 km/h) using the accelerator. Depress the RESUME button and the speed should climb to 50 mph (80.4 km/h).

Diagnosing No-Operation Concerns

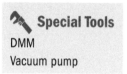

Special Tools
DMM
Vacuum pump

The first step in a verified no-operation concern is to check all fuses. Next, visually inspect the system for any obvious problems. If the visual inspection does not pinpoint the problem, perform the following steps:

1. Apply the brake pedal to observe proper brake light operation. If the brake lights do not operate, check the switch and circuit. Some brake switches have multiple internal switches; for example, one contact can be used for the brake lights, another contact used as an input to the PCM or TCM, and a third for the cruise control servo circuit. On these systems, just because the brake lights come on does not mean the brake switch is good. Use a scan tool to monitor the different brake switch inputs to confirm proper operation.
2. If the vehicle is equipped with a manual transmission, check to assure that the clutch deactivator switch is operating properly. Use an ohmmeter or voltmeter to test its operation.
3. Check for proper operation of the actuator lever and throttle linkage.
4. Disconnect the vacuum hose between the check valve and the servo (on the servo side of the check valve). Apply 18 inches of vacuum to the open end of the hose to test the check valve. It should hold the vacuum. If not, replace the check valve.
5. Check the vacuum dump valve for proper operation.
6. Test control switches and circuits following the procedure already learned. Use the circuit diagram and switch continuity charts to aid in testing.
7. Test servo operation.
8. Test speed sensor operation.
9. If all tests indicate proper operation, yet the system is not operational, replace the amplifier (controller).

Diagnosing Continuously Changing Speeds

If the vehicle speed changes up and down while the cruise control is on, use the following steps to locate the problem:

1. Check the actuator linkage for smooth operation.
2. Check the speedometer for proper routing and to make sure there are no kinks in the cable.
3. Test the servo.
4. Check the speed sensor.
5. Check the operation of the vacuum dump valve.
6. Check all electrical connections.
7. If none of these tests locate the fault, replace the amplifier (controller).

Diagnosing Intermittent Operation

Intermittent operation is usually caused by loose electrical or vacuum connections. If a visual inspection fails to locate the fault, test drive the vehicle and identify when the intermittent problem occurs. If the problem occurs during normal cruising, begin at step 1. If the problem occurs when operating the control buttons, or when the steering wheel is rotated, begin with step 3.

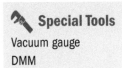

Special Tools
Vacuum gauge
DMM

1. Connect the vacuum gauge to the hose entering the servo. There should be at least 2.5 inches Hg (8.5 kPa) of vacuum.
2. Test the servo assembly.
3. Use the service information's switch continuity chart and system schematic to test switch operation. Turn the steering wheel through its full range while testing the switches. For example, using the Ford system shown in **Figure 13-37**, this test would be conducted by disconnecting the connector at the amplifier and connecting an ohmmeter between the terminal for circuit 151 and ground (with the ignition switch off). While rotating the steering wheel throughout its full range, make the following checks:

- Depress the OFF button; the ohmmeter reading should read between 0 and 1 ohm.
- Depress the SET/ACCEL button and check for a reading between 646 and 714 ohms.
- Depress the COAST button and the ohmmeter should read between 126 and 114 ohms.
- When the RESUME button is depressed, the reading should be between 2,310 and 2,090 ohms.

If the resistance values fluctuate while the steering wheel is being turned, the most likely cause is contamination on the slip rings or faulty clockspring. If the resistance values are above specifications, check the switches and ground circuit.

If the preceding tests (or the road test) fail to identify the fault, conduct a simulated road test while wiggling the electrical and vacuum connections. Monitor the brake switch input with a scan tool. A misadjusted or faulty brake switch can cause the system to shut off if it receives a brake applied input.

Component Testing

Testing of the safety switches and circuits is performed using normal testing procedures you have already learned. Testing of the servo assembly, dump valve, and speed sensor is included to familiarize you with these procedures.

Classroom Manual
Chapter 13, page 425

Servo Assembly. Actuator tests vary depending on design. Some manufacturers use vacuum servos and others use stepper motors. Be sure to follow the service procedures for the vehicle you are diagnosing. The **servo** controls the position of the throttle by receiving a controlled amount of vacuum. The following servo assembly test is a common test for Ford's cruise control system. Use the schematic in Figure 13-37 to perform the following test.

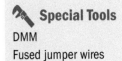

Special Tools
DMM
Fused jumper wires

Disconnect the eight-pin connector to the amplifier. Connect an ohmmeter between circuits 144 and 145. The resistance value should be between 40 and 125 ohms. Move the lead from circuit 145 to circuit 146. The resistance value should read between 60 and 190 ohms. If the resistance levels are out of specifications, check and repair the wiring between the amplifier and the servo. If the wiring is good, replace the servo.

If the resistance values are within specifications, leave the amplifier disconnected and start the engine. Jump 12 volts to circuit 144 and jump circuit 146 to ground. Momentarily jump circuit 145 to ground. The servo actuator arm should pull in and the engine speed should increase.

⚠ Caution
Be ready to abort the test by turning off the ignition, if engine rpm should rise to a level where internal damage may result.

⚡ WARNING Be sure to have the transmission in PARK or NEUTRAL. Block the wheels and set the parking brake before performing the servo test.

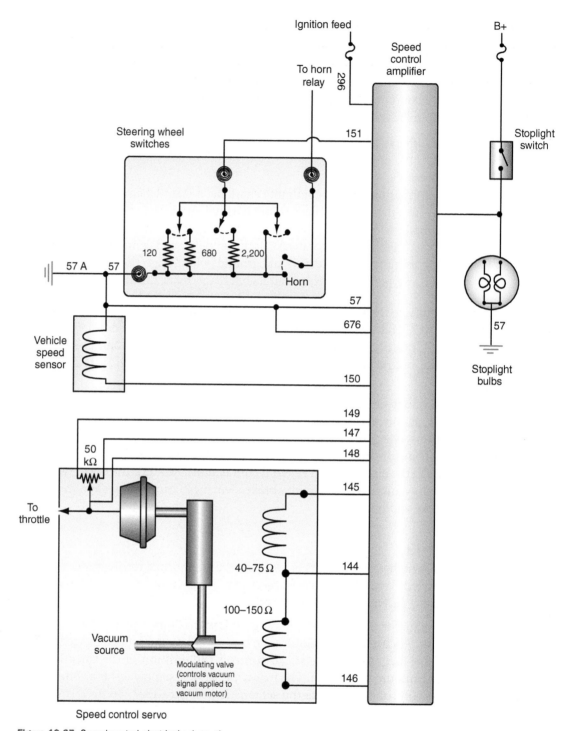

Figure 13-37 Speed control electrical schematic.

Remove the jumper to ground on circuit 146. The servo should release, and engine speed should return to idle. The servo must be replaced if it does not operate as described.

Special Tool

Hand vacuum pump

Dump Valve. The **dump valve** is a safety switch that releases vacuum to the servo when the brake pedal is pressed. A dump valve that is stuck open or leaks will cause a no-operation or an erratic-operation complaint. Failure of the dump valve to release vacuum may not, by itself, be noticed by the driver. It is part of a fail-safe system. If the dump valve does not release, the electrical switch signal is also used to disengage the cruise control system when the brakes are applied. It is good practice to test the dump valve.

To test the dump valve, disconnect the vacuum hose from the servo assembly to the dump valve. Connect a hand vacuum pump to the hose and apply vacuum to the dump valve. If vacuum cannot be applied, either the hose or the dump valve is defective.

If the valve holds vacuum, press the brake pedal. The vacuum should be released. If not, adjust the dump valve according to the service procedures. If the dump valve fails to release vacuum and it is properly adjusted, it must be replaced.

Speed Sensor. Disconnect the connector from the amplifier (Figure 13-37) and connect an ohmmeter between circuits 150 and 57A. The resistance should be approximately 200 ohms. If the resistance value is less than 200 ohms, check for a short in the circuits between the amplifier and the speed sensor. If there is no problem in the wiring, the coil in the sensor is shorted. If the resistance value is infinite, there is an open in the wires or in the sensor coil.

To test the sensor separate of the wiring harness, disconnect the wire connector from the sensor and connect the ohmmeter between the two terminals. This test should be used after testing at the amplifier connector to determine if there is a fault in the entire circuit.

> **Special Tool**
> DMM

Component Replacement

The two most common components to be replaced in the cruise control system are the servo and the switches. The following section covers replacement of these units.

Servo Assembly Replacement. Follow **Photo Sequence 31** to replace the servo assembly.

Reverse the procedure to install the servo assembly. To adjust the actuator cable, leave the cable adjusting clip off and pull the cable until all slack is removed. Maintain light pressure on the cable and install the adjusting clip. The clip must snap into place.

PHOTO SEQUENCE 31
Typical Procedure for Replacing the Cruise Control Servo Assembly

P31-1 Tools required to replace the servo assembly: fender covers, screwdriver set, combination wrench set, ratchet, and socket set.

P31-2 Remove the retaining screws attaching the speed control actuator cable to the accelerator cable bracket and intake manifold support bracket.

P31-3 Disconnect the cable from the brackets.

P31-4 Disconnect the speed control cable from the accelerator cable.

P31-5 Disconnect the electrical connection to the servo assembly.

P31-6 Remove the two retaining bolts that attach the servo assembly bracket to the shock tower.

PHOTO SEQUENCE 31 (CONTINUED)

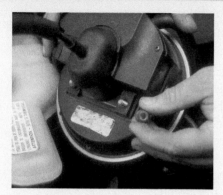

P31-7 Remove the two bolts that attach the servo assembly to the bracket.

P31-8 Remove the servo and cable assembly.

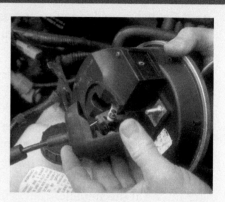

P31-9 Remove the two cable covers to servo assembly retaining bolt and pull off the cover.

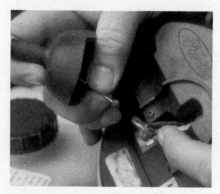

P31-10 Remove the cable from the servo assembly.

Switch Replacement. Switch removal differs depending on location. If the switch is a part of the multiple switch assembly on the turn signal stock, refer to the service information for removing this switch. The following is a common method of switch replacement for switches contained in the steering wheel.

⚠ WARNING **Follow the service information procedure for disarming the air bag system before performing this task. Failure to disarm the air bag system may result in accidental deployment and personal injury.**

With the air bag system properly disarmed, remove the air bag module. Disconnect the electrical connections to the switch assembly. Remove the screws that attach the switch assembly to the steering wheel. Then remove the switch.

To install the new switch assembly, position it into the steering wheel pad cover and attach the retaining screws. If the horn connectors have to be disconnected, attach them to the pad cover. Reinstall the air bag module and rearm the air bag system.

ADAPTIVE CRUISE CONTROL SYSTEM SERVICE

Classroom Manual
Chapter 13, page 428

The **adaptive cruise control (ACC)** system adjusts and maintains appropriate following distances between vehicles. The ACC system uses laser radar or infrared transceiver sensors to determine the vehicle-to-vehicle distances and relational speeds. The system maintains both the vehicle speed and a set distance between vehicles.

Figure 13-38 Usually the ACC sensor is located behind the front grill.

The laser radar or infrared sensor is mounted behind the front grill or fascia (**Figure 13-38**). The infrared system uses an integrated sensor and module assembly, while the laser radar sensor is usually remotely mounted from the distance module.

Hardwire circuits are diagnosed with the DMM using conventional methods. Since the system uses data from several different modules and also relies on several modules to activate separate functions of the system, diagnostics will require the use of a scan tool. You will need to access all of the modules involved with the ACC system and retrieve any DTCs they contain. The cause of these DTCs will need to be remedied. Also, communication errors will result in the ACC system not operating.

The following discusses the diagnosis of different faults associated with a system that uses infrared which is interrogated with the ACC module (**Figure 13-39**). The diagnostics will be similar for systems that use laser radar sensors.

If the voltage input to the ACC module drops below 9 volts while the engine is running, a "voltage low" DTC will be set. The module will disable the ACC system as long as the fault is active. This type of condition can result in intermittent ACC operation if due to loose or corroded terminals and connectors. Also, a charging system that has

Special Tools

Scan tool
DMM
Test light
Alignment screen
Masking tape
Grease pencil
Tape measure

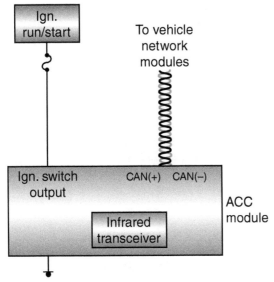

Figure 13-39 ACC sensor circuit.

intermittent problems can also cause this code to set. For intermittent conditions, use the wiring information to guide you through a wiggle test and inspection of all connectors and power and ground circuits.

> **CUSTOMER CARE** When servicing the ACC system, talk with the customer to find out the type of driving environment when the reported problem took place. Heavy rain, snow, ice, and fog can cause the ACC system to be deactivated. System shutdown for these reasons is considered normal and should be properly explained to the customer.

The voltage low fault can be caused by a faulty ignition feed circuit or a faulty ground circuit. Both of these circuits should be tested under load.

To test the ground circuit, disconnect the ACC module harness connector and use a test light connected to battery voltage. Probe the ground circuit with the test light. The test light should illuminate brightly. Use a voltmeter to read the voltage on the test light probe. If the circuit is good, the voltmeter should read close to 0 volt (**Figure 13-40**). If the test light does not illuminate, and 12 volts is seen on the voltmeter, the ground circuit is open. If the test light did illuminate and voltage above 300 mV is read on the voltmeter, there is resistance in the ground circuit.

The ignition feed circuit is tested by attaching the test light to a good ground and moving the probe to the ignition feed circuit terminal in the ACC module harness connector. The test light should illuminate brightly when the ignition switch is turned to the RUN position. A voltage reading at the test light probe should read battery voltage (**Figure 13-41**). If the light does not illuminate and the voltmeter reads 0 volt, there is an open in the circuit. If the test light is dim and less than battery voltage is read by the voltmeter, there is high resistance in the circuit.

If the ground and the ignition feed circuits test good (and the voltage low code is still active), the module is faulty and requires repair or replacement.

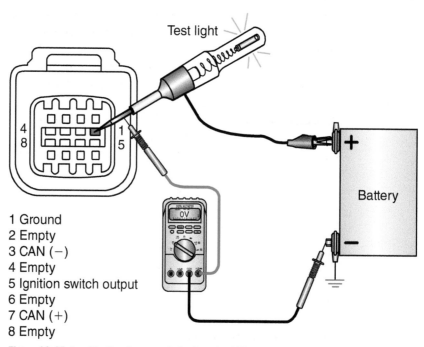

1 Ground
2 Empty
3 CAN (−)
4 Empty
5 Ignition switch output
6 Empty
7 CAN (+)
8 Empty

Figure 13-40 Load testing the ground circuit to the ACC module.

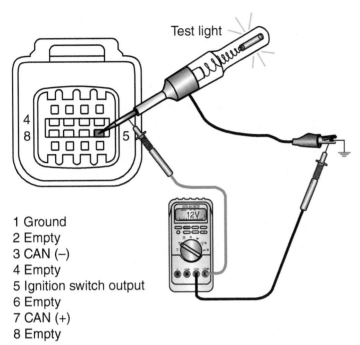

1 Ground
2 Empty
3 CAN (−)
4 Empty
5 Ignition switch output
6 Empty
7 CAN (+)
8 Empty

Figure 13-41 Load testing the power supply circuit to the ACC module.

If the voltage to the module exceeds 16 volts, a voltage high fault is detected by the module. The module will disable the ACC system and set a DTC. This fault can be caused by a charging system problem. Use the scan tool and check for any charging system–related DTCs stored in the PCM.

If the charging system operates properly, test the ignition feed and ground circuits as just described. If there is no problem in these circuits, the module is faulty.

If the ACC module detects that the sensor lens is damaged or missing, it will disable the ACC system and set a DTC. This DTC can be the result of dirt or debris on the lens, suspension modifications, installation of an aftermarket grill, or the installation of accessories in front of the sensor.

Begin by performing a thorough visual and physical inspection of the ACC sensor. Look for a missing or damaged sensor lens. If the lens is damaged or missing, the sensor/module will need to be replaced.

If the lens is not damaged, inspect the ACC module harness connector terminals for corrosion, damage, or terminal pushout. Also, use the wiring information as a guide to inspect the wiring harness and connectors. Carefully inspect all power and ground circuits. If all tests do not isolate a root cause for the fault, replace the module.

If the sensor/module assembly is replaced or removed and reinstalled in the vehicle, the sensor must be adjusted. If the module detects that the sensor is out of adjustment, it will disable the system and set a DTC. To perform this procedure, a special alignment screen is required (**Figure 13-42**). The screen is black with specifically located reflective tape strips. The tape is applied, so there is a single vertical stripe and two inverted "T" shapes.

To prepare the vehicle for the alignment procedure, the following must be checked and corrected if necessary:

1. There is no damage to the suspension, proper wheel alignment, damaged fascia, or broken grill.
2. Tire inflation pressures are correct.
3. The vehicle is level.
4. The ACC lens is clean and free from obstructions.
5. There is no load in the vehicle.

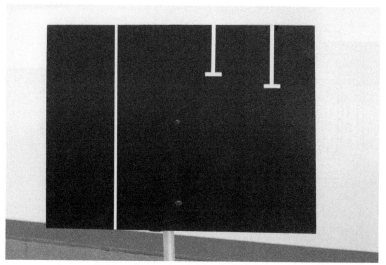

Figure 13-42 ACC alignment screen.

6. Sensor assembly mounting screws are properly torqued.
7. The fuel tank is FULL. Add 6.5 pounds (2.94 kg) of weight over the fuel tank for each gallon of missing fuel.
8. Close and latch the hood.

> **⚡ Caution**
>
> Ensure that there are no reflective surfaces or objects within 2 meters (6.56 feet) behind the aiming board. These objects may cause an error during the aiming process and be interpreted as a board misalignment.

Locate the centerline of the vehicle and mark the location on the windshield and rear window. The alignment screen is set 157.5 inches (4 meters) from the sensor (**Figure 13-43**). A sight notch on the top of the screen is used to align the screen with the center of the vehicle. The height of the screen is adjusted to specifications. The specifications are based on the measured distance from the ground to the center of the vertical adjustment ball screw (**Figure 13-44**). A chart is used to determine the height that the screen is adjusted to (**Table 13-1**). The screen height must be equal at both upper corners.

With the scan tool, access the "Align ACC sensor" function. The first step in the procedure will be to do a rough alignment. The vertical alignment cover is placed on the alignment screen to cover both of the alignment "T's." This forces the sensor to aim on only the vertical stripe. Use the adjusting screws to adjust the sensor (**Figure 13-45**) until the scan tool informs you that the rough alignment has been completed.

Next, remove the vertical alignment cover and perform the fine alignment procedure. This forces the sensor to locate all of the reflections from the tape. Once this step is completed, the scan tool will clear the internal compensation and inform you that the alignment procedure is completed.

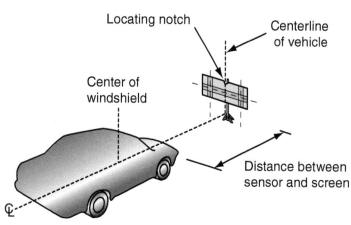

Figure 13-43 Vehicle and alignment screen set-up.

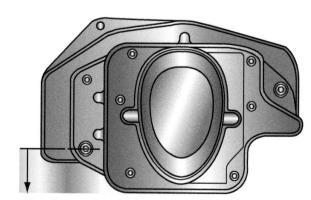

Figure 13-44 Sensor height measurement location.

TABLE 13-1 TABLE OF SPECIFICATIONS FOR ALIGNMENT SCREEN HEIGHT BASED ON THE HEIGHT OF THE SENSOR.

Sensor Height	Board Height
Height Specifications Measured in Millimeters	
651	853
652	854
653, 654	855
655, 656	856
657, 658	857
659	858
660, 661	859
662, 663	860
664	861
665, 666	862
667, 668	863
669	864
670, 671	865
672, 673	866
674	867
675, 676	868
677, 678	869
679, 680	870
681	871
682, 683	872
684, 685	873
686	874
687, 688	875
689, 690	876
691	877
692, 693	878
694, 695	879
696, 697	880
698	881
699, 700	882
701, 702	883
703	884
704, 705	885
706, 707	886
708	887
709, 710	888
711, 712	889
713	890
714, 715	891
716, 717	892
718, 719	893
720	894
721, 722	895
723, 724	896
725	897
726, 727	898
728, 729	899
730, 731	900
732, 733	901
734	902
735, 736	903
737, 738	904
739, 740	905
741	906
742	907

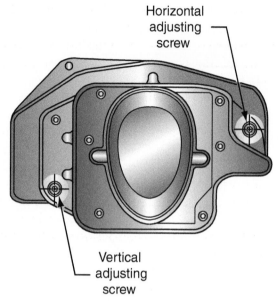

Figure 13-45 ACC sensor/module adjustment locations.

ELECTRONIC SUNROOF DIAGNOSIS

Classroom Manual
Chapter 13, page 430

Special Tools

Test light
DMM
Ammeter

Troubleshooting, the causes of slow, intermittent, or no sunroof operation, is a relatively simple procedure. Unlike many systems, the sunroof operation is not usually integrated with other systems. Because the system stands alone, diagnostics are generally performed in the same manner as testing any other motor-driven accessory. The following is the diagnostic procedure used to troubleshoot the electronic sunroof system shown (**Figure 13-46**). In addition, refer to the diagnostic charts at the end of this chapter to determine the causes of other system malfunctions.

Slow sunroof operation may be caused by excessive resistance in the circuit or motor. Excessive resistance can be determined by performing a voltage drop test. Obtain the correct schematic for the vehicle being diagnosed and follow through the circuit to locate

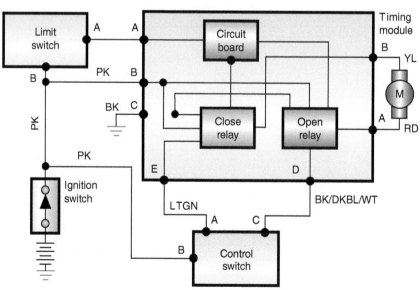

Figure 13-46 Electronic sunroof schematic.

the cause of the excessive resistance. Resistance can also occur inside the motor as a result of brush wear, bushing wear, and corroded connections.

Shorted armature or field coils can also result in slow motor operation. An ammeter can be used to test the current draw of the circuit to determine motor condition. If testing the motor and its circuits does not locate the cause of slow operation, then the problem is mechanical. Check the drive cable and tracks for any signs of wear or damage.

Intermittent problems may be the result of loose or corroded connections. To locate the cause of an intermittent fault, operate the system while wiggling the wires. This will assist in isolating the location of the poor connection. Some systems also use a circuit breaker to protect the motor. The circuit breaker may be overheating and tripping prematurely, or there may be resistance to window movement in the rails. The circuit breaker will trip, and then cool down and reset. Check that the glass is able to move easily in the rails. In addition, many intermittent problems are caused by a faulty control switch. Operate the control switch several times while performing the circuit test. Replace the switch if it fails at any time during the test.

Follow the procedures listed in the service information to locate an intermittent or no-operation fault within the circuit. Perform the usual visual inspections of the circuit before continuing. Be sure to check for proper system grounds.

CASE STUDY

The customer complains that although the fan blower motor operates well in low- and medium-speed positions, it runs slow in the high-speed position. A look at the wiring schematic in **Figure 13-47** for the circuit indicates that the resistor block does not control high-speed operation. Because the motor operates properly in low and medium speeds, the fault is not in the motor or the power supply feed. By examining the schematic, the only place that there could be resistance in the high-speed circuit is at the switch wiper or in the circuit between the switch and the resistor block. A voltage drop test confirms that the location of the resistance is the connector to the switch.

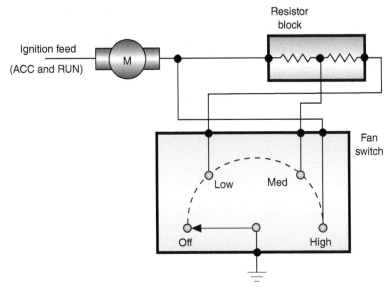

Figure 13-47 Three-speed blower motor schematic.

CASE STUDY

The owner of a 2016 Charger says the automatic door lock feature fails to operate. The technician checks all accessible connections.

Next, a scan tool is connected to the diagnostic tester and the output test is initiated. The door locks operate properly while in this test mode. The technician then initiates the switch test. The scan tool indicates that the driver's side door ajar switch is constantly reading open. The service information instructs the technician to disconnect the door ajar switch connector. When disconnected, the scan tool indicates that the switch is open. Next, a fused jumper wire is connected across the harness side connector. The scan tool still indicates that the switch is open. According to the service information, when the jumper wire is connected, the input state should indicate closed.

Voltage from the cabin compartment node (CCN) to the switch connector is tested. The voltage should be 12 volts, but 0 volt is measured. It is found that the wire between the CCN and the connector is good. According to the service information, the CCN must be replaced since it is not sending out the 12-volt reference signal to the left-front door switch.

The resistance of the terminal lead from the CCN to the switch is measured. Per the service manual specifications, the resistance value is too high and the CCN requires replacement.

ASE-STYLE REVIEW QUESTIONS

1. All of the following can cause the horn to sound continuously **EXCEPT:**
   - A. Circuit between relay and horn shorted to power.
   - B. Short to ground between relay and horn.
   - C. Horn switch contacts stuck closed.
   - D. Short to ground between the relay coil and the horn switch.

2. The two-speed windshield wiper operates in HIGH position only.
   Technician A says the low-speed brush may be worn.
   Technician B says the motor has a faulty ground connection.
   Who is correct?
   - A. A only
   - B. B only
   - C. Both A and B
   - D. Neither A nor B

3. What is the **LEAST LIKELY** cause of slow windshield wiper operation?
   - A. Binding mechanical linkage.
   - B. Excessive resistance in the motor's ground circuit.
   - C. Short to power at the low-speed brush.
   - D. Worn common brush.

4. Aligning the ACC sensor is being discussed.
   Technician A says the height of the screen is determined by the height of the sensor.
   Technician B says the screen is positioned directly in front of the sensor.
   Who is correct?
   - A. A only
   - B. B only
   - C. Both A and B
   - D. Neither A nor B

5. A customer states that the vehicle alarm will trip when there is no apparent attempt of entry.
   Technician A says the fault may be a loose lock cylinder.
   Technician B says a misadjusted jamb switch may be the cause.
   Who is correct?
   - A. A only
   - B. B only
   - C. Both A and B
   - D. Neither A nor B

6. The heater fan motor does not operate in high-speed position.
   Technician A says the cause is a faulty resistor block.
   Technician B says the motor is defective.
   Who is correct?
   - A. A only
   - B. B only
   - C. Both A and B
   - D. Neither A nor B

7. The grid of a rear window defogger removes only some areas of fog from the window.

   Technician A says the timer circuit is faulty.

   Technician B says the grid is damaged.

   Who is correct?

   A. A only   C. Both A and B

   B. B only   D. Neither A nor B

8. The passenger-side power window does not operate in either direction, despite which switch is used (master or window).

   Technician A says the problem is a faulty master switch.

   Technician B says the problem is a worn motor.

   Who is correct?

   A. A only   C. Both A and B

   B. B only   D. Neither A nor B

9. A sunroof is opening slower than normal and appears to be jerky in its operation. The most likely cause of this problem is:

   A. A shorted armature or field coil in the sunroof motor.

   B. A stuck open limit switch.

   C. A stuck closed limit switch.

   D. A stuck OPEN switch.

10. The power door locks will lock the door, but they do not unlock it.

    Technician A says the motor is faulty.

    Technician B says the unlock relay is faulty.

    Who is correct?

    A. A only   C. Both A and B

    B. B only   D. Neither A nor B

ASE CHALLENGE QUESTIONS

1. The driver-side heated seat shown in **Figure 13-48** fails to warm at any level request. The harness connector for the left heated seat is disconnected, the system activated, and voltage measured. The voltmeter reads 0 volt. What is the **MOST LIKELY** cause of this reading?

   A. The circuit between the HSM and the connector is open.

   B. The circuit between the HSM and the connector is shorted to ground.

   C. The HSM is faulty.

   D. Improper diagnostics of an HSD system.

2. A vehicle's cruise control system is inoperative. Which of the following will not cause this problem?

   A. Stuck closed brake pedal vacuum switch.

   B. Open stop lamp switch.

   C. Inoperative speedometer.

   D. Open cruise engagement switch.

3. The computer-controlled power window for the driver side does not operate in either direction. Measuring the voltage on the switch MUX input circuit shows 12 volts regardless of the position of the switch. Which of the following would be the **MOST LIKELY** cause?

   A. This is normal operation.

   B. Open in the MUX return circuit.

   C. A shorted reference voltage on the MUX circuit.

   D. A short to ground on the MUX return circuit.

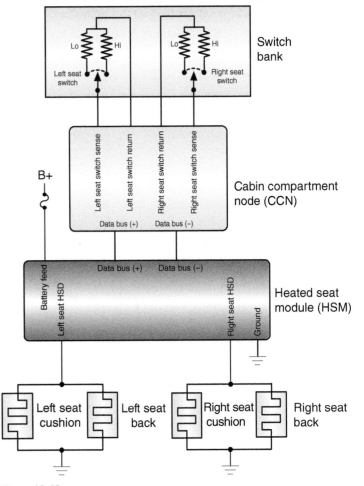

Figure 13-48

4. A completely inoperative GM automatic door lock system is being discussed; the door locks will not open or close manually or automatically.

Technician A says that there may be an open in the power feed circuit to the automatic door lock *controller*.

Technician B says that one or more doorjamb switches may be faulty.

Who is correct?

A. A only

B. B only

C. Both A and B

D. Neither A nor B

5. The blower motor in **Figure 13-49** operates at a high speed only. The **LEAST LIKELY** cause of this problem would be:

A. An open in the motor ground circuit.

B. An open in the control circuit.

C. A short to ground in the control circuit.

D. A short to ground in the motor ground circuit.

6. The horn of a vehicle equipped with a horn relay sounds continuously. Which of the following could be the cause of this problem?

A. Shorted horn relay coil.

B. Open horn switch.

C. Grounded wire on the switched side of the horn relay.

D. Open horn relay coil.

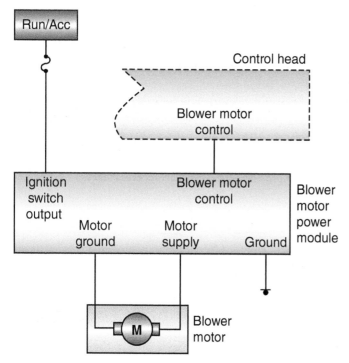

Figure 13-49

7. The high-speed position of a two-speed windshield wiper system is inoperative; the low-speed position is fine.

Which of the following could cause this problem?

A. High resistance on the ground side of the motor.

B. Open motor resistor.

C. Faulty park switch.

D. Worn motor brush.

8. An A/C blower motor turns slowly. A voltmeter that is connected across the power and ground terminals of the motor indicates 13.6 volts when the blower speed switch is in the HIGH position and the engine is running; an ammeter indicates that the motor is drawing about 3 amps.

Technician A says that the blower motor ground circuit may have excessive resistance.

Technician B says that the blower motor armature may be binding.

Who is correct?

A. A only

B. B only

C. Both A and B

D. Neither A nor B

9. The blower motor of a vehicle equipped with a ground-side-controlled motor circuit is running at high speed whenever the engine is running; the blower switch has no control of the motor speed.

Technician A says that ground side of the motor may be shorted to ground.

Technician B says that the blower switch contacts may have excessive resistance.

Who is correct?

A. A only

B. B only

C. Both A and B

D. Neither A nor B

10. A power window motor is inoperative. A voltmeter that is connected across the power and ground terminals of the motor indicates 0 volt when the window switch is moved to the DOWN position; when the ground lead of the voltmeter is connected to a chassis ground (and the switch is in the DOWN position), the voltmeter then indicates 12.6 volts.

Technician A says that the window motor has an open internal circuit.

Technician B says that the window switch may have an open contact.

Who is correct?

A. A only

B. B only

C. Both A and B

D. Neither A nor B

Name _____   Date _____

DIAGNOSING A WINDSHIELD WIPER CIRCUIT

Upon completion of this job sheet, you should be able to diagnose an inoperative or a poorly working windshield wiper system.

NATEF Correlation

This job sheet addresses the following **MLR** tasks:

A.3.	Use wiring diagrams to trace electrical/electronic circuits.
A.4.	Demonstrate proper use of a digital multimeter (DMM) when measuring source voltage, voltage drop (including grounds), current flow, and resistance.
A.5.	Demonstrate knowledge of the causes and effects from shorts, grounds, opens, and resistance problems in electrical/electronic circuits.
A.11.	Identify electrical/electronic system components and configuration.
E.8.	Verify windshield wiper and washer operation; replace wiper blades.

This job sheet addresses the following **AST** task:

G.4.	Describe operation of safety systems and related circuits (e.g., horn, air bags, seat belt pretensioners, occupancy classification, wipers, washers, speed control/collision avoidance, heads-up display, park assist, and back-up camera); determine needed repairs.

This job sheet addresses the following **AST/MAST** tasks:

A.3.	Demonstrate proper use of a digital multimeter (DMM) when measuring source voltage, voltage drop (including grounds), current flow, and resistance.
A.4.	Demonstrate knowledge of the causes and effects from shorts, grounds, opens, and resistance problems in electrical/electronic circuits.
A.7.	Use wiring diagrams during the diagnosis (troubleshooting) of electrical/electronic circuit problems.

This job sheet addresses the following **MAST** task:

G.4.	Diagnose operation of safety systems and related circuits (e.g., horn, air bags, seat belt pretensioners, occupancy classification, wipers, washers, speed control/collision avoidance, heads-up display, park assist, and back-up camera); determine needed repairs.

ASE NATEF

Tools and Materials

- A vehicle
- Wiring diagram for the chosen vehicle
- Service information for the chosen vehicle
- DMM

Describe the vehicle being worked on:

Year _____ Make _____ Model _____

VIN _____ Engine type and size _____

Procedure   **Task Completed**

1. Describe the general operation of the windshield wipers. Check the operation in all speeds and modes.

_____

_____

_____

_____

2. Check the mechanical linkages for evidence of binding or breakage. Record your findings.

_____

_____

_____

_____

3. Locate the wiring diagram for the windshield wiper circuit. Print out the diagram and ☐ highlight the power, grounds, and the circuit controls. Attach the printout to this job sheet.

4. Describe how the motor is controlled to operate at different speeds.

_____

_____

_____

_____

5. Connect the voltmeter across the ground circuit; energize the motor. What was your reading on the meter? What does this indicate?

_____

_____

_____

_____

6. Probe the power feed to the motor in the various switch positions. Observe your voltmeter readings. What were they? What is indicated by these readings?

_____

_____

_____

_____

7. Describe the general operation of the windshield wiper motor.

_____

_____

_____

_____

Instructor's Response

_____

_____

_____

Name _____   Date _____

TESTING THE WINDSHIELD WASHER CIRCUIT

Upon completion of this job sheet, you should be able to diagnose problems in the windshield washer circuit.

NATEF Correlation

This job sheet addresses the following **MLR** tasks:

A.3.	Use wiring diagrams to trace electrical/electronic circuits.
A.4.	Demonstrate proper use of a digital multimeter (DMM) when measuring source voltage, voltage drop (including grounds), current flow, and resistance.
A.5.	Demonstrate knowledge of the causes and effects from shorts, grounds, opens, and resistance problems in electrical/electronic circuits.
A.11.	Identify electrical/electronic system components and configuration.
E.8.	Verify windshield wiper and washer operation; replace wiper blades.

This job sheet addresses the following **AST** task:

G.4.	Describe operation of safety systems and related circuits (e.g., horn, air bags, seat belt pretensioners, occupancy classification, wipers, washers, speed control/collision avoidance, heads-up display, park assist, and back-up camera); determine needed repairs.

This job sheet addresses the following **AST/MAST** tasks:

A.3.	Demonstrate proper use of a digital multimeter (DMM) when measuring source voltage, voltage drop (including grounds), current flow, and resistance.
A.4.	Demonstrate knowledge of the causes and effects from shorts, grounds, opens, and resistance problems in electrical/electronic circuits.
A.7.	Use wiring diagrams during the diagnosis (troubleshooting) of electrical/electronic circuit problems.

This job sheet addresses the following **MAST** task:

G.4.	Diagnose operation of safety systems and related circuits (e.g., horn, air bags, seat belt pretensioners, occupancy classification, wipers, washers, speed control/collision avoidance, heads-up display, park assist, and back-up camera); determine needed repairs.

ASE CERTIFIED **NATEF**

Tools and Materials

- A vehicle
- Wiring diagram for the chosen vehicle
- DMM
- Hand tools
- Fused jumper wire

Describe the vehicle being worked on:

Year _____ Make _____ Model _____

VIN _____ Engine type and size _____

Procedure

1. Describe the general operation of the windshield wipers.

_____

_____

_____   **Task Completed**

_____

2. Check the fluid level in the washer fluid reservoir.   ☐

   Replenish the level, if necessary.

3. Remove the fluid output line from the washer pump. Activate the washer pump. Does fluid come out of the pump? _____

   What are your conclusions from this?

   _____

   _____

   _____

   _____

4. If fluid comes out of the pump but none sprays on the windshield, disconnect the fluid lines from the washer fluid nozzles. Activate the pump. Does fluid come out of the pump? _____

   What are your conclusions from this?

   _____

   _____

   _____

   _____

5. If the pump does not operate when switched on, run a fused jumper wire from the battery to the pump. What happened? What are your conclusions from this test?

   _____

   _____

   _____

   _____

6. Connect the voltmeter across the ground of the pump. Activate the pump. Describe what happened. What are your conclusions from this?

   _____

   _____

   _____

   _____

7. What are your recommendations about the windshield washer system?

   _____

   _____

   _____

   _____

Instructor's Response

_____

_____

_____

Name _____   **Date** _____

TESTING A REAR WINDOW DEFOGGER

Upon completion of this job sheet, you should be able to diagnose inoperative and poorly working rear window defogger units.

NATEF Correlation

This job sheet addresses the following **MLR** tasks:

A.3.   Use wiring diagrams to trace electrical/electronic circuits.

A.4.   Demonstrate proper use of a digital multimeter (DMM) when measuring source voltage, voltage drop (including grounds), current flow, and resistance.

A.5.   Demonstrate knowledge of the causes and effects from shorts, grounds, opens, and resistance problems in electrical/electronic circuits.

A.11.   Identify electrical/electronic system components and configuration.

This job sheet addresses the following **AST** tasks:

G.1.   Describe operation of comfort and convenience accessories and related circuits (such as: power windows, power seats, pedal height, power locks, trunk locks, remote start, moonroof, sunroof, sunshade, remote keyless entry, voice activation, steering wheel controls, back-up camera, park assist, cruise control, and auto dimming headlamps); determine needed repairs.

G.4.   Describe operation of safety systems and related circuits (e.g., horn, air bags, seat belt pretensioners, occupancy classification, wipers, washers, speed control/collision avoidance, heads-up display, park assist, and back-up camera); determine needed repairs.

This job sheet addresses the following **AST/MAST** tasks:

A.3.   Demonstrate proper use of a digital multimeter (DMM) when measuring source voltage, voltage drop (including grounds), current flow, and resistance.

A.4.   Demonstrate knowledge of the causes and effects from shorts, grounds, opens, and resistance problems in electrical/electronic circuits.

A.7.   Use wiring diagrams during the diagnosis (troubleshooting) of electrical/electronic circuit problems.

This job sheet addresses the following **MAST** task:

G.1.   Diagnose operation of comfort and convenience accessories and related circuits (such as: power windows, power seats, pedal height, power locks, trunk locks, remote start, moonroof, sunroof, sunshade, remote keyless entry, voice activation, steering wheel controls, back-up camera, park assist, cruise control, and auto dimming headlamps); determine needed repairs.

G.4.   Diagnose operation of safety systems and related circuits (e.g., horn, air bags, seat belt pretensioners, occupancy classification, wipers, washers, speed control/collision avoidance, heads-up display, park assist, and back-up camera); determine needed repairs.

Tools and Materials
- A vehicle with a rear window defogger
- Wiring diagram for the chosen vehicle
- DMM

Describe the vehicle being worked on:

Year _____ Make _____ Model _____

VIN _____ Engine type and size _____

Procedure   **Task Completed**

1. From the wiring diagram, identify the power feed to the rear window defogger. With the circuit activated, check for voltage at the power terminal to the defogger unit. Is battery voltage present? _____

   If no voltage is present, continue your testing at the control switch.   ☐

   If much less than battery voltage is present at the terminal, check for excessive   ☐
   resistance in the power circuit.

2. Identify the ground circuit for the defogger; measure the voltage drop across this part of the circuit. Your reading is _____. If it is more than 0.1 volt, check the wire, connectors, and ground connection for the cause of higher-than-normal resistance.

3. If all of the grids in the defogger circuit are good, applied voltage should be dropped across each grid. To test the condition of each grid, measure the voltage drop across each grid, starting from the top grid and moving down. List your findings.

   _____

   _____

   _____

   _____

   _____

4. If one or more of the grids do not drop applied voltage, continue your diagnosis by   ☐
   moving the negative meter lead toward the positive side of the grid and watch the
   meter. When battery voltage is measured, this indicates that there is an open after that
   point. The grid should be repaired.

5. On a working grid, connect the positive lead of the meter to the power feed terminal and move the negative lead to the following positions; record your measured voltage.

   3/4 of the way across the grid: _____

   1/2 of the way across the grid: _____

   1/4 of the way across the grid: _____

   Describe what is happening.

   _____

   _____

   _____

   _____

6. Turn off the circuit, measure the resistance across the following points of the grid, and record your resistance readings.

3/4 of the way across the grid: _____

1/2 of the way across the grid: _____

1/4 of the way across the grid: _____

Describe why you had these readings.

_____

_____

_____

_____

Instructor's Response

_____

_____

_____

Name _____   **Date** _____

PWM HIGH-SIDE-DRIVER-ACTUATED HEATED SEAT CIRCUIT TESTING

Upon completion of this job sheet, you will be able to determine proper operation of the heated seat circuit.

NATEF Correlation

This job sheet addresses the following **MLR** tasks:

A.3.   Use wiring diagrams to trace electrical/electronic circuits.

A.4.   Demonstrate proper use of a digital multimeter (DMM) when measuring source voltage, voltage drop (including grounds), current flow, and resistance.

A.5.   Demonstrate knowledge of the causes and effects from shorts, grounds, opens, and resistance problems in electrical/electronic circuits.

A.11.   Identify electrical/electronic system components and configuration.

This job sheet addresses the following **AST** tasks:

G.1.   Describe operation of comfort and convenience accessories and related circuits (such as: power windows, power seats, pedal height, power locks, trunk locks, remote start, moonroof, sunroof, sunshade, remote keyless entry, voice activation, steering wheel controls, back-up camera, park assist, cruise control, and auto dimming headlamps); determine needed repairs.

G.5.   Describe body electronic systems circuits using a scan tool; check for module communication errors (data bus systems); determine needed action.

This job sheet addresses the following **AST/MAST** tasks:

A.3.   Demonstrate proper use of a digital multimeter (DMM) when measuring source voltage, voltage drop (including grounds), current flow, and resistance.

A.4.   Demonstrate knowledge of the causes and effects from shorts, grounds, opens, and resistance problems in electrical/electronic circuits.

A.7.   Use wiring diagrams during the diagnosis (troubleshooting) of electrical/electronic circuit problems.

This job sheet addresses the following **MAST** tasks:

A.11.   Check electrical/electronic circuit waveforms; interpret readings and determine needed repairs.

G.1.   Diagnose operation of comfort and convenience accessories and related circuits (such as: power windows, power seats, pedal height, power locks, trunk locks, remote start, moonroof, sunroof, sunshade, remote keyless entry, voice activation, steering wheel controls, back-up camera, park assist, cruise control, and auto dimming headlamps); determine needed repairs.

G.5.   Diagnose body electronic systems circuits using a scan tool; check for module communication errors (data bus systems); determine needed action. ASE content area: *Accessories Diagnosis and Repair*; tasks: diagnose (troubleshoot) body electronic system circuits using a scan tool; determine necessary action.

Tools and Materials
- Scan tool
- Lab scope
- DMM (2)
- Service information
- Appropriate vehicle

Describe the vehicle being worked on:

Year _____ Make _____ Model _____

VIN _____ Engine type and size _____

Procedure   **Task Completed**

1. Using appropriate service information, identify the driver-side heated seat circuits.

   _____

   _____

2. Backprobe the control circuit terminal of the heater seat circuit at either the control   ☐
   module or the grid connectors. Connect two DMMs to the back-probing tools. Set
   one DMM to read voltage and the second DMM on percent duty cycle (+ trigger).

3. Connect a scan tool to the vehicle, and access the heated seat data.   ☐

4. With the ignition switch in the RUN position, record the voltage and heater duty cycle
   as shown on the scan tool.

   _____

   _____

5. Use the scan tool to start the actuation for the seat heater. Observe the voltage and
   percent duty-cycle reading on the voltmeters. Record these values when the actuator
   is started.

   _____

   _____

   _____

6. Observe the DMM volts and percent duty cycle, and record these values after the
   actuator test has been run for about 30 seconds.

   _____

   _____

   _____

7. What does this test prove?

   _____

   _____

   _____

   _____

8. Stop the activation and start the engine. Use the heated seat switch to turn the system onto the HI setting. Record the voltage and the percent duty cycle when the heater is first turned on.

_____

_____

_____

9. Record the voltage and the percent duty cycle after the voltages have stabilized.

_____

_____

_____

10. How is this duty cycle different from the duty cycle after the heater has been on for a short time?

_____

_____

_____

_____

Instructor's Response

_____

_____

_____

Name _____   Date _____

LAB SCOPE TESTING OF A BLOWER MOTOR POWER MODULE

Upon completion of this job sheet, you will be able to determine proper operation of the blower motor power module.

NATEF Correlation

This job sheet addresses the following **MLR** tasks:

A.3.   Use wiring diagrams to trace electrical/electronic circuits.

A.4.   Demonstrate proper use of a digital multimeter (DMM) when measuring source voltage, voltage drop (including grounds), current flow, and resistance.

A.5.   Demonstrate knowledge of the causes and effects from shorts, grounds, opens, and resistance problems in electrical/electronic circuits.

A.11.   Identify electrical/electronic system components and configuration.

This job sheet addresses the following **AST** tasks:

G.1.   Describe operation of comfort and convenience accessories and related circuits (such as: power windows, power seats, pedal height, power locks, trunk locks, remote start, moonroof, sunroof, sunshade, remote keyless entry, voice activation, steering wheel controls, back-up camera, park assist, cruise control, and auto dimming headlamps); determine needed repairs.

G.5.   Describe body electronic systems circuits using a scan tool; check for module communication errors (data bus systems); determine needed action.

This job sheet addresses the following **AST/MAST** tasks:

A.3.   Demonstrate proper use of a digital multimeter (DMM) when measuring source voltage, voltage drop (including grounds), current flow, and resistance.

A.4.   Demonstrate knowledge of the causes and effects from shorts, grounds, opens, and resistance problems in electrical/electronic circuits.

A.7.   Use wiring diagrams during the diagnosis (troubleshooting) of electrical/electronic circuit problems.

This job sheet addresses the following **MAST** tasks:

A.11.   Check electrical/electronic circuit waveforms; interpret readings and determine needed repairs.

G.1.   Diagnose operation of comfort and convenience accessories and related circuits (such as: power windows, power seats, pedal height, power locks, trunk locks, remote start, moonroof, sunroof, sunshade, remote keyless entry, voice activation, steering wheel controls, back-up camera, park assist, cruise control, and auto dimming headlamps); determine needed repairs.

G.5.   Diagnose body electronic systems circuits using a scan tool; check for module communication errors (data bus systems); determine needed action.

Tools and Materials

- Scan tool
- Lab scope
- DMM
- Service information
- Appropriate vehicle

Describe the vehicle being worked on:

Year _____ Make _____ Model _____

VIN _____ Engine type and size _____

Procedure   **Task Completed**

1. Use the appropriate service information and identify the input circuit from the control module to the power module and the output motor circuits.

    _____

    _____

    _____

2. Is the motor speed controlled by PWM of the power side or ground side of the motor? _____

3. Set up the lab scope for two channels. Both channels should be set to read 12-volt square-wave patterns.   ☐

4. Using back-probing tools, connect channel l red lead to the power module terminal for the input circuit. Connect channel 2 red lead to the blower motor output control circuit identified in step 2.   ☐

5. With the blower motor turned OFF, record the voltage on.

    Channel 1:_____

    Channel 2:_____

6. Slowly turn the blower switch from LOW to HI while observing the scope pattern. Describe your results.

    _____

    _____

    _____

7. Based on your observations, if the input circuit to the power module is shorted to ground, the blower speed is: ☐ HIGH   ☐ LOW   ☐ OFF

8. If the input circuit has an open, the blower operates only at what speed:
    ☐ HIGH   ☐ LOW   ☐ OFF

Instructor's Response

_____

_____

_____

Name _____   Date _____

TESTING THE ELECTRONIC CRUISE CONTROL SYSTEM

Upon completion of this job sheet, you should be able to test the electronic cruise control system using a scan tool or stand-alone diagnostic routines and determine needed repairs.

NATEF Correlation

This job sheet addresses the following **MLR** tasks:

A.3. Use wiring diagrams to trace electrical/electronic circuits.

A.4. Demonstrate proper use of a digital multimeter (DMM) when measuring source voltage, voltage drop (including grounds), current flow, and resistance.

A.5. Demonstrate knowledge of the causes and effects from shorts, grounds, opens, and resistance problems in electrical/electronic circuits.

A.11. Identify electrical/electronic system components and configuration.

This job sheet addresses the following **AST** tasks:

G.4. Describe operation of safety systems and related circuits (e.g., horn, air bags, seat belt pretensioners, occupancy classification, wipers, washers, speed control/collision avoidance, heads-up display, park assist, and back-up camera); determine needed repairs.

G.5. Describe body electronic systems circuits using a scan tool; check for module communication errors (data bus systems); determine needed action.

This job sheet addresses the following **AST/MAST** tasks:

A.3. Demonstrate proper use of a digital multimeter (DMM) when measuring source voltage, voltage drop (including grounds), current flow, and resistance.

A.4. Demonstrate knowledge of the causes and effects from shorts, grounds, opens, and resistance problems in electrical/electronic circuits.

A.7. Use wiring diagrams during the diagnosis (troubleshooting) of electrical/electronic circuit problems.

This job sheet addresses the following **MAST** tasks:

A.11. Check electrical/electronic circuit waveforms; interpret readings, and determine needed repairs.

G.4. Diagnose operation of safety systems and related circuits (e.g., horn, air bags, seat belt pretensioners, occupancy classification, wipers, washers, speed control/collision avoidance, heads-up display, park assist, and back-up camera); determine needed repairs.

G.5. Diagnose body electronic systems circuits using a scan tool; check for module communication errors (data bus systems); determine needed action.

ASE NATEF

Tools and Materials

- A vehicle equipped with electronic cruise control
- Scan tool
- Service information

Describe the vehicle being worked on:

Year _____ Make _____ Model _____

VIN _____ Engine type and size _____

Procedure

1. Does the system have a stand-alone computer, or is the system a part of the PCM functions? ☐ STAND-ALONE ☐ PCM

2. If possible, perform the simulated road test and record your results.

   _____

   _____

3. Are there any other related symptoms? ☐ Yes ☐ No

   If yes, describe the symptom.

   _____

   _____

4. Refer to proper service information to determine if a self-diagnostic routine can be performed. If so, describe how to enter diagnostics.

   _____

   _____

   _____

5. Is the self-diagnostic routine capable of displaying fault codes? ☐ Yes ☐ No

6. Perform the procedure listed in step 4 and record the results and any DTCs.

   _____

   _____

7. Use a scan tool and access the electronic cruise control system. Does the scan tool indicate that DTCs are present? ☐ Yes ☐ No

   If yes, record the DTCs.

   _____

   _____

8. Based on the results so far, what tests need to be performed to find the cause of the fault?

   _____

   _____

9. Perform the tests listed in step 8. What is your determination and recommendation?

   _____

   _____

Instructor's Response

_____

_____

_____

Name _____   Date _____

ADAPTIVE CRUISE CONTROL SERVICE

Upon completion of this job sheet, you will verify operation of the laser radar sensor used in adaptive cruise control systems.

NATEF Correlation

This job sheet addresses the following **MLR** tasks:

A.3.   Use wiring diagrams to trace electrical/electronic circuits.

A.4.   Demonstrate proper use of a digital multimeter (DMM) when measuring source voltage, voltage drop (including grounds), current flow, and resistance.

A.5.   Demonstrate knowledge of the causes and effects from shorts, grounds, opens, and resistance problems in electrical/electronic circuits.

A.6.   Check operation of electrical circuits using a test light.

A.11.   Identify electrical/electronic system components and configuration.

This job sheet addresses the following **AST** tasks:

G.4.   Describe operation of safety systems and related circuits (e.g., horn, air bags, seat belt pretensioners, occupancy classification, wipers, washers, speed control/collision avoidance, heads-up display, park assist, and back-up camera); determine needed repairs.

G.5.   Describe body electronic systems circuits using a scan tool; check for module communication errors (data bus systems); determine needed action.

This job sheet addresses the following **AST/MAST** tasks:

A.3.   Demonstrate proper use of a digital multimeter (DMM) when measuring source voltage, voltage drop (including grounds), current flow, and resistance.

A.4.   Demonstrate knowledge of the causes and effects from shorts, grounds, opens, and resistance problems in electrical/electronic circuits.

A.5.   Demonstrate proper use of a test light on an electrical circuit.

A.7.   Use wiring diagrams during the diagnosis (troubleshooting) of electrical/electronic circuit problems.

This job sheet addresses the following **MAST** tasks:

A.11.   Check electrical/electronic circuit waveforms; interpret readings; and determine needed repairs.

G.4.   Diagnose operation of safety systems and related circuits (e.g., horn, air bags, seat belt pretensioners, occupancy classification, wipers, washers, speed control/collision avoidance, heads-up display, park assist, and back-up camera); determine needed repairs.

G.5.   Diagnose body electronic systems circuits using a scan tool; check for module communication errors (data bus systems); determine needed action.

Tools and Materials

- Scan tool
- DMM
- Test light
- Alignment screen

- Tape
- Grease pencil
- Tape measure
- Wiring diagram
- Vehicle-equipped adaptive cruise control

Describe the vehicle being worked on:

Year _____ Make _____ Model _____

VIN _____ Engine type and size _____

Procedure

Task 1

1. Where is the sensor located?

   _____

   _____

   _____

2. Is the sensor integrated with the module? ☐ Yes ☐ No

   If no, where is the module? located?

   _____

   _____

   _____

3. What other modules are used by the ACC system for operation or inputs?

   _____

   _____

   _____

4. Identify the circuits to the sensors.

   _____

   _____

   _____

   _____

5. What supplies the voltage to the sensor? _____

6. Where is the ground for the sensor located? _____

7. Disconnect the sensor harness connector, and use a DMM to measure the voltage on the supply circuit with the ignition switch in the RUN position (engine off). What is your reading? _____

8. Using a test light, load the supply circuit and use the voltmeter to read the voltage at the test light probe._____

9. If the readings between steps 7 and 8 are different, explain why.

   _____

   _____

   _____

10. Describe how you will test the ground circuit.   **Task Completed**

_____

_____

_____

11. Test the ground circuit using the method described in step 9, and record your results.

_____

_____

_____

12. With the sensor harness disconnected, use the scan tool to retrieve fault codes.

_____

_____

_____

13. What are the observable symptoms of the ACC system with the sensor disconnected?

_____

_____

_____

14. Reconnect all connectors and clear any DTCs.   ☐

Task 2

   1. When is the sensor calibration procedure to be performed?

_____

_____

   2. What is the required height of the alignment screen?

_____

   3. How far from the sensor is the screen to be located?

_____

   4. Determine the centerline of the vehicle, and locate the screen in the proper position.

   5. What other preparations must be made prior to aligning the sensor?

_____

_____

_____

   6. Describe how the alignment procedure is to be performed.

_____

_____

_____

_____

7. Perform the alignment procedure. Was the procedure successful? ☐ Yes ☐ No

If no, what must be done to correct the problem?

_____

_____

Instructor's Response

_____

_____

_____

Name _____   **Date** _____

TESTING THE ANTITHEFT SYSTEM OPERATION

Upon completion of this job sheet, you should be able to test the antitheft system operation through the self-test diagnostic routine and determine needed repairs.

NATEF Correlation

This job sheet addresses the following **MLR** tasks:

A.3.	Use wiring diagrams to trace electrical/electronic circuits.
A.4.	Demonstrate proper use of a digital multimeter (DMM) when measuring source voltage, voltage drop (including grounds), current flow, and resistance.
A.5.	Demonstrate knowledge of the causes and effects from shorts, grounds, opens, and resistance problems in electrical/electronic circuits.
A.6.	Check operation of electrical circuits using a test light.
A.11.	Identify electrical/electronic system components and configuration.

This job sheet addresses the following **AST** tasks:

G.2.	Describe operation of security/antitheft systems and related circuits (such as: theft deterrent, door locks, remote keyless entry, remote start, and starter/fuel disable); determine needed repairs.
G.5.	Describe body electronic systems circuits using a scan tool; check for module communication errors (data bus systems); determine needed action.

This job sheet addresses the following **AST/MAST** tasks:

A.3.	Demonstrate proper use of a digital multimeter (DMM) when measuring source voltage, voltage drop (including grounds), current flow, and resistance.
A.4.	Demonstrate knowledge of the causes and effects from shorts, grounds, opens, and resistance problems in electrical/electronic circuits.
A.5.	Demonstrate proper use of a test light on an electrical circuit.
A.7.	Use wiring diagrams during the diagnosis (troubleshooting) of electrical/ electronic circuit problems.

This job sheet addresses the following **MAST** tasks:

G.2.	Diagnose operation of security/antitheft systems and related circuits (such as: theft deterrent, door locks, remote keyless entry, remote start, and starter/fuel disable); determine needed repairs.
G.5.	Diagnose body electronic systems circuits using a scan tool; check for module communication errors (data bus systems); determine needed action.

ASE **NATEF**

Tools and Materials

- A vehicle equipped with an antitheft system
- Scan tool
- Service information

Describe the vehicle being worked on:

Year _____ Make _____ Model _____

VIN _____ Engine type and size _____

Procedure

1. Can the system self-diagnostic test be performed stand-alone or through the use of a scan tool?   ☐ STAND-ALONE   ☐ SCAN TOOL

2. If it can be performed stand-alone, describe how the process is performed.

   _____

   _____

   _____

3. If a scan tool must be used, list the screen menu selections required to perform the self-test.

   _____

   _____

   _____

   _____

4. Enter the diagnostic self-test and record your results.

   _____

   _____

   _____

5. Based on the results so far, what tests need to be performed to find the cause of the fault?

   _____

   _____

6. Perform the tests listed in step 5. What is your determination and recommendation?

   _____

   _____

Instructor's Response

_____

_____

_____

DIAGNOSTIC CHART 13-1

PROBLEM AREA:	Horn system.
SYMPTOMS:	Horn sounds even when switch is not activated.
POSSIBLE CAUSES:	**1.** Faulty horn switch. **2.** Horn control circuit shorted to ground. **3.** Faulty horn relay.

DIAGNOSTIC CHART 13-2

PROBLEM AREA:	Horn system.
SYMPTOMS:	Intermittent horn operation.
POSSIBLE CAUSES:	**1.** Faulty horn switch. **2.** Poor clockspring contacts. **3.** Faulty horn relay. **4.** Poor ground connection at horn. **5.** Poor ground connection at switch.

DIAGNOSTIC CHART 13-3

PROBLEM AREA:	Horn system.
SYMPTOMS:	Horn fails to sound when the horn switch is activated.
POSSIBLE CAUSES:	**1.** Blown fuse. **2.** Faulty horn switch. **3.** Poor clockspring contacts. **4.** Faulty horn relay. **5.** Poor ground connection at horn. **6.** Poor ground connection at switch.

DIAGNOSTIC CHART 13-4

PROBLEM AREA:	Wiper system operation.
SYMPTOMS:	Wipers operate any time the ignition switch is in the RUN position.
POSSIBLE CAUSES:	**1.** Faulty wiper switch. **2.** Defective park switch or activation arm. **3.** Shorted control circuit.

DIAGNOSTIC CHART 13-5

PROBLEM AREA:	Wiper system operation.
SYMPTOMS:	Wipers operate some of the time.
POSSIBLE CAUSES:	**1.** Poor ground connection. **2.** Poor control circuit connection. **3.** Faulty switch. **4.** Worn motor contacts or brushes.

DIAGNOSTIC CHART 13-6

PROBLEM AREA:	Wiper system operation.
SYMPTOMS:	Wipers fail to function when switch is activated.
POSSIBLE CAUSES:	**1.** Blown fuse. **2.** Open in the control circuit. **3.** Short in the control circuit. **4.** Faulty wiper motor. **5.** Poor ground connection. **6.** Mechanical linkage binding. **7.** Faulty wiper switch.

DIAGNOSTIC CHART 13-7

PROBLEM AREA:	Wiper system operation.
SYMPTOMS:	**1.** Wipers operate only at low speed. **2.** Wipers operate only at high speed.
POSSIBLE CAUSES:	**1.** Faulty wiper switch. **2.** Worn brushes. **3.** Poor control circuit connections. **4.** Open in control circuit.

DIAGNOSTIC CHART 13-8

PROBLEM AREA:	Wiper system operation.
SYMPTOMS:	**1.** Wipers stop on the windshield when switch is turned off. **2.** Wipers remain on when wiper switch is turned off.
POSSIBLE CAUSES:	**1.** Faulty park switch. **2.** Activation arm broken or out of adjustment.

DIAGNOSTIC CHART 13-9

PROBLEM AREA:	Wiper system operation.
SYMPTOMS:	Wipers will not shut off.
POSSIBLE CAUSES:	**1.** Faulty switch. **2.** Faulty park switch or circuit.

DIAGNOSTIC CHART 13-10

PROBLEM AREA:	Intermittent wiper system operation.
SYMPTOMS:	No intermittent wiper operation, low and high speeds operate normally.
POSSIBLE CAUSES:	Faulty switch.

DIAGNOSTIC CHART 13-11

PROBLEM AREA:	Wiper motor operation.
SYMPTOMS:	Slow or no wiper operation.
POSSIBLE CAUSES:	**1.** Faulty wiper motor. **2.** Binding wiper linkage.

DIAGNOSTIC CHART 13-12

PROBLEM AREA:	Washer system operation.
SYMPTOMS:	Washer operates without switch activation.
POSSIBLE CAUSES:	**1.** Faulty switch. **2.** Short in control circuit.

DIAGNOSTIC CHART 13-13

PROBLEM AREA:	Washer system operation.
SYMPTOMS:	Intermittent washers operation.
POSSIBLE CAUSES:	**1.** Faulty switch. **2.** Faulty pump. **3.** Poor control circuit connections. **4.** Poor ground connection.

DIAGNOSTIC CHART 13-14

PROBLEM AREA:	Washer system operation.
SYMPTOMS:	Windshield washer fails to operate.
POSSIBLE CAUSES:	**1.** Blown fuse. **2.** Faulty switch. **3.** Faulty motor. **4.** Poor ground connection. **5.** Open or short in control circuit. **6.** Restriction in hoses.

DIAGNOSTIC CHART 13-15

PROBLEM AREA:	Power side window operation.
SYMPTOMS:	Power windows operate slower than normal.
POSSIBLE CAUSES:	**1.** Control circuit resistance. **2.** Faulty motor. **3.** Poor ground connection. **4.** Improperly adjusted regulators. **5.** Binding linkages.

DIAGNOSTIC CHART 13-16

PROBLEM AREA:	Power side window operation.
SYMPTOMS:	Intermittent power window operation.
POSSIBLE CAUSES:	**1.** Faulty switch. **2.** Faulty motor. **3.** Poor ground connections. **4.** Poor control circuit connections. **5.** Binding linkage. **6.** Faulty circuit breaker.

DIAGNOSTIC CHART 13-17

PROBLEM AREA:	Power side window operation.
SYMPTOMS:	Power windows fail to operate when switch is activated.
POSSIBLE CAUSES:	**1.** Faulty switch. **2.** Faulty motor. **3.** Poor ground connections. **4.** Poor control circuit connections. **5.** Binding linkage. **6.** Faulty circuit breaker.

DIAGNOSTIC CHART 13-18

PROBLEM AREA:	Power seat operation.
SYMPTOMS:	Power seats operate slower than normal.
POSSIBLE CAUSES:	**1.** Control circuit resistance. **2.** Faulty motor. **3.** Poor ground connection. **4.** Binding linkages. **5.** Faulty motor transmission.

DIAGNOSTIC CHART 13-19

PROBLEM AREA:	Power seat operation.
SYMPTOMS:	Intermittent power seat operation.
POSSIBLE CAUSES:	**1.** Faulty switch. **2.** Faulty motor. **3.** Poor ground connections. **4.** Poor control circuit connections. **5.** Binding linkage. **6.** Faulty circuit breaker. **7.** Faulty transmission.

DIAGNOSTIC CHART 13-20

PROBLEM AREA:	Power seat operation.
SYMPTOMS:	Power windows fail to operate when switch is activated.
POSSIBLE CAUSES:	**1.** Faulty switch. **2.** Faulty motor. **3.** Poor ground connections. **4.** Poor control circuit connections. **5.** Binding linkage. **6.** Faulty circuit breaker. **7.** Faulty transmission.

DIAGNOSTIC CHART 13-21

PROBLEM AREA:	Memory seat system operation.
SYMPTOMS:	Seat does not move to preset positions; power seat works normally.
POSSIBLE CAUSES:	**1.** Faulty motor position sensor. **2.** Faulty switch. **3.** Faulty control module. **4.** Open control module power feed circuit. **5.** Poor control module ground circuit. **6.** Improper park/neutral switch input. **7.** Bus communications error.

DIAGNOSTIC CHART 13-22

PROBLEM AREA:	Heated seat.
SYMPTOMS:	Implausible voltage on control circuit.
POSSIBLE CAUSES:	**1.** Control circuit shorted to voltage. **2.** Control circuit shorted to ground. **3.** Control circuit open or high resistance. **4.** Heater element. **5.** Control module.

DIAGNOSTIC CHART 13-23

PROBLEM AREA:	Heated seat.
SYMPTOMS:	Voltage on control circuit too low.
POSSIBLE CAUSES:	**1.** Control circuit shorted to ground. **2.** Control circuit open or high resistance. **3.** Heater element. **4.** Control module.

DIAGNOSTIC CHART 13-24

PROBLEM AREA:	Heated seat.
SYMPTOMS:	Voltage on control circuit high.
POSSIBLE CAUSES:	**1.** Control circuit shorted to voltage. **2.** Control circuit open or high resistance. **3.** Heater element. **4.** Control module.

DIAGNOSTIC CHART 13-25

PROBLEM AREA:	Seat fan.
SYMPTOMS:	Seat fails to cool properly.
POSSIBLE CAUSES:	**1.** Control circuit shorted to voltage. **2.** Control circuit open. **3.** Control circuit high resistance. **4.** Fan return circuit open. **5.** Faulty fan motor. **6.** Control module.

DIAGNOSTIC CHART 13-26

PROBLEM AREA:	Rear window defogger operation.
SYMPTOMS:	Rear window defogger fails to clear the window.
POSSIBLE CAUSES:	**1.** Open in the grid. **2.** Excessive circuit resistance. **3.** Faulty switch. **4.** Blown fuse. **5.** Open in control circuit. **6.** Grid connection loose. **7.** Poor ground connection.

DIAGNOSTIC CHART 13-27

PROBLEM AREA:	Rear window defogger operation.
SYMPTOMS:	Intermittent rear window defogger operation.
POSSIBLE CAUSES:	**1.** Faulty switch. **2.** Poor circuit control connection. **3.** Poor ground connection. **4.** Loose connection to grid.

DIAGNOSTIC CHART 13-28

PROBLEM AREA:	Power door lock system operation.
SYMPTOMS:	Power door locks fail to operate some of the time.
POSSIBLE CAUSES:	**1.** Faulty circuit breaker. **2.** Faulty switch. **3.** Poor control circuit connections. **4.** Poor ground connections. **5.** Faulty motor or solenoid. **6.** Faulty relay.

DIAGNOSTIC CHART 13-29

PROBLEM AREA:	Power door lock system operation.
SYMPTOMS:	All door locks do not operate in either direction.
POSSIBLE CAUSES:	**1.** Faulty circuit breaker. **2.** Faulty master switch. **3.** Open in control circuit. **4.** Short in control circuit. **5.** Poor ground connection. **6.** Faulty relay. **7.** Poor relay control circuit connections.

DIAGNOSTIC CHART 13-30

PROBLEM AREA:	Power door lock system operation.
SYMPTOMS:	One door lock does not operate in either direction.
POSSIBLE CAUSES:	**1.** Obstruction or binding of linkage. **2.** Open in control circuit. **3.** Short in control circuit. **4.** Poor ground connection. **5.** Faulty motor or solenoid.

DIAGNOSTIC CHART 13-31

PROBLEM AREA:	Power door lock system operation.
SYMPTOMS:	All locks work from one switch only.
POSSIBLE CAUSES:	**1.** Faulty switch. **2.** Open in control circuit. **3.** Short in control circuit.

DIAGNOSTIC CHART 13-32

PROBLEM AREA:	Cruise control operation.
SYMPTOMS:	Cruise control speed changes over or below set requests.
POSSIBLE CAUSES:	**1.** Faulty servo. **2.** Defective controller. **3.** Faulty speed sensor. **4.** Throttle linkage adjustment. **5.** Faulty amplifier. **6.** Faulty dump valve.

DIAGNOSTIC CHART 13-33

PROBLEM AREA:	Cruise control operation.
SYMPTOMS:	Cruise control fails to set.
POSSIBLE CAUSES:	**1.** Faulty switch or circuit. **2.** Faulty servo or circuit. **3.** Defective controller. **4.** Poor ground connection. **5.** Poor control circuit connections. **6.** Faulty relay or circuit. **7.** Faulty speed sensor. **8.** Faulty speed sensor circuit. **9.** Throttle linkage adjustment. **10.** Faulty or misadjusted brake switch.

DIAGNOSTIC CHART 13-34

PROBLEM AREA:	Adaptive cruise control.
SYMPTOMS:	Disabling of system.
POSSIBLE CAUSES:	**1.** Wiring harness, terminal, connector damage. **2.** Damaged sensor lens. **3.** Obstruction in front of sensor. **4.** Improper sensor alignment. **5.** High resistance in sensor voltage supply circuit. **6.** Open in sensor voltage supply circuit. **7.** High resistance in sensor ground circuit. **8.** Open in sensor ground circuit. **9.** Sensor supply circuit shorted to sensor ground circuit. **10.** Sensor supply circuit shorted to ground. **11.** Faulty sensor. **12.** Internal controller fault.

DIAGNOSTIC CHART 13-35

PROBLEM AREA:	Sunroof operation.
SYMPTOMS:	Sunroof opens or closes slower than normal.
POSSIBLE CAUSES:	**1.** Binding track. **2.** Excessive circuit resistance. **3.** Worn motor. **4.** Misadjusted linkage. **5.** Trim panel mispositioned. **6.** Cable guides mispositioned. **7.** Slipping motor clutch.

DIAGNOSTIC CHART 13-36

PROBLEM AREA:	Sunroof operation.
SYMPTOMS:	Intermittent sunroof operation.
POSSIBLE CAUSES:	**1.** Worn or defective motor. **2.** Faulty controller. **3.** Loose ground connection. **4.** Poor control circuit connection. **5.** Binding linkage and/or tracks.

DIAGNOSTIC CHART 13-37

PROBLEM AREA:	Sunroof operation.
SYMPTOMS:	Sunroof fails to move in either direction.
POSSIBLE CAUSES:	**1.** Worn or defective motor. **2.** Faulty controller. **3.** Loose ground connection. **4.** Poor control circuit connection. **5.** Binding linkage and/or tracks. **6.** Faulty circuit breaker or fuse. **7.** Trim panel mispositioned. **8.** Cable guides mispositioned. **9.** Slipping motor clutch.

DIAGNOSTIC CHART 13-38

PROBLEM AREA:	Antitheft system operation.
SYMPTOMS:	System fails to arm or to operate some of the time.
POSSIBLE CAUSES:	**1.** Faulty controller. **2.** Defective switches. **3.** Poor wire connections. **4.** Poor ground connections.

DIAGNOSTIC CHART 13-39

PROBLEM AREA:	Antitheft system operation.
SYMPTOMS:	System won't disarm.
POSSIBLE CAUSES:	**1.** Lock cylinder disarm switch loose. **2.** Open in the lock cylinder disarm switch circuit. **3.** Defective lock cylinder disarm switch. **4.** Defective lock cylinder. **5.** Faulty control module power and ground. **6.** Faulty control module.

DIAGNOSTIC CHART 13-40

PROBLEM AREA:	Antitheft system operation.
SYMPTOMS:	System trips and sounds alarm by itself.
POSSIBLE CAUSES:	**1.** Doorjamb, hood, or trunk switch loose. **2.** Loose connection in doorjamb switch, hood switch, or trunk switch circuit. **3.** Defective doorjamb, hood, or trunk switch. **4.** Faulty control module.

DIAGNOSTIC CHART 13-41

PROBLEM AREA:	Antitheft system operation.
SYMPTOMS:	System won't trip when door is opened.
POSSIBLE CAUSES:	**1.** Doorjamb switch circuit open. **2.** Faulty doorjamb switch. **3.** Faulty control module.

CHAPTER 14

SERVICING RADIO FREQUENCY AND INFOTAINMENT SYSTEMS

Upon completion and review of this chapter, you should be able to:

- Diagnose remote keyless entry systems.
- Program transmitters into the RKE system.
- Diagnose the immobilizer system.
- Program keys to the immobilizer system.
- Diagnose the tire pressure monitoring system.
- Diagnose the vehicle audio and video entertainment system for no operation.
- Diagnose the vehicle audio and video entertainment system for causes of radio noise.
- Diagnose the audio system for noise concerns.
- Service the speakers.
- Diagnose navigational system faults.

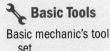

Basic Tools

Basic mechanic's tool set
Service information

Terms To Know

Antenna	Radiated noise	Remote keyless entry (RKE)
Conducted interference	Radio choke	Speakers
Immobilizer system	Radio interference	Tire pressure monitoring
PIN code	Radio noise	(TPM) system

INTRODUCTION

Usually when radio frequency (RF) is mentioned, the thought goes to audio systems. Although this is one use of RF, today's vehicles can have several systems that operate with RF. These include the remote keyless entry (RKE), remote start, immobilizer, and tire pressure monitoring (TPM) systems. This chapter discusses the diagnosis and servicing of the vehicle infotainment system as well as other radio frequency–operated systems.

REMOTE KEYLESS ENTRY AND REMOTE START

Many modern vehicles are equipped with a **remote keyless entry (RKE)** system that is used to lock and unlock the doors, turn on the interior lights, and release the trunk latch prior to approaching the vehicle. A small receiver is installed in the vehicle. A handheld key FOB is the transmitter assembly (**Figure 14-1**). Pressing a button on the FOB will allow operation of the RKE system, which is typically a function of the body control module (BCM) or cabin compartment node (CNN). The electronics of the RKE system are responsible for translating coded input from the key FOB into commands for module driver outputs. If the code is accepted, the control module generates a bus message command that is sent to other modules, or perform a direct output to the drivers to execute the requested function. **Figure 14-2** is a simple schematic of an RKE system.

Classroom Manual
Chapter 14, page 443

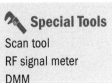

Special Tools

Scan tool
RF signal meter
DMM

Figure 14-1 RKE FOB transmitter.

Immobilizer		Messaging
RKE	Microprocessor	Audible outputs
Inputs		Driver outputs

Figure 14-2 RKE system schematic.

To diagnose an inoperative system, begin by confirming that the door locks, trunk latch, and interior lamps work normally when manually activated. If these systems check out fine, detailed diagnosis of the remote system is necessary. Follow the specific steps in the diagnostic procedures for the problem being experienced.

The transponder and receiver system operate at a fixed radio frequency. If the unit does not work from a normal distance, check for two conditions: weak batteries in the remote transmitter or a stronger radio transmitter close by (radio station, airport transmitter, etc.). The transponder operation can be tested using an RF signal meter (**Figure 14-3**). By operating the transponder while the RF signal meter is on, the strength of the signal is displayed by the light-emitting diodes (LEDs). If the transponder fails to indicate a signal, test the transponder batteries. Usually, two 3-volt (V) batteries are used. Also, be sure the batteries are installed correctly since they are polarity sensitive. If the transponders test good, confirm that they are properly programmed to the system.

Confirm that any supporting modules are receiving the bus communication messages from the control module. These will usually set "Loss of communication" fault codes. The cause can be bus circuit issues or faulty control modules. Always confirm proper power and ground circuit to the control module before condemning them.

Some perceived problems with the remote start feature may actually be normal operation. For example, most systems will only allow the engine to run for 5 to 10 minutes, and then the engine is shut off. The customer may think that the engine did not start if they approach the vehicle after the timeout has occurred. In addition, typically the engine can

Classroom Manual
Chapter 14, page 443

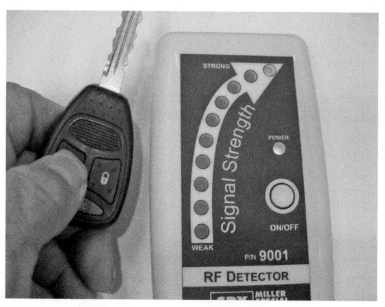

Figure 14-3 RF detector used to determine if FOB transmission signal is being sent.

only be started a couple of times without the ignition actually being placed in the RUN position. If the engine is started remotely too many times, the FOB request is denied, and the horn will double chirp to indicate the engine will not be started. In order to reactivate the remote start function, the ignition must be placed in the RUN position. The customer may mistake this operation of the system as an intermittent fault.

Other inputs to the remote start system include door ajar and hood switches. If any door is not properly shut, or the hood is open, the remote start function is disabled. If the remote start feature is not functional, confirm the operation of all inputs that can result in the system being disabled.

Diagnosing Systems with Aftermarket Kits

Diagnosing problems with the aftermarket installations of remote keyless entry and start systems can present some challenges. First, it is completely up to the installer to determine where the components are located in the vehicle. The control module can be located in the trunk, in one of the side kick panels, behind the dash, or under a seat. The same is true for the location of relays, diodes, program switches, power feeds, and grounds. The next challenge is determining what the function of each circuit to and from the control module is. Often a search on the Web will provide installation guides for the unit that is installed. In addition, copies of the owner's manual may be found that provide instruction of how to use the system and program remote transmitters. Also, with the system installed there may be more than one relay in a single circuit. Finally, what probably makes add-on kits the most challenging is the fact that the system is designed to work on several different applications, instead of just the vehicle it is installed in. This means there are different configurations of positive and ground circuits, relay output circuits, alarm disarm circuits, and switched inputs (**Figure 14-4**).

Classroom Manual
Chapter 14, page 444

Some systems will have diagnostic capabilities. If a fault is detected, they may flash an LED indicating a problem. By using the program switch to enter diagnostic mode, the LED will flash a trouble code. The process varies based on manufacturer.

Many of the control circuits operate bypass relays. These relays allow the remote keyless entry/remote start electronic control unit (ECU) to operate the systems by bypassing the factory systems. When diagnosing the systems first determine if the system works at all. For example, if the customer issue is that the dome lights do not come on when the UNLOCK button on the FOB is pressed; you need to determine if the dome lights operate at all. If they work when the door is opened or the dome light switch is turned on then the problem probably lies in the add-on system.

If the problem system does not operate at all, disconnect the aftermarket ECU and see if the problem goes away. For example, no, intermittent, or irrational power door lock operation may be caused by a problem with the add-on RKE system. If unplugging the module (or removing power to it) results in proper power door lock operation, the problem is within the add-on system.

Once it has been determined that the aftermarket system is causing the issue, use normal voltage and ground testing techniques to isolate the problem. If an insulation guide with the manufacturer's wiring diagram is not available, you should try to trace the circuits to and from the ECM and create your own schematic.

If you determine that the problem is within a FOB, refer to the owners guide to attempt reprogramming of the FOBs. This may require the use of a bitwrite program on some systems. The system may be equipped with a program button that, when activated according to instructions, will erase the FOBs and then allow for reprogramming them.

IMMOBILIZER SYSTEM DIAGNOSIS AND SERVICE

Classroom Manual
Chapter 14, page 450

The **immobilizer system** is designed to provide protection against unauthorized vehicle use by disabling the engine if an invalid key is used to start the engine or an attempt to hot-wire the ignition system is made. Most systems consist of an immobilizer module,

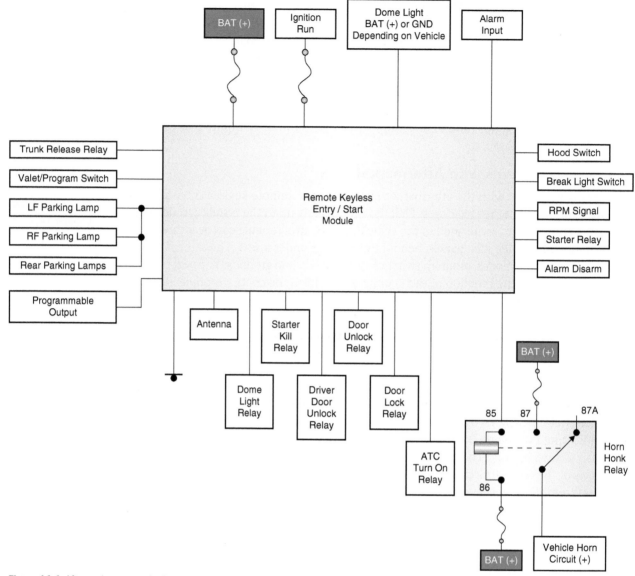

Figure 14-4 Aftermarket remote keyless entry and start systems can present some diagnostic challenges.

ignition key with transponder chip, the powertrain control module (PCM), and an indicator lamp. The special keys have a transponder chip under the covering (**Figure 14-5**). The immobilizer module radiates an RF signal through its antenna to the key. The transponder signals back to the antenna its identification number. If the message identifies the key as being valid, the immobilizer module sends a message over the data bus to the PCM indicating that the engine may be started.

Special Tools

Scan tool
DMM

AUTHOR'S NOTE The following service procedures are provided as an example of the system. Different manufacturers will have varying procedures and operation of their systems.

If the response received from the key transponder is missing or identifies the key as invalid, the immobilizer module sends an "invalid key" message to the PCM over the data bus. This prevents the engine from starting. In addition, the indicator lamp provides system status messages. The indicator lamp normally illuminates for 3 seconds for a bulb

Figure 14-5 Internal transponder is located under the cover of the key.

check when the ignition switch is first placed in the RUN position. After the bulb check is complete, the lamp should go out. If the lamp remains on after the bulb check, this indicates that the immobilizer module has detected a system malfunction or that the system has become inoperative. If the lamp flashes after the bulb check is completed, this indicates that an invalid key is detected or that a key-related fault is present.

AUTHOR'S NOTE The default condition in the PCM is "invalid key." If the PCM does not receive any messages from the immobilizer module, the engine is prevented from starting.

A common service to the immobilizer system is the addition of keys. Usually, the system can record several different key codes. However, the manufacturer may only provide two keys at the time of vehicle purchase. If additional keys are purchased, they will require programming into the system prior to use. Each key has a unique transponder identification code that is permanently programmed into it. When a key is programmed into the immobilizer module, the transponder identification code is then stored in the immobilizer's memory. In addition, the key learns the secret key code from the immobilizer module and programs it into its transponder memory. Have all previously programmed keys available when programming new keys. They may need to be reprogrammed also.

AUTHOR'S NOTE Systems differ on the customer programming procedure. The example provided here is only one method that is used. Always refer to the proper service information for the exact procedure of the vehicle you are servicing.

Most manufacturers provide a method the customer can follow to program new keys. This option is not available on all systems. The following is one method for customer programming of the keys.

In order to perform the customer programming method, two valid keys are used to enter the programming routine. The immobilizer module needs to see two unique key

IDs before it allows programming. One key cannot be used two times. Also, the key that is being programmed to the vehicle must be a blank key. If it has been programmed to another vehicle, it cannot be programmed again. If the sequence is not followed, the immobilizer will abort the programming process. Customer programming of a new key may be performed by the following procedure:

1. Insert one of the two valid keys into the ignition switch, and turn the ignition switch to the RUN position. Leave the ignition switch in the RUN position for at least 3 seconds but not more than 15 seconds.
2. Cycle the ignition switch to the OFF position and remove the key.
3. Within 15 seconds, insert the second valid key into the ignition switch and turn the ignition switch to the RUN position. Leave the ignition switch in the RUN position while observing the immobilizer warning lamp.
4. In about 10 seconds, the security indicator in the instrument cluster starts to flash and a single chime sounds. This indicates that the immobilizer system is in programming mode.
5. Within 60 seconds, turn the ignition switch to the OFF position and remove the key.
6. Insert the blank key into the ignition switch, and turn the ignition switch to the RUN position.
7. In about 10 seconds, an audible chime sounds and the security indicator stays illuminated solid for 3 seconds and then turns off. This indicates that the new key has been successfully programmed.
8. Cycle the ignition switch to the OFF position before attempting to start the engine.

If additional keys are to be programmed, the entire procedure must be performed for each key. The system automatically exits programming mode after a blank key is programmed.

If the vehicle owner loses a key, or a damaged key needs to be replaced, the new transponder IDs will need to be programmed. Unless there are still two additional keys that are programmed, the technician will need to use the secure method of programming the new key. This method requires the use of a diagnostic scan tool. Also, a unique **PIN code** that is programmed into the immobilizer module will need to be obtained. This is a secure code that must be obtained from the vehicle owner, from the original vehicle invoice, or from the vehicle manufacturer. **Photo Sequence 32** shows how additional keys can be programmed.

Problems with the immobilizer system can result in an engine no-start condition. Since the system uses many hardwired components and circuits, these can be diagnosed and tested using normal diagnostic tools and procedures. However, to reliably diagnose the electronic message inputs used to provide the electronic features of the immobilizer system requires the use of a diagnostic scan tool.

PHOTO SEQUENCE 32
Programming Additional Immobilizer Keys

P32-1 Connect the scan tool to access the immobilizer module, and locate the function for programming additional keys.

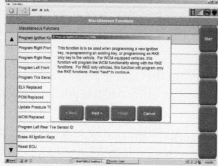

P32-2 Pressing the ENTER key on the scan tool will start the programming function.

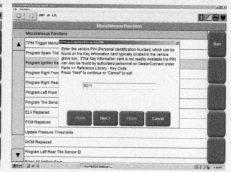

P32-3 Enter the PIN number when prompted.

PHOTO SEQUENCE 32 (CONTINUED)

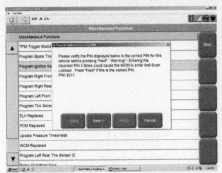

P32-4 Confirm that the PIN number entered is correct.

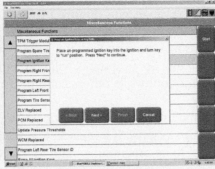

P32-5 Insert the blank key into the ignition switch, and turn to the RUN position.

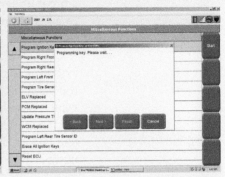

P32-6 Pressing the NEXT key on the scan tool will start the procedure for programming the key.

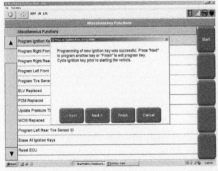

P32-7 A confirmation is provided that the key was successfully programmed, and you are asked if you wish to program another key.

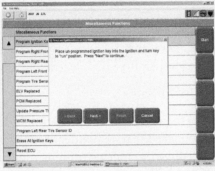

P32-8 If another key is to be programmed, the same steps will be repeated.

P32-9 When all keys have been programmed, turn the ignition to the OFF position before attempting to start the engine.

Most immobilizer system transponders do not use batteries. If the immobilizer transponder is located in the same FOB as the RKE system's transponder, the batteries only support the RKE functions. However, some systems do share the batteries. In this case if the batteries are weak, the engine will not start. These systems typically monitor the battery strength and display warning messages in the instrument panel or vehicle information center when replacement is required.

Systems that do not require batteries use a piezoelectric crystal that is "excited" when hit by the radio frequencies from the control module's antenna. The movement of the crystal generates a voltage to power the transponder.

If the diagnostics leads to replacement of the immobilizer module or the PCM, specific procedures may need to be followed. Usually, the replacement module requires initialization. This will require the use of a scan tool and access to the PIN. This process transfers the required data between modules so the system is operative and the existing keys can still be used. After the initialization procedure is performed, the keys may need to be programmed to the new immobilizer module. However, since the secret code data matches, the same keys can be reused.

CUSTOMER CARE If the customer has an issue with intermittent no-starts, one problem may be that he or she has more than one transponder on his or her key ring. If the transponders are close enough together they may send messages at the same time. The immobilizer control module will receive these corrupted messages and broadcast an invalid key message and set a DTC.

Classroom Manual
Chapter 14, page 447

TIRE PRESSURE MONITORING SYSTEMS

The **tire pressure monitoring (TPM) system** is a safety system that notifies the driver if a tire is underinflated or overinflated. Tire pressure sensors (TPSs) are attached to the rim by a sleeve nut on the valve stems (**Figure 14-6**). The TPS's internal transmitter broadcasts tire pressure and temperature information to a dedicated ECU.

When diagnosing a concern with the TPM system, you need to determine when the system is in operation. Some systems will only transmit tire inflation information when the vehicle is in motion, while others continue to monitor tire pressure even with the vehicle stationary.

The receiver module can set a diagnostic trouble code (DTC) when one of the transmitters fails to produce a signal. If a replacement transmitter is installed in the tire, the ID number may need to be programmed into the TPM ECU. However, some systems automatically learn the new transmitter when the vehicle is driven. This is done by the ECU receiving an unrecognized signal at the same transmission intervals as the recognized transmitters. When this occurs, the ECU stores the ID and begins monitoring that transmitter.

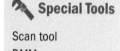

Special Tools

Scan tool
DMM
TPM analyzer
Training magnet

> **AUTHOR'S NOTE** It may be necessary to drive the vehicle up to 10 miles before the replacement transmitter ID is learned.

Some systems require that the ID location be trained into the receiving module. This allows the ECU to place the transponder number to the location on the vehicle. Location learning is usually done by entering training mode and placing a magnet around the tire's valve stem (**Figure 14-7**) in a specified order. When the magnet is placed around the valve stem, it pulls a reed switch closed and causes the TPS to send its data. As each location is learned, the ID is locked to that position. This procedure needs to be performed each time the tires are rotated to a different location on the vehicle. Systems that use transponders (**Figure 14-8**) relearn tire position each time the vehicle is in motion.

> **AUTHOR'S NOTE** Tire pressure sensor ID programming is different than location programming.

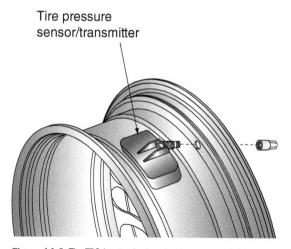

Tire pressure
sensor/transmitter

Figure 14-6 The TPS is attached to the rim as part of the valve stem.

Figure 14-7 The training magnet placed around the valve stem causes the sensor to transmit.

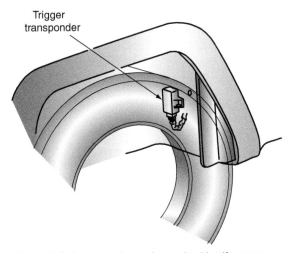

Figure 14-8 A transponder can be used to identify sensor location.

Testing of the ability of the tire pressure sensors to send their data can be done by a special TPS analyzer (**Figure 14-9**). The analyzer interacts with the tire pressure sensor without contact through wireless communication. The analyzer excites the TPS and captures its data for display (**Figure 14-10**). The display includes the transponder ID, the tire pressure reading, and the status of the sensor.

If there is no response from the TPS, the analyzer illuminates the FAIL light. This means either the TPS is faulty or the improper sensor for the system is installed.

Some analyzers will communicate with the scan tool to allow for transferring the tire pressure sensor identities and vehicle wheel position to the scan tool for programming into the TPM ECU.

If the system uses tire pressure transponders to excite the sensors, the transponders can be tested using the scan tool. Consider **Figure 14-11** of a transponder system. Only three transponders are used since the fourth position is inferred by the ECU. If a problem with the transponder is suspected, use the scan tool to initiate the TPM trigger module test (**Figure 14-12**). This test will verify the communications between the transponder and the TPM ECU. After the test is completed, the scan tool displays the results (**Figure 14-13**). Notice that the right front module did not acknowledge the ECU. Referring back to Figure 14-11, there is no transponder in this location.

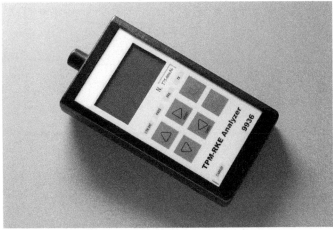

Figure 14-9 TPM analyzer.

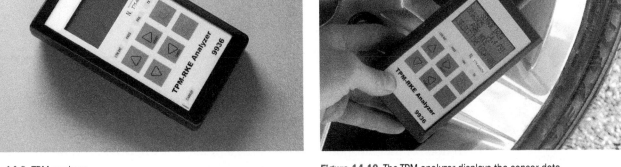

Figure 14-10 The TPM analyzer displays the sensor data.

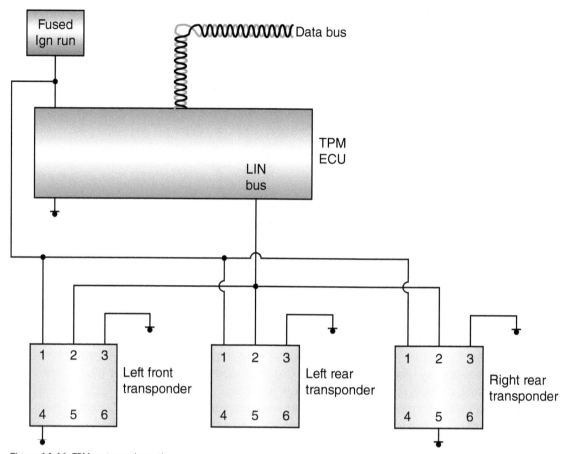

Figure 14-11 TPM system schematic.

Miscellaneous Functions	
Miscellaneous Functions	
▲ TPM Trigger Module Test	**Start**
Program Spare Tire Sensor ID	
Program Ignition Keys or Key FOBs	
Program Right Front Tire Sensor ID	
Program Right Rear Tire Sensor ID	
Program Left Front Tire Sensor ID	
Program Tire Sensor ID w/TPM Tool	
ELV Replaced	
PCM Replaced	
Update Pressure Thresholds	
WCM Replaced	
▼ Program Left Rear Tire Sensor ID	

Figure 14-12 Trigger module test used to verify communications between transponder and ECU.

If a transponder does not acknowledge the ECU, test the ignition run and the ground circuits. If these test to be good, test the local interconnect network (LIN) bus circuit. If all of these tests do not indicate a problem, the transponder is suspect.

If none of the transponders respond, the problem is likely to be a short to ground or to voltage in the LIN bus. If this is not the case, the ECU is faulty. Check powers and grounds to the ECU before condemning it.

Miscellaneous Functions

Miscellaneous Functions				Start
▲ TPM Trigger Module				
Program Spare Tire				
Program Ignition Key				
Program Right Front				
Program Right Rear				
Program Left Front T				
Program Tire Sensor II				
ELV Replaced				
PCM Replaced				
Update Pressure Thr				
WCM Replaced				
Program Left Rear Tire Sensor ID				
▼				

IPM Trigger Module Test ☒

Trigger Module Test complete. Press "Finish" to exit

Name	Value
Left Front Module Communication Acknowledge	Yes
Right Front Module Communication Acknowledge	No
Right Rear Module Communication Acknowledge	Yes
Left Rear Module Communication Acknowledge	Yes

[<Back] [Next>] [Finish] [Cancel]

Figure 14-13 Results of trigger module tests indicate if transponder responded.

CUSTOMER CARE Customers should be made aware that using the TPM display as a pressure gauge while inflating their tires may not work. Since many systems do not transmit air pressure data at a constant rate unless the vehicle is driving over a specified speed, the display will not change. This could result in excessively overinflated tires.

INFOTAINMENT SYSTEM SERVICE

Today's audio/video systems integrate several different functions. They may include AM/FM radio, satellite audio and video, CD player, DVD player, MP3 player, iPod connectivity, hands-free cell phone, navigational systems, cellular services, and Wi-Fi connectivity. Most are diagnosed using a scan tool since the system will set fault codes when a failure is determined. A central telematics module may act as a gateway for all of the individual systems that encompass the telematics system as a whole.

Vehicle Audio System Service

Repairs that are allowed to be performed for internal radio faults are dictated by warranty and shop policies. Radios and navigational monitors may be sent to specialized radio shops for repair. Keep in mind that repairs that involve replacing of backlighting bulbs and LEDs can often be done in-house. Just because you may not repair internal problems does not excuse you from doing proper diagnostics. Often the tendency is to remove the unit when the customer has described having a particular problem before performing a thorough prediagnosis routine. In many cases, the units show "NO TROUBLE FOUND" and are sent back to the dealership or shop. Most of the problems could have been solved without removing the unit.

Classroom Manual
Chapter 14, pages
455, 461

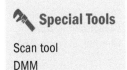
Special Tools

Scan tool
DMM

⚙ **SERVICE TIP** Do not blame the radio unless all other possible problem sources have been eliminated.

Before removing the radio/component, consider the following:

■ Test the vehicle's radio system outdoors and away from tall buildings.
■ Most radio noise can be located on weak amplitude modulation (AM) stations at the low-frequency end of the tuning band.

- When using a test antenna, the base must be grounded to the vehicle's body. Do not hold the antenna mast in your hands.
- Most noises heard in the radio system enter through the antenna.

External problems such as faulty speakers, broken wires, antenna issues, and so forth can be diagnosed and repaired by the automobile technician. The technician must determine the exact nature of the problem when performing a diagnosis. Determining whether the problem is intermittent or constant or whether the problem occurs when the vehicle is moving or stationary will help pinpoint the nature of the problem.

The diagnostic charts at the end of this chapter provide a guide to diagnosing poor radio performance complaints. Remember that the infotainment and telematics systems rely on data bus communications between individual modules. Be sure to confirm proper operation of the data bus communication system.

If the infotainment or telematics component does not appear to power on, the first step is to check for available voltage and good grounds. Battery voltage must be available to the radio, navigational module, hands-free module, and so forth on the power terminal. Test for voltage on the circuit while the circuit is loaded; do not trust open circuit testing. Most radios require two voltage feeds: direct battery and ignition run (**Figure 14-14**). Also, perform a voltage drop test of the ground circuit. If these circuits are good, an internal problem exists in the radio tuner.

> ⚙ **SERVICE TIP** Just because the clock works does not mean that power is being fed to the radio internal circuitry. Always check for a blown fuse or disconnected wire when diagnosing an inoperative radio.

Wiring diagrams for component systems are more complex. In addition to power and audio signal wires, note that some systems have a serial data wire for microprocessor communication between components for the controlling of unit functions.

Most vehicles that are equipped with an amplifier provide on-board diagnostics by use of a scan tool. On these systems speaker-related problems are diagnosed through the amplifier. For vehicles that are not equipped with an amplifier, speaker-related problems are diagnosed through the radio. Typical fault conditions that can be detected by the system are speaker circuits shorted to ground, shorted to voltage, shorted together, or open.

Classroom Manual
Chapter 14, page 456

Like the radio, the CD/DVD system components generate DTCs for faults that can be retrieved using the scan tool. Use the diagnostic charts at the end of this chapter to assist in identifying common customer complaints for poor CD system performance.

Classroom Manual
Chapter 14, page 461

The satellite radio system will also generate DTCs that can be read by the scan tool. The system will perform diagnostic system checks for audio output, presence of the antenna, antenna signal, and current subscription status. If there is no satellite radio audio output from the speakers, first check that the satellite radio subscription is initiated and that it has not expired. Also, check to confirm that the satellite antenna is not blocked from the satellite field of view. It is best to have the vehicle in an area that provides an unobstructed line of sight to the entire horizon when diagnosing a no-reception complaint. If the subscription is active and the antenna is not obstructed, a hardware failure is indicated. This will require repair or replacement of the satellite radio module, the satellite radio antenna, the vehicle's radio, the amplifier, or wire harness as directed in the service manual diagnostic procedures.

Classroom Manual
Chapter 14, page 458

Another concern may be the display of "Updating Channels" on the radio screen. This may indicate a hardware communication failure. Confirm that the satellite radio module is communicating on the data bus.

If there are no DTCs, begin diagnostics by inspecting the antenna connection. Disconnect the antenna from the telematics module and inspect for damage, corrosion, and proper fit.

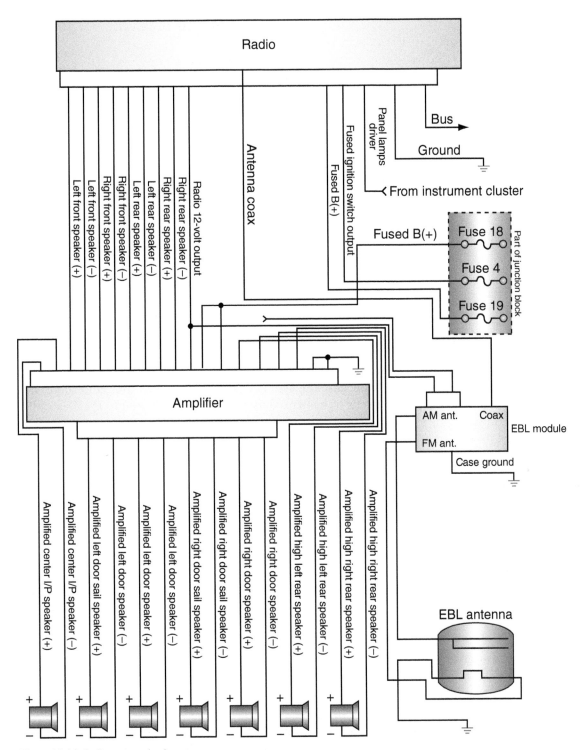

Figure 14-14 Audio system circuits.

If no problem was found with the antenna connection, use a test antenna (**Figure 14-15**) connected directly to the telematics module. Attempt to tune in several satellite stations. If the reception is still not available, or poor quality, replace the telematics module.

> 🌼 **SERVICE TIP** The antenna test kit includes a test antenna and different cable connection adapters. The test antenna must be properly grounded to the vehicle. If using the test antenna improves reception, the antenna or cable is faulty.

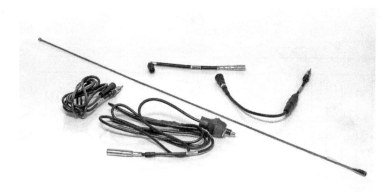

Figure 14-15 Test antenna kit.

If the system operates properly with the test antenna, work from the telematics module toward the antenna by connecting the test antenna at all antenna connections. This will isolate the location of the issue and determine if it is a cable or antenna fault.

Replacing Radios Equipped with HDDs

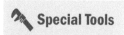

Special Tools

Scan tool
External hard drive
USB cable
Battery charger

A service concern that needs to be addressed is the replacement of a radio if it is equipped with an internal hard drive. Because the systems can hold over two thousand songs and other data, the hard drive needs to be backed up and the data restored to the new radio. Depending on the system, the backup procedure may be available only through the use of a scan tool. Hard drive backup is by using an external hard drive attached to the universal serial bus (USB) port of the radio. The following are the typical steps to backing up the HDD:

1. Connect a battery charger to the vehicle's battery and set to maintain 13 volts. The process of backing up the HDD can take several hours.
2. Turn the ignition to the RUN position with the engine OFF.
3. Power on the external HDD.
4. Connect the external HDD to the radio using the USB cable. The radio display will change to the "Manage My Files" screen.
5. Connect the scan tool and perform the steps to place the radio into "Dealership Mode."
6. Follow the on-screen instructions on the radio to back up the radio HDD.

Antenna Service

Classroom Manual
Chapter 14, pages
440, 442

Most complaints associated with the AM/FM radio are concerning poor reception. This is usually the result of an improper ground connection of the **antenna**. An antenna is used to collect AM and frequency modulation (FM) radio signal waves. The radio station's broadcast tower transmits electromagnetic energy through the air that induces an alternating current (AC) voltage in the antenna. The radio receiver processes this AC voltage signal and converts it to an audio output.

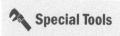

Special Tools

DMM
Jumper wires
Test antenna

The base of the antenna provides a path-to-chassis ground. Resistance in any portion of the vehicle's ground path can affect the overall performance of the audio system. The following conditions should be checked whenever diagnosing a poor radio reception concern:

- Loose or corroded battery cable terminals.
- High-resistance body and engine grounds.
- Loss of antenna and audio system grounds.

The radio is attached to the antenna lead by a coaxial cable that carries the signal through the center conductor. The outer conductor of the cable acts as a shield to protect the radio signal from interference. For the cable to be effective as a shield, it must be securely grounded at the body connection. A poor ground causes excess ignition noise in AM reception, or erratic sound. To check the antenna ground with a fixed antenna, do the following:

1. Tune the radio for a weak AM station or signal.
2. Unscrew and remove the antenna mast.
3. Connect a short jumper wire between the mast mounting stud and the cable lead.

If the radio no longer receives the radio station, the antenna ground is good. If the radio station is still received, the ground is poor or open.

For vehicles that have a power antenna, the ground can be checked by the following procedures:

1. Fully lower the antenna and then separate the motor harness connector. You may need to unplug the motor while it is running to keep the antenna about an inch from its base.
2. Connect an alligator clip near the top of the antenna. This will serve as the antenna.
3. Tune the radio to a weak AM station.
4. Remove the alligator clip and connect a short jumper between the mounting bracket and the top of the mast.

> **SERVICE TIP** The test antenna provides a reliable method for testing antenna grids that are in the front or rear window of the vehicle.

Like the test for the fixed mast antenna, if the radio no longer receives the radio station, the antenna ground is good. If the radio station is still received, the ground is poor or open.

Additional tests of the antenna include verification that there is continuity between the tip of the antenna mast to the tip of the radio connector. Also, verify there is no continuity between the antenna mast and chassis ground.

A test antenna kit is available for determining problems with the antenna or its cable (refer to Figure 14-15). The kit includes a test antenna and different cable connection adapters. The test antenna must be properly grounded to the vehicle. If using the test antenna improves reception, the antenna or cable is faulty.

Radio Noise Diagnostics

Radio noise is undesired interference or static (popping, clicking, or crackling) obstructing the normal sound of the radio station. Reception to the radio receiver from the antenna can be interfered with by noise resulting from radio frequency interference (RFI). This is especially true of the AM band. **Radio interference** generally refers to all undesirable signals that enter the radio and are heard as static or noise.

Classroom Manual
Chapter 14, page 458

Radiated noise results from an accessory that produces a signal that is detected by the antenna. The noise is picked up by the radio receiver and amplified through the audio circuits. The FM band is susceptible to electromagnetic interference (EMI), but usually is not as noticeable compared to AM. Control of RFI and EMI noise is done by proper ground connections at the radio antenna base, radio receiver, engine-to-body, and the heater core. In addition, resistor-type spark plugs and radio suppression secondary ignition wiring are used. The radio itself also has internal suppression devices, such as capacitors that shunt AC noise to ground and slow sudden changes of voltage in a circuit.

Common sources of radiated interference include the following:

- Defective ignition components.
- Motors such as heater blower, wiper, fan, power seats, and power windows.
- Electric fuel pumps.
- A loose ground strap or loose antenna mount.
- Noise-producing RFI interference that results from the operation of electrical circuit such as turn signals may cause a popping noise.

Some systems use a **radio choke**. The choke is a winding of wire. In a direct current (DC) circuit the choke acts like a short, but in an AC circuit it acts as a high resistance. The choke blocks the noisy AC current but allows the DC current to pass normally. The radio choke blocks **conducted interference**. Conductive interference is audible at very low volume levels and results from problems with the power circuit. The radio choke is installed in the ignition switch output circuit to prevent unwanted and random signals on the power circuit from reaching the radio. Sources of conducted interference can come from the generator, courtesy lamps, or control modules. In addition, radio static may be produced by nearby high-power transmission lines.

To inspect for the source of radio noise resulting from RFI or EMI, attempt to identify the component, which is the source. For example, an ignition system problem can cause radio noise that is engine speed related. Once the source is identified, the ground path and connections to that component should be checked. If a capacitor is used in the circuit, it must also be checked. All ground connections must be verified prior to replacing any components. Some of the grounds and connections that need to be inspected include the following:

- Radio antenna base ground.
- Radio receiver chassis ground.
- Generator.
- Engine-to-body ground straps.
- Electric fuel pump.
- Ignition module.
- Heater core ground strap.
- Wiper motors.
- Blower motor.
- Exhaust system ground straps.

In addition, spark plugs and spark plug wire routing and condition should be checked.

> **SERVICE TIP** If a noise suppression capacitor is suspected of being the cause of radio interference, a clip-on capacitor can be used to test the noise suppressor. Connect the test capacitor in place of the suppressor. If the radio static is reduced, a new suppressing capacitor is required.

A "diagnostic RF sniffer" tool can be made from an old piece of antenna lead-in from a mast or power antenna (**Figure 14-16**). This sniffer is used, along with the radio, to locate "hot spots" that are generating RFI noise. The noise can be found in wiring harnesses, in the upper part of the dash, or even between the hood and the windshield. When checking for noise on a wire, it is best to hold the sniffer parallel and close to the wire.

Classroom Manual
Chapter 14, page 454

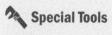

Special Tools

AA batteries
DMM

Speaker Service

Speakers turn the electrical energy from the radio receiver amplifier into acoustical energy. A speaker moves the air using a permanent magnet and an electromagnet (**Figure 14-17**). The electromagnet is energized when the amplifier delivers current to the voice coil at the speaker.

A faulty speaker distorts the sound of the radio or may be totally inoperative. A failure with the speaker requires its replacement since it is not serviceable. The exception would be if a terminal connection became loose or broken; then it may be soldered.

> **SERVICE TIP** Speakers that are "blown" may result from not being matched with the radio/amplifier. The speakers must be rated at the same or higher watts as the radio/amplifier. Speakers with lower ratings blow when the volume is turned up. Manufacturer-installed systems are matched; however, aftermarket-installed radios may not be.

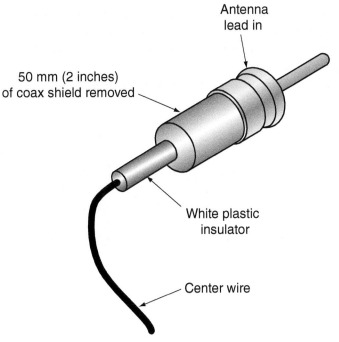

Figure 14-16 Making an RF sniffer from an old antenna cable.

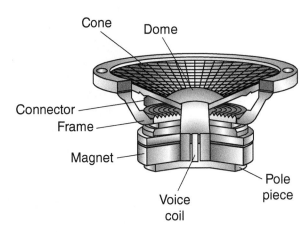

Figure 14-17 Speaker components.

To test an inoperative speaker, disconnect the harness connector and momentarily connect an AA battery across the speaker terminals. Make and break the connection quickly. The speaker should make a popping noise as the battery is connected and disconnected. If the speaker makes the popping noise, it passes, and the problem is in the wiring of the speaker. If no noise is made, the speaker is defective and must be replaced.

The resistance of the electromagnetic voice coil can be tested using an ohmmeter. Disconnect the harness connector to the speaker, and measure the resistance of the speaker. The ohmmeter should read the low resistance of the speaker voice coil (about 8 ohms [Ω]). Excessive, or infinite, readings indicate the speaker must be replaced.

TELEMATICS CONNECTIVITY ISSUES

Connectivity issues associated with cell phones and audio/video devices (IPod®, IPad®, and MP3 players) can be very difficult to diagnose. The problem lies in determining if the problem is within the vehicle's system or the customer's equipment. First, confirm that the device being paired to the telematics system is compatible. Manufacturers publish a list of compatible devices that will work with their system on their websites (**Figure 14-18**). Also, confirm the total number of devices currently paired to the telematics system. The telematics system has a limit to the number of components that can be paired to it.

Special multimedia test kits are available to help diagnose customer concerns with features of the telematics systems (**Figure 14-19**). The purpose of this tool is to simulate a known good cell phone or a known good media source. Using the test kit, it can be determined if the fault is related to the vehicle's system or the device being used.

The test kit can be used to diagnose the following:

- The AUX port.
- The USB port (both power and audio input).
- Bluetooth phone pairing issues.
- Bluetooth hands free calling.
- Bluetooth streaming audio.
- Bluetooth SMS messaging (send and receive).

Classroom Manual
Chapter 14, page 462

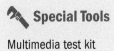 **Special Tools**

Multimedia test kit
Scan tool

Figure 14-18 Manufacturer websites provide information concerning device compatibility.

Figure 14-19 Multimedia test kit isolates if the fault is in the vehicle system or in the device.

The Bluetooth pairing test validates the ability of the telematics system to pair with a device by simulating the use of a known good compatible phone to test the system operation. **Photo Sequence 33** illustrates how this process is performed.

If the telematics system can be pair to the multimedia test tool, the connectivity function is working properly and the problem is with the customer's device. If pairing does not occur, confirm that the telematics software is up to date. Also, refer to the appropriate service information and Technical Service Bulletins (TSBs) for relevant diagnostic information.

If the customer states that the telematics system was working with their device but then stopped working, try to determine if the device was recently updated. Updates to device software or firmware may cause it to become out of sync or incompatible with the telematics system.

Next, clear the device pairing from the telematics system and reboot the customer's device. Attempt to pair the device to the telematics system again and check the operation. In addition, check to see if there are any software updates to the telematics system itself.

WI-FI CONNECTIVITY DIAGNOSTICS

Classroom Manual
Chapter 14, page 462

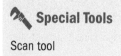

Special Tools

Scan tool
DMM

Like a home Wi-Fi system, the telematics Wi-Fi uses a wireless router to connect to an Internet service. The Hotspot feature of the Wi-Fi connection is limited to a radius of about 100 feet (30.48 meters) from the router. The hardwired circuits between components related to the Wi-Fi router and the Wi-Fi system may be diagnosed using a DMM and conventional procedures. Reference the appropriate wiring diagrams to determine test locations and expected test values. However, conclusive diagnosis of the telematics module and the bus communication between modules of the Wi-Fi system will require a scan tool.

PHOTO SEQUENCE 33
Performing a Bluetooth Pairing Test

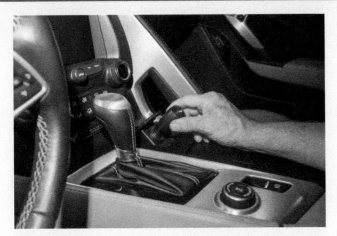

P33-1 Connect the power outlet power adapter into the 12-volt receptacle.

P33-2 Connect the USB cable into the power outlet power adapter to utilize a Bluetooth connection.

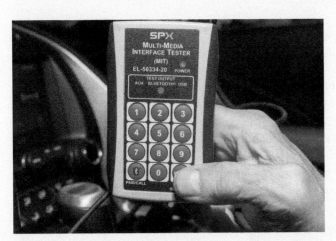

P33-3 Place the tool into Bluetooth mode.

P33-4 Use the vehicle controls to place the system into pairing mode.

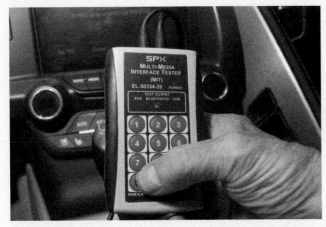

P33-5 Press the PAIR button on the test tool. The tool may identify the telematics system.

P33-6 Enter the pairing PIN number. The test tool should pair with the telematics system.

Usually faulty components of the Wi-Fi system are not serviceable. For example, the antenna may be internal to the telematics module and, if faulty, will require the module to be replaced. However, some systems use a separate antenna that can be replaced by itself. Also, the Wi-Fi antenna may be used as the cellular antenna. Determining any related symptoms between different features of the telematics system can help isolate the root cause of the customer's concern.

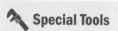

Classroom Manual
Chapter 14, page 463

Special Tools

Scan tool
DMM

Navigational System Service

The navigational system receives global positioning system (GPS) signals from up to eight satellites to display the position and direction of the vehicle. A gyro-sensor, along with the vehicle speed and steering angle sensors, enables the system to display the present vehicle position even in locations where GPS signals may be blocked.

Diagnosis of the navigational system is by the use of a scan tool and is DTC based. Using the DTC, refer to the appropriate service information for the diagnostic routine. All electrical components of the system must have confirmed power and ground circuits before replacement.

Components of the system are generally replaced and not serviced. A typical procedure for replacing the navigational system GPS antenna (**Figure 14-20**) is:

1. Disconnect and isolate the battery negative cable.
2. Remove the instrument panel top cover.
3. Remove the radio.
4. Remove the antenna mounting fasteners.
5. In some cases, each end of the cable is cut and the remaining portion of the cable is left in the instrument panel.
6. Position the new antenna in place and feed the wire harness through the opening by carefully pulling up on the instrument panel.
7. Install the antenna mounting fasteners.
8. Secure the antenna wire harness into place.
9. Install the radio.
10. Install the instrument panel top cover.
11. Reconnect the battery negative cable.

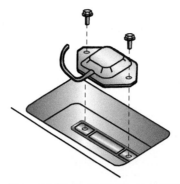

Figure 14-20 Navigational system GPS antenna.

CASE STUDY

A customer brought his or her vehicle into the service facility because of poor radio reception. The technician noticed interference noises coming from all of the speakers. He also observed that the noise appeared to be engine speed related.

While inspecting for the cause of the interference, he noticed that the coil-on-plug ignition system used on this engine had capacitors for each bank. The capacitor on the left bank was plugged in, but the capacitor for the right bank was not connected. After plugging the capacitor in, the radio noise ceased.

CASE STUDY

A customer brought her vehicle into the service facility because she was unable to pair their cell phone to the telematics system. Following the published procedure, the technician attempted to pair the phone. All attempts to pair the phone failed. Next, the technician confirmed that the cell phone was compatible with the telematics system. Finally, the technician checked for any software updates to the telematics system, there was none.

The technician connected the scan tool to determine how many devices were paired to the system. The scan tool reported that there were only three devices paired to the system. The service information indicated that the system can hold 12 paired devices. In addition, the scan tool did not report any DTCs recorded in the telematics system module.

To determine if the pairing issue is a fault in the telematics system of the customer's device, the technician connected the multimedia test kit. After performing the set-up steps and entering the PIN, the tool was not able to pair to the system. This isolated the fault to being in the telematics system and not the device. Following the diagnostic routine in the service information it was determined that the telematics module required replacement. After the module was replaced, proper system operation was confirmed and the vehicle was returned to the customer.

ASE-STYLE REVIEW QUESTIONS

1. The doors fail to lock or unlock when the RKE system is used; however, they will lock and unlock when the door switch is activated. Which is the **LEAST LIKELY** cause?

   A. FOB battery voltage too low.

   B. RKE antenna faulty.

   C. Faulty lock/unlock relay.

   D. Data bus communication circuit fault.

2. *Technician A* says most audio systems provide for speaker-related problems to be diagnosed with a scan tool through the amplifier.

   Technician B says if "Updating Channels" on the radio screen is constantly displayed, the subscription for satellite radio has expired.

   Who is correct?

   A. A only   C. Both A and B

   B. B only   D. Neither A nor B

3. *Technician A* says if the immobilizer module is replaced, new keys will be required.

   Technician B says procedures that copy data to the new immobilizer module may need to be performed for the vehicle to start.

   Who is correct?

   A. A only   C. Both A and B

   B. B only   D. Neither A nor B

4. A new tire pressure sensor has been installed onto a wheel.

   Technician A says the ID number may need to be programmed into the TPM ECU.

   Technician B says the system may learn the new transmitter ID when the vehicle is driven.

   Who is correct?

   A. A only   C. Both A and B

   B. B only   D. Neither A nor B

5. TPM systems that do not use a transponder are being discussed.

   Technician A says TPM sensors may require location training by use of a magnet.

   Technician B says the training magnet procedure needs to be performed whenever tire pressure is corrected.

   Who is correct?

   A. A only   C. Both A and B

   B. B only   D. Neither A nor B

6. A customer states that the engine starts and then dies almost immediately.

   Technician A says the immobilizer key may be faulty.

   Technician B says a data bus circuit may be open.

   Who is correct?

   A. A only   C. Both A and B

   B. B only   D. Neither A nor B

7. *Technician A* says the resistance between the antenna tip and chassis ground should be less than 1 ohm.

Technician B says the resistance between the terminals of a speaker should be very high.

Who is correct?

A. A only   C. Both A and B

B. B only   D. Neither A nor B

8. TPM systems that use transponders are being discussed.

Technician A says a fault code indicating no response from a transponder can be due to a faulty tire pressure sensor.

Technician B says the scan tool may support a transponder trigger test to isolate the problem.

Who is correct?

A. A only   C. Both A and B

B. B only   D. Neither A nor B

9. To program ignition keys into the immobilizer system, all of the following may be required **EXCEPT:**

A. PIN number.

B. Scan tool.

C. DMM.

D. All previously programmed keys.

10. Using a TPM analyzer results in the illumination of the FAIL light when attempting to retrieve data from the sensor. Which of the following is the **MOST LIKELY** cause?

A. Too low of tire pressure.

B. Transponder power supply circuit open.

C. Data circuit between transponder and TPM ECU defective.

D. Improper tire pressure sensor installed.

ASE CHALLENGE QUESTIONS

1. The RKE system fails to operate. An RF sniffer tool does not indicate a signal is being received from the FOB when a button is pushed. Which of the following is the **LEAST LIKELY** likely cause?

A. FOB not programmed.

B. Batteries too low.

C. Damaged FOB.

D. Batteries installed wrong.

2. On a vehicle equipped with an immobilizer system, the engine starts for a few seconds and then dies. What is the **LEAST LIKELY** likely cause of this?

A. An invalid key was used.

B. The immobilizer module has not been initialized.

C. Lost communication between immobilizer module and PCM.

D. A faulty starter relay.

3. *Technician A* says weak and fading AM signals could be the result of a bad or ungrounded antenna.

Technician B says most radio noise enters by way of the radio.

Who is correct?

A. A only   C. Both A and B

B. B only   D. Neither A nor B

4. *Technician A* says most complaints associated with poor reception from the AM/FM radio, satellite radio, navigational system, or cellular phone are usually the result of an improper ground connection of the antenna.

Technician B says resistance in any portion of the vehicle's ground path can affect the overall performance of the telematics system.

Who is correct?

A. A only   C. Both A and B

B. B only   D. Neither A nor B

5. A customer states that his or her engine starts and then dies. This seems to be intermittent. The customer makes arrangements to leave the vehicle for three days to be diagnosed and repaired. During these three days, the vehicle does not have any problems. The immobilizer module does have a stored "Invalid Key." What is the **MOST LIKELY** cause of this problem?

A. The battery in the key is low.

B. The customer has more than one transponder key on the ring.

C. The cut of the key blade is worn.

D. None of the above

Name _____   Date _____

ANALYZING THE RKE SYSTEM

Upon completion of this job sheet, you should be able to determine proper operation of the RKE system and perform diagnostics.

NATEF Correlation

This job sheet addresses the following **MLR** tasks:

A.3.	Use wiring diagrams to trace electrical/electronic circuits.
A.4.	Demonstrate proper use of a digital multimeter (DMM) when measuring source voltage, voltage drop (including grounds), current flow, and resistance.
A.5.	Demonstrate knowledge of the causes and effects from shorts, grounds, opens, and resistance problems in electrical/electronic circuits.
A.6.	Check operation of electrical circuits using a test light.
A.11.	Identify electrical/electronic system components and configuration.
E.6.	Describe the operation of keyless entry/remote-start systems.

This job sheet addresses the following **AST** tasks:

G.2.	Describe operation of security/antitheft systems and related circuits (e.g., theft deterrent, door locks, remote keyless entry, remote start, and starter/fuel disable); determine needed repairs.
G.5.	Describe body electronic systems circuits using a scan tool; check for module communication errors (data bus systems); determine needed action.

This job sheet addresses the following **AST/MAST** tasks:

A.3.	Demonstrate proper use of a digital multimeter (DMM) when measuring source voltage, voltage drop (including grounds), current flow, and resistance.
A.4.	Demonstrate knowledge of the causes and effects from shorts, grounds, opens, and resistance problems in electrical/electronic circuits.
A.7.	Use wiring diagrams during the diagnosis (troubleshooting) of electrical/electronic circuit problems.

This job sheet addresses the following **MAST** tasks:

A.11.	Check electrical/electronic circuit waveforms; interpret readings and determine needed repairs.
G.2.	Diagnose operation of security/antitheft systems and related circuits (e.g., theft deterrent, door locks, remote keyless entry, remote start, and starter/fuel disable); determine needed repairs.
G.5.	Diagnose body electronic systems circuits using a scan tool; check for module communication errors (data bus systems); determine needed action.

ASE **NATEF**

Tools and Materials

- Scan tool
- Lab scope
- DMM
- Extra FOBs
- Wiring diagram

Describe the vehicle being worked on:

Year _____ Make _____ Model _____

VIN _____ Engine type and size _____

Procedure   **Task Completed**

Task 1

This task involves programming of the RKE FOBs to the system using a scan tool.

1. Using the service information, determine the FOB programming function using the scan tool and describe the procedure.

   _____

   _____

   _____

2. Do all FOBs have to be programmed at the same time, or can individual FOBs be entered?

   _____

3. Connect the scan tool to the vehicle, access the RKE ECU, and select the function to   ☐
   program RKE.

4. Follow the instructions given by the scan tool to an additional FOB. How can you tell that the RKE module has been programmed successfully?

   _____

   _____

5. How many FOBs can be programmed to the system? _____

6. How many FOBs can be programmed to operate with the memory seat function?

   _____

7. When is it necessary to program the RKE module?

   _____

   _____

Task 2

The following task involves programming the RKE FOB without using the scan tool.

1. Look up the procedure for programming the RKE system without the use of the scan tool (customer programming mode). Describe the procedure.

   _____

   _____

   _____

2. Perform the procedure to program the additional FOB. How do you know if the RKE module has been programmed successfully?

   _____

   _____

Task Completed

Task 3

For this task, you will analyze the input and output signal transmission of the RKE module.

1. Determine all of the modules used for RKE operation and provide their function.

   _____

   _____

   _____

2. How does the RKE module communicate with the other modules?

   _____

   _____

3. Identify the circuits that are attached to the RKE module.

   _____

   _____

   _____

4. Set up the lab scope to view a 12-volt square wave at 0.4 s/Div, and connect the lead to   ☐
    the program circuit. Set the record trigger to falling edge.

   _____

5. Enter customer programming mode and record the results of the lab scope reading.

   _____

   _____

6. Move the lab scope lead to the RKE interface circuit, and change the time display to
    20 ms/Div. Cycle the ignition switch from OFF to RUN, and then press the Lock or
    Unlock button on the key FOB. Describe the lab scope sine wave.

   _____

7. Where is the voltage coming from for the two circuits just tested?

   _____

   _____

8. What would be the result if the interface circuit was opened?

   _____

   _____

Instructor's Response

_____

_____

_____

Name _____   Date _____

PROGRAMMING AN IMMOBILIZER KEY

Upon completion of this job sheet, you should be able to determine operating characteristics of the immobilizer system and properly program keys.

NATEF Correlation

This job sheet addresses the following **AST** tasks:

G.2.   Describe operation of security/antitheft systems and related circuits (e.g., theft deterrent, door locks, remote keyless entry, remote start, and starter/fuel disable); determine needed repairs.

G.5.   Describe body electronic systems circuits using a scan tool; check for module communication errors (data bus systems); determine needed action.

This job sheet addresses the following **MAST** tasks:

G.2.   Diagnose operation of security/antitheft systems and related circuits (e.g., theft deterrent, door locks, remote keyless entry, remote start, and starter/fuel disable); determine needed repairs.

G.5.   Diagnose body electronic systems circuits using a scan tool; check for module communication errors (data bus systems); determine needed action.

Tools and Materials

- Vehicle equipped with an immobilizer system
- Extra keys
- Scan tool
- PIN (if required)
- Service information

Describe the vehicle being worked on:

Year _____ Make _____ Model _____

VIN _____ Engine type and size _____

Procedure   **Task Completed**

Task 1

1. According to the service information, does the immobilizer system of the assigned vehicle support customer programming of extra keys?   ☐ Yes   ☐ No

2. If yes, describe the procedure.

_____

_____

_____

3. Obtain an extra key from your instructor, and attempt to start the vehicle with the new key. Describe the results.

_____

_____

4. Perform the customer programming method described in step 1 to program a   ☐
   new key.

Task Completed

5. How do you know the key was successfully programmed?

_____

_____

Task 2

1. Connect the scan tool to the DLC, and access the immobilizer system. _____ ☐

2. If possible, perform a Module ID and record the following items:

   Version: _____ Part Number: _____ Country Code: _____

3. Does the immobilizer system on the assigned vehicle require the use of a PIN?

   ☐ Yes   ☐ No

   If yes, obtain the PIN for the vehicle and record it: _____

4. How do you obtain the PIN?

   _____

   _____

5. If an incorrect PIN is entered or the correct procedure is not followed, are there any actions that must be done to reenter the programming mode?   ☐ Yes   ☐ No

   If yes, describe what must be done.

   _____

   _____

6. Is there a special procedure that must be followed if the PCM is replaced on the assigned vehicle?   ☐ Yes   ☐ No

   If yes, describe the procedure.

   _____

   _____

   _____

7. Is there a special procedure that must be followed if the immobilizer module is replaced on the assigned vehicle?   ☐ Yes   ☐ No

   If yes, describe the procedure.

   _____

   _____

8. Do the keys have to be reprogrammed after the immobilizer module is replaced?   ☐ Yes   ☐ No

9. Use the scan tool to erase all current ignition keys. ☐

10. Attempt to start the engine, and record the results.

_____

_____

11. Use the scan tool to read DTCs and record.

_____

_____

12. Using the scan tool, access the function to program ignition keys and follow the   ☐
instructions on the scan tool or in the service manual.

13. Does the vehicle start?   ☐ Yes   ☐ No   ☐

14. Clear any DTCs that were set.

Task 3

1. Wrap one of the programmed keys with aluminum foil, and attempt to start the
engine. Record your results.

_____

_____

2. Use the scan tool to read any DTCs in the system and record.

_____

_____

3. According to the service manual, what could cause the DTCs to set?

_____

_____

_____

Instructor's Response

_____

_____

_____

Name _____   Date _____

TIRE PRESSURE MONITORING SYSTEM TRAINING (MAGNET TYPE)

Upon completion of this job sheet, you should be able to train the TPM ECU to recognize the source locations of tire pressure sensor signals.

NATEF Correlation

This job sheet addresses the following **AST** tasks:

G.4.   Describe operation of safety systems and related circuits (e.g., horn, air bags, seat belt pretensioners, occupancy classification, wipers, washers, speed control/collision avoidance, heads-up display, park assist, and back-up camera); determine needed repairs.

G.5.   Describe body electronic systems circuits using a scan tool; check for module communication errors (data bus systems); determine needed action.

This job sheet addresses the following **MAST** tasks:

G.4.   Diagnose operation of safety systems and related circuits (e.g., horn, air bags, seat belt pretensioners, occupancy classification, wipers, washers, speed control/collision avoidance, heads-up display, park assist, and back-up camera); determine needed repairs.

G.5.   Diagnose body electronic systems circuits using a scan tool; check for module communication errors (data bus systems); determine needed action.

ASE NATEF

Tools and Materials

- Vehicle equipped with TPM system
- Scan tool
- Training magnet

Describe the vehicle being worked on:

Year _____ Make _____ Model _____

VIN _____ Engine type and size _____

Procedure

1. Locate the procedure for training the sensor location in the service information. What is the procedure for putting the TPM ECU into training mode?

   _____

   _____

   _____

2. What is the order of sensor location training?

   _____

   _____

3. Perform the procedure described in step 1 and initiate the training procedure. What is the first step the system requires you to do?

   _____

   _____

   _____

4. Set the first sensor that was identified in step 2 into learn mode by positioning the training magnet over the valve stem for at least 5 seconds. The sensor will transmit a message indicating to the ECU that it is in learn mode. Explain how you are notified that the sensor ID and position have been learned.

_____

_____

5. Continue to train the remaining sensors. How long do you have between each sensor training operation? _____

6. What happens if the time in step 5 elapses? _____

7. How do you know the training procedure was successful?

_____

_____

Instructor's Response

_____

_____

_____

Name _____   Date _____

TIRE PRESSURE MONITORING SYSTEM TESTING

Upon completion of this job sheet, you should be able to determine proper operation of the TPM system and determine the location of a faulty transmitter.

NATEF Correlation

This job sheet addresses the following **AST** tasks:

G.4.   Describe operation of safety systems and related circuits (e.g., horn, air bags, seat belt pretensioners, occupancy classification, wipers, washers, speed control/collision avoidance, heads-up display, park assist, and back-up camera); determine needed repairs.

G.5.   Describe body electronic systems circuits using a scan tool; check for module communication errors (data bus systems); determine needed action.

This job sheet addresses the following **MAST** tasks:

G.4.   Diagnose operation of safety systems and related circuits (e.g., horn, air bags, seat belt pretensioners, occupancy classification, wipers, washers, speed control/collision avoidance, heads-up display, park assist, and back-up camera); determine needed repairs.

G.5.   Diagnose body electronic systems circuits using a scan tool; check for module communication errors (data bus systems); determine needed action.

ASE NATEF

Tools and Materials
- Vehicle equipped with TPM system
- Scan tool

Describe the vehicle being worked on:

Year _____ Make _____ Model _____

VIN _____ Engine type and size _____

Procedure

Task Completed

1. Locate the pressure thresholds in the service information for the vehicle you are assigned.

   _____

   _____

2. Use a handheld tire pressure gauge to measure the actual tire pressures in the classroom vehicle and record your results.

   _____

   _____

3. Turn the ignition switch to the RUN position. Lower the windows in order to be able to hear the audible low tire pressure warning from outside the vehicle.   ☐

4. Quickly release air pressure from one of the tires until the low tire pressure warning sounds. Measure and record the tire pressure.

   _____

5. Why did the low tire pressure warning sound if the tire pressure is still above the low pressure ON threshold?

_____

_____

6. If a vehicle is displaying a Low Pressure warning message and all tire pressures check within the low- and high-pressure thresholds, how could this feature be used to locate the faulty transmitter?

_____

_____

Instructor's Response

_____

_____

_____

Name _____   Date _____

USING THE TPM ANALYZER

Upon completion of this job sheet, you should be able to determine proper operation of the TPM system and determine the location of a faulty transmitter.

NATEF Correlation

This job sheet addresses the following **AST** tasks:

G.4.   Describe operation of safety systems and related circuits (e.g., horn, air bags, seat belt pretensioners, occupancy classification, wipers, washers, speed control/collision avoidance, heads-up display, park assist, and back-up camera); determine needed repairs.

G.5.   Describe body electronic systems circuits using a scan tool; check for module communication errors (data bus systems); determine needed action.

This job sheet addresses the following **MAST** tasks:

G.4.   Diagnose operation of safety systems and related circuits (e.g., horn, air bags, seat belt pretensioners, occupancy classification, wipers, washers, speed control/collision avoidance, heads-up display, park assist, and back-up camera); determine needed repairs.

G.5.   Diagnose body electronic systems circuits using a scan tool; check for module communication errors (data bus systems); determine needed action.

Tools and Materials

- Vehicle equipped with TPM system
- Scan tool
- TPM analyzer

Describe the vehicle being worked on:

Year _____   Make _____   Model _____

VIN _____   Engine type and size _____

Procedure   Task Completed

1. With the analyzer's power turned on, select the model year and vehicle type from the selection screens.

2. Place the analyzer near the valve stem of one of the tires and start the procedure to retrieve the sensor information. Record the information displayed by the analyzer.

   _____

   _____

   _____

3. Lock the senor location into the analyzer.   ☐

4. Repeat for all sensors.

   _____

   _____

   _____

   _____

   _____

5. What could be the cause of the analyzer to display a fail condition when attempting to identify the sensor?

_____

_____

_____

6. Connect the analyzer to the scan tool using the USB port or wireless communications.   ☐

7. Select the function on the scan tool for programming the sensor location into the TPM ECU using the analyzer and follow the instructions on the screen. Describe the procedure.

_____

_____

_____

Instructor's Response

_____

_____

_____

Name _____   Date _____

TESTING THE AUDIO SYSTEM

Upon completion of this job sheet, you should be able to determine proper operation of the audio system.

NATEF Correlation

This job sheet addresses the following **MLR** tasks:

A.3.   Use wiring diagrams to trace electrical/electronic circuits.

A.4.   Demonstrate proper use of a digital multimeter (DMM) when measuring source voltage, voltage drop (including grounds), current flow, and resistance.

This job sheet addresses the following **AST/MAST** tasks:

A.3.   Demonstrate proper use of a digital multimeter (DMM) when measuring source voltage, voltage drop (including grounds), current flow, and resistance.

A.7.   Use wiring diagrams during the diagnosis (troubleshooting) of electrical/electronic circuit problems.

This job sheet address the following **AST** tasks:

G.3.   Describe operation of entertainment and related circuits (e.g., radio, DVD, remote CD changer, navigation, amplifiers, speakers, antennas, and voice-activated accessories); determine needed repairs.

G.5.   Describe body electronic systems circuits using a scan tool; check for module communication errors (data bus systems); determine needed action.

This job sheet address the following **MAST** tasks:

G.3.   Diagnose operation of entertainment and related circuits (e.g., radio, DVD, remote CD changer, navigation, amplifiers, speakers, antennas, and voice-activated accessories); determine needed repairs.

G.5.   Diagnose body electronic systems circuits using a scan tool; check for module communication errors (data bus systems); determine needed action.

ASE NATEF

Tools and Materials

- Scan tool
- DMM
- Test antenna kit
- Wiring diagram

Describe the vehicle being worked on:

Year _____ Make _____ Model _____

VIN _____ Engine type and size _____

Procedure

Task 1

For this task, you will identify the capabilities of the scan tool to diagnose and test audio system inputs and outputs.

1. Connect the scan tool to the assigned vehicle, and establish communication. Turn the radio on, perform a Module ID, and record the software version of the radio below:

_____

_____

2. Access the amplifier and connect a jumper wire between the power amplifier RR speaker input circuit from the radio and chassis ground.

3. Center the balance controls on the radio. What is the effect on the audio system?

_____

_____

4. Read radio DTCs. Are any detected? ☐ Yes ☐ No

5. Disconnect the short from the amplifier connector, and move it to the RR speaker output circuit and ground.   ☐

6. What is the effect on the audio system?

_____

_____

7. Cycle the ignition switch two times and read DTCs. Are any detected? ☐ Yes ☐ No

If yes, list the DTCs that were set.

_____

_____

Why did these DTCs set?

_____

_____

8. Explain your results.

_____

_____

_____

9. Restore all equipment to its original condition.   ☐

10. Is the vehicle equipped with a DVD player? ☐ Yes ☐ No

If yes, are there any tests available? ☐ Yes ☐ No

If yes, list the available tests.

_____

_____

Perform the tests and describe your results.

_____

_____

_____

11. What components can be tested using the actuator test function in the scan tool?

_____

_____

_____

12. What data information is available on the scan tool?

_____

_____

_____

Task 2

For this task, you will identify the capabilities of the scan tool to diagnose and test satellite radio system inputs and outputs.

1. Connect the scan tool to the assigned vehicle and establish communication with the   ☐
satellite radio system.

2. Are there any satellite radio system tests available?   ☐ Yes   ☐ No

If yes, list the available tests, perform the tests, and record the results.

_____

_____

_____

3. Clear any DTCs that are set.   ☐

4. At the satellite radio unit, disconnect one of the antenna cables.   ☐

5. Check for DTCs and list any that were set.

_____

_____

6. Reconnect the antenna.

7. Can the scan tool be used to verify that the signal is present?   ☐ Yes   ☐ No

8. Can the scan tool be used to verify that the subscription is present?   ☐ Yes   ☐ No

9. Clear any DTCs that were set.

10. Explain your results.

_____

_____

_____

Task 3

For this task, you will analyze radio reception performance, identify mast antenna faults, and perform antenna diagnostic tests.

1. Move the vehicle to a location that provides good reception.   ☐

2. Use a jumper wire to short the antenna mast to ground. When tuned to a strong AM station, what is the effect of the shorted antenna mast?

    _____

    _____

3. Switch to an FM station. What is the effect of the shorted antenna mast?

    _____

    _____

4. Remove the jumper wire and disconnect the antenna lead from the radio. What is the effect on FM and AM reception?

    _____

    _____

5. Locate the resistance specifications for the antenna in the service information.

    _____

    _____

6. Using a DMM, measure the resistance between the coax (outer conductor) connector and negative battery cable. What is the reading?

    _____

7. If resistance is excessive, what could be the cause?

    _____

    _____

8. Using a DMM, measure the resistance between the tip of the radio antenna connector (inner conductor) and antenna mast. What is the reading? _____

9. If resistance is excessive, what could be the cause?

    _____

    _____

10. Test the resistance between the tip of the radio antenna connector (inner conductor) and ground. What is the reading?

    _____

11. List some items that could cause radio frequency interference in the audio system.

    _____

    _____

    _____

Task 4

For this task, you will analyze speaker performance and diagnostics.

1. Using the service information, identify the circuits to the LR speaker.

   _____

2. Connect the scan tool to the DLC of the assigned vehicle and establish communication with the radio. ☐

3. With the ignition switch in the OFF position, disconnect the LR speaker connector and connect a jumper between the LR Speaker (+) circuit and ground. ☐

4. Turn the ignition switch to the RUN position and turn the radio on. What is the effect on the audio system?

   _____

   _____

Note: Cycle the ignition switch three times and leave the radio on. It may take some time for a code to set.

5. Record DTCs.

   _____

   _____

6. Remove the short and erase DTCs. ☐

7. Move the short to the speaker (–) circuit to ground and cycle the ignition key three times with the radio on. What is the effect on the audio system?

   _____

   _____

8. Are there any DTCs set?   ☐ Yes   ☐ No

   If yes, list them.

   _____

   _____

   If no, why not?

   _____

   _____

9. Remove the short and reconnect the speaker.

Instructor's Response

_____

_____

_____

Name _____   Date _____

USB INPUT/OUTPUT TEST

Upon completion of this job sheet, you should be able to validate the ability of the telematics system to play audio from a customer's device that has been connected to the USB port. The test will also determine if the USB port can provide power to a connected device.

NATEF Correlation

This job sheet addresses the following **MLR** tasks:

A.3.   Use wiring diagrams to trace electrical/electronic circuits.

A.4.   Demonstrate proper use of a digital multimeter (DMM) when measuring source voltage, voltage drop (including grounds), current flow, and resistance.

This job sheet addresses the following **AST/MAST** tasks:

A.3.   Demonstrate proper use of a digital multimeter (DMM) when measuring source voltage, voltage drop (including grounds), current flow, and resistance.

A.7.   Use wiring diagrams during the diagnosis (troubleshooting) of electrical/electronic circuit problems.

This job sheet addresses the following **AST** tasks:

G.3.   Describe operation of entertainment and related circuits (e.g., radio, DVD, remote CD changer, navigation, amplifiers, speakers, antennas, and voice-activated accessories); determine needed repairs.

G.5.   Describe body electronic systems circuits using a scan tool; check for module communication errors (data bus systems); determine needed action.

This job sheet addresses the following **MAST** tasks:

G.3.   Diagnose operation of entertainment and related circuits (e.g., radio, DVD, remote CD changer, navigation, amplifiers, speakers, antennas, and voice-activated accessories); determine needed repairs.

G.5.   Diagnose body electronic systems circuits using a scan tool; check for module communication errors (data bus systems); determine needed action.

Tools and Materials

- Vehicle equipped with a telematics system with USB port
- Multimedia test kit
- Scan tool
- Service information

Describe the vehicle being worked on:

Year _____ Make _____ Model _____

VIN _____ Engine type and size _____

Procedure

1. Use the service information and record the USB port's voltage and amperage output.

_____

_____

_____

2. What is the voltage and amperage requirement of the customer's device for operation and charging?

_____

_____

_____

3. Connect the multimedia tester's USB cable into the vehicle's USB port.   ☐

4. Does the tester power on?   ☐ Yes   ☐ No

5. Based on the results of step 4, what does this indicate?

_____

_____

_____

6. Place the multimedia test tool into USB mode.

   NOTE: If the system does not automatically enter USB mode, use the vehicle controls to place the telematics system into USB mode.

7. Does the telematics system play audio from the multimedia tester?

   Yes _____, go to step 8.

   No _____, go to step 12

8. What does the test confirm?

_____

_____

_____

_____

9. Use the service information or vehicle manufacturer's website and confirm the device is compatible with the telematics system. Is the device compatible?

   Yes   No

10. Are there any updates for the customer device?   ☐ Yes   ☐ No

11. Check the following items and record the results.

    Check the multimedia ports for damaged connection points or debris in the ports.

_____

_____

_____

    Check wiring circuits and connections to the USB port for concerns.

_____

_____

_____

Check for related TSBs.

_____

_____

_____

Check for vehicle-related system updates.

_____

_____

_____

12. What is your conclusion concerning the operation of the USB port?

_____

_____

_____

_____

Instructor's Response

_____

_____

_____

_____

_____

_____

_____

_____

DIAGNOSTIC CHART 14-1

PROBLEM AREA:	Keyless entry/remote start system operation.
SYMPTOMS:	Keyless entry fails to operate intermittently.
POSSIBLE CAUSES:	**1.** Defective key pad. **2.** Open circuit. **3.** Shorted circuit. **4.** Defective controller.

DIAGNOSTIC CHART 14-2

PROBLEM AREA:	Keyless entry/remote start system operation.
SYMPTOMS:	Keyless entry fails to operate.
POSSIBLE CAUSES:	**1.** Defective key pad. **2.** Open circuit. **3.** Shorted circuit. **4.** Defective controller.

DIAGNOSTIC CHART 14-3

PROBLEM AREA:	Immobilizer System.
SYMPTOMS:	Indicator lamp does not illuminate.
POSSIBLE CAUSES:	**1.** Faulty LED. **2.** Blown fuse. **3.** Faulty ground. **4.** Faulty battery feed. **5.** Faulty ignition feed.

DIAGNOSTIC CHART 14-4

PROBLEM AREA:	Immobilizer System.
SYMPTOMS:	Indicator lamp flashes during bulb check. Vehicle starts and dies.
POSSIBLE CAUSES:	**1.** Invalid key. **2.** Faulty key. **3.** Faulty antenna. **4.** Faulty module.

DIAGNOSTIC CHART 14-5

PROBLEM AREA:	Immobilizer System.
SYMPTOMS:	Indicator lamp stays on after bulb check.
POSSIBLE CAUSES:	**1.** Immobilizer fault has been detected. **2.** Immobilizer system inoperative.

DIAGNOSTIC CHART 14-6

PROBLEM AREA:	Tire pressure monitoring system operation.
SYMPTOMS:	Tire pressure light is on or display shows a low tire.
POSSIBLE CAUSES:	**1.** Low tire pressure. **2.** Faulty tire sensor. **3.** Faulty control module. **4.** Control module power feed circuit open or high resistance. **5.** Control module ground circuit fault. **6.** Bus communications error. **7.** Transponder failure. **8.** Display module fault.

DIAGNOSTIC CHART 14-7

PROBLEM AREA:	Audio System.
SYMPTOMS:	Radio will not turn on.
POSSIBLE CAUSES:	**1.** Open power battery feed circuit. **2.** Open ignition feed circuit. **3.** Poor radio ground circuit. **4.** Loss of bus communications.

DIAGNOSTIC CHART 14-8

PROBLEM AREA:	Audio System.
SYMPTOMS:	Radio will not produce sound.
POSSIBLE CAUSES:	**1.** Open power battery feed circuit. **2.** Open ignition feed circuit. **3.** Poor ground connection. **4.** Defective radio. **5.** Open speak circuit. **6.** Defective amplifier. **7.** Open power feed to amplifier. **8.** Poor amplifier ground circuit. **9.** Bus communications error. **10.** Stuck MUTE button.

DIAGNOSTIC CHART 14-9

PROBLEM AREA:	Audio System.
SYMPTOMS:	No sound in AM or FM mode. CD audio operates normally.
POSSIBLE CAUSES:	**1.** Faulty antenna connection. **2.** Poor antenna ground. **3.** Faulty radio.

DIAGNOSTIC CHART 14-10

PROBLEM AREA:	Audio System.
SYMPTOMS:	Excessive noise heard in AM audio.
POSSIBLE CAUSES:	**1.** Faulty antenna connection. **2.** Poor antenna ground. **3.** Faulty engine to chassis ground.

DIAGNOSTIC CHART 14-11

PROBLEM AREA:	Audio System.
SYMPTOMS:	Poor radio reception.
POSSIBLE CAUSES:	**1.** Faulty antenna connection. **2.** Poor antenna ground. **3.** Faulty radio.

DIAGNOSTIC CHART 14-12

PROBLEM AREA:	Audio CD System.
SYMPTOMS:	Radio will not go into CD mode.
POSSIBLE CAUSES:	**1.** CD not inserted into player. **2.** Stuck MODE button. **3.** Faulty player.

DIAGNOSTIC CHART 14-13

PROBLEM AREA:	Audio CD System.
SYMPTOMS:	CD will not eject.
POSSIBLE CAUSES:	**1.** CD adhesive label came loose. **2.** Warped CD. **3.** Faulty player.

DIAGNOSTIC CHART 14-14

PROBLEM AREA:	Audio CD System.
SYMPTOMS:	CD will not play.
POSSIBLE CAUSES:	**1.** Faulty power feed circuit. **2.** Poor ground connection. **3.** Defective player.

DIAGNOSTIC CHART 14-15

PROBLEM AREA:	Audio CD System.
SYMPTOMS:	CD will not insert.
POSSIBLE CAUSES:	**1.** Damaged edges of CD. **2.** CD surface defects. **3.** Wrong disc size. **4.** Faulty player.

DIAGNOSTIC CHART 14-16

PROBLEM AREA:	Audio CD System.
SYMPTOMS:	CD skips or jumps tracks.
POSSIBLE CAUSES:	**1.** CD surface defects. **2.** Excessive vehicle vibration. **3.** Loose mounting bolts. **4.** Faulty player.

DIAGNOSTIC CHART 14-17

PROBLEM AREA:	Video Display.
SYMPTOMS:	DVD screen does not power on.
POSSIBLE CAUSES:	**1.** Power and/or CAN circuit maybe open. **2.** Screen software issue.

DIAGNOSTIC CHART 14-18

PROBLEM AREA:	Video Display.
SYMPTOMS:	Green DVD screen.
POSSIBLE CAUSE:	Defective satellite video receiver module.

DIAGNOSTIC CHART 14-19

PROBLEM AREA:	Video Display.
SYMPTOMS:	Black DVD screen.
POSSIBLE CAUSES:	**1.** Defective satellite video receiver module. **2.** Video circuit damaged.

DIAGNOSTIC CHART 14-20

PROBLEM AREA:	Satellite Radio.
SYMPTOMS:	Poor or no satellite audio reception. Radio will not go into CD mode.
POSSIBLE CAUSES:	**1.** Satellite antenna. **2.** Satellite antenna connection. **3.** Open or shorted satellite antenna cable. **4.** Telematics module.

DIAGNOSTIC CHART 14-21

PROBLEM AREA:	Satellite Video.
SYMPTOMS:	Satellite video has no video.
POSSIBLE CAUSES:	**1.** Obstructed line of sight to the sky. **2.** Antenna or antenna cable inoperative.

DIAGNOSTIC CHART 14-22

PROBLEM AREA:	Satellite Video Display.
SYMPTOMS:	Screen does not power on.
POSSIBLE CAUSES:	**1.** Power and/or CAN circuit may be open. **2.** Screen software issue.

DIAGNOSTIC CHART 14-23

PROBLEM AREA:	Satellite Video Display.
SYMPTOMS:	Green screen.
POSSIBLE CAUSE:	Defective satellite video receiver module.

DIAGNOSTIC CHART 14-24

PROBLEM AREA:	Satellite Video Display.
SYMPTOMS:	Black screen.
POSSIBLE CAUSES:	**1.** Defective satellite video receiver module. **2.** Video circuit damaged.

DIAGNOSTIC CHART 14-25

PROBLEM AREA:	Satellite Video.
SYMPTOMS:	Satellite video has no video.
POSSIBLE CAUSES:	**1.** Obstructed line of sight to the sky. **2.** Antenna or antenna cable inoperative.

DIAGNOSTIC CHART 14-26

PROBLEM AREA:	Telecommunication System.
SYMPTOMS:	Failed phone pairing attempts.
POSSIBLE CAUSES:	**1.** Phone does not support Hands-Free Profile. **2.** Phone not Bluetooth enabled. **3.** PIN numbers between phone and system not matching. **4.** Phone has reached maximum number of allowed devices paired.

DIAGNOSTIC CHART 14-27

PROBLEM AREA:	Telecommunication System.
SYMPTOMS:	Poor phone audio quality.
POSSIBLE CAUSES:	**1.** Microphone failure. **2.** Rearview mirror not properly fixed to mounting button.

DIAGNOSTIC CHART 14-28

PROBLEM AREA:	Cellular service.
SYMPTOMS:	Poor or no cellular Reception.
POSSIBLE CAUSES:	**1.** Cellular antenna connection. **2.** Open or shorted cellular antenna cables. **3.** Open or shorted cellular antenna. **4.** Cellular antenna. **5.** Cellular module.

DIAGNOSTIC CHART 14-29

PROBLEM AREA:	Wi-Fi Hotspot.
SYMPTOMS:	Wi-Fi Hotspot inoperative.
POSSIBLE CAUSES:	**1.** Wi-Fi service registration incomplete. **2.** Wi-Fi subscription service is inactive, expired, or not installed. **3.** Wi-Fi capability is not enabled in user's device. **4.** Selected Wireless Access Point (WAP) is incorrect. **5.** Wireless Access Point (WAP) password or security settings incorrect. **6.** Cellular antenna.

DIAGNOSTIC CHART 14-30

PROBLEM AREA:	Navigation.
SYMPTOMS:	Poor or no GPS reception.
POSSIBLE CAUSES:	**1.** GPS antenna connection. **2.** Open or shorted GPS antenna cables. **3.** Open or shorted GPS antenna. **4.** GPS antenna. **5.** GPS module.

CHAPTER 15

SERVICING PASSIVE RESTRAINT AND VEHICLE SAFETY SYSTEMS

Upon completion and review of this chapter, you should be able to:

- Diagnose automatic seat belt systems.
- Service the air bag system safely.
- Use a scan tool to properly retrieve air bag system trouble codes.
- Replace the air bag module according to manufacturer's service information standards.
- Replace the clockspring assembly according to manufacturer's service information standards.
- Properly diagnose air bag system pressure sensors and their circuits.
- Properly diagnose air bag system accelerometer sensors.

- Properly diagnose the occupant classification system.
- Validate the occupant classification system.
- Properly diagnose the park assist system and demonstrate the proper use of DMMs and scan tools to diagnose the ultrasonic sensor.
- Properly diagnose the back-up camera system.
- Diagnose and service the park guidance system.
- Use a scan tool and DMM to diagnose the common sensors used by ESP and rollover mitigations systems.

Terms To Know

Automatic seat belt system

Live module

OCS validation test

OCS service kit

Park assist system (PAS)

INTRODUCTION

In this chapter, you will learn how to properly and safely service automatic passive restraint systems. Federal mandates concerning the equipping of these systems have assured that today's technician must be qualified to service them. The safety of the driver and/or passengers depends on the ability of the technician to properly diagnose and repair these systems. In this chapter, we include procedures for replacing the air bag module and the clockspring. In addition, there are typical procedures for validating the occupant classification system (OCS).

We also discuss diagnostics of vehicle safety systems such as lane change warning, park assist, advanced park guidance, and obstruction detection. It includes diagnostics of the various sensors used by these systems.

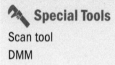

Classroom Manual
Chapter 15, page 470

Special Tools

Scan tool
DMM

AUTOMATIC SEAT BELT SERVICE

The **automatic seat belt system** automatically puts the shoulder and/or lap belt around the driver or occupant. The automatic seat belt is operated by direct current (DC) motors that move the belts by means of carriers on tracks. Even though the components used in the automatic seat belt system vary according to the manufacturer, the basic principles of locating and repairing the cause of a problem are similar. Refer to the service information to obtain a circuit diagram of the system. The circuit schematic, troubleshooting charts, and diagnostic charts will assist you in finding the fault.

In addition, most manufacturers provide a troubleshooting chart for the automatic seat belt system. Troubleshooting the circuits, switches, lamps, and motors as indicated in this chart requires only the skills described in previous chapters. In addition to the troubleshooting chart, an operational logic chart will provide helpful information when testing switches. Use a diagnostic chart to test the operation of the system and to determine faults with the module.

Some systems are capable of storing diagnostic trouble codes that can be retrieved by a scan tool. The scan tool is also used to perform a functional test of the system. To perform this test, turn the ignition switch to the RUN position and connect the scan tool to the diagnostic connector. Select the system test and follow the scan tool instructions. If there is a failure in the system, the scan tool will display the code.

Drive Motor Assembly Replacement

A faulty drive motor can cause slow or no operation of the automatic seat belts. If the technician diagnoses the component is faulty, the motor must be replaced. A typical procedure for motor assembly replacement is described here.

Begin by disconnecting the negative battery cable. If the vehicle is equipped with side air bag curtains, wait for the required amount of time specified by the manufacturer to allow the air bag system to discharge. Remove the B-pillar trim to gain access to the drive motor. Disconnect the electrical connector from the seat belt motor and locate the wiring harness out of the area. Remove the screws that attach the drive belt track to the motor. Next, remove the screw that attaches the vertical guide to the pillar. Slide the belt down far enough to disengage the sprockets from the motor drive gear teeth. Finally, remove the screws that attach the motor to the pillar.

AIR BAG SAFETY AND SERVICE WARNINGS

Classroom Manual
Chapter 15, page 473

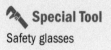

Special Tool

Safety glasses

Whenever working on the air bag system, it is important to follow some safety warnings. There are safety concerns with both deployed and live air bag modules. The air bag module is composed of the air bag and inflator assembly that are packaged in a single module. The module is mounted in the center of the steering wheel.

1. Wear safety glasses when servicing the air bag system.
2. Wear safety glasses when handling an air bag module.
3. Always disconnect the battery negative cable, isolate the cable end, and wait for the amount of time specified by the vehicle manufacturer before proceeding with the necessary diagnosis or service. The required amount of time may be as much as 15 minutes. Failure to observe this precaution may cause accidental air bag deployment and personal injury.
4. Handle all sensors with care. Do not strike or jar a sensor in such a manner that deployment may occur.
5. Replacement air bag system parts must have the same part number as the original. Replacement parts of lesser quality or questionable quality must not be used.

Improper or inferior components may result in improper air bag deployment and injury to the vehicle occupants.

6. Do not strike or jar a sensor or an air bag system occupant restraint controller (ORC). This may cause air bag deployment or the sensor to become inoperative.

7. Before an air bag system is powered up, all sensors and mounting brackets must be properly mounted and torqued to ensure correct sensor operation. If sensor fasteners do not have the proper torque, improper air bag deployment may result in injury to the vehicle occupants.

8. When carrying a **live module** that has not been deployed, face the trim and bag away from your body.

9. Do not carry the module by its wires or connector.

10. When placing a live module on a bench, face the trim and air bag up.

11. Deployed air bags may have a powdery residue on them. Sodium hydroxide is produced by the deployment reaction and is converted to sodium carbonate when it comes into contact with atmospheric moisture. It is unlikely that sodium hydroxide will still be present. However, wear safety glasses and gloves when handling a deployed air bag. Wash hands after handling.

12. A live air bag module must be deployed before disposal. Because the deployment of an air bag is an explosive process, improper disposal may result in injury and in fines. A deployed air bag should be disposed of according to EPA and manufacturer procedures.

13. Do not use a battery- or AC-powered voltmeter, ohmmeter, or any other type of test equipment not specified in the service information. Never use a test light to probe for voltage.

14. Never reach across the steering wheel to turn the ignition switch on.

15. Air bag circuits are usually located in a YELLOW conduit, use YELLOW wires, and/or have YELLOW connectors. Never work on or around these circuits without first disconnecting the battery and letting the system disarm.

DIAGNOSTIC SYSTEM CHECK

Before an air bag system is diagnosed, a system check is performed to avoid diagnostic errors. Always consult the manufacturer's specific information because the diagnostic system check may vary depending on the vehicle. The diagnostic system check involves observing the air bag warning light to determine if it is operating normally. A typical diagnostic system check follows:

1. Turn on the ignition switch and observe the air bag warning light. On some GM systems, this light should flash seven to nine times and then go out. On most other vehicles, the air bag warning light should be illuminated continually for 6 to 8 seconds and then go out. If the air bag warning light does not operate properly, further system diagnosis is necessary.

2. Observe the air bag warning light while cranking the engine. On many GM vehicles, this should cause the light to be illuminated continually. Always refer to the vehicle manufacturer's service information. During engine cranking, if the air bag warning light does not operate as specified by the vehicle manufacturer, a complete system diagnosis is required.

3. Observe the air bag warning light after the engine starts. The light should turn off a few seconds after the engine is started. If the air bag warning light remains off, there are no current diagnostic trouble codes (DTCs) in the air bag system module. If the air bag warning light remains on, obtain the DTCs with a scan tool or flash code method. Not all manufacturers provide for fault code retrieval by flash code methods—these vehicles will require a scan tool.

RETRIEVING FAULT CODES

Special Tool
Scan tool

Most air bag systems will store fault codes that can be retrieved by a scan tool. Usually the air bag system retains two types of fault codes: active and stored. Active fault codes turn the air bag warning lamp on. Stored codes are faults that are intermittent. Some manufacturers also display how long (in minutes) a code was active. Depending on the manufacturer, the fault codes may be stored in nonvolatile memory. These codes will not be erased if the battery is disconnected or damaged. The only way to erase these codes is by use of the scan tool.

To retrieve fault codes, connect the scan tool to the data link connector (DLC). From 1996 on, this is usually the J1962 connector. On earlier model vehicles, there may be a separate DLC for the air bag system. This connector could be located under the seat, in the glove box, or in a direct connection to the air bag control module. Always refer to the proper service information to determine the DLC location. Turn the ignition switch to the RUN position.

⚡ WARNING **When turning the ignition switch on, do not reach across the steering wheel. Make sure that your body is away from any air bag modules that may deploy.**

Follow the scan tool instructions to request supplemental restraint system (SRS) Code Display. Record all stored and active fault codes. Active codes cannot be erased—the cause of the problem must be corrected so the code can go to a stored state. Use the proper diagnostic chart to trace the cause of the fault.

Classroom Manual
Chapter 15, page 477

Some fault codes require that the ORC be replaced before any further diagnostics can be performed. For example, in some GM systems, a displayed code 52 indicates that enough accident information is stored in the EEPROM to fill its memory. In most systems, it takes four simultaneously closed arming and crash sensor events to fill the memory. This code requires that the ORC be replaced. A code 52 cannot be erased nor can any further diagnosis be performed until the ORC is replaced. If a code 71 is set, then an ORC failure is detected. A code 71 requires that the ORC be replaced before any other diagnostic procedures can be performed. A code 71 cannot be erased. If a code 52 or 71 is not displayed and there are other history codes, then the technician will be instructed to go to a diagnostic chart to locate the causes of the intermittent fault. If there are no history codes, diagnose remaining current codes from the lowest to the highest number.

General Motors called its control module the diagnostic energy reserve module (DERM).

AIR BAG SYSTEM TESTING

Classroom Manual
Chapter 15, page 463

Once the DTCs have been recorded, the technician will use the proper diagnostic chart to locate the fault. It is very important to follow *all* procedures listed. If the chart calls for the use of an ohmmeter to check a circuit, the technician would have been instructed to disconnect the harness from the air bag module in a prior step. If the technician ignores this step and connects the ohmmeter to the circuit, it is possible that the air bag(s) may deploy.

Since most systems use shorting bars at the connectors, the technician must remember to lift these up in order to properly diagnose the harness. Most manufacturers provide test connections or test harnesses to be plugged into the harness connectors (**Figure 15-1**). These test connectors will lift the shorting bar and provide test locations for connecting the digital multimeter (DMM). Before connecting the special wiring harness, remove and isolate the battery negative cable. Then wait for the time period specified by the vehicle manufacturer.

 **Special Tools**
DMM
Test harness

⚡ WARNING **Use only the vehicle manufacturer's recommended tools and equipment for air bag system service and diagnosis. Failure to observe this precaution may result in unwarranted air bag deployment and personal injury.**

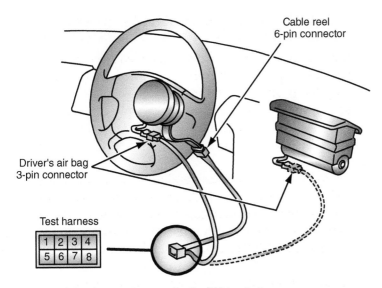

Figure 15-1 Test harness connected at the SRS terminals.

⚡ **WARNING** Do not use battery- or AC-powered voltmeters or ohmmeters except those meters specified by the vehicle manufacturer. Failure to observe this precaution may result in unwarranted air bag deployment and personal injury.

⚡ **WARNING** Do not use nonpowered probe-type test lights or self-powered test lights to diagnose the air bag system. Unwarranted air bag deployment and personal injury may result.

⚡ **WARNING** Follow the vehicle manufacturer's service and diagnostic procedures. Failure to observe these precautions may cause inaccurate diagnosis, unnecessary repairs, or unwarranted air bag deployment resulting in personal injury.

Air Bag Simulators

Most automotive manufacturers, and many after market suppliers, developed a method of performing air bag diagnostics by use of an air bag simulator (**Figure 15-2**). The simulator is connected in place of air bag components to represent a known good circuit and resistances. In addition, the simulator makes air bag service safer since the air bag module can be removed from the vehicle. The simulator allows for a safe test sequence that should identify any circuit problems. If a circuit problem does exist that would potentially cause the air bag to deploy, the simulator will indicate it and will remain in that mode until the problem is fixed.

🔧 **Special Tool**
Air bag load simulator

If an active code goes stored after the simulator is installed, the problem is with the disconnected component. For example, if there is an active DTC for "Driver's side squib circuit open," but when the air bag module is removed and the simulator installed in its place the DTC becomes stored, then the open is within the air bag module.

⚡ **WARNING** Never attempt to measure the resistance of an air bag module; deployment may occur.

⚡ **WARNING** Never attempt to repair an air bag module.

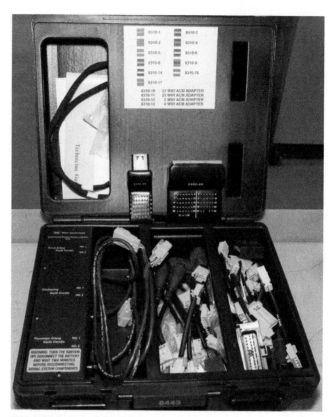

Figure 15-2 Air bag simulator tool.

INSPECTION AFTER AN ACCIDENT

Classroom Manual
Chapter 15, page 480

Any time the vehicle is involved in an accident, even if the air bag was not deployed, all air bag system components should be inspected. The wiring harness must be inspected for damage and repaired or replaced as needed. Any damaged or dented components must also be replaced. Do not attempt to repair any of the sensors or modules. Service is by replacement only.

In the event of deployment, the service information will provide a list of components that must be replaced. The list of components will vary depending on the manufacturer.

CLEANUP PROCEDURE AFTER DEPLOYMENT

Special Tools
Safety glasses
Particle dust mask
Protective clothing
   with long sleeves
Tape
Shop vacuum cleaner

If the air bag has been deployed, the residue inside the passenger compartment must be removed before entering the vehicle. Tape the air bag exhaust vents closed to prevent additional powder from escaping. Use the shop vacuum cleaner to remove any powder from the vehicle's interior. Work from the outside to the center of the vehicle. Vacuum the heater and air-conditioning (A/C) vents. Run the heater fan blower motor on low speed and vacuum any powder that is blown from the plenum.

COMPONENT REPLACEMENT

None of the sensors, air bag modules, or controllers used in the air bag system is repairable. In addition, seat belt pretensioners are not serviceable. If any of these components are found to be defective, then the component must be replaced.

If wiring repair is required, first refer to the proper service information. Some manufacturers do not recommend wire harness repair and others may have very specific procedures they require.

Air Bag Module Replacement

Classroom Manual
Chapter 15, page 474

 WARNING Before replacing any component of the air bag system, follow the service information procedure for disarming the system—even if the air bag is deployed.

 WARNING Wear safety glasses when working on the air bag system.

> ⚙ **SERVICE TIP** Anytime the steering column is to be removed from the vehicle, either remove the clockspring or lock the steering wheel so it cannot rotate. This will prevent the clockspring from being accidently extended beyond its travel limits.

Follow **Photo Sequence 34** as a guide to replacement of an air bag module. When removing the air bag module, the clockspring should be maintained in its correct index position at all times. Failure to do so can cause damage to the enclosure, wiring, or module. Any of these situations can cause the air bag system to default into a nonoperative mode. Always follow the service information for the vehicle you are working on.

The module bolts must be torqued to the specific value. Usually, new bolts are supplied with the air bag module.

Reverse the procedure to reinstall the module. To rearm the system, connect the yellow two-way electrical connector at the base of the steering wheel and install the connector position assurance (CPA). Turn the ignition switch to the RUN position; then replace the SRS fuse and connect the battery. This procedure is done to keep the technician out of the vehicle when the system is powered up.

With the ignition switch in the RUN position, observe the SRS warning light. It should perform its bulb test and then shut off. Perform the SRS diagnostic system check to confirm proper operation.

PHOTO SEQUENCE 34
Typical Procedure for Removing the Air Bag Module

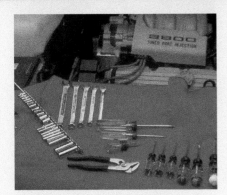

P34-1 Tools required to remove the air bag module include safety glasses, seat covers, screwdriver set, torx driver set, battery terminal pullers, battery pliers, assorted wrenches, ratchet and socket set, and service information.

P34-2 Place the seat and fender covers on the vehicle.

P34-3 Place the front wheels in the straight ahead position and turn the ignition switch to the lock position.

PHOTO SEQUENCE 34 (CONTINUED)

P34-4 Disconnect the negative battery cable.

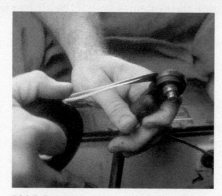

P34-5 Tape the cable terminal to prevent accidental connection with the battery post. Note: A piece of rubber hose can be substituted for the tape.

P34-6 Remove the SIR fuse from the fuse box. Wait for 10 minutes to allow the reserve energy to dissipate.

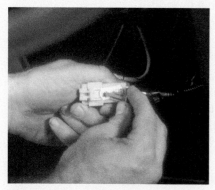

P34-7 Remove the connector position assurance (CPA) from the yellow electrical connector at the base of the steering column.

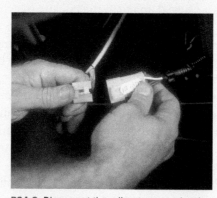

P34-8 Disconnect the yellow two-way electrical connector.

P34-9 Remove the four bolts that secure the module from the rear of the steering wheel.

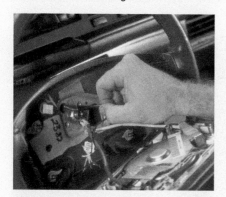

P34-10 Rotate the horn lead 1/4 turn and disconnect it.

P34-11 Disconnect the electrical connectors.

P34-12 Remove the module.

Classroom Manual
Chapter 15, page 476

 **Special Tools**

Steering wheel puller set

Clockspring Replacement

The clockspring (**Figure 15-3**) should be inspected any time the vehicle has been involved in an accident, even if the air bag was not deployed. In addition, the heat generated when an air bag is deployed may damage the clockspring. For this reason, it should be replaced whenever the air bag is deployed. Exact procedures vary according to the manufacturer.

Figure 15-3 Clockspring.

⚡ **WARNING** **Wear safety glasses when servicing the air bag system.**

In most cases (but not all), the front wheels must be located into the straight ahead position. Once the air bag system has been properly disarmed, the air bag module is removed. Mark the steering shaft and steering wheel with index marks for installation. In most cases, the steering wheel will need to be pulled off with a special puller after the lock nut is removed (**Figure 15-4**).

Once the upper and lower steering column shrouds are removed, the lower clock spring connector can be disconnected. If the clockspring is to be reused, tape it to prevent rotation. Remove the retaining fasteners and the clockspring.

Replacement of the clockspring is done in the reverse order. Some replacement clocksprings have a locking insert to prevent the rotation of the rotor. Do not remove this insert

Figure 15-4 Removing the steering wheel may require a puller.

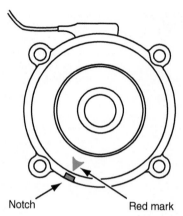

Figure 15-5 The index marks should align if the clockspring is properly centered.

until the clockspring is secured onto the column by the retaining screws. If the clockspring has been moved for some reason, then it must be re-centered. Usually this is done by gently rotating the rotor in one direction. Do not use excessive force, or damage to the clockspring may occur. Once the clockspring is in this full stop position, rotate the rotor in the opposite direction while counting the turns until the other stop is reached. Next, rotate the rotor back half the number of turns counted so it is in the middle of its travel. At this time, some form of index mark (**Figure 15-5**) or tabs should be aligned. Torque the new steering wheel attaching nut to specifications. When the steering wheel is replaced, rearm the system and perform the verification test.

DIAGNOSING AIR BAG SYSTEM PRESSURE SENSORS

All sensor diagnostics begin with a thorough inspection of the sensor and its related circuits. This includes looking for physical damage and proper electrical connection. Carefully examine the wiring harness and connectors between the sensor and the control module, including any dedicated ground circuits. All connectors and connector terminals must be tested for looseness, pullout, or damage. If the sensor has a hose connected to it, the condition of the hose must also be determined.

This section discusses the diagnostics of different types of pressure sensors. Examples of diagnostic routines will be provided. Although the diagnostics discussed may pertain to a specific function, the diagnostics of the circuits of most pressure sensors will be compatible.

Pressure sensors are generally a form of strain gauge that determines the amount of applied pressure by measuring the strain a material experiences when subjected to the pressure. Most strain gauge pressure sensors are a form of piezoresistive construction. A piezoresistive sensor changes in resistance value as the pressure applied to the sensing material changes. Common piezoresistive pressure sensors include Wheatstone bridges and capacitive discharge.

Diagnosing MEMS Pressure Sensors

MEMS pressure sensors are used to determine and confirm side impacts (**Figure 15-6**). Seldom is this the only input used to determine an impact event. Other inputs include vehicle motion, yaw and lateral sensor inputs, and accelerometer inputs. MEMS pressure sensors are not diagnosed in the same manner as most strain or pressure sensors since they have their own internal processing units and will usually communicate data to the ACM differently than a typical pressure sensor.

Caution

All restraint system sensors and modules must be properly fastened. Most sensors require proper orientation. In addition, all fasteners must be properly torqued. Often the manufacturer will discourage the reuse of fasteners.

Classroom Manual
Chapter 15, pages 479, 485

Special Tools

Scan tool
DMM
Jumper wires
Backprobing tools
Service information

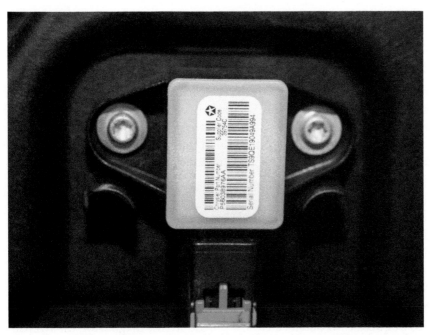

Figure 15-6 MEMS pressure sensor mounted inside the door to measure sudden pressure changes.

The MEMS sensor contains an electronic communication chip that allows it to communicate sensor status and sensor fault information to the ACM. The impact sensors perform their own self-test using power supplied by the ACM, and then communicate this status back to the ACM with periodic updates. The ACM microprocessor continuously monitors the electrical circuits to determine the system readiness. If the ACM detects a monitored system fault, it sets a DTC and illuminates the air bag indicator lamp. The impact sensor receives battery current and ground through dedicated sensor plus and minus circuits from the ACM (**Figure 15-7**). Communication between the impact sensors and the ACM is by modulating the current in the sensor plus circuit.

The hardwired circuits between the pressure-type impact sensors and the ACM may be diagnosed using conventional diagnostic tools and procedures. However, a scan tool is

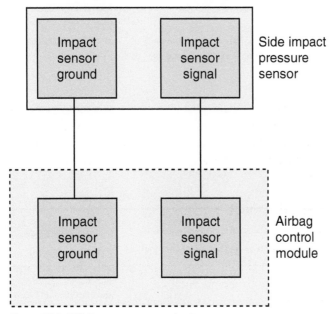

Figure 15-7 MEMS pressure sensor circuit.

the most common method for diagnosing the impact sensor, the electronic controls, or the communication between the sensor and the ACM.

Use the scan tool to access DTCs stored in the ACM. The following are some diagnostic examples for different DTCs that may be encountered.

Lost Communication with Side-Impact Sensor. The repair of this fault condition involves verifying the integrity of the wiring between the ACM and the sensor. The ACM monitors communication with the impact sensor continuously and will set this type of DTC when it no longer receives notification of the status update from the impact sensor. Possible causes for the setting of this DTC include the following:

- Side-impact sensor signal circuit open
- Side-impact sensor ground circuit open
- Side-impact sensor signal and ground circuit shorted together
- Side-impact sensor signal circuit shorted to ground
- Side-impact sensor signal circuit shorted to voltage
- Side-impact sensor ground circuit shorted to voltage
- Faulty impact sensor
- Faulty ACM

Photo Sequence 35 details a typical procedure for diagnosing the MEMS pressure sensor for this fault. The diagnostics used for this fault include those used for several other faults as well.

PHOTO SEQUENCE 35
Diagnosing the MEMS Pressure Sensor

P35-1 When retrieving the DTCs, pay attention to if they are active or stored faults. Also, record all of the DTCs to help determine if a common fault is responsible.

P35-2 If the DTC is active, turn the ignition OFF, disconnect the battery, and wait for the specified amount of time before proceeding.

P35-3 Disconnect the side-impact sensor connector. Check connectors and terminals. Clean and repair as necessary.

P35-4 Disconnect the ACM connector and inspect the connector and terminals for damage. Clean and repair as necessary.

P35-5 Connect the air bag simulator load tool adaptor to the ACM connectors.

P35-6 Measure the resistance of the side-impact sensor signal circuit between the side-impact sensor connector and the load tool adaptor. If the resistance is more than 1.0 ohm, the signal circuit has excessive resistance or an open condition that requires repair.

P35-7 Measure the resistance of the impact sensor ground circuit between the sensor connector and the load tool adapter. If the resistance is more than 1.0 ohm, the ground circuit has excessive resistance or an open condition that requires repair.

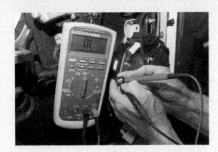

P35-8 Measure the resistance between the side-impact sensor signal circuit and the sensor ground circuit at the sensor connector. If the resistance is low, repair the condition that is causing the two circuits to be shorted together.

P35-9 Move the test leads to measure the resistance between the sensor signal circuit and chassis ground. If the resistance reading is below specifications, repair the signal circuit for a short to ground.

P35-10 Turn the ignition switch to the RUN position, and then reconnect the battery. Measure the voltage on the signal circuit. If voltage is present, repair the signal circuit for a short-to-voltage condition.

P35-11 Move the test leads to measure the voltage on the sensor ground circuit between the load tool adaptor and chassis ground. If voltage is present, repair the sensor ground circuit for a short-to-voltage condition.

P35-12 If all tests pass, turn the ignition to the OFF position and disconnect the battery. After waiting the specified amount of time, replace the impact sensor.

P35-13 Remove any special tools used and reconnect all disconnected components (except the battery).

P35-14 Turn the ignition to the RUN position, and reconnect the battery. Use the scan tool to clear all codes in the ACM.

P35-15 If needed, perform the module reset function with the scan tool.

P35-16 After waiting one minute, recheck for any active DTCs. If the same code for loss of communication returns, replace the ACM. If the code does not return, the repair is complete.

Side-Impact Sensor Configuration Mismatch. As discussed earlier, when the ACM is powered, it sends a test current to the impact sensors. During the test, the ACM is also looking to see if the impact sensors that it discovered agree with the preprogrammed vehicle configuration of components assigned to the ACM. The DTC sets when the ACM finds an impact sensor, while the configuration of the module does not indicate that the device should be present. Diagnosing this type of fault is centered on the confirmation that the correct ACM and wiring harness are installed. Usually, the fault is set after someone did some swapping of modules or attempted to install a used module. The fault can also set if the wiring harness has been changed and the improper one was installed.

Diagnosing Air Bag System Accelerometer Sensors

Classroom Manual
Chapter 15, page 479

Electro-mechanical accelerometers are tested by removing them from the system and measuring for open and shorted conditions. The circuits to the electro-mechanical sensors are tested for continuity, excessive resistance, and shorts as any other circuit would be. Be sure to follow all safety warnings and cautions associated with the test procedures.

MEMS-type accelerometers are diagnosed in much the same way as the MEMS pressure sensors just discussed. The MEMS accelerometer performs its own self-test routine upon receiving power from the ACM. The status of the diagnostic routine is then communicated to the ACM. The diagnostic routine is run on a constant basis when the ignition is turned on. If at any time the routine determines a fault, a DTC is set and the SRS warning lamp is illuminated.

The diagnostic process of checking circuit integrity is identical to that explored previously in Photo Sequence 35 for the MEMS pressure sensor. This code requires that the entire circuit be tested and will encompass most of the diagnostics that may be associated with other DTCs. Loss of communication between the accelerometer and the ACM can result from any of the following conditions:

- Impact sensor signal circuit open
- Impact sensor ground circuit open
- Impact sensor signal circuit shorted to sensor ground circuit
- Impact sensor signal circuit shorted to ground
- Impact sensor signal circuit shorted to voltage
- Impact sensor ground circuit shorted to voltage
- Faulty impact sensor
- Faulty ACM

OCCUPANT CLASSIFICATION SYSTEM SERVICE

Classroom Manual
Chapter 15, page 490

Diagnosing the electrical functions of the occupant classification system (OCS) is performed using a scan tool. DTCs will be recorded any time a fault is detected in the system. In addition, if a fault is detected, the passenger air bag disable lamp (PADL) is turned on, indicating to the driver that the air bags have been suppressed.

Wheatstone bridge is used in the OCS to determine the presence of an occupant and to classify them according to their weight. There are two main types of systems used: the bladder and the strain gauge system.

The bladder system uses a silicone-filled bladder positioned between the seat foam and the seat support (**Figure 15-8**). A pressure sensor is connected by a hose to the bladder (**Figure 15-9**). When the seat is occupied, pressure that is applied to the bladder

Special Tools
Scan tool
DMM

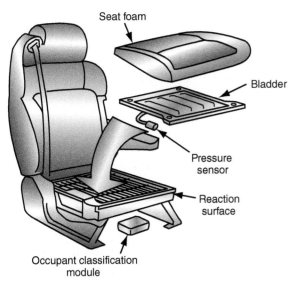

Seat foam

Bladder

Pressure
sensor

Reaction
surface

Occupant classification
module

Figure 15-8 Bladder-type occupant classification system.

Figure 15-9 The pressure sensor changes voltage signals as weight on the bladder changes.

disperses the silicone and the pressure sensor reads the increase in pressure. The pressure reading is input to the OCM (**Figure 15-10**).

Because pressure is used to determine seat occupation—and ultimately occupant classification—the system will correct for changes in atmospheric pressures. In addition, natural aging of the seat foam is also learned by monitoring gradual changes. When the seat is not occupied, the OCM compares the sensor voltage with the value stored in memory. The voltage values measured by the sensor will change as weight is added to the seat. Based on the change of voltage from the sensor, the OCM can determine the weight of the occupant.

The bladder system typically uses a strain gauge connected to the seat belt anchor (**Figure 15-11**). Because an infant seat that is securely strapped into the seat will cause an increase of downward pressures on the bladder, the belt tension sensor (BTS) is used to determine that the seat belt is tight around an object (about 24–26 lbs [11–12 kg] of force). The reading will indicate that the belt is tighter than what normally would be for a belt around a person. This would indicate that the belt is securing a child safety seat.

Circuit Performance Fault Code Diagnostics. Depending on the system, the ACM will usually sample the signal voltage from the sensor when the ignition is turned to the RUN position. If the voltage value indicates a pressure difference that is above or below an acceptable level, a DTC will be set and cause the air bag system warning lamp to illuminate. A circuit performance fault-type code is set if the circuit is over or under voltage for a specified length of time.

Although manufacturers that use a Wheatstone bridge–type sensor will have differing diagnosis of the sensors, if the sensor signal voltage is within the programmed value range of the air bag control module (ACM), a DTC may not set. For example, an open signal circuit will set a DTC, but resistance in the circuit that causes the voltage to be off by a few millivolts may not set a fault code, or it may trigger intermittent faults.

In some systems, testing the Wheatstone bridge sensor is a simple way of monitoring the sensor data output and voltage with a scan tool (**Figure 15-12**). With the ignition switch in the RUN position, record the pressure reading and voltage value. The pressure reading should represent static pressure (such as atmospheric pressure or no strain), and the voltage should confirm the reading. Compare this reading with the actual pressure and the two should be equal. A sensor that does not read correct static pressure cannot accurately read sudden pressure changes. If the reading is not correct, verify the condition

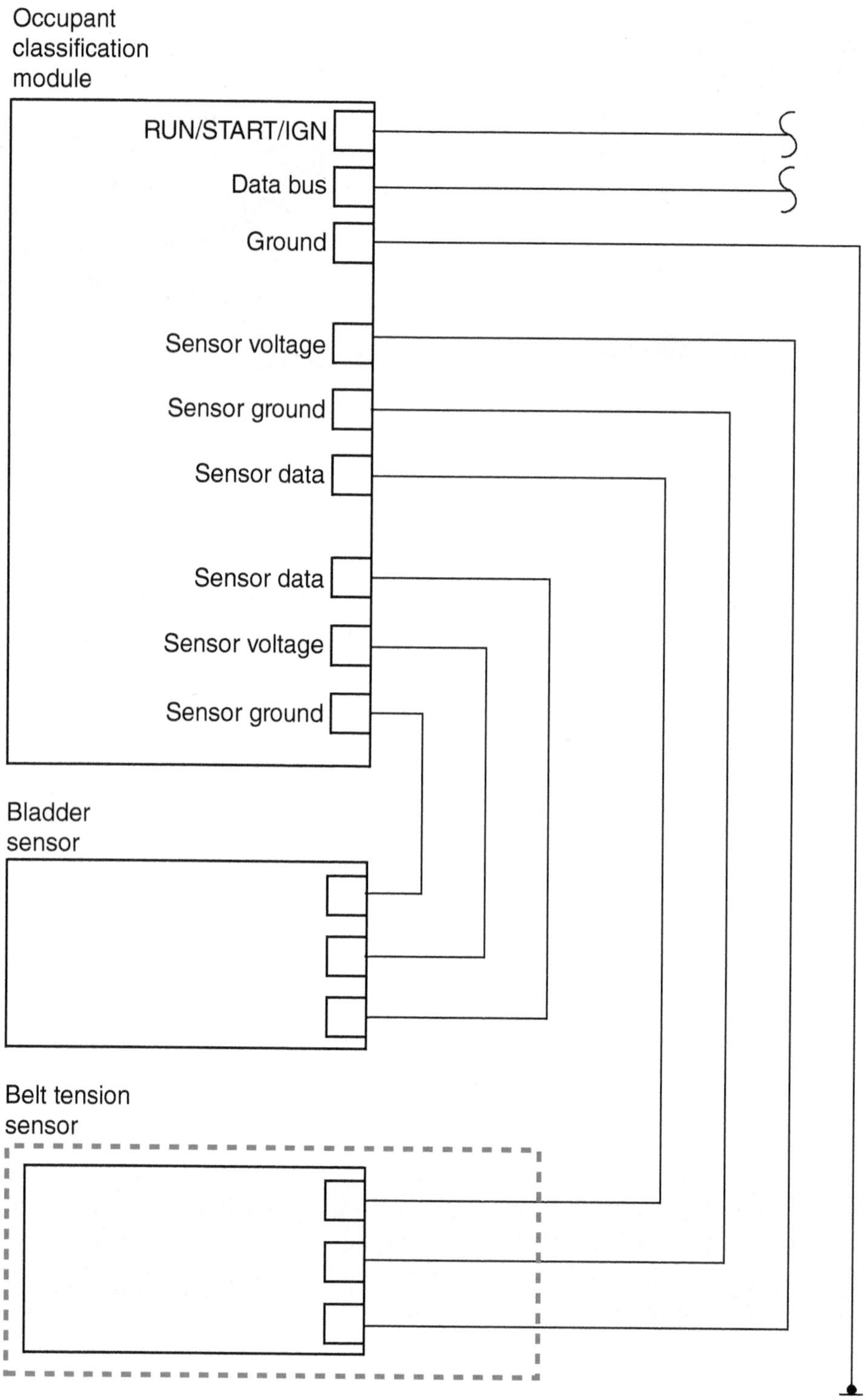

Figure 15-10 Electrical schematic of the bladder occupant classification system.

of all the circuits to the sensor, as discussed next. If the circuits are good, replace the sensor.

If a "Sensor Circuit High" fault is set, it is the result of the signal voltage being above a specified value. The circuit high fault is the result of any of the following conditions:

- Signal circuit open or high resistance
- Return circuit open or high resistance
- Signal circuit shorted to voltage

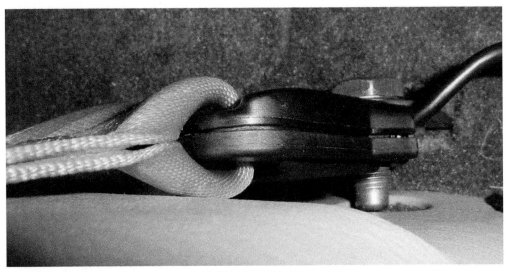

Figure 15-11 Belt tension sensor incorporated within the webbing loop mount.

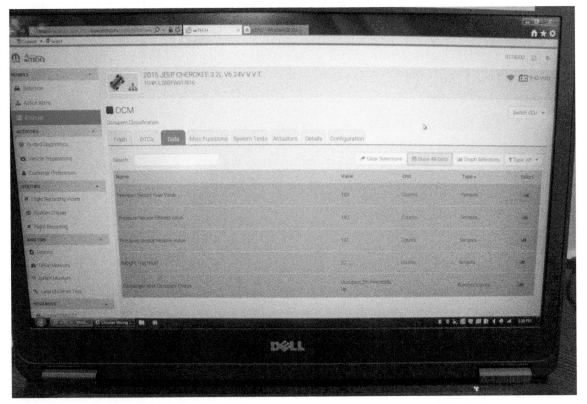

Figure 15-12 Strain gauge values as seen on a scan tool.

- 5-volt supply circuit shorted to voltage
- Faulty sensor
- Internal ACM fault

   To perform a quick test of the circuits, consider the schematic of a sensor circuit (**Figure 15-13**). If the scan tool indicates a voltage reading of 5 volts, this would be caused by any of the conditions listed above. Unplug the sensor and use a jumper wire to jump the signal circuit to ground. The scan tool should now read 0 volt. If it still reads 5 volts, the circuit is open between the ACM and the connector. If the voltage is above 0 volt, there is excessive resistance in the signal circuit between the ACM and the connector.

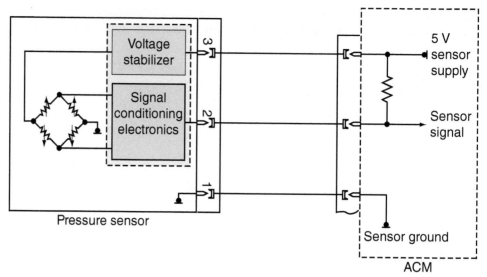

Figure 15-13 Typical pressure sensor circuit.

If the reading was 0 volt, move the jumper wire to short the signal circuit to the return circuit. The scan tool should now read 0 volt in a proper operating circuit. If the scan tool displays 5 volts, the return circuit is open. If the reading is above 0 volt, there is excessive resistance in the return circuit. If all of the readings are normal, the sensor is faulty.

> **⚙ SERVICE TIP** If the sensor return circuit was faulty within the ACM, all sensors that use this circuit would have a high voltage reading. If the return circuit of the sensor indicates an open circuit, check the return circuit from another input component. If this circuit shows good continuity, the problem is between a splice or in-line connector in the return circuit and the sensor connector. Use the wiring diagrams to locate these.

If a sensor circuit low fault is set, it is the result of the signal voltage being below a specified value. The circuit low fault is the result of any of the following conditions:

- 5-volt supply circuit open or high resistance
- 5-volt supply circuit shorted to ground
- 5-volt supply circuit shorted to sensor return circuit
- Signal circuit shorted to ground
- Signal circuit shorted to sensor return circuit
- Faulty sensor
- Internal ACM fault

If the scan tool reading with the ignition in the RUN position and sensor connected was 0 volt, this would be caused by any of the conditions listed above. If the voltage changes to 5 volts when the sensor is unplugged, the sensor is faulty. If the voltage remains at 0 volt when the sensor is unplugged, reconnect the sensor and backprobe the supply circuit. Use a voltmeter to check the supply circuit for 5 volts. It is important to test this voltage with the sensor connected to provide a load against the supply circuit. If the voltage is low, there is either high resistance, an open, or a fault in the ACM. These problems can be isolated by using a voltmeter to perform a voltage drop test of the circuit.

If the voltage is higher than 5 volts, this indicates a short to voltage condition. Isolate the offending circuit by removing fuses and relay one at a time while monitoring the voltage on the supply circuit. Once the voltage returns to normal after a fuse is pulled, use the wiring diagram to determine if any connectors are common between the two circuits. Next, inspect the wiring harness for damage or any condition that can cause the short.

> ⚙ **SERVICE TIP** Anytime you are using a scan tool to monitor a pressure reading, do not accept the value until you confirm the voltage. By correlating the signal voltage to the pressure reading, a fault in the circuit can be determined.

To test the sensor circuit shown in Figure 15-13 using a DMM, backprobe the 5-volt supply circuit wire (connector plugged to the sensor) with a voltmeter with the ignition in the RUN position. The voltage should be between 4.5 and 5.2 volts. If the voltage is too high, check for a short-to-voltage condition.

If the voltage is too low, backprobe the ACM connector to check the voltage on the circuit at the ACM. If the voltage is within specifications at the ACM but low at the sensor, repair the 5-volt supply circuit wire for high resistance. If this voltage is low at the computer, turn the ignition off, disconnect the battery, and wait for 2 minutes. Disconnect the ACM connector and use an ohmmeter to test for a short to ground. If the resistance is below specifications, this indicates that the circuit is shorted to chassis ground. If the reading is above the minimum specifications, use the ohmmeter to measure the resistance between the 5-volt supply circuit and the sensor return circuit. If the resistance is below specifications, locate the short between the two circuits in the wiring harness. If no problem is found, inspect the wiring and connectors between the sensor and the ACM while looking for any chafed, pierced, pinched, or partially broken wires. Inspect for broken, bent, pushed out, or corroded terminals. Be sure to review any technical service bulletins (TSBs) that may apply. If all of these fail to isolate the cause, replace or repair the ACM.

If the voltage supply circuit is good, test the sensor return (ground) circuit. With the ignition switch in the RUN position, backprobe the ground circuit terminal at the connector (with the connector plugged into the sensor). Connect the voltmeter from the sensor ground wire to the battery ground. If the voltage drop across this circuit exceeds specifications, test the voltage drop on the ACM's ground circuits. If this is good, repair the ground wire from the sensor to the computer.

If the first test of the 5-volt supply circuit indicates a voltage reading between 4.5 and 5.2 volts, the signal circuit will need to be tested for an open or high resistance. To test the sensor signal circuit, backprobe the sensor signal wire with the sensor connected and turn the ignition switch to the RUN position (engine off). Connect the DMM from the signal wire to ground. The voltage reading indicates the barometric pressure signal from the sensor to the ACM. If the signal circuit voltage, as compared to actual barometric pressure, confirms that the sensor's voltage supply and return circuits are good, replace the sensor.

The barometric pressure voltage signal varies depending on altitude and atmospheric conditions. Follow this calculation to obtain an accurate barometric pressure reading:

1. Determine present barometric pressure; for example, 29.85 inches.
2. Multiply your altitude by 0.001; for example, 600 feet times $0.001 = 0.6$.
3. Subtract the altitude correction from the present barometric pressure; for example, $29.85 - 0.6 = 29.79$.
4. Check the vehicle manufacturer's specifications to obtain the proper barometric pressure voltage signal in relation to the present barometric pressure.

> **AUTHOR'S NOTE** Some sensor signal circuits may have 0 volt when the connector is separated. Always confirm the expected voltage values with the proper service information.

If the 5-volt supply, signal, and return circuits all test good, reconnect the ACM harness connector and the battery. At the sensor harness connector, use a jumper wire to connect the signal circuit and the return circuit. With the ignition switch in the RUN position, view the sensor voltage displayed on the scan tool. It should read 0 volt and set a DTC for a circuit low code. If this test passes, replace the sensor. If this test does not pass, perform all of the inspections discussed earlier and then replace or repair the ACM.

Sensor Performance Fault Diagnostics. A fault with the performance or circuit of the pressure sensor will generate a DTC that can be retrieved by a scan tool. A sensor performance fault code is usually set when the voltage of the signal circuit is within acceptable parameters; however, the reading is not acceptable for the conditions. In addition, the ACM may determine that the ports of the sensor are blocked due to a reading that is different than expected. For example, if the pressure sensors in the two front doors do not indicate the same amount of pressure, the sensor that shows a change from the reading when the ignition was turned off as compared to when it was turned on again may set a performance code.

> ⚙ **SERVICE TIP** Be sure to properly identify the fault code. There is a difference between diagnostic routines for a *sensor* performance fault code and a *circuit* performance fault code.

> ⚠ **Caution**
>
> Use any special tools, such as breakout boxes, which the manufacturer calls for in the service information. Failure to follow these guidelines may result in damaged terminals.

The pressure sensor will usually set a performance fault if the sensor provides abnormal readings over a specified amount of time (**Figure 15-14**). To diagnose this type of fault, start with a visual inspection of the sensor, paying close attention to the sample port opening. If the visual inspection fails to isolate the cause of the performance fault code, the following steps can be used to determine the root cause. Refer to the proper service information to determine the conditions that cause the DTC to set and for the wiring diagram for the system. For this example, consider the wiring diagram in **Figure 15-15**.

With the ignition switch in the OFF position and the battery disconnected for two minutes, disconnect the pressure sensor connector and the ACM connector. Measure the resistance of each sensor circuit between the two connectors. If the resistance in any of

Absolute BARO reading	Lowest allowable voltage at -40°F	Lowest allowable voltage at 257°F	Lowest allowable voltage at 77°F	Designed output voltage	Highest allowable voltage at 77°F	Highest allowable voltage at 257°F	Highest allowable voltage at -40°F
31.0"	4.548 V	4.632 V	4.716 V	4.800 V	4.884 V	4.968 V	5.052 V
30.9"	4.531 V	4.615 V	4.699 V	4.783 V	4.867 V	4.951 V	5.035 V
30.8"	4.514 V	4.598 V	4.682 V	4.766 V	4.850 V	4.934 V	5.018 V
30.7"	4.497 V	4.581 V	4.665 V	4.749 V	4.833 V	4.917 V	5.001 V
30.6"	4.480 V	4.564 V	4.648 V	4.732 V	4.816 V	4.900 V	4.984 V
30.5"	4.463 V	4.547 V	4.631 V	4.715 V	4.799 V	4.883 V	4.967 V
30.4"	4.446 V	4.530 V	4.614 V	4.698 V	4.782 V	4.866 V	4.950 V
30.3"	4.430 V	4.514 V	4.598 V	4.682 V	4.766 V	4.850 V	4.934 V
30.2"	4.413 V	4,497 V	4.581 V	4.665 V	4.749 V	4.833 V	4.917 V
30.1"	4.396 V	4.480 V	4.564 V	4.648 V	4.732 V	4.816 V	4.900 V
30.0"	4.379 V	4.463 V	4.547 V	4.631 V	4.715 V	4.799 V	4.883 V

Figure 15-14 Barometric pressure voltage values at various pressures.

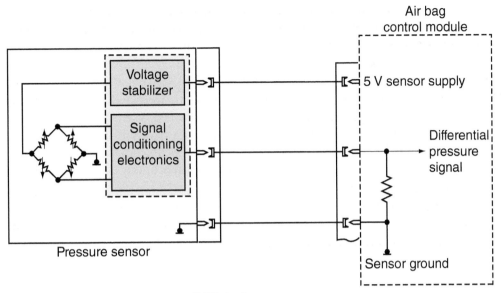

Figure 15-15 Differential pressure sensor (DPS) circuit.

the circuits is greater than that specified by the manufacturer, repair that circuit. Usually the maximum amount of resistance allowed will be about 1 ohm.

If the circuits all test within specifications, replace the sensor. Follow the service information requirements to perform a verification test of the system. If the performance fault code returns, the service information will generally lead you to replacing the ACM. Be sure that all of your testing up to this point has been complete and thorough. Before replacing the ACM, it must be determined to be the last possible cause of the problem.

Bladder-type systems produced by Delphi have some unique service requirements. If diagnostics indicate that the occupant classification module (OCM) requires replacement, this is possible only if the original seat is in the vehicle. Since the ORC stores the aging calibrations of the seat, if a new OCM is installed, the ORC will recognize the new module and will continue to use the old seat wear calibrations.

If diagnostics of a bladder-type system lead the technician to replace any of the following items, an **OCS service kit** must be installed:

- Bladder
- Sensor
- Seat foam
- Cloth seat cover

The OCS service kit consists of the seat foam, the bladder, the pressure sensor, the OCM, and the wiring. The service kit is calibrated as an assembly. The service kit OCM has a special identification data bit that is transmitted once it is connected. When the service kit is installed, the ORC will identify the service kit OCM and will clear all calibration data related to the system components.

The wiring of the service kit is hot glued to the module and the sensor to prevent separation of the components. A tag may also be located on the wiring harness, which identifies it as a service kit.

The service kit wiring uses a single connector that mates to the existing vehicle wiring harness. The connection on the vehicle harness that originally connected to the OCM now connects to the service kit harness. The connector on the vehicle harness that originally connected it to the pressure sensor will not be used with the service kit. Tie this wire to the harness to prevent it from getting caught in the seat tracks as the seat is moved.

Classroom Manual
Chapter 15, page 490

⚠ **Caution**
Since the ORC stores the seat-aging calibrations, swapping seats is not a good practice. The ORC will not have the correct calibrations for the replacement seat and the system will not work as intended. This may lead to air bag deployment when an infant is in the seat.

Classroom Manual
Chapter 15, page 491

Special Tools

Validation weight set
Scan tool

⚠ **Caution**

If the seat requires removal from the vehicle, the strain gauges are suscepti- ble to damage. Do not drop the seat or allow anyone to sit in it while it is out of the vehicle.

Strain gauge systems by TRW do not have these service requirements. However, any time a component is replaced on either system, the system must be validated before the vehicle is returned to the owner.

OCS Validation

Whenever the passenger-side front-seat retaining bolts are loosened, the seat is removed from the vehicle, the seat is replaced, the seat trim is replaced, a sensor is replaced, or the OCM is replaced, an **OCS validation test** must be performed. Virtually any service that is done to the passenger-side front seat will require the technician to validate OCS. The OCS validation test is done to confirm that the system can properly classify the occupant. This task usually requires the use of special weights (**Figure 15-16**).

The special weight set has three parts. The tool approximates federal standards for the weight of occupants:

1. The base weight of the tool weighs 37 pounds (17 kg) to validate for the classifica- tion of a rear-facing infant seat (RFIS) weight.
2. The addition of the 10-pound (4.5-kg) weight to the base validates for the weight classification of a child.
3. The addition of the 52-pound (24-kg) weight to the assembly validates for the weight classification of the fifth percentile female.

The total weight of the tool is 99 pounds (45 kg). Weights are added to the base in the proper order and placed in the correct position with the dowel pins. There are differences between the bladder-type and strain gauge—type systems that affect which weights will be used. It is imperative that the correct procedure be followed.

Photo Sequence 36 illustrates a typical procedure for validating the OCS.

Belt Tension Sensor Diagnostic Test

Classroom Manual
Chapter 15, page 491

The following is a typical procedure for testing the belt tension sensor (BTS) used on the Delphi bladder OCS. Since this sensor is used to determine if an infant seat is installed in the front passenger-side seat, it is critical that proper diagnostics be followed. Always refer to the service information for the vehicle you are working on.

Special Tools

Scan tool
Pull scale

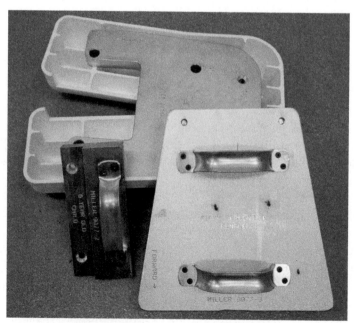

Figure 15-16 Special OCS weights used to validate the system.

PHOTO SEQUENCE 36
Occupant Classification Validation

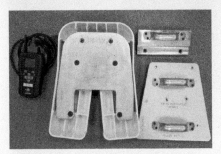

P36-1 Tools required to perform this task include a scan tool and special verification weights.

P36-2 Connect the scan tool and check for active battery voltage, OCM internal failure, or communication DTCs. Correct any of these conditions before continuing.

P36-3 Make sure the seat is empty and is in its full rearward position, with the back rest in a normal upright position.

P36-4 Use the scan tool to access the "OCM Verification" screen and follow the instructions to begin the test.

P36-5 Once the test has been started verify that the PADL is illuminated.

P36-6 When instructed, add the correct weight amount to the seat. Follow the scan tool prompts to complete this phase of the process. The scan tool should confirm that the phase was completed successfully.

P36-7 When instructed, add the additional weight and follow the prompts to complete this phase. The scan tool should confirm the phase was completed successfully.

P36-8 Confirm that the PADL is now off.

P36-9 Use the scan tool to access DTCs and check for any active codes. If active codes are present, this condition must be repaired and the system verification procedure repeated. Clear any stored codes.

With the scan tool connected, monitor the output of the BTS. This will usually be listed in counts (**Figure 15-17**). With the front passenger-side seat unoccupied and the BTS in its static state, the counts should be between 39 and 69. If the counts are outside of this range, the seat belt retractor and the BTS assembly must be replaced.

If the counts are within specified range, use a pull scale (**Figure 15-18**) to apply a load to the BTS. Connect the hook of the scale through the BTS webbing loop. Grab the pull

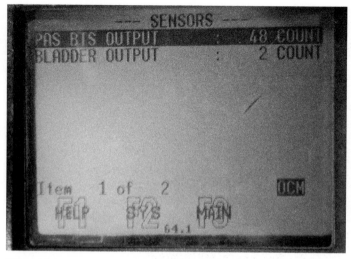

Figure 15-17 Scan tool will show the BTS counts.

Figure 15-18 Using the pull scale to obtain pressure against the BTS.

handle at the top of the spring scale and pull the spring scale straight up, keeping the horizontal bar level with the door sill and the spring scale in line with the BTS. Apply and hold 25 pounds (11.5 kg) of pressure on the sensor while monitoring the BTS counts on the scan tool. The counts should increase to between 153 and 204. If the counts are not within this range, replace the passenger seat belt retractor and BTS assembly.

If the counts are within range, release the pressure applied with the pull scale and monitor the BTS counts again. With no load applied to the BTS, the output should return to between 39 and 69 counts within 20 seconds. If the counts are not within this range, replace the passenger seat belt retractor and BTS assembly.

If all tests discussed pass, then the BTS is operating as intended. Confirm all electrical connections before returning the vehicle to the customer.

DIAGNOSING PARK ASSIST SYSTEMS

Classroom Manual
Chapter 15, page 501

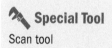

Special Tool

Scan tool

The **park assist system (PAS)** is a parking aid that alerts the driver to obstacles located in the path immediately behind the vehicle.

The **park assist system (PAS)** provides an example for diagnosing systems that use ultrasonic sensors (**Figure 15-19**). When an object is detected, the system uses an LED display and warning chimes to provide the driver with visual and audible warnings of the object's presence.

Ultrasonic transceiver sensors in the bumpers locate and identify the proximity of obstacles in the path of the vehicle. Each of the sensors receives battery voltage and ground from the park assist module. Each sensor has a dedicated serial bus communication circuit to the module (**Figure 15-20**).

If a customer states that his or her park assist system is not functioning properly, begin your testing by confirming that the sensors are not covered by dirt or other objects. Also, confirm that the parking brake is not applied and the parking brake switch is operating properly.

The microprocessor in the park assist module contains the park assist system logic circuits and uses on-board diagnostics to continuously monitor the entire park assist system electrical circuits and components. If the park assist module determines there is a problem in any of the park assist system circuits or components, it will store a DTC.

If a problem is detected, the hardwire circuits connecting components related to the park assist display may be diagnosed using a DMM and conventional diagnostic procedures using the proper wiring information. However, DMM diagnostic methods will not diagnose problems associated with the park assist display, the electronic controls, or communication between modules and other devices that provide some features of the

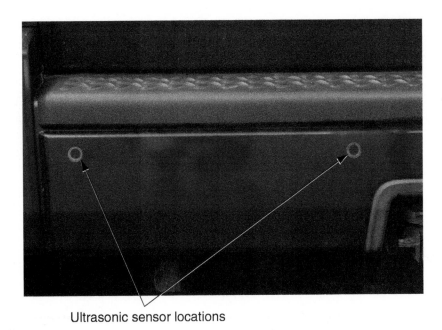

Ultrasonic sensor locations

Figure 15-19 Ultrasonic sensors used for back-up obstacle detection are usually located in the rear fascia/bumper.

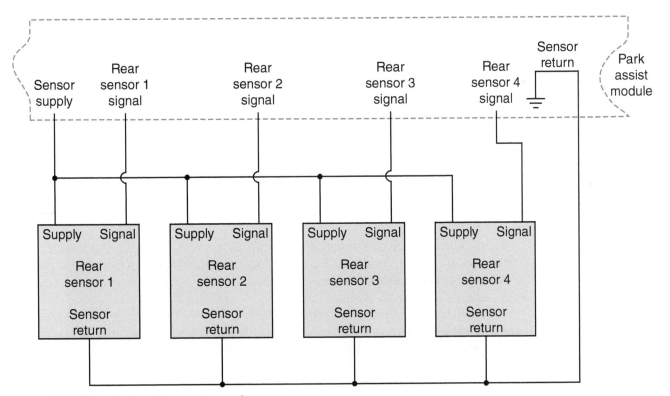

Figure 15-20 Wiring of park assist system transponders.

park assist system. Problems in these areas are more easily identified and repaired by use of a scan tool.

The park assist display is controlled by the park assist module. In addition, the display receives voltage and ground from the module (**Figure 15-21**). A dedicated serial bus provides for bidirectional communication between the display and the module.

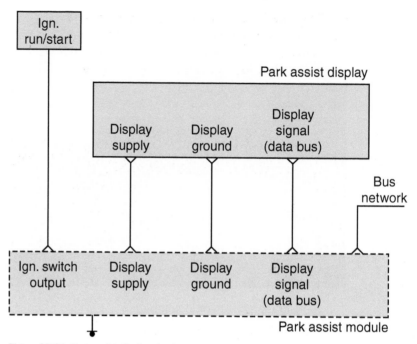

Figure 15-21 Park assist display circuit.

When diagnosing problems associated with the display, keep in mind that the microprocessor in the park assist module has complete control over the LED display. The module continually monitors the display status and will store a DTC if a fault is detected.

BACK-UP CAMERA SYSTEMS

Classroom Manual
Chapter 15, page 499

Special Tools
Scan tool
Service information

The back-up camera assists the driver in providing additional viewing of the area behind their vehicle. When the vehicle's transmission is put into reverse, the radio changes modes and uses the rear camera (**Figure 15-22**). The rear camera video is displayed on the radio

Figure 15-22 Rearview camera.

display (**Figure 15-23**). The camera is activated by a feed from the back-up lamp circuit (**Figure 15-24**). The camera sends a standard NTSC (TV) video signal to the radio. Most problems with the back-up camera system will result in a fault code being set. The DTCs can be accessed by the scan tool. Also, the scan tool may provide for activation functions to help isolate the problem area. **Table 15-1** is a diagnostic chart that directs the technician to possible causes for the condition.

Figure 15-23 Rearview camera display on the radio screen.

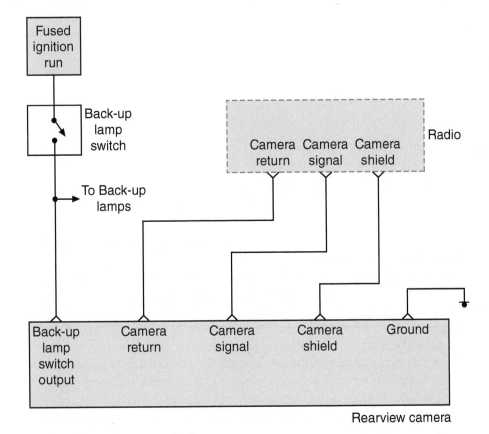

Figure 15-24 Rearview camera circuit.

Table 15-1 Diagnostic Chart for the Rear View Camera System

Condition	Possible Causes	Correction
Camera Inoperative	**1.** Fuse inoperative.	**1.** Check the back-up fuse. Replace fuse, if required, and repair cause of blown fuse.
	2. Wiring damaged.	**2.** Check for voltage at the camera connector for back-up lamp switch fee circuit. Repair wiring if required.
	3. Camera ground damaged.	**3.** Check for continuity between the ground wire at the camera connector and ground. Repair wiring if required.
	4. Video wiring damaged.	**4.** Check for continuity between the radio connector and the camera connector. There should be continuity.
	5. Camera inoperative.	**5.** Replace the camera. Refer to the appropriate Service information.
No Video Display	**1.** Fuse inoperative.	**1.** Check radio fuses. Replace fuses, if required, and repair cause of blown fuse.
	2. Radio connector damaged.	**2.** Check for loose or corroded radio connector. Repair, as necessary.
	3. Wiring damaged.	**3.** Check for battery voltage at radio connector. Repair wiring if required.
	4. Radio ground damaged.	**4.** Check for continuity between radio chassis and a known good ground. There should be continuity. Repair ground if required.
	5. Video wiring damaged.	**5.** Check for continuity between the radio connector and the camera connector. There should be continuity.
	6. Radio inoperative.	**6.** Refer to appropriate diagnostic service information for diagnosing the radio system.

Classroom Manual
Chapter 15, page 502

 **Special Tools**

Scan tool
60-mm diameter pole
Digital angle gauge
Tape measure

DIAGNOSING PARK GUIDANCE SYSTEMS

The park guidance system encompasses many different systems and inputs for proper operation. The ABS, electric steering, cameras, monitors, and radar sensors all make up the system. A fault in any of these systems or components can result in the parking guidance system shutting down.

The steering wheel angle sensor is used to monitor the rotation of the steering wheel during the parking maneuvers. If the battery is disconnected, the steering angle sensor may require initialization before the parking system will operate. Steering sensor initialization may require the use of a scan tool. However, many systems can be initialized in the following manner:

1. Turn the ignition to the RUN position.
2. Move the shift lever to the "R" position to confirm that the park guidance system is inoperative. "System initializing" may appear in the rearview monitor.
3. Move the shift lever to the "P" position and turn the steering wheel lock to lock.
4. Return the shift lever to the "R" position.

If the system is initialized, "Check surroundings for safety" or a similar message should be displayed indicating that the system is ready.

The park guidance system runs continuous diagnostics of the module logic, data bus, actuators, and inputs. Faults will set DTCs, thus most diagnostics are DTC driven. The scan tool can be used to monitor data PIDs and may also provide some bidirectional diagnostics. However, before condemning any modules or components confirm that the power and ground circuits are functioning properly.

To test the detection range of the sensors, the ignition switch is placed in the RUN position, and the guidance system switch is in the ON position. Different gear positions will need to be selected during the test, so be sure to choke the tires. For example, when measuring the front corner sensors the gear selection can be in any gear except PARK. The front center sensor can be tested in any gear position other than PARK or REVERSE. The rear corner and rear center sensors require the gear selection be in REVERSE.

To test the range detection of the corner sensors, move a 2.36-inch (60-mm) pole outward from the center of the sensor. Move the pole in the direction the sensor is pointing (**Figure 15-25**). The pole can be held either parallel or perpendicular to the ground, but not at a diagonal. The pole should be detected by the sensor when within the specified range. Move the pole left and right of the sensor's centerline to determine the span of the detectable range. The scan tool data PID will indicate the presence of the pole and the approximate range it is from the sensor. Alternately, the system display and warning buzzer should indicate that the sensors have detected the pole.

To test the detection range of the front and rear center sensors, use a wall or barrier (**Figure 15-26**). Be sure to test for proper sensor pattern width. Typically, the sensor pattern will extend further than the sides of the vehicle.

The ultrasonic sensors require adjustment anytime they are removed or replaced. The actual procedure will vary depending on the manufacturer of the park guidance system and the sensor used. The following procedure for the Toyota Prius is offered as a sample of the process.

First, the work area must be confirmed as being level and the vehicle is at the correct ride height. Next, measure the distance from the floor to each of the sensors and compare with specifications (**Figure 15-27**). If the measurement is not correct, confirm proper

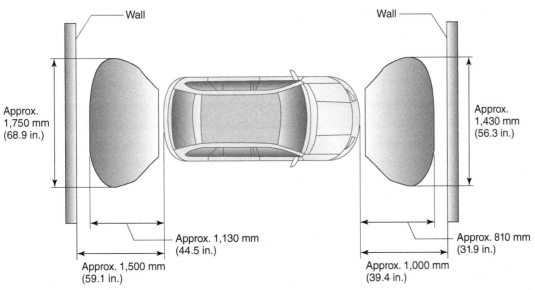

Figure 15-25 Range of sensor detection.

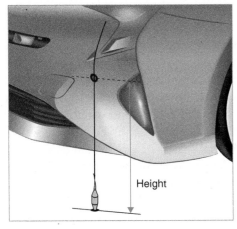

Figure 15-26 Measure the distance from the floor to the sensor.

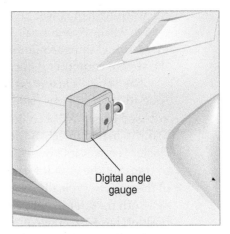

Figure 15-27 The digital angle gauge will display the orientation of the sensor.

Figure 15-28 If a fault is detected in the ESP system, the warning lamps are illuminated.

mounting. Some sensors can be adjusted slightly, but if the measurement and specification difference is sufficient, check for body damage and suspension component wear.

With proper sensor height confirmed, attach the digital angle gauge to the sensor (**Figure 15-28**). The angle gauge must read within the acceptable range as indicated in the service information. If the angle is not within tolerance, check for proper sensor mounting and body damage.

ESP AND ERM SYSTEM DIAGNOSIS

Classroom Manual
Chapter 15, pages
495, 504

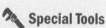

Special Tools

Scan tool
DMM
Backprobing tools
Jumper wires
Test light
Service information

The first indication to the vehicle operator that a problem is present in the ESP or the electronic rollover mitigation (ERM) system is the illumination of the warning light (**Figure 15-29**). This light may be lit along with the antilock brake warning light, the red brake warning light, and the brake assist system (BAS) warning light. If the red brake warning lamp is illuminated, it means there is a problem with the base brake system. In this case the other associated amber warning lights may illuminate also since the brake problem affects the entire system. The ABM, and other modules used within the systems, continuously monitors the circuits for continuity faults along with short activation tests of solenoids, pumps, and other actuators. If a fault is detected, the amber warning lamps illuminated means the system is disabled for the rest of that ignition cycle. When the vehicle is restarted and the self-test of the system determines that there are no faults, the warning lamps will be turned off after the bulb check process is completed and the system will be operational. If a fault is detected (or appears anytime during the drive cycle) the warning lamps are illuminated again and the system becomes nonoperational. Anytime the system computer detects a fault a DTC is retained.

The sensor cluster on ESP- and ERM-equipped vehicles contains both the lateral (side-to-side) and yaw (vehicle rotational) sensors. The data supplied by the sensor cluster are used by the ABM to determine how to control engine torque and which brake(s) to apply to maintain vehicle control during an ESP or rollover mitigation event. The sensor cluster receives its power and ground from the ABM. It also has a dedicated data circuit to transmit the data information (**Figure 15-30**). The sensor cluster is centrally located in the vehicle, typically under the center console (**Figure 15-31**).

Both the lateral and yaw sensors operate on an internal 5-volt reference. The neutral position has an average of 2.5 volts. **Figure 15-32** illustrates the typical voltage of a lateral acceleration sensor signal, and **Figure 15-33** illustrates that of a yaw sensor signal. The sensor cluster can be tested by removing its fasteners, observing the scan tool display

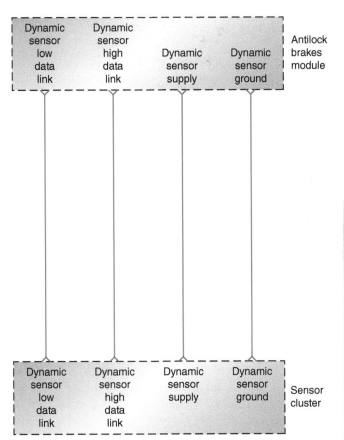

Figure 15-29 ESP sensor cluster circuit.

Figure 15-30 The sensor cluster is mounted on the centerline of the vehicle, usually under the console.

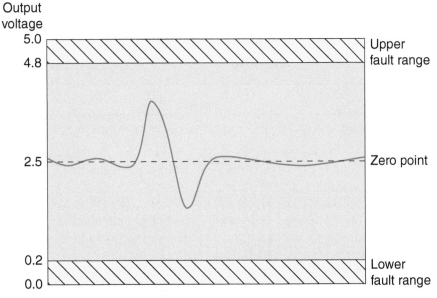

Figure 15-31 Voltage output of the lateral sensor.

while moving the sensor back and forth (**Figure 15-34**), and turning it around its center (**Figure 15-35**).

To transmit instantaneous data to the ABM, these signals are provided on a dedicated CAN C bus. Because of this data transmission procedure, it may not be possible to use a voltmeter or lab scope to monitor the voltage outputs. A scan tool is required to diagnose the sensor.

Output
voltage

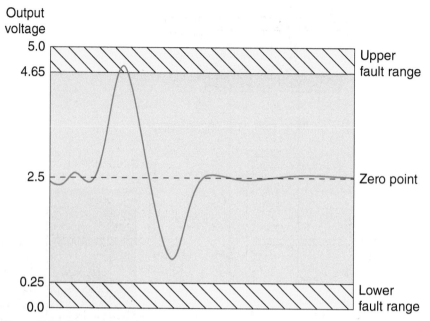

Figure 15-32 Voltage output of the yaw sensor.

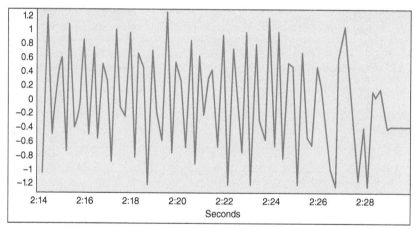

Figure 15-33 Scan tool display of the sensor cluster output as the sensor is moved back and forth.

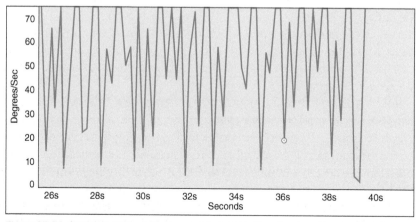

Figure 15-34 Scan tool display of the sensor cluster output as the sensor is rotated around its center.

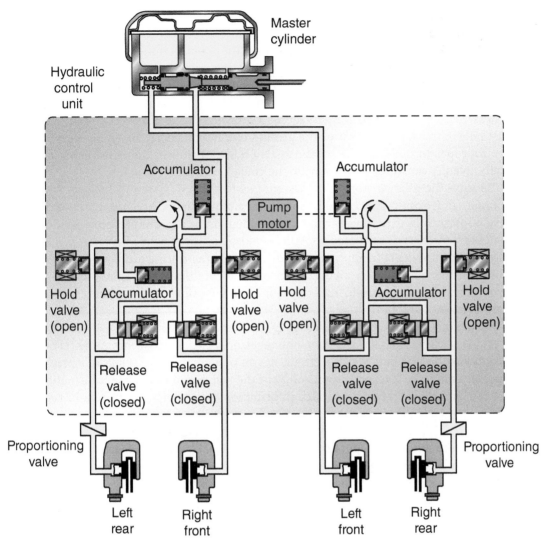

Figure 15-35 Hydraulic control solenoids used to control ABS and ESP functions.

The sensor is initialized when it is first installed. It is possible that it will go to an uninitialized state if the supply voltage drops below an acceptable level. If this occurs, the warning lamp will be illuminated and system operation may be inhibited. This can be caused by damage to the wires, terminals, or connectors. It can also be caused by high resistance in the supply and ground circuits. If there are no circuit failures, the sensor or the ABM is faulty.

To test the supply and ground circuits, disconnect the harness connector at the sensor and connect a voltmeter across the supply and ground circuit terminals. With the ignition switch in the RUN position, the voltage should be higher than 10.5 volts. If the voltage is correct, the sensor is faulty.

If the voltage reading was lower than 10.5 volts, disconnect the ABM harness connector (leave the sensor connector unplugged) and use an ohmmeter to check the supply circuit for high resistance. If no problem was found in the supply circuit, use the ohmmeter to measure the resistance of the ground circuit. If these do not indicate a circuit problem, it is possible that the supply circuit is shorted to another circuit in the harness. Use the ohmmeter to measure the resistance between the supply circuit and the other circuits in the harness. Any reading below 10K ohms indicates that the circuits are shorted together.

If the cause of the low-voltage reading is not isolated, the problem is a faulty ABM. Be sure to test power and ground circuits before performing repairs or replacing the ABM.

A DTC will be set for supply voltage high if the ABM detects voltage on the sensor cluster supply circuit when the power is turned off. This condition can be caused by a short to voltage in the wiring harness or connector damage. If the circuit does not have a defect, the ABM is faulty since it is not properly powering down the circuit.

Systems that use G-sensors will also set fault codes if the measured acceleration signal is higher than specifications for too long. This type of fault can be caused by wiring harness, terminal, or connector damage, so a thorough inspection is required. Also, high resistance in the supply circuit or the ground circuit can cause this fault. Keep in mind that improper torquing of the sensor fasteners may result in the ABM setting a code. Finally, a faulty sensor or ABM may be the cause.

Because the rollover mitigation system ties to the ESP system, which uses the ABS hydraulic control unit (HCU) and its solenoids, along with the ABS pump motor or active brake booster to initiate the applying of individual brake channels (**Figure 15-36**), diagnostics of the actuators are based on ABS diagnostic routines. The ERM system uses these same ABS components to apply an outside brake caliper during a sideways slide to work to prevent the vehicle from rolling. In some slide conditions, the solenoids are used to prevent pressurized hydraulic fluid from entering the caliper so the wheels will continue to roll instead of grabbing.

Failure of a solenoid circuit will set a DTC. Remember that the DTC directs the technician to the circuit that has the fault, not necessarily to the actual component. For example, a fault code that indicates the circuit has an open may be the result of a broken wire or a faulty solenoid. It is the responsibility of the technician to use the information from the DTC and the appropriate service information to properly diagnose and repair the condition.

The HCU solenoids can be tested with a lab scope. **Figure 15-37** illustrates a pattern obtained from a normally functioning solenoid valve. Using the solenoid activation test function in the scan tool and a lab scope, observe the voltage levels and inductive kicks. The inductive kicks indicate that the solenoid is working properly.

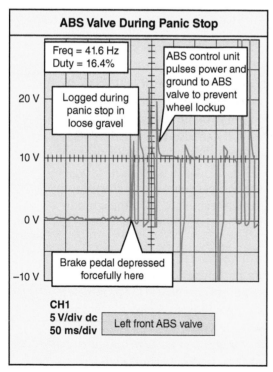

Figure 15-36 Lab scope trace pattern as ABS solenoid is being activated during wheel lockup.

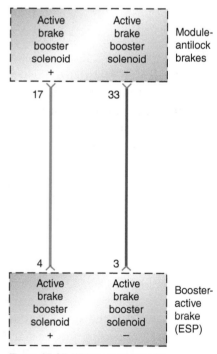

Figure 15-37 Wiring schematic of the command circuit between the ABM and the active booster solenoid.

AUTHOR'S NOTE Obtaining a lab scope pattern of the solenoid activation is only possible if there is a harness between the ABM and the HCU. If the ABM and HCU are integrated, you will not be able to access the solenoid circuit.

When the ABM determines that the active brake booster solenoid circuit has failed the diagnostic test, a DTC is set. This can be the result of any of the following conditions:

- Wiring harness, terminal, or connector damage
- Active brake booster solenoid (1) circuit and active brake booster solenoid (2) circuit shorted to together
- Active brake booster solenoid (1) circuit shorted to ground, voltage, or open
- Active brake booster solenoid (2) circuit shorted to ground, voltage, or open
- Faulty active brake booster solenoid
- Faulty ABM

Consider **Figure 15-38** as an example of testing the active booster solenoid. Using the scan tool, read and record all DTCs and any freeze frame or snapshot data. If the code is active, inspect the connectors of the ABM and the active booster solenoid (**Figure 15-39**) for broken, bent, pushed-out, or corroded terminals and repair any problems found.

To test for activation of the solenoid, disconnect the harness connector at the ABM (be sure the solenoid harness is connected). With the ignition in the RUN position, connect a jumper wire between the solenoid (2) and chassis ground at the ABM harness connector. Connect a jumper wire for several seconds between the solenoid (1) circuit and a 12-volt supply at the ABM harness connector. There should be brake application when these connections are made. If there is, the ABM is defective and needs to be replaced. If there is no activation, the circuits and the solenoid need to be diagnosed.

The resistance of the solenoid can be tested with an ohmmeter. Usually, these solenoids have a low resistance value that is within the range of 1 to 2 ohms. Always confirm the specified resistance value with the proper service information. If the resistance value is out of specifications, replace the solenoid.

With the solenoid connector disconnected, connect a voltmeter between the voltage supply circuit and chassis ground. Turn the ignition switch to the RUN position and observe the reading. If the voltage is 0 volt, there is an open, a short to ground, or the two circuits are shorted together. Use normal ohmmeter test procedures to determine the type of fault and to locate the cause. In addition, a test light connected to 12 volts can be used to determine if the circuit is shorted to ground.

To check for an open circuit, disconnect the solenoid and the ABM harness connectors. Connect a jumper wire between the solenoid (1) circuit and ground. Using a 12-volt

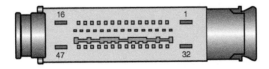

Module-antilock brakes

Booster-active brake (ESP)

Figure 15-38 Example of connector terminal locations for active brake booster system.

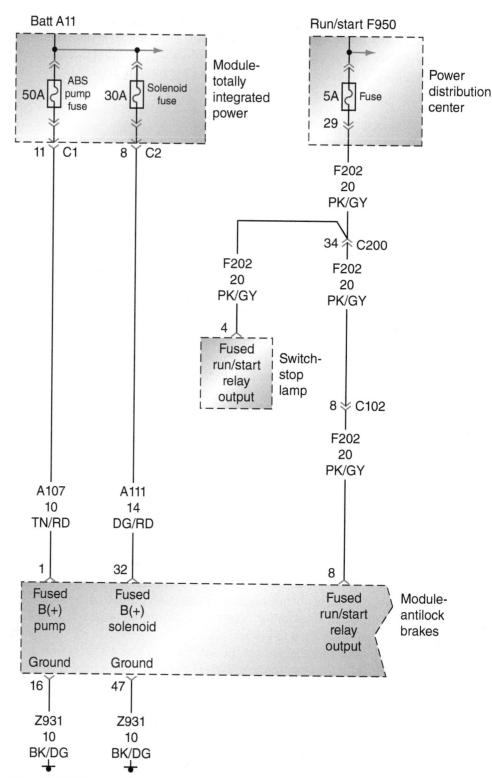

Figure 15-39 Wiring diagram of power and ground circuits for the ABM.

test light connected to 12 volts, check the solenoid (1) circuit. If the test light illuminates, replace the ABM. If it does not, the circuit has an open.

If a voltage reading was recorded in the initial voltage test, remember that this was an open circuit test and is not conclusive. Load the supply circuit with a test light, and measure the voltage again. If the voltage is less than battery volts, there is high resistance in the circuit. If the supply circuit checks good, test the ground circuit for shorts and opens. With both the ABM and solenoid harness connectors disconnected, use a test light connected to 12 volts to check the solenoid (2) circuit for a short to ground. If the test light illuminates, repair the short-to-ground condition.

If the test light does not illuminate, test the circuit for a short to voltage by connecting the test light to chassis ground and probing the solenoid (2) circuit. If the light comes on, repair the short-to-voltage condition.

To test for an open, disconnect the ABM and the solenoid harness connectors. Connect a jumper wire between the solenoid (2) circuit and chassis ground. Using a test light connected to 12 volts, check the solenoid (2) circuit. If the test light does not illuminate, there is an open in the circuit.

Faults within the ABS pump motor circuit can be the result of any of the following conditions:

- ABS pump motor fuse open
- Wiring harness, terminal, or connector damage
- Fused B(1) circuit shorted to ground or open
- Ground circuit open
- Faulty ABM

Use the scan tool to read and record all DTCs. Reference any freeze frame or snapshot data that are available. Erase the DTCs. Next, drive the vehicle above 25 mph (40 Km/h) and then park it and cycle the ignition switch. Use the scan tool again to read and record any DTCs and other information associated with them. If DTCs return, follow the service information for the diagnostic routine for the specific DTC. The following is a typical procedure for diagnosing the ABS pump circuits.

Begin by using the scan tool to perform an actuation test of the pump motor. If the pump motor actuates, but you just recorded DTCs for the circuit, the problem is intermittent. Carefully examine all connectors, terminals, and wiring for damage.

If the pump motor fails to actuate, check the voltage supply circuit (**Figure 15-40**). First confirm that the fuse is not open. If the fuse is blown, locate and repair the cause of the excessive amperage. Since the pump motor is usually an integral component of the HCU,

Figure 15-40 To accurately test the ground circuit, it should be tested under load.

there are internal circuits that may not be easily tested. If the pump motor is a stand-alone component, be sure to test all of the circuits from the ABM to the motor.

If the fuse is good, check for proper voltage applied to the fused B(1) circuit. Disconnect the ABM harness connector, and load the circuit with a test light. The voltage should be close to battery voltage. If it is less than battery voltage, there is resistance in the circuit. If the voltage reading was 0 volt, the circuit is open (a short to ground has already been eliminated by checking the fuse).

If there is proper voltage on the supply circuit, check the ground circuit next. With the ABM harness connector disconnected, use a test light attached to a 12-volt source and test the ground circuit(s). If the test light illuminates, measure the loaded voltage on the circuit. If the voltmeter reads more than 100 mV, there is excessive resistance on the ground circuit(s). If the test light does not illuminate, locate and repair the open in the ground circuit.

AUTHOR'S NOTE See *Today's Technician: Automotive Suspension and Steering Systems* Shop Manual for descriptions of diagnosing and testing the various automatic suspension system actuators that may be used by ESP and ERM systems.

CASE STUDY

A customer brings their vehicle to the repair shop because they have observed that the air bag warning light is always illuminated. The technician confirms the warning lamp operation by performing the diagnostic system check. Next, the technician uses the scan tool to retrieve fault codes. The scan tool displays a current code for an open driver-side air bag squib circuit. Following the service information procedure, the technician removes the air bag module and tests the clockspring circuit. Since the clockspring circuit proves to be in good condition, the technician replaces the air bag module. Upon completion of installing the new module, the diagnostic system test is run again to confirm that the repair was successful.

The air bag module was replaced due to the process of elimination. Since the clockspring circuit proved to be good, the air bag module had to be faulty. The resistance of the squib in the module was not tested because this can cause the air bag to deploy and the service information provided several warnings against it.

CASE STUDY

A customer brought her 2014 Jeep Grand Cherokee into the shop because the "ESP" light was always on. The technician verified the condition and used his scan tool to retrieve DTCs. A code C2114 was set for "dynamic sensor supply voltage low." After consulting the service information, the technician accessed the sensor cluster and the ABM. While taking out the center console to gain access to the sensor cluster, he saw that some additional electrical accessories were installed. These included a power outlet for a DVD player and a hands-free module. Following the diagnostic procedures, he tested the circuits and found there was resistance in the sensor ground circuit. Upon closer inspection of the circuit wire, he located a damaged section. Apparently, during the installation of the extra accessories this wire was damaged and not repaired. After cutting out the damaged section and soldering in a new wire, the system fault was no longer active. After test driving the vehicle to confirm the repair, the vehicle was returned to its owner.

CASE STUDY

A customer has brought his vehicle into the shop because the park guidance system does not operate. When he activates the system, a warning buzzer sounds and the message center displays, "Park system not available." The technician confirms the customer's concern and connects the scan tool to check for DTCs. DTC C1694 for "Steering angle initialization incomplete" was reported. The technician performed the calibration procedure by turning the steering wheel from full left to full right. The DTC was cleared and system confirmation routine performed. However, the DTC reappeared and the system would not initiate. Following the service information system procedures, the technician performed the "Steering center memorize" function of the scan tool. Next, she performed the "Max steering angle memorize" function. The system verification was performed and was successful. When the customer returned to pick up his vehicle, he informed the technician that another shop replaced the battery the day before the park guidance system stopped working. With today's vehicles, the technician must reference all applicable service information when performing even the most routine of service procedures, including battery replacement.

ASE-STYLE REVIEW QUESTIONS

1. The occupant classification system must be validated whenever:

   A. The seat has been removed from the vehicle.

   B. The OCM has been replaced.

   C. A new sensor has been installed.

   D. All of the above.

2. The park assist guidance system fails to activate.
   Technician A says this can be due to a steering angle sensor that is not initialized.
   Technician B says this can be caused by a sensor that is covered by debris.
   Who is correct?

   A. A only

   B. B only

   C. Both A and B

   D. Neither A nor B

3. The PADL is off when the seat is empty. This may indicate:

   A. A faulty OCM.

   B. Normal operation.

   C. The BTS is indicating the seat belt is cinched.

   D. All of the above.

4. Replacement of an air bag module is being discussed.
   Technician A says to follow the service information procedure for disarming the system.
   Technician B says to wear safety glasses.
   Who is correct?

   A. A only

   B. B only

   C. Both A and B

   D. Neither A nor B

5. *Technician A* says the clockspring should be inspected any time the vehicle has been involved in an accident, even if the air bag was not deployed.
   Technician B says the heat that is generated when an air bag is deployed may damage the clockspring.
   Who is correct?

   A. A only

   B. B only

   C. Both A and B

   D. Neither A nor B

6. *Technician A* says air bag residue should be swept from the vehicle's interior using a whisk broom.
   Technician B says whenever a vehicle is involved in an accident the air bag control module must be replaced.
   Who is correct?

   A. A only

   B. B only

   C. Both A and B

   D. Neither A nor B

7. Air bag system service is being discussed.
   Technician A says before an air bag system component is replaced, the negative battery cable should be disconnected and the technician should wait the specified time advised in the service information.
   Technician B says this waiting period is necessary to dissipate the reserve energy in the air bag system computer.
   Who is correct?

   A. A only

   B. B only

   C. Both A and B

   D. Neither A nor B

8. Air bag sensor service is being discussed.

   Technician A says incorrect torque on air bag sensor fasteners may cause improper air bag deployment.

   Technician B says the arrow on an air bag sensor must face toward the driver's side of the vehicle.

   Who is correct?

   A. A only
   B. B only
   C. Both A and B
   D. Neither A nor B

9. Air bag system diagnosis is being discussed.

   Technician A says on some Ford products, the air bag computer prioritizes faults and flashes the code representing the highest priority fault.

   Technician B says on some air bag systems, the air bag computer disarms the system if a fault occurs that could result in an unwarranted air bag deployment.

   Who is correct?

   A. A only
   B. B only
   C. Both A and B
   D. Neither A nor B

10. The purpose of the air bag load tool is being discussed.

    Technician A says the load tool is used to load the air bag system circuits to prevent accidental deployment.

    Technician B says the load tool is used to simulate known good resistances of the circuit.

    Who is correct?

    A. A only
    B. B only
    C. Both A and B
    D. Neither A nor B

ASE CHALLENGE QUESTIONS

1. A customer states that the "Pass Air Bag Off" light comes on whenever they lay their briefcase on the front passenger seat.

   Technician A says the OCS is too sensitive and needs to be validated.

   Technician B says this is normal since the weight of the briefcase is matching the weight of a small child.

   Who is correct?

   A. A only
   B. B only
   C. Both A and B
   D. Neither A nor B

2. The customer is concerned about the "Pass Air Bag Off" light not illuminating when their eight-year-old child sits on the seat. The vehicle is equipped with a bladder-type system.

   Technician A says this is normal operation.
   Technician B says a DTC will set for this condition.

   Who is correct?

   A. A only
   B. B only
   C. Both A and B
   D. Neither A nor B

3. A customer states that while driving the vehicle, the air bag warning lamp illuminates intermittently.

   Technician A says this can be caused by a loose connection to one of the system's sensors.

   Technician B says this may indicate a defect that will set a trouble code.

   Who is correct?

   A. A only
   B. B only
   C. Both A and B
   D. Neither A nor B

4. The parking assist guidance system does not perform a parallel parking maneuver with a displayed message that the parking space is not of sufficient size. However, the selected parking spaces are large enough to park the vehicle.

   Technician A says this can be caused by a misaligned ultrasonic corner sensor.

   Technician B says this can be caused by a disconnected rear center sensor.

   Who is correct?

   A. A only
   B. B only
   C. Both A and B
   D. Neither A nor B

5. Side-impact air bags are being discussed.

   Technician A says most systems have a control module or sensor located in the B-pillar.

   Technician B says the side air bags only deploy when the front air bags deploy.

   Who is correct?

   A. A only
   B. B only
   C. Both A and B
   D. Neither A nor B

Name _____ Date _____

WORKING SAFELY AROUND AIR BAGS

Upon completion of this job sheet, you should be able to work safely around and with air bag systems.

NATEF Correlation ⎯⎯⎯⎯⎯⎯⎯⎯⎯⎯⎯⎯⎯⎯⎯⎯⎯⎯⎯⎯⎯⎯⎯⎯⎯⎯⎯⎯⎯⎯

This job sheet addresses the following **MLR** tasks:

A.4.   Demonstrate proper use of a digital multimeter (DMM) when measuring source voltage, voltage drop (including grounds), current flow, and resistance.

E.4.   Disable and enable supplemental restraint system (SRS); verify indicator lamp operation.

This job sheet addresses the following **AST/MAST** task:

A.3.   Demonstrate proper use of a digital multimeter (DMM) when measuring source voltage, voltage drop (including grounds), current flow, and resistance.
This job sheet addresses the following AST tasks:

G.4.   Describe operation of safety systems and related circuits (e.g., horn, air bags, seat belt pretensioners, occupancy classification, wipers, washers, speed control/collision avoidance, heads-up display, park assist, and back-up camera); determine needed repairs.

G.5.   Describe body electronic systems circuits using a scan tool; check for module communication errors (data bus systems); determine needed action.

This job sheet addresses the following **MAST** tasks:

G.4.   Diagnose operation of safety systems and related circuits (e.g., horn, air bags, seat belt pretensioners, occupancy classification, wipers, washers, speed control/collision avoidance, heads-up display, park assist, and back-up camera); determine needed repairs.

G.5.   Diagnose body electronic systems circuits using a scan tool; check for module communication errors (data bus systems); determine needed action.

ASE NATEF

Tools and Materials
- A vehicle with air bags
- Service information for the chosen vehicle
- Component locator for the chosen vehicle
- Safety glasses
- A DMM

Describe the vehicle being worked on:

Year _____ Make _____ Model _____

VIN _____ Engine type and size _____

Procedure

1. Locate the information about the air bag system in the service information. How are the critical parts of the system identified in the vehicle?

_____

_____

_____

_____

_____

_____

2. List the main components of the air bag system and describe their locations.

_____

_____

_____

_____

_____

3. There are some very important guidelines to follow when working with and around air bag systems. These are listed here with some key words left out. Read through these and fill in the blanks with the correct words.

a. Wear _____ _____ when servicing an air bag system and when handling an air bag module.

b. Wait at least _____ minutes after disconnecting the battery before beginning any service. The reserve _____ module is capable of storing enough energy to deploy the air bag for up to _____ minutes after battery voltage is lost.

c. Always handle all _____ and other components with extreme care. Never strike or jar a sensor, especially when the battery is connected; this can cause deployment of the air bag.

d. Never carry an air bag module by its _____ or _____, and, when carrying it, always face the trim and air bag _____ from your body. When placing a module on a bench, always face the trim and air bag.

e. _____ air bags may have a powdery residue on them. _____ is produced by the deployment reaction and is converted to_____when it comes in contact with the moisture in the atmosphere.

Although it is unlikely that harmful chemicals will still be on the bag, it is wise to wear _____ _____ and _____ when handling a deployed air bag. Wash your hands immediately after handling a deployed air bag.

f. A live air bag must be _____ before it is disposed of. A deployed air bag should be disposed of in a manner consistent with the _____ and the manufacturer's recommended procedures.

g. Never use a battery- or A/C-powered _____, _____, or any other type of test equipment in the system unless the manufacturer specifically says to. Never probe with a _____ _____ for voltage.

Instructor's Response

_____

_____

_____

Name _____   Date _____

DIAGNOSING "DRIVER SQUIB CIRCUIT OPEN" FAULT

Upon completion of this job sheet, you should be able to diagnose the cause of a "driver squib circuit open" fault in an air bag system and determine needed repairs.

NATEF Correlation

This job sheet addresses the following **MLR** tasks:

A.3.	Use wiring diagrams to trace electrical/electronic circuits.
A.5.	Demonstrate knowledge of the causes and effects from shorts, grounds, opens, and resistance problems in electrical/electronic circuits.
A.7.	Using fused jumper wires, check operation of electrical circuits.
A.11.	Identify electrical/electronic system components and configuration.
E.4.	Disable and enable supplemental restraint system (SRS); verify indicator lamp operation.

This job sheet addresses the following **AST** tasks:

G.4.	Describe operation of safety systems and related circuits (e.g., horn, airbags, seat belt pretensioners, occupancy classification, wipers, washers, speed control/collision avoidance, heads-up display, park assist, and back-up camera); determine needed repairs.
G.5.	Describe body electronic systems circuits using a scan tool; check for module communication errors (data bus systems); determine needed action.

This job sheet addresses the following **AST/MAST** tasks:

A.4.	Demonstrate knowledge of the causes and effects from shorts, grounds, opens, and resistance problems in electrical/electronic circuits.
A.6.	Use fused jumper wires to check operation of electrical circuits.
A.7.	Use wiring diagrams during the diagnosis (troubleshooting) of electrical/electronic circuit problems.

This job sheet addresses the following **MAST** tasks:

G.4.	Diagnose operation of safety systems and related circuits (e.g., horn, air bags, seat belt pretensioners, occupancy classification, wipers, washers, speed control/collision avoidance, heads-up display, park assist, and back-up camera); determine needed repairs.
G.5.	Diagnose body electronic systems circuits using a scan tool; check for module communication errors (data bus systems); determine needed action.

ASE CERTIFIED **NATEF**

Tools and Materials

A vehicle equipped with a driver-side air bag (bugged by instructor prior to performing this task)

- Service information for chosen vehicle
- Safety glasses
- Battery terminal puller
- Fused jumper wires
- Scan tool

Describe the vehicle being worked on:

Year _____ Make _____ Model _____

VIN _____ Engine type and size _____

Procedure   **Task Completed**

1. Describe the normal conditions in which the air bag warning lamp should operate.

   _____

   _____

   _____

2. Perform the diagnostic system check. How did the air bag warning light respond?

   _____

   _____

3. Follow all safety warnings and cautions listed in the service information and connect the scan tool. Record all fault codes the scan tool displays.

   Active: _____

   _____

   Stored: _____

   _____

4. Is the code for "Driver-Side Squib Circuit" or equivalent active?   ☐ Yes ☐ No
   If no, consult your instructor.

5. List all possible causes that can set this fault code.

   _____

   _____

   _____

   _____

6. Disconnect the negative battery cable and isolate it.   ☐

7. How long does the service information say you must wait before proceeding?

   _____

   _____

8. Follow the service information instructions to remove the driver-side air bag module   ☐
   from the steering wheel.

9. Connect a jumper wire across the upper connector of the clockspring.   ☐

10. With the ignition switch in the RUN position, connect the battery ground cable.   ☐

11. Use the scan tool and record the active fault code.

   _____

   _____

12. Does the fault code indicate the circuit is shorted?   ☐ Yes ☐ No   **Task Completed**

13. If you answered yes to question 12, what is the faulty component?

_____

_____

14. If you answered no to question 12, locate the lower clockspring connector. Describe this connector's location.

_____

_____

_____

15. Disconnect the negative battery cable and isolate it. Wait the recommended amount of time before proceeding.   ☐

16. Disconnect the lower clockspring connector and connect a jumper wire across the control module side of the harness connector.   ☐

17. With the ignition switch in the RUN position, reconnect the battery ground cable.   ☐

18. Use the scan tool to retrieve active fault codes and record them.

_____

19. Does the fault code indicate the circuit is shorted?   ☐ Yes ☐ No

20. If you answered yes to question 19, what is the faulty component?

_____

21. If you answered no to question 19, locate the connector to the air bag control module. Describe this connector's location.

_____

_____

_____

22. Disconnect the negative battery cable and isolate it. Wait for the recommended amount of time before proceeding.   ☐

23. Disconnect the air bag control module connector and locate the wires of the driver-side squib circuit. What are the color codes of the wires?

_____

_____

_____

24. Use an ohmmeter to test the wire harness between the air bag control module connector and the lower clockspring connector. Record your results.

_____

_____

25. Based on your results, what have you determined to be the location of the fault?

_____

_____

Instructor's Response

_____

_____

_____

Name _____   Date _____

DIAGNOSING AIR BAG PRESSURE SENSOR CIRCUIT PERFORMANCE FAULT

Upon completion of this job sheet, you should be able to diagnose the cause of a pressure sensor circuit fault code.

NATEF Correlation

This job sheet addresses the following **MLR** tasks:

A.4.   Demonstrate proper use of a digital multimeter (DMM) when measuring source voltage, voltage drop (including grounds), current flow, and resistance.

A.5.   Demonstrate knowledge of the causes and effects from shorts, grounds, opens, and resistance problems in electrical/electronic circuits.

A.7.   Using fused jumper wires, check operation of electrical circuits.

This job sheet addresses the following **AST/MAST** tasks:

A.3.   Demonstrate proper use of a digital multimeter (DMM) when measuring source voltage, voltage drop (including grounds), current flow, and resistance.

A.4.   Demonstrate knowledge of the causes and effects from shorts, grounds, opens, and resistance problems in electrical/electronic circuits.

A.6.   Use fused jumper wires to check operation of electrical circuits.

This job sheet addresses the following **AST** task:

G.4.   Describe operation of safety systems and related circuits (e.g., horn, air bags, seat belt pretensioners, occupancy classification, wipers, washers, speed control/collision avoidance, heads-up display, park assist, and back-up camera); determine needed repairs.

This job sheet addresses the following **MAST** task:

G.4.   Diagnose operation of safety systems and related circuits (e.g., horn, air bags, seat belt pretensioners, occupancy classification, wipers, washers, speed control/collision avoidance, heads-up display, park assist, and back-up camera); determine needed repairs.

ASE NATEF

Tools and Materials

Vehicle equipped with air bag system pressure sensor, such as the strain gauges used for OCS (bugged by instructor prior to performing the task). Note: Although it would be ideal to use a vehicle with pressure sensors associated with the air bag system, this task can be performed on most transducer-type pressure sensor circuits if needed.

- Service information
- Fused jumper wire
- Backprobing tools
- Scan tool

Describe the vehicle being worked on:

Year _____ Make _____ Model _____

VIN _____ Engine type and size _____

Procedure

1. Read and record any fault codes in the air bag system.

   Active: _____

   Stored: _____

2. Monitoring the sensor data output and voltage with a scan tool with the ignition switch in the RUN position, record the pressure reading and voltage value.

   _____

3. What do these values represent?

   _____

   _____

   _____

4. Do the readings in step 2 indicate actual conditions or are they substitute values?

   _____

5. What can cause the "Sensor Circuit High" fault to set?

   _____

   _____

   _____

6. Unplug the sensor and use a fused jumper wire to jump the signal circuit to ground. What does the scan tool indicate for the voltage value?

   _____

   _____

   _____

7. Based on this reading, can you determine the location of the fault?

   _____

8. If the voltage is above 0 volt, what would this indicate?

   _____

   _____

   _____

9. If the reading in step 6 is 0 volt, what would be your next diagnostic step?

   _____

   _____

   _____

10. Based on your readings, what is the fault?

    _____

    _____

    _____

Instructor's Response

_____

_____

_____

Name _____   Date _____

TESTING THE MEMS SENSOR

Upon completion of this job sheet, you should be able to diagnose the cause of a MEMS sensor loss of communication fault code.

NATEF Correlation

This job sheet addresses the following **MLR** tasks:

A.4.	Demonstrate proper use of a digital multimeter (DMM) when measuring source voltage, voltage drop (including grounds), current flow, and resistance.
E.5.	Demonstrate knowledge of the causes and effects from shorts, grounds, opens, and resistance problems in electrical/electronic circuits.
E.4.	Disable and enable supplemental restraint system (SRS); verify indicator lamp operation.
E.5.	Remove and reinstall door panel.

This job sheet addresses the following **AST/MAST** tasks:

E.3.	Demonstrate proper use of a digital multimeter (DMM) when measuring source voltage, voltage drop (including grounds), current flow, and resistance.
E.4.	Demonstrate knowledge of the causes and effects from shorts, grounds, opens, and resistance problems in electrical/electronic circuits.

This job sheet addresses the following **AST** tasks:

G.4.	Describe operation of safety systems and related circuits (e.g., horn, air bags, seat belt pretensioners, occupancy classification, wipers, washers, speed control/collision avoidance, heads-up display, park assist, and back-up camera); determine needed repairs.
E.5.	Describe body electronic systems circuits using a scan tool; check for module communication errors (data bus systems); determine needed action.

This job sheet addresses the following **MAST** tasks:

G.4.	Diagnose operation of safety systems and related circuits (e.g., horn, air bags, seat belt pretensioners, occupancy classification, wipers, washers, speed control/collision avoidance, heads-up display, park assist, and back-up camera); determine needed repairs.
G.5.	Diagnose body electronic systems circuits using a scan tool; check for module communication errors (data bus systems); determine needed action.

ASE CERTIFIED **NATEF**

Tools and Materials

Vehicle equipped with a MEMS pressure or accelerometer sensor (bugged by instructor prior to performing task)

- Service information
- Air bag simulator load tool and adapters
- Scan tool
- DMM
- Safety glasses
- Battery terminal puller

Describe the vehicle being worked on:

Year _____ Make _____ Model _____

VIN _____ Engine type and size _____

Procedure   **Task Completed**

1. Record DTCs

   Active: _____

   Stored: _____

2. Disconnect the battery and wait for the specified amount of time before proceeding.   ☐

3. Follow the procedure in the service information to remove the door panel (if necessary).   ☐

4. Disconnect the connector of the sensor identified by the DTC. Check the condition of the connectors and terminals.   ☐

5. Disconnect the ACM connector and inspect the connector and terminals for damage.   ☐

6. Connect the air bag simulator load tool adaptor to the ACM connectors, and measure the resistance of the sensor signal circuit between the sensor connector and the load tool adaptor.

   _____

7. If the resistance is too high, what would this indicate?

   _____

   _____

   _____

8. Measure the resistance of the sensor ground circuit between the sensor connector and the load tool adapter.

   _____

9. If the resistance is too high, what would this indicate?

   _____

   _____

   _____

10. Measure the resistance between the sensor signal circuit and the sensor ground circuit at the sensor connector.

    _____

11. What would a low reading in step 9 indicate?

    _____

    _____

    _____

12. Move the test leads to measure the resistance between the sensor signal circuit and chassis ground.

    _____

13. If the resistance reading is below specifications, what would this indicate?

    _____

    _____

    _____

14. Turn the ignition switch to the RUN position, and then reconnect the battery. Measure the voltage on the signal circuit.

_____

15. If voltage is present, what is the next step?

_____

_____

_____

16. Move the test leads to measure the voltage on the sensor ground circuit between the load tool adaptor and chassis ground. Record your results.

_____

17. If voltage is present, what is the next step?

_____

_____

_____

18. If all tests pass, what is the likely problem?

_____

_____

_____

_____

Instructor's Response

_____

_____

_____

Name _____   Date _____

SIDE-IMPACT SENSOR REPLACEMENT

Upon completion of this job sheet, you should be able to properly replace the side-impact sensor.

NATEF Correlation

This job sheet addresses the following **MLR** task:

E.5.   Remove and reinstall door panel.

This job sheet addresses the following **AST** task:

G.4.   Describe operation of safety systems and related circuits (e.g., horn, air bags, seat belt pretensioners, occupancy classification, wipers, washers, speed control/collision avoidance, heads-up display, park assist, and back-up camera); determine needed repairs.

This job sheet addresses the following **MAST** task:

G.4.   Diagnose operation of safety systems and related circuits (e.g., horn, air bags, seat belt pretensioners, occupancy classification, wipers, washers, speed control/collision avoidance, heads-up display, park assist, and back-up camera); determine needed repairs.

ASE CERTIFIED NATEF

Tools and Materials

- Vehicle equipped with side-impact sensors
- Service information
- Scan tool
- Torque wrench
- Safety glasses
- Battery terminal puller

Describe the vehicle being worked on:

Year _____ Make _____ Model _____

VIN _____ Engine type and size _____

Procedure

Task Completed

1. Locate the fastener torque specifications in the service information and record.

   _____

2. Are there any safety warnings or cautions listed in the service information concerning the installation procedures of the sensor?   ☐ Yes ☐ No

3. If yes, describe.

   _____

   _____

   _____

4. Follow the service information procedures to remove the door panel.   ☐

5. Is there a requirement for proper orientation of the sensor?   ☐ Yes ☐ No

   If yes, describe how to determine proper orientation.

   _____

   _____

   _____

6. Is a gasket used between the sensor and the mounting surface area?   ☐ Yes ☐ No

7. Is the gasket in good condition?   ☐ Yes ☐ No

8. Install the sensor.

Instructor's Response

_____

_____

_____

_____

Name _____   Date _____

CLOCKSPRING REPLACEMENT AND CENTERING

Upon completion of this job sheet, you should be able to remove, replace, and properly center the clockspring used in an air bag system.

NATEF Correlation

This job sheet addresses the following **MLR** task:

E.4.   Disable and enable supplemental restraint system (SRS); verify indicator lamp operation.

This job sheet addresses the following **AST** tasks:

G.4.   Describe operation of safety systems and related circuits (e.g., horn, air bags, seat belt pretensioners, occupancy classification, wipers, washers, speed control/collision avoidance, heads-up display, park assist, and back-up camera); determine needed repairs.

G.5.   Describe body electronic systems circuits using a scan tool; check for module communication errors (data bus systems); determine needed action.

This job sheet addresses the following **MAST** tasks:

G.4.   Diagnose operation of safety systems and related circuits (e.g., horn, air bags, seat belt pretensioners, occupancy classification, wipers, washers, speed control/collision avoidance, heads-up display, park assist, and back-up camera); determine needed repairs.

G.5.   Diagnose body electronic systems circuits using a scan tool; check for module communication errors (data bus systems); determine needed action.

Tools and Materials

- A vehicle equipped with a driver-side air bag
- Service information for chosen vehicle
- Safety glasses
- Battery terminal puller
- Steering wheel puller

Describe the vehicle being worked on:

Year _____ Make _____ Model _____

VIN _____ Engine type and size _____

Procedure

Task Completed

1. According to the service information, in what position must the front wheels be before beginning?

   _____

   _____

2. Place the front wheels in the position described in step 1.   ☐

3. Disconnect the battery negative cable and isolate it.   ☐

4. According to the service information, how long must you wait before proceeding?

   _____

   _____

5. Remove the air bag module from the steering wheel.   ☐

6. Remove the steering wheel attaching bolt or nut. Can this fastener be reused?
    ☐ Yes ☐ No

7. Mark the shaft and steering wheel with index marks for reinstallation.   ☐

8. Use a steering wheel puller to remove the steering wheel from the shaft.   ☐

9. Remove the upper and lower steering column shrouds.   ☐

10. Disconnect the clockspring connector from the steering column harness.   ☐

11. Remove the retaining screws (or release the locking tabs) and the clockspring.
                                                                                                  ☐
12. If the same clockspring is to be reinstalled, describe the procedure for centering the
    clockspring.

    _____

    _____

    _____

13. In what position must the front wheel and steering column be to install the
    clockspring?

    _____

14. Install all components. What is the torque specification for the clockspring fasteners
    (if used)?

    _____

    What is the torque specification for the steering wheel fastener?

    _____

    What is the torque specification for the air bag module fasteners?

    _____

Instructor's Response

_____

_____

_____

Name _____   Date _____

OCCUPANT CLASSIFICATION SYSTEM VALIDATION PROCEDURE

Upon completion of this job sheet, you should be able to properly perform the OCS validation procedure.

NATEF Correlation

This job sheet addresses the following **AST** tasks:

G.4.   Describe operation of safety systems and related circuits (e.g., horn, air bags, seat belt pretensioners, occupancy classification, wipers, washers, speed control/collision avoidance, heads-up display, park assist, and back-up camera); determine needed repairs.

G.5.   Describe body electronic systems circuits using a scan tool; check for module communication errors (data bus systems); determine needed action.

This job sheet addresses the following **MAST** tasks:

G.4.   Diagnose operation of safety systems and related circuits (e.g., horn, air bags, seat belt pretensioners, occupancy classification, wipers, washers, speed control/collision avoidance, heads-up display, park assist, and back-up camera); determine needed repairs.

G.5.   Diagnose body electronic systems circuits using a scan tool; check for module communication errors (data bus systems); determine needed action.

ASE NATEF

Tools and Materials

- Vehicle with OCS
- Scan tool
- Validation weight set
- Battery charger

Describe the vehicle being worked on:

Year _____ Make _____ Model _____

VIN _____ Engine type and size _____

Procedure   **Task Completed**

1. Connect a battery charger to the vehicle's battery and set so 13 volts is read across the terminals.   ☐

2. Is the system on the assigned vehicle a strain gauge or bladder system?

   ☐ Strain gauge   ☐ Bladder system

3. Connect the scan tool to the DLC and access the OCM module. Provide the following information (if available):

   OCM part number: _____

   Software version: _____

4. Make sure the passenger front seat is empty. What is the state of the PADL?

   _____

5. Sit in the seat and describe the state of the PADL.

   _____

6. Use the scan tool to access the sensor display function. Record the values while remaining in the seat.

   _____

   _____

7. Move out of the seat and record the values.

_____

_____

8. Assure that the seat is empty and navigate the scan tool to the validation function. What instructions are provided on the scan tool screen when the test is started (if any)?

_____

_____

_____

_____

9. According to the scan tool, what is the first step that needs to be performed?

_____

_____

10. What is the first weight amount to be placed on the seat?

_____

11. During the validation procedure, what is the state of the PADL?

_____

_____

12. Complete the validation procedure while listing the required weight amounts.

_____

_____

_____

13. How do you know that the procedure was completed?

_____

14. What is the state of the PADL once the procedure is completed?

_____

15. Clear any DTCs that may have been set.   ☐

16. Disconnect a sensor and record any DTCs.

_____

_____

17. Reconnect the sensor. Did the fault go from active to stored?

_____

18. What needs to be done to clear the DTC?

_____

19. Perform the required task list in step 18.   ☐

Instructor's Response

_____

_____

_____

Name _____   Date _____

ANALYZING THE PARK ASSIST SYSTEM

Upon completion of this job sheet, you will be able to verify operation of the ultrasonic sensor.

NATEF Correlation

This job sheet addresses the following **MLR** tasks:

A.3.   Use wiring diagrams to trace electrical/electronic circuits.

A.4.   Demonstrate proper use of a digital multimeter (DMM) when measuring source voltage, voltage drop (including grounds), current flow, and resistance.

A.5.   Demonstrate knowledge of the causes and effects from shorts, grounds, opens, and resistance problems in electrical/electronic circuits.

A.11.   Identify electrical/electronic system components and configuration.

This job sheet addresses the following **AST** tasks:

G.4.   Describe operation of safety systems and related circuits (e.g., horn, air bags, seat belt pretensioners, occupancy classification, wipers, washers, speed control/collision avoidance, heads-up display, park assist, and back-up camera); determine needed repairs.

G.5.   Describe body electronic systems circuits using a scan tool; check for module communication errors (data bus systems); determine needed action.

This job sheet addresses the following **AST/MAST** tasks:

A.3.   Demonstrate proper use of a digital multimeter (DMM) when measuring source voltage, voltage drop (including grounds), current flow, and resistance.

A.4.   Demonstrate knowledge of the causes and effects from shorts, grounds, opens, and resistance problems in electrical/electronic circuits.

A.7.   Use wiring diagrams during the diagnosis (troubleshooting) of electrical/electronic circuit problems.

This job sheet addresses the following **MAST** tasks:

G.4.   Diagnose operation of safety systems and related circuits (e.g., horn, air bags, seat belt pretensioners, occupancy classification, wipers, washers, speed control/collision avoidance, heads-up display, park assist, and back-up camera); determine needed repairs.

G.5.   Diagnose body electronic systems circuits using a scan tool; check for module communication errors (data bus systems); determine needed action.

Tools and Materials

- Scan tool
- DMM
- Wiring diagram
- Vehicle equipped with a park assist system

Describe the vehicle being worked on:

Year _____ Make _____ Model _____

VIN _____ Engine type and size _____

Procedure   **Task Completed**

1. Where are the ultrasonic transceiver sensors located on the assigned vehicle?

   _____

   _____

   _____

2. Identify the circuits to the sensors.

   _____

   _____

   _____

3. What supplies the voltage to the sensor?

   _____

4. Where is the ground for the sensor located?

   _____

5. Block the wheels and set the parking brake.   ☐

6. Disconnect a sensor harness connector, and use a DMM to measure the voltage on
   the supply circuit with the ignition switch in the RUN position (engine off) and the
   transmission in reverse. What is your reading?

   _____

7. Reconnect the sensor, backprobe the connector for the voltage supply circuit, and use
   the voltmeter to measure voltage with the ignition switch in the RUN position (engine
   off) and the transmission in reverse. What is your reading?

   _____

8. If the readings between steps 6 and 7 are different, explain why.

   _____

   _____

   _____

9. Describe how you will test the ground circuit.

   _____

   _____

   _____

10. Test the ground circuit using the method described in step 9, and record your results.

    _____

11. With the sensor harness disconnected, use the scan tool to retrieve fault codes.   ☐

12. Was the module capable of determining the sensor circuit open?   ☐ Yes   ☐ No
    If so, explain how this can be done.

    _____

    _____

    _____

13. What are the observable symptoms with the park assist system with the sensor disconnected?

_____

_____

_____

Instructor's Response

_____

_____

_____

_____

Name _____   Date _____

PERFORMING STEERING ANGLE SETTING PROCEDURE (PARK GUIDANCE SYSTEM)

Upon completion of this job sheet, you should be able to properly perform the steering angle setting procedure on a vehicle equipped with the park assist guidance system.

NATEF Correlation

This job sheet addresses the following **AST** tasks:

G.4.   Describe operation of safety systems and related circuits (e.g., horn, air bags, seat belt pretensioners, occupancy classification, wipers, washers, speed control/collision avoidance, heads-up display, park assist, and back-up camera); determine needed repairs.

G.5.   Describe body electronic systems circuits using a scan tool; check for module communication errors (data bus systems); determine needed action.

This job sheet addresses the following **MAST** tasks:

G.4.   Diagnose operation of safety systems and related circuits (e.g., horn, air bags, seat belt pretensioners, occupancy classification, wipers, washers, speed control/collision avoidance, heads-up display, park assist, and back-up camera); determine needed repairs.

G.5.   Diagnose body electronic systems circuits using a scan tool; check for module communication errors (data bus systems); determine needed action.

Tools and Materials

- Vehicle equipped with park assist guidance
- Scan tool
- Service information

Describe the vehicle being worked on:

Year _____ Make _____ Model _____

VIN _____ Engine type and size _____

Procedure

For this job sheet task, you will perform the steering angle sensor initialization, center memorization, and maximum angle memorization routines.

1. Using the service information system, locate the steering angle sensor initialization procedure. List the steps below:

   _____

   _____

   _____

2. How do you confirm that the steering angle sensor has been properly initialized?

   _____

   _____

   _____

3. Use the scan tool to access the "Steering angle setting" function.   **Task Completed**

4. Center the steering wheel. What is the maximum tolerance the steering wheel can be off-center?

_____

5. Select "Steering center memorized."   ☐

6. Turn the steering wheel to the left lock position and then to the right lock position and select "Max steering memorize."   ☐

7. How is proper steering angle adjustment confirmed?

_____

_____

_____

Instructor's Response

_____

_____

_____

Name _____   **Date** _____

PARK ASSIST GUIDANCE SYSTEM OPERATION CHECK

Upon completion of this job sheet, you should be able to properly perform an operation check of the park assist guidance system and recognize fault detection characteristics.

NATEF Correlation ————————————————————————————

This job sheet addresses the following **AST** tasks:

G.4.   Describe operation of safety systems and related circuits (e.g., horn, air bags, seat belt pretensioners, occupancy classification, wipers, washers, speed control/collision avoidance, heads-up display, park assist, and back-up camera); determine needed repairs.

G.5.   Describe body electronic systems circuits using a scan tool; check for module communication errors (data bus systems); determine needed action.

This job sheet addresses the following **MAST** tasks:

G.4.   Diagnose operation of safety systems and related circuits (e.g., horn, air bags, seat belt pretensioners, occupancy classification, wipers, washers, speed control/collision avoidance, heads-up display, park assist, and back-up camera); determine needed repairs.

G.5.   Diagnose body electronic systems circuits using a scan tool; check for module communication errors (data bus systems); determine needed action.

NATEF

Tools and Materials

- Vehicle equipped with park assist guidance.
- Scan tool
- 60-mm diameter bar
- Tape measure
- Service information

Describe the vehicle being worked on:

Year _____ Make _____ Model _____

VIN _____ Engine type and size _____

Procedure

For this job sheet task, you will perform an operational check of the park assist guidance system and observe system fault detection.

1. Unplug the right front corner sensor.   **Task Completed**

2. Attempt to activate the system to perform a parking maneuver. What methods are used to indicate a fault in the system?

_____

_____

_____

3. Record any DTCs that were set.

_____

_____

_____

4. Reconnect the sensor and clear any DTCs. Confirm the system is operating properly.   ☐

5. Cover a rear corner sensor.   ☐

6. Attempt to activate the system to perform a parking maneuver. What methods are used to indicate a fault in the system?

_____

_____

_____

7. Record any DTCs that were set.

_____

_____

_____

8. Uncover the sensor and clear any DTCs. Confirm that the system is operating properly. ☐

9. Place the ignition in the RUN position and make sure that the guidance system is activated.

10. Record the gear selection required to measure the detection distance for each sensor:

Left front: _____

Right front: _____

Center front: _____

Left rear: _____

Right rear: _____

Center rear: _____

11. Move a 2.36-inch (60-mm) diameter pole near each corner sensor to measure its detection range. Record your results.

Left front: _____

Right front: _____

Left rear: _____

Right rear: _____

12. Move the pole to determine the sensor span for each of the corner sensors. Record your results.

Left front: _____

Right front: _____

Left rear: _____

Right rear: _____

13. Are the corner sensor detection ranges within specifications?   ☐ Yes ☐ No

Instructor's Response

_____

_____

_____

DIAGNOSTIC CHART 15-1

PROBLEM AREA:	Air bag system operation.
SYMPTOMS:	Air bag warning lamp illuminated.
POSSIBLE CAUSES:	**1.** Squib circuit shorted to ground. **2.** Defective clockspring. **3.** Squib circuit open. **4.** Sensor circuit open. **5.** Sensor circuit shorted. **6.** Faulty module. **7.** Poor battery feed circuit to control module. **8.** Poor ignition feed circuit to control module. **9.** Poor control module ground circuit. **10.** Loss of bus communications. **11.** Faulty instrument cluster lamp circuit.

DIAGNOSTIC CHART 15-2

PROBLEM AREA:	Loss of communication between MEMS sensor and the ACM.
SYMPTOMS:	Air bag lamp illuminated.
POSSIBLE CAUSES:	**1.** Sensor signal circuit open. **2.** Sensor ground circuit open. **3.** Sensor signal circuit shorted to sensor ground circuit. **4.** Sensor signal circuit shorted to ground. **5.** Sensor signal circuit shorted to voltage. **6.** Sensor ground circuit shorted to voltage. **7.** Faulty impact sensor. **8.** Faulty ACM.

DIAGNOSTIC CHART 15-3

PROBLEM AREA:	Side curtain squib circuit open.
SYMPTOMS:	Air bag lamp illuminated.
POSSIBLE CAUSES:	**1.** Squib line 1 circuit open. **2.** Squib line 2 circuit open. **3.** Failure of side curtain air bag squib. **4.** Faulty control module.

DIAGNOSTIC CHART 15-4

PROBLEM AREA:	Squib circuit shorted to voltage.
SYMPTOMS:	Air bag lamp illuminated.
POSSIBLE CAUSES:	**1.** Squib line 1 circuit shorted to voltage. **2.** Squib line 2 circuit shorted to voltage. **3.** Failure of side curtain air bag squib. **4.** Faulty control module.

DIAGNOSTIC CHART 15-5

PROBLEM AREA:	Squib circuit shorted to ground.
SYMPTOMS:	Air bag lamp illuminated.
POSSIBLE CAUSES:	**1.** Squib line 1 circuit shorted to ground. **2.** Squib line 2 circuit shorted to ground. **3.** Failure of side curtain air bag squib. **4.** Faulty control module.

DIAGNOSTIC CHART 15-6

PROBLEM AREA:	Squib circuits shorted together.
SYMPTOMS:	Air bag lamp illuminated.
POSSIBLE CAUSES:	**1.** Squib line 1 circuit shorted to squib line 2. **2.** Failure of side curtain air bag squib. **3.** Faulty control module.

DIAGNOSTIC CHART 15-7

PROBLEM AREA:	Ultrasonic sensor circuit failure.
SYMPTOMS:	Disabling of system.
POSSIBLE CAUSES:	**1.** Wiring harness, terminal, connector damage. **2.** High resistance in sensor voltage supply circuit. **3.** High resistance in sensor ground circuit. **4.** Sensor signal circuit shorted to ground. **5.** Signal circuit shorted to sensor ground circuit. **6.** Sensor supply circuit shorted to sensor ground circuit. **7.** Sensor supply circuit shorted to ground. **8.** Sensor supply circuit shorted to voltage. **9.** Faulty sensor. **10.** Internal controller fault.

DIAGNOSTIC CHART 15-8

PROBLEM AREA:	Ultrasonic module.
SYMPTOMS:	Disabling of system.
POSSIBLE CAUSES:	**1.** Wiring harness, terminal, connector damage. **2.** High resistance in module voltage supply circuit. **3.** High resistance in module ground circuit. **4.** Charging system failure. **5.** Internal controller fault.

DIAGNOSTIC CHART 15-9

PROBLEM AREA:	Infrared sensor circuit failure.
SYMPTOMS:	Disabling of system.
POSSIBLE CAUSES:	**1.** Wiring harness, terminal, connector damage. **2.** Damaged lens. **3.** Obstruction in front of sensor. **4.** Improper sensor alignment. **5.** High resistance in sensor voltage supply circuit. **6.** Open in sensor voltage supply circuit. **7.** High resistance in sensor ground circuit. **8.** Open in sensor ground circuit. **9.** Sensor supply circuit shorted to sensor ground circuit. **10.** Sensor supply circuit shorted to ground. **11.** Faulty sensor. **12.** Internal controller fault.

DIAGNOSTIC CHART 15-10

PROBLEM AREA:	Camera system.
SYMPTOMS:	Disabling of system. Lack of image display.
POSSIBLE CAUSES:	**1.** Wiring harness, terminal, connector damage. **2.** Damaged lens. **3.** Obstruction in front of camera. **4.** Improper camera calibration/alignment. **5.** High resistance in voltage supply circuit. **6.** Open in voltage supply circuit. **7.** High resistance in ground circuit. **8.** Open in ground circuit. **9.** Damaged camera video wiring. **10.** Faulty camera.

DIAGNOSTIC CHART 15-11

PROBLEM AREA:	Park assist.
SYMPTOMS:	When reverse (R) has been selected, an image of the area behind the vehicle is not displayed.
POSSIBLE CAUSES:	**1.** Parking assist ECU. **2.** Power source circuit. **3.** Camera feed circuit. **4.** Reverse signal circuit.

DIAGNOSTIC CHART 15-12

PROBLEM AREA:	Park assist.
SYMPTOMS:	When the target parking position is set, the initial display position of the target parking space is very far from the actual parking space.
POSSIBLE CAUSES	**1.** Ultrasonic sensor. **2.** Front bumper cover. **3.** Parking assist ECU.

DIAGNOSTIC CHART 15-13

PROBLEM AREA:	Park assist.
SYMPTOMS:	The vehicle width extension lines do not overlap with the predicted path lines when the steering wheel is centered.
POSSIBLE CAUSES:	**1.** Steering angle setting. **2.** Improper mounting of the rear camera assembly. **3.** Height control sensor/vehicle height difference setting. **4.** Rear camera optical axis adjustment. **5.** Parking assist ECU.

DIAGNOSTIC CHART 15-14

PROBLEM AREA:	Park assist.
SYMPTOMS:	"System initializing" is displayed.
POSSIBLE CAUSES:	**1.** Steering angle setting. **2.** Spiral cable with sensor subassembly. **3.** Parking assist ECU.

DIAGNOSTIC CHART 15-15

PROBLEM AREA:	Park assist.
SYMPTOMS:	Assist operation stops frequently.
POSSIBLE CAUSES:	**1.** Inspect the tires (tire size, tire pressure, and tire wear). **2.** Rear television camera optical axis adjustment. **3.** Parking assist ECU.

Upon completion and review of this chapter, you should be able to:

- Demonstrate proper safety precautions associated with servicing the hybrid electric vehicle.
- Properly remove and install the high-voltage service plug.
- Access and interpret DTCs, information codes, freeze frame data, and history data.

- Determine the cause of HV battery system failures.
- Determine failures of the inverter/converter assembly.
- Determine the cause of system main relay failures.
- Replace the system main relays.

Terms To Know

Battery ECU	High-voltage electronic control unit (HV ECU)	Information codes
Converter		Inverter
Freeze frame data	High-voltage service plug	System main relay (SMR)
	History data	

INTRODUCTION

This chapter discusses some of the service procedures for a common hybrid system. At the present time, specially trained dealership technicians perform most of the service of the hybrid system. Because of this, the main focus of this chapter will be on safety concerns associated with the high-voltage system.

The hybrid electric vehicle (HEV) system combines the operating characteristics of an internal combustion engine and an electric motor. In addition, the system can use regenerative braking to recover energy that normally would be lost to heat and use it to supplement the power of the engine. The sample HEV used to describe the service procedures in this chapter includes the following components (**Figure 16-1**):

- Hybrid transaxle that integrates the MG1, the MG2, and the planetary gear unit.
- Inverter assembly.
- HV ECU.
- ECM.
- High-voltage (HV) battery.
- Battery ECU.
- Service plug.
- The system main relay (SMR).
- Auxiliary battery.

Classroom Manual
Chapter 16, page 512

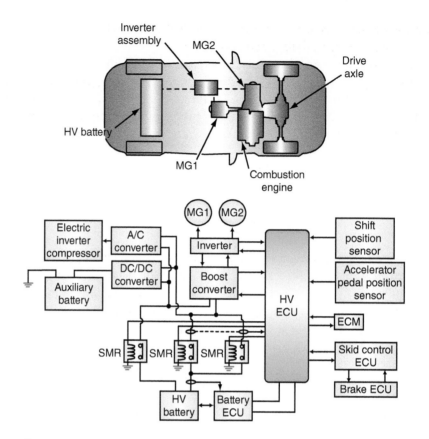

Figure 16-1 Components of an HEV system.

SAFETY PRECAUTIONS

Since the hybrid system can use voltages in excess of 500 volts (both DC and AC), it is vital that the service technician be familiar with, and follow, all safety precautions. Failure to perform the correct procedures can result in electrical shock, battery leakage, or an explosion. The following are the safety procedures that must be followed whenever servicing the HEV's high-voltage systems:

- Remove the key from the ignition.
- Disconnect the negative (−) terminal of the auxiliary (12-volt) battery. Always disconnect the auxiliary battery prior to removing the high-voltage service plug.
- Remove the high-voltage service plug and put it where it cannot be accidentally reinstalled by someone else.
- Cover the high-voltage service plug receptacle with insulation tape.
- Do not attempt to test or service the system for 5 minutes after the high-voltage service plug is removed. At least 5 minutes is required to discharge the high-voltage condenser inside the inverter.
- Test the integrity of the insulating gloves prior to use.
- Wear high-voltage insulating gloves when disconnecting the service plug.
- Never cut the orange high-voltage power cables. The wire harnesses, terminals, and connectors of the high-voltage system are identified by orange. In addition, high-voltage components may have a "High Voltage" caution label attached to them.
- Cover the terminals of a disconnected connector with insulation tape.
- Never open high-voltage components.
- Use a digital multimeter (DMM) to confirm that high-voltage circuits have 0 volt before performing any service procedure.
- Use insulated tools when available.

- Do not wear metallic objects that may cause electrical shorts.
- Follow the service manual diagnostic procedures.
- Wear protective safety goggles when inspecting the HV battery.
- Before touching any of the high-voltage system wires or components, wear insulating gloves, make sure the power switch is off, and disconnect the auxiliary battery.
- Turn the power switch to the OFF position prior to performing a resistance check.
- Turn the power switch to the OFF position prior to disconnecting or reconnecting any connectors or components.
- Isolate with insulation tape any high-voltage wires that have been removed.
- Properly torque the high-voltage terminals.

Insulating Glove Integrity Test

The insulating gloves that the technician wears for protection while servicing the high-voltage system must be tested for integrity before use. If there is a leak in the gloves, high-voltage electricity can travel through the hole to the technician's body. To test a glove, blow air into it and then fold it at the base to seal the air inside. Slowly roll the base of the glove toward the fingers. If the glove holds pressure, its insulating properties are intact. If any leaks are detected, discard the glove.

HIGH-VOLTAGE SERVICE PLUG

The HEV is equipped with a **high-voltage service plug** that disconnects the HV battery from the system. Usually, this plug is located near the battery (**Figure 16-2**). Prior to disconnecting the high-voltage service plug, the vehicle must be turned off and the negative terminal of the auxiliary battery must be disconnected. Once the high-voltage service plug is removed, the high-voltage circuit is shut off at the intermediate position of the HV battery.

The high-voltage service plug assembly contains a safety interlock reed switch. The reed switch is opened when the clip on the high-voltage service plug is lifted. The open reed switch turns off power to the SMR. The main fuse for the high-voltage circuit is inside the high-voltage service plug assembly.

However, never assume that the high-voltage circuits are off. The removal of the high-voltage service plug does not disable the individual HV batteries. Use a DMM to verify that 0 volt is in the system before beginning service. When testing the circuit for voltage, set the voltmeter to the 400 VDC scale.

Special Tools

DMM capable of reading 400 VDC
Insulating gloves
Insulating tape

Caution

Once the high-voltage service plug is removed, do not operate the power switch. Doing so may damage the hybrid vehicle control ECU.

Figure 16-2 The high-voltage service plug is usually located near the HV battery.

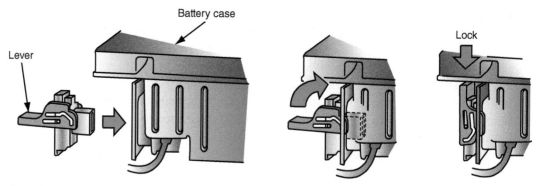

Figure 16-3 To install the service plug, make sure the lever is down and then fully locked once installed.

After the high-voltage service plug is removed, a minimum of 5 minutes must pass before beginning service on the system. This is required to discharge the high voltage from the condenser in the inverter circuit.

To install the high-voltage service plug, make sure the lever is locked in the DOWN position (**Figure 16-3**). Slide the plug into the receptacle, and lock it in place by lifting the lever upward. Once it is locked in place, it closes the reed switch.

> ⚙️ **SERVICE TIP** DTCs will be erased once the batteries are disconnected. Prior to disconnecting the system, be sure to check and record DTCs.

SELF-DIAGNOSIS CAPABILITIES

Classroom Manual
Chapter 16, page 522

Special Tool
Scan tool

The DTCs associated with the high-voltage system may include both SAE codes and manufacturer codes. SAE codes must be set as prescribed by the SAE, while manufacturer codes can be set by a manufacturer.

The **high-voltage electronic control unit (HV ECU)** controls the two motor generators (MG1 and MG2) and the engine based on torque demand. Also, these units are controlled based on regenerative brake control and the HV battery's state of charge (SOC). It is the responsibility of the HV ECU to provide reliable circuit shutdown in the event of a malfunction. The HV ECU uses three relays housed in the SMR assembly to connect and disconnect the high-voltage circuit. If a malfunction is detected, the HV ECU will use the relays to control the system based on programmed instructions stored in its memory. If the system is determined to be malfunctioning, the HV ECU will illuminate the master warning lamp in the instrument cluster. In addition, it may illuminate the HV system warning, the HV battery warning, or the discharge warning lamps.

Diagnostic trouble codes (DTCs) are set when the fault occurs. To access the DTCs, a scan tool with the proper interface module (if needed) is connected to the data link connector (DLC). The scan tool will also provide information codes, freeze frame data, and history data.

Information codes are additional codes that provide more information and freeze frame data concerning the DTCs. Information codes along with the DTC indicate more precisely the location of the fault. These codes are accessed using the scan tool while in the HV ECU system screen (**Figure 16-4**).

The **freeze frame data** is a recording of the driving condition when the malfunction occurred. This is useful for determining how the vehicle was operating at the time and for locating any input or output values that are out of range.

History data information can be useful for determining if a customer's concern is actually a problem with the system. It provides a means of determining if the vehicle owner's driving habits may be the cause of the problem. The data will display information such as if the gear shift lever was moved before the vehicle was ready, if the transmission was shifted into park while the vehicle was moving, if the accelerator pedal was depressed while in the NEUTRAL position, and so on. Use the proper service manual for information on using this data.

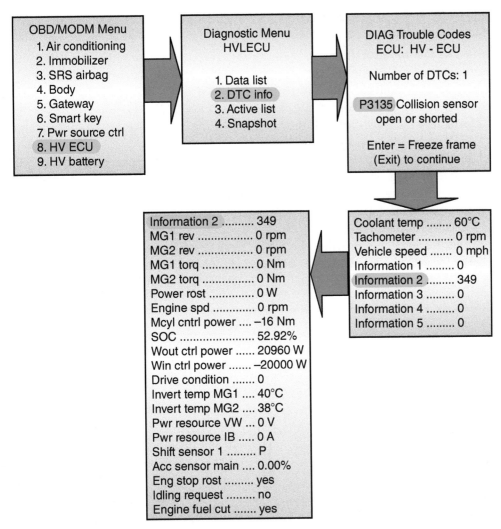

Figure 16-4 Accessing information codes.

In addition, the scan tool may provide activation tests of the HV ECU. Different tests are available. Using the service manual, determine what test is required for the specific test.

The following are typical mode functions:

- Mode 1. This mode runs the engine continuously in the PARK position and to check HV ECU operation. It can also be used to disable traction control so the speedometer test can be performed.
- Mode 2. Cancels the traction control that is affected when the rotational difference between the front and rear wheels is excessive.
- Inverter stop. Keeps the inverter power transistor on to determine if there is an internal leak in the inverter or the HV control ECU.
- Cranking request. Activates the motor generator continuously to crank the engine in order to measure the compression.

CUSTOMER CARE Since the hybrid vehicle is still new on the market, the chances are the vehicle owner has never driven one before. Take the time to explain the proper operation of the vehicle to prevent any misunderstandings of the system function.

HIGH-VOLTAGE BATTERY SERVICE

Classroom Manual
Chapter 16, page 512

A **battery ECU** monitors the condition of the HV battery assembly. The battery ECU determines the SOC of the HV battery by monitoring voltage, current, and temperature. The battery ECU collects data and transmits it to the HV ECU to be used for proper charge and discharge control.

The battery ECU also controls the operation of the battery blower motor to maintain proper HV battery temperature.

Special Tools

DMM
Scan tool

The HV battery stores power generated by MG1 and recovered by MG2 during regenerative braking (**Figure 16-5**). The HV battery must also supply power to the electric motor when the vehicle is first started from a stop or when additional power is needed. A typical HV battery uses several nickel-metal-hydride modules and can provide over 270 volts (**Figure 16-6**).

When the vehicle is moving, the HV battery is subjected to repetitive charge and discharge cycles. The HV battery is discharged by MG2 during acceleration mode and then is recharged by regenerative braking. An amperage sensor (**Figure 16-7**) is used so

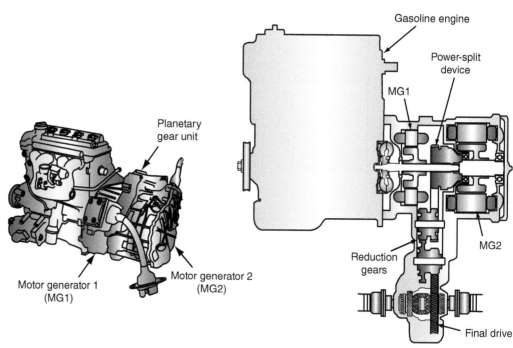

Figure 16-5 Layout of the HEV transaxle.

Figure 16-6 The HV battery modules.

Figure 16-7 The amperage sensor.

the battery ECU can transmit requests to the HV ECU to maintain the SOC of the HV battery. The battery ECU attempts to keep the SOS at 60%. The battery ECU also monitors delta SOC to determine if it is capable of maintaining acceptable levels of charge. The normal, low-to-high SOC delta is 20%.

If the battery ECU sends abnormal messages to the HV ECU, the HV ECU illuminates the warning light and enters fail-safe control. DTCs and informational codes are set along with freeze frame data. Fail-safe control can result in the battery ECU restricting or stopping the charging and discharging of the HV battery.

If there is a leak in the high-voltage system insulation that may seriously harm a person, the system will enter fail-safe control and set DTCs. This will occur if the battery ECU determines the insulation resistance of the power cable to be 100 kΩ, or less.

Whenever an HV battery malfunction occurs, use the scan tool to view the "HV Battery Data List." This provides all HV battery system information.

> ⚙ **SERVICE TIP** If inspection and testing fail to locate the leak, then it is possible that water has entered into the battery assembly or into the converter/inverter assembly.

High-Voltage Battery Charging

If the SOC of the HV battery is too low to allow the engine to run, the HV battery will need to be recharged. On some vehicles it requires the use of a special HV battery charger (**Figure 16-8**). In addition, some manufacturers will allow only specially trained people to recharge the battery. Some manufacturers will not even supply the charging equipment to the dealer; a representative of the company performs the task of recharging the HV battery.

HV battery recharging must be performed outside. The correct cable is connected between the vehicle and the charger (**Figure 16-9**). When using the charger, the immediate area must be secured and marked with warning tape. It requires about 3 hours to recharge the battery to an SOC of about 50%.

> ⚡ **WARNING** Do not attempt to recharge the HV battery with a standard 12-volt battery charger. HV battery chargers are now available from aftermarket suppliers that will recharge most HV batteries.

The best method for recharging the HV battery is to allow the HV system to replenish the battery. This is done by simply driving the vehicle while providing several opportunities

MG1 functions as the control element for the planetary gear set. It recharges the HV battery and supplies electrical power to drive MG2. MG1 also functions as the starter for the engine. MG2 is used for power at low speeds and for supplemental power when needed at higher speeds.

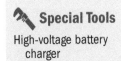
Special Tools
High-voltage battery charger
Warning tape

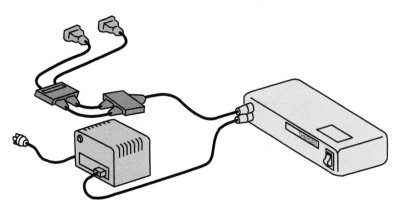

Figure 16-8 HV battery charger.

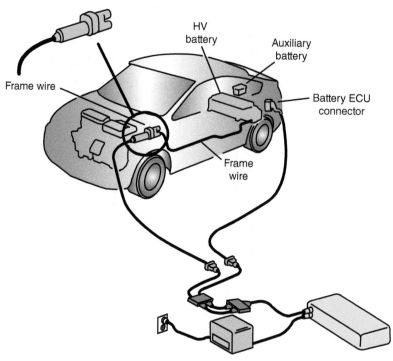

Figure 16-9 HV battery charger connection.

for regenerative braking. Monitor the SOC of the HV battery while driving the vehicle to assure that the SOC is increasing after each regenerative braking event.

Inverter/Converter Assembly

The **inverter** (**Figure 16-10**) controls the current flow between MG1, MG2, and the HV battery. The inverter converts HV battery DC voltage into three-phase alternating current (AC) for MG1 and MG2. It also converts (rectifies) high-voltage AC from MG1 and MG2 to DC voltage to charge the HV battery. The HV ECU controls the activation of the power transistors to perform these functions (**Figure 16-11**).

The **converter** is a DC/DC transformer (**Figure 16-12**). It converts the voltage from 270 VDC to 14 VDC to recharge the auxiliary battery and to power 12-volt electrical

Special Tool

Scan tool

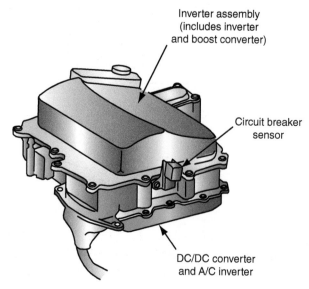

Figure 16-10 The inverter assembly.

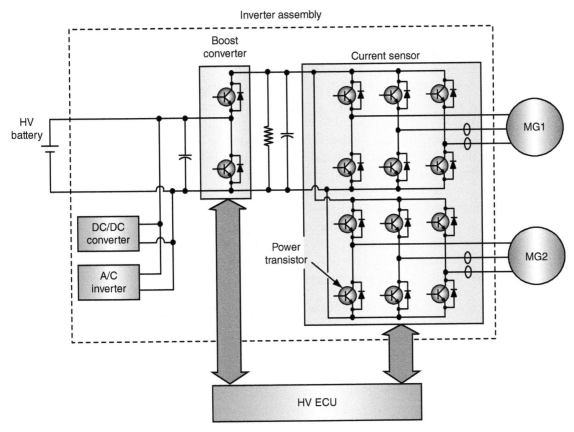

Figure 16-11 Inverter assembly internal electrical circuit.

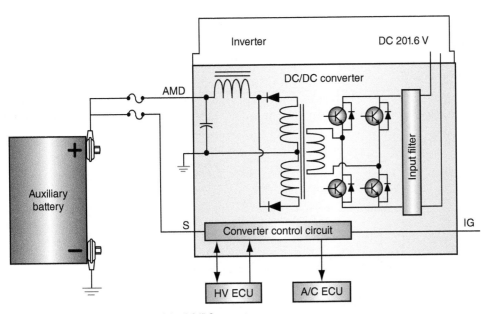

Figure 16-12 The internal circuit of the DC/DC converter.

components. If the DC/DC converter malfunctions, the auxiliary battery voltage will drop until it is no longer possible to drive the vehicle. The HV ECU monitors operation of the DC/DC converter, illuminates the warning lamp, and sets a DTC if a failure is determined.

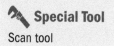

Special Tool

Scan tool

SYSTEM MAIN RELAY

The HV ECU controls the operation of the **system main relay (SMR)** to connect and disconnect the power source of the high-voltage system. Three relays are used. One is for the negative side, while the other two are for the positive side (**Figure 16-13**).

Initially SMR1 and SMR3 are turned on to energize the circuit for the HV battery. This provides the needed current for the motor generator to start the engine. Then SMR2 is energized and SMR1 is turned off. This makes the current from the generator flow through a resistor, thus controlling the amount of current flow. This protects the circuit from excessive initial current from the generator. Finally, SMR2 is turned on and SMR1 is turned off to allow free flow of current in the circuit.

During shutdown, SMR2 is turned off first, then SMR3. This provides a means for the HV ECU to verify that the relays have been properly turned off.

The HV ECU checks that the system main relay is operating normally. If a fault is detected, DTCs and information data will be stored. The following is a typical procedure to be used if diagnosis leads to replacement of one of the SMRs:

1. Secure the proper service manual, tested insulating gloves, and insulating tape.
2. Gain access to the auxiliary battery and disconnect the negative terminal.
3. Wearing the insulated gloves, remove the high-voltage service plug.
4. Use insulating tape to cover the terminals of the plug receptacle.
5. After gaining access to the battery carrier panel, remove the junction terminal. Be sure to wear your insulating gloves.
6. Remove the ground wire, SMR2 cover, and disconnect the relay connector to remove SMR2.
7. Remove SMR3 by disconnecting the main battery cable, disconnecting the connector, and removing the fasteners.
8. Disconnect the connector and the ground terminal for SMR1.
9. Remove the fasteners and SMR1.

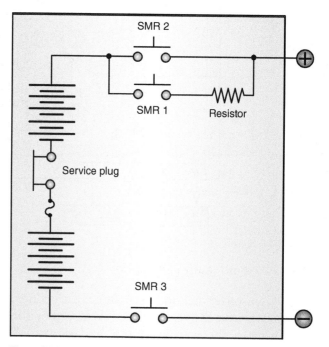

Figure 16-13 Schematic of the SMR.

CASE STUDY

An owner of a 2005 Prius has brought their vehicle into the claiming that the master warning light comes on every once in a while but goes out after a couple of starts. The technician retrieves the DTCs from the HV ECU and finds a "Shift before ready" code. Upon investigation of the information code, the technician also refers to history data. Here it is determined that the cause of the fault was that the customer was not waiting for the ready light to stop flashing before placing the transmission into drive. The warning light would go out after three starts if the same condition did not reoccur. The technician took the time to go over the proper startup sequence with the vehicle owner so future problems could be avoided.

ASE-STYLE REVIEW QUESTIONS

1. All of the following statements concerning hybrid high-voltage system safety is true **EXCEPT**:

   A. Disconnect the motor generators prior to turning the ignition off.

   B. Disconnect the negative (−) terminal of the auxiliary battery before removing the service plug.

   C. Do not attempt to test or service the system for 5 minutes after the high-voltage service plug is removed.

   D. Turn the power switch to the OFF position prior to performing a resistance check.

2. *Technician A* says HEV batteries can provide over 270 volts.

   Technician B says the HEV high voltage from the MG1 and MG2 to the inverter/converter can be more than 500 volts.

   Who is correct?

   A. A only   C. Both A and B

   B. B only   D. Neither A nor B

3. When working on the high-voltage system, which of the following should be done?

   A. Always place the high-voltage service plug where someone will not accidentally reinstall it.

   B. Before servicing, use a voltmeter set on 400 VDC to determine if the high-voltage system voltage is at 0 volt.

   C. Test the integrity of the insulating gloves prior to use.

   D. All of the above.

4. *Technician A* says the main system relay should be removed before disconnecting the service plug.

   Technician B says the high-voltage components are usually identified with a warning label.

   Who is correct?

   A. A only   C. Both A and B

   B. B only   D. Neither A nor B

5. To test the integrity of the insulating gloves:

   A. Fill the gloves with water to see if there is a leak.

   B. Fill the gloves with air and submerge in water to see if air bubbles arise from any leaks.

   C. Shine a flashlight into the glove and see if light escapes.

   D. None of the above.

6. The high-voltage service plug:

   A. Disconnects the inverter/converter from the motor generators.

   B. Disconnects the auxiliary battery from the HV battery.

   C. Disconnects the HV battery from the system.

   D. Provides a connection for the battery charger.

7. *Technician A* says once the service plug is disconnected, there is no high voltage in the vehicle systems.

   Technician B says prior to disconnecting the high-voltage service plug, the vehicle must be turned off and the negative terminal of the auxiliary battery must be disconnected.

   Who is correct?

   A. A only
   B. B only
   C. Both A and B
   D. Neither A nor B

8. *Technician A* says the HV ECU can shut down the high-voltage system if a fault is detected.

   Technician B says if the auxiliary battery voltage goes low, the HV ECU will direct regenerative braking energy to the auxiliary battery.

   Who is correct?

   A. A only
   B. B only
   C. Both A and B
   D. Neither A nor B

9. *Technician A* says the HV battery is charged with a conventional flooded battery charger set at 3.5 amps.

   Technician B says the HV battery can only be charged if it is removed from the vehicle.

   Who is correct?

   A. A only
   B. B only
   C. Both A and B
   D. Neither A nor B

10. *Technician A* says information codes provide more specific indications of the fault location.

    Technician B says the freeze frame data is a recording of the driving condition when the malfunction occurred.

    Who is correct?

    A. A only
    B. B only
    C. Both A and B
    D. Neither A nor B

ASE CHALLENGE QUESTIONS

1. *Technician A* says if an intermittent fault code is being set in the high-voltage system, water may be entering a connector.

   Technician B says to locate the cause of current leakage through the high-voltage cable insulation, spray the cable with water.

   Who is correct?

   A. A only
   B. B only
   C. Both A and B
   D. Neither A nor B

2. *Technician A* says if the vehicle fails to start, the system main relay may have a fault.

   Technician B says if the vehicle fails to start, the HV battery may be too low.

   Who is correct?

   A. A only
   B. B only
   C. Both A and B
   D. Neither A nor B

3. *Technician A* says a faulty reed switch in the service plug may cause the vehicle not to start.

   Technician B says a water cooler pump failure may result in a "HV battery too hot" fault code.

   Who is correct?

   A. A only
   B. B only
   C. Both A and B
   D. Neither A nor B

4. *Technician A* says the engine is started by the auxiliary battery if the HV SOC is below 15%.

   Technician B says if the engine fails to start, the inverter may have malfunctioned.

   Who is correct?

   A. A only
   B. B only
   C. Both A and B
   D. Neither A nor B

5. *Technician A* says if the vehicle operates slowly from a stop, the engine may require a tune-up.

   Technician B says the HEV system does not provide a method of performing a compression test on the engine.

   Who is correct?

   A. A only
   B. B only
   C. Both A and B
   D. Neither A nor B

Name _____   Date _____

HYBRID SAFETY

Upon completion of this job sheet, you should be familiar with the critical safety procedures involved in servicing a high-voltage hybrid system.

NATEF Correlation

This job sheet addresses the following **MLR/AST/MAST** task:

B.7.   Identify safety precautions for high-voltage systems on electric or hybrid electric, and diesel vehicles.

ASE CERTIFIED **NATEF**

Tools and Materials

- HEV
- Service manual
- Insulating gloves
- Eye protection

Describe the vehicle being worked on:

Year _____ Make _____ Model _____

VIN _____ Engine type and size _____

Procedure

Task Completed

1. Use the service manual information and determine the location on the vehicle for the high-voltage service plug.

   _____

   _____

2. What must be done prior to disconnecting the service plug?

   _____

   _____

3. How long must you wait after the plug is disconnected before servicing the system?

   _____

   _____

4. Access the 12-volt auxiliary battery and remove the negative terminal.   ☐

5. Test the insulating gloves for leaks. Are the gloves safe to use?   ☐ Yes ☐ No

   If No, inform your instructor.

6. Put on the insulating gloves and eye protection.   ☐

7. Remove the service plug. What device(s) are integrated into the service plug assembly?

   _____

   _____

8. Reinstall the service plug.   ☐

9. Review your observations with your instructor.   ☐

Instructor's Response

_____

_____

_____

Name _____   Date _____

HYBRID SYSTEM DTCS

Upon completion of this job sheet, you should be able to diagnose hybrid faults by retrieving DTCs, information codes, and data values.

NATEF Correlation

This job sheet addresses the following **AST** task:

G.5.   Describe body electronic systems circuits using a scan tool; check for module communication errors (data bus systems); determine needed action.

This job sheet addresses the following **MAST** task:

G.5.   Diagnose body electronic systems circuits using a scan tool; check for module communication errors (data bus systems); determine needed action.

ASE CERTIFIED   NATEF ®

Tools and Materials

- HEV
- Scan tool
- Service manual

Describe the vehicle being worked on:

Year _____ Make _____ Model _____

VIN _____ Engine type and size _____

Procedure

1. With the vehicle started, observe if any warning lamps are illuminated. If so, which ones?

   _____

   _____

2. Connect the scan tool and check for DTCs in the PCM and the HV ECU. Record any DTCs displayed.

   _____

   _____

3. Access and record any information codes that are displayed.

   _____

   _____

4. Use the service manual to determine what system component is affected. Record your results.

   _____

   _____

5. Access the "HV ECU Data List" screen and record any values that are out of range.

   _____

   _____

6. Based on your results, what would be the next logical approach to take?

_____

_____

_____

Instructor's Response

_____

_____

_____

_____

DIAGNOSTIC CHART 16-1

PROBLEM AREA:	High-voltage system.
SYMPTOMS:	Engine will not start. Warning lamp illumination.
POSSIBLE CAUSES:	**1.** Auxiliary battery low. **2.** HV battery low SOC. **3.** Faulty motor generator. **4.** Faulty HV ECU. **5.** Poor electrical connections in HV cable. **6.** Service plug not installed.

DIAGNOSTIC CHART 16-2

PROBLEM AREA:	HV battery.
SYMPTOMS:	Low SOC. Warning lamp illumination.
POSSIBLE CAUSES:	**1.** Poor electrical connections. **2.** Faulty motor generator or circuits. **3.** Faulty HV ECU or circuits. **4.** Faulty battery ECU or circuits. **5.** Faulty amperage sensor or circuits. **6.** Faulty HV battery.

1. The current draw of a window motor is being measured.
   Technician A says the ammeter can be connected on the power supply side of the motor.
   Technician B says the ammeter can be connected on the ground side of the motor.
   Who is correct?

   A. A only

   B. B only

   C. Both A and B

   D. Neither A nor B

2. The digital readout of an auto-ranging DVOM that is in the "volts" position displays "13.7." Which of the following statements is true about this measurement?

   A. The actual value is 13.7 volts.

   B. The actual value is 13.7 mV.

   C. The actual value is 137 mV.

   D. More information is needed in order to determine the actual value.

3. A voltmeter that is connected across the input and output terminals of an instrument cluster illumination lamp rheostat indicates 12.6 volts, with the switch in the minimum brightness position and the engine off. Which of the following statements is true?

   A. The voltage available at the lamps will be about 12.6 volts.

   B. The voltage available at the lamps will be 0.0 volt.

   C. The rheostat is operating correctly.

   D. More information is needed in order to determine whether the lamps are operating correctly.

4. A voltmeter indicates 0.45 volt. Which of the following represents this measurement?

   A. 0.450 mV.

   C. 4.5 mV.

   B. 450 mV.

   D. 0.045 volt.

5. The following information about a solenoid control signal has been gathered using a lab scope: Frequency is 10 hertz and pulse width is 5 ms.
   Technician A says this means that the solenoid is being turned on and off 10 times per second and that the length of time the injector is open during each "on" pulse is 5 ms.

   Technician B says this means that the solenoid is being turned on and off 10 times per second and that the length of time the solenoid is closed during each "off" cycle is 95 ms.
   Who is correct?

   A. A only

   B. B only

   C. Both A and B

   D. Neither A nor B

6. Lab scope testing is being discussed.
   Technician A says the preferred method to use when observing the output of a potentiometer is to select an external trigger.
   Technician B says that when analyzing the output of a low-voltage computer input sensor, a high trigger level should be selected.
   Who is correct?

   A. A only

   B. B only

   C. Both A and B

   D. Neither A nor B

7. The left-rear and right-rear taillights and the left-rear brake light of a vehicle turn on dimly whenever the brake pedal is depressed; however, the right-rear brake light operates at the correct brightness.
   Technician A says the left-rear taillight and brake light may have a poor ground connection.
   Technician B says the brake light switch may have excessive resistance.
   Who is correct?

   A. A only

   B. B only

   C. Both A and B

   D. Neither A nor B

8. The horn of a vehicle equipped with a horn relay sounds weak and distorted whenever it is applied. Which of the following is the **LEAST LIKELY** cause of this problem?

   A. High resistance in the relay load circuit.

   B. High resistance in the horn ground circuit.

   C. Excessive voltage drop between the relay load contact and the horn.

   D. A 12-volt drop across the relay coil winding.

9. The circuit breaker that protects an electric window circuit opens whenever an attempt is made to lower the window.
Technician A says the internal resistance of the motor is too high.
Technician B says the window regulator may be sticking.
Who is correct?

A. A only
B. B only
C. Both A and B
D. Neither A nor B

10. The relay coil resistance of an inoperative electric antenna circuit is 0.5 kΩ.
Technician A says this could prevent the proper operation of the antenna motor.
Technician B says this could result in the antenna rising but not retracting.
Who is correct?

A. A only
B. B only
C. Both A and B
D. Neither A nor B

NOTE: Question 11 refers to the horn circuit on page 863.

11. The horn circuit wiring diagram is being discussed.
Technician A says the wire that provides the ground path for the high-note horn is an 18-gauge black wire that is part of circuit Z1.
Technician B says the wire that provides power to the horn relay coil is a violet wire that is connected to terminal 87 of the relay.
Who is correct?

A. A only
B. B only
C. Both A and B
D. Neither A nor B

NOTE: Question 12 refers to the rear wiper/washer on page 864.

12. The wiring diagram is being discussed.
Technician A says the wire that activates the rear wiper motor module originates at terminal G of the rear window/wiper switch.
Technician B says there are only three wires connected to the accessory switch panel of this vehicle.
Who is correct?

A. A only
B. B only
C. Both A and B
D. Neither A nor B

13. Domestic and import wire sizes are being discussed.
Technician A says a 14-gauge wire is larger than a 16-gauge wire.

Technician B says a 0.8 mm² wire is smaller than a 1.0 mm² wire.
Who is correct?

A. A only
B. B only
C. Both A and B
D. Neither A nor B

14. What is the expected total current flow through a 12-volt circuit with two 12-ohm resistors connected in parallel?

A. 0.5 amp
B. 1 amp
C. 2 amps
D. 32 amps

15. A hydrometer that is being used to measure the specific gravity of a battery indicates 1.240.
Technician A says if the ambient temperature is 70°F, the corrected specific gravity reading will be 1.236.
Technician B says if the battery temperature is 60°F, the corrected specific gravity reading will be 1.232.
Who is correct?

A. A only
B. B only
C. Both A and B
D. Neither A nor B

16. The specific gravity of each of the cells of a battery is as follows: 1.220, 1.245, 1.190, 1.205, 1.210, and 1.215. Which of the following procedures should be performed?

A. Charge the battery and then retest the specific gravity.
B. Perform a battery capacity test.
C. Perform a 3-minute charge test.
D. Replace the battery.

17. A battery has an open circuit voltage of 12.1 volts.
Technician A says a battery capacity test should now be performed.
Technician B says the battery should now be charged.
Who is correct?

A. A only
B. B only
C. Both A and B
D. Neither A nor B.

18. Battery test series is being discussed.
Technician A says to perform the battery capacity test first.
Technician B says the state of charge test is always the last test performed.
Who is correct?

A. A only
B. B only
C. Both A and B
D. Neither A nor B

19. All of the following could cause a slow cranking condition **EXCEPT**:

    A. Overadvanced ignition timing.

    B. Shorted neutral safety switch.

    C. Misaligned starter mounting.

    D. Low-battery state of charge.

20. A vehicle owner states that occasionally he is unable to start his car: The engine starts cranking at normal speed and then all of a sudden he hears a "whee" sound and at that point the engine stops cranking. After four or five attempts, the engine finally starts.
    Technician A says the starter drive gear may be slipping.
    Technician B says there may be excessive voltage drop across the starter-solenoid contacts.
    Who is correct?

    A. A only   C. Both A and B

    B. B only   D. Neither A nor B

21. A vehicle with a no-crank condition is being tested. When the starter-solenoid battery and start terminals are connected with a jumper wire, the starter begins to crank the engine.
    Technician A says the starter solenoid may be faulty.
    Technician B says the ignition switch may be faulty.
    Who is correct?

    A. A only   C. Both A and B

    B. B only   D. Neither A nor B

22. A vehicle is being tested for a slow-crank condition. One lead of a voltmeter is connected to the positive battery post and the other lead is connected to the motor terminal of the starter-solenoid. With the ignition key on, the voltmeter indicates 12 volts; when the engine is cranked, the voltmeter indicates 0.2 volt.
    Technician A says the positive side of the starter-solenoid load circuit is OK.
    Technician B says the negative side of the starter circuit may have excessive voltage drop.
    Who is correct?

    A. A only   C. Both A and B

    B. B only   D. Neither A nor B

23. A technician performs an output test on an AC generator (alternator) rated at 100 amps with an internal regulator. During the test, it produces 30 amps. The cause of the low output may be:

    A. Shorted diode.

    B. Open sense circuit to the regulator.

    C. Worn brushes.

    D. Faulty capacitors.

24. A charging system voltage output test reveals the following information:

    1. Base voltage: 12.6 volts.
    2. At 1,500 rpm: 13.3 volts.
    3. At 2,000 rpm with loads on: 13.0 volts.

    Technician A says the alternator output voltage is lower than normal.
    Technician B says there may be excessive resistance on the insulated side of the charging circuit.
    Who is correct?

    A. A only   C. Both A and B

    B. B only   D. Neither A nor B

25. Charging system voltage and amperage on a vehicle equipped with an external voltage regulator is 12.1 volts and 0 amp at 1,500 rpm with all loads applied. When the alternator is full-fielded under the same conditions, the values rise to 13.8 volts and 95 amps. All of the following statements concerning these tests are correct **EXCEPT**:

    A. The alternator may have an open field circuit.

    B. The voltage regulator's voltage limiter contacts may have excessive resistance.

    C. The voltage limiter may have an open coil winding.

    D. The alternator is capable of charging correctly.

26. Alternator diode testing is being discussed.
    Technician A says an AC voltmeter can be used to check for faulty diodes.
    Technician B says an ammeter can be used to check for faulty diodes.
    Who is correct?

    A. A only   C. Both A and B

    B. B only   D. Neither A nor B

27. *Technician A* says that a short to ground in the PCI bus network circuit will result in a total bus network failure.
    Technician B says an open PCI bus circuit wire to the BCM will result in total bus network failure.
    Who is correct?

    A. A only   C. Both A and B

    B. B only   D. Neither A nor B

28. *Technician A* says that the normal at-rest voltage for the PCI bus is 0 volt.

Technician B says that the normal active voltage on the PCI bus is 12 volts.
Who is correct?

A. A only	C. Both A and B
B. B only	D. Neither A nor B

29. *Technician A* says that the location of an open CAN B bus circuit can be located by shorting the other circuit to ground while observing the module communications on the scan tool.
Technician B says if the fault code indicated that a CAN C circuit is shorted, this would cause a total failure of the CAN C bus.
Who is correct?

A. A only	C. Both A and B
B. B only	D. Neither A nor B

30. *Technician A* says the normal voltmeter reading on the CAN B(+) circuit with the ignition key in the RUN position is between 280 and 920 mV.
Technician B says normal CAN C bus termination is 60 ohms.
Who is correct?

A. A only	C. Both A and B
B. B only	D. Neither A nor B

31. How much total resistance is in a 12-volt circuit that is drawing 4 amps?

A. 0.333 ohm	C. 4.8 ohms
B. 3 ohms	D. 48 ohms

32. A replacement halogen bulb for a composite lamp was replaced but it lasted for a very short period of time; all of the other bulbs on the car are working fine. Which of the following could account for the premature failure of the bulb?

A. Excessive charging system voltage.

B. Fingerprints on the bulb.

C. Excessive voltage drop in the power feed circuit to the bulb.

D. Excessive resistance in the ground circuit of the bulb.

33. The turn signals of a vehicle are inoperative. The green indicator bulbs that are supposed to flash when the turn signal switch is moved to either the left- or the right-turn position do not turn on at all.
Technician A says the turn signal flasher contacts could have fused closed.

Technician B says the circuit from the turn signal flasher to the turn signal switch may be open.
Who is correct?

A. A only	C. Both A and B
B. B only	D. Neither A nor B

NOTE: Question 34 refers to the headlight switch on page 865.

34. Referring to the wiring diagram: the lighting circuit of this vehicle has the wire that is connected to the headlamp dimmer switch shorted to ground. Which of the following statements concerning this problem is true?

A. The 10-amp fuse will blow when the headlights are turned on.

B. The 15-amp fuse will blow immediately.

C. The 10-amp fuse will blow immediately.

D. The circuit breaker will open the circuit when the headlights are turned on.

NOTE: Question 35 refers to the multifunction switch on page 866.

35. Which of the following statements about the circuit is true?

A. The retract relay-up coil power feed comes from terminal 106 of the multifunction switch.

B. The power feed to the multifunction switch comes from the theft warning subcontrol amp.

C. The retract relay-up load circuit power feed comes from terminal 111 of the multifunction switch.

D. Terminal 106 of the multifunction switch is connected to ground.

36. *Technician A* says a voltmeter connected to the input wire of an IVR should indicate a fluctuating voltage.
Technician B says an open IVR on a single-gauge system could prevent the proper operation of the coolant temperature warning light.
Who is correct?

A. A only	C. Both A and B
B. B only	D. Neither A nor B

37. The fuel gauge of a vehicle equipped with an electromagnetic gauge cluster indicates a full tank, regardless of the actual amount of fuel in the tank. Which of the following could be the cause of this problem?

   A. Open IVR.

   B. Open fuel gauge sender unit.

   C. An open circuit between the gauge and the sending unit.

   D. A grounded circuit between the fuel gauge and the fuel gauge sender unit.

   NOTE: Question 38 refers to the charging system wiring diagram on page 867.

38. Referring to the wiring diagram: the charge indicator light of this vehicle does not come on when the ignition switch is in the RUN position and the engine is turned off. Which of the following statements concerning this problem is true?

   A. The 10-ohm resistor inside the voltage regulator may be open.

   B. The field relay contacts may be stuck closed.

   C. The rotor winding may be shorted.

   D. The 50-ohm resistor may be open.

   NOTE: Question 39 refers to the two-speed wiper circuit on page 868.

39. The low-speed position of the windshield wiper system is inoperative; the high-speed position is working fine.
   Technician A says circuit 58 may be open.
   Technician B says circuit 63 may be open.
   Who is correct?

   A. A only   C. Both A and B

   B. B only   D. Neither A nor B

   NOTE: Question 40 refers to the blower motor circuit on page 869.

40. The medium-high speed of the blower motor circuit is inoperative; the rest of the blower speeds are fine.
   Technician A says circuit 752 may be open.
   Technician B says the middle resistor in the blower motor resistor assembly may be open.
   Who is correct?

   A. A only   C. Both A and B

   B. B only   D. Neither A nor B

   NOTE: Question 41 refers to the power window circuit on page 869.

41. The power window motor in the wiring diagram is completely inoperative. With the master window switch placed in the DOWN position, the following voltages are measured at each terminal:

Terminal Number	Voltage
1	12
2	0
3	12
4	0
5	12
6	12
7	0
8	12
9	0
10	12
11	0

   Which of the following statements represents the cause of this problem?

   A. The master switch is faulty.

   B. The window switch is faulty.

   C. The motor is faulty.

   D. There is a poor ground in the circuit.

   NOTE: Question 42 refers to the body computer on page 870.

42. The "liftgate ajar" lamp is remaining on even after the liftgate is closed. Voltage at BCM terminal J2-4 is 0.0 volt.
   Technician A says that the BCM may be faulty.
   Technician B says that the liftgate switch may be stuck closed.
   Who is correct?

   A. A only   C. Both A and B

   B. B only   D. Neither A nor B

43. A Chrysler multiplex system is being discussed.
   Technician A says a voltmeter connected to the "Bus +" of any multiplexed component should indicate 12 volts with the ignition switch in the ON position.

Technician B says a 12-volt test light can be used to test for power at various multiplexed components.
Who is correct?

A. A only

C. Both A and B

B. B only

D. Neither A nor B

44. Sweep-testing a potentiometer with a voltmeter and an ohmmeter is being discussed.
Referring to a voltmeter, *Technician A* says that the positive lead should be connected to the feedback terminal and that the negative lead should be connected to the ground terminal.
Referring to an ohmmeter, *Technician B* says that the positive lead should be connected to the feedback terminal and that the negative lead should be connected to the reference terminal.
Who is correct?

A. A only

C. Both A and B

B. B only

D. Neither A nor B

NOTE: Question 45 refers to the diagnostic chart on page 871.

45. The diagnostic chart is being used to troubleshoot a GM vehicle that has a current diagnostic trouble code 24 stored in memory. The ECC display panel indicates the correct vehicle speed. Which of the following represents the probable cause of the trouble code?

A. Open circuit 437.

B. Faulty speedometer gear.

C. Faulty vehicle speed sensor.

D. Possible ECM problem.

NOTE: Question 46 refers to the Ford-illuminated entry system on page 872.

46. The illuminated entry system is inoperative; neither the right nor the left outer door handle activates the system. The courtesy lights and each lock cylinder LED operates when the courtesy lamp switch is turned on. Which of the following represents the possible cause of this problem?

A. Open in circuit 54.

B. Short to ground in circuit 465.

C. Open in circuit 57.

D. Short to ground in circuit 54.

NOTE: Question 47 refers to the instrument cluster schematic on page 873.

47. The electronic instrument cluster is totally inoperative; none of the display segments illuminate. *Technician A* says there may be an open in circuit 389.
Technician B says circuit 151 may be open.
Who is correct?

A. A only

C. Both A and B

B. B only

D. Neither A nor B

48. The ambient temperature sensor of a vehicle equipped with an SATC A/C system is being tested. The resistance of the sensor is about 1 kΩ.
Technician A says the sensor is out of range and needs to be replaced.
Technician B says the temperature of the sensor needs to be known before it can be accurately tested.
Who is correct?

A. A only

C. Both A and B

B. B only

D. Neither A nor B

49. Air bag module replacement is being discussed.
Technician A says the negative battery cable should be disconnected at the beginning of the repair.
Technician B says the reserve energy should be allowed to dissipate after battery power is removed.
Who is correct?

A. A only

C. Both A and B

B. B only

D. Neither A nor B

NOTE: Question 50 refers to the GM electronic sunroof circuit on page 874.

50. The electric sunroof of the vehicle will close but not open. All of the following could be the cause of this problem **EXCEPT:**

A. Open wire between the "open" relay and the motor.

B. Faulty "open" relay load circuit contacts.

C. Faulty control switch.

D. Open wire between the control switch and the "open" relay.

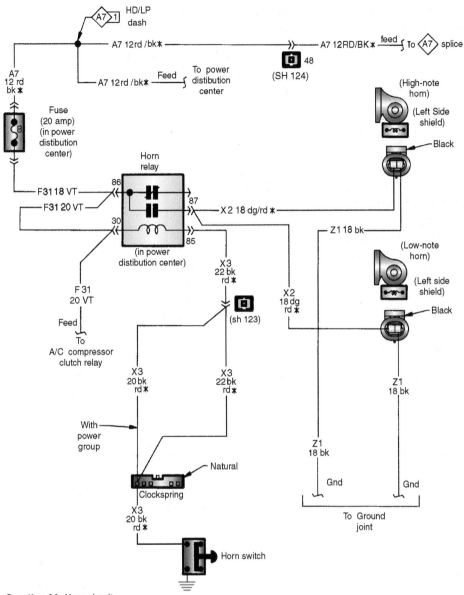

Question 11 Horn circuit.

Rear Wiper/Washer

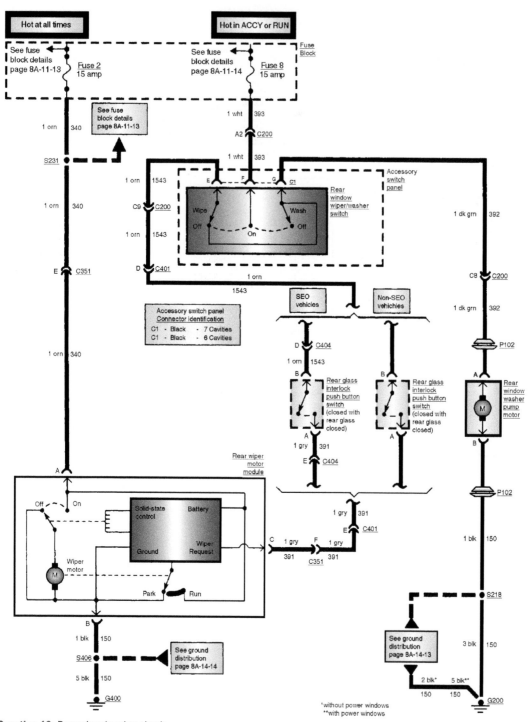

Question 12 Rear wiper/washer circuit.

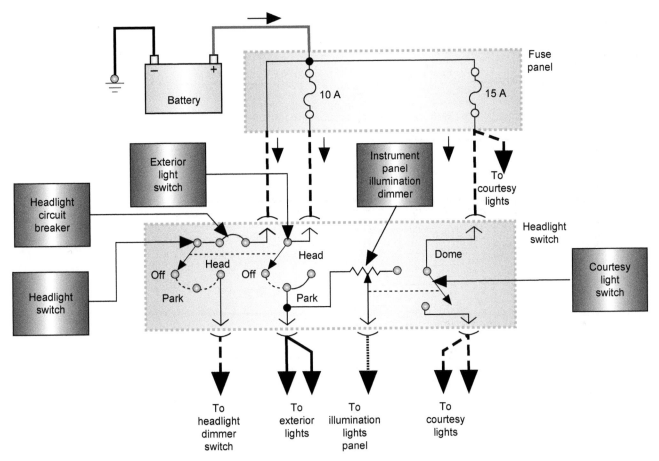

Question 34 Headlight switch circuit.

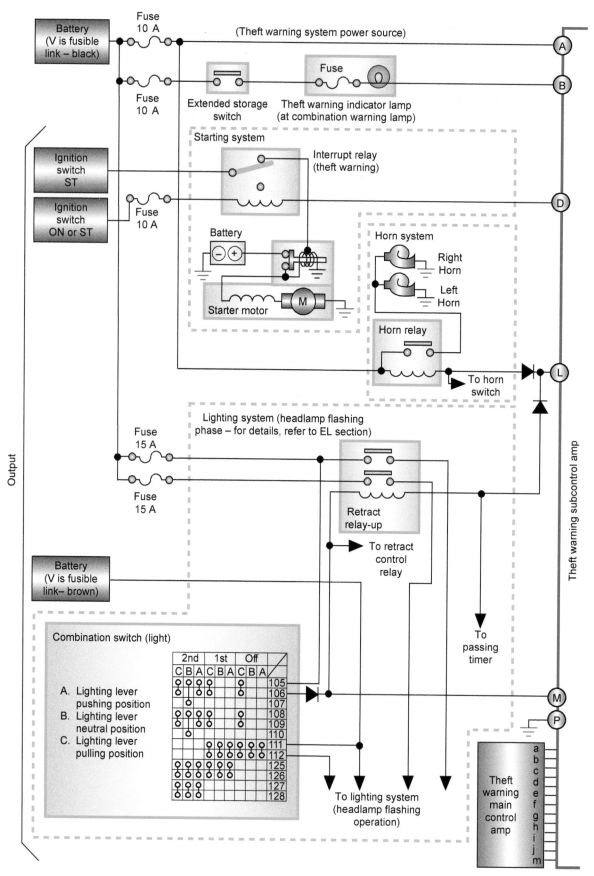

Question 35 Multifunction switch schematic.

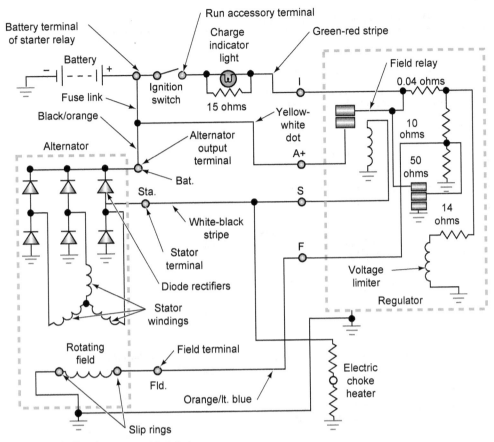

Question 38 Charging system wiring diagram.

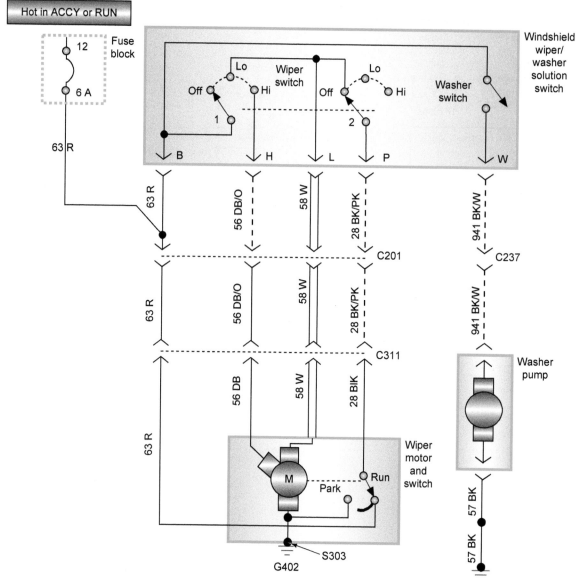

Question 39 Two-speed wiper circuit

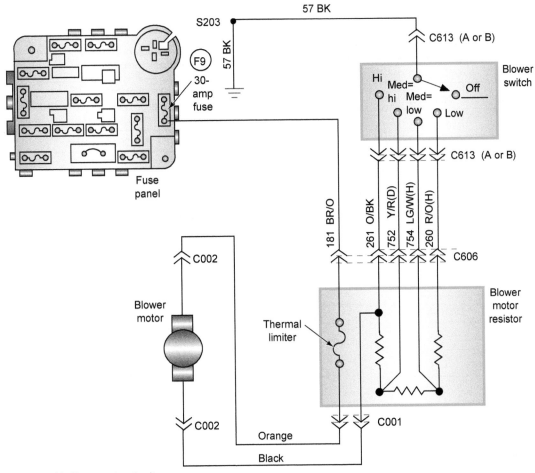

Question 40 Blower motor circuit.

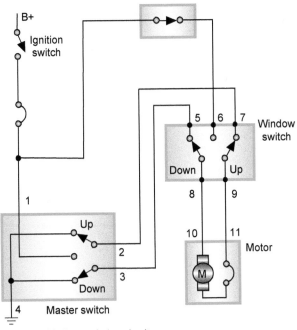

Question 41 Power window circuit.

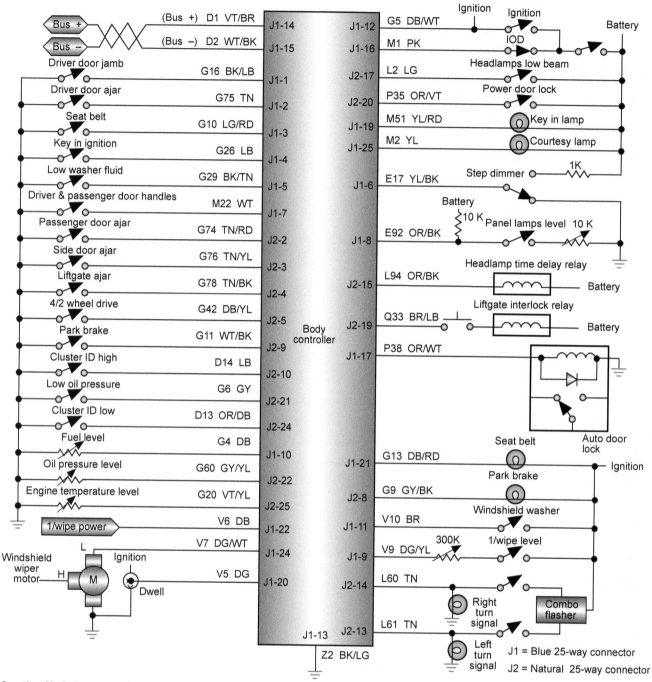

Question 42 Body computer inputs and outputs.

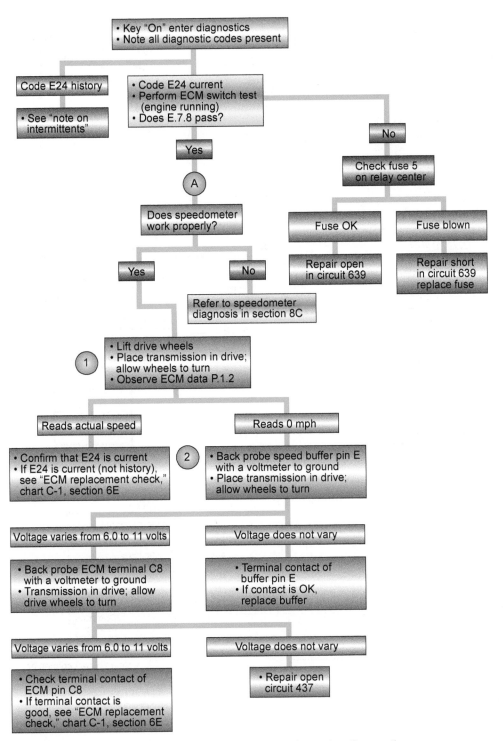

When all diagnoses and repairs are completed, clear codes and verify operation.

Question 45 Diagnostic chart.

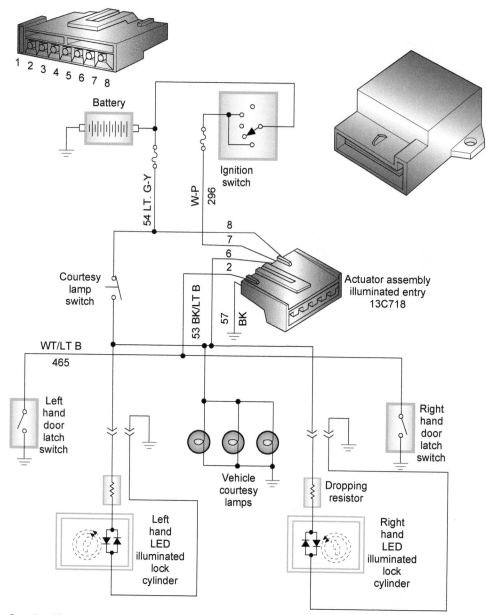

Question 46 Ford-illuminated entry system diagram.

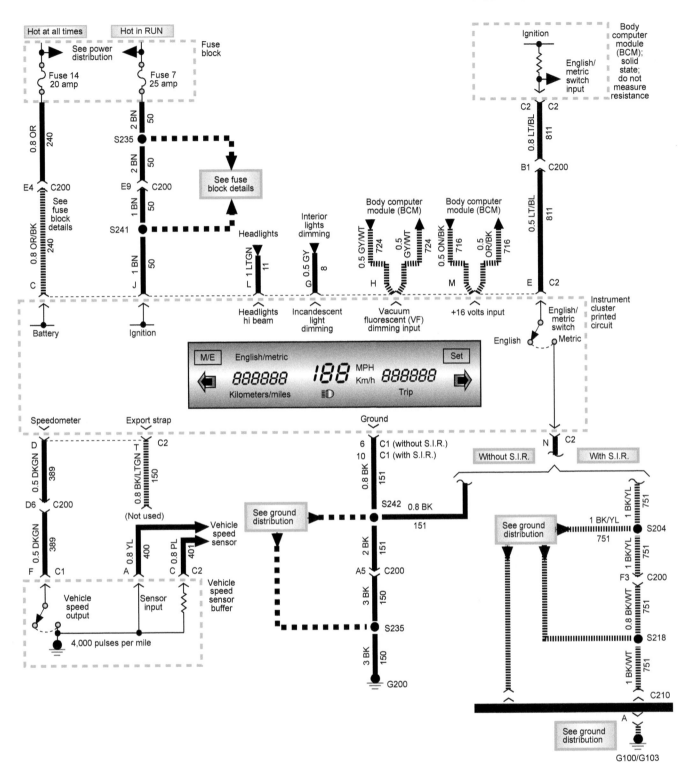

Question 47 Instrument cluster schematic.

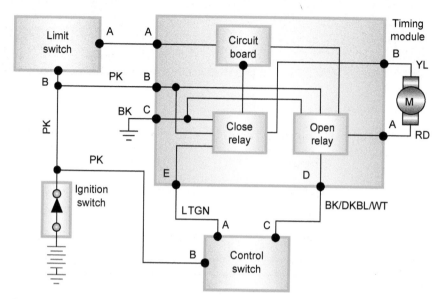

Question 50 GM electronic sunroof diagram.

APPENDIX B
METRIC CONVERSIONS

to convert these	to these	multiply by
TEMPERATURE		
Centigrade degrees	Fahrenheit degrees	1.8 then +32
Fahrenheit degrees	Centigrade degrees	0.556 after −32
LENGTH		
Millimeters	Inches	0.03937
Inches	Millimeters	25.4
Meters	Feet	3.28084
Feet	Meters	0.3048
Kilometers	Miles	0.62137
Miles	Kilometers	1.60935
AREA		
Square centimeters	Square inches	0.155
Square inches	Square centimeters	6.45159
VOLUME		
Cubic centimeters	Cubic inches	0.06103
Cubic inches	Cubic centimeters	16.38703
Cubic centimeters	Liters	0.001
Liters	Cubic centimeters	1,000
Liters	Cubic inches	61.025
Cubic inches	Liters	0.01639
Liters	Quarts	1.05672
Quarts	Liters	0.94633
Liters	Pints	2.11344
Pints	Liters	0.47317
Liters	Ounces	33.81497
Ounces	Liters	0.02957

to convert these	to these	multiply by
WEIGHT		
Grams	Ounces	0.03527
Ounces	Grams	28.34953
Kilograms	Pounds	2.20462
Pounds	Kilograms	0.45359
WORK		
Centimeter kilograms	Inch pounds	0.8676
Inch pounds	Centimeter kilograms	1.15262
Meter kilograms	Foot pounds	7.23301
Foot pounds	Newton meters	1.3558
PRESSURE		
Kilograms/sq. cm	Pounds/sq. inch	14.22334
Pounds/sq. inch	Kilograms/sq. cm	0.07031
Bar	Pounds/sq. inch	14.504
Pounds/sq. inch	Bar	0.06895

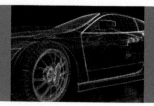
Alltest Instruments, Inc.
Farmingdale, NJ

Baum Tools Unlimited, Inc.
Sarasota, FL

Carquest Corp.
Raleigh, NC

Fluke Corp.
Everett, WA

Kent-Moore, Div. of SPX Corp.
Charlotte, NC

Mac Tools
Westerville, OH

Matco Tools
Stow, OH

NAPA Hand/Service Tools
Atlanta, GA

OTC, Div. of Bosch Automotive Service Solutions
Warren, MI

Parts Plus, Division of Automotive Distribution Network
Germantown, TN

Pico Technology
Tyler, TX

Snap-On Tools Corp.
Kenosha, WI

APPENDIX D
J1930 TERMINOLOGY LIST

NEW TERM	NEW ACRONYMS/ ABBREVIATIONS	OLD ACRONYMS/ TERM	NEW TERM	NEW ACRONYMS/ ABBREVIATIONS	OLD ACRONYMS/ TERM
Accelerator pedal	AP	Accelerator	Diagnostic test mode	DTM	Self-test mode
Air cleaner	ACL	Air cleaner	Diagnostic trouble code	DTC	Self-test code
Air conditioning	A/C	A/C Air conditioning	Distributor ignition	DI	CBD DS TFI Closed bowl distributor Duraspark ignition Thick film ignition
Barometric pressure	BARO	BP Barometric pressure			
Battery positive voltage	B+	BATT+ Battery positive	Early fuel evaporation	EFE	EFE Early fuel evaporation
Camshaft position	CMP	Camshaft sensor	Electrically erasable programmable read-only memory	EEPROM	E2PROM
Carburetor	CARB	CARB Carburetor			
Charge air cooler	CAC	After cooler Intercooler	Electronic ignition	EI	DIS EDIS Distributorless ignition system Electronic distributorless ignition system
Closed loop	CL	EEC			
Closed throttle position	CTP	CTP Closed throttle position			
Clutch pedal position	CPP	CES CIS Clutch engage switch Clutch interlock switch	Engine control module	ECM	ECM Engine control module
			Engine coolant level	ECL	Engine coolant level
Continuous fuel injection	CFI	Continuous fuel injection	Engine coolant temperature	ECT	ECT
Continuous trap oxidizer	CTOX	CTO			Engine coolant temperature
Crankshaft position	CKP	CPS VRS Variable reluctance sensor	Engine speed	RPM	RPM Revolutions per minute
Data link connector	DLC	Self-test connector			

NEW TERM	NEW ACRONYMS/ ABBREVIATIONS	OLD ACRONYMS/ TERM	NEW TERM	NEW ACRONYMS/ ABBREVIATIONS	OLD ACRONYMS/ TERM
Erasable programmable read-only memory	EPROM	EPROM Erasable programmable read-only memory	Indirect fuel injection	IFI	IDFI Indirect fuel injection
Evaporative emission	EVAP	EVP sensor EVR solenoid	Inertia fuel shutoff	IFS	Inertia switch
Exhaust gas recirculation	EGR	EGR Exhaust gas recirculation	Intake air temperature	IAT	ACT Air charge temperature
Fan control	FC	EDF Electro-drive fan	Knock sensor	KS	KS Knock sensor
Flash electrically erasable programmable read-only memory	FEEPROM	FEEPROM Flash electrically erasable programmable read-only memory	Malfunction indicator lamp	MIL	CEL "CHECK ENGINE" light "SERVICE ENGINE SOON" light
Flash-erasable programmable read-only memory	FEPROM	FEPROM Flash-erasable programmable read-only memory	Manifold absolute pressure	MAP	MAP Manifold absolute pressure
Flexible fuel	FF	FCS FFS FFV Fuel compensation sensor Flex fuel sensor	Manifold differential pressure	MDP	MDP Manifold differential pressure
			Manifold surface temperature	MST	MST Manifold surface temperature
Fourth gear	4GR	Fourth gear	Manifold vacuum zone	MVZ	MVZ Manifold vacuum zone
Fuel pump	FP	FP Fuel pump	Mass airflow	MAF	MAF Mass airflow
Generator	GEN	ALT Alternator	Mixture control	MC	Mixture control
Ground	GND	GND Ground	Multiport fuel injection	MFI	EFI Electronic fuel injection
Heated oxygen sensor	HO_2S	HEGO Heated exhaust gas oxygen sensor	Nonvolatile random access memory	NVRAM	NVM Nonvolatile memory
Idle air control	IAC	IAC Idle air bypass control	On-board diagnostic	OBD	Self-test On-board diagnostic
Idle speed control	ISC	Idle speed control	Open loop	OL	OL Open loop
Ignition control module	ICM	DIS module EDIS module TFI module	Oxidation catalytic converter	OC	COC Conventional oxidation catalyst

NEW TERM	NEW ACRONYMS/ ABBREVIATIONS	OLD ACRONYMS/ TERM	NEW TERM	NEW ACRONYMS/ ABBREVIATIONS	OLD ACRONYMS/ TERM
Oxygen sensor	O2S	EGO	Secondary air injection	AIR	AM CT MTA Air management Conventional thermactor Managed thermactor air Thermactor
PARK/NEUTRAL position	PNP	NDS NGS TSN Neutral drive switch Neutral gear switch Transmission select neutral			
Periodic trap oxidizer	PTOX	PTOX Periodic trap oxidizer	Sequential multiport fuel injection	SFI	SEFI Sequential electronic fuel injection
Power steering pressure	PSP	PSPS Power steering pressure switch	Service reminder indicator	SRI	SRI Service reminder indicator
Powertrain control module	PCM	ECA ECM ECU EEC processor Engine control assembly Engine control module Engine control unit	Smoke puff limiter	SPL	SPL Smoke puff limiter
			Supercharger	SC	SC Supercharger
			Supercharger bypass	SCB	SCB Supercharger bypass
Programmable read-only memory	PROM	PROM Programmable read-only memory	System readiness test	SRT	—
Pulsed secondary air injection	PAIR	MPA PA Thermactor II Managed pulse air Pulse air	Thermal vacuum valve	TVV	Thermal vacuum switch
			Third gear	3GR	Third gear
Random access memory	RAM	RAM Random access memory	Three-way catalytic converter	TWC	TWC Three-way catalytic converter
Read-only memory	ROM	ROM Read-only memory	Three-way + oxidation catalytic converter	TWC + OC	TWC and COC Dual bed Three-way catalyst and conventional oxidation catalyst
Relay module	RM	RM Relay module			
Scan tool	ST	GST NGS Generic scan tool New-generation STAR tester Enhanced scan tool OBD II ST	Throttle body	TB	TB Throttle body
			Throttle body fuel injection	TBI	CFI Central fuel injection EFI
			Throttle position	TP	TP Throttle position

NEW TERM	NEW ACRONYMS/ ABBREVIATIONS	OLD ACRONYMS/ TERM
Torque converter clutch	TCC	CCC CCO MCCC Converter clutch control Converter clutch override Modulated converter clutch control
Transmission control module	TCM	4EAT module
Transmission range	TR	—
Turbocharger	TC	TC Turbocharger
Vehicle speed sensor	VSS	VSS Vehicle speed sensor

NEW TERM	NEW ACRONYMS/ ABBREVIATIONS	OLD ACRONYMS/ TERM
Voltage regulator	VR	VR Voltage regulator
Volume airflow	VAF	VAF Volume airflow
Warm-up oxidation catalytic converter	WU-OC	WU-OC Warm-up oxidation catalytic converter
Warm-up three-way catalytic converter	WU-TWC	WU-TWC Warm-up three-way catalytic converter
Wide-open throttle	WOT	Full throttle WOT Wide-open throttle

GLOSSARY

Note: **Terms are highlighted in bold,** followed by Spanish translation in color.

A circuit A generator circuit that uses an external grounded field circuit. The regulator is on the ground side of the field coil.

Circuito A Circuito regulador del generador que utiliza un circuito inductor externo puesto a tierra. En el circuito A, el regulador se encuentra en el lado a tierra de la bobina inductora.

Activation test A bidirectional scan tool function that allows the technician to command an actuator to test its operation.

Prueba de activación Una función de la herramienta de exploración bidireccional que le permite al técnico comandar un actuador para probar su operación.

Active headlight system (AHS) A vehicle safety feature that is designed to enhance nighttime safety by providing drivers additional time to act or react to approaching road hazards by predicting upcoming corners or turns in the road.

Sistema de faros activos (AHS) Una característica de seguridad del vehículo diseñada para mejorar la seguridad nocturna, proporcionando a los conductores tiempo adicional para actuar o reaccionar ante los peligros de la carretera al acercarse a las esquinas o virajes próximos en la carretera.

Actuators Devices that perform the actual work commanded by the computer. They can be in the form of a motor, relay, switch, or solenoid.

Accionadores Dispositivos que realizan el trabajo efectivo que ordena la computadora. Dichos dispositivos pueden ser un motor, un relé, un conmutador o un solenoide.

Adaptive cruise control (ACC) An additional function of the cruise control system that uses laser radar or infrared transceiver sensors to determine the vehicle-to-vehicle distances and relational speeds. The system maintains both the vehicle speed and a set distance between vehicles.

Control de crucero adaptable (ACC) Una función adicional del sistema de control de crucero que utiliza un radar láser o sensores de transmisores infrarrojos para determinar las distancias entre vehículos y las velocidades de relación. El sistema mantiene tanto la velocidad del vehículo como una distancia fija entre vehículos.

Air bag module Composed of the air bag and inflator assembly that is packaged into a single module.

Unidad del Airbag Formada por el conjunto del Airbag y el inflador. Este conjunto se empaqueta en una sola unidad.

Air bag system A supplemental restraint that deploys a bag out of the steering wheel or passenger-side dash panel to provide additional protection against head and face injuries during an accident.

Sistema de Airbag Resguardo complementario que expulsa una bolsa del volante o del panel de instrumentos del lado del pasajero para proveer protección adicional contra lesiones a la cabeza y a la cara en caso de un accidente.

Air gap The space between the trigger wheel teeth and the sensor.

Entrehierro El espacio entre los dientes de la rueda disparadora y el sensor.

Ambient temperature The temperature of the outside air.

Temperatura ambiente Temperatura del aire ambiente.

Ambient temperature sensor Thermistor used to measure the temperature of the air entering the vehicle.

Sensor de temperatura ambiente Termistor utilizado para medir la temperatura del aire que entra al vehículo.

Ammeter A test meter used to measure current draw.

Amperímetro Instrumento de prueba utilizado para medir la intensidad de una corriente.

Amperes *See* current.

Amperios *Véase* corriente.

Analog A voltage signal that is infinitely variable or can be changed within a given range.

Señal analógica Señal continua y variable que debe traducirse a valores numéricos discontinuos para poder ser trataba por una computadora.

Analog signal Varying voltage with infinite values within a defined range.

Señal análoga Voltaje variable con infinidad de valores dentro de un límite definido de velocidad.

Antenna A wire or a metal stick that increases the amount of metal the transmitter's waves can interact with to convert radio frequency electrical currents into electromagnetic waves and vice versa.

Antena Un cable o varilla de metal que aumenta la cantidad de metal con la que pueden interactuar las ondas del transmisor para convertir las corrientes eléctricas de radio frecuencia en ondas electromagnéticas y viceversa.

Antenna trimmer screw An adjusting screw located near the antenna lead connector on the radio tuner that tunes the radio circuit to match the resistance of the antenna.

Tornillo compensador de antena Un tornillo de ajuste ubicado cerca del conector del cable de plomo de la antena en el sintonizador de radio que ajusta el circuito de radio para que coincida con la resistencia de la antena.

Antilock brake system (ABS) A brake system that automatically pulsates the brakes to prevent wheel lock-up under panic stop and poor traction conditions.

Frenos antibloqueo Sistema de frenos que pulsa los frenos automáticamente para impedir el bloqueo de las ruedas en casos de emergencia y de tracción pobre.

Antitheft device A device or system that prevents illegal entry or driving of a vehicle. Most are designed to deter entry.

Dispositivo a prueba de hurto Un dispositivo o sistema quepreviene la entrada o conducción ilícita de un vehículo. Lamayoría se diseñan para detener la entrada.

A-pillar The pillar in front of the driver or passenger that supports the windshield.

Soporte A Soporte enfrente del conductor o del pasajero que sostiene el parabrisas.

Arming sensor A device that places an alarm system into "ready" to detect an illegal entry.

Sensor de armado Un dispositivo que pone "listo" un sistema dealarma para detectar una entrada ilícita.

Asbestos *Asbestos* is the name given to a set of six naturally occurring silicate minerals that are resistant to heat and corrosion. The mineral fibers include chrysotile, amosite, crocidolite, tremolite, anthophyllite, and actinolite.

Asbesto Asbestoes el nombre dado a un conjunto de seis minerales de silicato que son resistentes al calor y la corrosión de origen natural. Las fibras minerales incluyen; crisotilo, amosita, crocidolita, tremolita, antofilita, actinolita.

Aspirator Tubular device that uses a Venturi effect to draw air from the passenger compartment over the in-car sensor. Some manufacturers use a suction motor to draw the air over the sensor.

Aspirador Dispositivo tubular que utiliza un efecto venturi para extraer aire del compartimiento del pasajero sobre el sensor dentro del vehículo. Algunos fabricantes utilizan un motor de succión para extraer el aire sobre el sensor.

Audio system The sound system for a vehicle; can include radio cassette player, CD player, amplifier, and speakers.

Sistema de audio El sistema de sonido de un vehículo; puedeincluir el radio, el tocacaset, el toca discos compactos, el amplificador, y las bocinas.

Automatic door locks A system that automatically locks all doors by activating one switch.

Cerraduras automáticas de puerta Un sistema que cierra todas laspuertas automaticamente al activar un solo conmutador.

Automatic seat belt system Automatically puts the shoulder and/or lap belt around the driver or occupant. The automatic seat belt is operated by DC motors that move the belts by means of carriers on tracks.

Sistema automático de correas de asiento Funciona automáticamente, poniendo el cinturón de seguridad sobre el hombro y el pecho del chofer y el pasajero del automóvil. La correa automática del asiento trabaja por medio de motores DC, o sea corriente directa, que da movimiento a las correas transportadoras.

Automatic traction control A system that prevents slippage of one of the drive wheels. This is done by applying the brake at that wheel and/or decreasing the engine's power output.

Control automático de tracción Un sistema que previene el patinaje de una de las ruedas de mando. Esto se efectúa aplicando el freno en esa rueda y/o disminuyendo la salida de potencia del motor.

Average responding A method used to read AC voltage.

Respuesta media Un método que se emplea para leer la tensión decorriente alterna.

B circuit A generator regulator circuit that is internally grounded. In the B circuit, the voltage regulator controls the power side of the field circuit.

Circuito B Circuito regulador del generador puesto internamente a tierra. En el circuito B, el regulador de tensión controla el lado de potencia del circuito inductor.

Backprobe A term used to mean that a test is being performed on the circuit while the connector is still connected to the component. The test probes are inserted into the back of the wire connector.

Sonda exploradora de retorno Término utilizado para expresar que se está llevando a cabo una prueba del circuito mientras el conectador sigue conectado al componente. Las sondas de prueba se insertan a la parte posterior del conectador de corriente.

BAT Terminal of a generator, starter, or solenoid that has direct battery feed connected to it.

Forma de aproximación del balancín Terminal de un generador, encendedor o solenoide conectado directamente a la batería.

Battery ECU Used to monitor the condition of the HV battery assembly in a hybrid electric vehicle (HEV). The battery ECU determines the SOC of the HV battery by monitoring voltage, current, and temperature.

UCE de la batería Se utiliza para monitoreard del ensamblado de la batería del VH en un VHE. La UCE de la batería determina el EDC de la batería del VH al monitorear el voltaje, la corriente y la temperatura.

Battery holddowns Brackets that secure the battery to the chassis of the vehicle.

Portabatería Los sostenes que fijan la batería al chasis del vehículo.

Battery leakage test Used to determine if current is discharging across the top of the battery case.

Prueba de pérdida de corriente de la batería Prueba utilizada para determinar si se está descargando corriente a través de la parte superior de la caja de la batería.

Battery terminal test Checks for poor electrical connections between the battery cables and terminals; uses a voltmeter to measure voltage drop across the cables and terminals.

Prueba del borne de la batería Verifica si existen conexiones eléctricas pobres entre los cables y los bornes de la batería. Utiliza un voltímetro para medir caídas de tensión entre los cables y los bornes.

B-codes DTCs that are assigned to the vehicle's body systems and control modules.

Códigos B Instrucciones de transmisión digital (DTC) que se asignan a los sistemas de la carrocería del vehículo y a los módulos de control.

Bench test A term used to indicate that the unit is to be removed from the vehicle and tested.

Prueba de banco Término utilizado para indicar que la unidad será removida del vehículo para ser examinada.

Bendix drive A type of starter drive that uses the inertia of the spinning starter motor armature to engage the drive gear to the gears of the flywheel. This type of starter drive was used on early models of vehicles and is rarely seen today.

Acoplamiento Bendix Un tipo del acoplamiento del motor de arranque que usa la inercia de la armadura del motor de arranque giratorio para endentar el engranaje de mando con los engranajes del volante. Este tipo de acoplamiento del motor de arranque se usaba en los modelos vehículos antiguos y se ven raramente.

Bezel The retaining trim around a component.

Bisel El resto del decorado alrededor de un componente.

Bimetallic gauges Simple dial and needle indicators that transform the heating effect of electricity into mechanical movement.

Calibradores bimetálicos Un cuadrante simple y agujas indicadoras que transforman el efecto del calor de la electricidad en un movimiento mecánico.

Binary numbers Strings of zeroes and ones, or on and off, which represent a numeric value.

Números binarios Franjas que indican ceros (0) y unos (1), de prendido/apagado, representando el valor numérico.

B-Pillar The pillar located over the shoulder of the driver or passenger.

Soporte B Soporte ubicado sobre el hombro del conductor o del pasajero.

Breakout box Allows the technician to test circuits, sensors, and actuators by providing test points.

Caja de interruptores Le permite al técnico probar los circuitos, los monitores y actuadores al indicar los puntos de prueba.

Brushes Electrically conductive sliding contacts, usually made of copper and carbon.

Escobillas Contactos deslizantes de conducción eléctrica, por lo general hechos de cobre y de carbono.

Bucking coil One of the coils in a three-coil gauge. It produces a magnetic field that bucks or opposes the low reading coil.

Bobina compensadora Una de las bobinas de un calibre de tres bobinas. Produce un campo magnético que es contrario o en oposición a la bobina de baja lectura.

Buffer A buffer cleans up a voltage signal. These are used with PM generator sensors to change the AC voltage to a digitalized signal.

Separador Un separador aguza una señal del tensión. Estos se usan con los sensores generadores PM para cambiar la tensión de corriente alterna a una señal digitalizado.

Bulkhead connector A large connector that is used when many wires pass through the bulkhead or firewall.

Conectador del tabique Un conectador que se usa al pasar muchos alambres por el tabique o mamparo de encendidos.

Bus bar A common electrical connection to which all of the fuses in the fuse box are attached. The bus bar is connected to battery voltage.

Barra colectora Conexión eléctrica común a la que se conectan todos los fusibles de la caja de fusibles. La barra colectora se conecta a la tensión de la batería.

Capacitance discharge sensor A form of piezo sensor that uses the discharge from a variable capacitor.

Sensor de descarga de capacitancia Un tipo de sensor que utiliza la descarga de un capacitor variable.

Capacitor An electrical device made from layers of conductors and dielectric used to store and release electrical energy; also used to absorb voltage changes in the circuit.

Capacitor Un dispositivo eléctrico que se fabrica de capas de conductores y material dieléctrico y se utiliza para almacenar y liberar energía eléctrica; también se utiliza para absorber los cambios de voltaje en el circuito.

Capacity test The part of the battery test series that checks the battery's ability to perform when loaded.

Prueba de capacidad Parte de la serie de prueba de la batería que verifica la capacidad de funcionamiento de la batería cuando está cargada.

Captured signal A signal stored in a DSO's memory.

Señal de captura La señal restauradora se mantiene en la memoria indicada con las letras DSO.

Carbon monoxide An odorless, colorless, and toxic gas that is produced as a result of combustion.

Monóxido de carbono Gas inodoro, incoloro y tóxico producido como resultado de la combustión.

Carbon tracking A condition where paths of carbon will allow current to flow to points that are not intended. This condition is most commonly found inside distributor caps.

Rastreo de carbón Una condición en la cual las trayectorias del carbón permiten fluir el corriente a los puntos no indicados. Esta condición se encuentra comúnmente dentro de las tapas del distribuidor.

Cartridge fuses *See* maxi-fuse.

Fusibles cartucho *Véase* maxifusible.

Cathode ray tube Similar to a television picture tube. It contains a cathode that emits electrons and an anode that attracts them. The screen of the tube will glow at points that are hit by the electrons.

Tubo de rayos catódicos Parecidos a un tubo de pantalla de televisor. Contiene un cátodo que emite los electrones y un ánodo que los atrae. La pantalla del tubo iluminará en los puntos en donde pegan los electrones.

Caustic Chemicals that have the ability to destroy or eat through something and that are extremely corrosive.

Cáusticos Químicos que tienen la habilidad de destruir o carcomer algo, y que son extremadamente corrosivos.

Cell element The assembly of a positive and a negative plate in a battery.

Elemento de pila La asamblea de una placa positiva y negativa en una batería.

Charge To pass an electric current through a battery in an opposite direction than during discharge.

Cargar Pasar una corriente eléctrica por la batería en una dirección opuesta a la usada durante la descarga.

Charge rate The speed at which the battery can safely be recharged at a set amperage.

Indicador de carga eléctrica La velocidad a la cual la batería puede ser recargada seguramente a un amperaje establecido.

Charging system requirement test Diagnostic test used to determine the total electrical demand of the vehicle's electrical system.

Prueba del requisito del sistema de carga Prueba diagnóstica utilizada para determinar la exigencia eléctrica total del sistema eléctrico del vehículo.

CHMSL The abbreviation for center high-mounted stop light, often referred to as the third brake light.

CHMSL La abreviación para el faro de parada montada alto en el centro que suele referirse como el faro de freno tercero.

Circuit The path of electron flow consisting of the voltage source, conductors, load component, and return path to the voltage source.

Circuito Trayectoria del flujo de electrones, compuesto de la fuente de tensión, los conductores, el componente de carga y la trayectoria de regreso a la fuente de tensión.

Clamping diode A diode that is connected in parallel with a coil to prevent voltage spikes from the coil reaching other components in the circuit.

Diodo de bloqueo Un diodo que se conecta en paralelo con una bobina para prevenir que los impulsos de tensión lleguen a otros componentes en el circuito.

Clockspring Maintains a continuous electrical contact between the wiring harness and the air bag module.

Muelle de reloj Mantiene un contacto eléctrico continuo entre el cableado preformado y la unidad del Airbag.

Closed circuit A circuit that has no breaks in the path and allows current to flow.

Circuito cerrado Circuito de trayectoria ininterrumpida que permite un flujo continuo de corriente.

Cold cranking amps (CCA) Rating that indicates the battery's ability to deliver a specified amount of current to start an engine at low ambient temperatures.

Amperaje de arranque en frío (CCA, por su sigla en inglés) Régimen que indica la capacidad de la batería para proporcionar una cantidad de corriente específica, capaz de hacer arrancar un motor a temperatura ambiente baja.

Cold soak A key-off condition that allows the engine coolant temperature to equal with ambient temperature.

Amperios de arranque en frío Tasa indicativa de la capacidad de la batería para producir una cantidad específica de corriente para arrancar un motor a bajas temperaturas ambiente.

Color codes Used to assist in tracing the wires. In most color codes, the first group of letters designates the base color of the insulation and the second group of letters indicates the color of the tracer.

Códigos de colores Utilizados para facilitar la identificación de los alambres. Típicamente, el primer alfabeto representa el color base del aislamiento y el segundo representa el color del indicador.

Common connector A connector that is shared by more than one circuit and/or component.

Conector común Un conector que se comparte entre más de un circuito y/o componente.

Commutator A series of conducting segments located around one end of the armature.

Conmutador Serie de segmentos conductores ubicados alrededor de un extremo de la armadura.

Component locator Service manual used to find where a component is installed in the vehicle. The component locator uses both drawings and text to lead the technician to the desired component.

Manual para indicar los elementos componentes Manual de servicio utilizado para localizar dónde se ha instalado un componente en el vehículo. En dicho manual figuran dibujos y texto para guiar al mecánico al componente deseado.

Composite bulb A headlight assembly that has a replaceable bulb in its housing.

Bombilla compuesta Una asamblea de faros cuyo cárter tiene una bombilla reemplazable.

Compound motor A motor that has the characteristics of a series-wound and a shunt-wound motor.

Motor compuesta Un motor que tiene las características de un motor exitado en serie y uno en derivación.

Computer An electronic device that stores and processes data and is capable of operating other devices.

Computadora Dispositivo electrónico que almacena y procesa datos y que es capaz de ordenar a otros dispositivos.

Conductance A measurement of the battery's plate surface that is available for chemical reaction, determining how much power the battery can supply.

Conductancia Medida de la superficie de la placa de la batería que está lista para la reacción química, determinando así cuánta potencia puede suplir la batería.

Conducted interference An electromagnetic interference (EMI) caused by the physical contact of the conductors.

Interferencia conducida Una interferencia electromagnética (EMI) provocada por el contacto físico de los conductores.

Conductor A material in which electrons flow or move easily.

Conductor Una material en la cual los electrones circulen o se mueven fácilmente.

Continuity Refers to the circuit being continuous with no opens.

Continuidad Se refiere al circuito ininterrumpido, sin aberturas.

Control assembly Provides for driver input into the automatic temperature control microprocessor. The control assembly is also referred to as the control panel.

Montaje de control Un microprocesador automático que le da al Chofer la información necesaria sobre el control de temperatura. El montaje de control es llamado también panel de control.

Converter A DC/DC transformer that converts the voltage from 270 volts DC to 14 volts DC to recharge the auxiliary battery and to power 12-volt electrical components on an HEV.

Convertidor Transformador CD-CD que convierte el voltaje de 270 voltios a 14 voltios de CD para recargar la batería auxiliary y para darles potencia a los componentes eléctricos de 12 voltios en un VHE.

Corona effect A condition where high voltage leaks through a wire's insulation and produces a light or an illumination; worn insulation on spark plug wires causes this.

Efecto corona Una condición en la cual la alta tensión se escapa por la insulación del alambre y produce una luz o una iluminación; esto se causa por la insulación desgastada en los alambres de las bujías.

Cowl The top portion of the front of the automobile body that supports the windshield and dashboard.

Capucha Esta es la parte principal de la carrocería en el frente del automóvil y es la que sostiene el parabrisas y el tablero de instrumentos.

Crash sensor Normally open electrical switch designed to close when subjected to a predetermined amount of jolting or impact.

Sensor de impacto Un conmutador normalmente abierto diseñado a cerrarse al someterse a un sacudo de una fuerza predeterminada o un impacto.

Crimping The process of bending, or deforming by pinching, a connector so that the wire connection is securely held in place.

Engarzado Proceso a través del cual se curva o deforma un conectador mediante un pellizco para que la conexión de alambre se mantenga firme en su lugar.

Crimping tool Has different areas to perform several functions. This single tool cuts a wire, strips the insulation, and crimps the connector.

Herramienta prensadora Tiene diferentes áreas para poder ofrecer varias funciones. Esta simple herramienta cortará el cable, aislará y prensará el conectador.

Crocus cloth Used to polish metals. While polishing, the cloth removes very little metal.

Paño de color azafrán Se usa para pulir los metales. Su acción es suave y al pulir remueve pequeñas partículas de metal.

Cross-fire The undesired firing of a spark plug that results from the firing of another spark plug. This is caused by electromagnetic induction.

Encendido transversal El encendido no deseable de una bujía que resulta del encendido de otra bujía. Esto se causa por la inducción electromagnética.

CRT The common acronym for a cathode ray tube.

CRT La sigla común de un tubo de rayos catódicos.

Curb height The height of the vehicle when it has no passengers or loads, and normal fluid levels and tire pressure.

Altura del contén La altura del vehículo cuando no lleva pasajeros ni cargas, y los niveles de los fluidos y de la presión de las llantas son normales.

Current The aggregate flow of electrons through a wire. One ampere represents the movement of 6.25 billion billion electrons (or one coulomb) past one point in a conductor in one second.

Corriente Flujo combinado de electrones a través de un alambre. Un amperio representa el movimiento de 6,25 mil millones de mil millones de electrones (o un colombio) que sobrepasa un punto en un conductor en un segundo.

Current-draw test Diagnostic test used to measure the amount of current that the starter draws when actuated. It determines the electrical and mechanical condition of the starting system.

Prueba de la intensidad de una corriente Prueba diagnóstica utilizada para medir la cantidad de corriente que el arrancador tira cuando es accionado. Determina las condiciones eléctricas y mecánicas del sistema de arranque.

Current output test Diagnostic test used to determine the maximum output of the AC generator.

Prueba de la salida de una corriente Prueba diagnóstica utilizada para determinar la salida máxima del generador de corriente alterna.

Cycle One set of changes in a signal that repeats itself several times.

Ciclo Una serie de cambios en una señal que se repite varias veces.

Darlington A special type of configuration usually consisting of two transistors fabricated on the same chip or mounted in the same package.

Darlington Tipo especial de configuración que generalmente consiste en 2 transistores fabricados en el mismo chip o montados en el mismo paquete.

D'Arsonval gauge A gauge design that uses the interaction of a permanent magnet and an electromagnet and the total field effect to cause needle movement.

Calibrador D'Arsonval Calibrador diseñado para utilizar la interacción de un imán permanente y de un electroimán, y el efecto inductor total para generar el movimiento de la aguja.

D'Arsonval movement A small coil of wire mounted in the center of a permanent horseshoe-type magnet. A pointer or needle is mounted to the coil.

Movimiento D'Arsonval Se refiere a una pequeña bobina de cable montada en el centro de un permanente imán diseñado como la herradura de caballo. El puntero o la aguja está montada sobre la bobina.

Deep cycling Discharging the battery completely before recharging it.

Operación cíclica completa La descarga completa de la batería previo al recargo.

Demonstration mode SmartBeam™ function that allows operation of the automatic high beams and high-beam indicator while the vehicle is stationary and under any ambient lighting conditions.

Mando de demostración Función del rayo inteligente de marca registrada que permite la operación de rayos altos automáticos y del indicador de rayo alto cuando el vehículo está parado y bajo cualesquiera condiciones de iluminación ambientales.

DERM Designed to provide an energy reserve of 36 volts to assure deployment for a few seconds when vehicle voltage is low or lost. The DERM also maintains constant diagnostic monitoring of the electrical system. It will store a code if a fault is found and provide driver notification by illuminating the warning light.

DERM El significado de este término se refiere al diseño que mantiene una reserva de energía de 36 voltios la que asegura, por unos pocos segundos, la salida de la corriente cuando el voltaje del vehículo es bajo o se ha perdido. El DERM mantiene también un constante diagnóstico monitor del sistema eléctrico. Guarda también una clave en caso de que se encuentre alguna falla y le da al chofer una alerta iluminando las señales.

Diagnostic module Part of an electronic control system that provides self-diagnostics and/or a testing interface.

Módulo de diagnóstico Parte de un sistema controlado electronicamente que provee autodiagnóstico y/o una interfase de pruebas.

Diagnostic trouble codes (DTCs) Fault codes that represent a circuit failure in a monitored system.

Códigos de destello Códigos de fallas de diagnóstico (CFD o DTC) que se muestran por medio de los destellos de una lámpara o diodo luminiscente.

Diaphragm A thin, flexible, circular plate that is held around its outer edge by the horn housing, allowing the middle to flex.

Diafragma Es una fina placa circular flexible que es sostenida alrededor de su borde externo por el cuerno del embrague, permitiendo que el centro se doble.

Digital A voltage signal is either on-off, yes-no, or high-low.

Digital Una señal de tensión está Encendida-Apagada, es Sí-No o Alta-Baja.

Digital multimeter (DMM) Displays values using liquid crystal displays instead of a swinging needle. They are basically computers that determine the measured value and display it for the technician.

Multímetro digital Exhibe valores usando cristal líquido para desplegar, en cambio de una aguja giratoria. Son básicamente computadores que determinan el valor medido y lo muestran al técnico.

Digital signal A voltage value that has two states. The states can be on-off or high-low.

Señal digital El valor del voltaje que tiene dos ventajas. Una señal muestra prendido/apagado y la otra alta/baja.

Dimmer switch A switch in the headlight circuit that provides the means for the driver to select either high-beam or low-beam operation, and to switch between the two. The dimmer switch is connected in series within the headlight circuit and controls the current path for high and low beams.

Conmutador reductor Conmutador en el circuito para faros delanteros que le permite al conductor elegir la luz larga o la luz corta, y conmutar entre las dos. El conmutador reductor se conecta en serie dentro del circuito para faros delanteros y controla la trayectoria de la corriente para la luz larga y la luz corta.

Diode An electrical one-way check valve that will allow current to flow only in one direction.

Diodo Válvula eléctrica de retención, de una vía, que permite que la corriente fluya en una sola dirección.

Diode rectifier bridge A series of diodes that are used to provide a reasonably constant DC voltage to the vehicle's electrical system and battery.

Puente rectificador de diodo Serie de diodos utilizados para proveerles una tensión de corriente continua bastante constante al sistema eléctrico y a la batería del vehículo.

Diode trio Used by some manufacturers to rectify the stator of an AC generator current so that it can be used to create the magnetic field in the field coil of the rotor.

Trío de diodos Utilizado por algunos fabricantes para rectificar el estátor de la corriente de un generador de corriente alterna y poder así utilizarlo para crear el campo magnético en la bobina inductora del rotor.

Direct drive A situation where the drive power is the same as the power exerted by the device that is driven.

Transmisión directa Una situación en la cual el poder de mando es lo mismo que la potencia empleada por el dispositivo arrastrado.

Discriminating sensors Part of the air bag circuitry; these sensors are calibrated to close with speed changes that are great enough to warrant air bag deployment. These sensors are also referred to as crash sensors.

Sensores discriminadores Una parte del conjunto de circuitos de Airbag; estos sensores se calibran para cerrar con los cambios de la velocidad que son bastante severas para justificar el despliegue del Airbag. Estos sensores también se llaman los sensores de impacto.

Drive coil A hollowed field coil used in a positive engagement starter to attract the movable pole shoe of the starter.

Bobina de excitación Una bobina inductora hueca empleada en un encendedor de acoplamiento directo para atraer la pieza polar móvil del encendedor.

DSO A common acronym for a digital storage oscilloscope.

DSO Una sigla común del osciloscopio de almacenamiento digital.

Dual ramping A sensor circuit that provides a more precise method of determining the temperature. This circuit starts with a high-voltage value when cold and decreases as temperature increases. When the voltage reaches a predetermined value, the PCM switches to lower resistance circuit causing the sense reading to go high again and provides a second set of inputs.

Voltaje de diente de sierra doble Un circuito de sensor que brinda un método más preciso de determinar la temperatura. Este circuito comienza con un valor de voltaje alto cuando está frío y desciende a medida que la temperatura se incrementa. Cuando el voltaje alcanza un valor predeterminado, el PCM pasa a un circuito de resistencia más baja, lo que hace que la lectura del sensor vuelva a subir y proporciona un segundo conjunto de información entrante.

Dump valve A safety switch that releases vacuum to the servo when the brake pedal is pressed.

Válvula de descargue Un interruptor de seguridad que facilita el vacío al servo cuando se presiona el pedal del freno.

Duty cycle The percentage of on time to total cycle time.

Ciclo de trabajo Porcentaje del trabajo efectivo a tiempo total del ciclo.

Eddy currents Small induced currents.

Corriente de Foucault Pequeñas corrientes inducidas.

Electrical load The working device of the circuit.

Carga eléctrica Dispositivo de trabajo del circuito.

Electrochemical The chemical action of two dissimilar materials in a chemical solution.

Electroquímico Acción química de dos materiales distintos en una solución química.

Electrolysis The production of chemical changes by passing electrical current through an electrolyte.

Electrólisis La producción de los cambios químicos al pasar un corriente eléctrico por un electrolito.

Electrolyte A solution of 64% water and 36% sulfuric acid.

Electrolito Solucion de un 64 percent de agua y un 36 percent de ácido sulfúrico.

Electromagnetic gauge Gauge that produces needle movement by magnetic forces.

Calibrador electromagnético Calibrador que genera el movimiento de la aguja mediante fuerzas magnéticas.

Electromagnetic induction The production of voltage and current within a conductor as a result of relative motion within a magnetic field.

Inducción electrómagnética Producción de tension y de corriente dentro de un conductor como resultado del movimiento relativo dentro de un campo magnético.

Electromagnetic interference (EMI) An undesirable creation of electromagnetism whenever current is switched on and off.

Interferencia electromagnética Fenómeno de electromagnetismo no deseable que resulta cuando se conecta y se desconecta la corriente.

Electromagnetism A form of magnetism that occurs when current flows through a conductor.

Electromagnetismo Forma de magnetismo que ocurre cuando la corriente fluye a través de un conductor.

Electromechanical A device that uses electricity and magnetism to cause a mechanical action.

Electromecánico Un dispositivo que causa una acción mecánica por medio de la electricidad y el magnetismo.

Electromotive force (EMF) *See* voltage.

Fuerza electromotriz *Véase* tensión.

Electronic level controller (ELC) The computer that controls the automatic suspension system.

Controler de nivel electrónico Este es el computador que controla el sistema automático de suspensión.

Electrostatic discharge (ESD) straps Ground your body to prevent static discharges that may damage electronic components.

Correas electroestáticas de descargue Estas aislan su cuerpo para prevenir descargas estáticas que puedan dañar los componentes electrónicos.

EMI Electromagnetic interference.

EMI La interferencia electromagnética.

Environmental data A saved snapshot of the conditions when the fault occurred.

Datos ambientales Una instantánea guardada de las condiciones en que se produjo el fallo.

Excitation current Current that magnetically excites the field circuit of the AC generator.

Corriente de excitación Corriente que excita magnéticamente al circuito inductor del generador de corriente alterna.

Face shields Clear plastic shields that protect the entire face.

Careta de soldador Caretas de plástico claro que protegen toda la cara.

Fail safe Computer substitution of a fixed input value if a sensor circuit should fail. This provides for system operation, but at a limited function; also referred to as the "limp-in" mode.

Falla Sustitución por la computadora de un valor fijo de entrada en caso de que ocurra una falla en el circuito de un sensor. Esto asegura el funcionamiento del sistema, pero a una capacidad limitada.

False wipes Unnecessary wipes that occur when the control knob of the multifunction switch is in one of the five automatic wiper sensitivity positions and no rain or moisture is apparent within the wipe pattern on the windshield glass.

Limpieza de parabrisas falsa Limpieza de parabrisas innecesaria que se produce cuando la perilla de control del interruptor multifunción se encuentra en una de las cinco posiciones de sensibilidad del limpiaparabrisas, y no hay lluvia ni humedad aparente dentro del patrón ni del cristal del parabrisas.

Fast charging Battery charging using a high amperage for a short period of time.

Carga rápida Carga de la batería que utiliza un amperaje máximo por un corto espacio de tiempo.

Feedback 1. Data concerning the effects of the computer's commands is fed back to the computer as an input signal; used to determine if the desired result has been achieved. 2. A condition that can occur when electricity seeks a path of lower resistance, but the alternate path operates another component than that intended. Feedback can be classified as a short.

Realimentación 1. Datos referentes a los efectos de las órdenes de la computadora se suministran a la misma como señal de entrada. La realimentación se utiliza para determinar si se ha logrado el resultado deseado. 2. Condición que puede ocurrir cuando la electricidad busca una trayectoria de menos resistencia, pero la trayectoria alterna opera otro componente que aquel deseado. La realimentación puede clasificarse como un cortocircuito.

Fiber optics A medium of transmitting light through polymethyl methacrylate plastic that keeps the light rays parallel even if there are extreme bends in the plastic.

Transmisión por fibra óptica Técnica de transmisión de luz por medio de un plástico de polimetacrilato de metilo que mantiene los rayos de luz paralelos aunque el plástico esté sumamente torcido.

Field current-draw test Diagnostic test that determines if there is current available to the field windings.

Prueba de la intensidad de una corriente inductora Prueba diagnóstica que determina si se está generando corriente a los devanados inductores.

Field relay The relay that controls the amount of current going to the field windings of a generator. This is the main output control unit for a charging system.

Relé inductor El relé que controla la cantidad del corriente a los devanados inductores de un generador. Es la unedad principal de potencia de salida de un sistema de carga.

Fire blanket A safety device consisting of a sheet of fire retardant material that is thrown on top of a small fire in order to smother it.

Manta contra incendios Un dispositivo de seguridad que consiste en una sábana de material retardante de fuego que se lanza sobre un pequeño incendio para reprimirlo.

Fire extinguisher A portable apparatus that contains chemicals, water, foam, or special gas that can be discharged to extinguish a small fire.

Extinctor de incendios Aparato portátil que contiene elementos químicos, agua, espuma o gas especial que pueden descargarse para extinguir un incendio pequeño.

Flammable A substance that supports combustion.

Inflamable Sustancia que promueve la combustión.

Flash To remove the existing programming and overwrite it with new software.

Limpiar Así se remueve el programa existente y lo reemplaza con programas nuevos.

Flash codes Diagnostic trouble codes (DTCs) that are displayed by flashing a lamp or LED.

Códigos intermitentes Códigos de fallas de diagnóstico (CFD) que se muestran en una lámpara intermitente o LED.

Flat rate A pay system in which technicians are paid for the amount of work they do. Each job has a flat rate time. Pay is based on that time, regardless of how long it takes to complete the job.

Tarifa bloque Sistema de pago en el que se le paga al técnico por cada hora de trabajo realizado. Cada trabajo tiene un tiempo de tarifa bloque. El pago se basa en ese tiempo sin importar cuánto tiempo se lleve para terminarlo.

Floor jack A portable hydraulic tool used to raise and lower a vehicle.

Gato de pie Herramienta hidráulica portátil utilizada para levantar y bajar un vehículo.

Forward bias A positive voltage that is applied to the P-type material and a negative voltage to the N-type material of a semiconductor.

Polarización directa Tensión positiva aplicada al material P y tensión negativa aplicada al material N de un semiconductor.

Free speed test Diagnostic test that determines the free rotational speed of the armature. This test is also referred to as the no-load test.

Prueba de velocidad libre Prueba diagnóstica que determina la velocidad giratoria libre de la armadura. A dicha prueba se le llama prueba sin carga.

Freeze frame An OBD II requirement which specifies that engine conditions be recorded when an emissions-related fault is first detected.

Cuadro congelado Un requisito OBD II que especifica que las condiciones del motor deben registrarse en cuanto se detecta una falla relacionada con las emisiones.

Freeze frame data A recording of the driving condition when the malfunction occurred. This is useful for determining how the vehicle was operating at the time and for locating any input or output values that are out of range.

Datos de trama fija Archivo de las condiciones de manejo cuando sucede una falla. Es conveniente para determinar cómo operaba el vehículo en ese momento y para localizar cualesquier valores de entrada y salida que estén fuera de banda.

Frequency The number of complete oscillations that occur during a specific time, measured in hertz.

Frecuencia El número de oscilaciones completas que ocurren durante un tiempo específico, medidas en Hertz.

Full field Field windings that are constantly energized with full battery current. Full fielding will produce maximum AC generator output.

Campo completo Devanados inductores que se excitan constantemente con corriente total de la batería. EL campo completo producirá la salida máxima de un generador de corriente alterna.

Full field test Diagnostic test used to isolate if the detected problem lies in the AC generator or the regulator.

Prueba de campo completo Prueba diagnóstica utilizada para determinar si el problema descubierto se encuentra en el generador de corriente alterna o en el regulador.

Functional test Checks the operation of the system as the technician observes the different results.

Prueba funcional Verifica la operación del sistema mientras el técnico observa los diferentes resultados.

Fuse A replaceable circuit protection device that will melt should the current passing through it exceed its rating.

Fusible Dispositivo reemplazable de protección del circuito que se fundirá si la corriente que fluye por el mismo excede su valor determinado.

Fuse box A term used to indicate the central location of the fuses contained in a single holding fixture.

Caja de fusibles Término utilizado para indicar la ubicación central de los fusibles contenidos en un solo elemento permanente.

Fusible link A wire made of meltable material with a special heat-resistant insulation. When there is an overload in the circuit, the link melts and opens the circuit.

Cartucho de fusible Alambre hecho de material fusible con aislamiento especial resistente al calor. Cuando ocurre una sobrecarga en el circuito, el cartucho se funde y abre el circuito.

Gassing The conversion of a battery's electrolyte into hydrogen and oxygen gas.

Burbujeo La conversión del electrolito de una batería al gas de hidrógeno y oxígeno.

Gauge 1. A device that displays the measurement of a monitored system by the use of a needle or pointer that moves along a calibrated scale. 2. The number that is assigned to a wire to indicate its size. The larger the number, the smaller the diameter of the conductor.

Calibrador 1. Dispositivo que muestra la medida de un sistema regulado por medio de una aguja o indicador que se mueve a través de una escala calibrada. 2. El número asignado a un alambre indica su tamaño. Mientras mayor sea el número, más pequeño será el diámetro del conductor.

Gauss gauge A meter that is sensitive to the magnetic field surrounding a wire conducting current. The gauge needle will fluctuate over the portion of the circuit that has current flowing through it. Once the ground has been passed, the needle will stop fluctuating.

Calibador gauss Instrumento sensible al campo magnético que rodea un alambre conductor de corriente. La aguja del calibrador se moverá sobre la parte del circuito a través del cual fluye la corriente. Una vez se pasa a tierra, la aguja dejará de moverse.

Glitches Unwanted voltage spikes that are seen on a voltage trace. These are normally caused by intermittent opens or shorts.

Irregularidades espontáneos Impulsos de tensión no deseables que se ven en una traza de tensión. Estos se causan normalmente por las aberturas o cortos intermitentes.

Grids A series of horizontal, ceramic, silver-compounded lines that are baked into the surface of the window.

Rejillas Una serie de líneas horizontales, de cerámica plateada, combinadas y endurecidas (horneadas) en la superficie de la ventana.

Ground The common negative connection of the electrical system that is the point of lowest voltage.

Tierra Conexión negativa común del sistema eléctrico. Es el punto de tensión más baja.

Ground circuit test A diagnostic test performed to measure the voltage drop in the ground side of the circuit.

Prueba del circuito a tierra Prueba diagnóstica llevada a cabo para medir la caída de tensión en el lado a tierra del circuito.

Ground side The portion of the circuit that is from the load component to the negative side of the source.

Lado a tierra Parte del circuito que va del componente de carga al lado negativo de la fuente.

Ground strap A length of wire or small cable used to connect two components together so current can flow through both components; commonly used to provide a ground path from the engine block to the chassis.

Correa de tierra Un trozo de alambre o un cable pequeño que se utiliza para conectar dos componentes entre sí para que la corriente pueda fluir entre ellos; se utiliza comúnmente para suministrar una puesta a tierra del bloque del motor al chasis.

Grounded circuit An electrical defect that allows current to return to ground before it has reached the intended load component.

Circuito puesto a tierra Falla eléctrica que permite el regreso de la corriente a tierra antes de alcanzar el componente de carga deseado.

Growler Test equipment used to test starter armatures for shorts and grounds. It produces a very strong magnetic field that is capable of inducing a current flow and magnetism in a conductor.

Indicador de cortocircuitos Equipo de prueba utilizado para localizar cortociruitos y tierra en armaduras de arranque. Genera un campo magnético sumamente fuerte, capaz de inducir flujo de corriente y magnetismo en un conductor.

Hall-effect switch A sensor that operates on the principle that if a current is allowed to flow through a thin conducting material exposed to a magnetic field, another voltage is produced.

Conmutador de efecto Hall Sensor que funciona basado en el principio de que si se permite el flujo de corriente a través de un material conductor delgado que ha sido expuesto a un campo magnético, se produce otra tensión.

Halogen The term used to identify a group of chemically related nonmetallic elements. These elements include chlorine, fluorine, and iodine.

Halógeno Término utilizado para identificar un grupo de elementos no metálicos relacionados químicamente. Dichos elementos incluyen el cloro, el flúor y el yodo.

Hand tools Tools that use only the force generated from the body to operate.

Herramientas manuales Herramientas que sólo utilizan la fuerza que genera el cuerpo para manejarlas.

Hard codes Failures that were detected the last time the BCM tested the circuit.

Códigos indicadores de dureza Fallas que fueron detectadas la última vez que el funcionamiento indicado con las letras BCM probó el circuito.

Hard-shell connector An electrical connector that has a hard plastic shell which holds the connecting terminals of separate wires.

Conectador de casco duro Conectador eléctrico con casco duro de plástico que sostiene separados los bornes conectadores de alambres individuales.

Hazard Communication Standard The original bases of the right-to-know laws.

Normalización de Comunicado sobre Riesgos Las bases originales de las leyes de derecho de información.

Hazardous materials Materials that can cause illness, injury, or death or that pollute water, air, or land.

Materiales peligrosos Materiales que pueden causar enfermedades, daños, o la muerte, o que contaminan el agua o la tierra.

Hazardous waste A waste that is on the Environmental Protection Agency (EPA) list of known and harmful materials.

Desecho peligroso Desecho que está en la lista de materiales conocidos y peligrosos de la Agencia de Protección Ambiental (EPA).

Heat shrink tube A hollow insulation material that shrinks to an airtight fit over a connection when exposed to heat.

Tubería contraída térmicamente Material aislante hueco que se contrae para acomodarse herméticamente sobre una conexión cuando se encuentra expuesto al calor.

Hertz A measurement of frequency.

Hertzios Es una unidad de frecuencia.

High-efficiency particulate arresting (HEPA) A special vacuum cleaner used to capture asbestos dust.

Detención de partículas de alto rendimiento (HEPA) Una aspiradora especial que se utiliza para capturar polvo de asbesto.

High-voltage electronic control unit (HV ECU) In an HEV, this unit controls the motor generators and the engine based on torque demand.

Unidad de control electrónico de alto voltaje (UCE AV) En un VHE controla los generadores del motor y el motor basándose en la demanda del par motor.

High-voltage service plug The HEV is equipped with a high-voltage service plug that disconnects the HV battery from the system.

Bujía de servicio de alto voltaje El VHE está equipado con una bujía de servicio de alto voltaje que desconecta la batería del VH del sistema.

History data Data that is stored when a fault is set in an HEV, which can be used to determine if a customer's concern is actually a problem with the system.

Historial de datos Los datos que se archivan cuando ocurre una falla en un VHE y que pueden usarse para determinar si la preocupación del cliente es en realidad un problema con el sistema.

Hoist A lift that is used to raise an entire vehicle.

Elevador Montacargas utilizado para elevar el vehículo en su totalidad.

Hold-in winding A winding that holds the plunger of a solenoid in place after it moves to engage the starter drive.

Devanado de retención Un devanado que posiciona el núcleo móvil de un solenoide después de que mueva para accionar el acoplamiento del motor de arranque.

Hydrometer A test instrument used to check the specific gravity of the electrolyte to determine the battery's state of charge.

Hidrómetro Instrumento de prueba utilizado para verificar la gravedad específica del electrolito y así determinar el estado de carga de la batería.

Hydrostatic lock A condition where noncompressible liquid that is trapped in the combustion chamber of the engine prevents the piston from moving upward.

Bloqueo hidrostático Una condición en la cual el líquido no comprimible que queda atrapado en la cámara de combustión del motor evita que el pistón se mueva hacia arriba.

Igniter A combustible device that converts electrical energy into thermal energy to ignite the inflator propellant in an air bag system.

Ignitor Un dispositivo combustible que convierte la energía eléctrica a la energía termal para encender el propelente inflador en un sistema Airbag.

Illuminated entry actuator Contains a printed circuit and a relay.

Actuador iluminado de entradas Contiene un circuito impreso y un relevador.

Immobilizer system Designed to provide protection against unauthorized vehicle use by disabling the engine if an invalid key is used to start the vehicle or if an attempt to hot-wire the ignition system is made.

Sistema inmovilizador Diseñado para brindar protección contra uso no autorizado de vehículos al desactivar el motor si se utiliza una llave no válida para arrancar el vehículo o si hay un intento de hacerle un puente al sistema de encendido.

Impedance The combined opposition to current created by the resistance, capacitance, and inductance of a test meter or circuit.

Impedancia Oposición combinada a la corriente generada por la resistencia, la capacitancia y la inductancia de un instrumento de prueba o de un circuito.

Inductive reactance The result of current flowing through a conductor and the resultant magnetic field around the conductor that opposes the normal flow of current.

Reactancia inductiva El resultado de un corriente que circule por un conductor y que resulta en un campo magnético alrededor del conductor que opone el flujo normal del corriente.

Information codes HEV codes that provide more information and freeze frame data concerning the DTCs. Information codes along with the DTC indicate more precisely the location of the fault.

Códigos de información Códigos del VHE que proporcionan mayor información y datos de trama fija que concierne a los códigos de diagnóstico de fallas (CDF). Los códigos de información junto con los CDF indican con más precisión la localización de la falla.

Instrument voltage regulator (IVR) Provides a constant voltage to the gauge regardless of the voltage output of the charging system.

Instrumento regulador de tensión Le provee tensión constante al calibrador, sin importar cual sea la salida de tensión del sistema de carga.

Insulated circuit resistance test A voltage drop test that is used to locate high resistance in the starter circuit.

Prueba de la resistencia de un circuito aislado Prueba de la caída de tensión utilizada para localizar alta resistencia en el circuito de arranque.

Insulated side The portion of the circuit from the positive side of the source to the load component.

Lado aislado Parte del circuito que va del lado positivo de la fuente al componente de carga.

Insulator A material that does not allow electrons to flow easily through it.

Aislador Una material que no permite circular fácilmente los electrones.

Integrated circuit (IC chip) A complex circuit of thousands of transistors, diodes, resistors, capacitors, and other electronic devices that are formed into a small silicon chip. As many as 30,000 transistors can be placed on a chip that is 1/4 inch (6.35 mm) square.

Circuito integrado (Fragmento CI) Circuito complejo de miles de transistores, diodos, resistores, condensadores, y otros dispositivos electrónicos formados en un fragmento pequeño de silicio. En un fragmento de 1/4 de pulgada (6,35 mm) cuadrada, pueden colocarse hasta 30.000 transistores.

Intermittent codes Those that have occurred in the past but were not present during the last BCM test of the circuit.

Códigos intermitentes   Son los que han ocurrido en el pasado pero no estuvieron presentes durante la última prueba del circuito, efectuada de acuerdo con las especificaciones de las letras BCM.

Inverter   Controls current flow between the motor generators and the HV battery in an HEV; converts HV battery DC voltage into three-phase alternating current for the motor generators and also converts high-voltage AC from the motor generators to DC voltage to charge the HV battery.

Inversor   Controla el flujo de la corriente entre los generadores del motor y la batería del VH en un VHE. Convierte el voltaje de la batería del VH en corriente alterna de tres fases para los generadores del motor, y también convierte la corriente alterna de alto voltaje de los generadores del motor a voltaje de corriente directa para cargar la batería del VH.

J1962 breakout box (BOB)   A special tool that provides a passthrough test point which connects in series between the DLC and the scan tool. This provides easy testing of voltages and resistance of any of the DLC circuits without risk of damage to the DLC.

Caja de conexiones J1962   Herramienta especial que proporciona un paso mediante el punto de prueba que se conecta en series entre el circuito del control del enlace de datos (DLC) y el instrumento de exploración. Esto proporciona una prueba fácil de los voltajes y la resistencia en cualquiera de los circuitos del DLC sin riesgo de dañar el DLC.

Jump assist   A feature of the HEV to provide for the 12-volt auxiliary battery to recharge the HV battery.

Puente auxiliar   Una característica del HEV para que la batería auxiliar de 12 voltios recargue la batería HV.

Jumper wire   A wire used in diagnostics that is made up of a length of wire with a fuse or circuit breaker and has alligator clips on both ends.

Cable conector   Una alambre empleado en los diagnósticos que se comprende de un trozo de alambre con un fusible o un interruptor y que tiene una pinza de conexión en ambos lados.

Keyless entry   A lock system that allows for locking and unlocking of a vehicle with a touch keypad instead of a key.

Entrada sin llave   Un sistema de cerradura que permite cerrar y abrir un vehículo por medio de un teclado en vez de utilizar una llave.

Knock   The spontaneous auto-ignition of the remaining air-fuel mixture in the engine combustion chamber that occurs after normal combustion has started causing the formation of standing ultrasonic waves.

Autoencendido   El encendido automático espontáneo de la mezcla de combustible/aire restante en la cámara de combustión del motor que se produce después de que comienza la combustión normal, y que causa la formación de ondas ultrasónicas estacionarias.

Knock sensor   Measures engine knock, or vibration, and converts the vibration into a voltage signal.

Sensor de autoencendido   Mide el autoencendido del motor, o vibración, y convierte la vibración en una señal de voltaje.

KOEO   Key ON, Engine OFF.

KOEO   Llave puesta, motor apagado.

KOER   Key ON, Engine RUNNING.

KOER   Llave puesta, motor en marcha.

kV   Kilovolt or 1,000 volts.

kV   Kilovolito o 1000 voltios.

Lamination   The process of constructing something with layers of materials that are firmly connected.

Laminación   El proceso de construir algo de capas de materiales unidas con mucha fuerza.

Lamp   A device that produces light as a result of current flow through a filament. The filament is enclosed within a glass envelope and is a type of resistance wire that is generally made from tungsten.

Lámpara   Dispositivo que produce luz como resultado del flujo de corriente a través de un filamento. El filamento es un tipo de alambre de resistencia hecho por lo general de tungsteno, que es encerrado dentro de una bombilla.

Lamp sequence check   Used to determine problems by observing the operation of the warning lights under different conditions.

Lámpara verificadora de secuencias   Usada para determinar problemas al observar el funcionamiento de las luces de alerta bajo condiciones diferentes.

Light-emitting diode (LED)   A gallium-arsenide diode that converts the energy developed when holes and electrons collide during normal diode operation into light.

Diodo emisor de luz   Diodo semiconductor de galio y arseniuro que convierte en luz la energía producida por la colisión de agujeros y electrones durante el funcionamiento normal del diodo.

Limit switch   A switch used to open a circuit when a predetermined value is reached. Limit switches are normally responsive to a mechanical movement or temperature changes.

Disyuntor de seguridad   Un conmutador que se emplea para abrir un circuito al alcanzar un valor predeterminado. Los disyuntores de seguridad suelen ser responsivos a un movimiento mecánico o a los cambios de temperatura.

Liquid crystal display (LCD)   A display that sandwiches electrodes and polarized fluid between layers of glass. When voltage is applied to the electrodes, the light slots of the fluid are rearranged to allow light to pass through.

Visualizador de cristal líquido   Visualizador digital que consta de dos láminas de vidrio selladas, entre las cuales se encuentran los electrodos y el fluido polarizado. Cuando se aplica tensión a los electrodos, se rompe la disposición de las moléculas para permitir la formación de carácteres visibles.

Live module   An ABS module that has not been deployed.

Módulo activo   El módulo que no ha sido desplegado.

Logic probe   A test instrument used to detect a pulsing signal.

Sonda lógica   Un instrumento de prueba que se emplea para detectar una señal pulsante.

Magnetic induction sensors   Sensors that use the principle of inducing a voltage into a winding by using a moving magnetic field.

Sensores de inducción magnética   Sensores que utilizan el principio de inducir un voltaje en un devanado mediante un campo magnético móvil.

Magnetic pulse generator   Sensor that uses the principle of magnetic induction to produce a voltage signal. Magnetic pulse generators are commonly used to send data concerning the speed of the monitored component to the computer.

Generador de impulsos magnéticos   Sensor que funciona según el principio de inducción magnética para producir una señal de tensión. Los generadores de impulsos magnéticos se utilizan comúnmente para transmitir datos a la computadora relacionados a la velocidad del componente regulado.

Magnetism   An energy form resulting from atoms aligning within certain materials, giving the materials the ability to attract other metals.

Magnetismo   Forma de energía que resulta de la alineación de átomos dentro de ciertos materiales y que le da a éstos la capacidad de atraer otros metales.

Magnetoresistive (MR) sensors   Consist of the magnetoresistive sensor element, a permanent magnet, and an integrated signal conditioning circuit to make use of the magnetoresistive effect.

Sensores de movimiento magneto resistivos (MR)   Están compuestos por el elemento sensor de movimiento magneto resistivo, un imán permanente y un circuito de condicionamiento de señales integrado para hacer uso del efecto magneto resistivo.

Material safety data sheets (MSDS)   Contain information about each hazardous material in the workplace.

Hojas de datos de seguridad del material (HDSM)   Contienen información sobre cada material peligroso en el lugar de trabajo.

Maxi-fuse   A circuit protection device that looks similar to blade-type fuses except they are larger and have a higher amperage capacity. Maxi-fuses are used because they are less likely to cause an underhood fire when there is an overload in the circuit. If the fusible link burns in two, it is possible that the hot side of the fuse could come into contact with the vehicle frame and the wire could catch on fire.

Maxifusible   Dispositivo de protección del circuito parecido a un fusible de tipo de cuchilla, pero más grande y con mayor capacidad de amperaje. Se utilizan maxifusibles porque existen menos probabilidades de que ocasionen un incendio debajo de la capota cuando ocurra una sobrecarga en el circuito. Si el cartucho de fusible se quemase en dos partes, es posible que el lado "cargado" del fusible entre en contacto con el armazón del vehículo y que el alambre se encienda.

Metri-pack connector Special wire connectors used in some computer circuits. They seal the wire terminals from the atmosphere, thereby preventing corrosion and other damage.

Conector metri-pack Los conectores de alambres especiales que se emplean en algunos circuitos de computadoras. Impermealizan los bornes de los alambres, así previniendo la corrosión y otros daños.

Millisecond Equals 1/100th of a second.

Milesegundos Equivale a 1/100 de un segundo.

Molded connector An electrical connector that usually has one to four wires that are molded into a one-piece component.

Conectador moldeado Conectador eléctrico que por lo general tiene hasta un máximo cuatro alambres que se moldean en un componente de una sola pieza.

MSDS Material safety data sheet.

MSDS Hojas de Dato de Seguridad de los Materiales.

Multifunction switch Can have a combination of any of the following switches in a single unit: headlights, turn signal, hazard, dimmer, horn, and flash to pass.

Interruptor de múltiples funciones Este permite lograr una combinación de cualquiera de los siguientes interruptores: faroles, señal de doblaje, peligro, interruptor reducidor de luz, cuerno, señal de pasada a otro carril.

Multimeter A test instrument that measures more than one electrical property.

Multímetro Un instrumento diagnóstico que mide más de una propiedad eléctrica.

Multiplying coil Made of 10 wraps of wire. This multiplies the ammeter reading so that the tester's scale can be used to read lower amperage. For example, if the needle is pointing to 25 amps when using the multiplying coil, the actual reading is 2.50 amps.

Bobina múltiple Está hecha de diez (10) cables enroscados. Esto multiplica la lectura del amperímetro y así la escala de prueba puede ser usada para ver si el amperaje está bajo. Por ejemplo, si la aguja está indicando 25 amperajes cuando se está usando la bobina múltiple la lectura actual es de 2.50 amperajes.

Negative temperature coefficient (NTC) Thermistors reduce their resistance as the temperature increases.

Coeficiente de temperatura negativa (NTC, por su sigla en inglés) Termistores cuya estructura de cristal disminuye el valor de la resistencia elecétrica a medida que aumenta la temperatura.

Neutral safety switch A switch used to prevent the starting of an engine unless the transmission is in PARK or NEUTRAL.

Disyuntor de seguridad en neutral Un conmutador que se emplea para prevenir que arranque un motor al menos de que la transmisión esté en posición PARK r NEUTRAL.

No crank A term used to mean that when the ignition switch is placed in the START position, the starter does not turn the engine.

Sin arranque Término utilizado para expresar que cuando el botón conmutador de encendido está en la posición START, el arrancador no enciende el motor.

No-crank test Diagnostic test performed to locate any opens in the starter or control circuits.

Prueba sin arranque Prueba diagnóstica llevada a cabo para localizar aberturas en los circuitos de arranque o de mando.

Noise An unwanted voltage signal that rides on a signal. Noise is usually the result of radio frequency interface (RFI) or electromagnetic induction (EMI).

Ruido Una señal indeseada del voltaje que aparece montada en una señal. El ruido es generalmente el resultado de la radio frecuencia de contacto (RFI) o de la inducción electromagnética (EMI).

Normally closed (NC) switch A switch designation denoting that the contacts are closed until acted upon by an outside force.

Conmutador normalmente cerrado Nombre aplicado a un conmutador cuyos contactos permanecerán cerrados hasta que sean accionados por una fuerza exterior.

Normally open (NO) switch A switch designation denoting that the contacts are open until acted upon by an outside force.

Conmutador normalmente abierto Nombre aplicado a un conmutador cuyos contactos permanecerán abiertos hasta que sean accionados por una fuerza exterior.

NTC Negative temperature coefficient.

NTC Las iniciales NTC representan la temperatura coeficiente negativa.

OBD II Stands for on-board diagnostics, second generation.

OBD II Las letras OBD II se refieren a los diagnósticos del tablero, segunda generación.

Occupational safety glasses Eye protection that is designed with special high-impact lens and frames, and side protection.

Gafas de protección para el trabajo Gafas diseñadas con cristales y monturas especiales resistentes y provistas de protección lateral.

OCS service kit An OCS part set that consists of the seat foam, the bladder, the pressure sensor, the occupant classification module (OCM), and the wiring. The service kit is calibrated as an assembly. The service kit OCM has a special identification data bit that is transmitted once it is connected.

Kit de servicio SCO Juego de partes SCO que consiste en la huleespuma del asiento, el depósito, el sensor de presión, el módulo de clasificación del ocupante (MCO) y el alambrado. El kit de servicio está calibrado como un ensamblado. El kit de servicio SCO tiene un bit de datos de identificación especial que se transmiten cuando se haya conectado.

OCS validation test Special procedure that confirms the OCS can properly classify the occupant. This task usually requires the use of special weights.

Prueba de revalidación del SCO Procedimiento especial que confirma que el SCO puede clasificar apropiadamente al ocupante. Esta tarea generalmente requiere el uso de pesas especiales.

Odometer A mechanical counter in the speedometer unit that indicates total miles accumulated on the vehicle.

Odómetro Aparato mecánico en la unidad del velocímetro con el que se cuentan las millas totales recorridas por el vehículo.

Offset Placed off center. Refers to the number of degrees a timing light or meter should be set to provide accurate ignition timing readings.

Desviación Ubicado fuera de lo central. Se refiere al número de grados que se debe ajustar una luz de temporización o un medidor para proveer las lecturas exactas del tiempo de encendido.

Ohm Unit of measure for resistance. One ohm is the resistance of a conductor such that a constant current of one ampere in it produces a voltage of one volt between its ends.

Ohmio Unidad de resistencia eléctrica. Un ohmio es la resistencia de un conductor si una corriente constante de 1 amperio en el conductor produce una tensión de 1 voltio entre los dos extremos.

Ohmmeter A test meter used to measure resistance and continuity in a circuit.

Ohmiómetro Instrumento de prueba utilizado para medir la resistencia y la continuidad en un circuito.

Ohm's law Defines the relationship between current, voltage, and resistance.

Ley de Ohm Define la relación entre la corriente, la tensión y la resistencia.

One-hand rule A service procedure used on high-voltage EV and HEV vehicles that requires working with only one hand to prevent an electric shock from the high-voltage system passing through your body.

Regla de una mano Un procedimiento de servicio que se utiliza en vehículos EV y HEV de alto voltaje, el cual se opera con una sola mano para evitar que pase una descarga electrostática del sistema de alto voltaje a través del cuerpo.

Open An electrical term used to indicate that a circuit is not complete or is broken.

Abierto Es un término de electricidad, usado para indicar que el circuito no está completo o está roto.

Open circuit A term used to indicate that current flow is stopped. By opening the circuit, the path for electron flow is broken.

Circuito abierto Término utilizado para indicar que el flujo de corriente ha sido detenido. Al abrirse el circuito, se interrumpe la trayectoria para el flujo de electrones.

Open circuit testing Testing for voltage by disconnecting a connector and then measuring for applied voltage.

Prueba de circuito abierto Prueba de voltaje en donde se desconecta un conector y luego se mide el voltaje aplicado.

Open circuit voltage test Used to determine the battery's state of charge. It is used when a hydrometer is not available or cannot be used.

Prueba de la tensión en un circuito abierto Sirve para determinar el estado de carga de la batería. Esta prueba se lleva a cabo cuando no se dispone de un hidrómetro o cuando el mismo no puede utilizarse.

Optics test   Test of the automatic high-beam system to test the ability to recognize ambient light through the lens and the windshield.

Prueba óptica   Prueba del sistema automático de luces altas para averiguar la habilidad de reconocer las luces ambientales mediante la lente y el parabrisas.

Overload   Excess current flow in a circuit.

Sobrecarga   Flujo de corriente superior a la que tiene asignada un circuito.

Overrunning clutch   A clutch assembly on a starter drive used to prevent the engine's flywheel from turning the armature of the starter motor.

Embrague de sobremarcha   Una asamblea de embrague en un acoplamiento del motor de arranque que se emplea para prevenir que el volante del motor dé vueltas al armazón del motor de arranque.

Oxygen sensor   A voltage-generating sensor that measures the amount of oxygen present in an engine's exhaust.

Sensor de oxígeno   Un sensor generador de tensión que mide la cantidad del oxígeno presente en el gas de escape de un motor.

Parallel circuit   A circuit that provides two or more paths for electricity to flow.

Circuito en paralelo   Circuito que provee dos o más trayectorias para que circule la electricidad.

Parasitic drains   Constant drains of the battery due to accessories that draw small amounts of current.

Drenaje parásita   Los constantes drenajes en la batería son causados debido a que los accesorios atraen pequeñas cantidades de corriente.

Parasitic loads   Electrical loads that are still present when the ignition switch is in the OFF position.

Cargas parásitas   Cargas eléctricas que todavía se encuentran presentes cuando el botón conmutador de encendido está en la posición OFF.

Park assist system (PAS)   A parking aid that alerts the driver to obstacles that are located in the path immediately behind the driver.

Sistema de ayuda al aparcamiento (PAS)   Ayuda al aparcamiento que avisa al conductor de los obstáculos que se encuentran en la trayectoria inmediatamente detrás del conductor.

Park switch   Contact points located inside the wiper motor assembly that supply current to the motor after the wiper control switch has been turned to the PARK position. This allows the motor to continue operating until the wipers have reached their PARK position.

Conmutador PARK   Puntos de contacto ubicados dentro del conjunto del motor del frotador que le suministran corriente al motor después de que el conmutador para el control de los frotadores haya sido colocado en la posición PARK. Esto permite que el motor continue su funcionamiento hasta que los frotadores hayan alcanzado la posición original.

Passive seat belt system   Seat belt operation that automatically puts the shoulder and/or lap belt around the driver or occupant. The automatic seat belt is moved by DC motors that move the belts by means of carriers on tracks.

Sistema pasivo de cinturones de seguridad   Función de los cinturones de seguridad que automáticamente coloca el cinturón superior y/o inferior sobre el conductor o pasajero. Motores de corriente continua accionan los cinturones automáticos mediante el uso de portadores en pistas.

Peak   The highest voltage value in one cycle of an AC voltage sine wave.

Máxima eficiencia   El valor más alto del voltaje de un ciclo de corriente alterna (AC) en un voltaje de onda senoidal.

Peak-to-peak voltage   The total voltage measured between the peaks of an AC sine wave.

Voltaje de máximas eficiencias   La medida total del voltaje cuando los puntos son de máxima eficiencia en una onda senoidal de corriente alterna (AC).

Photo cell   A variable resistor that uses light to change resistance.

Fotocélula   Resistor variable que utiliza luz para cambiar la resistencia.

Photo cell resistance assembly   A technician-made test tool that replaces the photo cell to produce predictable results.

Montaje de resistencia fotocélula   Una herramienta técnica de prueba que remplaza la fotocélula para producir resultados visibles.

Phototransistor   A transistor that is sensitive to light.

Fototransistor   Transistor sensible a la luz.

Photovoltaic diodes   Diodes capable of producing a voltage when exposed to radiant energy.

Diodos fotovoltaicos   Diodos capaces de generar una tensión cuando se encuentran expuestos a la energía de radiación.

Pickup coil   The stationary component of the magnetic pulse generator consisting of a weak permanent magnet that is wound around by fine wire. As the timing disc rotates in front of it, the changes of magnetic lines of force generate a small voltage signal in the coil.

Bobina captadora   Componente fijo del generador de impulsos magnéticos compuesta de un imán permanente débil devanado con alambre fino. Mientras gira el disco sincronizador enfrente de él, los cambios de las líneas de fuerza magnética generan una pequeña señal de tensión en la bobina.

Piezoelectric sensors   Produce a proportional voltage output resulting from deformation of the element as pressure is applied.

Sensores piezoeléctricos   Producen una salida de voltaje proporcional que es el resultado de una deformación del elemento a medida que se aplica presión.

Piezoelectric transducer   A sensor that is capable of measuring the pressures associated with ultrasonic waves.

Transductor piezoeléctrico   Un sensor capaz de medir las presiones relacionadas con las ondas ultrasónicas.

Piezoresistive   Resistance that changes in value as the pressure applied to the sensing material changes.

Piezoresistencia   Resistencia que cambia su valor a medida que varía la presión que se aplica al material de detección.

Piezoresistive sensor   A sensor that is sensitive to pressure changes.

Sensor piezoresistivo   Sensor susceptible a los cambios de presión.

PIN code   A personal identification code used by the manufacturer to allow reprogramming and other services to the immobilizer system.

Código NIP   Un código de identificación personal utilizado por el fabricante para permitir la reprogramación y otros servicios en el sistema inmovilizador.

Pinion gear   A small gear; typically refers to the drive gear of a starter drive assembly or the small drive gear in a differential assembly.

Engranaje de piñón   Un engranaje pequeño; tipicamente se refiere al engranaje de arranque de una asamblea de motor de arranque o al engranaje de mando pequeño de la asamblea del diferencial.

Pinpoint test   A specific component test.

Prueba de precisión   Una prueba hecha con un componente específico.

Plate straps   Metal connectors used to connect the positive or negative plates in a battery.

Abrazaderas de la placa   Los conectores metálicos que sirven para conectar las placas positivas o negativas de una batería.

Plates   The basic structure of a battery cell; each cell has at least one positive plate and one negative plate.

Placas   La estructura básica de una celula de batería; cada celula tiene al menos una placa positiva y una placa negativa.

PMGR   An abbreviation for permanent magnet gear reduction.

PMGR   Una abreviación de desmultiplicación del engranaje del imán permanente.

Pneumatic tools   Power tools that are powered by compressed air.

Herramientas neumáticas   Herramientas mecánicas accionadas por aire comprimido.

Polarizers   Glass sheets that make light waves vibrate in only one direction. This converts light into polarized light.

Polarizadores   Las láminas de vidrio que hacen vibrar las ondas de luz en un sólo sentido. Esto convierte la luz en luz polarizada.

Polarizing   The process of light polarization or of setting one end of a field as a positive or negative point.

Polarizadora   El proceso de polarización de la luz o de establecer un lado de un campo como un punto positivo o negativo.

Polymer   Compound that consists of a plastic with carbon grains.

Polímero   Compuesto fabricado a partir de un plástico con granos de carbón.

Positive engagement starter   A type of starter that uses the magnetic field strength of a field winding to engage the starter drive into the flywheel.

Acoplamiento de arranque positivo   Un tipo de arrancador que utilisa la fuerza del campo magnético del devanado inductor para accionar el acoplamiento del arrancador en el volante.

Positive temperature coefficient (PTC)   Thermistors that increase their resistance as the temperature increases.

Coeficiente de temperatura positiva (PTC, por su sigla en inglés) Termistores que aumentan su resistencia a medida que aumenta la temperatura.

Potential The ability to do something; typically voltage is referred to as the potential. If you have voltage, you have the potential for electricity.

Potencial La capacidad de efectuar el trabajo; típicamente se refiere a la tensión como el potencial. Si tiene tensión, tiene la potencial para la electricidad.

Potentiometer A voltage divider circuit that is used to measure linear or rotary movement.

Potenciómetro Un circuito divisor de voltaje que se utiliza para medir el movimiento lineal o rotativo.

Power formula A formula used to calculate the amount of electrical power a component uses. The formula is $P = I \times E$, where P stands for power (measured in watts), I stands for current, and E stands for voltage.

Formula de potencia Una formula que se emplea para calcular la cantidad de potencia eléctrica utilizada por un componente. La formula es $P = I \times E$, en el que el P quiere decir potencia (medida en wats), I representa el corriente y el E representa la tensión.

Power tools Tools that use forces other than those generated from the body. They can use compressed air, electricity, or hydraulic pressure to generate and multiply force.

Herramientas mecánicas Herramientas que utilizan fuerzas distintas a las generadas por el cuerpo. Dichas fuerzas pueden ser el aire comprimido, la electricidad, o la presión hidráulica para generar y multiplicar la fuerza.

Powertrain control module (PCM) The computer that controls the engine operation.

Módulo de control de potencia El computador que controla la operación del motor.

Pressure control solenoid A solenoid used to control the pressure of a fluid, commonly found in electronically controlled transmissions.

Solenoide de control de la presión Un solenoide que controla la presión de un fluido, suele encontrarse en las transmisiones controladas electronicamente.

Primary wiring Conductors that carry low voltage and low current. The insulation of primary wires is usually thin.

Hilos primarios Hilos conductores de tensión y corriente bajas. El aislamiento de hilos primarios es normalmente delgado.

Program A set of instructions that the computer must follow to achieve desired results.

Programa Conjunto de instrucciones que la computadora debe seguir para lograr los resultados deseados.

PROM (programmable read only memory) Memory chip that contains specific data which pertains to the exact vehicle the computer is installed in. This information may be used to inform the CPU of the accessories that are equipped on the vehicle.

PROM (memoria de sólo lectura programable) Fragmento de memoria que contiene datos específicos referentes al vehículo particular en el que se instala la computadora. Esta información puede utilizarse para informar a la UCP sobre los accesorios de los cuales el vehículo está dotado.

Protection device Circuit protector that is designed to turn off the system that it protects. This is done by creating an open to prevent a complete circuit.

Dispositivo de protección Protector de circuito diseñado para "desconectar" el sistema al que provee protección. Esto se hace abriendo el circuito para impedir un circuito completo.

Prove-out circuit A function of the ignition switch that completes the warning light circuit to ground through the ignition switch when it is in the START position. The warning light is on during engine cranking to indicate to the driver that the bulb is working properly.

Circuito de prueba Función del boton conmutador de encendido que completa el circuito de la luz de aviso para que se ponga a tierra a través del botón conmutador de encendido cuando éste se encuentra en la posición START. La luz de aviso se encenderá durante el arranque del motor para avisarle al conductor que la bombilla funciona correctamente.

PTC Positive temperature coefficient.

PTC Iniciales identificadoras de la temperatura positiva coeficiente.

Pull to seat A method used to install the terminals into the connector.

Tire el asiento Un método usado para instalar los terminales dentro del conectador.

Pulse width The length of time in milliseconds that an actuator is energized.

Duración de impulsos Espacio de tiempo en milisegundos en el que se excita un accionador.

Pulse width modulation On-off cycling of a component. The period of time for each cycle does not change; only the amount of on time in each cycle changes.

Modulación de duración de impulsos Modulación de impulsos de un componente. El espacio de tiempo de cada ciclo no varía; lo que varía es la cantidad de trabajo efectivo de cada ciclo.

Push to seat A method used to install the terminals into the connector.

Empuje el asiento Un método usado para instalar los terminales dentro del conectador.

Quartz swing needle Displays that are similar in design to the aircore electromagnetic gauges used in conventional analog instrument panels.

Aguja de oscilación del cuarzo Mostrarios que son similares en diseño al del núcleo de aire de los calibres electromagnéticos usados convencionalmente en los paneles de instrumentos análogos.

Quick test Isolates the problem area and determines whether the starter motor, solenoid, or control circuit is at fault.

Prueba rápida Aisla el área con problemas y determina si el arrancador del motor, el solenoid, y el circuito de control están fallando.

Radial grid A type of battery grid that has its patterns branching out from a common center.

Rejilla radial Un tipo de rejilla de bateria cuyos diseños extienden de un centro común.

Radiated noise Signals that are generated from an electrical accessory and are detected by the antenna.

Ruido radiado Señales que se generan desde un accesorio eléctrico y que son detectadas por la antena.

Radio choke Absorbs voltage spikes and prevents static in the vehicle's radio.

Impedancia del radio Absorba los impulsos de la tensión y previene la presencia del estático en el radio del vehículo.

Radio frequency interface (RFI) Produced when electromagnetic radio waves of sufficient amplitude escape from a wire or connector.

Radio frecuencia de contacto (RFI) Se produce cuando las ondas de radio eletromagnéticas y de suficiente amplitud se escapan de un cable o conector.

Radio interference All undesirable signals that enter the radio and are heard as static or noise.

Interferencia de radio Toda señal indeseable que entra a través de la radio y que se escucha en forma de estática o ruido.

Radio noise An undesired interference or static noise that obstructs the normal sound of the radio station.

Ruido de radio Interferencia o ruido estático indeseado que obstaculiza la emisión normal de las estaciones de radio.

Ratio A mathematical relationship between two or more things.

Razón Una relación matemática entre dos cosas o más.

Rectification The conversion of AC current to DC current.

Rectificación Proceso a través del cual la corriente alterna es transformada en una corriente continua.

Refractometer A special meter used to measure the specific gravity of a liquid by the refraction of light.

Refractómetro Herramienta especial que usa para medir la gravedad específica de un líquido por la refracción de la luz.

Relative compression testing A test method for determining if a cylinder has lower compression relative to other cylinders by using current draw during cranking.

Prueba de compresión relativa Método de prueba para determinar si un cilindro tiene compresión más baja comparada a los otros cilindros al utilizar llamada de corriente durante la desencoladura.

Relay A device that uses low current to control a high-current circuit. Low current is used to energize the electromagnetic coil, while high current is able to pass over the relay contacts.

Relé Dispositivo que utiliza corriente baja para controlar un circuito de corriente alta. La corriente baja se utiliza para excitar la bobina electromagnética, mientras que la corriente alta puede transmitirse a través de los contactos del relé.

Remote keyless entry (RKE)   A radio frequency system that allows the vehicle to be unlocked or locked from a distance.

Entrada remota sin llave (RKE)   Un sistema de radiofrecuencia que permite el vehículo para ser abierta o cerrada desde una distancia.

Reserve-capacity rating   An indicator, in minutes, of how long a vehicle can be driven with the headlights on, if the charging system should fail. The reserve-capacity rating is determined by the length of time, in minutes, that a fully charged battery can be discharged at 25 amps before battery cell voltage drops below 1.75 volts per cell.

Clasificación de capacidad en reserva   Indicación, en minutos, de cuánto tiempo un vehículo puede continuar siendo conducido, con los faros delanteros encendidos, en caso de que ocurriese una falla en el sistema de carga. La clasificación de capacidad en reserva se determina por el espacio de tiempo, en minutos, en el que una batería completamente cargada puede descargarse a 25 amperios antes de que la tensión del acumulador de la batería disminuya a un nivel inferior de 1,75 amperios por acumulador.

Resistance   Opposition to current flow.

Resistencia   Oposición que presenta un conductor al paso de la corriente eléctrica.

Resistance wire   A special type of wire that has some resistance built into it. These typically are rated by ohms per foot.

Alambre de resistencia   Un tipo de alambre especial que por diseño tiene algo de resistencia. Estos tipicamente tienen un valor nominal de ohm por pie.

Resistive shorts   Shorts to ground that pass through a form of resistance first.

Cortocircuitos resistivos   Cortocircuitos a tierra que primero pasan por una forma de resistencia.

Resistor block   A series of resistors with different values.

Bloque resistor   Serie de resistores que tienen valores diferentes.

Resource Conservation and Recovery Act (RCRA)   Law that makes hazardous waste generators responsible for the waste from the time it becomes a waste material until proper waste disposal is completed.

Ley de la Conservación y Recuperación de los Recursos   Ley que hace responsables a los que generan residuos peligrosos por su desecho desde el momento en que se convierte en material de desecho hasta que se completa su destrucción apropiada de desechos.

Reversed bias   A positive voltage is applied to the N-type material and a negative voltage is applied to the P-type material of a semiconductor.

Polarización inversa   Tensión positiva aplicada al material N y tensión negativa aplicada al material P de un semiconductor.

RFI   Common acronym for radio frequency interference.

RFI   Una sigla común de la interferencia de radiofrecuencia.

Rheostat   A two-terminal variable resistor used to regulate the strength of an electrical current.

Reostático   Un resistor variable de dos terminales usado para regular la potencia de un circuito eléctrico.

Right-to-know laws   Laws concerning hazardous materials and wastes that protect every employee in a workplace by requiring employers to provide a safe working place as it relates to hazardous materials. The right-to-know laws state that employees have a right to know when the materials they use at work are hazardous.

Leyes de derecho de información   Leyes contra materiales peligrosos y desechos que protegen a cada empleado en un lugar de trabajo al requerir que los patrones proporcionen un lugar de trabajo segura contra los materiales peligrosos. Las leyes de derecho de información declaran que los empleados tienen derecho a saber cuando y qué materiales que usan en el trabajo son peligrosos.

RMS   Root mean square; a method for measuring AC voltage.

RMS   Raíz de la media de los cuadrados; un método para medir la tensión del corriente alterna.

Root mean square (RMS)   Meters convert AC signal to a comparable DC voltage signal.

Corriente efectiva (RMS)   Los medidores convierten la señal de corriente alterna (AC) a una señal de voltaje comparable (DC).

Rotor   The component of the AC generator that is rotated by the drive belt and creates the rotating magnetic field of the AC generator.

Rotor   Parte rotativa del generador de corriente alterna accionada por la correa de transmisión y que produce el campo magnético rotativo del generador de corriente alterna.

Safety goggles   Eye protection device that fits against the face and forehead to seal off the eyes from outside elements.

Gafas de seguridad   Dispositivo protector que se coloca delante de los ojos para preservarlos de elementos extraños.

Safety stands   Support devices used to hold the vehicle off the floor after it has been raised by a floor jack.

Soportes de seguridad   Dispositivos de soporte utilizados para sostener el vehículo sobre el suelo después de haber sido levantado con el gato de pie.

Scan tool   A microprocessor designed to communicate with the BCM. It accesses trouble codes and runs system operation, actuator, and sensor tests.

Herramienta analizadora   Un microprocesador diseñado para comunicarse usando la información de las iniciales BCM. Así se encuentran los códigos que indican problemas, cómo funciona el sistema de conducción, los actuadores y monitores de prueba.

Scanner   A diagnostic test tool that is designed to communicate with the vehicle's on-board computer.

Dispositivo de exploración   Herramienta de prueba diagnóstica diseñada para comunicarse con la computadora instalada en el vehículo.

Sealed-beam headlight   A self-contained glass unit that consists of a filament, an inner reflector, and an outer glass lens.

Faro delantero sellado   Unidad de vidrio que contiene un filamento, un reflector interior y una lente exterior de vidrio.

Secondary wiring   Conductors, such as battery cables and ignition spark plug wires, that are used to carry high voltage or high current. Secondary wires have extra-thick insulation.

Hilos secundarios   Conductores, tales como cables de batería e hilos de bujías del encendido, utilizados para transmitir tensión o corriente alta. Los hilos secundarios poseen un aislamiento sumamente grueso.

Self-test input (STI)   The single pigtail connector located next to the self-test connector.

Autoprueba de entrada (STI)   Simple cable flexible de conección localizado junto al conector de autoprueba.

Semiconductors   An element that is neither a conductor nor an insulator. Semiconductors are materials that conduct electric current under certain conditions, yet will not conduct under other conditions.

Semiconductores   Elemento que no es ni conductor ni aislante. Los semiconductores son materiales que transmiten corriente eléctrica bajo ciertas circunstancias, pero no la transmiten bajo otras.

Sender unit   The sensor for the gauge. It is a variable resistor that changes resistance values with changing monitored conditions.

Unidad emisora   Sensor para el calibrador. Es un resistor variable que cambia los valores de resistencia según cambian las condiciones reguladas.

Sensitivity controls   A potentiometer that allows the driver to adjust the sensitivity of the automatic dimmer system to surrounding ambient light conditions.

Controles de sensibilidad   Un potenciómetro que permite que el conductor ajusta la sensibilidad del sistema de intensidad de iluminación automático a las condiciones de luz ambientales.

Sensor   Any device that provides an input to the computer.

Sensor   Cualquier dispositivo que le transmite información a la computadora.

Sensor dropout   When an induction sensor stops producing an output voltage due to slow speeds of the target.

Desenganche del sensor   Cuando un sensor de inducción deja de producir voltaje de salida debido a velocidades bajas del objetivo.

Series circuit   A circuit that provides a single path for current flow from the electrical source through all the circuit's components, and back to the source.

Circuito en serie   Circuito que provee una trayectoria única para el flujo de corriente de la fuente eléctrica a través de todos los componentes del circuito, y de nuevo hacia la fuente.

Series–parallel circuit   A circuit that has some loads in series and some in parallel.

Circuito en series paralelas   Circuito que tiene unas cargas en serie y otras en paralelo.

Series-wound motor   A type of motor that has its field windings connected in series with the armature. This type of motor develops its maximum torque output at the time of initial start. Torque decreases as motor speed increases.

Motor con devanados en serie Un tipo de motor cuyos devanados inductores se conectan en serie con la armadura. Este tipo de motor desarrolla la salida máxima de par de torsión en el momento inicial de ponerse en marcha. El par de torsión disminuye al aumentar la velocidad del motor.

Servo Controls the position of the throttle by receiving a controlled amount of vacuum.

Control servo Este sirve para controlar la posición de la válvula reguladora al recibir una cantidad controlada de vacío.

Servomotor An electrical motor that produces rotation of less than a full turn. A feedback mechanism is used to position itself to the exact degree of rotation required.

Servomotor Motor eléctrico que genera rotación de menos de una revolución completa. Utiliza un mecanismo de realimentación para ubicarse al grado exacto de la rotación requerida.

Short An unwanted electrical path; sometimes this path goes directly to ground.

Corto Una trayectoria eléctrica no deseable; a veces este trayectoria viaja directamente a tierra.

Short to ground A condition that allows current to return to ground before it has reached the intended load component.

Corto a la tierra Su función permite que la corriente regrese a la tierra, antes de que haya llegado a un componente intencionalmente cargado.

Shorted circuit Allows current to bypass part of the normal path.

Circuito corto Este circuito permite que la corriente pase por una parte del recorrido normal.

Shunt-wound motor A type of motor whose field windings are wired in parallel to the armature. This type of motor does not decrease its torque as speed increases.

Motor con devanados en derivación Un tipo de motor cuyos devanados inductores se cablean paralelos a la armadura. Este tipo de motor no disminuya su par de torsión al aumentar la velocidad.

Shutter wheel A metal wheel consisting of a series of alternating windows and vanes. It creates a magnetic shunt that changes the strength of the magnetic field from the permanent magnet of the Hall-effect switch or magnetic pulse generator.

Rueda obturadora Rueda metálica compuesta de una serie de ventanas y aspas alternas. Genera una derivación magnética que cambia la potencia del campo magnético, del imán permanente del conmutador de efecto Hall o del generador de impulsos magnéticos.

Sine wave A waveform that shows voltage changing polarity.

Onda senoidal Una forma de onda que muestra un cambio de polaridad en la tensión.

Single-phase voltage The sine wave voltage induced in one conductor of the stator during one revolution of the rotor.

Tensión monofásica La tensión en forma de onda senoidal inducida en un conductor del estator durante una revolución del rotor.

Sinusoidal A waveform that is a true sine wave.

Senoidal Una forma de onda que es una onda senoidal verdadera.

Slow charging Battery charging rate between 3 and 15 amps for a long period of time.

Carga lenta Índice de carga de la batería de entre 3 y 15 amperios por un largo espacio de tiempo.

Slow cranking A term used to mean that the starter drive engages the ring gear, but the engine turns too slowly to start.

Arranque lento Término utilizado para expresar que el mecanismo de transmisión de arranque engrana la corona, pero que el motor se enciende de forma demasiado lenta para arrancar.

Soft codes Codes those that have occurred in the past, but were not present during the last BCM test of the circuit.

Códigos suaves Códigos que han ocurrido en el pasado, pero que no estaban presentes durante la última prueba BCM del circuito.

Soldering The process of using heat and solder (a mixture of lead and tin) to make a splice or connection.

Soldadura Proceso a través del cual se utiliza calor y soldadura (una mezcla de plomo y de estaño) para hacer un empalme o una conexión.

Solderless connectors Hollow metal tubes that are covered with insulating plastic. They can be butt connectors or terminal ends.

Conectadores sin soldadura Tubos huecos de metal cubiertos de plástico aislante. Pueden ser extremos de conectadores o de bornes.

Solenoid An electromagnetic device that uses movement of a plunger to exert a pulling or holding force.

Solenoide Dispositivo electromagnético que utiliza el movimiento de un pulsador para ejercer una fuerza de arrastre o de retención.

Solenoid circuit resistance test Diagnostic test used to determine the electrical condition of the solenoid and the control circuit of the starting system.

Prueba de la resistencia de un circuito solenoide Prueba diagnóstica utilizada para determinar la condición eléctrica del solenoide y del circuito de mando del sistema de arranque.

Speakers Turn electrical energy provided by the radio receiver amplifier into acoustical energy using a permanent magnet and electromagnet to move air to produce sound.

Altavoces Convierten la energía eléctrica que suministra el amplificador receptor de radio en energía acústica mediante el uso de un imán permanente y un electroimán, los cuales mueven el aire para producir sonido.

Specific gravity The weight of a given volume of a liquid divided by the weight of an equal volume of water.

Gravedad específica El peso de un volumen dado de líquido dividido por el peso de un volumen igual de agua.

Speedometer An instrument panel gauge that indicates the speed of the vehicle.

Velocímetro Calibrador en el panel de instrumentos que marca la velocidad del vehículo.

Splice The joining of single wire ends or the joining of two or more electrical conductors at a single point.

Empalme La unión de los extremos de un alambre o la unión de dos o más conductores eléctricos en un solo punto.

Splice clip A special connector used along with solder to assure a good connection. The splice clip is different from solderless connectors in that it does not have insulation.

Grapa para empalme Conectador especial utilizado junto con la soldadura para garantizar una conexión perfecta. La grapa para empalme se diferencia de los conectadores sin soldadura porque no está provista de aislamiento.

Square waves Identified by having straight vertical sides and a flat top.

Ondas cuadradas Para saber cuáles son, deben tener lados rectos verticales y una punta plana.

Stabilize Removing the surface charge of the battery by placing a large load on the battery for 15 seconds.

Estabilizar Para hacerlo hay que remover la superficie cargada de la batería, colocando en la batería, por 15 segundos, una carga grande.

Starter drive The part of the starter motor that engages the armature to the engine flywheel ring gear.

Transmisión de arranque Parte del motor de arranque que engrana la armadura a la corona del volante de la máquina.

State of charge The condition of a battery's electrolyte and plate materials at any given time.

Estado de carga Condición del electrolito y de los materiales de la placa de una batería en cualquier momento dado.

Stator The stationary coil of the AC generator in which current is produced.

Estátor Bobina fija del generador de corriente alterna donde se genera corriente.

Stator neutral junction The common junction of Wye stator windings.

Unión de estátor neutral La unión común de los devanados de un estátor Y.

Stepped resistor A resistor that has two or more fixed resistor values.

Resistor de secciones escalonadas Resistor que tiene dos o más valores de resistencia fija.

Stepper motor An electrical motor that contains a permanent magnet armature with two or four field coils; can be used to move the controlled device to the desired location. By applying voltage pulses to selected coils of the motor, the armature turns a specific number of degrees. When the same voltage pulses are applied to the opposite coils, the armature rotates the same number of degrees in the opposite direction.

Motor paso a paso Motor eléctrico que contiene una armadura magnética fija con dos o cuatro bobinas inductoras. Puede utilizarse para mover el dispositivo regulado a cualquier lugar deseado. Al aplicárseles impulsos de tensión a ciertas bobinas del motor, la armadura girará un número específico de grados. Cuando estos mismos impulsos de tensión se aplican a las bobinas opuestas, la armadura girará el mismo número de grados en la dirección opuesta.

Sulfation   A chemical action within the battery that interferes with the ability of the cells to deliver current and accept a charge.

Sulfatado   Acción química dentro de la batería que interfiere con la capacidad de los acumuladores de transmitir corriente y recibir una carga.

System main relay (SMR)   Used to connect and disconnect the power source of the high-voltage system in an HEV.

Relé principal del sistema del (SMR o RPS)   Se usa para conectar y desconectar la fuente de potencia del sistema de alto voltaje en un VHE.

Tachometer   An instrument that measures the speed of the engine in revolutions per minute (rpm).

Tacómetro   Instrumento que mide la velocidad del motor en revoluciones por minuto (rpm).

Test light   Checks for electrical power in a circuit.

Luz de prueba   Verifica la potencia eléctrica en un circuito.

Thermistor   A solid-state variable resistor made from a semiconductor material that changes resistance as a function of temperature.

Termistor   Un resistor variable de estado sólido, fabricado a partir de material semiconductor que cambia la resistencia en función de la temperatura.

Three-minute charge test   A reasonably accurate method for diagnosing a sulfated battery for use on conventional batteries.

Prueba de carga de tres minutos   Método bastante preciso en baterías convencionales para diagnosticar una batería sulfatada.

Throttle position sensor (TPS)   Sensor used to modify the pulse width calculation based on position of the throttle plate and the rate of change of the throttle plate.

Sensor de posición del acelerador (TPS, por sus siglas en inglés)   Sensor que se utiliza para modificar el cálculo de la amplitud del pulso sobre la base de la posición de la placa del acelerador y el índice de cambio de la placa del acelerador.

Timer control   A potentiometer that is part of the headlight switch in some systems. It controls the amount of time the headlights stay on after the ignition switch is turned off.

Control temporizador   Un potenciómetro que es parte del conmutador de los faros en algunos sistemas. Controla la cantidad del tiempo que quedan prendidos los faros después de apagarse la llave del encendido.

Tin   Process of applying solder to the tip of the soldering iron to provide better heating control.

Estaño   El proceso de aplicar soldadura en la punta de un hierro soldador para dar mejor control al calor.

Tire pressure monitoring (TPM) system   A safety system that notifies the driver if a tire is underinflated or overinflated.

Sistema de monitoreo de presión de las llantas (TPM, por su sigla en inglés)   Un sistema de seguridad que informa al conductor si una llanta está muy inflada o poco inflada.

Torque converter   A hydraulic device found on automatic transmissions. It is responsible for controlling the power flow from the engine to the transmission; works like a clutch to engage and disengage the engine's power to the drive line.

Convertidor de par   Un dispositivo hidráulico en las transmisiones automáticas. Se encarga de controlar el flujo de la potencia del motor a la transmisión; funciona como un embrague para embragar y desembragar la potencia del motor con la flecha motríz.

Trimotor   A three-armature motor.

Trimotor   Es un motor de tres armaduras.

Trouble codes   Output of the self-diagnostics program in the form of a numbered code that indicates faulty circuits or components. Trouble codes are two or three digital characters that are displayed in the diagnostic display if the testing and failure requirements are both met.

Códigos indicadores de fallas   Datos del programa autodiagnóstico en forma de código numerado que indica los circuitos o los componentes defectuosos. Dichos códigos se componen de dos o tres carácteres digitales que se muestran en el visualizador diagnóstico si se llenan los requisitos de prueba y de falla.

Troubleshooting   The diagnostic procedure of locating and identifying the cause of the fault. It is a step-by-step process of elimination using cause and effect.

Detección de fallas   Procedimiento diagnóstico a través del cual se localiza e identifica la falla. Es un proceso de eliminación que se lleva a cabo paso a paso por medio de causa y efecto.

TVRS   An abbreviation for television-radio-suppression cable.

TVRS   Una abreviación del cable de supresíon del televisión y radio.

Two-button diagnostics   The buttons on the instrument panel that, when pressed in the right combination, place the module into self-test mode.

Diagnósticos de dos botones   Estos son los botones que aparecen en el panel de instrumentos, los cuales al presionarlos en la combinación correcta, ponen el módulo en estado de autoprueba.

Two-coil gauge   A gauge design that uses the interaction of two electromagnets and the total field effect upon an armature to cause needle movement.

Calibrador de dos bobinas   Calibrador diseñado para utilizar la interacción de dos electroimanes y el efecto inductor total sobre una armadura para generar el movimiento de la aguja.

U-codes   Diagnostic trouble codes assigned to the vehicle communication network.

Códigos en U   Códigos de fallo de diagnóstico asignados a la red de comunicación del vehículo.

Vacuum distribution valve   A valve used in vacuum-controlled concealed headlight systems. It controls the direction of vacuum to various vacuum motors or to vent.

Válvula de distribución al vacío   Válvula utilizada en el sistema de faros delanteros ocultos controlado al vacío. Regula la dirección del vacío a varios motores al vacío o sirve para dar salida del sistema.

Vacuum fluorescent display (VFD)   A display type that uses anode segments coated with phosphor and bombarded with tungsten electrons to cause the segments to glow.

Visualización de fluorescencia al vacío   Tipo de visualización que utiliza segmentos ánodos cubiertos de fósforo y bombardeados de electrones de tungsteno para producir la luminiscencia de los segmentos.

Valve body   A unit that consists of many valves and hydraulic circuits. This unit is the central control point for gear shifting in an automatic transmission.

Cuerpo de la válvula   Una unedad que consiste de muchas válvulas y circuitos hidráulicos. Esta unedad es el punto central de mando para los cambios de velocidad en una transmisión automática.

Variable resistor   A resistor that provides for an infinite number of resistance values within a range.

Resistor variable   Resistor que provee un número infinito de valores de resistencia dentro de un margen.

Vehicle identification number (VIN)   A number that is assigned to a vehicle for identification purposes. The identification plate is usually located on the cowl, next to the left-upper instrument panel.

Número de identificación del vehículo   Número asignado a cada vehículo para fines de identificación. Por lo general, la placa de identificación se ubica en la bóveda, al lado del panel de instrumentos superior de la izquierda.

Vehicle lift points   The areas that the manufacturer recommends for safe vehicle lifting. They are the areas that are structurally strong enough to sustain the stress of lifting.

Puntos para elevar el vehículo   Áreas específicas que el fabricante recomienda para sujetar el vehículo a fin de lograr una elevación segura. Son las áreas del vehículo con una estructura suficientemente fuerte para sostener la presión de la elevación.

Vehicle module scan   A special function of the scan tools that queries all of the modules on the bus to respond and then lists those that did reply.

Explorador del módulo del vehículo   Función especial de los instrumentos de exploración que cuestionarán a todos los módulos en el bus para que respondan, y luego harán una lista de aquellos que respondieron.

Volatile   A substance that vaporizes or explodes easily.

Volátil   Sustancia que se evapora o explota con facilidad.

Volatility   The tendency of a fluid to evaporate quickly or pass off in the form of vapor.

Volatilidad   La tendencia de un fluido a evaporarse rápidamente o disiparse en forma de vapor.

Volt   The unit used to measure the amount of electrical force.

Voltio   Unidad práctica de tensión para medir la cantidad de fuerza eléctrica.

Voltage   The difference or potential that indicates an excess of electrons at the end of the circuit the farthest from the electromotive force. It is the electrical pressure that causes electrons to move through a circuit. One volt is the amount of pressure required to move one amp of current through one ohm of resistance.

Tensión Diferencia o potencial que indica un exceso de electrones al punto del circuito que se encuentra más alejado de la fuerza electromotriz. La presión eléctrica genera el movimiento de electrones a través de un circuito. Un voltio equivale a la cantidad de presión requerida para mover un amperio de corriente a través de un ohmio de resistencia.

Voltage drop A resistance in the circuit that reduces the electrical pressure available after the resistance. The resistance can be the load component, the conductors, any connections, or an unwanted resistance.

Caída de tensión Resistencia en el circuito que disminuye la presión eléctrica disponible después de la resistencia. La resistencia puede ser el componente de carga, los conductores, cualquier conexión o resistencia no deseada.

Voltage drop test Determines if the battery, regulator, and AC generator are all operating at the same potential.

Prueba de la caída de tensión en el voltaje Determina si la batería, el regulador, y el generador de corriente alterna AC están todos funcionando al mismo potencial.

Voltage limiter Connected through the resistor network of a voltage regulator. It determines whether the field will receive high, low, or no voltage. It controls the field voltage for the required amount of charging.

Limitador de tensión Conectado por el red de resistores de un regulador de tensión. Determina si el campo recibirá alta, baja o ninguna tensión. Controla la tensión de campo durante el tiempo indicado de carga.

Voltage output test Used to make a quick determination about whether the charging system is working properly. If the charging system is operating correctly, then check for battery drain.

Prueba de salida del voltaje Se usa para determinar en forma rápida si el sistema de carga está trabajando correctamente. Si el sistema de carga está funcionando correctamente, entonces revíselo todo para saber si hay algún drenaje en la batería.

Voltage regulator Used to control the output voltage of the AC generator, based on charging system demands, by controlling field current.

Regulador de tensión Dispositivo cuya función es mantener la tensión de salida del generador de corriente alterna, de acuerdo a las variaciones en la corriente de carga, controlando la corriente inductora.

Voltmeter A test meter used to read the pressure behind the flow of electrons.

Voltímetro Instrumento de prueba utilizado para medir la presión del flujo de electrones.

Warning light A lamp that illuminates to warn the driver of a possible problem or hazardous condition.

Luz de aviso Lámpara que se enciende para avisarle al conductor sobre posibles problemas o condiciones peligrosas.

Watt The unit of measure of electrical power, which is the equivalent of horsepower. One horsepower is equal to 746 watts.

Watio Unidad de potencia eléctrica, equivalente a un caballo de vapor. 746 watios equivalen a un caballo de vapor (CV).

Wattage A measure of the total electrical work being performed per unit of time.

Vataje Medida del trabajo eléctrico total realizado por unidad de tiempo.

Waveform The electronic trace that appears on a scope; it represents voltage over time.

Forma de onda La trayectoria electrónica que aparece en un osciloscopio; representa la tensión a través del tiempo.

Weather-pack connector An electrical connector that has rubber seals on the terminal ends and on the covers of the connector half to protect the circuit from corrosion.

Conectador resistente a la intemperie Conectador que tiene sellos de caucho en los extremos de los bornes y en las cubiertas de la parte del conectador para proteger el circuito contra la corrosión.

Wet fouling A condition of a spark plug in which it is wet with oil.

Engrase húmedo Una condición de la bujía en la cual se moja de aceite.

Wheatstone bridge A series–parallel arrangement of resistors between an input terminal and ground. Flexing of the disc on which the resistors are laid changes their value.

Puente Wheatstone Una disposición en serie paralela de resistores entre un terminal de entrada y la tierra. La flexión del disco en el que están colocados los resistores cambia su valor.

Wiring diagram An electrical schematic that shows a representation of actual electrical or electronic components and the wiring of the vehicle's electrical systems.

Esquema de conexiones Esquema en el que se muestran las conexiones internas de los componentes eléctricos o electronicos reales y las de los sistemas eléctricos del vehículo.

Wiring harness A group of wires enclosed in a conduit and routed to specific areas of the vehicle.

Cableado preformado Conjunto de alambres envueltos en un conducto y dirigidos hacia áreas específicas del vehículo.

Work order A legal document representing an agreement for a repair facility to provide services to a customer and give the shop the authorization to perform the service at the quoted estimate. The work order contains customer, vehicle, and shop information and describes issues customers experience with their vehicle. It is also used by the technician to track diagnostic and repair times.

Orden de trabajo Un documento legal que representa un acuerdo para que una instalación de reparación preste servicios a un cliente y le dé autorización al taller para realizar el servicio según la estimación cotizada. La orden de trabajo contiene información del cliente, vehículo y taller, y describe los problemas que experimentan los clientes con su vehículo. También la utiliza el técnico para rastrear los diagnósticos y tiempos de reparación.

Workplace hazardous materials information systems (WHMIS) Canadian equivalent to MSDS.

Sistemas de información acerca de los materiales peligrosos en el área de trabajo (SIMPAT o WHMIS) Equivalente canadiense a las hojas de datos de la seguridad de un material.

Worm gear A type of gear whose teeth wrap around the shaft. The action of the gear is much like that of a threaded bolt or screw.

Engranaje de tornillo sin fin Un tipo de engranaje cuyos dientes se envuelven alrededor del vástago. El movimiento del engranaje es muy parecido a un perno enroscado o una tuerca.

Wye connection A type of stator winding in which one end of the individual windings is connected at a common point. The structure resembles the letter Y.

Conexión Y Un tipo de devanado estátor en el cual una extremidad de los devanados individuales se conectan en un punto común. La estructura parece la letra "Y."

INDEX